Chimpanzees in Context

Chimpanzees in Context

A Comparative Perspective on Chimpanzee Behavior, Cognition, Conservation, and Welfare

EDITED BY LYDIA M. HOPPER
AND STEPHEN R. ROSS

With a Foreword by Jane Goodall

The University of Chicago Press
Chicago and London

The University of Chicago Press, Chicago 60637
The University of Chicago Press, Ltd., London

Published 2020
Printed in the United States of America

29 28 27 26 25 24 23 22 21 20   1 2 3 4 5

ISBN-13: 978-0-226-72784-4 (cloth)
ISBN-13: 978-0-226-72798-1 (paper)
ISBN-13: 978-0-226-72803-2 (e-book)
DOI: https://doi. org/10.7208/chicago/9780226728032.001.0001

Library of Congress Cataloging-in-Publication Data

Names: Hopper, Lydia M., editor. | Ross, Stephen R., editor. | Goodall, Jane, 1934- writer of foreword. | Lester E. Fisher Center for the Study and Conservation of Apes, host institution, organizer.
Title: Chimpanzees in context : a comparative perspective on chimpanzee behavior, cognition, conservation, and welfare / edited by Lydia M. Hopper and Stephen R. Ross ; with a foreword by Jane Goodall.
Description: Chicago ; London : The University of Chicago Press, 2020. | Papers from a conference of the same name organized and hosted by the Lester E. Fischer Center for the Study and Conservation of Apes, Lincoln Park Zoo, Chicago in 2016. | Includes bibliographical references and index.
Identifiers: LCCN 2020017311 | ISBN 9780226727844 (cloth) | ISBN 9780226727981 (paperback) | ISBN 9780226728032 (ebook)
Subjects: LCSH: Chimpanzees—Behavior—Congresses. | Chimpanzees—Conservation—Congresses. | Chimpanzees—Ecology—Congresses. | Chimpanzees—Research—Congresses. | Social behavior in animals—Congresses. | Cognition in animals—Congresses. | Animal welfare—Congresses. | LCGFT: Conference papers and proceedings.
Classification: LCC QL737.P94 C56 2020 | DDC 599.885—dc23
LC record available at https://lccn.loc.gov/2020017311

♾ This paper meets the requirements of ANSI/NISO Z39.48-1992 (Permanence of Paper).

Contents

PART 3 Studying Chimpanzees

PART 4 Communication

PART 5 Cooperation

PART 6 Tool Use, Cognition, and Culture

Foreword

JANE GOODALL

To really understand the complex life cycle and social behavior of chimpanzee, you need time. Chimpanzees, like the other great apes (bonobos, gorillas, orangutans, and humans), mature very slowly—infants continue to share their mother's nest, ride her back, and suckle (though less frequently) until about five years old when the next offspring is often born, and even then they remain emotionally connected to their mother, traveling with her for several years, strengthening family bonds that may persist throughout life. A female will not have her first infant until she is between 10 and 13 years old, and, like humans and other great apes, she has a long gestation period (about 8 months for chimpanzees). And there is an interval averaging five years between births. Chimpanzees can live to be over 50 years in the wild (and up to 70 years in captivity). As a result of this long life, and the fact that each individual has his or her own distinct personality, any worthwhile behavioral study of chimpanzees must be long term and must concentrate on individual life histories.

So in 1986, when Paul Heltne, then director of the Chicago Academy of Sciences, approached me with the idea of organizing a conference to coincide with the publication of my book *The Chimpanzees of Gombe: Patterns of Behaviour*, I had already been studying the Gombe chimpanzees for a long time—over two decades. By then other scientists had established field stations in different parts of the chimpanzee range across Africa and a few of these studies had also been operating for decades (the longest was started by Japanese primatologists at Mahale Mountains National Park in 1966). In other words, we were just at the start of the era of long-term studies on known individuals. The first infant chimpanzees at Gombe whose life histories were recorded from birth were born in 1964. And yet, even by 1986, these individuals were only just taking their place as adults in the complex fusion-fission society of their community. By that time, there was a research station at Gombe and a number of individual chimpanzees were known extremely well—such as the famous female Flo and her family—adult son Faben, adolescent son Figan, and juvenile Fifi, sister to little Flint born in 1964.

When Paul discussed with me the plan for a conference, to be hosted by the Chicago Academy of Sciences, he initially suggested we invite speakers to discuss not only chimpanzees and bonobos (then known as pygmy chimpanzees), but also the third African ape, the gorilla—and to invite paleontologists to discuss early hominids. He felt that such a conference would help to set the great apes in an evolutionary perspective. But I begged him to organize, for once, a conference *just* for the chimpanzees (the extent to which bonobos differed from chimpanzees, biologically and behaviorally, was not fully realized at the time). Finally, Paul agreed. It would be a meeting to specifically discuss the behavior of chimpanzees, *Pan troglodytes.*

At that first conference in 1986 there were not so many of us "chimp people." Everyone accepted the invitation to attend and it was the first time the Japanese primatologists had brought their findings to the Western world. We also invited some researchers working with captive colonies. It was a very exciting meeting. The underlying goal was to explore similarities and differences in the behavior of chimpanzees at the different study sites, to begin to understand how the environment might affect behavior and whether (as I believed) there were also cultural differences passed from one generation to the next through observation and imitation. But there was also a session to discuss conservation issues, and another on conditions in some captive situations—such as the cruel training of chimpanzees for entertainment and their treatment in medical research laboratories. For me the experience in Chicago was life changing, as it was then that I realized, reluctantly, that I had to leave the research at Gombe to others, and work instead on conservation issues and the welfare of captive chimpanzees worldwide.

The conference was so successful that in 1991 a second meeting was organized in Chicago. This time the stated aim was to discuss chimpanzee cultural diversity. Several new field sites had been established across Africa by then and the researchers were able to compare and begin to analyze the different cultural behaviors observed at these various study sites. I believe that it was because of this conference that behavioral scientists across the board began to accept that cultural behavior was not unique to humans. In fact, it was becoming clear that cultural behavior was also to be found in other long-lived animals—such as elephants, whales, and dolphins. It was exciting to find that more researchers were not only prepared to share their knowledge with their peers, but also eager to do so because of the new light it shed on chimpanzee adaptability.

About a decade later, my good friend Professor Tetsuro Matsuzawa of Kyoto University, made the suggestion to convene the third of these chimpanzee meetings. He suggested Steve Ross and Elizabeth Lonsdorf of Chicago's Lincoln Park Zoo host the meeting and together they chose the theme of "The Mind of the Chimpanzee." This third meeting convened in 2006 and participants explored many aspects of chimpanzee intelligence, learning, and emotions. Some of the papers compared chimpanzee cognitive performance with that of other species, including humans, though again, the focus was primarily on chimpanzees.

The fourth, and most recent, of these conferences was held in 2016, again organized by Lincoln Park Zoo scientists, with the theme "Chimpanzees in Context." In some ways, the meeting would break new ground, as species other than chimpanzees were included in the discussions. Alongside the many experienced chimpanzee scientists—such as Tetsuro Matsuzawa, Bill McGrew, John Mitani, Anne Pusey, Frans de Waal, Andy Whiten, and Richard Wrangham, scientists studying orangutans, dolphins, and ravens also participated. At that meeting it was exciting to hear from scientists young and old, about their important work studying and protecting not only chimpanzees, but also a whole variety of other primate species around the world. I was especially pleased to realize that more and more researchers, particularly some of the younger ones, were passionately involved in conservation and welfare issues, and were increasingly working with local communities so as to help improve the future for chimpanzees

Thus, the simple gathering of 1986 has grown over the years and today our discussion of chimpanzees intersects with a broader view of all the other species that make up the natural world around us. The collective contribution of researchers studying different primates has greatly increased our understanding of primate evolution. The growing realization of the need to protect primates and their habitats, develop sanctuaries for infants orphaned by the bushmeat and live animal trades, and improve their lives in zoos around the world, has generated a determination to move with some urgency to advance all such initiatives. It has highlighted the shocking fact that many species, including all the apes, are threatened with extinction.

Our increased understanding of primate behavior has helped us to better understand ourselves. We are not (as was commonly believed in the early 1960s) the only species able to use and make tools, or to have personalities, minds, and emotions. There is, after all, no difference in *kind* between us and other animals. Knowing that our closest living relatives are the great apes and studying ways in which our behavior is so similar to theirs, also helps us appreciate the main *difference*—the explosive development of the human intellect. How strange that the most intellectual species is destroying our only home, Planet Earth. We are in the middle of the sixth great extinction; one that is caused by our irresponsible behavior and the absurd belief that there can be unlimited economic development and human population growth on a planet of finite natural resources.

As I suggested earlier, it takes considerable time to study the many facets of a chimpanzee's life. But we don't have much time left if we are to do something to help the survival of our closest living relatives. Now it is time to use our intellect to start healing the harm we have inflicted, to protect the habitats of our primate relatives (along with biodiversity) before it is quite too late. There is so much more to learn about the chimpanzees and other primates, and I pray that in another ten years or so, the next of this series of conferences may bring with it, signs that humankind is acting now and taking the necessary steps to save chimpanzees and other species from extinction.

Preface: Understanding Chimpanzees in Context

LYDIA M. HOPPER AND STEPHEN R. ROSS

The contents of this book reflect themes discussed at Chimpanzees in Context, a meeting hosted by Lincoln Park Zoo's Lester E. Fisher Center for the Study and Conservation of Apes in August 2016. This meeting was the fourth in the *Understanding Chimpanzees* series, each of which resulted in an edited volume (Heltne and Marquardt 1989; Lonsdorf, Ross, and Matsuzawa 2010; Wrangham et al. 1996). Therefore, this volume is also the fourth in a series; a collection of works that have brought together contemporary observations and research related to chimpanzees in order to create a thorough picture of our closest living relative, *Pan troglodytes*.

The previous *Understanding Chimpanzees* meetings and volumes focused tightly on chimpanzees, examining their behavior and cognition, both in the wild and in captivity. While they allowed for detailed cross-site comparisons of chimpanzee behavior, revealing their cultural differences or intra-species comparisons, there was little comparison of chimpanzees to other primate species, or even non-primate species. By expanding the focus to include other relevant taxa, our aim with the Chimpanzees in Context meeting we hosted, and this resultant volume, was to put chimpanzees in context; the context of theory and practice, as well as the context of what we know for other species, disciplines, approaches, and questions.

But what do we mean by "in context"? This could be interpreted in a number of ways, and, indeed, it was our hope that it would be. The Merriam-Webster dictionary defines "context" as 1) the parts of a discourse that surround a word or passage and can throw light on its meaning and 2) interrelated conditions in which something exists or occurs. By seeking different perspectives, we hope to interpret the lives and experiences of chimpanzees in a more comprehensive manner and to form a richer understanding as to the ways in which chimpanzees' unique experiences shape their behavior.

We invited authors to share contemporary insights about chimpanzee behavior and to set them against what we know for other species. Beyond species comparisons, we also encouraged the authors to demonstrate how different methodologies might provide different insights, how different cultural experiences might influence our perspectives of chimpanzees, and how the different ecologies in which chimpanzees live might affect the expression of their behavior. Thus, by "in context" we mean more than simply "as compared to other species," and it is our hope that this volume provides a comprehensive and varied perspective on chimpanzees through the eyes of those who study and work with them, as well as from those who focus on other species altogether.

Perhaps the most obvious interpretation of comparative research is studies in which a single methodology is deployed and applied to more than one species, revealing a deeper understanding of the phylogeny of a behavior or physiological process. For example, we can compare life history characteristics across apes (Knott and Harwell, chapter 1 this volume) and the physiology that underpins developmental differences across chimpanzees and bonobos (Behringer et al., chapter 2 this volume). By comparing species, we can also learn how different social systems shape and regulate their relationships (Taglialatela et al., chapter 4 this volume; Wittig et al., chapter 5 this volume) or the ways in which they communicate (Hobaiter, chapter 10 this volume; Townsend, Watson, and Slocombe, chapter 11 this volume). To emphasize the impact of taking a comparative approach, we have included chapters in this volume that provide perspectives on chimpanzees from researchers studying a wide range of species including other apes (Clay, chapter 12 this volume; Rosenbaum, Santymire, and Stoinski, chapter 6 this volume), monkeys (Luncz and van de Waal, chapter 18 this volume; Vale and Brosnan, chapter 15 this volume), as well as non-primate species, such as dolphins (Mann, Stanton, and Murray, chapter 3 this volume) and corvids (Massen, Schaake, and Bugnyar, chapter 16 this volume).

We also highlight the importance that the context of the experimental protocol, study site, or social grouping can have when interpreting chimpanzee behavior and cognition. Chapters in this volume shed light not only on how to design methods that facilitate effective comparative research, but also on how these methods have been refined. Specifically, they reveal the importance of designing protocols that can be tested with multiple species with different physiologies and from different ecological niches (Massen, Schaake, and Bugnyar, chapter 16 this volume) or that can be applied in different contexts, such as with wild and captive populations (Hopper and Carter, chapter 7 this volume; Yamamoto, chapter 14 this volume). In recognition of the importance of a strong yet flexible methodological approach, many of the contributing authors highlight the evolution of methodological techniques (Beran, Perdue, and Parrish, chapter 20 this volume; Duguid et al., chapter 13 this volume; Martin and Adachi, chapter 8 this volume) or the refinement of theoretical approaches (Tennie, Hopper, and van Schaik, chapter 19 this volume) over time. Beyond the development of an experimental technique over time, the authors also underscore how the context of the experimental protocol may impact the results obtained and, importantly, our interpretation of them (Rosenbaum, Santymire, and Stoinski, chapter 6 this volume; Vale and Brosnan, chapter 15 this volume).

Beyond the experimental approach, it is vital to consider the context of the environment—both physical and social—in which chimpanzees live. The chimpanzees' social group, and their standing within it, will influence their social relationships and potentially the skills they possess or can acquire (Luncz and van de Waal, chapter 18 this volume). The authors discuss not only how the social elements of a chimpanzees' world affect them, but also how this compares to the social bonds and dominance hierarchies formed by members of other species (Wittig et al., chapter 5 this volume). Beyond the social environment, the physical environment is also important to consider when interpreting chimpanzees' behavior; we cannot begin to understand chimpanzees without studying how they respond to different environments both in the wild (Pruetz, Bogart, and Lindshield, chapter 17 this volume) and in captive settings (Ross, chapter 24 this volume) or even between these two environments (Cronin and Ross, chapter 22 this volume; Yamamoto, chapter 14 this volume).

Perhaps the most important context facing chimpanzees this century, is the context of human influence. It is important to recognize that all the environments in which all chimpanzees exist today, are in some way under the influence of human activity. Whether they are housed and managed in a captive setting (Herrelko, Vick, and Buchanan-Smith, chapter 23 this volume; Ross, chapter 24 this volume) or facing anthropogenic threats in the wild (Chapman et al., chapter 25 this volume; Hartel et al., chapter 26 this volume), the lives of chimpanzees are inextricably intertwined with our own species for better or for worse. The threats chimpanzees face are varied, but all relate to anthropogenic pressure and understanding those threats and studying ways to mitigate or ameliorate them are vital (Hartel et al., chapter 26 this volume; Morgan et al., chapter 27 this volume). Concurrently, it is our duty to provide the best possible care to chimpanzees in managed settings such as zoos, laboratories, and sanctuaries. By understanding the reasons that chimpanzees have come to be housed in captive settings (Hirata et al., chapter 9 this volume) as well as those factors that enhance or limit their well-being in captivity (Bloomsmith et al., chapter 21 this volume) we can best devise and evaluate the most efficacious strategies for their care (Cronin and Ross, chapter 22 this volume; Ross, chapter 24 this volume).

While we have framed our book into eight thematic sections, we encourage readers to seek links across them. Indeed, the very theme of "context" provides a network of cross-connections throughout the chapters we have included in this book. Collectively, the chapters remind us of the importance of considering the social, ecological, and cognitive contexts of chimpanzee behavior, and how these contexts shape our interpretation of our understanding of chimpanzees. Only by leveraging these powerful perspectives, do we stand a chance at improving how we understand, care for, and protect this species.

References

Heltne, P. G., and L. A. Marquardt. 1989. *Understanding Chimpanzees*. Cambridge, MA: Harvard University Press.

Lonsdorf, E. V., S. R. Ross, and T. Matsuzawa. 2010. *The Mind of the Chimpanzee: Ecological and Experimental Perspectives*. Chicago: University of Chicago Press.

Wrangham, R. W., W. C. McGrew, F. B. M. de Waal, and P. G. Heltne. 1996. *Chimpanzee Cultures*. Cambridge, MA: Harvard University Press.

Acknowledgments

The contents of this volume reflect topics discussed at Chimpanzees in Context, the fourth meeting of the Understanding Chimpanzees series. Chimpanzees in Context was hosted by Lincoln Park Zoo's Lester E. Fisher Center for the Study and Conservation of Apes in August 2016 and held at Lincoln Park Zoo, Chicago, USA. We are extremely grateful to the staff and volunteers at Lincoln Park Zoo, without whom this meeting would not have been possible. In particular, we wish to thank Kevin Bell, Megan Ross, Lisa Faust, Josh Rupp, Ashley Bedore, Jillian Braun, and Maureen Leahy for administrative and logistical support. The success of the meeting itself was largely the result of exceptional work by Fisher Center staff and volunteers, including Anne Kwiatt, Crystal Egelkamp, and Sarah Jacobson and was financially supported by the Arcus Foundation and the David Bohnett Foundation, and with in-kind support provided by Lincoln Park Zoo. Thanks are also due to the Leo S. Guthman Foundation, who provided core support to the Fisher Center. We are also grateful to the conference steering committee: Kristin Bonnie, Sarah Brosnan, Katherine Cronin, Brian Hare, Elizabeth Lonsdorf, Tetsuro Matsuzawa, John Mitani, David Morgan, Frans de Waal, Andrew Whiten, and Roman Wittig. Considering the production of this volume, we thank all the contributing authors and the Fisher Center staff and interns who assisted us in compiling the materials, notably Sarah Huskisson, Jesse Leinwand, Christina Doelling, Samantha Earl, and Rachel Nelson. Finally, we acknowledge the literally hundreds of individual chimpanzees that continue to inspire us to work on their behalf.

PART ONE

Life Histories and Developmental Milestones

1

Ecological Risk and the Evolution of Great Ape Life Histories

CHERYL D. KNOTT AND FAYE S. HARWELL

Introduction

Great apes exemplify selection for slow life histories, characterized by late age at first reproduction, long inter-birth intervals, and long lifespans. However, they also show substantial variation in these features between and within genera, species, and populations. Although differences in life history characteristics between ape genera have been noted by researchers, it has often been within the context of understanding why humans differ from the great apes (Hill et al. 2001; Robson, van Schaik, and Hawkes 2006). There has been less exploration of what accounts for the differences *between* the great apes.

Documenting great ape life history differences is important for several reasons. First, we can understand each species' particular adaptations and evolutionary history better if we isolate what features distinguish it from other species (Gould and Lewontin 1979). Second, it provides data to test hypotheses about the relationship between ecology, behavior, morphology, and life history among the apes, and more broadly, within primates. Third, understanding what features have shaped life history within our closest relatives allows us to see what is unique about humans, and thus gain a better understanding of our own evolution. Lastly, it allows for more accurate reconstructions of the life history of extinct hominins, which is crucial for understanding when particular human life history features emerged (Hill et al. 2001; Robson, van Schaik, and Hawkes 2006). For additional comparative perspectives on great ape life histories, see Behringer et al., chapter 2 this volume, and Mann, Stanton, and Murray, chapter 3 this volume.

There are many hypotheses about which factors select for a species' life history, but there is not a single explanation accounting for all variation observed at the genus, species, and individual levels. Variables explaining broad differences between groups often do not shed light on individual differences in life history within a population. Here, we focus specifically on which explanations can most inform our understanding of life history differences between the great apes. We call particular attention to differences in environmental variability. We argue that fluctuations in energy availability associated with ecological risk provide the best explanation for differences in the evolution of great ape life histories. For each great ape species, we demonstrate how energy availability relates to their life history features. Lastly, we compare wild and captive settings, which provide vastly different environmental conditions.

Approach

WHAT ARE THE DIFFERENCES IN GREAT APE LIFE HISTORIES?

Life history refers to the timing of major "life events" such as menarche, first birth, subsequent births, menopause, and death for an individual. The life history profile of a population or species is constructed by recording these life events longitudinally for many individuals. Accordingly, determining the life history parameters of a particular population must come from long-term study sites where the age of numerous individuals is known and individuals are followed throughout their lifetime. The longest great ape studies are those on the chimpanzee populations of Gombe and Mahale, both in Tanzania, which have been monitored for over 50 years (Goodall 1968; Nishida 1968). This means that the longest-running wild sites are only just beginning to document the full lives of individual apes. Furthermore, many sites experience periods of disruption, due to political instability or illegal logging, sometimes leading to key life events of individuals being missed (Gruen, Fultz, and Pruetz 2013; Riley and Bezanson 2018). Additionally, it can be difficult to study certain aspects of life history because of the behavioral ecology of a species. For instance, female chimpanzees often emigrate from their natal group prior to their first birth and may be difficult to track outside the core study area. Likewise, habituation and home range size may hamper consistent monitoring of some ape populations. Despite these limitations, we now have life history data for hundreds of individual wild and captive apes. Table 1.1 displays the latest great ape life history variables available in the published literature grouped by genera, species, and population. It should be noted that the newest great ape species, the Tapanuli orangutan (*Pongo tapanuliensis*), is not included in this chapter as there is little known of its life history at this time (Nater et al. 2017).

TABLE 1.1. Reproductive parameters of wild and captive apes. Means are given, followed by medians in parentheses. Ranges and sample sizes, when available, are provided in parentheses below (*n* is the number of samples or intervals, *f* refers to females and *m* to males). Where multiple sources from one site exist, the most recent publication was cited.

Species	*Site*	*Infant Mortality*	*Age at Menarche*	*First Sexual Behavior*	*Age at First Birth (Female)*	*Interbirth Interval*	*Age at Last Birth (Female)*	*Life Expectancy at Birth*	*Life Expectancy at Maturity*	*Age at Death*	*Maximum Lifespan*
Orangutan (*Pongo* overall)	All Wild	0.07[46,l,e]	—	—	14.4[46,h,l] (12.7–16.3, n=13)	7.8[46,j,l,q] (5.5–9.3, n=49)	—	—	—	—	—
Orangutan (Bornean- *P. pygmaeus*)	Danum Valley	0.13[46] (n=16)	—	—	—	7.0 (7.0)[46,j,q] (n=1)	—	—	—	—	—
Orangutan (Bornean- *P. pygmaeus*)	Gunung Palung	0.04[46] (n=28)	—	12.5[23] (n=1)	15.1[46,h] (14.4–16.4, n=3)	7.6 (7.6)[46,j,q] (6.8–8.3, n=11)	—	—	—	—	—
Orangutan (Bornean- *P. pygmaeus*)	Kinabatangan	0.12[46] (n=17)	—	—	—	7.4 (7.5)[46,j,q] (6.5–8.0, n=4)	—	—	—	—	—
Orangutan (Bornean- *P. pygmaeus*)	Sebaganau	0.05[46] (n=19)	—	—	13.3[46,h] (12.7–13.8, n=2)	6.8 (7.0)[46,j,q] (5.5–7.8, n=4)	—	—	—	—	—
Orangutan (Bornean- *P. pygmaeus*)	Tanjung Puting		—	10–11[11,h] (n=1)	15.7 (16.0)[45] (15–16, n=3)	7.7 (7.7)[12,h,j,k] (5.2–10.4, n=23, f=11)	—	—	—	—	—
Orangutan (Bornean- *P. pygmaeus*)	Tuanan	0.09[46] (n=35)	—	—	14.1[46,h] (13.3–15.0, n=2)	7.0 (7.0)[46,j,q] (5.9–8.3, n=13)	—	—	—	—	—
Orangutan (Sumatran- *P. abelii*)	Ketambe	0.0[46] (n=32)	—	—	13.7[46,h] (13.3–14.1, n=7)	7.7 (7.25)[46,j,q] (6.3–9.3, n=14)	41[50,g,h]	28.8*f*, 33.7*m*[50,h]	—	—	53*f*, 58*m*[50,h] (n=13*f*, 15*m*)

(continues)

TABLE 1.1. *(continued)*

Species	Site	Infant Mortality	Age at Menarche	First Sexual Behavior	Age at First Birth (Female)	Interbirth Interval	Age at Last Birth (Female)	Life Expectancy at Birth	Life Expectancy at Maturity	Age at Death	Maximum Lifespan
Orangutan (Sumatran- *P. abelii*)	Suaq	0.15[46] (n=27)	—	—	14.7[46,h] (13.7–15.7, n=2)	7.5 (7.2)[46,j,q] (6.3–9.3, n=8)	—	—	—	—	—
Orangutan (*Pongo* overall)	Captive		7.7 (7.6)[2,28,29,a,d] (4.5–11.1, n=8)	5.5–9.0 [2,27]	11.7[5] (7–33, n=314)	(5.5) [39,51] (2.1–6.9, n=12)	41[39,g] (n=1234)	—	—	—	58[39,g] (n=2566)
Orangutan (Bornean- *P. pygmaeus*)	Captive	0.21[1,e] (n=440)	—	—	15.5[1] (n=158)	6.3[1] (n=27)	29.0[1] (n=69)	—	37.1*f*, 30.6*m*[1,k] (n=23*f*, 15*m*)	(27.3*m*)[43]	33.7*m*[1,g] (n=122*m*)
Orangutan (Sumatran- *P. abelii*)	Captive	0.22[1,e] (n=504)	—	—	16.4[1] (n=183)	5.8[1] (n=27)	26.2[1] (n=67)	24.6*f*, 16.7*m*[51,i] (n=329*f*, 292*m*)	33.0*f*, 29.3*m*[1,k] (n=33*f*, 48*m*)	(32.8*f*, 25.2*m*)[43]	40.8*m*[1,g] (n=109*m*)
Gorilla (Mountain- *Gorilla beringei beringei*)	Karisoke	0.27[35,f] (n=181)	7.3[15,b] (6.5–7.9, n=3)	(6.3)[49] (5.8–7.1)	9.84[36] (8.0–13.6, n=31)	4.0[35,j] (2.3–7.25, n=88, f=66)	—	33*f*, 23*m*[4,i] (31–35*f*, 22–29*m*)	—	34.6*f*[35,h] (21.3–43.7, n=14)	39*f*, 35*m*[4,g]
Gorilla (Grauer's– *G. b. graueri*)	Kahuzi-Biega	0.26[53]	—	—	—	4.6[53] (3.4–6.6, n=9)	—	—	—	—	—
Gorilla (Western Lowland– *G. gorilla gorilla*)	Lossi	0.22[34,o] (n=12)	—	—	—	5 [34,o,p] (n=3)	—	—	—	—	—

Gorilla (Western Lowland– *G. g. gorilla*)	Mbeli Bai	0.65[34,o] (n=28)	—	—	—	5.2[34,o] (n=3)	—	—	—	—	—
Gorilla (Western Lowland- *G. g. gorilla*)	Captive	—	—	—	9.3[6] (7.5–10.4, n=8)	4.2 (4.0)[40] (2.3–6.4, n=16, f=13)	17.1[33] (5.5–35.0)	23.3[24] (n=868)	—	(38.3*f*, 31.7*m*)[43]	54[20]
Bonobo (*Pan paniscus*)	Wamba	0.18[10, e] (n=22)	8–9[25]	—	14.2 (14.0)[25] (13.0–15.0, n=6)	4.8[10,j] (n=28, f=17)	—	—	—	—	—
Bonobo (*P. paniscus*)	Captive	—	8.2 (7.7)[44] (6.0–11.2, n=9)	—	10.8 (10.0)[25,44,d] (7.7–20.0, n=20)	3.6[44] or 5.1 (5.2)[16] (1.9–7.6, n=21, f=14) [16]	—	—	—	(31.5*m*)[43]	60[26,g]
Chimpanzee (*Pan troglodytes* overall)	All Wild	—	—	—	13.3[37]	(5.9)[8,j,q] (1.81–11.0, n=173, f=165)	55.0[8,g,h] (n=165)	14.6*f*, 11.2*m*[18] (n=148*f*, 156*m*)	30.4*f*, 29.2*m*[18] (n=148*f*, 156*m*)	—	55*f*, 46*m*[18,g,h] (n=148*f*, 156*m*)
Chimpanzee (Western- *P. t. versus*)	Bossou	0.20*f*, 0.20*m*[42]	8.5[9, c] (8.0–9.0, n=5)	—	10.9[42,h] (9.7–13.9, n=9, f=5)	5.3[9] (6.5)[8,j,q] (1.9–10.7, n=21, f=10)	39.5[8,g,h] (n=10)	—	—	—	—
Chimpanzee (Eastern- *P. t. schweinfurthii*)	Budongo	—	—	—	—	5.2[9] (5.3)[8,j,q] (4.8–7.0, n=13, f=17)	40.6[8,g,h] (n=17)	—	—	—	—

(continues)

TABLE 1.1. *(continued)*

Species	Site	Infant Mortality	Age at Menarche	First Sexual Behavior	Age at First Birth (Female)	Interbirth Interval	Age at Last Birth (Female)	Life Expectancy at Birth	Life Expectancy at Maturity	Age at Death	Maximum Lifespan
Chimpanzee (Western rehabilitants- *P. t. versus*)	Gambia	—	—	—	—	(6.1)[8,j,q] (1.80–11.0, n=43, f=25)	32.7[8,g,h] (n=25)	—	—	—	—
Chimpanzee (Eastern- *P. t. schweinfurthii*)	Gombe	0.24*f*, 0.34*m*[48] (n= 37*f*, 45*m*)	10.8[48,c] (8.5–13.5, n=8)	—	13.3[48] (11.1–17.2, n=4)	5.5[13](6.1)[8,j,q] (4.0–6.5, n=21, f=13)[13] (3.3–8.1, n=74, f=41)[8]	49.2[8,g,h] (n=41)	20*f*, 15*m*[52,h]	—	—	55[52,g,h] (n=35)
Chimpanzee (Eastern- *P. t. schweinfurthii*)	Kibale (Kanywara)	0.29[9,e]	11.1[22,c] (n=1)	11.1[22] (n=1)	15.4 (15.0)[22] (14–18, n=5)	6.6[9] (5.9)[8,j,q] (2.4–8.2, n=21, f=17)[8]	55.0[8,g,h] (n=17)	21.6*f*, 17.1*m*[30,h] (n=123)	—	—	64[52,g,h] (n=39)
Chimpanzee (Eastern- *P. t. schweinfurthii*)	Kibale (Ngogo)	0.23*f*, 0.18*m*[52,e] (n= 86*f*, 80*m*)	—	—	—	(5.9)[8,j,q] (2.4–8.2, n=21, f=17)	55.0[8,g,h] (n=17)	35.8*f*, 29.6*m*[52,h,k] (n=86*f*, 80*m*)	—	—	66[52,g,h] (n=66)
Chimpanzee (Eastern- *P. t. schweinfurthii*)	Mahale	0.60*f*, 0.60*m*[31,e] (n= 40*f*, 44*m*)	10.2 (10.3)[32,c,d] (n=10)	10.9 (10.7)[31] (9.1–13.4, n=20)	15.6 (13.2)[32] (12–23, n=5, f=3)	6.0[9](5.75)[8,j,q] (3.7–11.0, n=116, f=62)	44.0[8,g,h] (n=62)	—	—	—	48[32,g] (n=152)
Chimpanzee (*P. t. versus*)	Taï	0.4*f*, 0.4*m*[3]	10[41,c]	—	13.7 (13.75)[3,h] (12.5–18.5, n=8, f=7)	5.75 (5.4)[3,q] (4.0–10.0, n=33, f=19)	43.0[41,g,h]	7*f*, 8*m*[52,h]	—	—	46[52,g,h] (n=15)

Chimpanzee (Eastern- *P. t. schweinfurthii*)	Mahale	0.60*f*, 0.60*m*[31,e] (n= 40*f*, 44*m*)	10.2 (10.3)[32,c,d] (n=10)	10.9 (10.7)[31] (9.1–13.4, n=20)	15.6 (13.2)[32] (12–23, n=5, f=3)	6.0[9](5.75)[8,j,q] (3.7–11.0, n=116, f=62)	44.0[8,g,h] (n=62)	—	—	—	48[32,g] (n=152)
Chimpanzee (*P. t. versus*)	Taï	0.4*f*, 0.4*m*[3]	10[41,c]	—	13.7 (13.75)[3,h] (12.5–18.5, n=8, f=7)	5.75 (5.4)[3,q] (4.0–10.0, n=33, f=19)	43.0[41,g,h]	7*f*, 8*m*[52,h]	—	—	46[52,g,h] (n=15)
Chimpanzee (*P. troglodytes* overall)	Captive	—	—	—	11.7[38] (n=214)	2.1[38] (n=1253)	—	29.4*f*, 20.6*m*[7]	—	(37.4*f*, 31.7*m*)[43]	53.4[20,g]
Human (*Homo sapiens*)	Ache	—	14.0[47]	—	17.7[17]	3.1[21]	42.1[21]	37[14]	50.4*f*, 48.7*m*[47] (n=971)	—	85[17,g]
Human (*H. sapiens*)	Hazda	—	16.0[47]	—	19.0[47]	3.2[17]	45.0[17]	34[14] (n=706)	44.0[47]	—	—
Human (*H. sapiens*)	Hiwi	—	12.6[47]	—	20.5 [47]	3.7[21]	37.8[21]	27[14] (n=375)	36.3[47]	—	—
Human (*H. sapiens*)	!Kung	—	15.5[19]	—	19.2[19]	3.4[21]	37[21]	36[14]	38.1[14]	—	—

Sources: [1]Anderson et al. 2008; [2]Asano 1967; [3]Boesch and Boesch-Achermann 2000; [4]Bronikowski et al. 2011; [5]Cocks 2007; [6]Dixson 1981; [7]Dyke et al. 1995; [8]Emery Thompson et al. 2007; [9]Emery Thompson 2013; [10]Furuichi et al. 1998; [11]Galdikas 1995; [12]Galdikas and Wood 1990; [13]Goodall 1986; [14]Gurven & Kaplan 2007; [15]Harcourt, Stewart, and Fossey 1981; [16]Harvey 1997; [17]Hill & Hurtado 1996 (Ache); [18]Hill et al. 2001; [19]Howell 1979; [20]Judge & Carey 2000; [21]Kaplan et al. 2000; [22]Knott 2001 citing Wrangham; [23]Knott, Emery Thompson, and Wich 2009; [24]Kohler, Preston, and Lackey 2006; [25]Kuroda 1989; [26]Lowenstine, McManamon, and Terio 2016; [27]Lippert 1977; [28]Markham 1990; [29]Markham 1995; [30]Muller & Wrangham 2014; [31]Nishida, Takasaki, and Takahata 1990; [32]Nishida et al. 2003; [33]Ogden & Wharton 1997; [34]Robbins et al. 2004; [35]Robbins et al. 2006; [36]Robbins et al. 2009; [37]Robson, van Schaik, and Hawkes 2006; [38]Roof et al. 2005; [39]Shumaker, Wich, and Perkins 2008; [40]Sievert, Karesh, and Sunde 1991; [41]Stumpf 2011; [42]Sugiyama 2004; [43]Survival Species Library 2018; [44]Thompson-Handler 1990; [45]Tilson et al. 1993; [46]van Noordwijk et al. 2018; [47]Walker et al. 2006; [48]Wallis 1997; [49]Watts 1991; [50]Wich et al. 2004; [51]Wich, de Vries, et al. 2009; [52]Wood et al. 2017; [53]Yamagiwa & Kahekwa 2001.

Notes: [a] Age at first menstruation, [b] Age at first swelling, [c] Age at first *full* swelling, [d] The mean and median were calculated by pooling all the original raw data from multiple studies, [e] Infant mortality before 4 years of age, [f] Infant mortality before 3 years of age, [g] Oldest known record in study, [h]Includes individuals with age estimates, [i]Survival age of 50% of individuals, [j]Excluding data from intervals with a known miscarriage or removal of infant, [k] Computed using a Kaplan-Meier analysis with censored and uncensored intervals, [l] Includes all field sites except Tanjung Puting, [o]Includes data at sites where monitoring was not continuous, [p]Age estimated to the nearest year, [q]Calculated using estimated birth dates.

Results

JUVENILE GROWTH RATE

For mammals, the first stage of life is growing to adult body size and developing primary and secondary sexual characteristics necessary for reproduction (Arendt 1997). The timing of puberty in human and nonhuman primates is intimately linked with both genetic predispositions and environmental factors, as well as being regulated by energetics (Ebling 2005). In the wild, juvenile great ape growth rate is difficult to monitor as most field studies employ noninvasive methods. However, new advances in laser photogrammetry methods (Bergeron 2007; Galbany et al. 2016; Rothman, Chapman, et al. 2008) as well as hormonal measurements (Emery Thompson et al. 2016) hold promise for future investigations of juvenile growth.

In captivity, gorillas have the largest neonates, followed by chimpanzees, orangutans, and lastly bonobos (Leigh and Shea 1996; Seitz 1969). Postnatally, gorillas grow faster than the other African apes and both sexes have earlier growth spurts than the other species. Male gorillas achieve their much larger body size by growing for longer than do either of the *Pan* species (chimpanzees and bonobos), or female gorillas. Female gorillas cease growth earlier than chimpanzees but later than bonobos. Chimpanzees and bonobos differ from each other in both the rate and duration of growth, with different patterns observed for each sex (Leigh and Shea 1996). Leigh and Shea (1995) found that captive female orangutans keep growing until an estimated 18 years of age, whereas male orangutans appear to have indeterminate growth (Leigh 1994). In contrast, female gorillas stop growing at approximately 9.5 years while males continue until 12.5–13 years. In captive female bonobos, growth ceases between 8–10 years whereas males grow until the age of 11–12 years old. In chimpanzees, females are reported to grow until 11.5–12 years and males until 12–13 years (Leigh and Shea 1995).

AGE AT PUBERTY

In mammals, puberty is defined as the period when juveniles achieve sexual maturity. More specifically, it indicates a shift in energy allocations from growth to reproduction (Charnov and Berrigan 1993; Ellison et al. 2012). The best-documented marker for the onset of puberty in human and nonhuman primates is age of menarche in females. However, menarche is less apparent and often difficult to detect in great apes in the wild. Chimpanzee and bonobo females have conspicuous sexual swellings around the time of ovulation (Wrangham 1993). From captive studies, it is understood that *Pan* females start to have sexual swellings 1–2 years prior to ovulation and menarche (Earnhardt et al. 2003). Thus, the first sexual swelling serves as a proxy for the onset of puberty in these females. However, chimpanzee and bonobo females normally emigrate to new communities after menarche. Thus, at most sites, researchers do not know when individual adult females reached menarche, as these individuals transferred into the community after menarche occurred. Nulliparous female gorillas also have sexual swellings at the time of ovulation, but theirs are significantly less prominent than those of either *Pan* species and are hard to see from any distance in the field (Watts 1991). Female orangutans do not have sexual swellings during ovulation, nor do they have copious menstrual bleeding (Ogden and Wharton 1997; Sodaro 1997). Thus, we also report on age at first sexual behavior as another comparative measure of age at puberty (table 1.1).

Timing of puberty in both male and female apes likely has effects on long-term reproductive success. Among great apes, females experience adolescent subfecundity between the onset of puberty and their first birth (Bercovitch and Ziegler 2002; Dixson 1998; Knott 2001). Within a population, individuals undergoing early development likely start ovulating earlier, conceive their first offspring earlier, and have a longer reproductive career (Ellis 2004; Emery Thompson et al. 2007; Hawkes et al. 1998). Across female great apes (table 1.1), mountain gorillas have the earliest age at menarche, averaging 7.3 years at the Karisoke research site, Rwanda (Harcourt, Stewart, and Fossey 1981), which aligns with them having the earliest reported age at first sexual activity (6.3 years) (Harcourt, Stewart, and Fossey 1981; Watts 1991). In chimpanzees, age at menarche ranges from 8.5 to 11.1 years. Bonobos have an estimated age of menarche of 8–9 years (Kuroda 1989). Age at menarche has been even more difficult to determine in wild orangutans, because of the lack of a sexual swelling, but first sexual behavior starts around 10–13 years of age (Galdikas 1995; Knott, Emery Thompson, and Wich 2009).

Less is known about the variation in the onset of puberty in male apes, although there are accounts of changes in sexual behavior and first age to sire offspring. Among great apes, male puberty is most distinguishable in gorillas. Development of the silver back and sagittal crest are two obvious secondary sexual characteristics that start developing at puberty (Robbins 1995). Behaviorally, male mountain gorillas in Karisoke, Rwanda, emigrate around 13.5 years of age and achieve dominance at an average age of 15 years (Robbins 1995). Male orangutans exhibit a rare phenomenon called bi-maturism where individuals develop secondary sexual characteristics at variable time points or not at all (Emery Thompson, Zhou, and Knott 2012). The process of developing these secondary sexual characteristics is called flanging and involves the attainment of cheek pads (flanges), a throat sac, the ability to long call, increased body size, and longer, darker hair (Knott and Kahlenberg 2011; Kuze, Malim, and Kohshima 2005). Despite variability in the timing of flange development, all males likely attain sexual maturity around the same time (Maggioncalda, Sapolsky, and Czekala 1999). Thus, puberty in orangutans cannot be ascertained from secondary sexual characteristics. Male chimpanzees and bonobos also lack observable secondary sexual characteristics (Gavan 1953). Future research on puberty in apes could highlight whether either sex experiences greater variability in the timing of puberty.

AGE AT FIRST BIRTH

Age at first birth marks the beginning of the reproductive stage of life, which continues until death in great apes. Reports of age at first birth have become more accurate with the increase in birth data at long-term great ape field sites. Additionally, genetic testing is now used to determine paternity, which will eventually allow for estimates of first successful fertilization among males (Goossens et al. 2006; Nsubuga et al. 2008; Vigilant et al. 2001). In females, this life history trait is affected by the duration of puberty, menstrual status (ovulating vs. amenorrheic), and mating opportunities. For males, age at which fertilization is possible is affected by the timing of spermatogenesis and availability of mating opportunities for young adult males. Thus, variation in age at first birth emphasizes both social and physiological challenges that individuals face when starting the reproductive stage of life.

Among great apes, age at first birth for females follows the same pattern as other development variables (table 1.1). Gorillas have the earliest births, at about 9.8 years of age in the wild (Harcourt, Stewart, and Fossey 1981; Watts 1991). Although female gorillas normally transfer groups at puberty, these data come from Karisoke, Rwanda, where females of known birthdate often transfer to another habituated group, thus ages are likely quite accurate (Robbins et al. 2009). Orangutan females are philopatric, so first birth can be ascertained accurately for some individuals, averaging 14.4 years across wild sites (van Noordwijk et al. 2018). Age at first birth for chimpanzees and bonobos is less certain as these species typically transfer at puberty. Robson, van Schaik, and Hawkes (2006) estimate the average age at first birth for chimpanzees across most sites to be 13.3 years, although the range is quite broad (table 1.1). Kuroda (1989) estimates the average age at first birth for bonobos at the Wamba research site, Democratic Republic of Congo (DRC), as 14.2 years.

INTER-BIRTH INTERVAL

Inter-birth interval (IBI) is one of the better-documented life history variables for great ape field sites, as it can be recorded for all mothers regardless of their chronological age being known. IBI refers to the timing between sequential births and is generally the duration of time invested in a single offspring along with the time needed to return to positive energy balance and resume ovulation (Emery Thompson, Muller, and Wrangham 2012). Orangutans have a mean IBI of 7.8 years, the longest of any great ape as well as any mammal (Galdikas and Wood 1990; van Noordwijk et al. 2018). Although earlier reports suggested that Sumatran orangutans had significantly longer IBIs than Bornean orangutans (Wich et al. 2004), the addition of more data reveals no significant inter-island difference (van Noordwijk et al. 2018). In contrast, the average IBI for mountain gorillas at Karisoke, Rwanda, is 4.0 years, nearly half that of orangutans (Robbins et al. 2006). Data from western gorillas are limited, but initial reports based on six observed IBIs from the Lossi and Mbeli Bai sites, both in the Republic of Congo, suggest slightly longer IBIs of around 5 years (Robbins et al. 2004). Chimpanzees and bonobos have intermediate IBI values, with an average of 5.9 years across all chimpanzee sites and 4.8 years for bonobos in Wamba, DRC (table 1.1). Sample sizes for orangutans, chimpanzees, and mountain gorillas are now quite large and thus there is confidence that this variation reflects real species

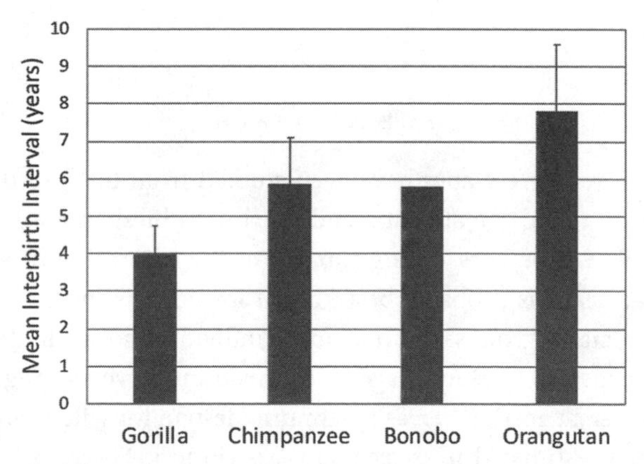

FIGURE 1.1. Interbirth intervals, in years, of great apes in the wild. Means are shown for gorillas (Robbins et al. 2006), bonobos (Furuichi et al. 1998), and orangutans (van Noordwijk et al. 2018), medians for chimpanzees (Emery Thompson et al. 2007). Standard deviations are shown where reported.

differences (fig. 1.1). Gestation length does not appear to greatly differ between ape genera though (Ogden and Wharton 1997; Sodaro 1997). Thus, variation in IBIs is likely derived from differences in the duration of lactation, and associated amenorrhea, as well as maternal energetic status related to their feeding ecology (Knott, Emery Thompson, and Wich 2009).

AGE AT LAST BIRTH

Menopause is considered a life history trait of humans that is rarely experienced by nonhuman great apes. Reproductive viability is apparently sustained for a greater proportion of the lifespan for ape females compared to human females (Walker and Herndon 2008), although evidence from the wild is still limited. Age at last birth has yet to be reported for gorillas, bonobos, or most orangutan sites. One Sumatran orangutan site, Ketambe, estimates last birth at 41 years of age (Wich et al. 2004). In captive orangutans, ovarian hormones showed no decrease in females over the age of 35 years, including two females who were sampled 15 years earlier (Durgavich, Knott, and Emery Thompson 2012). In chimpanzees, age at last birth has been estimated to be as old as 55 years of age (Emery Thompson et al. 2007). Analysis of ovarian hormone function shows that wild female chimpanzee fertility declines in pace with general health and that healthy females are able to maintain high birth rates late into life, showing no evidence of menopause (Emery Thompson et al. 2007). However, recent reports from the Ngogo research site, in Kibale Forest, Uganda, suggest that at least six female chimpanzees have lived 10 years or more after the birth of their last offspring, without having reproduced again (Wood et al. 2017).

AGE AT DEATH

Relatively few wild great apes have been studied from birth to death. Thus, life expectancy, mean age at death, and maximum lifespan are largely based on individuals whose ages were estimated at the beginning of a long-term study. Age at death is probably best known for gorillas in the wild because they live in tractable groups. The maximum individual gorilla lifespan at Karisoke was recorded as 39 years of age for a female and 35 years of age for a male (Bronikowski et al. 2011), whereas maximum lifespan for wild orangutans and chimpanzees is estimated to exceed 50 years (Emery Thompson et al. 2007; Wich et al. 2004; Wood et al. 2017). Thus, in the wild, gorillas appear to have shorter lifespans than the other great apes. In wild orangutans, Wich et al.

(2004) report a maximum lifespan at the Ketambe site in Sumatra of 53 years for a female and 58 years for a male. But these are estimated ages, particularly for males (Wich et al. 2004), and it is highly unlikely that average male lifespan is longer than female lifespan, in this highly dimorphic species with high rates of male-male competition. Thus, data are still too limited for orangutans and bonobos to determine whether mean or maximum lifespans in the wild differ from the better-studied chimpanzees.

In captivity, some individual apes, including gorillas, can reach ages of over 50 years. Animals in captivity are protected from predation, have carefully monitored diets, and receive regular medical care, all factors that would be expected to increase longevity. However, captivity also introduces other risk factors, such as a propensity toward obesity for some species and less exercise than they would get in the wild (Wich, Shumaker, et al. 2009). Cardiac disease is a commonly reported cause of morbidity and mortality in all great apes in captivity (Strong et al. 2016). Also, stress may play an important role in captive ape health (although wild apes, of course, also experience stress) (Hosey 2008; Morgan and Tromborg 2007). Thus, various factors may account for the lower mean life expectancy of orangutans and gorillas in captivity compared to the wild (table 1.1). These data, however, reflect a variety of husbandry practices over the last 80 or more years, and modern and improved care in captivity have led to longer lifespans during the more recent decades (Wich, Shumaker, et al. 2009).

Discussion

All of the life history variables we have described point to a grade shift between the different genera of great apes. Gorillas have the fastest life history consisting of the fastest growth rates, earliest ages at maturity, shortest IBIs, and earliest deaths. Orangutans stand out for their exceptionally long life history with slow development, late age at puberty, late age at first birth, longest IBIs, and a maximum lifespan of over 50 years. Chimpanzees and bonobos fall in between gorillas and orangutans in their life history pace, with the possible exception of maximum lifespan, which seems to be similar to orangutans. Data on wild bonobos is very limited, thus differences between the two *Pan* species are hard to assess. Thus far, bonobo life history parameters seem to be at the faster end of the chimpanzee distribution. Below, we present additional ecological data to argue that this grade shift between the great apes reflects differences in energy availability in their environments that have led to evolved differences in life histories.

EXTRINSIC ADULT MORTALITY

The dominant life history paradigm states that the length of the juvenile period is set by the level of extrinsic (or unavoidable) adult mortality (Charnov 1991, 1993; Charnov and Berrigan 1990, 1993). Charnov and Berrigan (1993) demonstrate that species with long life expectancies can "afford" a long juvenile period. Species with high extrinsic mortality are predicted to have early ages at reproduction, invest less energy in maintenance, senesce faster, and have a shorter maximum lifespan (Kirkwood and Austad 2000; Shattuck and Williams 2010). This explanation seems to explain why primates, as a taxonomic group, have delayed life histories compared to other mammals (Charnov and Berrigan 1993). However, differences in extrinsic adult mortality do not seem to explain life history differences amongst great apes.

In the wild, gorillas have the shortest life expectancies and fastest overall life history. Charnov's model (1993) would suggest that gorillas should have greater extrinsic mortality than do the other great apes, followed by chimpanzees, bonobos, and orangutans, respectively. Three sources of mortality have been suggested for great apes: infant mortality, predation (including hunting), and disease (Fay et al. 1995; Ryan and Walsh 2011; Watts 1989; Wilkie, Sidle, and Boundzanga 1992). Chimpanzees and gorillas show similar levels of infant mortality with most sites ranging 20–29%, but there are several outliers (65% for gorillas in Mbeli Bai, 60% for chimpanzees in Mahale, and 40% for chimpanzees in Taï) (table 1.1). Bonobos have a reported infant mortality of 18% at Wamba (Furuichi et al. 1998). In contrast, orangutans stand out for having extremely low infant mortality at all sites (van Noordwijk et al. 2018).

Due to their large body sizes, great apes are protected against many sources of predation experienced by smaller animals. The African leopard has been known to kill gorillas, chimpanzees, and bonobos with some frequency (D'Amour, Hohmann, and Fruth 2006; Robbins et al. 2004). In Borneo and Sumatra, there are only two suspected cases of the much smaller Sunda clouded leopard, weighing 11–16 kg (Grassman et al. 2005), attacking a wild orangutan: a juvenile at one site (Kanamori et al. 2012) and an adult female at another site, who was traveling on the ground after habitat loss (Marzec et al. 2016). Sumatran orangutans seem to avoid their other big cat predator, the Sumatran tiger, by being fully arboreal (van Schaik 1999). Thus, orangutans appear to be more protected from predation due to the lack of large predators and their arboreal lifestyle.

Disease is often categorized as a negative effect of sociality (Nunn et al. 2008). Disease outbreaks have killed a substantial number of chimpanzees and gorillas at different time points (Leendertz, Pauli, et al. 2006, see also

Chapman et al., chapter 25 this volume), with many of these cases being associated, either directly or indirectly, with human contact (Köndgen et al. 2008; Dunay et al. 2018). In chimpanzees, deaths have been shown to be caused by respiratory viruses (Emery Thompson et al. 2018; Kaur et al. 2008; Lonsdorf et al. 2018; Negrey et al. 2019; Patrono et al. 2018; Scully et al. 2013; Williams et al. 2008), Ebola (Formenty et al. 1999; Georges et al. 1999; Leroy et al. 2004), anthrax (Leendertz et al. 2004), and possibly polio (Goodall 1983). Similarly, for gorillas, respiratory viruses (Grutzmacher et al. 2016), Ebola (Walsh et al. 2003), scabies (Kalema-Zikusoka, Kock, and Macfie 2002), and anthrax (Leendertz, Lankester, et al. 2006) have been verified as causes of death, while measles (Ferber 2000) has been suspected. Less is known about the prevalence of disease among bonobos (Inogwabini and Leader-Williams 2012), but respiratory viruses are also a known cause of mortality (Grutzmacher et al. 2018). Such disease outbreaks have not yet been reported for any wild orangutan population. Thus, there is evidence that mortality resulting from disease is greater in African apes, who are more social, compared to orangutans (Wich, Shumaker, et al. 2009). Orangutans lead semi-solitary lives, which likely results in the lowest disease exposure among the great apes, although ex-captive populations, living at higher densities, can experience higher level of disease (Kilbourn et al. 2003).

With the exception of orangutans, differences in predation, disease, and infant mortality do not appear to account for the differences in the pace of life history between the great apes (van Schaik and Isler 2012; Watts 2012). Orangutans, with their high rates of survivorship in the wild, do appear to have lower extrinsic mortality than the African apes. However, the data do not support the conclusion that gorillas have higher extrinsic mortality than do chimpanzees or bonobos.

ECOLOGICAL VARIABILITY AND RISK AVERSION

In animals, such as great apes, where extrinsic mortality is low and many individuals die of old age, alternative explanations are needed to understand what factors are most important in determining life history. Whereas most sources of extrinsic mortality (i.e., predation, disease, infant mortality) are thought to select for faster life histories, one type of extrinsic mortality, the risk of starvation, has been shown to extend lifespan (Kirkwood and Austad 2000). Periods of famine, or experimentally induced low food intake, lead to a longer lifespan in a variety of animals (Austad 1993; Kirkwood and Austad 2000; Shanley and Kirkwood 2000; Weindruch and Walford 1988), often accomplished by metabolic changes. We propose that the relative risk of starvation is a major factor in determining great ape life history pace as a form of ecological risk aversion (Leigh and Shea 1995, 1996; Watts 2012).

Janson and Van Schaik (1993) proposed the "Juvenile Risk Aversion" hypothesis predicting that animals inhabiting environments of high ecological risk have been selected to have slower growth rates to avoid the risk of starvation and death during low food periods. Juvenile risk aversion predicts slower growth rates, and a slower life history, when ecological risk is high, as slower growth is less metabolically costly. Developing juveniles may buffer their ecological risk by having slower growth rates (Janson and van Schaik 1993). This concept can be extended to explain more than just the duration of juvenility, though, and can help us also understand trade-offs between reproduction and maintenance during other life history stages. For example, Knott (2001) proposed the concept of "ecological energetics" to understand what determines variability in reproductive parameters between, as well as within, the apes. Energetics, which includes energy intake and expenditure, is the pathway through which ecological risk is manifested in an animal's physiology. Adults, who are no longer growing, can buffer this risk by lowering their reproductive costs and spreading their reproductive investment over a longer period. In females, this means lengthening IBI by delaying the next conception through lowering ovarian function when ecological risk is high (Ellison et al. 1993; Knott 2001).

How then can we quantify ecological risk? Janson and van Schaik (1993) argue that folivores may grow faster than frugivores because they face less ecological risk, as leaves are typically more abundant and predictable than fruit. Leigh (1994) has also shown that folivorous primates grow faster than frugivorous ones. The diets of the great apes largely consist of fruit and leaves as well as other resources like insects, bark, small prey, and extracted plant parts. However, the variation in fruit consumption among the great apes is appreciable. Eastern and western lowland gorillas consistently eat the most herbaceous diets, albeit there are differences in fruit consumption between the two species (Doran-Sheehy et al. 2009; Watts 1996). Orangutans usually incorporate fruit in their diet, but these values can be as low as 10% (Morrogh-Bernard et al. 2009). Chimpanzee and bonobo diets are generally considered higher quality in comparison to other great apes as a result of the greater percentage of fruit they consume (Watts 2012; Wrangham, Conklin-Brittain, and Hunt 1998). As the most folivorous great ape species, gorillas do indeed have the overall fastest life history. However, the much more frugivorous *Pan* species do *not* have slower life histories than orangutans. Thus, the folivory-frugivory distinction is incomplete as an explanation for life history variation between the great apes.

We propose that a more informative way to quantify differences in ecological risk is the degree of monthly variation in caloric and nutrient intake.

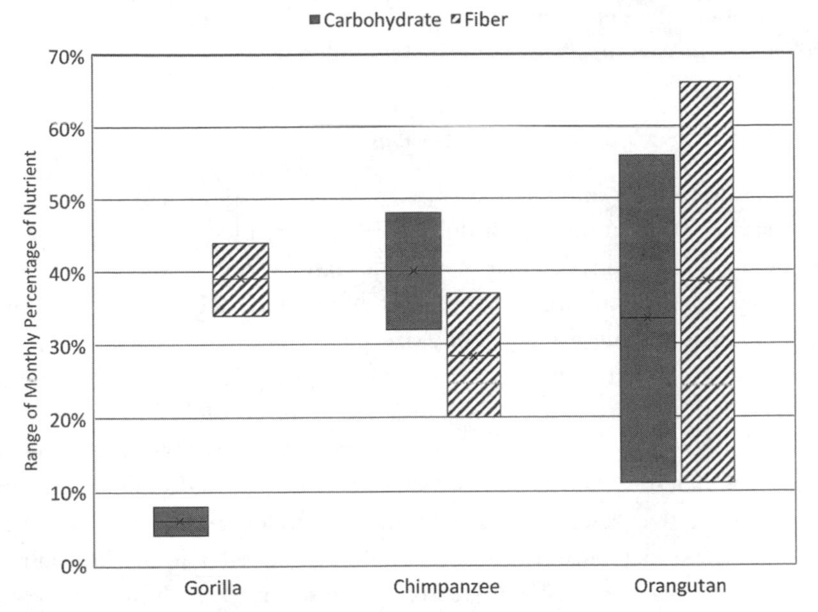

FIGURE 1.2. Range of the monthly percentage of carbohydrates (solid) and fiber (hashed) in wild gorilla diet (Rothman, Dierenfeld, et al. 2008) and chimpanzee and orangutan diets (Conklin-Brittain et al. 2006).

Figure 1.2 shows comparative data from three great ape sites on variation in monthly dietary intake of fiber and carbohydrates for gorillas, chimpanzees, and orangutans. Fiber is normally considered an indicator of a low-quality diet and low-calorie intake, whereas carbohydrates are normally indicative of a high quality, high calorie diet (Marshall and Wrangham 2007). Data on nutrient intake in these species reveal that gorillas show little monthly variation in the energetic and nutritional content of their diet, with fiber ranging 34–44% and carbohydrate ranging 4–8% (Rothman, Dierenfeld, et al. 2008). Orangutans, in contrast, show dramatic fluctuations in diet, with carbohydrate intake ranging 11–56% and fiber ranging 11–66% (Conklin-Brittain, Knott, and Wrangham 2006). Orangutans, especially Bornean orangutans, likely experience the greatest fluctuation in fruit availability, with fruit comprising anywhere from 10 to 100% of their monthly diet (Knott 1998; Morrogh-Bernard et al. 2009). Chimpanzees show intermediate levels of nutrient and caloric variation, with fiber and carbohydrates ranging 20–37% and 32–48%, respectively (Conklin-Brittain, Knott, and Wrangham 2006). These data on the degree of fluctuations in nutrient and energy intake correspond to the pace of life history observed in the great apes. Below, we examine what is known about the degree of ecological risk in each of the great ape taxa and consider how well this predicts life history differences.

Gorillas

Gorillas, particularly mountain gorillas, live in an environment with little ecological risk (Watts 2012). This is due to their dietary niche of eating a largely ubiquitous herbaceous diet that shows little seasonal fluctuation (Watts 1984). Leigh and Shea (1996) argue that the fast growth rates seen in gorillas allow them to rapidly achieve a large body size and the gut maturation needed for fermentation of a high fiber diet. Fast juvenile growth leads to earlier age at puberty and first birth. Fast growth may also be responsible for earlier mortality in wild gorillas. Within-species comparisons show that faster growth leads to a trade-off of lower viability later on in life (Metcalfe and Monaghan 2003). Genes that favor faster growth may have pleiotropic effects that result in earlier senescence (Kirkwood and Rose 1991; Metcalfe and Monaghan 2003). Thus, heavy investment in early growth seems to result in more rapid aging, or less investment in maintenance, and thus earlier death.

Gorillas can also afford to have faster IBIs than the other apes. Gorillas are not often food limited and likely have greater regularity in their daily diet compared to the other great apes. We predict the habitat of gorillas is able to provide enough food resources for females to optimize their reproductive output and spend as little time as possible amenorrheic between births. The shorter IBIs found in gorillas, compared to the other apes, also fits well with predictions based on "ecological energetics." In mountain gorillas, Robbins et al. (2006) found that females had significantly longer IBIs for their first birth (4.4 years) compared to their second (3.6 years). Robbins et al. (2006) attribute this to longer time needed to recoup maternal condition after birth for primiparous mothers, some of whom may not have been fully mature at the time of conception. Interestingly, preliminary data indicates that western gorillas, with greater fruit consumption, have longer IBIs on average (Robbins et al. 2004). Because of the incorporation of more fruit, western gorillas should face greater fluctuations in resource availability and thus may be expected to have somewhat slower life history than do mountain gorillas, with a possible later age at first birth and longer IBIs (Doran and McNeilage 2001; Robbins et al. 2004).

Orangutans

Many rainforests where orangutans live are characterized by mast fruiting, dramatic fluctuations in flower and fruit availability (Marshall et al. 2009).

Orangutans seem to have a number of adaptations to cope with these extreme fluctuations in energy and nutrient availability. These include fat storage during fruit-rich times to subsist on during fruit-poor times, large body size as a buffer against starvation, and the lowest metabolic rate of any mammal (Knott 1998; Pontzer et al. 2010; Wheatley 1987). Thus, orangutans live in an environment of pronounced ecological risk, which we argue has selected for a suite of slow life history features.

As shown by Leigh and Shea (1995), orangutan juveniles have the slowest growth of any primate. This leads them to have lower energetic requirements than other apes, an additional mechanism that buffers them from fluctuations in energy intake. The extended juvenile period observed in orangutans may be necessary for ecological risk aversion (Janson and van Schaik 1993; van Noordwijk and van Schaik 2005). The long IBI of orangutans could also be interpreted as an energy-saving adaptation that buffers maternal energy reserves. Ovarian function in orangutans responds to changes in energy balance, with orangutans more likely to conceive when they are in positive energy balance (Knott, Emery Thompson, and Wich 2009). Due to variability in food availability and the cost of arboreal locomotion, orangutans may have evolved a longer period of juvenile dependence and a slower overall life history to avoid risks associated with their lifestyle and habitat (van Noordwijk and van Schaik 2005; van Noordwijk et al. 2018).

Chimpanzees and Bonobos

Chimpanzees and bonobos show intermediate levels of ecological risk compared to gorillas and orangutans, and have an intermediate paced life history. Chimpanzees are probably the most specialized in eating ripe fruit, spending on average more than 50% of their feeding time on fruits (Kaplan et al. 2000; Wrangham, Conklin-Brittain, and Hunt 1998). Between and within chimpanzee populations we know that food availability is a major determinant of IBI and reproductive success (Emery Thompson et al. 2004; Pusey, Williams, and Goodall 1997). In comparison to chimpanzees, bonobos incorporate more leaves and pith into their daily diet (Malenky et al. 1994). Between the two species, bonobos likely experience less ecological risk due to their greater reliance on terrestrial herbaceous vegetation (Chapman, White, and Wrangham 1994). Bonobos are also predicted to have larger average group sizes than do chimpanzees as a result of lowered foraging competition (Chapman, White, and Wrangham 1994).

Comparing the life history pace of the two species is difficult. Because of political turmoil in the Democratic Republic of Congo, sufficient data have been collected only on wild bonobos at Wamba, even though other sites exist. Among chimpanzee sites, there are many inter-population differences, which may reflect disparities in ecological risk. For instance, the Kanyawara and Mahale populations of chimpanzees, both in Tanzania, have the oldest averages for age at first birth and longest averages for IBIs (table 1.1). These two populations have seemingly "slower" life histories overall than the bonobos at Wamba, Democratic Republic of Congo, whereas other chimpanzee populations, such as Bossou, exhibit a "faster" pace than Wamba bonobos (table 1.1). While we have characterized both *Pan* species as having intermediate-paced life histories, future studies on food availability and updates to life history trait values will likely be insightful in discerning differences between species.

CAPTIVE GREAT APE COMPARISONS AND SELECTION FOR PLASTICITY

Captive apes benefit from having a reliable and more than adequate food supply, little need to expend energy, low disease load, and regular veterinary care when compared to their wild counterparts. Thus, they are relatively free from ecological risk. Many life history parameters in captivity are highly influenced by captive management (Anderson et al. 2008; Shumaker, Wich, and Perkins 2008). Nevertheless, studies of captive apes reveal the degree of plasticity in some life history features. Apes have faster growth rates in captivity, and when it can be determined, reach menarche at an earlier age. IBIs are highly controlled by captive management, yet when babies are allowed to stay with their mothers and contraception is not administered to the mother, they still have shorter IBIs than in the wild (Shumaker, Wich, and Perkins 2008). Overall, orangutans exhibit the greatest variance in these life history features when comparing captive and wild data, whereas gorillas show the least variation. Bonobos and chimpanzees are intermediate between these two extremes in the amount of variation they display between wild and captive settings (table 1.1).

Species differences in the degree of plasticity are probably best demonstrated by age at first birth for female great apes. Orangutans have their first births, on average, at 11.7 years of age in captivity (Cocks 2007; Lippert 1974), but individuals have been known to reproduce as early as 7 years of age (Shumaker, Wich, and Perkins 2008). This is compared to the wild where the mean age of first birth is 14.4 years with a range of 12.7 to 16.3 years (van Noordwijk et al. 2018). Age at first birth in captivity has been calculated as 11.7 years for chimpanzees (Roof et al. 2005) compared to 13.3 years in the wild (Robson, van Schaik, and Hawkes 2006). Bonobos have a reported age at first birth of

10.8 years in captivity (Kuroda 1989; Thompson-Handler 1990) as opposed to 14.2 years in the wild (Kuroda 1989). Thus, captivity appears to decrease age at first birth for all great ape species, except possibly gorillas (Knott 2001). Captive female gorillas have a mean age at first birth of 9.3 years (Dixson 1981), remarkably similar to the mean of 9.8 years at the Karisoke field site, Rwanda (Robbins et al. 2009). This may be because wild gorillas are already reproducing as early as they can in the wild (Knott 2001). The earlier age at first birth in captivity of chimpanzees, bonobos, and especially orangutans indicates a high degree of phenotypic plasticity. Although age at first birth in captivity is impacted by captive management decisions (Shumaker, Wich, and Perkins 2008), the captive data demonstrate that these ape species are *capable* of reproducing at much earlier ages in captivity than they are able to in the wild, indicating that they mature physically at a much earlier age, including having earlier ages of menarche. This supports the interpretation that under conditions of abundant food, life history for individuals is accelerated. Furthermore, it aligns with the general prediction that animals experiencing highly variable environments will evolve greater plasticity, allowing them to adapt their growth and reproduction depending on the circumstances (Kirkwood and Austad 2000; Stearns and Koella 1986).

HUMANS

Human life history presents an anomaly. In other mammals, including great apes, the rate of offspring production reflects the overall species growth rates; species with faster growth rates having shorter IBIs (Charnov 1991; Charnov and Berrigan 1993). This is not the case for humans, who have the slowest growth rates, latest ages at first reproduction, and longest life spans, yet have the shortest IBIs among the great apes (table 1.1). Thus, additional factors are needed to understand the evolution of human life history. One factor selecting for an overall slower life history may be the energetic demands of a large brain. Isler and van Schaik (2009) propose that in humans the requirements of having a large brain may have selected for tradeoffs in growth and reproduction. Thus, based on their other life history variables, we would have expected humans to have even longer IBIs than orangutans, instead of the shortest IBIs amongst the great apes. This shortening of the human IBI can be attributed to both changes in human social structure and the advent of food sharing, as well as the way we obtain and process our food (Hrdy 2009; Wrangham 2009). These two changes have allowed humans to decrease their maternal energetic burdens, leading to shorter inter-birth intervals in humans compared to the other great apes (Knott 2001).

Future Directions

Our review emphasizes the importance of ecology in structuring a species' life history, which points to the problem of using single factors, such as body size or phylogenetic relatedness, to extrapolate about extinct hominins. All the living great apes occupy a unique socio-ecological niche with a suite of life history adaptations. Body size alone, for example, does not tell us about diet and life history in these closely related species. The life history of chimpanzees is often used to predict and model life history traits throughout the hominin lineage (de León et al. 2008; Hawkes et al. 1998; Kaplan et al. 2000; Leigh 2004). While the human-chimpanzee comparison provides insight regarding the uniqueness of human life history, the ecology of each species likely plays a more important role in shaping life history than does their phylogenetic position. As we obtain a better understanding of the relationships between diet, resource availability, body size, mortality, and life history in these living species it will allow for more informed models about the past lives of extinct hominins. The relationship that we argue here between environmental variability, ecological risk aversion, and phenotypic plasticity in the great apes lends support to the hypothesis that ecological variability selected for increased phenotypic plasticity during human evolution (Potts 1998; Vrba 1985).

Despite the accumulation of values for table 1.1, there are numerous life history traits still unknown in captive and wild settings alike. First, we found that several variables do not have a standard for how they are recorded or analyzed. We were particularly interested in life expectancy at maturity, but there was not a standard way to report this, which is compounded by variation amongst species in the age at which maturity is achieved. Thus, we reported on life expectancy at birth, since this was often included in findings from life tables. Similarly, we chose not to report data on age at weaning as there was not a clear consensus on how this is defined. Some studies recorded age at weaning as the mother's first rejection of the offspring while others looked at the complete cessation of breastfeeding or separation of the mother-offspring pair (Lee 1996). Additionally, puberty has not been thoroughly investigated in great apes. For humans, Marshall and Tanner (1969, 1970) described the sex-specific stages of puberty and the age ranges for when these developmental changes take place. Comparable studies have not been conducted for great ape species, likely as a result of the difficult nature of the research. Continued hormonal studies documenting the development of ovarian function in both captive and wild juveniles and adolescent apes will help elucidate the impact of energetics on age at puberty, separate from any human manipulation of breeding. Studies on the age of puberty in males are especially sparse. One way to record age of puberty in male great apes is to determine testosterone level from urine samples. Such hormonal studies can be time consuming and often difficult to execute but will lead to a greater understanding of life history trait variation among the great apes. Additionally, as genetic analyses add data on paternity, we will be able to determine the age at which individual males conceive their first offspring. These gaps in knowledge will guide the direction of future captive and field great ape research on life histories. Both the continuation of data collection at long-term field sites and captive environments, as well as the development of new methods to examine unanswered questions, are necessary to refine our understanding of great ape life histories.

Future comparative studies of wild and captive great apes should investigate the impact that food availability may have on growth trajectories. Captive great apes usually experience puberty at younger ages, possibly indicative of faster growth resulting from greater food availability (Ogden and Wharton 1997; Shumaker, Wich, and Perkins 2008). Captive and wild inter-species comparisons can also reveal not only how these variables differ between species, but how phenotypically plastic they are within each species (Knott, Emery Thompson, and Wich 2009). Hormonal measurements of the onset of puberty can document differences in the approximate duration of juvenility, but investigations of actual growth rates are necessary to establish species differences in the timing of peak growth velocity and sensitivity to food availability in the environment. For long-term great ape field sites, it will be important to establish or continue methodologies, such as laser photogrammetry and hormonal analysis, that allow for longitudinal studies of growth. As data collection becomes more standardized and technology improves the accuracy of data, we expect that the findings presented here will change slightly. We anticipate the ever-growing accumulation of data at long-term great ape field sites will allow for more complete findings relating to *all* life stages and more informed predictions pertaining to variation in life histories.

References

Anderson, H. B., M. Emery Thompson, C. D. Knott, and L. Perkins. 2008. "Fertility and mortality patterns of captive Bornean and Sumatran orangutans: Is there a species difference in life history?" *Journal of Human Evolution* 54 (1): 34–42.

Arendt, J. D. 1997. "Adaptive intrinsic growth rates: An integration across taxa." *Quarterly Review of Biology* 72 (2): 149–77.

Asano, M. 1967. "A note on the birth and rearing of an orang-utan *Pongo pygmaeus* at Tama Zoo, Tokyo." *International Zoo Yearbook* 7 (1): 95–96.

Austad, S. N. 1993. "Retarded senescence in an insular population of opossums." *Journal of Zoology* 229: 695–708.

Bercovitch, F. B, and T. E. Ziegler. 2002. "Current topics in primate socioendocrinology." *Annual Review of Anthropology* 31 (1): 45–67.

Bergeron, P. 2007. "Parallel lasers for remote measurements of morphological traits." *Journal of Wildlife Management* 71 (1): 289–92.

Boesch, C., and H. Boesch-Achermann. 2000. *The Chimpanzees of the Tai Forest: Behavioural Ecology and Evolution*. New York: Oxford University Press.

Bronikowski, A. M., J. Altmann, D. K. Brockman, M. Cords, L. M. Fedigan, A. Pusey, T. Stoinski, W. F. Morris, K. B. Strier, and S. C. Alberts. 2011. "Aging in the natural world: Comparative data reveal similar mortality patterns across primates." *Science* 331 (6022): 1325–28.

Chapman, C. A., F. J. White, and R. W. Wrangham. 1994. "Party size in chimpanzees and bonobos." In *Chimpanzee Cultures*, edited by R. W. Wrangham, W. C. McGrew, F. B. M. de Waal, and P. G. Heltne, 41–58. Cambridge, MA: Harvard University Press.

Charnov, E. L. 1991. "Evolution of life history variation among female mammals." *Proceedings of the National Academy of Sciences* 88: 1134–37.

Charnov, E. L. 1993. *Life History Invariants*. Oxford: Oxford University Press.

Charnov, E. L., and D. Berrigan. 1990. "Dimensionless numbers and life history evolution: Age of maturity versus the adult lifespan." *Evolutionary Ecology* 4 (3): 273–75.

Charnov, E. L., and D. Berrigan. 1993. "Why do female primates have such long lifespans and so few babies? Or life in the slow lane." *Evolutionary Anthropology* 1: 191–94.

Cocks, L. 2007. "Factors affecting mortality, fertility, and well-being in relation to species differences in captive orangutans." *International Journal of Primatology* 28 (2): 421–28.

Conklin-Brittain, N. L., C. D. Knott, and R. W. Wrangham. 2006. "Energy intake by wild chimpanzees and orangutans. Methodological considerations and a preliminary comparison." In *Feeding Ecology in Apes and Other Primates*, edited by G. Hohmann, M. Robbins, and C. Boesch, 445–71. Cambridge: Cambridge University Press.

D'Amour, D. E., G. Hohmann, and B. Fruth. 2006. "Evidence of leopard predation on bonobos (*Pan paniscus*)." *Folia Primatologica* 77 (3): 212–17.

de León, M. S. P, L. Golovanova, V. Doronichev, G. Romanova, T. Akazawa, O. Kondo, H. Ishida, and C. P. E. Zollikofer. 2008. "Neanderthal brain size at birth provides insights into the evolution of human life history." *Proceedings of the National Academy of Sciences* 105 (37): 13764–68.

Dixson, A. 1981. *The Natural History of the Gorilla*. New York: Columbia University Press.

Dixson, A. 1998. *Primate Sexuality: Comparative Studies of the Prosimian, Monkeys, Apes, and Human Beings*. New York: Oxford University Press.

Doran, D., and A. McNeilage. 2001. "Subspecific variation in gorilla behavior: The influence of ecological and social factors." In *Mountain Gorillas: Three Decades of Research at Karisoke*, 123–49. Cambridge: Cambridge University Press.

Doran-Sheehy, D., P. Mongo, J. Lodwick, and N. L. Conklin-Brittain. 2009. "Male and female western gorilla diet: Preferred foods, use of fallback resources, and implications for ape versus Old World monkey foraging strategies." *American Journal of Physical Anthropology* 140 (4): 727–38.

Dunay, E., K. Apakupakul, S. Leard, J. L. Palmer, and S. L. Deem. 2018. "Pathogen transmission from humans to great apes is a growing threat to primate conservation." *EcoHealth* 15 (1): 148–62.

Durgavich, L. S., C. D. Knott, and M. Emery Thompson. 2012. "Captive female orangutans do not exhibit hormonal signs of age-related reproductive decline." *American Journal of Physical Anthropology* 147 (54): 136.

Dyke, B., T. B. Gage, P. L. Alford, B. Swenson, and S. Williams-Blangero. 1995. "Model life table for captive chimpanzees." *American Journal of Primatology* 37 (1): 25–38.

Earnhardt, J. M., S. R. Ross, E. V. Lonsdorf, and A. E. Pusey. 2003. "A demographic comparison of wild chimpanzees from Gombe and a managed population from North American zoos." *American Journal of Primatology* 60: 62–63.

Ebling, F. J. P. 2005. "The neuroendocrine timing of puberty." *Reproduction* 129 (6): 675–83.

Ellis, B. J. 2004. "Timing of pubertal maturation in girls: An integrated life history approach." *Psychological Bulletin* 130 (6): 920.

Ellison, P. T., C. Panter-Brick, S. F. Lipson, and M. T. O'Rourke. 1993. "The ecological context of human ovarian function." *Human Reproduction* 8 (12): 2248.

Ellison, P. T., M. W. Reiches, H. Shattuck-Faegre, A. Breakey, M. Konecna, S. Urlacher, and V. Wobber. 2012. "Puberty as a life history transition." *Annals of Human Biology* 39 (5): 352–60.

Emery Thompson, M. 2013. "Comparative reproductive energetics of human and nonhuman primates." *Annual Review of Anthropology* 42: 287–304.

Emery Thompson, M., J. H. Jones, A. E. Pusey, S. Brewer-Marsden, J. Goodall, D. Marsden, T. Matsuzawa, T. Nishida, V. Reynolds, Y. Sugiyama, and R. W. Wrangham. 2007. "Aging and fertility patterns in wild chimpanzees provide insights into the evolution of menopause." *Current Biology* 17 (24): 2150–56.

Emery Thompson, M., Z. P. Machanda, E. J. Scully, D. K. Enigk, E. Otali, M. N. Muller, T. L. Goldberg, C. A. Chapman, R. W. Wrangham. 2018. "Risk factors for respiratory illness in a community of wild chimpanzees (*Pan troglodytes schweinfurthii*)." *Royal Society Open Science* 5 (9): 180840.

Emery Thompson, M., M. N. Muller, K. Sabbi, Z. P. Machanda, E. Otali, and R. W. Wrangham. 2016. "Faster reproductive rates trade off against offspring growth in wild chimpanzees." *Proceedings of the National Academy of Sciences* 113 (28): 7780–85.

Emery Thompson, M., M. N. Muller, and R. W. Wrangham. 2012. "The energetics of lactation and the return to fecundity in wild chimpanzees." *Behavioral Ecology* 23 (6): 1234–41.

Emery Thompson, M., R. W. Wrangham, V. Reynolds, and A. E. Pusey. 2004. "Natural variation in ovarian function in East African chimpanzees (*Pan troglodytes schweinfurthii*): Noninvasive hormonal assessment in females from three study populations." *American Journal of Primatology* 62 (S1): 122.

Emery Thompson, M., A. Zhou, and C. D. Knott. 2012. "Low testosterone correlates with delayed development in male orangutans." *PLoS One* 7 (10): 1–7.

Fay, J. M., R. Carroll, J. C. Kerbis Peterhans, and D. Harris. 1995. "Leopard attack on and consumption of gorillas in the Central African Republic." *Journal of Human Evolution* 29 (1): 93–99.

Ferber, D. 2000. "Human diseases threaten great apes." *Science* 289 (5483): 1277–78.

Formenty, P., C. Boesch, M. Wyers, C. Steiner, F. Donati, F. Dind, F. Walker, and B. Le Guenno. 1999. "Ebola virus outbreak among wild chimpanzees living in a rain forest of Côte d'Ivoire." *Journal of Infectious Diseases* 179 (S1): 120–26.

Furuichi, T., G. Idani, H. Ihobe, S. Kuroda, K. Kitamura, A. Mori, T. Enomoto, N. Okayasu, C. Hashimoto, and T. Kano. 1998. "Population dynamics of wild bonobos (*Pan paniscus*) at Wamba." *International Journal of Primatology* 19 (6): 1029–43.

Galbany, J., T. S. Stoinski, D. Abavandimwe, T. Breuer, W. Rutkowski, N. V. Batista, F. Ndagijimana, and S. C. McFarlin. 2016. "Validation of two independent photogrammetric techniques for determining body measurements of gorillas." *American Journal of Primatology* 78 (4): 418–31.

Galdikas, B. M. F. 1995. "Social and reproductive behavior of wild adolescent female orangutans." In *The Neglected Ape*, edited by R. D. Nadler, B. F. M. Galdikas, L. K. Sheeran, and N. Rosen, 163–82. New York: Plenum Press.

Galdikas, B. M. F., and J. W. Wood. 1990. "Birth spacing patterns in humans and apes." *American Journal of Physical Anthropology* 83 (2): 185–91.

Gavan, J. A. 1953. "Growth and development of the chimpanzee: A longitudinal and comparative study." *Human Biology* 25: 93–143.

Georges, A.-J., E. M. Leroy, A. A. Renaut, C. T. Benissan, R. J. Nabias, M. T. Ngoc, P. I. Obiang, J. P. M. Lepage, E. J. Bertherat, and D. D. Bénoni. 1999. "Ebola hemorrhagic fever outbreaks in Gabon, 1994–1997: Epidemiologic and health control issues." *Journal of Infectious Diseases* 179 (S1): 65–75.

Goodall, J. 1968. "The behaviour of free-living chimpanzees in the Gombe Stream Reserve." *Animal Behaviour Monographs* 1: 161–311.

Goodall, J. 1983. "Population dynamics during a 15-year period in one community of free-living chimpanzees in the Gombe National Park, Tanzania." *Z. Tierpsychol.* 61: 1–60.

Goodall, J. 1986. *The Chimpanzees of Gombe: Patterns of Behavior.* Cambridge, MA: Harvard University Press.

Goossens, B., J. M. Setchell, S. S. James, S. M. Funk, L. Chikhi, A. Abulani, M. Ancrenaz, I. Lackman-Ancrenaz, and M. W. Bruford. 2006. "Philopatry and reproductive success in Bornean orang-utans (*Pongo pygmaeus*)." *Molecular Ecology* 15 (9): 2577–88.

Gould, S. J., and R. C. Lewontin. 1979. "The spandrels of San Marco and the panglossian paradigm: A critique of the adaptationist programme." *Proceedings of the Royal Society of London B* 205: 581–98.

Grassman, L. I., M. E. Tewes, N. J. Silvy, and K. Kreetiyutanont. 2005. "Ecology of three sympatric felids in a mixed evergreen forest in north-central Thailand." *Journal of Mammalogy* 86 (1): 29–38.

Gruen, L., A. Fultz, and J. Pruetz. 2013. "Ethical issues in African great ape field studies." *Institute for Laboratory Animal Research Journal* 54 (1): 24–32.

Grutzmacher, K. S., V. Keil, S. Metzger, L. Wittiger, I. Herbinger, S. Calvignac-Spencer, K. Mätz-Rensing, O. Haggis, L. Savary, S. Köndgen, F. H. Leendertz. 2018. "Human respiratory syncytial virus and Streptococcus pneumoniae infection in wild bonobos." *EcoHealth* 15 (2): 462–66.

Grutzmacher, K. S., S. Köndgen, V. Keil, A. Todd, A. Feistner, I. Herbinger, K. Petrzelkova, T. Fuh, S. A. Leendertz, S. Calvignac-Spencer, and F. H. Leendertz. 2016. "Codetection of respiratory syncytial virus in habituated wild western lowland gorillas and humans during a respiratory disease outbreak." *EcoHealth* 13 (3): 499–510.

Gurven, M., and H. Kaplan. 2007. "Longevity among hunter-gatherers: A cross-cultural examination." *Population and Development Review* 33 (2): 321–65.

Harcourt, A. H., K. J. Stewart, and D. Fossey. 1981. "Gorilla reproduction in the wild." In *Reproductive Biology of the Great Apes*, edited by C. E. Graham, 265–79. New York: Academic Press.

Harvey, N. C. 1997. "Gestation, parturition, interbirth intervals, and lactational recovery in bonobos." In *The Care and Management of Bonobos in Captive Environments*, edited by

J. Mills, G. Reinartz, H. De Bois, L. Van Elsacker, and B. Van Puijenbroeck. Milwaukee: Zoological Society of Milwaukee County.

Hawkes, K., J. F. O'Connell, N. G. Blurton Jones, H. Alvarez, and E. L. Charnov. 1998. "Grandmothering, menopause, and the evolution of human life histories." *Proceedings of the National Academy of Sciences* 95: 1336–39.

Hill, K., C. Boesch, J. Goodall, A. E. Pusey, J. Williams, and R. W. Wrangham. 2001. "Mortality rates among wild chimpanzees." *Journal of Human Evolution* 40: 437–50.

Hill, K., and A. Magdalena Hurtado. 1996. *Ache Life History: The Ecology and Demography of a Foraging People*. New York: Aldine de Gruyter.

Hosey, G. 2008. "A preliminary model of human–animal relationships in the zoo." *Applied Animal Behaviour Science* 109 (2–4): 105–27.

Howell, N. 1979. *Demography of the Dobe !Kung*. New York: Academic Press.

Hrdy, S. B. 2009. *Mothers and Others: The Evolutionary Origins of Mutual Understanding*. Cambridge, MA: Harvard University Press.

Inogwabini, B.-I., and N. Leader-Williams. 2012. "Effects of epidemic diseases on the distribution of bonobos." *PLoS One* 7 (12): e51112.

Isler, K., and C. P. van Schaik. 2009. "The expensive brain: A framework for explaning evolutionary changes in brain size." *Journal of Human Evolution* 57 (4): 392–400.

Janson, C. H., C. P. van Schaik. 1993. "Ecological risk aversion in juvenile primates: Slow and steady wins the race." In *Juvenile Primates: Life History, Development, and Behaviour*, edited by M. E. Pereira and L. A. Fairbanks, 57–74. New York: Oxford University Press.

Judge, D. S., and J. R. Carey. 2000. "Postreproductive life predicted by primate patterns." *Journals of Gerontology Series A: Biological Sciences and Medical Sciences* 55 (4): B201–9.

Kalema-Zikusoka, G., R. A. Kock, and E. J. Macfie. 2002. "Impenetrable National Park, Uganda." *Veterinary Record* 150: 12–15.

Kanamori, T., N. Kuze, H. Bernard, T. P. Malim, and S. Kohshima. 2012. "Fatality of a wild Bornean orangutan (*Pongo pygmaeus morio*): Behavior and death of a wounded juvenile in Danum Valley, North Borneo." *Primates* 53 (3): 221–26.

Kaplan, H., K. Hill, J. Lancaster, and A. M. Hurtado. 2000. "A theory of human life history evolution: Diet, intelligence, and longevity." *Evolutionary Anthropology* 9 (4): 156–84.

Kaur, T., J. Singh, S. Tong, C. Humphrey, D. Clevenger, W. Tan, B. Szekely, Y. Wang, Y. Li, E. A. Muse, M. Kiyono, S. Hanamura, E. Inoue, M. Nakamura, M. A. Huffman, B. Jian, T. Nishida. 2008. "Descriptive epidemiology of fatal respiratory outbreaks and detection of human-related metapneumovirus in wild chimpanzees (*Pan troglodytes*) at Mahale Mountains National Park, western Tanzania." *American Journal of Primatology* 70: 755–65.

Kilbourn, A. M., W. B. Karesh, N. D. Wolfe, E. J. Bosi, R. A. Cook, and M. Andau. 2003. "Health evaluation of free-ranging and semi-captive orangutans (*Pongo pymaeus pygmaeus*) in Sabahh, Malaysia." *Journal of Wildlife Diseases* 39 (1): 73–83.

Kirkwood, T. B., and S. N. Austad. 2000. "Why do we age?" *Nature* 408 (6809): 233.

Kirkwood, T. B., and M. R. Rose. 1991. "Evolution of senescence: Late survival sacrificed for reproduction." *Philosophical Transactions of the Royal Society of London B* 332 (1262): 15–24.

Knott, C. D. 1998. "Changes in orangutan caloric intake, energy balance, and ketones in response to fluctuating fruit availability." *International Journal of Primatology* 19 (6): 1061–79.

Knott, C. D. 2001. "Female reproductive ecology of the apes: Implications for human evolution." In *Reproductive Ecology and Human Evolution*, edited by P. T. Ellison, 429–63. New York: Aldine de Gruyter.

Knott, C. D., M. E. Emery Thompson, and S. A. Wich. 2009. "The ecology of reproduction in wild orangutans." In *Orangutans: Geographic Variation in Behavioral Ecology and Conservation*, edited by S. A. Wich, S. S. Utami, T. Mitra Setia, and C. van Schaik, 171–88. Oxford: Oxford University Press.

Knott, C. D., and S. Kahlenberg. 2011. "Orangutans: Understanding forced copulations." In *Primates in Perspective*, edited by C. J. Campbell, A. Fuentes, K. C. MacKinnon, S. Bearder, and R. M. Stumpf, 290–305. New York: Oxford University Press.

Kohler, I. V., S. H. Preston, and L. B. Lackey. 2006. "Comparative mortality levels among selected species of captive animals." *Demographic Research* 15: 413–34.

Köndgen, S., H. Kühl, P. K. N'Goran, P. D. Walsh, S. Schenk, N. Ernst, R. Biek, P. Formenty, K. Mätz-Rensing, B. Schweiger, S. Junglen, H. Ellerbrok, A. Nitsche, T. Briese, W. I. Lipkin, G. Pauli, C. Boesch, and F. H. Leendertz. 2008. "Pandemic human viruses cause decline of endangered great apes." *Current Biology* 18 (4): 260–64.

Kuroda, S. 1989. "Developmental retardation and behavioral characteristics of pygmy chimpanzees." In *Understanding Chimpanzees*, edited by P. G. Heltne and L. A. Marquardst, 184–93. Cambridge, MA: Harvard University Press.

Kuze, N., T. P. Malim, and S. K. Kohshima. 2005. "Developmental changes in the facial morphology of the Borneo Orangutan (*Pongo pygmaeus*): Possible signals in visual communication." *American Journal of Primatology* 65 (4): 353–76.

Lee, P. C. 1996. "The meanings of weaning: Growth, lactation, and life history." *Evolutionary Anthropology* 5 (3): 87–98.

Leendertz, F. H., H. Ellerbrok, C. Boesch, E. Couacy-Hymann, K. Mätz-Rensing, R. Hakenbeck, C. Bergmann, P. Abaza, S. Junglen, and Y. Moebius. 2004. "Anthrax kills wild chimpanzees in a tropical rainforest." *Nature* 430 (6998): 451.

Leendertz, F. H., F. Lankester, P. Guislain, C. Néel, O. Drori, J. Dupain, S. Speede, P. Reed, N. Wolfe, and S. Loul. 2006. "Anthrax in Western and Central African great apes." *American Journal of Primatology* 68 (9): 928–33.

Leendertz, F. H., G. Pauli, K. Maetz-Rensing, W. Boardman, C. Nunn, H. Ellerbrok, S. A. Jensen, S. Junglen, and C. Boesch. 2006. "Pathogens as drivers of population declines: The importance of systematic monitoring in great apes and other threatened mammals." *Biological Conservation* 131 (2): 325–37.

Leigh, S. R. 1994. "Ontogenetic correlates of diet in anthropoid primates." *American Journal of Physical Anthropology* 94: 499–522.

Leigh, S. R. 2004. "Brain growth, life history, and cognition in primate and human evolution." *American Journal of Primatology* 62 (3): 139–64.

Leigh, S. R., and B. T. Shea. 1995. "Ontogeny and the evolution of adult body size dimorphism in apes." *American Journal of Primatology* 36: 36–60.

Leigh, S. R., and B. T. Shea. 1996. "Ontogeny of body size variation in African apes." *American Journal of Physical Anthroplogy* 99: 43–65.

Leroy, E. M., P. Rouquet, P. Formenty, S. Souquière, A. Kilbourne, J.-M. Froment, M. Bermejo, S. Smit, W. Karesh, and R. Swanepoel. 2004. "Multiple Ebola virus transmission events and rapid decline of central African wildlife." *Science* 303 (5656): 387–90.

Lippert, W. 1974. "Beobachtungen zum Schwangerschafts- und Geburtsverhalten beim Orang-Utan (*Pongo pygmaeus*) im Tierpark Berlin." *Folia Primatalogica* 21: 108–34.

Lippert, W. 1977. Erfahrungen bei der Aufzucht von Orang-Utans (*Pongo pygmaeus*) im Tierpark Berlin." *Zoologische Garten* 47: 209–25.

Lonsdorf, E. V., T. R. Gillespie, T. M. Wolf, I. Lipende, J. Raphael, J. Bakuza, C. M. Murray, M. L. Wilson, S. Kamenya, D. Mjungu, D. A. Collins, I. C. Gilby, M. A. Stanton, K. A. Terio, H. J. Barbian, M. Ramirez, A. Krupnick, E. Seidl, J. Goodall, B. H. Hahn, A. E. Pusey, and D. A. Travis. 2018. "Socioecological correlates of clinical signs in two communities of wild chimpanzees (*Pan troglodytes*) at Gombe National Park, Tanzania." *American Journal of Primatology* 80 (1): e22562.

Lowenstine, L. J., R. McManamon, and K. A. Terio. 2016. "Comparative pathology of aging great apes: Bonobos, chimpanzees, gorillas, and orangutans." *Veterinary Pathology* 53 (2): 250–76.

Maggioncalda, A. N., R. M. Sapolsky, and N. M. Czekala. 1999. "Reproductive hormone profiles in captive male orangutans: Implications for understanding developmental arrest." *American Journal of Physical Anthropology* 109 (1): 19–32.

Malenky, R. K., S. Kuroda, E. O. Vineberg, and R. W. Wrangham. 1994. "The significance of terrestrial herbaceous foods for bonobos, chimpanzees and gorillas." In *Chimpanzee Cultures*, edited by R. W. Wrangham, W. C. McGrew, F. B. M. de Waal, and P. G. Heltne, 59–75. Cambridge, MA: Harvard University Press.

Markham, R. J. 1990. "Breeding orangutans at Perth zoo: Twenty years of appropriate husbandry." *Zoo Biology* 9: 171–82.

Markham, R. J. 1995. "Doing it naturally: Reproduction in captive orangutans (*Pongo pygmaeus*)." In *The Neglected Ape*, edited by R. D. Nadler, B. F. M. Galdikas, L. K. Sheeran, and N. Rosen, 273–378. New York: Plenum Press.

Marshall, A. J., M. Ancrenaz, F. Q. Brearley, G. M. Fredriksson, N. Ghaffar, M. Heydon, S. J. Husson, M. Leighton, K. R. McConkey, H. C. Morrogh-Bernard, J. Proctor, C. P. van Schaik, C. P. Yeager, and S. A. Wich. 2009. "The effects of forest phenology and floristics on populations of Bornean and Sumatran orangutans." In *Orangutans: Geographic Variation in Behavioral Ecology and Conservation*, edited by S. A. Wich, S. S. Utami, T. Mitra Setia, and C. van Schaik, 97–118. Oxford: Oxford University Press.

Marshall, A. J., and R. W. Wrangham. 2007. "Evolutionary consequences of fallback foods." *International Journal of Primatology* 28 (6): 1219–35.

Marshall, W. A., and J. M. Tanner. 1969. "Variations in pattern of pubertal changes in girls." *Archives of Disease in Childhood* 44 (235): 291.

Marshall, W. A., and J. M. Tanner. 1970. "Variations in the pattern of pubertal changes in boys." *Archives of Disease in Childhood* 45 (239): 13–23.

Marzec, A. M., J. A. Kunz, S. F. Falkner, S. S. U. Atmoko, S. E. Alavi, A. M. Moldawer, E. R. Vogel, C. Schuppli, C. P. van Schaik, and M. A. van Noordwijk. 2016. "The dark side of the red ape: Male-mediated lethal female competition in Bornean orangutans." *Behavioral Ecology and Sociobiology* 70 (4): 459–66.

Metcalfe, N. B., and P. Monaghan. 2003. "Growth versus lifespan: Perspectives from evolutionary ecology." *Experimental Gerontology* 38 (9): 935–40.

Morgan, K. N., and C. T. Tromborg. 2007. "Sources of stress in captivity." *Applied Animal Behaviour Science* 102 (3–4): 262–302.

Morrogh-Bernard, H. C., S. J. Husson, C. D. Knott, S. A. Wich, C. P. van Schaik, M. A. van Noordwijk, I. Lackman-Ancrenaz, A. J. Marshall, T. Kanamori, N. Kuze, and R. bin Sakong. 2009. "Orangutan activity budgets and diet." In *Orangutans: Geographic Variation in Behavioral Ecology and Conservation*, edited by S. A. Wich, S. S. Utami, T. Mitra Setia, and C. van Schaik, 119–34. Oxford: Oxford University Press.

Muller, M. N., and R. W. Wrangham. 2014. "Mortality rates among Kanyawara chimpanzees." *Journal of Human Evolution* 66: 107–14.

Nater, A., M. P. Mattle-Greminger, A. Nurcahyo, M. G. Nowak, M. de Manuel, T. Desai, C. Groves, M. Pybus, T. B. Sonay, C. Roos, A. R. Lameira, S. A. Wich, J. Askew, M. Davila-Ross, G. Fredriksson, G. de de Valles, F. Casals, J. Prado-Martinez, B. Goossens, E. J. Verschoor, K. Warren, S. I. Singleton, D. A. Marques, J. Pamungkas, D. Perwitasari-Farajallah, P. Rianti, A. Tuuga, I. G. Gut, M. Gut, P. Orozco-terWengel, C. P. van Schaik, J. Bertranpetit, M. Anisimova, A. Scally, T. Marques-Bonet, E. Meijaard, and M. Krutzen. 2017. "Morphometric, behavioral, and genomic evidence for a new orangutan species." *Current Biology* 27 (22): 3487–98.

Negrey, J. D., R. B. Reddy, E. J. Scully, S. Phillips-Garcia, L. A. Owens, K. E. Langergraber, J. C. Mitani, M. Emery Thompson, R. W. Wrangham, M. N. Muller, E. Otali, Z. Machanda, D. Hyeroba, K. A. Grindle, T. E. Pappas, A. C. Palmenberg, J. E. Gern, and T. L. Goldberg. 2019. "Simultaneous outbreaks of respiratory disease in wild chimpanzees caused by distinct viruses of human origin." *Emerging Microbes and Infections* 8 (1): 139–49.

Nishida, T. 1968. "The social group of wild chimpanzees in the Mahale Mountains." *Primates* 9 (3): 167–224.

Nishida, T., N. Corp, M. Hamai, T. Hasegawa, M. Hiraiwa-Hasegawa, K. Hosaka, K. D. Hunt, N. Itoh, K. Kawanaka, and A. Matsumoto-Oda. 2003. "Demography, female life history, and reproductive profiles among the chimpanzees of Mahale." *American Journal of Primatology* 59 (3): 99–121.

Nishida, T., H. Takasaki, and Y. Takahata. 1990. "Demongraphy and reproductive profiles." In *The Chimpanzees of the Mahale Mountains: Sexual and Life History Strategies*, edited by T. Nishida, 63–97. Tokyo: University of Tokyo Press.

Nsubuga, A. M., M. R. Robbins, C. Boesch, and L. Vigilant. 2008. "Patterns of paternity and group fission in wild multimale mountain gorilla groups." *American Journal of Physical Anthropology* 135 (3): 263–74.

Nunn, C. L., P. H. Thrall, K. Stewart, and A. H. Harcourt. 2008. "Emerging infectious diseases and animal social systems." *Evolutionary Ecology* 22 (4): 519–43.

Ogden, J., and D. Wharton. 1997. *The management of gorillas in captivity: Husbandry manual of the Gorilla Species Survival Plan (SSP®)*. Atlanta: The Gorilla SSP and the Atlanta/Fulton County Zoo, Inc.

Patrono, L.V., L. Samuni, V. M. Corman, L. Nourifar, C. Röthemeier, R. M. Wittig, C. Drosten, S. Calvignac-Spencer, and F. H. Leendertz. 2018. "Human coronavirus OC43 outbreak in wild chimpanzees, Côte d Ivoire, 2016." *Emerging Microbes and Infections* 7 (1): 118.

Pontzer, H., D. A. Raichlen, R. W. Shumaker, C. Ocobock, and S. A. Wich. 2010. "Metabolic adaptation for low energy throughput in orangutans." *Proceedings of the National Academy of Sciences* 107 (32): 14048–52.

Potts, R. 1998. "Variability selection in hominid evolution." *Evolutionary Anthropology* 7: 81–96.

Pusey, A., J. Williams, and J. Goodall. 1997. "The influence of dominance rank on the reproductive success of female chimpanzees." *Science* 277: 828–31.

Riley, E. P., and M. Bezanson. 2018. "Ethics of primate fieldwork: Toward an ethically engaged primatology." *Annual Review of Anthropology* 47: 493–512.

Robbins, A. M., M. M. Robbins, N. Gerald-Steklis, and H. D. Steklis. 2006. "Age-related patterns of reproductive success among female Mountain Gorillas." *American Journal of Physical Anthropology* 131 (4): 511–21.

Robbins, A. M., T. S. Stoinski, K. A. Fawcett, and M. M. Robbins. 2009. "Does dispersal cause reproductive delays in female mountain gorillas?" *Behaviour* 146 (4/5): 525–49.

Robbins, M. M. 1995. "A demographic analysis of male life history and social structure of mountain gorillas." *Behaviour* 132 (1): 21–47.

Robbins, M. M., M. Bermejo, C. Cipolletta, F. Magliocca, R. J. Parnell, E. Stokes. 2004. "Social structure and life-history patterns in western gorillas (*Gorilla gorilla gorilla*)." *American Journal of Primatology* 64 (2): 145–59.

Robson, S. L., C. P. van Schaik, and K. Hawkes. 2006. "The derived features of human life history." In *The Evolution of Human Life History*, edited by K. Hawkes and R. R. Paine, 17–44. Santa Fe: School of American Research Press.

Roof, K. A., W. D. Hopkins, M. K. Izard, M. Hook, and S. J. Schapiro. 2005. "Maternal age, parity, and reproductive outcome in captive chimpanzees (*Pan troglodytes*)." *American Journal of Primatology* 67 (2): 199–207.

Rothman, J. M., C. A. Chapman, D. Twinomugisha, M. D. Wasserman, J. E. Lambert, and T. L. Goldberg. 2008. "Measuring physical traits of primates remotely: The use of parallel lasers." *American Journal of Primatology* 70: 1–5.

Rothman, J. M., E. S. Dierenfeld, H. F. Hintz, and A. N. Pell. 2008. "Nutritional quality of gorilla diets: Consequences of age, sex, and season." *Oecologia* 155: 111–22.

Ryan, S. J., and P. D. Walsh. 2011. "Consequences of non-intervention for infectious disease in African great apes." *PLoS One* 6 (12): e29030.

Scully, E. J., S. Basnet, R. W. Wrangham, M. N. Muller, W. Otali, D. Hyeroba, K. A. Grindle, T. E. Pappas, M. Emery Thompson, Z. Machanda, K. E. Watters, A. C. Palmenberg, J. E. Gern, and T. L. Goldberg. 2013. "Lethal respiratory disease associated with human rhinovirus C in wild chimpanzees, Uganda." *Emerging Infectious Disease* 24 (2): 267–74.

Seitz, A. 1969. "Notes on the body weights of new-born and young orang-utans *Pongo pygmaeus*." *International Zoo Yearbook* 9 (1): 81–84.

Shanley, D. P., T. B. L. Kirkwood. 2000. "Calorie restriction and aging: A life history analysis." *Evolution* 54: 740–50.

Shattuck, M. R., and S. A. Williams. 2010. "Arboreality has allowed for the evolution of increased longevity in mammals." *Proceedings of the National Academy of Sciences* 107 (10): 4635–39.

Shumaker, R., S. A. Wich, and L. Perkins. 2008. "Reproductive life history traits of female orangutans (*Pongo* spp.)." In *Primate Reproductive Aging*, edited by S. Atsalis, S. W. Margulis, and P. R. Hof, 147–61. Basel: Karger.

Sievert, J., W. B. Karesh, and V. Sunde. 1991. "Reproductive intervals in captive female western lowland gorillas with a comparison to wild mountain gorillas." *American Journal of Primatology* 24: 227–34.

Sodaro, C. 1997. *Orangutan Species Survival Plan Husbandry Manual*. Chicago: Chicago Zoological Park.

Species Survival Library. Accessed from www.aza.org/species-survival-statistics.

Stearns, S. C., and J. C. Koella. 1986. "The evolution of phenotypic plasticity in life-history traits: Predictions of reaction norms for age and size at maturity." *Evolution* 40 (5): 893–913.

Strong, V. J., D. Grindlay, S. Redrobe, M. Cobb, and K. White. 2016. "A systematic review of the literature relating to captive great ape morbidity and mortality." *Journal of Zoo and Wildlife Medicine* 47 (3): 697–710.

Stumpf, R. M. 2011. "Chimpanzees and bonobos." In *Primates in Perspective*, edited by C. J. Campbell, A. Fuentes, K. C. MacKinnon, S. Bearder, and R. M. Stumpf, 340–56. New York: Oxford University Press.

Sugiyama, Y. 2004. "Demographic parameters and life history of chimpanzees at Bossou, Guinea." *American Journal of Physical Anthropology* 124 (2): 154–65.

Thompson-Handler, N. E. 1990. "The pygmy chimpanzee: sociosexual behavior, reproductive biology and life history." PhD diss., Yale University.

Tilson, R., U. S. Seal, K. Soemarna, W. Ramono, E. Sumardja, S. Poniran, C. van Schaik, M. Leighton, H. Rijksen, and A. Eudey. 1993. "Orangutan population and habitat viability analysis report." Orangutan population and habitat viability analysis workshop, Medan, North Sumatra, Indonesia.

van Noordwijk, M. A., S. S. U. Atmoko, C. D. Knott, N. Kuze, H. C. Morrogh-Bernard, F. C. Oram, C. Schuppli, C. P. van Schaik, and E. P. Willems. 2018. "The slow ape: High infant survival and long inter-birth intervals in orangutans." *Journal of Human Evolution* 125: 38–49.

van Noordwijk, M. A., and C. P. van Schaik. 2005. "Development of ecological competence in Sumatran orangutans." *American Journal of Physical Anthropology* 127 (1): 79–94.

van Schaik, C. P. 1999. "The socioecology of fission-fusion sociality in Orangutans." *Primates* 40 (1): 69–86.

van Schaik, C. P., and K. Isler. 2012. "Life history evolution in primates." In *The Evolution of Primate Societies*, edited by J. C. Mitani, J. Call, P. M. Kappeler, R. A. Palombit, and J. B. Silk, 220–44. Chicago: University of Chicago Press.

Vigilant, L., M. Hofreiter, H. Siedel, and C. Boesch. 2001. "Paternity and relatedness in wild chimpanzee communities." *Proceedings of the National Academy of Sciences* 98 (23): 12890–95.

Vrba, E. S. 1985. "Ecological and adaptive changes associated with early hominid evolution." In *Ancestors: The Hard Evidence*, edited by E. Delson, 63–71. New York: Liss.

Walker, M. L., J. G. Herndon. 2008. "Menopause in nonhuman primates?" *Biology of Reproduction* 79 (3): 398–406.

Walker, R., M. Gurven, K. Hill, A. Migliano, N. Chagnon, R. De Souza, G. Djurovic, R. Hames, A. M. Hurtado, and H. Kaplan. 2006. "Growth rates and life histories in twenty-two small-scale societies." *American Journal of Human Biology* 18 (3): 295–311.

Wallis, J. 1997. "A survey of reproductive parameters in the free-ranging chimpanzees of Gombe National Park." *Journal of Reproduction and Fertility* 109: 121–54.

Walsh, P. D., K. A. Abernethy, M. Bermejo, R. Beyers, P. De Wachter, M. E. Akou, B. Huijbregts, D. I. Mambounga, A. K. Toham, and A. M. Kilbourn. 2003. "Catastrophic ape decline in western equatorial Africa." *Nature* 422 (6932): 611.

Watts, D. P. 1984. "Composition and variability of mountain gorilla diets in the central Virungas." *American Journal of Primatology* 7: 323–56.

Watts, D. P. 1989. "Infanticide in mountain gorillas: New cases and a reconsideration of the evidence." *Ethology* 81 (1): 1–18.

Watts, D. P. 1991. "Mountain gorilla reproduction and sexual behavior." *American Journal of Primatology* 24: 211–225.

Watts, D. P. 1996. "Comparative socio-ecology of gorillas." In *Great Ape Societies*, edited by W. C. McGrew, L. F. Marchant, and T. Nishida, 16–28. New York: Cambridge University Press.

Watts, D. P. 2012. "The apes: Taxonomy, biogeography, life histories, and behavioral ecology." In *The Evolution of Primate Societies*, edited by J. C. Mitani, J. Call, P. M. Kappeler, R. A. Palombit, and J. B. Silk, 113–42. Chicago: University of Chicago Press.

Weindruch, R., and R. L. Walford. 1988. *The Retardation of Aging and Disease by Dietary Restriction*. Springfield, IL: Thomas.

Wheatley, B. P. 1987. "The evolution of large body size in orangutans: A model for hominoid divergence." *American Journal of Primatology* 13 (3): 313–24.

Wich, S. A., R. W. Shumaker, L. Perkins, H. de Vries. 2009. "Captive and wild orangutan (*Pongo* sp.) survivorship: A comparison and the influence of management." *American Journal of Primatology* 71 (8): 680–86.

Wich, S. A., S. S. Utami-Atmoko, T. M. Setia, H. Rijksen, C. Schürmann, J. Van Hooff, and C. P. van Schaik. 2004. "Life history of wild Sumatran orangutans (*Pongo abelii*)." *Journal of Human Evolution* 47 (6): 385–98.

Wich, S. A., H. de Vries, M. Ancrenaz, L. Perkins, R. W. Shumaker, A. Suzuki, and C. P. van Schaik. 2009. "Orangutan life history variation." In *Orangutans: Geographic Variation in Behavioral Ecology and Conservation*, edited by S. A. Wich, S. S. Utami, T. Mitra Setia, and C. P. van Schaik, 65–76. Oxford: Oxford University Press.

Wilkie, D. S., J. G. Sidle, and G. C. Boundzanga. 1992. "Mechanized logging, market hunting, and a bank loan in Congo." *Conservation Biology* 6 (4): 570–80.

Williams, J. M., E. V. Lonsdorf, J. Schumacher-Stankey, J. Goodall, and A. E. Pusey. 2008. "Causes of death in the Kasekela chimpanzees of Gombe National Park, Tanzania." *American Journal of Primatology* 70 (8): 766–77.

Wood, B. M., K. E. Langergraber, J. C. Mitani, and D. P. Watts. 2017. "Menopause is common among wild female chimpanzees in the Ngogo community." Presentation given at the 86th Annual Meeting of the American Association of Physical Anthropologists, Austin, TX.

Wrangham, R. W. 1993. "The evolution of sexuality in chimpanzees and bonobos." *Human Nature* 4 (1): 47–79.

Wrangham, R. W. 2009. *Catching Fire: How Cooking Made Us Human*. New York: Basic Books.

Wrangham, R. W., N. L. Conklin-Brittain, and K. D. Hunt. 1998. "Dietary response of chimpanzees and cercopithecines to seasonal variation in fruit abundance. I. Antifeedants." *International Journal of Primatology* 19 (6): 949–70.

Yamagiwa, J., and J. M. Kahekwa. 2001. "Dispersal patterns, group structure, and reproductive parameters of eastern lowland gorillas at Kahuzi in the absence of infanticide." In *Mountain Gorillas: Three Decades of Research at Karisoke*, edited by M. M. Robbins, P. Sicotte, and K. J. Stewart, 89–122. Cambridge: Cambridge University Press.

2

Growing Up: Comparing Ontogeny of Bonobos and Chimpanzees

VERENA BEHRINGER, JEROEN M. G. STEVENS, TOBIAS DESCHNER, AND GOTTFRIED HOHMANN

Introduction

For a long time, there was no evidence that chimpanzees lived south of the Congo River, but in the 1920s indications emerged for a chimpanzee subspecies (*Pan satyrus paniscus*) living in that area of Africa (Schwarz 1929). Coolidge (1933) later assigned a distinct species status to these apes: the bonobo (*Pan paniscus*). At the time, given the similarities between chimpanzees and bonobos, there was much discussion about the taxonomic classification of the bonobo, namely whether it should be considered a subspecies of the chimpanzee (*Pan troglodytes*) or a distinct species (Groves 1986; Horn 1979). Indeed, many aspects of the two sister-species' morphology, diet, and social organization are similar, which may explain why, for such a long time, they were not clearly distinguished.

Socially, both chimpanzees and bonobos live in large multi-male, multi-female communities that exhibit a flexible fission-fusion grouping pattern (Boesch and Boesch-Achermann 2000; Furuichi 1989; Goodall 1986; Kano 1992; Nishida et al. 1990). Males of both species are philopatric (Boesch and Boesch-Achermann 2000; Eriksson et al. 2006; Furuichi et al. 2012; Gilby et al. 2013), while females transfer once or multiple times (Furuichi 1989; Gerloff et al. 1999; Idani 1991a; Kahlenberg et al. 2008; Langergraber et al. 2007). Physically, adult bonobos and chimpanzees are similar in body weight (Shea 1983a) and both species show a moderate sexual dimorphism with males being heavier than females (Shea 1985), though the degree of dimorphism is even smaller in bonobos than in chimpanzees (Wrangham and Pilbeam 2002; Zihlman and Cramer 1978).

Nevertheless, since the discovery of the bonobo, studies have also emphasized differences between the two *Pan* species (Badrian and Badrian 1977; Groves 1986; Kuroda 1980; see also Knott and Harwell, chapter 1 this volume;

Taglialatela et al., chapter 4 this volume). To characterize chimpanzee behavior, researchers have used terms such as "demonic," "political," and "cultural" (de Waal 2007; Wrangham and Peterson 1996; Wrangham 1996). In contrast, bonobo behavior is typically characterized as being "gentle," "peaceful," and "tolerant" (de Waal 2001; de Waal and Lanting 1998; Furuichi 1997; Wrangham and Peterson 1996). This popular dichotomy reflects mainly differences of male behavior, since adult male chimpanzees form strong bonds with other males, and cooperate in the context of hunting, mate competition, and territorial behavior (Amsler 2010; Gilby et al. 2013; Mitani and Watts 2001; Watts 1998). Male chimpanzees employ aggressive and cooperative tactics to enhance their dominance, and use conditioning aggression toward females to increase the effectiveness of mate-guarding (Muller et al. 2011). The use of aggressive mating tactics by male chimpanzees enhances individual mating success and gives males dominance over females (Muller, Kahlenberg, and Wrangham 2009; Watts 1998). In bonobos, however, male bonding is rare and restricted to close kin (Furuichi and Ihobe 1994). They engage neither in cooperative hunting nor in border patrols, and there is no evidence for cooperative mate competition. Instead, male bonobos associate with females rather than other males, and inter-sexual relations are characterized by affiliative behavior (Furuichi 1989; Hohmann et al. 1999; Surbeck et al. 2017). Moreover, adult male bonobos engage in life-long bonds with their mothers (Furuichi and Ihobe 1994; Hohmann et al. 1999; Kano 1992), mothers support sons in conflicts with other males, and by doing so enhance the mating success of their sons (Furuichi 1997; Surbeck, Mundry, and Hohmann 2011). Male aggression is mainly restricted to assert intra-sexual dominance relations, but is not used in the context of mating, which leads to a system of co-dominance between the sexes (Furuichi 1997; Kuroda 1980; Vervaecke, de Vries, and van Elsacker 2000a).

Another dichotomy between bonobos and chimpanzees is found in female social relationships. Although variations between chimpanzee populations have become apparent through several long-term field studies, chimpanzee females are generally less gregarious than males, especially during non-fertile periods. Female chimpanzees invest less in social relationships with other females, and show clear aversion to females from other communities (Goodall 1986; Nishida 1989; Pusey and Schroepfer-Walker 2013; Wrangham 1986; Wrangham and Smuts 1980). In bonobos, however, females are the gregarious sex (Furuichi 2011, 1989; Kano 1980; Kuroda 1980); immigrating females engage in differentiated social and affiliative relations with residents (Furuichi 2011; Idani 1991a).

Another aspect contributing to the differentiated characterizations of the two *Pan* species, concerns intergroup encounters. Chimpanzees are highly territorial: males engage in boundary patrols, and both sexes participate in violent aggression when encountering strangers (Amsler 2010; Mitani and Watts 2001, 2005; Mitani, Watts, and Amsler 2010; Watts et al. 2006), a behavioral trait that appears to be entirely absent in bonobos. At the beginning of a bonobo intergroup encounter, tension can be high, resulting in many sociosexual interactions. However, antagonistic interactions rarely occur. Moreover, females engage in affinitive interactions like grooming. Males only infrequently show any interaction with members of other groups, but appeasement behavior between males, like mounting, occasionally occurs (Furuichi and Ihobe 1994; Hohmann and Fruth 2002; Idani 2003, 1991b, 1990).

While many studies have focused on the ultimate explanations for these interspecies differences, in terms of fitness and evolutionary past, the more proximate causes of the dichotomy remain understudied. Specifically, questions include (1) at what stage during the ontogeny do these species differences occur and (2) which proximate mechanisms are responsible for the emergence of the different phenotypes? We examine these questions here.

It has been argued that some of the morphological characteristics of the two *Pan* species reflect differences in their development (Corruccini and McHenry 1979). Specifically, bonobos have been characterized as being paedomorphic (retaining juvenile features as adults), referring to their less intense skull development in relation to body mass, derived from a more robust ancestor via "heterochrony" (Coolidge 1933; Coolidge and Shea 1982; McHenry and Corruccini 1981; Tuttle 1975; Vinicius and Lahr 2003). The term heterochrony refers to modifications in the rate and timing of specific constituents of ontogenetic trajectories, such as onset and offset of rates of growth, maturation, and development (Gould 1988; McKinney and McNamara 1991; Zelditch and Fink 1996). In an evolutionary context, heterochrony connects phylogeny and ontogeny (Arthur 2002). While the term of heterochrony was originally used to describe phenomenological characterizations of morphological changes during development (Berge and Penin 2004; Gould 1977), it was later extended to include cognitive, cellular, and molecular developmental scales (Klingenberg 1998; Langer 2006).

Comparative ontogenetic studies on bonobos and chimpanzees have explored to what extent interspecific differences reflect temporal differences in development (Berge and Penin 2004). Early studies were limited by the scarcity of information from living bonobos, and focused on material of museum collections. From this research, bonobos were described as paedomorphic (Coolidge 1933; Cramer 1977; Shea 1983a, 1983b), a finding that continues to gain support (Lieberman et al. 2007). More recent evidence from dental development shows that the eruption of permanent teeth happens later in bonobos than in chimpanzees (Bolter and Zihlman 2011), while a comparison of mandibular development in the two *Pan* species revealed that ontogenetic trajectories of changes of mandibular shape are parallel (Boughner and Dean 2008). Further evidence for bonobos' paedomorphic traits includes the high-pitched calls of adults; the persistence of immature patterns of hair coloration into adulthood, particularly the white tail tuft (Groves 1986); and the similarity in female cycles between adult bonobos and adolescent chimpanzees (Dahl 1986). In terms of behavioral development, it has been suggested that bonobo infants are paedomorphic and show delayed motor and social development compared with chimpanzees (Kuroda 1989; Wobber 2012; Wobber, Wrangham, and Hare 2010, but see de Lathouwers and Van Elsacker 2006 for contrasting results).

Approach

Endocrinological data can provide a novel and interesting perspective on the degree of heterochrony in the two *Pan* species, but relatively few such data exist. Physiological markers, like hormones, are considered to be mediators of the transition of life stages (Hochberg 2012; Jacobs and Wingfield 2000). To compare endocrine patterns during ontogeny in chimpanzees and bonobos, we measured several hormone markers in different life stages, to identify presence or absence of heterochrony during development in the two species (i.e., whether the two species showed different rates of development).

We selected four physiological markers that show specific patterns during ontogeny in humans, and allow for the identification of developmental milestones, which may also differ between the sexes as well as across species (table 2.1). The first two markers we studied, total triiodothyronine (T3) and insulin-like growth factor-binding protein (IGFBP-3), relate to somatic growth during the period between birth and puberty; dehydroepiandrosteron-sulphate (DHEA-S) is associated with adrenarche; while an increase of testosterone initiates the start of puberty in both males and females.

Patterns of somatic growth are well studied in chimpanzees (Gavan 1971; Hamada et al. 2003), where males have a faster growth rate than females (Leigh and Shea 1996). For bonobos, fewer data exist, but it has been suggested that the growth processes in chimpanzees and bonobos are heterochronic, and that weight differences between the sexes in bonobos are a result of longer growth duration in males than females. These different trajectories call for more elaborate work, using physiological markers in living subjects, to understand the underlying mechanism.

Thyroid hormones are an ideal candidate to measure somatic growth because they are involved in the regulation of metabolic activity and they

TABLE 2.1. Physiological markers representing events during ontogeny with their pattern, sex-specific differences in humans, and the corresponding sample size information in our study with bonobos and chimpanzees. Sample size does not correspond to the number of individuals because for some of them, more than one sample was analyzed (longitudinal and cross-sectional data sets).

					Bonobos		*Chimpanzees*	
Marker	*Event*	*Broad Pattern*	*Sex-Specific Pattern*	*Total Samples*	*Females*	*Males*	*Females*	*Males*
Creatinine	Muscle growth	Increases with age	Males have higher levels	256	62	47	86	58
Total T3	Skeletal and brain growth	Decreases after growth spurt	Earlier decline in females	235	53	47	64	36
IGFBP-3	Somatic growth	Increases from birth onwards, decreases after puberty	Females have higher levels	279	60	44	81	45
DHEA	Adrenarche	Low, increases until age 20, then decreases	None, individual differences	268	61	46	96	60
Testosterone	Onset of puberty	Low levels after the first year of life, increases with puberty	Males have higher levels	265	64	49	98	53

Note: DHEA, dehydroepiandrosterone; IGFBP-3, insulin-like growth factor-binding protein-3; total T3, total triiodothyronine.

control brain development and somatic growth (Bassett and Williams 2016; Venturi and Begin 2010). In humans, levels of the biologically active thyroid hormone T3 increases after birth, plateaus within the first week, remains at this level until puberty, and decreases after the adolescent growth spurt (Eworo et al. 2015; Michaud et al. 1991; Ryness 1972). Corresponding with a sex-difference in the adolescent growth spurt in humans, the decline in T3 levels occurs earlier in girls than boys (Garcia-Bulnes et al. 1977; Parra et al. 1980). In a previous study, we analyzed a urinary thyroid hormone (total T3) in bonobos and chimpanzees and found evidence for heterochrony. In both species, total T3 levels were higher in immatures than adults, but the timing of the decline of total T3 occurred many years later in bonobos compared to chimpanzees (Behringer et al. 2014b). To further explore differences in T3 levels between species and sex, we conducted analyses of a larger sample set and here we report, for the first time, on measures of long-term changes in urinary total T3 levels in bonobos.

The somatotropic axis consists of several growth factors, such as growth hormone, which is correlated with T3 concentrations (Grunfeld et al. 1988), and insulin-like growth factor (IGF), which is also influenced by thyroid hormones and insulin-like growth factor binding proteins (IGFBPs). In humans, changes in the somatotropic axis mark shifts between infancy and childhood (Hochberg 2012) and it is known that insulin-like growth factor binding protein 3 (IGFBP-3) declines over the course of life, and is lower in men than in women (DeLellis et al. 2004; Janssen et al. 1998). Our previous analyses of urinary IGFBP-3 levels in adult bonobos and chimpanzees also revealed an age-related decline, and one that was steeper in females than in males (Behringer et al. 2016). Here, we provide additional information on urinary IGFBP-3 levels of bonobos and chimpanzees during ontogeny, including individuals from the first year of life to 55 years old.

Adrenarche is an early stage in sexual maturation and represents another milestone of ontogeny. Adrenarche is also termed the "puberty of the adrenal glands" (Ibáñez et al. 2000) because it is characterized by changes in the release from the adrenal glands of dehydroepiandrosterone (DHEA) and its sulphate (DHEA-S). In humans, the *zona reticularis*, an area within the adrenal gland, secretes DHEA and DHEA-S from birth until the first year of life. Thereafter, the *zona reticularis* disappears and concentrations of the two hormones decrease. At three to five years old, the *zona reticularis* is reestablished and levels of DHEA and its sulphate increase again until the age of 25 years; afterward they decrease again with increasing age (de Peretti and Forest 1978). A similar developmental pattern was found in chimpanzees (Blevins et al. 2013; Collins, Nadler, and Preedy 1981; Copeland et al. 1985; Cutler et al. 1978; Smail et al. 1982) and bonobos (Behringer et al. 2012). In adult humans and chimpanzees, DHEA and DHEA-S levels decline with increasing age (Blevins et al. 2013), but the pattern in bonobos is unknown. The data we present here complement previous work (Behringer et al. 2012) by analyzing changes of urinary DHEA levels in bonobos and further from chimpanzees, and we include longitudinal data for some males.

Sex steroid hormones, such as testosterone, can be used to measure the maturation of ovaries and testes, which in humans indicates the onset of adult reproductive capacity (i.e., puberty) (Bogin 1997). In humans, sex steroid hormones are elevated during the first three months of life, called "mini puberty." Afterward sex steroids decline and stay low for the following 10–12 years, a stage that is called "juvenile pause" (Grumbach 2002). Then, maturation of testes and ovaries starts, and levels of sex steroid hormones increase (Byrd et al. 1998; Dorn and Biro 2011; Frasier, Gafford, and Horton 1969; Grumbach

2002; Lalwani, Reindollar, and Davis 2003; Styne 1994). In an earlier analysis, we found that urinary testosterone levels increase in female bonobos around 4–5 years of age but in female chimpanzees they did not increase until after ten years of age (Behringer et al. 2014a). In contrast, we also found that in males of both species, urinary testosterone levels increased around eight years of age (Behringer et al. 2014a). Our previous work is expanded by presenting results based on a larger data set; we include analyses of changes in testosterone levels in blood samples; and provide the first long-term data on urinary testosterone changes in male bonobo subjects.

SUBJECTS, SAMPLE COLLECTION, AND MEASUREMENT

To evaluate and compare developmental patterns in bonobos and chimpanzees, we collected urine and serum samples and measured four biological markers that can be related to different components of growth and development (table 2.1). Laboratory-based work was conducted at the endocrinology lab of the Max Planck Institute for Evolutionary Anthropology in Leipzig and the Justus-Liebig-University Gießen, Germany. All samples came from zoo-housed apes (see acknowledgments). Analytical methods used in this study have been validated, and detailed information about the methods used can be found in Hauser, Deschner, and Boesch (2008), Behringer et al. (2014a), and Behringer et al. (2016). More information on sample extraction dilution factors and other details of the laboratory protocol can also be found in these publications.

STATISTICAL ANALYSES

We ran general linear mixed models (GLMMs [Baayen 2008]), to examine changes of the biological markers with chronological age. All models were run in R (R version 3.2.5, R Core Team 2017) using the lmer function provided in the package lme4 (Bates et al. 2013).

To examine changes in levels of a given marker across lifespan, we built models that included a three-way interaction of the predictor variables age (z-transformed), sex, and species. To approximate a normal distribution, all response variables, total T3, IGFBP-3, DHEA, and testosterone, were log-transformed. The three-way interaction was included in all full models as markers were expected to differ between sexes and species. Our general assumption was that in both species, levels of developmental markers decline with age. Based on results of earlier studies comparing developmental changes of the two *Pan* species, we expected to see marker-specific differences between the species and/or the sexes in terms of the onset and steepness of this decline. For DHEA we expected an inverse u-shape, and therefore included age also as a squared term in the models. In all models, we included the z-transformed time of sample collection as a control variable, and random intercepts for ID and zoo to control for a possible influence of different individuals and for animal husbandry conditions. As the distribution of individuals of different ages varied between zoos, we included random slopes for age and sampling time in zoo, as well as for age and sampling time in ID.

Model assumptions (normal distribution and homogeneity of residuals) were assessed by visual inspections of a histogram, a q–q plot of the residuals, and by plotting residuals against fitted values. All model assumptions were met for all models. We examined Variance Inflation Factors (VIF [Field 2009]) using the function vif of the R-package car (Fox 2011) applied to a standard linear model excluding random effects and slopes. These indicated that collinearity was not an obvious issue (maximum VIF varied between 1.05 [T3 model] and 1.15 [IGFBP-3 model]).

To investigate the significance of the predictors age, species, and sex, and their interactions as a whole, we compared the full model with a null model (excluding predictor variables and interactions, but retaining the random effects and slopes) using a likelihood ratio test (Dobson and Barnett 2008; R function ANOVA). Significance for all tests was set at the $p = 0.05$ level.

For each urinary hormone and metabolite, we compared the full model with the null model. These comparisons revealed significance for all tested models (total T3 $\chi^2 = 24.86$, DF = 7, $p < 0.001$; IGFBP-3 $\chi^2 = 43.13$, DF = 7, $p < 0.001$; DHEA $\chi^2 = 37.22$, DF = 8, $p < 0.001$; testosterone $\chi^2 = 192.72$, DF = 7, $p < 0.001$), indicating that full model explained data variation better than the null model.

Results

URINARY TOTAL T3 LEVELS

For the urinary total T3 level, the three-way interaction of age with species and sex was not significant (Estimate = 0.06; SE = 0.186; p = 0.755), and nor were any of the two-way interactions in the reduced model (age *sex: Estimate = 0.021; SE = 0.092; p = 0.827; age*species: Estimate = 0.04; SE = 0.17; p = 0.715; sex*species: Estimate = −0.184; SE = 0.18; p = 0.311). Therefore, these interactions were removed from the model. The final model revealed that in both species urinary total T3 levels were significantly higher in males than in

TABLE 2.2. Results of the final model of the impact of age, species, and sex on urinary creatinine levels, urinary total T3 levels, urinary IGFBP-3 levels, and urinary DHEA levels during lifetime in bonobos and chimpanzees.

Response Variable	*Term*	*Estimate*	*SE*	*DF*	*P value*
Urinary creatinine	Intercept	−1.132	0.124		
	Daytime (a)	−0.129	0.083	1	0.132
	Species	0.093	0.134	1	0.516
	Age at sampling^2	−0.133	0.045	1	0.006
	Age at sampling*sex	0.208	0.093	1	0.034
Urinary total T3	Intercept	0.58	0.099		
	Daytime (a)	0.07	0.06	1	0.25
	Species	0.361	0.119	1	0.01
	Sex	0.203	0.088	1	0.023
	Age at sampling	−0.21	0.054	1	< 0.001
Urinary IGFBP-3	Intercept	1.916	0.107		
	Daytime (a)	−0.085	0.05	1	0.145
	Sex	−0.158	0.09	1	0.084
	Age at sampling * species	−0.221	0.094	1	0.033
Urinary DHEA	Intercept	3.401	0.233		
	Daytime (a)	−0.238	0.105	1	0.023
	Species	0.247	0.247	1	0.322
	Sex	0.358	0.173	1	0.044
	Age at sampling^2	−0.365	0.111	1	0.001

Note: DHEA, dehydroepiandrosterone; IGFBP-3, insulin-like growth factor-binding protein-3; total T3, total triiodothyronine.

(a) control predictor, * indicates an interaction.

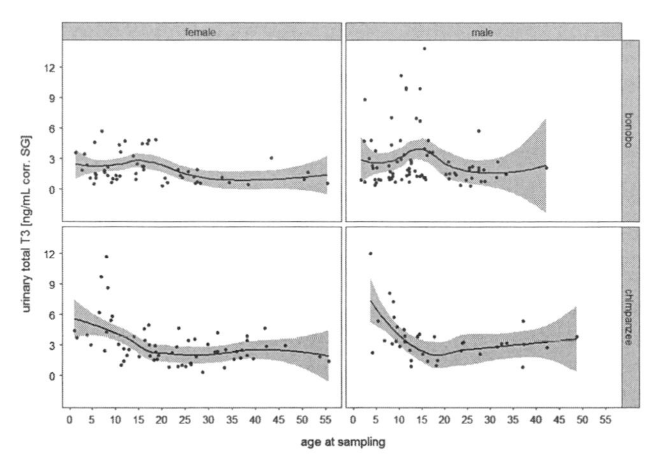

FIGURE 2.1. Measures of urinary total T3 levels in relation to age. Showing data from 100 bonobos (53 females, 47 males) and 100 chimpanzees (64 females, 36 males), collectively contributing 235 urine samples.

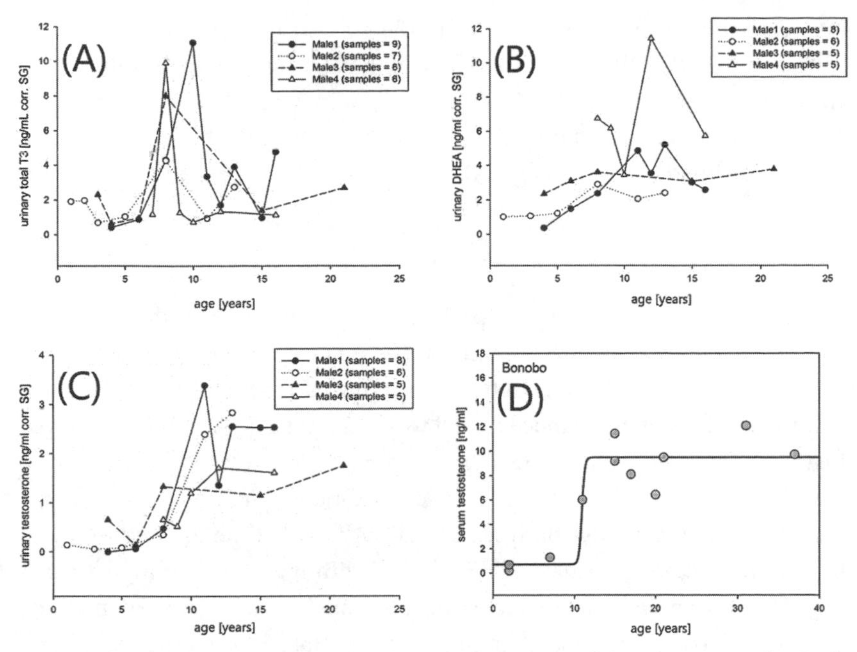

FIGURE 2.2. Changes in four male bonobos during the first 20 years of life in A) urinary total T3 levels, B) urinary DHEA levels, and C) urinary testosterone levels. D) Measures of eleven serum testosterone levels for male bonobos in relation to age.

females, and that they declined with age (table 2.2). Species-specific differences exist in the pattern of age-related changes: in chimpanzees, T3 levels decreased with age from birth onward until 10 to 15 years of age, after which they plateaued. In bonobos, urinary total T3 levels increased from birth until 10–15 years of age, after which they showed a decrease with age (table 2.2 and fig. 2.1). For a subset of four male bonobos, we plotted urinary total T3 level levels measured repeatedly during their first 20 years of life and found the same pattern as the cross-sectional data model, with peaks in T3 around 7–10 years and a decrease between 10 and 15 years (fig. 2.2a).

URINARY IGFBP-3 LEVELS

Variation in urinary IGFBP-3 levels was neither explained by the three-way interaction of age with species and sex (Estimate = -0.012; SE = 0.191; p = 0.952), nor by the two-way interactions of age with sex (Estimate = 0.125; SE = 0.092; p = 0.185) or sex with species (Estimate = 0.051; SE = 0.182; p = 0.783) in the reduced model. The final model revealed that the decline in urinary IGFBP-3 levels with age differed significantly between the two species, but not between males and females (table 2.2 and fig. 2.3). From birth onward, urinary IGFBP-3 levels were higher in chimpanzees than in bonobos. In chimpanzees, the IGFBP-3 levels showed an inverse relationship to age. In contrast, in bonobos urinary IGFBP-3 levels were stable in the first 15 years of life and declined thereafter (fig. 2.3).

URINARY DHEA LEVELS

The variance in urinary DHEA levels was not explained by the three-way interaction of age with species and sex (Estimate = -0.079; SE = 0.446; p = 0.861) and therefore this interaction was removed from the model. All two-way interactions of the reduced model were also not significant and removed (age*sex: Estimate = 0.088; SE = 0.226; p = 0.702; age*species: Estimate = 0.266; SE = 0.221; p = 0.231; sex*species: Estimate = -0.407; SE = 0.35; p = 0.248). In the final model, both sex and age at sample time squared were significant predictors of urinary DHEA levels, with males having higher levels than females, and with levels increasing after birth, reaching a peak between 15 and 25 years of age and declining afterward (table 2.2 and fig. 2.4). For a subset of four bonobo males, for whom we had longitudinal data, we found that individual urinary DHEA patterns reflected the cross-sectional data pattern (fig. 2.2b) and the individual variation in DHEA levels identified previously by Behringer et al. (2012).

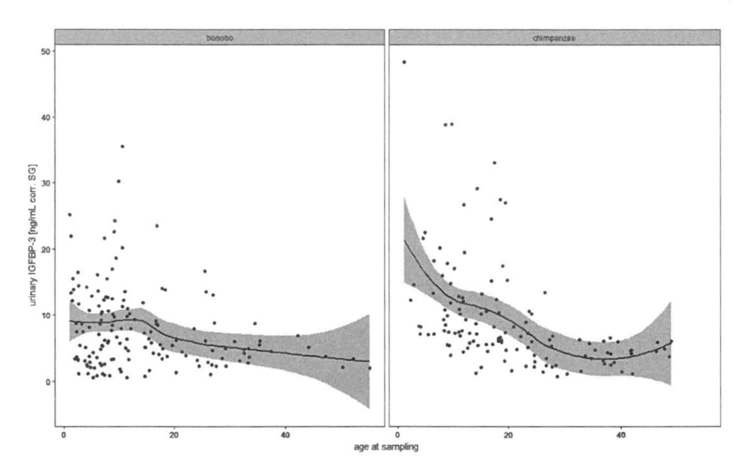

FIGURE 2.3. Measures of urinary IGFBP-3 levels in relation to age. Showing data from 104 bonobos (60 females, 44 males) and 126 chimpanzees (81 females, 45 males), collectively contributing 279 urine samples.

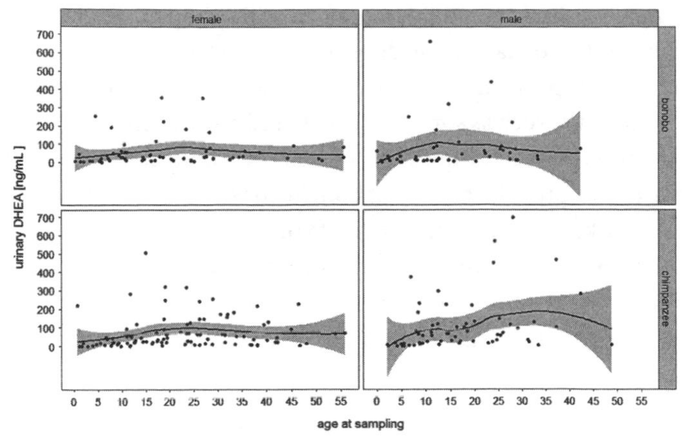

FIGURE 2.4. Measures of urinary DHEA levels in relation to age. Showing data from 107 bonobos (61 females, 46 males) and 156 chimpanzees (96 females, 60 males), collectively contributing 268 urine samples.

URINARY TESTOSTERONE LEVELS

Confirming our earlier results (Behringer et al. 2014a), the analyses of the larger data set showed that an increase in urinary testosterone occurs in female bonobos around the age of five, in male bonobos and chimpanzees around the age of eight, and in female chimpanzees at the age of ten years. Plotting the measures of the four male bonobos confirms the increase of urinary testosterone with age found in the cross-sectional study (fig. 2.2c). Furthermore, with a small sample set of serum samples (11 male bonobos) we were able to replicate the urinary testosterone pattern for males (fig. 2.2d).

Discussion

In comparison to other nonhuman primates, all great ape species grow slowly, experience brain growth later in life, and delay puberty and reproductive maturation (see also Knott and Harwell, chapter 1 this volume). These traits are shared with humans, and it is particularly data from bonobos and chimpanzees that are often considered in evolutionary models tracing ancestral traits of human development (Ponce de León and Zollikofer 2008; Rosati et al. 2007; Wrangham and Pilbeam 2002). Previous work has been strongly biased to somatic growth and to morphometric measures of skeletons and teeth. While this approach provides valuable information about age-related changes in size and shape of body parts, it provides only a snapshot taken at a particular developmental stage of a given individual. In contrast, we analyzed physiological markers from living bonobos and chimpanzees, facilitating the monitoring of the dynamics of developmental changes over time. Having measures of multiple markers, we were now able to study different developmental trajectories, such as somatic growth and reproductive maturation. Focusing attention to the key question of heterochronic developmental differences, our results provide three clear conclusions: (a) species differences exist in the two markers that are indicative of somatic growth (IGFBP-3 and total T3); (b) there are sex differences in markers related to puberty (testosterone) that also differ across the two species; and (c) sex differences in the markers signaling adrenarche (DHEA and DHEA-S) are similar in the two species.

SOMATIC GROWTH

Our data show that bonobos and chimpanzees differ from each other in their development, as assessed by physiological markers. The earliest differentiation in a marker is in urinary IGFBP-3 levels, with chimpanzees of both sexes having higher levels than bonobos, suggesting pronounced differences in somatic growth during infancy and juvenility, with perhaps a higher growth velocity in chimpanzees compared to bonobos. In both species, urinary IGFBP-3 levels converged at age 20, a time when somatic growth is thought to terminate. The differences in urinary IGFBP-3 levels obtained in our study indicate a marked difference in the life history of the two species, such as a different investment in growth during infancy and juvenility. Species differences in somatic growth have been related to the alternating effects of environmental pressures. The ecological risk aversion hypothesis proposes that, at the periphery of the group, juveniles are exposed to mortality risk from predators, while those in the center of the group are safer but face more intense resource competition. In this context, slow somatic growth is considered as an adaption to reduce negative energy balances due to resource competition with more efficient group members (Janson and van Schaik 2002). However, the data for testing this hypothesis for bonobos and chimpanzees are not yet available.

In chimpanzees, urinary total T3 levels declined from birth onward, whereas in bonobos total T3 levels increased from birth to adulthood (10–15 years) and declined thereafter. Although our cross-sectional data set shows substantial noise, the overall pattern of urinary IGFBP-3 and total T3 levels were corroborated by the longitudinal data for four male bonobos. It remains to be explored whether bonobos reach the highest T3 levels later in life than chimpanzees, which would reflect paedomorphism in somatic growth in bonobos, or whether bonobos have an increase in T3, which is absent in chimpanzees.

Another potential factor selecting for differences in somatic growth is mortality risk. If mortality risk increases at certain age classes, investments will shift in order to reach a less risky age class earlier (Stearns 2000). In wild chimpanzees, infant and juvenile mortality is higher than in bonobos (Hill et al. 2001; Muller and Wrangham 2014; Wood et al. 2017), which appears to reflect the impact of male infanticide (Townsend et al. 2007; Watts et al. 2006; Wilson et al. 2014). In wild bonobos, survival probability is unknown, but the few data on infant mortality suggest that this is very low when compared with chimpanzees (Furuichi et al. 1998, unpublished data of G. Hohmann from Lomako and LuiKotale). Both in captivity and in the wild, a greater proportion of chimpanzee infants die before the age of five than bonobos (de Lathouwers and Van Elsacker 2004; Furuichi et al. 1998). Therefore, chimpanzees would increase perhaps their survival probability if immature chimpanzees invest more in somatic growth velocity to terminate the critical period of high mortality risk faster than bonobos.

It is also possible that the differences in somatic growth are influenced by differences in maternal investment (Hochberg and Albertsson-Wikland 2008). Low levels of hormones influencing growth and the inferred delay of somatic growth are considered to be adaptations of immature individuals to enhance parental investment (Haig 2010). Wild female chimpanzees optimize reproductive output by investing in the number of offspring at the expense of offspring body mass (Emery Thompson et al. 2016), a strategy that would turn the potential benefits from slow somatic growth into a risk. In bonobos, females wean their offspring at an older age than chimpanzees do (de Lathouwers and Van Elsacker 2006). Furthermore, female bonobos may care simultaneously for two offspring (Furuichi et al. 1998), and mothers are tolerant when older offspring recruit maternal care such as transport and nipple contact (G. Hohmann, unpublished data from Lomako and LuiKotale). Taken together, these findings suggest extended maternal care to older offspring in bonobos. The trade-off of this strategy is lower reproductive output of mothers that would predict female bonobos have longer inter-birth intervals but increasing the survival of their offspring. However, the limited data that are available suggest that wild female bonobos at Wamba tend to have shorter inter-birth intervals than female chimpanzees (Furuichi et al. 1998). Whether this is due to food provisioning remains to be explored. However, reports from the same site show that female bonobos are able to care simultaneously for two offspring. In modern human societies inter-birth intervals are shorter than in hominoid primates and mothers care for multiple dependent offspring simultaneously; a reproductive strategy that requires that the energetic costs of infant care are shared with others (Sear and Mace 2008). The bonobo equivalent to the communal infant care of humans, may be the high rate of food sharing, the dominant role of females in the division of high-quality food, and the unconstrained access of immatures to divisible food (Hohmann and Fruth 1996).

ONSET OF MATURATION

Previous studies suggested that, because of the differences in male aggression, adult male bonobos may have lower levels of circulating testosterone, or may be less sensitive to testosterone, than male chimpanzees (Muller and Wrangham 2001). Another study predicted differences in male reproductive strategies may be associated with differences in the age-related patterns of testosterone production (Wobber et al. 2013). Those authors reported that testosterone levels of male chimpanzees declined from infancy to juvenility but then increased markedly during adolescence, while, in contrast, testosterone levels in bonobos were stable throughout this developmental period. The data obtained in our study do not confirm these findings: males of both species showed low urinary testosterone levels during infant and juvenile period and a significant rise in urinary testosterone levels after 8–9 years. The fact that the significant rise in testosterone in samples from adolescent males also appeared in both the longitudinal data and in the serum samples, indicates that the result is not an artifact of the cross-sectional data design or of the urine matrices. Moreover, the correspondence in androgen levels of the two species also applies to measures of DHEA signaling, the first postnatal surge of adrenal activity at the end of infancy, which happened in males of both species at the same time.

In contrast to similarities in age-related changes in testosterone levels in males, we found that female bonobos showed an increase in testosterone around five years, whereas female chimpanzees showed a corresponding peak at the age of ten. Provided that the rise in testosterone indicates the onset of puberty and that the onset of puberty triggers female transfer to neighboring groups, one would expect the endocrine differentiation to be associated with behavioral differences (see also Rosenbaum, Santymire, and Stoinski, chapter 6 this volume). In the wild, female bonobos start visiting neighboring groups at five to six years old, which is long before they reach adult size (Furuichi 2011). The few data that are available indicate that there is little or no resistance against immigrants and those new females are able to establish affiliative bonds with resident females (Furuichi 2011; Idani 1991a). This suggests that in bonobos the costs associated with female transfer are low. This is different for female chimpanzees, where immigrants face resistance of resident females and need support of resident males (Kahlenberg et al. 2008). The aggression against immigrant female chimpanzees has been related to the increase in feeding competition (Kahlenberg, Emery Thompson, and Wrangham 2008). Interspecific variation in migratory behavior can be explained with differences in resource abundance, and also by the benefits deriving from alliance formation and bonding among female bonobos (Moscovice, Deschner, and Hohmann 2015; Parish, de Waal, and Haig 2006; Vervaecke, de Vries, and van Elsacker 2000b).

Conclusion

In conclusion, the investment in early somatic growth for chimpanzees seems beneficial because it perhaps terminates the developmental stage with high mortality risk at a younger age. At least in female chimpanzees, large body size may promote dominance rank (Pusey et al. 2005) that, in turn, may increase reproductive success (Pusey and Schroepfer-Walker 2013). In males, dominance status and reproductive success appear to be influenced by social parameters such as cooperation and competition (Foerster et al. 2016). In bonobos, the mortality risk of immatures is low, frequent food sharing reduces the energetic costs of mothers, and the importance of body size for social dominance is reduced. Apart from mortality and maternal investment, the females of the two *Pan* species differ in terms of the impact of body size on social dominance. Female chimpanzees' reproductive success is predicted by access to resources, and the social integration of young females into a new group starts only after they have reached final adult size. In bonobos, females have a large time window to integrate into another group, establish bonds with unrelated females to improve their status, and support each other in conflicts with males. In comparison to the social system of chimpanzees, the bonobo society promotes in females early investment in social bonds, and the somatic and behavioral attributes of immature life stages may actually enhance the motivation of adult females to invest in social relations with immigrants. Under such conditions, a longer period of somatic growth would facilitate social integration, female bonding, and gregariousness.

Through our study we wished to ask at what stage during chimpanzee and bonobo ontogeny do differences in development occur and which proximate mechanisms (in this case hormones) may be responsible for the emergence of the different phenotypes. Our results show clear differences concerning markers of somatic growth (e.g., total T3, IGFBP-3) that are most obvious early in life, a pattern that supports the idea that development of the two species is heterochronic, and that bonobos are paedomorphic when compared with chimpanzees. Yet, looking at other markers, it turns out that heterochrony is not a general feature; the pattern of temporal differentiation varies, with bonobos being delayed in some traits (e.g., somatic growth) while chimpanzees are delayed in others (e.g., female sexual maturation). One dimension of species-differences in development that has received little attention concerns age-related changes in hormone levels between males and females and differences between females of the two species. Longitudinal monitoring of maturation will provide the context for those traits separating humans and primates as well as those that we share.

Acknowledgments

This research would not have been possible without the combined help from care takers, curators, veterinarians, and directors of the zoos contributing samples for hormone measures and other data. Given spatial constraints we are unable to list the names of all the people to whom we owe gratitude and the following institutions must represent all those who have been involved: Aalborg Zoo (Denmark), Allwetterzoo Münster (Germany), Apenheul (Netherlands), Artis (Netherlands), Badoca Safari Park (Portugal), Belfast Zoo (UK), Bergzoo Halle (Germany), Bioparc Fuengirola (Spain), Bioparc Valencia (Spain), Borås Djurpark (Sweden), Burgers' Zoo (Netherlands), Columbus Zoo and Aquarium (USA), Copenhagen Zoo (Denmark), Dudley Zoological Gardens (UK), Furuviksparken (Sweden), Givskud Zoo (Denmark), Kittenberger Kálmán Zoo and Botanical Garden (Hungary), Kristiansand Dyrepark (Norway), La Vallée des Singes (France), Ljubljana Zoo (Slovenia), Milwaukee County Zoo (USA), Ogród Zoologiczny w Warszawie (Poland), Parco Natura Viva (Italy), Planckendael Wild Animal Park (Belgium), San Diego Wild Animal Park (USA), San Diego Zoo (USA), Tierpark Hellabrunn (Germany), Tisch Family Zoological Gardens (Israel), Twycross Zoo (UK), Walter Zoo (Switzerland), Wilhelma Zoologisch-Botanischer Garten, Stuttgart (Germany), Zoo Antwerpen (Belgium), Zoo am Meer (Germany), Zoo Augsburg (Germany), Zoo-Aquarium de la Casa de Campo de Madrid (Spain), Zoo Berlin (Germany), Zoo Cologne (Germany), Zoo Frankfurt (Germany), Zoo Gdańsk (Poland), Zoo Heidelberg (Germany), Zoo Leipzig (Germany), Zoo Magdeburg (Germany), Zoologicka Zahrada Bratislava (Slovakia), ZOOM Erlebniswelt Gelsenkrichen (Germany), Zoo Wuppertal (Germany). The authors are grateful to Lydia Hopper and Steve Ross for the invitation to the conference and for their editorial input on earlier drafts of this chapter.

References

Amsler, S. J. 2010. "Energetic costs of territorial boundary patrols by wild chimpanzees." *American Journal of Primatology* 72: 3–103.

Arthur, W. 2002. "The emerging conceptual framework of evolutionary developmental biology." *Nature* 415: 757–64.

Baayen, R. H. 2008. *Analyzing Linguistic Data*. Cambridge: Cambridge University Press.

Badrian, A., and N. Badrian. 1977. "Pygmy chimpanzees." *Oryx* 13: 463.

Bassett, J. H. D., and G. R. Williams. 2016. "Role of thyroid hormones in skeletal development and bone maintenance." *Endocrine Reviews* 37: 135–87.

Bates, D., M. Maechler, B. Bolker, and S. Walker. 2013. "lme4: Linear mixed-effects models using Eigen and S4." arXiv preprint. arXiv:1406.5823.

Behringer, V., T. Deschner, C. Deimel, J. M. G. Stevens, and G. Hohmann. 2014a. "Age-related changes in urinary testosterone levels suggest differences in puberty onset and divergent life history strategies in bonobos and chimpanzees." *Hormones and Behavior* 66: 525–33.

Behringer, V., T. Deschner, R. Murtagh, J. M. G. Stevens, and G. Hohmann. 2014b. "Age-related changes in thyroid hormone levels of bonobos and chimpanzees indicate heterochrony in development." *Journal of Human Evolution* 66: 83–88.

Behringer, V., G. Hohmann, J. M. G. Stevens, A. Weltring, and T. Deschner. 2012. "Adrenarche in bonobos (*Pan paniscus*): Evidence from ontogenetic changes in urinary dehydroepiandrosterone-sulfate levels." *Journal of Endocrinology* 214: 55–65.

Behringer, V., S. A. Wudy, W. F. Blum, J. M. G. Stevens, T. Remer, C. Boesch, and G. Hohmann. 2016. "Sex differences in age-related decline of urinary insulin-like growth factor-binding protein-3 levels in adult bonobos and chimpanzees." *Frontiers in Endocrinology* 7: 1–10.

Berge, C., and X. Penin. 2004. "Ontogenetic allometry, heterochrony, and interspecific differences in the skull of African apes, using tridimensional Procrustes analysis." *American Journal of Physical Anthropology* 124: 124–38.

Blevins, J. K., J. E. Coxworth, J. G. Herndon, and K. Hawkes. 2013. "Brief communication: Adrenal androgens and aging: Female chimpanzees (*Pan troglodytes*) compared with women." *American Journal of Physical Anthropology* 151: 643–48.

Boesch, C., and H. Boesch-Achermann. 2000. *The Chimpanzees of the Taï Forest: Behavioural Ecology and Evolution*. Oxford: Oxford University Press.

Bogin, B. 1997. "Evolutionary hypotheses for human childhood." *Yearbook of Physical Anthropology* 40: 63–89.

Bolter, D. R., and A. L. Zihlman. 2011. "Brief communication: Dental development timing in captive *Pan paniscus* with comparisons to *Pan troglodytes*." *American Journal of Physical Anthropology* 145: 647–52.

Boughner, J. C., and M. C. Dean. 2008. "Mandibular shape, ontogeny and dental development in bonobos (*Pan paniscus*) and chimpanzees (*Pan troglodytes*)." *Evolutionary Biology* 35: 296–308.

Byrd, W., M. J. Bennett, B. R. Carr, Y. Dong, F. Wians, and W. Rainey. 1998. "Regulation of biologically active dimeric inhibin A and B from infancy to adulthood in the male." *Journal of Clinical Endocrinology and Metabolism* 83: 2849–54.

Collins, D. C., R. D. Nadler, and J. R. K. Preedy. 1981. "Adrenarche in the great apes." Presented at the Fourth Meeting of the American Society of Primatologists, San Antonio, TX.

Coolidge, H. J. 1933. "*Pan paniscus.* Pigmy chimpanzee from south of the Congo River." *American Journal of Physical Anthropology* 18: 1–59.

Coolidge, H. J., and B. T. Shea. 1982. "External body dimensions of *Pan paniscus* and *Pan troglodytes* chimpanzees." *Primates* 23: 245–51.

Copeland, K. C., J. W. Eichberg, C. R. Parker, and A. Bartke. 1985. "Puberty in the chimpanzee: Somatomedin-C and its relationship to somatic growth and steroid hormone concentrations." *Journal of Clinical Endocrinology and Metabolism* 60: 1154–60.

Corruccini, R., and H. McHenry. 1979. "Morphological affinities of *Pan paniscus*." *Science* 204: 1341–43.

Cramer, D. L. 1977. "Craniofacial morphology of *Pan paniscus.* A morphometric and evolutionary appraisal." *Contributions to Primatology* 10: 1–64.

Cutler, G. B., M. Glenn, M. Bush, G. D. Hodgen, C. E. Graham, and D. L. Loriaux. 1978. "Adrenarche: A survey of rodents, domestic animals, and primates." *Endocrinology* 103: 2112–18.

Dahl, J. F. 1986. "Cyclic perineal swelling during the intermenstrual intervals of captive female pygmy chimpanzees (*Pan paniscus*)." *Journal of Human Evolution* 15: 369–85.

de Lathouwers, M., and L. Van Elsacker. 2004. "Comparing maternal styles in bonobos (*Pan paniscus*) and chimpanzees (*Pan troglodytes*)." *American Journal of Primatology* 64: 411–23.

de Lathouwers, M., and L. Van Elsacker. 2006. "Comparing infant and juvenile behavior in bonobos (*Pan paniscus*) and chimpanzees (*Pan troglodytes*): A preliminary study." *Primates* 47: 287–93.

DeLellis, K., S. Rinaldi, R. J. Kaaks, L. N. Kolonel, B. Henderson, and L. Le Marchand. 2004. "Dietary and lifestyle correlates of plasma insulin-like growth factor-I (IGF-I) and IGF binding protein-3 (IGFBP-3): The multiethnic cohort." *Cancer Epidemiology, Biomarkers, and Prevention* 13: 1444–51.

de Peretti, E., and M. G. Forest. 1978. "Pattern of plasma dehydroepiandrosterone sulfate levels in humans from birth to adulthood: Evidence for testicular production." *Journal of Clinical Endocrinology and Metabolism* 47: 572–77.

de Waal, F. B. M. 2001. "Apes from Venus: Bonobos and human social evolution." In *Tree of Origin: What Primate Behavior Can Tell Us about Human Social Evolution*, edited by F. B. M. de Waal, 41–68. Cambridge, MA: Harvard University Press.

de Waal, F. B. M. 2007. *Chimpanzee Politics: Power and Sex among Apes.* 25th anniversary ed. Baltimore: Johns Hopkins University Press.

de Waal, F. B. M., and F. Lanting. 1998. *Bonobo: The Forgotten Ape.* Berkeley: University of California Press.

Dobson, A. J., and A. G. Barnett. 2008. *An Introduction to Generalized Linear Models.* 3rd ed. Boca Raton: CRC Press.

Dorn, L. D., and F. M. Biro. 2011. "Puberty and its measurement: A decade in review." *Journal of Research on Adolescence* 21: 180–95.

Emery Thompson, M., M. N. Muller, K. Sabbi, Z. P. Machanda, E. Otali, and R. W. Wrangham. 2016. "Faster reproductive rates trade off against offspring growth in wild chimpanzees." *Proceedings of the National Academy of Sciences* 113 (28): 7780–85.

Eriksson, J., H. Siedel, D. Lukas, M. Kayser, A. Erler, C. Hashimoto, G. Hohmann, C. Boesch, and L. Vigilant. 2006. "Y-chromosome analysis confirms highly sex-biased dispersal and suggests a low male effective population size in bonobos (*Pan paniscus*)." *Molecular Ecology* 15: 939–49.

Eworo, R., A. Enosakhare, Z. Okhormhe, and A. Udoh. 2015. 'Thyroid and growth hormones interdependence: A 'reciprocal potentiation,' the synergy for growth and development at childhood." *British Journal of Medicine and Medical Research* 9: 1–9.

Field, A. P. 2009. *Discovering Statistics Using SPSS*. 3rd ed. Los Angeles: SAGE Publications.

Foerster, S., M. Franz, C. M. Murray, I. C. Gilby, J. T. Feldblum, K. K. Walker, and A. E. Pusey. 2016. "Chimpanzee females queue but males compete for social status." *Scientific Reports* 6: 35404.

Fox, J. 2011. *An R Companion to Applied Regression*. 2nd ed. Thousand Oaks, CA: SAGE Publications.

Frasier, S. D., F. Gafford, and R. Horton. 1969. "Plasma androgens in childhood and adolescence." *Journal of Clinical Endocrinology and Metabolism* 29: 1404–8.

Furuichi, T. 1989. "Social interactions and the life history of female *Pan paniscus* in Wamba, Zaïre." *International Journal of Primatology* 10: 173–97.

Furuichi, T. 1997. "Agonistic interactions and matrifocal dominance rank of wild bonobos (*Pan paniscus*) at Wamba." *International Journal of Primatology* 18: 855–75.

Furuichi, T. 2011. "Female contributions to the peaceful nature of bonobo society." *Evolutionary Anthropology* 20: 131–42.

Furuichi, T., G. Idani, H. Ihobe, C. Hashimoto, Y. Tashiro, T. Sakamaki, M. N. Mulavwa, K. Yangozene, and S. Kuroda. 2012. "Long-term studies on wild bonobos at Wamba, Luo Scientific Reserve, D.R. Congo: Towards the understanding of female life history in a male-philopatric species." In *Long-Term Field Studies of Primates*, edited by P. M. Kappeler and D. P. Watts, 413–33. Berlin: Springer.

Furuichi, T., G. Idani, H. Ihobe, S. Kuroda, K. Kitamura, A. Mori, T. Enomoto, N. Okayasu, C. Hashimoto, and T. Kano. 1998. "Population dynamics of wild bonobos (*Pan paniscus*) at Wamba." *International Journal of Primatology* 19: 1029–43.

Furuichi, T., and H. Ihobe. 1994. "Variation in male relationships in bonobos and chimpanzees." *Behaviour* 130: 211–28.

Garcia-Bulnes, G., C. Cervantes, M. A. Cerbon, H. Tudon, R. M. Argote, and A. Parra. 1977. "Serum thyrotropin triiodothyronine and thyroxine levels by radioimmunoassay during childhood and adolescence." *Acta Endocrinologica* 86: 742–53.

Gavan, J. A. 1971. "Longitudinal, postnatal growth in chimpanzee." In: *Behavior, Growth, and Pathology of Chimpanzees*, edited by G. H. Bourne, 47–103. Baltimore: University Park Press.

Gerloff, U., B. Hartung, B. Fruth, G. Hohmann, and D. Tautz. 1999. "Intracommunity relationships, dispersal pattern and paternity success in a wild living community of bonobos (*Pan paniscus*) determined from DNA analysis of faecal samples." *Proceedings of the Royal Society of London B* 266: 1189–95.

Gilby, I. C., L J. N. Brent, E. E. Wroblewski, R. S. Rudicell, B. H. Hahn, J. Goodall, and A. E. Pusey. 2013. "Fitness benefits of coalitionary aggression in male chimpanzees." *Behavioral Ecology and Sociobiology* 67: 373–81.

Goodall, J. 1986. *The Chimpanzees of Gombe: Patterns of Behavior*. Cambridge, MA: Belknap Press of Harvard University Press.

Gould, S. J. 1977. *Ontogeny and Phylogeny*. Cambridge, MA: Belknap Press of Harvard University Press.

Gould, S. J. 1988. "The uses of heterochrony." In *Heterochrony in Evolution*, edited by M. L. McKinney, 1–13. Boston: Springer.

Groves, C. P. 1986. "Systematics of the great apes." In *Comparative Primate Biology, Vol. 1: Systematics, Evolution and Anatomy*, 187–217. New York: Liss.

Grumbach, M. M. 2002. "The neuroendocrinology of human puberty revisited." *Hormone Research* 57: 2–14.

Grunfeld, C., B. M. Sherman, and R. R. Cavalieri. 1988. "The acute effects of human growth hormone administration on thyroid function in normal men." *Journal of Clinical Endocrinology and Metabolism* 67: 1111–14.

Haig, D. 2010. "Transfers and transitions: Parent-offspring conflict, genomic imprinting, and the evolution of human life history." *Proceedings of the National Academy of Sciences* 107: 1731–35.

Hamada, Y., K. Chatani, T. Udono, Y. Kikuchi, and H. Gunji. 2003. "A longitudinal study on hand and wrist skeletal maturation in chimpanzees (*Pan troglodytes*), with emphasis on growth in linear dimensions." *Primates* 44: 259–71.

Hauser, B., T. Deschner, and C. Boesch. 2008. "Development of a liquid chromatography–tandem mass spectrometry method for the determination of 23 endogenous steroids in small quantities of primate urine." *Journal of Chromatography B* 862: 100–112.

Hill, K., C. Boesch, J. Goodall, A. Pusey, J. Williams, and R. Wrangham. 2001. "Mortality rates among wild chimpanzees." *Journal of Human Evolution* 40: 437–50.

Hochberg, Z. 2012. *Evo-Devo of Child Growth: Treatise on Child Growth and Human Evolution*. Hoboken: Wiley-Blackwell.

Hochberg, Z., and K. Albertsson-Wikland. 2008. "Evo-devo of infantile and childhood growth." *Pediatric Research* 64: 2–7.

Hohmann, G., and B. Fruth. 1996. "Food sharing and status in unprovisioned bonobos." In *Food and the Status Quest: An Interdisciplinary Perspective*, edited by P. W. Wiessner and W. Schiefenhövel, 47–68. Providence: Berghahn Books.

Hohmann, G., and B. Fruth. 2002. "Dynamics and social organization of bonobos (*Pan paniscus*)." In *Behavioural Diversity in Chimpanzees and Bonobos*, edited by C. Boesch, G. Hohmann, and L. F. Marchant. Cambridge: Cambridge University Press.

Hohmann, G., U. Gerloff, D. Tautz, and B. Fruth. 1999. "Social bonds and genetic ties: Kinship, association and affiliation in a community of bonobos (*Pan paniscus*)." *Behaviour* 136: 1219–35.

Horn, A. D. 1979. "The taxonomic status of the bonobo chimpanzee." *American Journal of Physical Anthropology* 51: 273–81.

Ibáñez, L., J. DiMartino-Nardi, N. Potau, and P. Saenger. 2000. "Premature adrenarche-normal variant or forerunner of adult disease?" *Endocrine Reviews* 21: 671–96.

Idani, G. 1990. "Relations between unit-groups of bonobos at Wamba, Zaire: Encounters and temporary fusions". *African Study Monographs* 11: 153–86.

Idani, G. 1991a. "Social Relationships between Immigrant and Resident Bonobo (*Pan paniscus*) Females at Wamba." *Folia Primatologica* 57: 83–95.

Idani, G. 1991b. "Case of inter-unit group encounters in pygmy chimpanzees at Wamba, Zaire." In *Primatology Today: Proceedings of the XIIIth Congress of the International Primatological Society, Nagoya and Kyoto, 18–24, July 1990*, edited by A. Ehara, A. Kimura, O. Takenaka, and M. Iwamoto, 231–34. Amsterdam: Elsevier.

Idani, G. 2003. "The unitgroups of wild bonobos; by instances of inter-unitgroup encounters." *Primate Research* 19: 23–31.

Jacobs, J. D., and J. C. Wingfield. 2000. "Endocrine control of life-cycle stages: A constraint on response to the environment?" *Condor* 102: 35–51.

Janson, C. H., and C. P. van Schaik. 2002. "Ecological risk aversion in juvenile primates: Slow and steady wins the race." In *Juvenile Primates: Life History, Development, and Behavior*, edited by M. E. Pereira and L. A. Fairbanks, 57–76. Chicago: University of Chicago Press.

Janssen, J. A. M. J. L., R. P. Stolk, H. P. A. Pols, D. E. Grobbee, F. H. de Jong, and S. W. J. Lamberts. 1998. "Serum free IGF-I, total IGF-I, IGFBP-1 and IGFBP-3 levels in an elderly population: Relation to age and sex steroid levels." *Clinical Endocrinology* 48: 471–78.

Kahlenberg, S. M., M. Emery Thompson, M. N. Muller, and R. W. Wrangham. 2008. "Immigration costs for female chimpanzees and male protection as an immigrant counterstrategy to intrasexual aggression." *Animal Behaviour* 76: 1497–1509.

Kahlenberg, S. M., M. Emery Thompson, and R. W. Wrangham. 2008. "Female competition over core areas in *Pan troglodytes schweinfurthii*, Kibale National Park, Uganda." *International Journal of Primatology* 29: 931–47.

Kano, T. 1980. "Social behavior of wild pygmy chimpanzees (*Pan paniscus*) of Wamba: A preliminary report." *Journal of Human Evolution* 9: 243–60.

Kano, T. 1992. *The Last Ape: Pygmy Chimpanzee Behavior and Ecology*. Stanford: Stanford University Press.

Klingenberg, C. P. 1998. "Heterochrony and allometry: The analysis of evolutionary change in ontogeny." *Biological Reviews* 73: 79–123.

Kuroda, S. 1980. "Social behavior of the pygmy chimpanzees." *Primates* 21: 181–97.

Kuroda, S. 1989. "Developmental retardation and behavioral characteristics of pygmy chimpanzees." In *Understanding Chimpanzees*, edited by P. G. Heltne and L. A. Marquardt, 184–93. Cambridge, MA: Harvard University Press.

Lalwani, S., R. H. Reindollar, and A. J. Davis. 2003. "Normal onset of puberty." *Obstetrics and Gynecology Clinics of North America* 30: 279–86.

Langer, J. 2006. "The heterochronic evolution of primate cognitive development." *Biological Theory* 1: 41–43.

Langergraber, K. E., H. Siedel, J. C. Mitani, R. W. Wrangham, V. Reynolds, K. Hunt, and L. Vigilant. 2007. "The genetic signature of sex-biased migration in patrilocal chimpanzees and humans." *PLoS One* 2: e973.

Leigh, S. R., and B. T. Shea. 1996. "Ontogeny of body size variation in African apes." *American Journal of Physical Anthropology* 99: 43–65.

Lieberman, D. E., J. Carlo, M. Ponce de León, and C. P. E. Zollikofer. 2007. "A geometric morphometric analysis of heterochrony in the cranium of chimpanzees and bonobos." *Journal of Human Evolution* 52: 647–62.

McHenry, H. M., and R. S. Corruccini. 1981. "*Pan paniscus* and human evolution." *American Journal of Physical Anthropology* 54: 355–67.

McKinney, M. L., and K. McNamara. 1991. *Heterochrony: The Evolution of Ontogeny*. New York: Plenum Press.

Michaud, P., A. Foradori, J. A. Rodriguez-Portales, E. Arteaga, J. M. Lopez, and R. Tellez. 1991. "A prepubertal surge of thyrotropin precedes an increase in thyroxine and 3,5,3'-triiodothyronine in normal children." *Journal of Clinical Endocrinology and Metabolism* 72: 976–81.

Mitani, J. C., and D. P. Watts. 2001. "Boundary patrols and intergroup encounters in wild chimpanzees." *Behaviour* 138: 299–327.

Mitani, J. C., and D. P. Watts. 2005. "Correlates of territorial boundary patrol behaviour in wild chimpanzees." *Animal Behaviour* 70: 1079–86.

Mitani, J. C., D. P. Watts, and S. J. Amsler. 2010. "Lethal intergroup aggression leads to territorial expansion in wild chimpanzees." *Current Biology* 20: R507–8.

Moscovice, L. R., T. Deschner, and G. Hohmann. 2015. "Welcome back: Responses of female bonobos (*Pan paniscus*) to fusions." *PLoS One* 10: e0127305.

Muller, M. N., S. M. Kahlenberg, and R. W. Wrangham. 2009. "Male aggression against females and sexual coercion in chimpanzees." In *Sexual Coercion in Primates and Humans: An Evolutionary Perspective on Male Aggression against Females*, edited by M. N. Muller and R. W. Wrangham, 184–217. Cambridge, MA: Harvard University Press.

Muller, M. N., M. E. Thompson, S. M. Kahlenberg, and R. W. Wrangham. 2011. "Sexual coercion by male chimpanzees shows that female choice may be more apparent than real." *Behavioral Ecology and Sociobiology* 65: 921–33.

Muller, M. N., and R. W. Wrangham. 2001. "The reproductive ecology of male hominoids." In *Reproductive Ecology and Human Evolution, Evolutionary Foundations of Human Behavior*, edited by P. T. Ellison, 397–428. New York: Aldine de Gruyter.

Muller, M. N., and R. W. Wrangham. 2014. "Mortality rates among Kanyawara chimpanzees." *Journal of Human Evolution* 66: 107–14.

Nishida, T. 1989. "Social interactions between resident and immigrant female chimpanzees." In *Understanding Chimpanzees*, edited by P. G. Heltne and L. A. Marquardt, 68–89. Cambridge, MA: Harvard University Press.

Nishida, T., H. Takasaki, and Y. Takahata, Y. 1990. "Demography and reproductive profiles." In *The Chimpanzees of the Mahale Mountains: Sexual and Life History Strategies*, edited by T. Nishida. Tokyo: University of Tokyo Press.

Parish, A. R., F. B. M. de Waal, and D. Haig. 2006. "The other 'closest living relative': How bonobos (*Pan paniscus*) challenge traditional assumptions about females, dominance, intra- and intersexual interactions, and hominid evolution." *Annals of the New York Academy of Sciences* 907: 97–113.

Parra, A., S. Villalpando, E. Junco, B. Urquieta, S. Alatorre, and G. Garcia-Bulnes. 1980. "Thyroid gland function during childhood and adolescence." *European Journal of Endocrinology* 93: 306–14.

Ponce de León, M. S., and C. P. E. Zollikofer. 2008. "Neanderthals and modern humans—chimps and bonobos: Similarities and differences in development and evolution." In *Neanderthals Revisited: New Approaches and Perspectives*, edited by J. J. Hublin, K. Harvati, and T. Harrison, 71–88. Dordrecht: Springer.

Pusey, A. E., G. W. Oehlert, J. M. Williams, and J. Goodall. 2005. "Influence of ecological and social factors on body mass of wild chimpanzees." *International Journal of Primatology* 26: 3–31.

Pusey, A. E., and K. Schroepfer-Walker. 2013. "Female competition in chimpanzees." *Philosophical Transactions of the Royal Society of London B* 368: 20130077.

R Core Team, 2016. "R: A language and environment for statistical computing." R foundation for statistical computing. Vienna, Austria.

Rosati, A. G., J. R. Stevens, B. Hare, and M. D. Hauser. 2007. "The evolutionary origins of human patience: Temporal preferences in chimpanzees, bonobos, and human adults." *Current Biology* 17: 1663–68.

Ryness, J. 1972. "The measurement of serum thyroxine in children." *Journal of Clinical Pathology* 25: 726–29.

Schwarz, E. 1929. "Das Vorkommen des Schimpansen auf dem linken Kongo-Ufer." *Revue de Zoologie et de Botanique Africaines* 16: 424–26.

Sear, R., and R. Mace. 2008. "Who keeps children alive? A review of the effects of kin on child survival." *Evolution and Human Behavior* 29: 1–18.

Shea, B. T. 1983a. "Allometry and heterochrony in the African apes." *American Journal of Physical Anthropology* 62: 275–89.

Shea, B. T. 1983b. "Paedomorphosis and neoteny in the pygmy chimpanzee." *Science* 222: 521–22.

Shea, B. T. 1985. "The ontogeny of sexual dimorphism in the African apes." *American Journal of Primatology* 8: 183–88.

Smail, P. J., C. Faiman, W. C. Hobson, G. B. Fuller, and J. S. D. Winter. 1982. "Further studies on adrenarche in nonhuman primates." *Endocrinology* 111: 844 48.

Stearns, S. C. 2000. "Life history evolution: Successes, limitations, and prospects." *Naturwissenschaften* 87: 476–86.

Styne, D. M. 1994. "Physiology of puberty." *Hormone Research* 41 (2): 3–6.

Surbeck, M., C. Girard-Buttoz, C. Boesch, C. Crockford, B. Fruth, G. Hohmann, K. E. Langergraber, K. Zuberbühler, R. M. Wittig, and R. Mundry. 2017. "Sex-specific association patterns in bonobos and chimpanzees reflect species differences in cooperation." *Royal Society Open Science* 4: 161081.

Surbeck, M., R. Mundry, and G. Hohmann. 2011. "Mothers matter! Maternal support, dominance status and mating success in male bonobos (*Pan paniscus*)." *Proceedings of the Royal Society of London B* 278: 590–98.

Townsend, S. W., K. E. Slocombe, M. Emery Thompson, and K. Zuberbühler. 2007. "Female-led infanticide in wild chimpanzees." *Current Biology* 17: R355–56.

Tuttle, R. 1975. "Parallelism, brachiation, and hominoid phylogeny." In *Phylogeny of the Primates*, edited by W. P. Luckett and F. S. Szalay, 447–80. Boston: Springer.

Venturi, S., and M. E. Begin. 2010. "Thyroid hormone, iodine and human brain evolution." In *Human Brain Evolution: The Influence of Freshwater and Marine Food Resources*, edited by S. C. Cunnane and K. M. Stewart. Hoboken: John Wiley and Sons.

Vervaecke, H., H. de Vries, and L. van Elsacker. 2000a. "Dominance and its behavioral measures in a captive group of bonobos (*Pan paniscus*)." *International Journal of Primatology* 21: 47–68.

Vervaecke, H., H. de Vries, and L. van Elsacker. 2000b. "Function and distribution of coalitions in captive bonobos (*Pan paniscus*)." *Primates* 41: 249–65.

Vinicius, L., and M. M. Lahr. 2003. "Morphometric heterochrony and the evolution of growth." *Evolution* 57: 2459–68.

Watts, D. P. 1998. "Coalitionary mate guarding by male chimpanzees at Ngogo, Kibale National Park, Uganda." *Behavioral Ecology and Sociobiology* 44: 43–55.

Watts, D. P., M. Muller, S. J. Amsler, G. Mbabazi, and J. C. Mitani. 2006. "Lethal intergroup aggression by chimpanzees in Kibale National Park, Uganda." *American Journal of Primatology* 68: 161–80.

Wilson, M. L., C. Boesch, B. Fruth, T. Furuichi, I. C. Gilby, C. Hashimoto, C. L. Hobaiter, G. Hohmann, N. Itoh, K. Koops, J. N. Lloyd, T. Matsuzawa, J. C. Mitani, D. C. Mjungu, D. Morgan, M. N. Muller, R. Mundry, M. Nakamura, J. Pruetz, A. E. Pusey, J. Riedel, C. Sanz, A. M. Schel, N. Simmons, M. Waller, D. P. Watts, F. White, R. M. Wittig, K. Zuberbühler, and

R. W. Wrangham. 2014. "Lethal aggression in *Pan* is better explained by adaptive strategies than human impacts." *Nature* 513: 414–17.

Wobber, V. E. 2012. "Comparative cognitive development and endocrinology in *Pan* and *Homo*." PhD diss., Harvard University.

Wobber, V. E., B. Hare, S. Lipson, R. Wrangham, and P. Ellison. 2013. "Different ontogenetic patterns of testosterone production reflect divergent male reproductive strategies in chimpanzees and bonobos." *Physiology and Behavior* 116–17: 44–53.

Wobber, V. E., R. Wrangham, and B. Hare. 2010. "Bonobos exhibit delayed development of social behavior and cognition relative to chimpanzees." *Current Biology* 20: 226–30.

Wood, B. M., D. P. Watts, J. C. Mitani, and K. E. Langergraber. 2017. "Favorable ecological circumstances promote life expectancy in chimpanzees similar to that of human hunter-gatherers." *Journal of Human Evolution* 105: 41–56.

Wrangham, R. W. 1986. "Ecology and social relationships in two species of chimpanzee." In *Ecological Aspects of Social Evolution: Birds and Mammals*, 333–52. Princeton, NJ: Princeton University Press.

Wrangham, R. W., W.C. McGrew, F.B.M. de Waal, and P.G. Heltne. 1996. *Chimpanzee Cultures*. Cambridge, MA: Harvard University Press.Wrangham, R. W., and D. Peterson. 1996. *Demonic Males: Apes and the Origins of Human Violence*. Boston: Houghton Mifflin.

Wrangham, R. W., and D. Pilbeam. 2002. "African apes as time machines." In *All Apes Great and Small*, edited by B. M. F. Galdikas, N. E. Briggs, L. K. Sheeran, G. L. Shapiro, and J. Goodall, 5–17. Boston: Kluwer Academic Publishers.

Wrangham, R. W., and B. B. Smuts. 1980. "Sex differences in the behavioural ecology of chimpanzees in the Gombe National Park, Tanzania." *Journal of Reproduction and Fertility* 28: 13–31.

Zelditch, M. L., and W. L. Fink. 1996. "Heterochrony and heterotopy: Stability and innovation in the evolution of form." *Paleobiology* 22: 241–54.

Zihlman, A. L., and D. L. Cramer. 1978. "Skeletal differences between pygmy (*Pan paniscus*) and common chimpanzees (*Pan troglodytes*)." *Folia Primatologica* 29: 86–94.

3

Dolphins and Chimpanzees: A Case for Convergence?

JANET MANN, MARGARET A. STANTON, AND CARSON M. MURRAY

Introduction

Bottlenose dolphins (Indo-Pacific *Tursiops aduncus* and common *T. truncatus*) are often compared with chimpanzees and bonobos (*Pan troglodytes* and *P. paniscus*) because of similarities in life history, social structure, and key behavioral traits such as fission-fusion dynamics, alliance formation, and sociocognitive abilities (tables 3.1 and 3.2). Convergence in these traits is even more striking when we consider the vast separation in their evolutionary histories and ecological differences. The order Primates and the superorder Cetartiodactyla, which includes cetaceans and even-toed ungulates, have a common ancestor approximately 90 million years ago (MYA). Cetaceans diverged from terrestrial artiodactyls about 50–53 MYA and divergence between *T. aduncus* and *T. truncatus* probably occurred at least 11–15 MYA (McGowen, Spaulding, and Gatesy 2009; Steeman et al. 2009). Phylogeny within the subfamily delphininae is not resolved, nor is it known whether the *Tursiops* genus is monophyletic or polyphyletic (Amaral et al. 2012; Jedensjö, Kemper, and Krützen 2017). In the case of the genus *Pan*, despite the close genetic relationship and some shared behavioral traits, there are notable differences between the two species in terms of intrasexual bonds, territoriality, and tool use, which have diverged since the two species split approximately 1–2 MYA (Prüfer et al. 2012).

Approach

In this chapter, we compare dolphins with chimpanzees across a range of traits and consider how they inform our understanding of the evolution of complex social systems (table 3.2, Gruber and Clay 2016; White and Wrangham 1988). We focus particularly on females because of our expertise, but also because the socioecology of females drives social structure (Kappeler and van Schaik

2002). Our understanding of dolphins and chimpanzees is shaped by inherent phylogenetic differences between species, demography, and local ecology. As dolphins and chimpanzees are long-lived, we focus almost exclusively on longitudinal field studies spanning 10 years or more.

Results

LIFE HISTORY

Life history theory provides a comparative framework for understanding reproductive and survival strategies, including the temporal patterning of development and reproduction across the lifespan (see also Knott and Harwell, chapter 1 this volume; Behringer et al., chapter 2 this volume). The theory can be applied to between-species comparisons, as well as within-species comparisons since local ecology and intrinsic factors may alter the tradeoffs for each individual differently (e.g., Altmann and Alberts 2003, 2005; Emery Thompson et al. 2016). Table 3.1 summarizes major life history variables for the two genera. Notably, there is complete overlap between the two. The

TABLE 3.1. Summary of chimpanzee and dolphin life history parameters by study site.

Milestone	*Species*	*Study Site*	*Average and/or Range (years)*	*References*
Weaning[1]	*P. troglodytes*	Gombe, Tanzania	3–5	Clark 1977; Pusey 1983; Van de Rijt-Plooij and Plooij 1987
	P. troglodytes	Kanyawara, Uganda	5.2	Machanda et al. 2015
	P. paniscus	Wamba, Democratic Republic of Congo	4–5	Kuroda 1989
	T. aduncus	Shark Bay, Australia	4; 2.3–8.3	Karniski, Krzyszczyk, and Mann 2018; Mann et al. 2000
	T. aduncus	Mikura Island, Japan	3.5; 3–5	Kogi et al. 2004
	T. truncatus	Sarasota, Florida, USA	1.5–2	Wells and Scott 2009
Sexual maturation[2]	*P. troglodytes*	Gombe, Tanzania	11.5	Walker et al. 2018
	P. troglodytes	Mahale, Tanzania	10.7	Emery Thompson 2013[3]
	P. troglodytes	Bossou, Guinea	8.5	Emery Thompson 2013
	T. aduncus	Shark Bay, Austaralia	10	Mann 2019; Mann et al. 2000
	T. truncatus	Sarasota, Florida, USA	5–10	Wells 2003
	T. truncatus	Moray Firth, Scotland	5–9	Robinson et al. 2017

TABLE 3.1. *(continued)*

Milestone	*Species*	*Study Site*	*Average and/or Range (years)*	*References*
Female dispersal	*P. troglodytes*	Kanyawara, Uganda	12.9	Emery Thompson 2013
	P. troglodytes	Gombe, Tanzania	12.5	Emery Thompson 2013
	P. troglodytes	Mahale, Tanzania	11.0	Emery Thompson 2013
	T. aduncus	Shark Bay, Australia	N/A	Tsai and Mann 2013
	T. truncatus	Sarasota, Florida, USA	N/A	Wells 2003
Age at first birth	*P. troglodytes*	Gombe, Tanzania	14.9 philopatric females; 16.2 dispersing females	Walker et al. 2018
	P. troglodytes	Mahale, Tanzania	13.2	Emery Thompson 2013
	P. troglodytes	Bossou, Guinea	10.9	Emery Thompson 2013
	P. troglodytes	Taï, Côte d'Ivoire	13.7	Emery Thompson 2013
	T. aduncus	Shark Bay, Australia	Mean=13.0; 9–20	Mann 2019
	T. truncatus	Sarasota, Florida, USA	5–10	Wells 2003
	T. truncatus	Moray Firth, Scotland	5–9	Robinson et al. 2017
	T. truncatus	SW Atlantic, Brazil	8–10	Fruet et al. 2015
Interbirth[4] interval (IBI)	*P. troglodytes*	Kanyawara, Uganda	6.6	Emery Thompson 2013
	P. troglodytes	Gombe, Tanzania	5.6	Emery Thompson 2013
	P. troglodytes	Mahale, Tanzania	6.0	Emery Thompson 2013
	P. troglodytes	Budongo, Uganda	5.2	Emery Thompson 2013
	P. troglodytes	Bossou, Guinea	5.3	Emery Thompson 2013
	P. troglodytes	Ngogo, Uganda	5.3	Watts 2012
	P. troglodytes	Tai, Côte d'Ivoire	5.8	Emery Thompson 2013
	T. aduncus	Shark Bay, Australia	4.6; 1–15 (surviving offspring average IBI is 4.9, minimum is 3)	Karniski, Krzyszczyk, and Mann 2018; Mann et al. 2000
	T. aduncus	Mikura Island, Japan	3.4	Kogi et al. 2004
	T. truncatus	Sarasota, Florida, USA	4.0; 2–6	Wells 2003
	T. truncatus	SW Atlantic, Brazil	3.0 (with surviving offspring, IBI range is 2–6 years)	Fruet et al. 2015
	T. truncatus	Moray Firth, Scotland	3.8; 2–9	Robinson et al. 2017

(continues)

TABLE 3.1. *(continued)*

Milestone	*Species*	*Study Site*	*Average and/or Range (years)*	*References*
Max lifespan (females)	*P. troglodytes*	Gombe, Tanzania	54.0	Bronikowski et al. 2011
	T. aduncus	Shark Bay, Australia	47+	Karniski et al. 2018; Mann et al. 2000
	T. truncatus	Sarasota, Florida, USA	65	Wells 2003, unpublished
Max lifespan (males)	*P. troglodytes*	Gombe, Tanzania	44.0	Bronikowski et al. 2011
	T. aduncus	Shark Bay, Australia	47+	Mann et al. 2000, unpublished
	T. truncatus	Sarasota, Florida, USA	40+	Wells 2003, *Naples Daily News* 2015

[1] Weaning is based on the age when suckling was last observed but it should be viewed with caution since comfort suckling does not necessary indicate milk transfer and night nursing could still be an important source of nutrition, even when day suckling is not observed. For Kanywara chimpanzees, we calculated an average from the maximum estimate in Machanda et al. 2015.

[2] For chimpanzees, sexual maturation was assumed as the age at first maximum swelling. For dolphins, it is based on first known conception.

[3] Emery Thompson (2013) provides a thorough review of female reproductive parameters so we refer to it and have only added more recent references where appropriate.

[4] Interbirth intervals include all intervals, regardless of infant survival, unless otherwise stated.

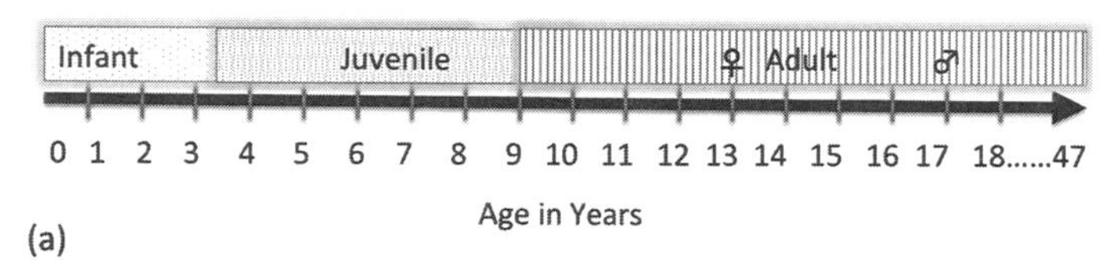

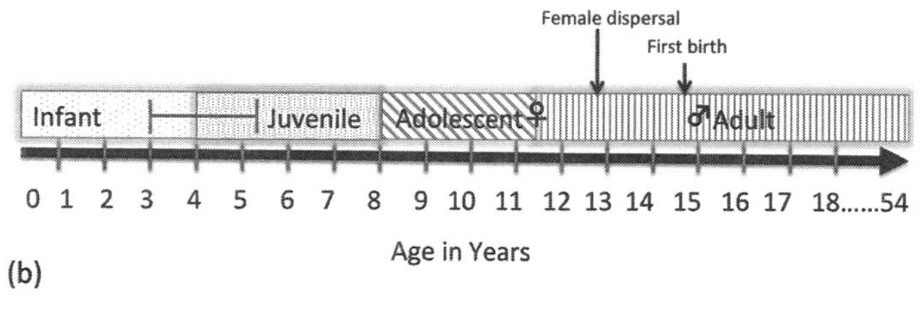

FIGURE 3.1. Typical life history parameters for a) wild dolphins and b) wild chimpanzees. Some dolphin calves continue to nurse as late as 8 years of age. An adolescent phase has not been characterized in dolphins. Female dolphins have their first calf, on average, at age 13 and males achieve their earliest paternities in their late teens. Maximum lifespan is not known but some dolphins are in their late 40s at present. See table 3.1 for more details. For chimpanzees, we show the average life history variables from the Kasekela community at Gombe National Park, Tanzania, but we include range bars for the end of infancy at other study sites and demonstrate the variation in weaning and interbirth interval. The end of female adolescence corresponds to the average of sexual maturation. Males in this and other populations have been categorized as adults at age 15. See table 3.1 for more details.

convergence between dolphin and chimpanzee life history parameters, as shown in figure 3.1, is superficially surprising given the extreme differences in habitat and energetic costs (discussed later in this chapter).

MATERNAL BEHAVIOR AND OFFSPRING DEVELOPMENT

Bottlenose dolphins and chimpanzees all have extended life histories characterized by long, slow periods of growth, delayed sexual maturation, and long lifespans (Charnov and Berrigan 1993; Harvey, Martin, and Clutton-Brock 1987; Mann et al. 2000; Whitehead and Mann 2000). As in most mammals, the mother-infant relationship is primary in these species; however, differences between the species in terms of physical precociality have important consequences for the mother-infant relationship and offspring development (table 3.2). Dolphin neonates are physically precocious as they must surface to breathe and swim to follow their mothers immediately after birth, an ability more similar to ungulate calves than primate infants (Mann and Smuts 1998; Ralls, Lundrigan, and Kranz 1987). Dolphin calves begin by swimming close and parallel to their mothers in echelon position, and soon move to infant position under the mother (fig. 3.2). Both the echelon and infant positions have been equated to infant carrying in primates (Miketa et al. 2018; fig. 3.3), for the hydrodynamic benefits that both of these positions provide to the calf and the energetic costs to the mother (Noren et al. 2008). The precocial locomotion of bottlenose dolphin calves allows individuals as young as 2–12 weeks of age to actively join or leave other individuals, including their mothers (Mann and Smuts 1998; Mann and Watson-Capps 2005). Interestingly, these temporary mother-calf separations continue throughout infancy and appear to emulate the fission-fusion dynamics (i.e., subgroups that change size and composition frequently) exhibited by adults of the species (Galezo, Krzyszczyk, and Mann 2018), while providing calves with the opportunity for self-socialization (Gibson and Mann 2008a; Mann and Watson-Capps 2005; Stanton, Gibson, and Mann 2011). Conversely, chimpanzee infants are in almost constant contact with their mothers for the first 4–6 months of life (Goodall 1986). Indeed, young chimpanzees in the Kasekela community, Tanzania, do not travel predominantly under their own power until around 3.5 years old on average (Lonsdorf et al. 2014a).

Given prolonged offspring dependency, both dolphin and chimpanzee mothers have considerable influence over their offspring's development; however, we have only recently begun to understand the variation in maternal behavior and mother-offspring relationships in wild populations. Despite

TABLE 3.2. Summary of key similarities and differences between bottlenose dolphins and chimpanzees

Trait	*Key similarities*	*Key differences*
Foraging ecology	• Reliance on foods that are heterogeneously distributed in space and time. • Low predation risk in most populations.	• Predominantly hunting in dolphins, gathering in chimpanzees. • Dolphins do not share food, while chimpanzees do. • Extractive foraging with tools common in chimpanzees but rare or nonexistent in dolphins. • Specialization possibly more common in dolphins.
Social organization	• Fission-fusion dynamics in both species.	• Larger and more open communities in dolphins than chimpanzees. • Male philopatry in chimpanzees, bisexual philopatry in dolphins, but association biased towards matrilineal kin. • Stronger sex segregation in dolphins compared to chimpanzees.
Social relationships	• Intrasexual long-term bonds present in both species. • Aggression and alliances are important features of social life but the scope and context varies between the species. • Infanticide reported among common chimpanzees and some dolphin populations.	• Nested-hierarchical alliances in dolphins, but chimpanzees have only first-order alliances. • Clear dominance hierarchies detectable in chimpanzees, but not in dolphins.
Life history	• Very similar life histories marked by a long dependency period.	• Chimpanzees have conspicuous sexual swellings, while dolphins have concealed ovulation and it is common for pregnancy to overlap with last year of nursing the current calf.
Maternal behavior	• Females invest intensively in each offspring. • Males do not care for offspring. • Offspring have protracted developmental periods during which extensive observational learning is possible.	• Very young dolphin calves frequently break contact and travel hundreds of meters from their mothers, whereas chimpanzee infants rarely break contact in the first six months. • Dolphins begin catching their own fish in the first months of life while chimpanzees do not forage independently until years later. Food sharing between mother and offspring absent in dolphins but common in chimpanzees. • Juvenile dolphins, particularly sons, are on their own post-weaning, but chimpanzee juveniles typically maintain strong association with their mothers.

FIGURE 3.2. Tool-using female dolphin, Torrent, wearing a marine sponge, and her offspring, Monsoon, in Shark Bay, Australia. Monsoon is swimming in infant position, under the mother. It is from this position that all nursing takes place. Like carrying in primates, this infant position exerts some hydrodynamic costs on the mother. Photo by Ewa Krzyszczyk, monkeymiadolphins.org.

precocial locomotion and opportunities for self-socialization, in Shark Bay, Australia, there is a relationship between dolphin calf sociality and maternal behavior. For example, a calf's number of associates is positively related to number of maternal associates, but male calves will seek out more associates during separations than female calves, particularly if the mother tends to be solitary (Gibson and Mann 2008a). Furthermore, maternal time spent foraging negatively correlates with gregariousness for mother and calf (Gibson and Mann 2008a), and maternal foraging behavior predicts the development of more difficult foraging tactics in their offspring, such that calves, particularly females, "do what mother does" (Mann et al. 2008; Sargeant and Mann 2009).

Like bottlenose dolphins, chimpanzees live in societies characterized by a high level of fission-fusion dynamics; however, compared to dolphin calves, chimpanzee infants are not physically precocious, and their social partners are confined to those individuals with whom the mother associates, particularly early in life. Interestingly, in the Kasekela community at Gombe, Tanzania, it has been found that chimpanzee mothers with male infants are more gregarious than mothers with female infants, raising the intriguing possibility that mothers are providing social opportunities for their sons that facilitate the development of sex-appropriate behaviors (Murray et al. 2014). However,

FIGURE 3.3. Photo of chimpanzee mother, Gremlin (born in 1970), and her infant Grendal (born in 2015). Grendal is riding "jockey style." Gombe Stream Reserve, Tanzania. Photo by M. A. Stanton.

how these differences in early socialization relate to adult outcomes is the topic of ongoing investigation. Indeed, sex differences in social behavior appear early in this species. For example, infant males travel under their own power earlier, increase distance from their mother earlier, have more social partners, and interact more with adult males compared to infant females (Lonsdorf et al. 2014a, 2014b). Conversely, young female chimpanzees learn to termite fish earlier compared to young males (Lonsdorf, Eberly, and Pusey 2004). These early differences foreshadow sex differences in chimpanzee adult behavior.

Compared to chimpanzees, socialization and social development during the infant period may be particularly important for bottlenose dolphins. During the first year of life, Shark Bay calves spend an increasing amount of time socializing, at which point it appears to peak (Mann and Watson-Capps 2005; Gibson and Mann 2008a). As they approach weaning, calves of both sexes increase the amount of time separated from their mothers; however, there are sex differences in behavior during separations, with female calves increasing their time spent foraging and decreasing their time in groups (Gibson and Mann 2008a). After weaning, bottlenose dolphin mother-offspring association decreases with daughters maintaining stronger bonds with their mothers than sons (Krzyszczyk et al. 2017; Mann et al. 2000; McHugh et al. 2011; Tsai and Mann 2013). Thus, these newly weaned juveniles must navigate their ecological and social surroundings in the absence of direct maternal care, and spend, on average, 42% of their time alone (Krzyszczyk et al. 2017; Mann et al. 2000; McHugh et al. 2011; Tsai and Mann 2013). Notably, mother-calf party formation during the infant period appears to enable calves, particularly males, to develop social skills (Gibson and Mann 2008a), and those males who are less socially integrated as calves are subsequently less likely to survive the juvenile period (Stanton and Mann 2012).

In terms of maternal investment, both taxa exhibit terminal investment and/or reproductive senescence. Older Shark Bay dolphin mothers wean calves late, at 5–8 years of age, compared to young mothers, who wean calves at 3–4 years age. Older females have longer interbirth intervals than younger (Karniski, Krzyszczyk, and Mann 2018), a pattern also documented in older chimpanzee females in both the Kanyawara community in Uganda (Emery Thompson et al. 2016) and Kasekela community in Tanzania (Jones et al. 2010). Furthermore, at the start of reproductive life, first-time Kasekela chimpanzee mothers behaviorally invest more in their infants (via grooming, playing, and nursing) compared to experienced mothers and, contrary to the pattern observed in most primates, firstborn chimpanzee infants are no more likely to die than laterborns (Stanton et al. 2014). Interestingly, disproportionately high levels of firstborn mortality were recently reported in the Ngogo chimpanzee community in Uganda (Wood et al. 2017), but data on the behavior of mothers based on parity in that community are not yet available. In Shark Bay, dolphin calf survival decreases with maternal age in a linear fashion (Karniski, Krzyszczyk, and Mann 2018). In contrast, first-born dolphin mortality is high in Sarasota, Florida, due to offloading of organochlorine compounds in milk (Wells et al. 2005). The long juvenile period, high social tolerance between females, and precociousness of calves might lessen the demand of dolphin maternal experience with first-borns. How experience with calves during the juvenile period shapes maternal behavior and/or calf survival with first-borns has not yet been determined.

Unlike bottlenose dolphins, juvenile chimpanzees continue to travel with their mothers and younger sibling after weaning for an additional 4–5 years until adolescence (Goodall 1986; Pusey 1983, 1990). Recent work indicates that chimpanzee mothers can continue to have considerable influence on offspring outcomes even after infancy. For example, in Mahale, Tanzania, orphaned male chimpanzees face significantly lower odds of survival than non-orphaned individuals, even if they lose their mothers after weaning (Nakamura et al. 2014). In the Kasekela community in Tanzania, natal males "inherit" their mother's core areas, continuing to concentrate space use to the area in which their mother ranged even after her death (Murray et al. 2008). Furthermore, Kasekela females who have a mother alive in the community, "jump the queue" and enter the adult female dominance hierarchy at a higher position than those who do not have a mother alive and present in the community (Foerster et al. 2016). Additionally, mothers and siblings appear to gain some benefit from continued association with juvenile offspring through the additional social interaction and/or bond. In Kasekela, juveniles remain strong grooming partners with their mothers (Pusey 1983, 1990; Stanton et al. 2017; Watts and Pusey 2002) and offspring who have an older sibling tend to be more likely to survive than those without an older sibling (Stanton et al. 2017).

SOCIAL ORGANIZATION

Bottlenose dolphins and chimpanzees share many aspects of their social organization, including multi-male, multi-female groups, promiscuous mating, alliance formation, and a fission-fusion social system in which subgroups or "parties" change in size and composition over the course of minutes or hours (table 3.2, see also Boesch 1996; Connor et al. 2000; Galezo, Krzyszczyk, and Mann 2018; Goodall 1986; Nishida 1968; Tsai and Mann 2013). These fission-fusion dynamics provide a useful lens to examine the fitness consequences of individual differences in sociality and social preferences, since individuals can be found alone or in subgroups of differing composition.

Bottlenose dolphin community size ranges from around 60 to well over 600 individuals depending on how community boundaries are defined. Subgroup size varies widely between resident populations from 3.8 (Campbell, Bilgre, and Defran 2002) to 17.2 (Lusseau et al. 2003). In Shark Bay, where there are over 600 individuals in a connected network (Mann et al. 2012), the average subgroup size is 4.8 individuals; however, the size and composition of these subgroups is likely dependent on social context (Galezo, Krzyszczyk, and Mann 2018; Gero et al. 2005; Smolker et al. 1992). Chimpanzee community size ranges from fewer than 20 individuals to over 200 and their average subgroup size varies substantially as well (recent community sizes from the smallest Bossou, Guinea, and Tai North, Côte d'Ivoire, communities: Fujisawa et al. 2016; Luncz and Boesch 2015; and the largest community of Ngogo, Uganda: Langergraber et al. 2017). Chimpanzee sites have reported subgroup size averages that range from three in the Republic of Congo (Morgan and Sanz 2003) to 15 in Senegal (Pruetz and Bertolani 2009).

While dolphin and chimpanzee societies share many similarities given their fission-fusion dynamics, there are noteworthy differences in terms of community openness and territoriality. Most bottlenose dolphin communities are "open," composed of a mix of residential and temporary individuals, with influxes of migratory members for periods of time (e.g., Cheney et al. 2013; Lusseau et al. 2003; Nekolny et al. 2017; Smolker et al. 1992; Wells 2003). However, bottlenose dolphin communities can vary from small discrete and closed communities, to those larger, open, and mixed communities that fluctuate seasonally. For example, the common bottlenose dolphin community in Doubtful Sound, New Zealand, is small and closed within a fjord, even though they have access to dolphins outside of it (Lusseau et al. 2003), while the Sarasota Bay community in Florida has a residential community of about 160 members, with non-residents from the Gulf of Mexico occasionally sighted in the Bay (Wells 2003). In Shark Bay, there are no well-defined community boundaries among Indo-Pacific bottlenose dolphins, just overlapping subcommunities (Mann et al. 2012; Smolker et al. 1992). Open communities tend to be large, where individuals might have hundreds of associates (Mann et al. 2012). All bottlenose dolphin societies show some degree of modularity, with stronger ties amongst subcommunity members than between.

Open communities have never been reported for chimpanzees. Male chimpanzees actively patrol boundaries (Goodall 1979; Watts and Mitani 2001), which carries energetic costs incurred by traveling more and feeding less than when not on patrol (Amsler 2010), and cooperatively defend boundaries through vocalizations and physical aggression (Boesch et al. 2008; Wilson and Wrangham 2003). Intergroup interactions can be lethal (reviewed in Wilson et al. 2014) and result in the extermination of a neighboring community (Goodall 1986), or expansion of territorial boundaries (Mitani, Watts, and Amsler 2010) and increased access to food (Williams et al. 2004; Wrangham 1999). Peaceful interactions between community members have never been reported for any chimpanzee study site. In contrast, aggressive intercommunity interactions have not been reported for bottlenose dolphins, although *interspecific* interactions between bottlenose dolphins and spotted dolphins

(*Stenella frontalis*) are often aggressive (Cusick and Herzing 2014). The cause(s) for such aggression are not known.

Another key difference between dolphins and chimpanzees surrounds dispersal patterns. The large sizes and fluid dynamics of many bottlenose dolphin communities (ranging usually from hundreds to thousands) may be one reason that bottlenose dolphins maintain bisexual philopatry (Tsai and Mann 2013). By comparison, in chimpanzees there is female-biased dispersal, a rare phenomena in mammals, including primates (Strier, Lee, and Ives 2014), but important for reducing inbreeding. In the Kasekela chimpanzee community, where ~50% of females do not disperse, some incestuous matings occur, but studies have demonstrated avoidance between maternal siblings (Pusey 1980; van Lawick-Goodall 1968; but see Stumpf et al. 2009) while another study reported surprisingly high rates of mating between fathers and daughters (Wroblewski 2010). Nonetheless, a recent study reported relatively few cases of inbreeding between close relatives, defined as having a degree of relatedness > 0.25; Walker et al. (2017) found that only four of 68 known paternities were between close relatives. Bottlenose dolphins do have some inbreeding as well, with deleterious effects (Frère et al. 2010a), but mothers and their adult sons generally avoid one another, especially when the mother is cycling (Wallen, Krzyszczyk, and Mann 2017).

SOCIAL BONDS

Like most mammals, dolphins and chimpanzees exhibit sex differences in social behavior that reflect species-specific reproductive strategies (table 3.2). In both species, males are considered the more gregarious sex and tend to form the strongest intrasexual social bonds. Enduring male-male bonds are associated with alliance formation in bottlenose dolphins; however, there is intrapopulational variability in the presence and prevalence of male alliances. In Shark Bay, males form first-order alliances composed of two to three strongly bonded individuals who cooperatively consort with females, as well as second-order alliances described as moderate bonds between members of first-order alliances, with additional evidence for a third level (Connor and Krützen 2015). Alliance membership is essential for mating success in Shark Bay (Krützen et al. 2004). Multilevel alliance formation has also been reported in the St. Johns River, Florida, while only first-order alliances exist in the Bahamas (Parsons et al. 2003), Port Stephens, Australia (Wiszniewski, Brown, and Möller 2012), and Sarasota, Florida (Owen, Wells, and Hofmann 2002). Interestingly, male-male alliances are absent in two well-studied populations where sexual dimorphism is substantial: Moray Firth, Scotland (Eisfeld and Robinson 2004), and Doubtful Sound, New Zealand (Lusseau 2007). Male-male bonds among chimpanzees, particularly eastern chimpanzees, can be enduring (Mitani 2009), carry fitness benefits (Gilby et al. 2013; Mitani 2009), and also function in cooperative hunting and territorial defense (Muller and Mitani 2005). In contrast to dolphins, mate-guarding coalitions of male chimpanzees have only been observed in the unusually large Ngogo community (Watts 1998), raising the possibility that socio-demographic factors, like group size and operational sex ratio, predict the occurrence of consorting alliances in both species.

Female dolphins and chimpanzees are broadly considered the less gregarious sex, particularly when anestrous; however, female sociality varies substantially both within and between populations of the same species or subspecies. This variation is expected based on socio-ecological theory with females responding to differing food distributions and predation pressures. The prevailing evidence suggests that social bonds between unrelated females in chimpanzees are far less prevalent than between males; however, chimpanzees at Taï National Forest, Côte d'Ivoire, are subject to higher predation risks than many eastern chimpanzee populations and females are more gregarious than their eastern counterparts. This population is considered "bisexually bonded" since males and females associate at high levels (Boesch 1996; Lehmann and Boesch 2005; Wittiger and Boesch 2013). Similarly, the Doubtful Sound dolphin population lives in a low-productivity habitat at the extreme southern edge of the species range and this small community features a greater proportion of strong, long-lasting associations within and between sexes, including between females (Lusseau et al. 2003). Even in communities where female sociability is lower than males, bonds persist between mother and daughter (chimpanzee: Goodall 1986, dolphin: Tsai and Mann 2013) and among unrelated female dyads or social cliques (chimpanzee: Fawcett 2000; Gilby and Wrangham 2008; Langergraber, Mitani, and Vigilant 2009; Wakefield 2013; dolphin: Frère et al. 2010b; Mann et al. 2012). Within populations, differences in female sociability are related to factors including reproductive state (chimpanzees: Goodall 1986; Matsumoto-Oda et al. 1998; dolphins: Smolker et al. 1992; Wallen, Krzyszczyk, and Mann 2017), presence, age, and sex of infants (chimpanzees: Foerster et al. 2015; Murray et al. 2014; Otali and Gilchrist 2006; dolphins: Gibson and Mann 2008a, 2008b; Mann and Watson-Capps 2005), presence of kin (chimpanzees: Foerster et al. 2015; dolphins: Frère et al. 2010b), and foraging tactics (chimpanzees: Lonsdorf 2006; dolphins: Mann et al. 2008, 2012; Smolker et al. 1997).

Interestingly, no dolphin field studies have reported direct indicators of female competition or dominance hierarchies. As fish resources are not alienable, there is little for female dolphins to directly compete over, especially since hunting is a solitary activity (Galezo, Krzyszczyk, and Mann 2018; Gibson and Mann 2008a). Agonism between females is remarkably rare (Scott et al. 2005). In Shark Bay, the one place agonism does occur is at a provisioning site where up to five adult females accept fish handouts from humans (Foroughirad and Mann 2013). While the fission-fusion social organization may mitigate female competition to some extent if females avoid each other to minimize competition, intrasexual competition is still a pervasive feature of female chimpanzee social life in some populations. Several chimpanzee populations have now characterized female dominance hierarchies as categorical (Pusey, Williams, and Goodall 1997; Wrangham, Clark, and Isabirye-Basuta 1992) and cardinal (Kahlenberg et al. 2008; Murray 2007; Newton-Fisher 2006). At Taï National Forest, Côte d'Ivoire, where females chimpanzees are highly gregarious, Wittig and Boesch (2003) found that female hierarchies were linear and that higher-ranking females won contests over food resources. The only other site to suggest linear hierarchies is Gombe National Park, Tanzania (Kasekela community: Foerster et al. 2016), where females are less gregarious and prefer to forage in differentiated core areas. At this site, high-ranking females had higher reproductive success (Pusey, Williams, and Goodall 1997) and significantly more productive core areas (Murray, Eberly and Pusey 2006), a result also found in the Kanyawara community in Kibale National Forest, Uganda (Emery Thompson et al. 2007). While the presence of dominance hierarchies should attenuate direct physical aggression, it is noteworthy that female chimpanzees can be severely aggressive to each other in two contexts: female-led infanticide and severe aggression toward immigrant females (Pusey et al. 2008; reviewed in Pusey and Schroepfer-Walker 2013). Researchers have suggested that these behaviors reflect long-term competition.

Intersexual conflict is an important feature of both dolphin and chimpanzee social life, often viewed through the lens of sexual coercion. Sexual coercion is a pervasive feature of bottlenose dolphin society at most sites, although the degree of allied coercion varies. For example, in Doubtful Sound, New Zealand, males do not form alliances and, although sexually dimorphic, intersexual aggression has not been reported (Lusseau 2007). Males compete, and even head butt, in Doubtful Sound, but aggression is not directed at females (Lusseau 2007). In contrast, sexual conflict and allied aggression toward females is common at several sites, most notably Shark Bay (Connor and Krützen 2015; Scott et al. 2005; Wallen, Patterson, and Krzyszczyk 2016), but also in southeast Australia (Möller et al. 2001), the Bahamas (Parsons et al. 2003) and, to a lesser degree, in Sarasota, Florida (Owen, Wells, and Hofmann 2002). Consortships of (usually cycling) females by the male dolphin alliances typically feature biting, hitting, chasing, and threat displays (Connor and Krützen 2015).

Male chimpanzees frequently direct aggression toward females both during estrous and in non-estrous periods, but studies have found an inconsistent relationship between aggression and mating behavior (Muller et al. 2007; Stumpf and Boesch 2006, 2010). Nonetheless, more coercive males in the Kasekela community at Gombe National Park, Tanzania, were more likely to sire offspring than less coercive males (Feldblum et al. 2014). As noted above, mate-guarding coalitions of males have only been observed in the large Ngogo chimpanzee community (Watts 1998), while consortships between an individual male and female appear to be more common in the Kasekela community than at other study sites (Boesch and Boesch-Achermann 2000; Reynolds 2005), where they are most often used by low-ranking males with less desirable females (Wroblewski et al. 2009) and result in approximately 20% of the conceptions (Constable et al. 2001; Wroblewski et al. 2009).

Male aggression toward females also occurs in the context of intercommunity interactions in chimpanzees and infanticide in both dolphins and chimpanzees (chimpanzees: reviewed in Wilson et al. 2014). Reports of infanticide vary among populations of common bottlenose dolphins (Moray Firth, Scotland: Patterson et al. 1998; and Virginia, USA: Dunn et al. 2002), but appears to be rare in Indo-Pacific bottlenose dolphins and possibly other sites where males and females are long-term residents (Wallen, Krzyszczyk, and Mann 2017). Overall, differences in male-female aggression between and within species are likely related to multiple factors such as familiarity and stability within communities, kinship, paternity confidence, and access to females when they resume cycling post-infanticide (table 3.2).

FORAGING ECOLOGY

Given the vastly different ecosystems that dolphins and chimpanzees inhabit, it is unsurprising that they differ dramatically in their foraging ecology (table 3.2). While both species rely on foods that are heterogeneous in space and time, this is more pronounced for dolphins, which rely exclusively on mobile prey. Dolphins are piscivores, although they also eat cephalopods (squid, octopus) and crustaceans (shrimp). Essentially, dolphins are hunters who use both echolocation and vision to locate prey and must then accelerate to chase and capture individual prey. Although large fish schools can attract substantial aggregations of dolphins, bottlenose dolphins tend to hunt alone or in small groups in most locations. Cooperative hunting has been reported at several locations (Duffy-Echevarria, Connor, and St Aubin 2008; Torres and Read 2009) and even human-dolphin cooperative fishing occurs, a cultural phenomenon that has lasted for more than a century in Laguna, Brazil (Daura-Jorge et al. 2012). Dolphins are catholic feeders, enjoying a wide diversity of prey. Not all fish are edible or easy prey, as some are poisonous (e.g., blowfish), have spines (e.g., flathead, lionfish), or are difficult to swallow whole or break up due to their size. As dolphins do not chew their prey, large prey items are typically broken into smaller pieces using the seafloor. Unfortunately, human observers can rarely see fish catches and the type of prey dolphins consume as fish are swallowed head first, instantly. Only large fish are brought to the surface. Dietary information is occasionally available through collection of scat, or stomach analysis of otoliths (ear bones from fish) from stranded animals. Fatty acids in blubber also provide some clues as long as fatty acid signatures are also known for relevant prey species.

Like dolphins, chimpanzees rely on food resources that are heterogeneously distributed. They are ripe-fruit specialists and fruit consistently contributes 60–70% of the diet at all long-term chimpanzee study sites (reviewed in Conklin-Brittain, Knott, and Wrangham 2001). Behavioral adaptations, facilitated by fission-fusion dynamics, and a reliance on fallback foods allow chimpanzees to maintain a relatively consistent diet of fruit despite seasonal scarcity and pronounced variation in chimpanzee habitats, which range from dense forests to more open environments. For example, Fongoli chimpanzees in Senegal live in a semi-arid habitat but fruit still comprises 63% of their diet (Pruetz 2006; Pruetz, Bogart, and Lindshield, chapter 17 this volume). Like dolphins, female chimpanzees in some populations minimize competition by foraging alone or with dependent offspring. For example, mothers of the Kasekela community in Gombe National Park, can spend up to 60% of their time in small family groups (Williams et al. 2002). However, females and males of the western chimpanzee subspecies are more gregarious, which is thought to reflect differences in predation pressure, food patch size, and the presence of fallback foods. In addition to fruit, chimpanzees also consume leaves, pith, nuts acquired through tool use at some study sites, and fauna through insectivory that is sometimes tool-assisted (see below, and also Pruetz. Bogart, and Lindshield, chapter 17 this volume; Luncz and van de Waal, chapter 18 this volume) and cooperative hunting for arboreal monkeys (Boesch 2002; Mitani, Merriwether and Zhang 2000; Vale and Brosnan, chapter 15 this volume). Within chimpanzees, faunivory patterns vary by sex as females consume a larger percentage of insects while cooperative hunting is predominantly a male activity (but see Gilby et al. 2017; Pruetz et al. 2015; and also Pruetz, Bogart, and Lindshield, chapter 17 this volume).

One of the most interesting aspects of dolphin foraging behavior is that they have developed a range of hunting tactics for specific prey. In Shark Bay over 20 distinct foraging tactics used by dolphins to aid in prey capture have been identified (Mann and Patterson 2013; unpublished data; Mann and Sargeant 2003; Sargeant et al. 2005); however, most adult females become specialists in one or two foraging tactics, although the same tactic might be used to acquire multiple prey species. Similar patterns have been reported in common bottlenose dolphins studied in Sarasota, Florida (Rossman et al. 2015). Possibly, habitat heterogeneity (Sargeant et al. 2007) and the evasive talents of fish have favored specialization. In addition, learning the habits of a limited range of fish might be easier than trying to catch what one can. This specialization has important implications for female dolphin range, as many foraging tactics are habitat-specific (Sargeant et al. 2007). As such, females form home ranges that overlap extensively with both males and other females (Tsai and Mann 2013), but likely have preferred core areas within those ranges. Relative to females, male dolphins tend to be generalists, likely because they must have large ranges in order to find and consort with cycling females (Randić et al. 2012; Wallen, Patterson, and Krzyszczyk 2016). While individual foraging specialization has not been reported for chimpanzees, overlapping core areas are also characteristic of eastern chimpanzees living in the Kasekela community in Tanzania and the Kanyawara community in Uganda (Emery Thompson et al. 2007; Murray, Mane, and Pusey 2007; Williams et al. 2002).

Given specialization, range size, solitary hunting, and prey mobility, direct competition between dolphin females over food is low. In Shark Bay, stealing prey from another dolphin is exceedingly rare. Although most fish are swallowed whole quickly, and large fish attract younger dolphins to observe (Mann, Sargeant, and Minor 2007), others will rarely touch the fish that another has caught. Even where cooperative hunting occurs (e.g., Gazda et al. 2005), bottlenose dolphins do not share their fish. Offspring must catch their own from a very young age (Mann and Smuts 1999), and appear to learn extensively from observation (including echolocation) (Sargeant and Mann 2009). Some foraging tactics take many years to perfect (Patterson, Krzyszczyk, and Mann 2015; Sargeant et al. 2005). In contrast, food sharing is more common among chimpanzees (see also Taglialatela et al., chapter 4 this volume; Yamamoto, chapter 14 this volume). Meat sharing is well documented in chimpanzees (e.g., Boesch and Boesch 1989; Nishida et al. 1992; Stanford et al. 1994; Mitani and Watts 2001; Gilby 2006), with some researchers suggesting that meat sharing may garner reproductive access and/or promote social bonds (Gomes and Boesch 2009; Nishida et al. 1992; but see Gilby et al. 2010; Mitani and Watts 2001). Sharing of plant foods is less common among chimpanzees, but has also been reported, particularly between mother and offspring (Goodall 1986).

TOOL USE

Tool use is rare in bottlenose dolphins (Mann and Patterson 2013) but ubiquitous in chimpanzees (Whiten et al. 1999). Chimpanzees use tools in a variety of contexts, especially extractive foraging, but dolphin tool use is almost exclusively for foraging. There are some critical similarities, particularly in the mode of transmission, which appears to be almost exclusively vertical, from mother to offspring, in both wild dolphins and chimpanzees (chimpanzee: Lonsdorf 2006, although see Luncz and van de Waal, chapter 18 this volume; dolphin: Mann et al. 2008). Another stark similarity is the female bias in tool use reported for dolphins and chimpanzees: females are more avid tool users than males (Boesch and Boesch 1990; Gruber, Clay, and Zuberbühler 2010; Mann et al. 2008; Pruetz, Bogart, and Lindshield, chapter 17 this volume). Lonsdorf, Eberly, and Pusey (2004) demonstrated that this sex bias begins early in life as young female chimpanzees start termite fishing and gain proficiency earlier in life than young males.

While chimpanzee tool use was first reported in the 1960s (Goodall 1986, with earlier reports from captivity, Köhler 1925; Yerkes 1925), sponge tool-use by dolphins was discovered in 1984 (Smolker et al. 1997) and "sponger" dolphins have been studied ever since. To date, regular sponge-tool use is found only in Shark Bay, Australia, although other sites around Australia have reported one-off sightings of dolphins carrying sponges (Mann and Patterson 2013). About four percent of the females in Shark Bay regularly use basket sponges that they detach from the seafloor, to ferret prey from a cluttered substrate in 7–16 m channels (Mann et al. 2008). The sponge is worn over the beak (fig. 3.2) and protects the beak from sharp shell and rocks below. It enables the dolphin to scare up fish that lack swimbladders and are thus "inaudible" via echolocation and are well-camouflaged bottom-dwellers (Patterson and Mann 2011). Strong vertical learning of sponging is evident, particularly from mother to daughter (>90% of daughters of "spongers," sponge-using dolphins, become regular spongers themselves) but also from mother to son (59%) (Mann and Patterson 2013; Mann et al. 2008). Spongers spend more time hunting than other dolphins, and their predominant foraging tactic is sponging—a clear specialization (Mann et al. 2008). Each sponge is used for multiple dives and multiple foraging bouts but is typically discarded in about an hour. Expertise in sponging improves with age, up until age 20—in that older females use their tools for longer and spend less time searching for tools (Patterson, Krzyszczyk, and Mann 2015). Developmentally, sponging appears later than most other hunting tactics, as calves do not begin sponging until age two or three years (Mann and Sargeant 2003; Sargeant and Mann 2009). Of the five dolphins observed sponging in 1984 and 1985, all continued to sponge for the rest of their lives.

Physiologically, chimpanzees and other primates have distinct hand morphology that allows for complex object manipulation (Aversi-Ferreira et al. 2010). Dolphins are at the other extreme, streamlined for effective movement through the marine environment and not built for object manipulation. This is likely why few study sites have reported tool use of any kind (Mann and Patterson 2013). So while chimpanzees are phenotypically biased for tool use, bottlenose dolphins are biased against it (Meulman, Seed, and Mann 2013). As a result, it is not surprising that while chimpanzees have been reported to use over 40 different tool types (Sanz and Morgan 2007), only one or two tool forms have been reported for each bottlenose dolphin species (Mann and Patterson 2013). Furthermore, sponging dolphins might need to inhibit their "natural" tendencies since they are tasked with searching for prey in an entirely new way. They typically rely on echolocation, which might not be efficient with fish that lack swimbladders. It also seems likely they would need to inhibit a predisposed resistance to putting something over the beak and face, which might interfere with echolocation and grasping prey. Nevertheless, Shark Bay bottlenose dolphins have found a way to exploit an otherwise empty niche (Patterson and Mann 2011).

Conclusions and Future Directions

The taxa focused on this chapter, bottlenose dolphins and chimpanzees, exhibit remarkable convergence in terms of social system, social relationships, and life history despite divergent morphological, ecological, and phylogenetic features. While members of both taxa feed on heterogeneously distributed resources, chimpanzees are primarily gatherers, while dolphins are exclusively hunters. This fundamental difference in foraging behavior suggests that food distribution is an unlikely candidate driving convergence between members of the two genera. However, social dynamics and extensive social learning continually prove to be important.

Figure 3.4 provides a snapshot of the social structure of the Shark Bay

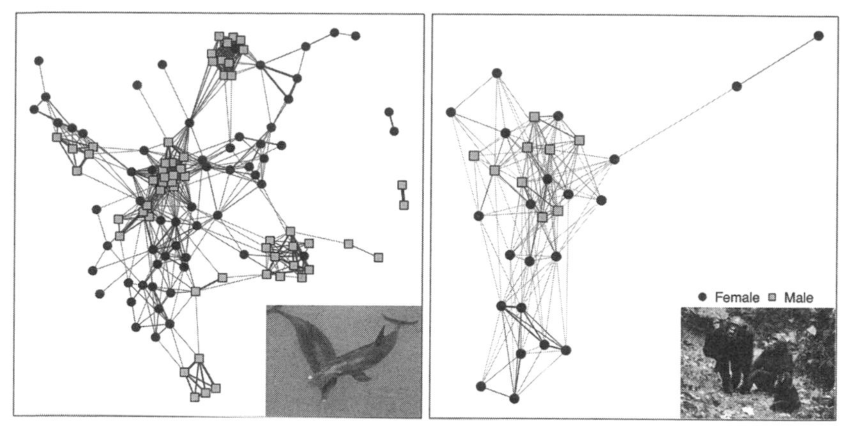

FIGURE 3.4. The social networks of Shark Bay bottlenose dolphins (eastern gulf, Australia) from 2012 to 2014 and Kasekela chimpanzees (Tanzania) from 2013 to 2014 generated using the igraph package (Csardi and Nepusz 2006) for R (version 3.3.3; R Core Development Team 2016). Nodes are individuals, and connections between individuals represent half-weight indices (HWI) of association based on shared group membership in dolphins and joint arrivals in chimpanzees. Dolphins sighted on more than 10 days and chimpanzees with more than 20 joint arrivals were included in their respective networks. Line thickness is proportionate to the HWI of that dyad and only those associations greater than the mean are represented. Photos by M. A. Stanton.

dolphin and Kasekela chimpanzee communities where differences, possibly driven in part by community size and openness, are evident. The strength of male allies and tight female clusters are more evident in the dolphins than chimpanzees. Higher modularity within a community might be more important when communities are large and not bounded. Notably, the accumulation of long-term data on different populations of both dolphins and chimpanzees referenced throughout this chapter is revealing interesting and sometimes extensive inter-population variability in foraging behavior, social behavior, and life history traits. Thus, we note that one of the most important similarities between these taxa is the capacity for behavioral flexibility. In addition to interpopulational differences, members of both genera exhibit substantial interindividual differences in foraging behavior (including tool use in extractive foraging) and sociability. Overall, understanding whether similar factors underlie within-species variation at both the population and individual levels will help reveal those pressures, or releases from constraint, responsible for this remarkable evolutionary convergence.

Longitudinal study at multiple sites is paramount for understanding ecological, social, and phylogenetic factors shaping large brains, slow life histories, and complex social dynamics. Here, we highlighted some of the similarities and differences between taxa with markedly different evolutionary histories and ecology, while also considering the notable variation within each species and even within populations. Continued dialogue between scientists and a comparative framework will help us move the field forward and ultimately inform our understanding of the largest brains on land and in the sea.

References

Altmann, J., and S. C. Alberts. 2003. "Variability in reproductive success viewed from a life history perspective in baboons." *American Journal of Human Biology* 15: 401–9.

Altmann, J., and S. C. Alberts. 2005. "Growth rates in a wild primate population: Ecological influences and maternal effects." *Behavioral Ecology and Sociobiology* 57 (5): 490–501.

Amaral, A. R., J. A. Jackson, L. M. Möller, L. B. Beheregaray, M. Manuela Coelho, 2012. "Species tree of a recent radiation: The subfamily Delphininae (Cetacea, Mammalia)." *Molecular Phylogenetics and Evolution* 64 (1): 243–53.

Amsler, S. J. 2010. "Energetic costs of territorial boundary patrols by wild chimpanzees." *American Journal of Primatology* 72 (2): 93–103.

Aversi-Ferreira, T. A., R. Diogo, J. M. Potau, G. Bello, J. F. Pastor, and M. Ashraf Aziz. 2010. "Comparative anatomical study of the forearm extensor muscles of *Cebus libidinosus* (Rylands *et al.*, 2000; Primates, *Cebidae*), modern humans, and other primates, with comments on primate evolution, phylogeny, and manipulatory behavior." *Anatomical Record* 293: 2056–70.

Boesch, C. 1996. "Social grouping in Tai chimpanzees." *Great Ape Societies* 101–13.

Boesch, C. 2002. "Cooperative hunting roles among Tai chimpanzees." *Human Nature* 13 (1), 27–46.

Boesch, C., and H. Boesch. 1989. "Hunting behavior of wild chimpanzees in the Tai National Park." *American Journal of Physical Anthropology* 78 (4): 547–73.

Boesch, C., and H. Boesch. 1990. "Tool use and tool making in wild chimpanzees." *Folia Primatologica* 54: 86–99.

Boesch, C., and H. Boesch-Achermann. 2000. *The Chimpanzees of the Taï Forest: Behavioural Ecology and Evolution*. New York: Oxford University Press.

Boesch, C., C. Crockford, I. Herbinger, R. Wittig, Y. Moebius, and E. Normand. 2008. "Intergroup conflicts among chimpanzees in Taï National Park: Lethal violence and the female perspective." *American Journal of Primatology* 70 (6): 519–32.

Bronikowski, A. M., J. Altmann, D. K. Brockman, M. Cords, L. M. Fedigan, A. Pusey, T. Stoinski, W. F. Morris, K. B. Strier, and S. C. Alberts. 2011. "Aging in the natural world: Comparative data reveal similar mortality patterns across primates." *Science* 331 (6022): 1325–28.

Campbell, G. S., B. A. Bilgre, and R. H. Defran. 2002. "Bottlenose dolphins (*Tursiops truncatus*) in Tuneffe Atoll, Belize: Occurrence, site fidelity, group size, and abundance." *Aquatic Mammals* 28 (2): 170–80.

Charnov, E. L., and D. Berrigan. 1993. "Why do female primates have such long lifespans and so few babies? Or life in the slow lane." *Evolutionary Anthropology* 1 (6): 191–94.

Cheney, B., P. M. Thompson, S. N. Ingram, P. S. Hammond, P. T. Stevick, J. W. Durban, R. M. Culloch, S. H. Elwen, L. Mandleberg, V. M. Janik, N. J. Quick, V. Islas-Villanueva, K. P. Robinson, M. Costa, S. M. Eisfeld, A. Walters, C. Phillips, C. R. Weir, P. G. H. Evans, P. Anderwald, R. J. Reid, J. B. Reid, and B. Wilson. 2013. "Integrating multiple data sources to assess the distribution and abundance of bottlenose dolphins *Tursiops truncatus* in Scottish waters." *Mammal Review* 43: 71–88.

Clark, C. B. 1977. "A preliminary report on weaning among chimpanzees of the Gombe National Park, Tanzania." In *Primate Biosocial Development: Biological, Social and Ecological Determinants*, edited by S. Chevalier-Skolinkoff and F. Poirer, 235–60. New York: Garland.

Conklin-Brittain, N. L., C. D. Knott, and R. W. Wrangham. 2001. "The feeding ecology of apes." In *The Apes: Challenges for the 21st Century, Conference Proceedings of the Brookfield Zoo*, 167–74. Chicago: Chicago Zoological Society.

Connor, R. C., and M. Krützen. 2015. "Male dolphin alliances in Shark Bay: Changing perspectives in a 30-year study." *Animal Behaviour* 103: 223–35.

Connor, R. C., R. Wells, J. Mann, and A. Read. 2000. "The bottlenose dolphin, *Tursiops* sp.: Social relationships in a fission-fusion society." In *Cetacean Societies: Field Studies of Dolphins and Whales*, edited by J. Mann, R. Connor, P. Tyack, and H. Whitehead, 91–126. Chicago: University of Chicago Press.

Constable, J. L., M. V. Ashley, J. Goodall, and A. E. Pusey. 2001. "Noninvasive paternity assignment in Gombe chimpanzees." *Molecular Ecology* 10 (5): 1279–1300.

Csardi, G., and T. Nepusz. 2006. "The igraph software package for complex network research." *InterJournal, Complex Systems* 1695.

Cusick, J. A., and D. L. Herzing. 2014. "The dynamic of aggression: How individual and group factors affect the long-term interspecific aggression between two sympatric species of dolphin." *Ethology* 120: 287–303.

Daura-Jorge, F. G., M. Cantor, S. Ingram, D. Lusseau, and P. C. Simões-Lopes. 2012. "The structure of a bottlenose dolphin society is coupled to a unique foraging cooperation with artisanal fishermen." *Biology Letters* 8: 702–5.

Duffy-Echevarria, E. E., R. C. Connor, and D. J. St Aubin. 2008. "Observations of strand-feeding behavior by bottlenose dolphins (*Tursiops truncatus*) in Bull Creek, South Carolina." *Marine Mammal Science* 24: 202–6.

Dunn, D. G., S. G. Barco, D. A. Pabst, and W. A. McLellan. 2002. "Evidence for infanticide in bottlenose dolphins of the Western North Atlantic." *Journal of Wildlife Diseases* 38 (3): 505–10.

Eisfeld, S. M., and K. P. Robinson. 2004. "The sociality of bottlenose dolphins in the outer southern Moray Firth, NE Scotland: Implications for current management proposals?" *European Research on Cetaceans* 18: 60–63.

Emery Thompson, M. 2013. "Reproductive ecology of female chimpanzees." *American Journal of Primatology* 75 (3): 222–37.

Emery Thompson, M., S. M. Kahlenberg, I. C. Gilby, and R. W. Wrangham. 2007. "Core area quality is associated with variance in reproductive success among female chimpanzees at Kibale National Park." *Animal Behaviour* 73 (3): 501–12.

Emery Thompson, M., M. N. Muller, K. Sabbi, Z. P. Machanda, E. Otali, and R. W. Wrangham. 2016. "Faster reproductive rates trade off against offspring growth in wild chimpanzees." *Proceedings of the National Academy of Sciences* 113 (28): 7780–85.

Fawcett, K. A. 2000. "Female relationships and food availability in a forest community of chimpanzees." PhD diss., University of Edinburgh.

Feldblum, J. T., E. E. Wroblewski, R. S. Rudicell, B. H. Hahn, T. Paiva, M. Cetinkaya-Rundel, A. E. Pusey, and I. C. Gilby. 2014. "Sexually coercive male chimpanzees sire more offspring." *Current Biology* 24 (23): 2855–60.

Foerster, S., M. Franz, C. M. Murray, I. C. Gilby, J. T. Feldblum, K. K. Walker, and A. E. Pusey. 2016. "Chimpanzee females queue but males compete for social status." *Scientific Reports* 6: 35404.

Foerster, S., K. McLellana, K. Schroepfer-Walker, C. M. Murray, C. Krupenyea, I. C. Gilby, and A. E. Pusey. 2015. "Social bonds in the dispersing sex: Partner preferences among adult female chimpanzees." *Animal Behaviour* 105: 139–52.

Foroughirad, V., and J. Mann. 2013. "Human fish provisioning has long-term impacts on the behaviour and survival of bottlenose dolphins." *Biological Conservation* 160: 242–49.

Frère, C. H., M. Krützen, A. M. Kopps, P. Ward, J. Mann, and W. B. Sherwin. 2010a. "Inbreeding tolerance and fitness costs in wild bottlenose dolphins." *Proceedings of the Royal Society of London B* 277 (1694): 2667–73.

Frère, C. H., M. Krützen, J. Mann, R. C. Connor, L. Bejder, and W. B. Sherwin. 2010b. "Social and genetic interactions drive fitness variation in a free-living dolphin population." *Proceedings of the National Academy of Sciences* 107 (46): 19949–54.

Fruet, P. F., R. Cezar Genoves, L. M. Möller, S. Botta, and E. R. Secchi. 2015. "Using mark-recapture and stranding data to estimate reproductive traits in female bottlenose dolphins (*Tursiops truncatus*) of the Southwestern Atlantic Ocean." *Marine Biology* 162 (3): 661–73.

Fujisawa, M., K. J. Hockings, A. G. Soumah, and T. Matsuzawa. 2016. "Placentophagy in wild chimpanzees (*Pan troglodytes verus*) at Bossou, Guinea." *Primates* 57 (2): 175–80.

Galezo, A., E. Krzyszczyk, and J. Mann. 2018. "Sexual segregation in Indo-Pacific bottlenose dolphins is driven by female avoidance of males." *Behavioral Ecology* 29 (2): 377–86.

Gazda, S. K., R. C. Connor, R. K. Edgar, and F. Cox. 2005. "A division of labour with role specialization in group-hunting bottlenose dolphins (*Tursiops truncatus*) off Cedar Key, Florida." *Proceedings of the Royal Society of London B* 272: 135–40.

Gero, S., L. Bejder, H. Whitehead, J. Mann, and R. C. Connor. 2005. "Behaviourally specific preferred associations in bottlenose dolphins, *Tursiops* spp." *Canadian Journal of Zoology* 83: 1566–73.

Gibson, Q. A., and J. Mann. 2008a. "Early social development in wild bottlenose dolphins: Sex differences, individual variation, and maternal influence." *Animal Behaviour* 76: 375–87.

Gibson, Q. A., and J. Mann. 2008b. "The size, composition, and function of wild bottlenose dolphin (*Tursiops* sp.) mother-calf groups in Shark Bay, Australia." *Animal Behaviour* 76: 389–405.

Gilby, I. C. 2006. "Meat sharing among the Gombe chimpanzees: Harassment and reciprocal exchange." *Animal Behaviour* 71 (4): 953–96.

Gilby, I. C., L. J. N. Brent, E. E. Wroblewski, R. S. Rudicell, B. H. Hahn, J. Goodall, and A. E. Pusey. 2013. "Fitness benefits of coalitionary aggression in male chimpanzees." *Behavioral Ecology and Sociobiology* 67 (3): 373–81.

Gilby, I. C., M. Emery Thompson, J. D. Ruane, and R. W. Wrangham. 2010. "No evidence of short-term exchange of meat for sex among chimpanzees." *Journal of Human Evolution* 59 (1): 44–53.

Gilby, I. C., Z. P. Machanda, R. C. O'Malley, C. M. Murray, E. V. Lonsdorf, K. Walker, D. C. Mjungu, E. Otali, M. N. Muller, M. Emery Thompson, A. E. Pusey, and R. W. Wrangham. 2017. "Predation by female chimpanzees: Toward an understanding of sex differences in meat acquisition in the last common ancestor of Pan and Homo." *Journal of Human Evolution* 110: 82–94.

Gilby, I. C., and R. W. Wrangham. 2008. "Association patterns among wild chimpanzees (*Pan troglodytes schweinfurthii*) reflect sex differences in cooperation." *Behavioral Ecology and Sociobiology* 62 (11): 1831.

Gomes, C. M., and C. Boesch. 2009. "Wild chimpanzees exchange meat for sex on a long-term basis." *PLoS One* 4: e5116.

Goodall, J. 1979. "Intercommunity interactions in the chimpanzee population of the Gombe National Park." In *The Great Apes*, edited by D. A. Hamburg and E. R. McCown, 13–53. Menlo Park, CA: Benjamin/Cummings.

Goodall, J. 1986. *The Chimpanzees of Gombe: Patterns of Behavior*. Cambridge, MA: Harvard University Press.

Gruber, T., and Z. Clay. 2016. "A comparison between bonobos and chimpanzees: A review and update." *Evolutionary Anthropology* 25 (5): 239–52.

Gruber, T., Z. Clay, and K. Zuberbühler. 2010. "A comparison of bonobo and chimpanzee tool use: Evidence for a female bias in the *Pan* lineage." *Animal Behaviour* 80 (6): 1023–1103.

Harvey, P. H., R. D. Martin, and T. Clutton-Brock. 1987. "Life histories in comparative perspective." In *Primate Societies*, edited by B. B. Smuts, D. L. Cheney, R. M. Seyfarth, and R. W. Wrangham, 181–96. Chicago: University of Chicago Press.

Jedensjö, M., C. M. Kemper, and M. Krützen. 2017. "Cranial morphology and taxonomic resolution of some dolphin taxa (Delphinidae) in Australian waters, with a focus on the genus *Tursiops*." *Marine Mammal Science* 33 (1): 187–205.

Jones, J. H., M. L. Wilson, C. Murray, and A. Pusey. 2010. "Phenotypic quality influences fertility in Gombe chimpanzees." *Journal of Animal Ecology* 79 (6): 1262–69.

Kahlenberg, S. M., M. Emery Thompson, M. N. Muller, and R. W. Wrangham. 2008. "Immigration costs for female chimpanzees and male protection as an immigrant counterstrategy to intrasexual aggression." *Animal Behaviour* 76 (5): 1497–1509.

Kappeler, P. M., and C. P. van Schaik. 2002. "Evolution of primate social systems." *International Journal of Primatology* 23 (4): 707–40.

Karniski, C., E. Krzyszczyk, and J. Mann. 2018. "Senescence impacts reproduction and maternal investment in bottlenose dolphins." *Proceedings of the Royal Society of London B* 285 (1883): 20181123.

Kogi, K., T. Hishii, A. Imamura, T. Iwatani, and K. M. Dudzinski. 2004. "Demographic parameters of Indo-Pacific bottlenose dolphins (*Tursiops Aduncus*) around Mikura Island, Japan." *Marine Mammal Science* 20: 510–26.

Köhler, W. 1925. *The Mentality of Apes*. Translated from the second revised edition by Ella Winter. New York: Liveright.

Krützen, M., L. M. Barre, R. C. Connor, J. Mann, and W. B. Sherwin. 2004. "O father: where art thou?—Paternity assessment in an open fission-fusion society of wild bottlenose dolphins (*Tursiops* sp.) in Shark Bay, Western Australia." *Molecular Ecology* 13: 1975–90.

Krzyszczyk, E., E. M. Patterson, M. A. Stanton, and J. Mann. 2017. "The transition to independence: Sex differences in social and behavioral development of wild bottlenose dolphins." *Animal Behaviour* 129: 43–59.

Kuroda, S. 1989. "Developmental retardation and behavioral characteristics of pygmy chimpanzees." In *Understanding Chimpanzees*, edited by P. G. Heltne and L. A. Marquardt, 184–94. Cambridge, MA: Harvard University Press.

Langergraber, K. E., J. Mitani, and L. Vigilant. 2009. "Kinship and social bonds in female chimpanzees (*Pan troglodytes*)." *American Journal of Primatology* 71 (10): 840–51.

Langergraber, K. E., D. P. Watts, L. Vigilant, and J. C. Mitani. 2017. "Group augmentation, collective action, and territorial boundary patrols by male chimpanzees." *Proceedings of the National Academy of Sciences* 114 (28): 7337–42.

Lehmann, J., and C. Boesch. 2005. "Bisexually bonded ranging in chimpanzees (*Pan troglodytes verus*)." *Behavioral Ecology and Sociobiology* 57 (6): 525–35.

Lonsdorf, E. V. 2006. "What is the role of mothers in the acquisition of termite-fishing behaviors in wild chimpanzees (*Pan troglodytes schweinfurthii*)?" *Animal Cognition* 9 (1): 36–46.

Lonsdorf, E. V., K. E. Anderson, M. A. Stanton, M. Shender, M. R. Heintz, J. Goodall, and C. M. Murray. 2014a. "Boys will be boys: Sex differences in wild infant chimpanzee social interactions." *Animal Behaviour* 88: 79–83.

Lonsdorf, E. V., L. Eberly, and A. E. Pusey. 2004. "Sex differences in learning in chimpanzees." *Nature* 428: 715–16.

Lonsdorf, E. V., A. C. Markham, M. R. Heintz, K. E. Anderson, D. J. Ciuk, J. Goodall, and C. M. Murray. 2014b. "Sex differences in wild chimpanzee behavior emerge during infancy." *PLoS One* 9 (6): e99099.

Luncz, L. V., and C. Boesch. 2015. "The extent of cultural variation between adjacent chimpanzee (*Pan troglodytes verus*) communities; A microecological approach." *American Journal of Physical Anthropology* 156 (1): 67–75.

Lusseau, D. 2007. "Why are male social relationships complex in the Doubtful Sound bottlenose dolphin population?" *PLoS One* 2 (4): e348.

Lusseau, D., K. Schneider, O. J. Boisseau, P. Haase, E. Slooten, and S. M. Dawson. 2003. "The bottlenose dolphin community of Doubtful Sound features a large proportion of long-lasting associations: Can geographic isolation explain this unique trait?" *Behavioral Ecology and Sociobiology* 54 (4): 396–405.

Machanda, Z., N. F. Brazeau, A. B. Bernard, R. M. Donovan, A. M. Papakyrikos, R. Wrangham, and T. M. Smith. 2015. "Dental eruption in East African wild chimpanzees." *Journal of Human Evolution* 82: 137–44.

Mann, J. 2019. "Maternal care and calf development in odontocetes." In *Ethology and Behavioral Ecology of Toothed Whales and Dolphins*, edited by B. Würsig, 95–116. Cham: Springer.

Mann, J., R. C. Connor, L. M. Barre, and M. R. Heithaus. 2000. "Female reproductive success in bottlenose dolphins (*Tursiops* sp.): Life history, habitat, provisioning, and group size effects." *Behavioral Ecology* 11: 210–19.

Mann, J., and E. M. Patterson. 2013. "Tool use by aquatic animals." *Philosophical Transactions of the Royal Society of London B* 368 (1630): 20120424.

Mann, J., and B. L. Sargeant. 2003. "Like mother, like calf: The ontogeny of foraging traditions in wild Indian Ocean bottlenose dolphins (*Tursiops* sp.)." In *The Biology of Traditions: Models and Evidence*, edited by D. M. Fragaszy and S. Perry, 236–66. Cambridge: Cambridge University Press.

Mann, J., B. L. Sargeant, and M. Minor. 2007. "Calf inspection of fish catches: Opportunities for oblique social learning?" *Marine Mammal Science* 23 (1): 197–202.

Mann, J., B. L. Sargeant, J. J. Watson-Capps, Q. A. Gibson, M. R. Heithaus, R. C. Connor, and E. Patterson. 2008. "Why do dolphins carry sponges?" *PLoS One* 3 (12): e3868.

Mann, J., and B. B. Smuts. 1998. "Natal attraction: Allomaternal care and mother-infant separations in wild bottlenose dolphins." *Animal Behaviour* 55 (5): 1097–1113.

Mann, J., and B. B. Smuts. 1999. "Behavioral development in wild bottlenose dolphin newborns (*Tursiops* sp.)." *Behaviour* 136: 529–66.

Mann, J., M. A. Stanton, E. M. Patterson, E. J. Bienenstock, and L. O. Singh. 2012. "Social networks reveal cultural behaviour in tool using dolphins." *Nature Communications* 3: 980.

Mann, J., and J. J. Watson-Capps. 2005. "Surviving at sea: Ecological and behavioural predictors of calf mortality in Indian Ocean bottlenose dolphins, sp." *Animal Behaviour* 69 (4): 899–909.

Matsumoto-Oda, A., K. Hosaka, M. A. Huffman, and K. Kawanaka. 1998. "Factors affecting party size in chimpanzees of the Mahale mountains." *International Journal of Primatology* 19 (6): 999–1011.

McGowen, M. R., M. Spaulding, and J. Gatesy. 2009. "Divergence date estimation and a comprehensive molecular tree of extant cetaceans." *Molecular Phylogenetics and Evolution* 53: 891–906.

McHugh, K. A., J. B. Allen, A. A. Barleycorn, and R. S. Wells. 2011. "Natal philopatry, ranging behavior, and habitat selection of juvenile bottlenose dolphins in Sarasota Bay, Florida." *Journal of Mammalogy* 92 (6): 1298–1313.

Meulman, E. J. M., A. Seed, and J. Mann. 2013. "If at first you don't succeed . . . Studies of ontogeny shed light on the cognitive demands of habitual tool use." *Philosophical Transactions of the Royal Society of London B* 368 (1630): 20130050.

Miketa, M. L., E. M. Patterson, E. Krzyszczyk, V. Foroughirad, and J. Mann. 2018. "Calf age and sex affects maternal diving behavior in Shark Bay bottlenose dolphins." *Animal Behaviour* 137: 107–17.

Mitani, J. C. 2009. "Male chimpanzees form enduring and equitable social bonds." *Animal Behaviour* 77 (3): 633–40.

Mitani, J. C., D. A. Merriwether, and C. Zhang. 2000. "Male affiliation, cooperation and kinship in wild chimpanzees." *Animal Behaviour* 59: 885–93.

Mitani, J. C., and D. P. Watts. 2001. "Why do chimpanzees hunt and share meat?" *Animal Behaviour* 61 (5): 915–24.

Mitani, J. C., D. P. Watts, and S. J. Amsler. 2010. "Lethal intergroup aggression leads to territorial expansion in wild chimpanzees." *Current Biology* 20 (12): R507–8.

Möller, L. M., L. B. Beheregaray, R. G. Harcourt, and M. Krützen. 2001. "Alliance membership and kinship in wild male bottlenose dolphins (*Tursiops aduncus*) of southeastern Australia." *Proceedings of the Royal Society of London B* 268: 1941–47.

Morgan, D., and C. Sanz. 2003. "Naïve encounters with chimpanzees in the Goualougo Triangle, Republic of Congo." *International Journal of Primatology* 24 (2): 369–81.

Muller, M. N., S. M. Kahlenberg, M. Emery Thompson, and R. W. Wrangham. 2007. "Male coercion and the costs of promiscuous mating for female chimpanzees." *Proceedings of the Royal Society of London B* 274 (1612): 1009–14.

Muller, M. N., and J. C. Mitani. 2005. "Conflict and cooperation in wild chimpanzees." *Advances in the Study of Behavior* 35: 275–331.

Murray, C. M. 2007. "Method for assigning categorical rank in female *Pan troglodytes schweinfurthii* via the frequency of approaches." *International Journal of Primatology* 28 (4): 853–64.

Murray, C. M., L. E. Eberly, and A. E. Pusey. 2006. "Foraging strategies as a function of season and rank among wild female chimpanzees (*Pan troglodytes*)." *Behavioral Ecology* 17 (6): 1020–28.

Murray, C. M., I. C. Gilby, S. V. Mane, and A. E. Pusey. 2008. "Adult male chimpanzees inherit maternal ranging patterns." *Current Biology* 18 (1): 20–24.

Murray, C. M., E. V. Lonsdorf, M. A. Stanton, K. R. Wellens, J. A. Miller, J. Goodall, and A. E. Pusey. 2014. "Early social exposure in wild chimpanzees: Mothers with sons are more gregarious than mothers with daughters." *Proceedings of the National Academy of Sciences* 111 (51): 201409507.

Murray, C. M., S. V. Mane, and A. E. Pusey. 2007. "Dominance rank influences female space use in wild chimpanzees, *Pan troglodytes*: Towards an ideal despotic distribution." *Animal Behaviour* 74 (6): 1795–1804.

Nakamura, M., H. Hayaki, K. Hosaka, N. Itoh, and K. Zamma. 2014. "Brief communication: Orphaned male chimpanzees die young even after weaning." *American Journal of Physical Anthropology* 153 (1): 139–43.

Naples Daily News. 2015. "2 of Sarasota Bay's oldest dolphins found dead." October 22, 2015. https://archive.naplesnews.com/news/environment/2-of-sarasota-bays-oldest-dolphins-found-dead-ep-1332535661-337621841.html.

Nekolny, S. R., M. Denny, G. Biedenbach, E. M. Howells, M. Mazzoil, W. N. Durden, L. Moreland, J. D. Lambert, and Q. A. Gibson. 2017. "Effects of study area size on home range estimates of common bottlenose dolphins *Tursiops truncatus*." *Current Zoology* 63 (6): 693–701.

Newton-Fisher, N. E. 2006. "Female coalitions against male aggression in wild chimpanzees of the Budongo Forest." *International Journal of Primatology* 27 (6): 1589–99.

Nishida, T. 1968. "The social group of wild chimpanzees in the Mahali Mountains." *Primates* 9 (3): 167–224.

Nishida, T., T. Hasegawa, H. Hayaki, Y. Takahata, and S. Uehara. 1992. "Meat-sharing as a coalition strategy by an alpha male chimpanzee." In *Topics in Primatology: Human Origins*, edited by T. Nishida, W. C. McGrew, P. Marler, M. Pickford, and F. B. M. de Waal, 159–74. Tokyo: University of Tokyo Press.

Noren, S. R., G. Biedenbach, J. V. Redfern, and E. F. Edwards. 2008. "Hitching a ride: The formation locomotion strategy of dolphin calves." *Functional Ecology* 22: 278–83.

Otali, E., and J. S. Gilchrist. 2006. "Why chimpanzee (*Pan troglodytes schweinfurthii*) mothers are less gregarious than nonmothers and males: The infant safety hypothesis." *Behavioral Ecology and Sociobiology* 59 (4): 561–70.

Owen, E. C. G., R. S. Wells, and S. Hofmann. 2002. "Ranging and association patterns of paired and unpaired adult male Atlantic bottlenose dolphins, *Tursiops truncatus*, in Sarasota, Florida, provide no evidence for alternative male strategies." *Canadian Journal of Zoology* 80 (12): 2072–89.

Parsons, K. M., J. W. Durban, D. E. Claridge, K. C. Balcomb, L. R. Noble, and P. M. Thompson. 2003. "Kinship as a basis for alliance formation between male bottlenose dolphins, *Tursiops truncatus*, in the Bahamas." *Animal Behaviour* 66: 185–94.

Patterson, E. M., E. Krzyszczyk, and J. Mann. 2015. "Age-specific foraging performance and reproduction in tool-using wild bottlenose dolphins." *Behavioral Ecology* 27 (2): 401–10.

Patterson, E. M., and J. Mann. 2011. "The ecological conditions that favour tool use and innovation in wild bottlenose dolphins (*Tursiops* sp.)." *PLoS One* 6 (7): e22243.

Patterson, I. A. P., R. J. Reid, B. Wilson, K. Grellier, H. M. Ross, and P. M. Thompson. 1998. "Evidence for infanticide in bottlenose dolphins: An explanation for violent interactions with harbour porpoises?" *Proceedings of the Royal Society of London B* 265: 1167–70.

Pruetz, J. D. 2006. "Feeding ecology of savanna chimpanzees (*Pan troglodytes verus*) at Fongoli, Senegal." In *Feeding Ecology in Apes and Other Primates*, edited by G. Hohmann, M. M. Robbins, and C. Boesch, 161–82. Cambridge: Cambridge University Press.

Pruetz, J. D., and P. Bertolani. 2009. "Chimpanzee (*Pan troglodytes verus*) behavioral responses to stresses associated with living in a savanna-mosaic environment: Implications for hominin adaptations to open habitats." *PaleoAnthropology*: 252–62.

Pruetz, J. D., P. Bertolani, K. Boyer Ontl, S. Lindshield, M. Shelley, and E. G. Wessling. 2015. "New evidence on the tool-assisted hunting exhibited by chimpanzees (*Pan troglodytes verus*) in a savannah habitat at Fongoli, Sénégal." *Royal Society Open Science* 2 (4): 140507.

Prüfer, K., K. Munch, I. Hellmann, K. Akagi, J. R. Miller, B. Walenz, S. Koren, G. Sutton, C. Kodira, R. Winer, J. R. Knight, J. C. Mullikin, S. J. Meader, C. P. Ponting, G. Lunter, S. Higashino, A. Hobolth, J. Dutheil, E. Karakoç, C. Alkan, S. Sajjadian, C. R. Catacchio, M. Ventura, T. Marques-Bonet, E. E. Eichler, C. André, R. Atencia, L. Mugisha, J. Junhold, N. Patterson, M. Siebauer, J. M. Good, A. Fischer, S. E. Ptak, M. Lachmann, D. E. Symer, T. Mailund, M. H. Schierup, A. M. Andrés, J. Kelso, and S. Pääbo. 2012. "The bonobo genome compared with the chimpanzee and human genomes." *Nature* 486 (7404): 527–31.

Pusey, A. E. 1980. "Inbreeding avoidance in chimpanzees." *Animal Behaviour* 28 (2): 543–52.

Pusey, A. E. 1983. "Mother-offspring relationships in chimpanzees after weaning." *Animal Behaviour* 31 (2): 363–77.

Pusey, A. E. 1990. "Behavioural changes at adolescence in chimpanzees." *Behaviour* 115 (3): 203–46.

Pusey, A. E., C. Murray, W. Wallauer, M. Wilson, E. Wroblewski, and J. Goodall. 2008. "Severe aggression among female *Pan troglodytes schweinfurthii* at Gombe National Park, Tanzania." *International Journal of Primatology* 29 (4): 949.

Pusey, A. E., and K. Schroepfer-Walker. 2013. "Female competition in chimpanzees." *Philosophical Transactions of the Royal Society of London B* 368 (1631): 20130077.

Pusey, A. E., J. Williams, and J. Goodall. 1997. "The influence of dominance rank on the reproductive success of female chimpanzees." *Science* 277 (5327): 828–31.

Ralls, K., B. Lundrigan, and K. Kranz. 1987. "Mother-young relationships in captive ungulates: Behavioral changes over time." *Ethology* 75: 1–14.

Randić, S., R. C. Connor, W. B. Sherwin, and M. Krützen. 2012. "A novel mammalian social structure in Indo-Pacific bottlenose dolphins (*Tursiops* sp.): Complex male alliances in an open social network." *Proceedings of the Royal Society of London B* 279 (1740): 3083–90.

Reynolds, V. 2005. *The Chimpanzees of the Budongo Forest: Ecology, Behaviour and Conservation*. Oxford: Oxford University Press.

van de Rijt-Plooij, H. H. C., and F. X. Plooij. 1987. "Growing independence, conflict and learning in mother-infant relations in free-ranging chimpanzees." *Behaviour* 101 (1): 1–86.

Robinson, K. P., T. M. C. Sim, R. M. Culloch, T. S. Bean, I. Cordoba Aguilar, S. M. Eisfeld, M. Filan, G. N. Haskins, G. Williams, and G. J. Pierce. 2017. "Female reproductive success and calf survival in a North Sea costal bottlenose dolphin (*Tursiops truncatus*) population." *PLoS One* 12 (9): e0185000.

Rossman, S., P. H. Ostrom, M. Stolen, N. B. Barros, H. Gandhi, C. A. Stricker, and R. S. Wells. 2015. "Individual specialization in the foraging habits of female bottlenose dolphins living in a trophically diverse and habitat rich estuary." *Oecologia* 178: 415.

Sanz, C. M., and D. B. Morgan. 2007. "Chimpanzee tool technology in the Goualougo Triangle Republic of Congo." *Journal of Human Evolution* 52: 420–33.

Sargeant, B. L., and J. Mann. 2009. "Developmental evidence for foraging traditions in wild bottlenose dolphins." *Animal Behaviour* 78 (3): 715–21.

Sargeant, B. L., J. Mann, P. Berggren, and M. Krützen. 2005. "Specialization and development of beach hunting, a rare foraging behavior, by wild bottlenose dolphins (*Tursiops* sp.)." *Canadian Journal of Zoology* 83: 1400–1410.

Sargeant, B. L., A. J. Wirsing, M. R. Heithaus, and J. Mann. 2007. "Can environmental heterogeneity explain individual foraging variation in wild bottlenose dolphins (*Tursiops* sp.)?" *Behavioral Ecology and Sociobiology* 61: 679–88.

Scott, E., J. Mann, J. J. Watson-Capps, B. L. Sargeant, and R. C. Connor. 2005. "Aggression in bottlenose dolphins: Evidence for sexual coercion, male-male competition, and female tolerance through analysis of tooth-rake marks and behaviour." *Behaviour* 142: 21–44.

Smolker, R. A., A. F. Richards, R. C. Connor, J. Mann, and P. Berggren. 1997. "Sponge-carrying by Indian Ocean bottlenose dolphins: Possible tool-use by a delphinid." *Ethology* 103: 454–65.

Smolker, R. A., A. F. Richards, R. C. Connor, and J. W. Pepper. 1992. "Sex differences in patterns of association among Indian Ocean bottlenose dolphins." *Behaviour* 123 (1/2): 38–69.

Stanford, C. B., J. Wallis, E. Mpongo, and J. Goodall. 1994. "Hunting decisions in wild chimpanzees." *Behaviour* 131 (1): 1–18.

Stanton, M. A., Q. A. Gibson, and J. Mann. 2011. "When mum's away: A study of mother and calf ego networks during separations in wild bottlenose dolphins (*Tursiops* sp.)." *Animal Behaviour* 82 (2): 405–12.

Stanton, M. A., E. V. Lonsdorf, A. E. Pusey, J. Goodall, and C. M. Murray. 2014. "Maternal behavior by birth order in wild chimpanzees (*Pan troglodytes*)." *Current Anthropology* 55 (4): 483–89.

Stanton, M. A., E. V. Lonsdorf, A. E. Pusey, and C. M. Murray. 2017. "Do juveniles help or hinder? Influence of juvenile offspring on maternal behavior and reproductive outcomes in wild chimpanzees (*Pan troglodytes*)." *Journal of Human Evolution* 111: 152–62.

Stanton, M. A., and J. Mann. 2012. "Early social networks predict survival in wild bottlenose dolphins." *PLoS One* 7 (10): e47508.

Steeman, M. E., M. B. Hebsgaard, R. E. Fordyce, S. Y. W. Ho, D. L. Rabosky, R. Nielsen, C. Rahbek, H. Glenner, M. V. Sørensen, and E. Willerslev. 2009. "Radiation of extant cetaceans driven by restructuring of the oceans." *Systematic Biology* 58 (6): 573–85.

Strier, K. B., P. C. Lee, and A. R. Ives. 2014. "Behavioral flexibility and the evolution of primate social states." *PLoS One* 9 (12): e114099.

Stumpf, R. M., and C. Boesch. 2006. "The efficacy of female choice in chimpanzees of the Taï Forest, Côte d'Ivoire." *Behavioral Ecology and Sociobiology* 60 (6): 749–65.

Stumpf, R. M., and C. Boesch. 2010. "Male aggression and sexual coercion in wild West African chimpanzees, *Pan troglodytes verus*." *Animal Behaviour* 79 (2): 333–42.

Stumpf, R. M., M. Emery Thompson, M. N. Muller, and R. W. Wrangham. 2009. "The context of female dispersal in Kanyawara chimpanzees." *Behaviour* 146 (4): 629–56.

Torres, L. G., and A. J. Read. 2009. "Where to catch a fish? The influence of foraging tactics on the ecology of bottlenose dolphins (*Tursiops truncatus*) in Florida Bay, Florida." *Marine Mammal Science* 25: 797–815.

Tsai, Y.-J. J., and J. Mann. 2013. "Dispersal, philopatry and the role of fission-fusion dynamics in bottlenose dolphins." *Marine Mammal Science* 29 (2): 261–79.

van Lawick-Goodall, J. 1968. "The behaviour of free-living chimpanzees in the Gombe Stream Reserve." *Animal Behaviour Monographs* 1: 161–311.

Wakefield, M. L. 2013. "Social dynamics among females and their influence on social structure in an East African chimpanzee community." *Animal Behaviour* 85 (6): 1303–13.

Walker, K. K., R. S. Rudicell, Y. Li, B. H. Hahn, E. Wroblewski, and A. E. Pusey. 2017. "Chimpanzees breed with genetically dissimilar mates." *Open Science* 4 (1): 160422.

Walker, K. K., C. S. Walker, J. Goodall, and A. E. Pusey. 2018. "Maturation is prolonged and variable in female chimpanzees." *Journal of Human Evolution* 114: 131–140.

Wallen, M. M., E. Krzyszczyk, and J. Mann. 2017. "Mating in a bisexually philopatric society: Bottlenose dolphin females associate with adult males but not adult sons during estrous." *Behavioral Ecology and Sociobiology* 71 (10): 153.

Wallen, M. M., E. M. Patterson, E. Krzyszczyk, and J. Mann. 2016. "Ecological costs to females in a system with allied sexual coercion." *Animal Behaviour* 115: 227–36.

Watts, D. P. 1998. "Coalitionary mate guarding by male chimpanzees at Ngogo, Kibale National Park, Uganda." *Behavioral Ecology and Sociobiology* 44 (1): 43–55.

Watts, D. P. 2012. "Long-term research on chimpanzee behavioral ecology in Kibale National Park, Uganda." In *Long-Term Field Studies of Primates*, edited by P. M. Kappeler and D. P. Watts, 313–38. Berlin: Springer.

Watts, D. P., and J. C. Mitani. 2001. "Boundary patrols and intergroup encounters in wild chimpanzees." *Behaviour* 138 (3): 299–327.

Watts, D. P., and A. E. Pusey. 2002. "Behavior of juvenile and adolescent great apes." In *Juvenile Primates: Life History, Development, and Behavior*, edited by M. E. Pereira and L. A. Fairbanks, 148–67. Chicago: University of Chicago Press.

Wells, R. S. 2003. "Dolphin social complexity: Lessons from long-term study and life history." In *Animal Social Complexity: Intelligence, Culture, and Individualized Societies*, edited by F. B. M. de Waal and P. L. Tyack, 32–56. Cambridge, MA: Harvard University Press.

Wells, R. S., and M. D. Scott. 2009. "Common bottlenose dolphin (*Tursiops truncatus*)." In *Encyclopedia of Marine Mammals*, edited by W. F. Perrin, B. Würsig, and J. G. M. Thewissen, 249–55. San Diego: Elsevier.

Wells, R. S., V. Tornero, A. Borrell, A. Aguilar, T. K. Rowles, H. L. Rhinehart, S. Hofmann, W. M. Jarman, A. A. Hohn, and J. C. Sweeney. 2005. "Integrating life-history and reproductive success data to examine potential relationships with organochlorine compounds for bottlenose dolphins (*Tursiops truncatus*) in Sarasota Bay, Florida." *Science of the Total Environment* 349: 106–19.

White, F. J., and R. W. Wrangham. 1988. "Feeding competition and patch size in the chimpanzee species *Pan paniscus* and *Pan troglodytes*." *Behaviour* 105 (1): 148–64.

Whitehead, H., and J. Mann. 2000. "Female reproductive strategies in cetaceans." In *Cetacean Societies: Field Studies of Dolphins and Whales*, edited by J. Mann, R. C. Connor, P. Tyack, and H. Whitehead, 219–46. Chicago: University of Chicago Press.

Whiten, A., J. Goodall, W. C. McGrew, T. Nishida, V. Reynolds, Y. Sugiyama, C. E. Tutin, R. W. Wrangham, and C. Boesch. 1999. "Cultures in Chimpanzees." *Nature* 399: 682–85.

Williams, J. M., A. E. Pusey, J. V. Carlis, B. P. Farm, and J. Goodall. 2002. "Female competition and male territorial behaviour influence female chimpanzees' ranging patterns." *Animal Behaviour* 63 (2): 347–60.

Wilson, M. L., C. Boesch, B. Fruth, T. Furuichi, I. C. Gilby, C. Hashimoto, C. L. Hobaiter, G. Hohmann, N. Itoh, K. Koops, J. N. Lloyd, T. Matsuzawa, J. C. Mitani, D. C. Mjungu, D. Morgan, M. N. Muller, R. Mundry, M. Nakamura, J. Pruetz, A. E. Pusey, J. Riedel, C. Sanz, A. M. Schel, N. Simmons, M. Waller, D. P. Watts, F. White, R. M. Wittig, K. Zuberbühler, and R. W. Wrangham. 2014. "Lethal aggression in *Pan* is better explained by adaptive strategies than human impacts." *Nature* 513 (7518): 414–17.

Wilson, M. L., and R. W. Wrangham. 2003. "Intergroup relations in chimpanzees." *Annual Review of Anthropology* 32 (1): 363–92.

Williams, J. M., G. W. Oehlert, J. V. Carlis, and A. E. Pusey. 2004. "Why do male chimpanzees defend a group range?" *Animal Behaviour* 68 (3): 523–32.

Wiszniewski, J., C. Brown, and L. M. Möller. 2012. "Complex patterns of male alliance formation in a dolphin social network." *Journal of Mammalogy* 93 (1): 239–50.

Wittig, R. M., and C. Boesch. 2003. "Food competition and linear dominance hierarchy among female chimpanzees of the Tai National Park." *International Journal of Primatology* 24 (4): 847–67.

Wittiger, L., and C. Boesch. 2013. "Female gregariousness in Western Chimpanzees (*Pan troglodytes verus*) is influenced by resource aggregation and the number of females in estrus." *Behavioral Ecology and Sociobiology* 67 (7): 1097–1111.

Wood, B. M., D. P. Watts, J. C. Mitani, and K. E. Langergraber. 2017. "Favorable ecological circumstances promote life expectancy in chimpanzees similar to that of human hunter-gatherers." *Journal of Human Evolution* 105: 41–56.

Wrangham, R. W. 1999. "Evolution of coalitionary killing." *American Journal of Physical Anthropology* 110 (S29): 1–30.

Wrangham, R. W., A. P. Clark, and G. Isabirye-Basuta. 1992. "Female social relationships and social organization of the Kibale Forest chimpanzees. In *Topics in Primatology: Human Origins*, edited by T. Nishida, W. C. McGrew, P. Marler, M. Pickford, and F. B. M. de Waal, 81–98. Tokyo: University of Tokyo Press.

Wroblewski, E. E. 2010. "Paternity and father-offspring relationships in wild chimpanzees, *Pan troglodytes schweinfurthii*." PhD diss., University of Minnesota.

Wroblewski, E. E., C. M. Murray, B. F. Keele, J. C. Schumacher-Stankey, B. H. Hahn, and A. E. Pusey. 2009. "Male dominance rank and reproductive success in chimpanzees, *Pan troglodytes schweinfurthii*." *Animal Behaviour* 77 (4): 73–885.

Yerkes, Robert M. 1925. *Almost Human*. New York: Century Company.

PART TWO

A Social Species

4

Social Behavior and Social Tolerance in Chimpanzees and Bonobos

JARED P. TAGLIALATELA, SARA A. SKIBA, ROBERT E. EVANS, STEPHANIE L. BOGART, AND NATALIE G. SCHWOB

Introduction

The adaptive value of living in increasingly large social groups remains an area of key importance for those interested in the evolutionary origins of human social behavior and language. A number of avian, cetacean, and primate species show a positive association between group size, communicative complexity, and social diversity (Freeberg and Harvey 2008; Krams et al. 2012; Marino 2002). Although living socially may be inherently costly due to increased competition for resources, species living in large social groups gain fitness benefits by experiencing increased protection from predators, increased mating opportunities and genetic diversity, improved access to consistent food resources, and occasionally, direct benefits to reproductive fitness (Engh et al. 2000; Henzi and Barrett 2003; Silk 2007a,b). In birds, for example, living socially is associated with reduced predation risk and increased reproductive success. Among the variety of nesting patterns exhibited by Brewer's blackbirds (*Euphagus cyanocephalus*), predation risk was found to be highest amongst nest sites that were widely dispersed, and lowest among clumped nest colonies (Horn 1968). Similarly, individual birds living in clumped nest colonies had higher reproductive success than individuals living in linearly dispersed nest sites (Horn 1968). Additionally, there is evidence that increased social diversity may be a predictor of reproductive fitness in mammals. For example, in bottlenose dolphins (*Tursiops* sp.), social diversity explained 44% of the variation in calving success (Frère et al. 2010). In other words, female dolphins living in larger, more diverse social groups produced more viable offspring than those in smaller groups.

Among primates, there is increasing evidence that there are direct reproductive benefits to living in relatively large, socially complex groups. For instance, the sociality of adult female baboons (*Papio* sp.) (as defined by the proportion of time spent in physical proximity, receiving grooming, or grooming conspecific[s]) is positively associated with infant survival, and is independent of individual rank or other environmental factors (Silk, Alberts, and Altmann 2003). Notwithstanding, higher-ranking female baboons do have shorter inter-birth intervals and produce offspring that are more likely to survive past the first year as compared to lower-ranking females (Bulger and Hamilton 1987). Therefore, it is possible that an individual's sociality is important for counterbalancing hierarchy discrepancies in large social groups, and that complex socio-communicative behavior facilitates this sociality (Henzi and Barrett 2003; Silk 2007b).

Human sociality and language represent the most complex socio-communicative system in the animal kingdom. However, the evolutionary origins of human social behavior, and the cognitive capacities associated with it, remain poorly understood. Comparing the social behavior of chimpanzees (*Pan troglodytes*) and bonobos (*Pan paniscus*) provides an opportunity to make inferences about the origins of our own sociality. Chimpanzees and bonobos are the only two extant species of the genus *Pan*, sharing a common ancestor with humans approximately 6 million years ago (MYA). Chimpanzees and bonobos subsequently diverged from a common ancestor as recently as 1.5 MYA (Becquet et al. 2007; Hey 2010). This relatively short period of time since their divergence is reflected in both the physiological and behavioral similarities between the two species (Prüfer et al. 2012). In regard to physical appearance, chimpanzees and bonobos are so similar that they were not identified as individual species until the early 1930s (de Waal 1988). Despite this physical similarity, bonobos are generally smaller and more gracile than their larger more robust chimpanzee counterparts (Goodall 1986). Both species exhibit similarly complex socio-communicative behavior, have the largest foraging party sizes and vocal repertoire sizes of all nonhuman apes, and have the highest rates of grooming among all ape species (Stanford 1998; McComb and Semple 2005). Furthermore, both species live in highly social, multi-male, multi-female groups, and have been observed to use socio-communicative behaviors to influence overall social standing (Gruber and Clay 2016; Liebal et al. 2014).

Despite these similarities, and their close phylogenetic relatedness, the two species display notable differences in behavior (Behringer et al., chapter 2 this volume; Clay, chapter 12 this volume; Yamamoto, chapter 14 this volume). Bonobos are regarded as more tolerant than chimpanzees, are matriarchal, and participate in high levels of socio-sexual behavior (Parish, de Waal, and Haig 2000; Stanford 1998). On the contrary, chimpanzees are considered to be territorial, exhibit higher levels of intra- and inter-specific aggression as compared to bonobos, and typically have a single male sitting atop the group hierarchy (de Waal 2008; Stanford 1998). Bonobos travel in larger feeding parties and live in larger social groups than chimpanzees, which has led to the common perspective that bonobos are more social than chimpanzees (Doran et al. 2002; Stanford 1998).

It has been hypothesized that differences in the habitats occupied by chimpanzees and bonobos may explain, at least in part, these behavioral differences. Both species are native to sub-Saharan Africa, but whereas chimpanzees live in a variety of habitats across the continent, bonobos are restricted in their range to a region of jungle on the south bank of the Congo River, in the Democratic Republic of Congo. Chimpanzees and gorillas live sympatrically within large portions of their habitat (Head et al. 2011; Stanford and Nkurunungi 2003), whereas bonobos live allopatrically to both species. Bonobos' geographic isolation is hypothesized as a driving factor in the differential evolutionary trajectories between chimpanzees and bonobos (Malenky and Wrangham 1994; Tutin et al. 1991).

Although ripe fruit is a major part of the bonobo diet, terrestrial herbaceous vegetation (THV) consumption appears to constitute a greater proportion of their total diet (33%) compared to chimpanzees' diet (7%; Rubenstein and Wrangham 1986). Chimpanzees rely on THV mostly during times of fruit scarcity, while bonobos consume THV at a relatively constant rate throughout the year (Chapman, White, and Wrangham 1994). The competitive exclusion principle, which states that two species cannot occupy the same ecological niche at the same time, may explain the dietary specializations observed in chimpanzees (frugivory) and the lack of specialization for either THV or ripe fruit by bonobos (Volterra 1928). In short, chimpanzees have faced selective pressures since their phylogenetic split with bonobos to expand their habitat use, specialize their diet, and rely more heavily on ripe fruit as a result of competition with gorillas, whereas bonobos have not. These ecological pressures may similarly have led to the differences in social behavior that are observed between the two species in the present day.

Approach and Results

Comparative research on chimpanzees and bonobos has been conducted only relatively recently, and putative differences in social behaviors between the two species have yet to be fully investigated (de Waal 1988; Goodall 1986; Malenky and Wrangham 1994; Tutin et al. 1991). This chapter presents both observational and experimental data from three novel studies directly comparing aspects of social behavior between captive bonobos and chimpanzees.

Specifically, we first compare social behavior by examining social proximity to conspecifics and grooming rates between bonobos and chimpanzees in captive settings. Second, we directly compare prosocial behavior during an experimental feeding and foraging task with both species. These data provide a number of new insights into the two species most closely related to humans.

SOCIAL PROXIMITY AND GROOMING

We sought to collect behavioral data that would allow us to compare sociality within the genus *Pan*. In order to assess sociality, we measured inter-individual distance (social proximity) between a focal individual and the nearest neighboring conspecific, as well as grooming rates (Sibbald et al. 2005). We created four categories of social proximity so that we could determine how much time is spent at varying social distances (see table 4.1 for details). In addition to social proximity, we collected data on any grooming in which the focal individual participated. Grooming is one of the most crucial mechanisms for maintaining social relationships in primates and it is seen as a reciprocal social behavior, especially in species with shallow dominance hierarchies (Stevens et al. 2005; Watts 2000). However, few studies have measured grooming in chimpanzees and bonobos directly, making it difficult to compare grooming rates between the two species. In addition, most comparisons that have been made between the social behaviors of these apes have been based on independent studies of the two species in the wild (Muroyama and Sugiyama 1994). Although data collected from wild apes is crucial to understanding behavioral variation in natural ecological conditions, captive studies permit

TABLE 4.1. Social proximity definitions for Social Proximity and Grooming study of chimpanzees and bonobos.

Proximity	*Description*
Close/ touching	Focal individual is in physical contact with a conspecific or close enough to touch a conspecific without reaching (~<1.5 meters)
Socially close	Focal individual is ~1.5–3.0 meters from the nearest conspecific
Alone	Focal individual is ~3.0–5.0 meters from the nearest conspecific
Isolated	Focal individual is >5.0 meters from the nearest conspecific

comparisons while controlling for the potentially confounding effects caused by variation in ecological conditions. In captivity, resource availability, group size and composition, and range size can be controlled for, allowing for direct comparisons of sociality between the two species.

In terms of sociality, we hypothesized that chimpanzee and bonobo social proximity and grooming rates would differ. Based on observations of these two species in the wild (McComb and Semple 2005; Stanford 1998), we predicted that bonobos would be in closer social proximity to conspecifics than chimpanzees in captive settings. Additionally, we predicted that the relative amount of time spent engaged in grooming activities would be higher for bonobos than it would be for chimpanzees.

Subjects and Housing

We collected data on 47 chimpanzees and bonobos at four zoos in North America accredited by the Association of Zoos and Aquariums between September 2015 and June 2016 (see table 4.2). Of the 24 bonobos, we collected data on 15 individuals (5 males; 10 females, aged 3–49 years) housed at the Milwaukee County Zoo located in Milwaukee, Wisconsin. Our observations at Milwaukee County Zoo were completed while the bonobos were in their primary indoor enclosure. This space was approximately 10,000 sq. ft. and included a variety of climbing structures, including a three-story mesh wall. The remaining nine bonobos (3 males; 6 females, aged 3–47 years) lived at the Jacksonville Zoo and Gardens in Jacksonville, Florida. We collected data at Jacksonville Zoo while the bonobos were in their 6,940 sq. ft. outdoor yard with a large and complex wooden climbing structure.

We included 23 chimpanzees in this study from the North Carolina Zoological Society in Asheboro, North Carolina, and the Chattanooga Zoo in Chattanooga, Tennessee. We collected data from 16 chimpanzees at North Carolina Zoological Society (4 males; 12 females, aged 3–45 years) from the visitor viewing area of the chimpanzees' large outdoor yard, which included tall grasses, a climbing structure, and a dry moat. The remaining seven chimpanzees (2 males; 5 females, aged 24–31 years) had recently been relocated to Chattanooga Zoo from the Yerkes National Primate Research Center in June 2015, three months prior to the beginning of our data collection period (September 2015–June 2016). At Chattanooga Zoo, we collected data from the visitor viewing area where all seven chimpanzees were given daily access to the approximately 5,000 sq. ft. enclosure. See table 4.2 for more information on the subjects and study sites.

TABLE 4.2. Subjects and study sites for Social Proximity and Grooming study

Species	*Zoo*	*Subjects, Housing, and Data Collection*
Bonobos	Milwaukee County Zoo, Milwaukee, Wisconsin	We collected data on 5 males and 10 females ranging in age from 3 to 49 years old. The bonobo enclosure is composed of two areas that are used seasonally. The primary indoor enclosure is ~10,000 sq. ft., and we conducted our observations from glass windows on the ground level of this enclosure. Within their enclosure the bonobos have access to a variety of climbing structures, including a mesh wall at the rear of their enclosure, which extends all the way to the roof three stories above. They are able to climb this mesh wall, and there are several common locations where the bonobos often congregate to groom or sit close to skylights. The second area to which the bonobos are given access in the summer months is a series of expansive tunnels with varying levels, creating a looped path with visual and auditory access to the majority of the enclosure.
	Jacksonville Zoo and Gardens, Jacksonville, Florida	We collected data on 3 males and 6 females ranging in age from 3 to 47 years old. The group of bonobos that we observed varied on a daily basis, simulating the species' natural fission-fusion social structure. There were 4–6 individuals per group, where some days one group was given access to the yard for the entire day, while on other days the groups were changed at noon. When this occurred, individuals from the morning group would occasionally be included in the afternoon group. The bonobo enclosure was ~6,940 sq. ft., surrounded by a ~2 ft deep water moat. Within their yard, the bonobos had a two-story waterfall and small pool along with a three-story climbing structure. There were ropes and hammocks distributed across the enclosure. Observations of these bonobos were conducted from visitor decks two stories above the yard.
Chimpanzees	North Carolina Zoological Society, Asheboro, North Carolina	We collected data on 4 males and 12 females ranging in age from 3 to 45 years old. These chimpanzees were separated into two groups, but the males were never changed between groups, and there were only 3–4 females that were frequently exchanged between groups. This resulted in frequent groupings of 1 male and 4–6 females or 2 males and 2–6 females. Access to their approximately 1-acre outdoor yard was given to one of these groups until 2:00 pm, when the other group was given access to the yard until 2:00 pm the following day. We conducted observations on these chimpanzees from 2 visitor viewing areas, which allowed the viewing of chimpanzees at ground level through glass walls.
	Chattanooga Zoo, Chattanooga, Tennessee	We collected data on 2 males and 5 females ranging in ages from 24 to 31 years old. In June 2015, these chimpanzees were transferred to the zoo from the Yerkes National Primate Research Center, and data collection was conducted between 2015 and 2016, beginning three months after the chimpanzees' arrival to the zoo. All subjects were given access to a ~5,000 sq. ft. outdoor enclosure every day, weather permitting. We observed the chimpanzees from visitor areas, which allowed us to view the chimpanzees at ground level from behind glass walls. The enclosure included a waterfall and small pool in addition to several climbing rocks and ropes that provide access to hammocks and additional climbing structures.

Observational Data Collection

We collected observational data in the form of 10-minute focal follows with 30-second instantaneous sampling intervals. During each 10-minute focal follow, a data point was collected on social proximity and grooming for the focal individual every 30 seconds. We recorded social proximity categorically (see table 4.1), along with the identity of the nearest neighbor at the determined social proximity. Additionally, if the focal individual was participating in grooming, we recorded the directionality of grooming (give, receive, or mutual) and the number and identity of grooming partners. Focal subjects were at least 3 years of age, and in a group of at least five individuals. Any interactions by the focal individual with infants two years of age or younger were not recorded.

Statistical Analysis

To test the hypothesis that distinct patterns of sociality are exhibited between chimpanzees and bonobos, we conducted two separate univariate analyses of variance (ANOVA). For the first analysis (see fig. 4.1a), we collapsed the categories of social proximity so that comparisons could be made between the amounts of time the two species spend outside of the close/touching social proximity category. This proportion was calculated as the time spent "alone" for each individual:

$$\frac{(\#\text{ of data points SC} + \#\text{ data points AL} + \#\text{ data points IS})}{(\text{total }\#\text{ data points} - \text{any OOV})}$$

where SC = socially close, AL = alone, IS = isolated and OOV = out of view.

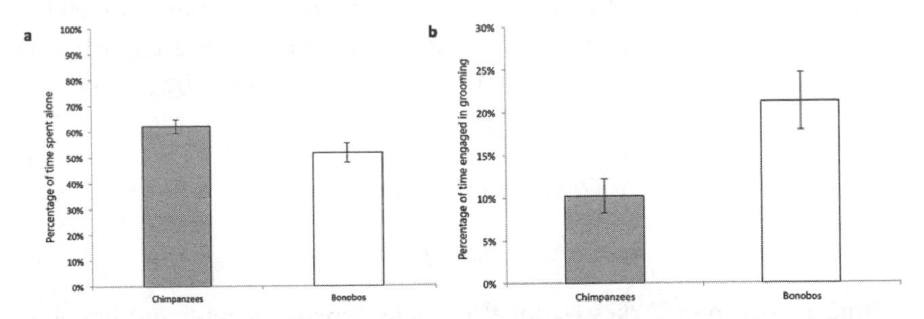

FIGURE 4.1. A) Percentage of time spent "alone," in a social proximity further than 1.5 m from nearest conspecific, between chimpanzees and bonobos (error bars indicate standard error). B) Percentage of time engaged in any grooming activity for chimpanzees and bonobos (error bars indicate standard error).

We screened the data for outliers using the boxplot function in IBM SPSS (version 22) using species as the category and proportion of time spent alone as the variable of interest. For the second analysis, we combined all the categories of grooming directionality so that we could create a value for each individual that reflects the proportion of time that the individual was engaged in any grooming activity. We similarly screened the grooming data for outliers using species as the category and proportion of time grooming as the variable of interest. All analyses were conducted in IBM SPSS (version 22).

Results

A total of 9,744 data points were collected for chimpanzees and bonobos (4,809 on chimpanzees and 4,935 on bonobos; 10, 10-minute focal-follows per individual). There were three bonobos for which we were unable to collect sufficient data and one chimpanzee who died during data collection. Removal of these individuals from analyses caused a reduction of 231 data points (chimpanzee: 42; bonobo: 189). We identified one bonobo as an outlier for spending time alone; she had recently given birth and was rarely engaged with group members. She was removed from all subsequent analyses. This resulted in 22 chimpanzees and 20 bonobos that we included in our data set.

Figure 4.1a depicts the proportion of time spent "alone" for both chimpanzees and bonobos. We performed a univariate ANOVA for percentage of time alone, with the fixed factors sex and species, and group size included as a covariate. This revealed a main effect for species, no main effect for sex, and no interaction between the two factors. Chimpanzees spent significantly more time alone (62.1%) than bonobos (51.6%), $F(1,42) = 6.04$, $p = 0.019$.

We carried out a second univariate ANOVA to evaluate the proportion of time spent grooming between the two species with the fixed factors, sex and species, and group size included as a covariate. This analysis also revealed a main effect for species, no main effect for sex, and no interaction between the two factors. The data indicate that bonobos spend roughly twice as much time grooming (21%) as do chimpanzees (11%), $F(1,42) = 5.15$, $p = 0.029$ (fig. 4.1b).

PROSOCIAL BEHAVIOR

Subjects

Chimpanzees from Yerkes National Primate Research Center and bonobos from Jacksonville Zoo and Milwaukee County Zoo participated in this study. Thirty-two subjects were tested on a social foraging task designed to replicate a clumped, but divisible, food dispersal situation and 34 subjects participated in a food sharing task simulating a shareable food dispersal situation. To reduce the potential effects of subject characteristics, we used a matched design pairing each bonobo to a chimpanzee subject based on sex and age.

Social Foraging Task

In this study, we gave pairs of chimpanzees and bonobos simultaneous access to a single food pile and we recorded their interactions. We tested subjects in pairs and each subject had two test sessions, each with a different social partner. For this task, we matched 16 bonobos (9 females; 7 males, aged 7–44 years, mean ± s.e.m. = 20.69 ± 2.85 years) with 16 chimpanzee subjects (9 females; 7 males, aged 16–57 years, mean ± s.e.m. = 22.75 ± 2.45 years). Social partners were never forced together and the pairing of a highly dominant individual with a low-ranking individual was avoided. Out of 64 total trials, 32 (16 chimpanzee and 16 bonobo trials) occurred with same sex pairs and 32 (16 chimpanzee and 16 bonobo trials) with mixed sex pairs.

Just prior to testing, each dyad (plus any dependent bonobo offspring) was separated from their home group and into an empty cage. A forage mixture consisting of a pile of shredded paper approximately 1 meter wide was mixed with small pieces of food and placed in an empty enclosure adjacent to where the pair was waiting. The forage mixture contained one chopped vegetable (e.g., sweet potato or green pepper) and a chopped fruit (e.g., apple or pear), as well as cereal filling a 946 mL container. The subjects were given access to the forage mixture simultaneously (fig. 4.2a).

We collected data for eight minutes using instantaneous sampling (Altmann 1974) every 15 seconds on the body orientations of both subjects and their proximity to each other. Body orientation of the subjects was recorded as either oriented away, such that their backs directed toward one another (back-back), or oriented toward one another, so that each individual's face was directed toward the other (face-face). If body orientation could not be categorized as either back-back or face-face, we considered the subjects mixed. For example, if the subjects were side by side or one subject was facing the other but the other was facing away, they were categorized as mixed orientation. Proximity was determined to be very close (within 0.5 meters), close/touching (both subjects could reach out a limb and touch the other, approximately between 0.5 to 1.5 m), or not proximate (subjects could not touch, > 1.5 m distance; see table 4.3). Inter-observer reliability on body orientation and proximity data was consistently >95% in agreement ($N = 5$ sessions for each category) and was re-checked frequently throughout the study.

FIGURE 4.2. Examples of the prosocial tasks include the (a) social foraging (clumped-dispersible) conducted with dyads of both species and the (b) food sharing task performed in social groups of chimpanzees and bonobos.

TABLE 4.3. Social proximity definitions for the Prosocial study of chimpanzees and bonobos

Proximity	*Description*
Very close	Subjects are within 0.5 meters of each other
Close/ touching	Both subjects can reach out a limb and touch the other (~0.5—1.5 meters)
Not proximate	Subjects could not touch (>1.5 meters)

Additionally, latency to "splitting the pile" was recorded when one or both subjects picked up any portion of the pile and moved the pile and themselves away from the other individual to a distance greater than 1.5 m.

Food Sharing Task

In this task, a single ape (the "owner") was given a monopolizable food reward and we recorded their interactions with group members. For this task we matched 17 bonobos (10 females; 7 males, aged 7–44 years, mean ± s.e.m. = 21.06 ± 2.66 years) with 17 chimpanzees (10 females; 7 males, aged 15–37 years, mean ± s.e.m. = 21.76 ± 1.58 years). Bonobo groups ranged from 2–5 (including owner) individuals (mean number of conspecifics present during owner's session = 2.18 ± 0.154). Chimpanzee groups ranged from 2–11 (including owner) individuals (mean number of group members present during owner's session = 3.35 ± 0.380).

Each of the 34 subjects (owners) received one session. Both the chimpanzees and bonobos were tested between their morning and afternoon/evening feeding sessions, with normal enrichment given during the afternoon/evening feeding. Therefore, no changes to daily feeding or enrichment schedules occurred. The food resource consisted of carrots broken in half and placed upright in a double-bagged brown paper bag with the top folded down three times. Carrots were placed in the bag to create a tight fit, so that if turned upside down none would fall out. The bag of carrots was given to a focal subject (owner) and the session lasted until all carrots were consumed or until the owner was no longer in possession of any carrots (duration ranged between six and 22 minutes [mean ± s.e.m. = 13.74 ± 0.669 min]; fig. 4.2b).

The bag was given to the owner by either tossing it directly to the owner or by placing it in a feeder box to allow access to the owner. If these methods were unavailable due to housing constraints, shifting of the animals was done so that the intended owner entered an empty enclosure containing the bag, then the rest of his or her social group was let into the same enclosure once the owner had possession of the bag.

One experimenter operated the video camcorder and narrated any important events ad libitum (e.g., how a group member acquired a carrot), while another collected proximity data using instantaneous focal recording every 30 seconds. Social proximity was recorded as the number and identity of any group members within reach of the owner, such that if each individual stretched out an arm or leg, they could touch one another (i.e., approximately 1.5 meters or less, see table 4.3). All observers collecting data on proximity were in agreement on number and identification >95% of the time (N = 8 sessions), and reliability was continually checked throughout data collection.

TABLE 4.4. Food sharing task definitions of possible methods by which a social partner obtained a carrot

Share	Active donation (give)	Owner willingly offers carrot to a proximal individual as opposed to the other taking the carrot.
	Tolerated theft	A conspecific takes carrot that is in contact (i.e., hand, foot, or in the bag they were holding) with the owner, without resistance.
Unattended	A conspecific obtains a carrot from the ground or other surface that is not in direct contact with the owner.	

Videos were coded to determine every instance a conspecific received a carrot and categorized it as either "a share" or "an unattended" (table 4.4). Following other researchers (Feistner and McGrew 1989; Jaeggi, Stevens, and van Schaik 2010), we liberally defined sharing as food being transferred with no resistance by the owner of the monopolized resource and the food had to be in the owner's "possession" (e.g., in their hand or foot, or in the bag they were holding; see table 4 in Jaeggi, Stevens, and van Schaik 2010). Previous research utilized food transfer subcategories distinguishing between different levels of requesting and resistance behaviors by both the owner and a social partner (Boesch and Boesch 1989; de Waal 1989, 1992; Jaeggi, Stevens, and van Schaik 2010). Here, we broadly defined food transfer as a share, including all types of tolerated theft and active donations. Tolerated theft occurred when the conspecific took the food from the owner. Active donations (gives) occurred when the owner willingly offered food to a social partner. Since active donations occurred only among chimpanzees, we lumped it with tolerated theft to define "share" in this study (table 4.4). If a conspecific obtained a carrot from the ground or anywhere other than from the owner, we coded it as an "unattended" carrot (table 4.4; previously labeled as "recovery" in other studies, see Boesch and Boesch 1989).

Statistical Analyses

Since the chimpanzee groups did not contain infants or juveniles, bonobos less than five years of age were excluded from all analyses. Additionally, given the presumed differences in motivation in mother-offspring sharing versus sharing among unrelated individuals, we excluded all mother-offspring and sibling dyads during the social foraging task and mother-offspring shares were excluded from analyses on the food sharing task.

Social foraging: Combining both test sessions, we calculated the proportion of each subject's body orientation and proximity to others. We conducted a separate ANOVA for each category (body orientation and proximity) using species as the fixed factor. If the subject split the foraging pile on *both* of their trials, it was scored at "split+"; and it was scored "split-" if during at least one of the trials splitting never occurred. We used a chi-square to determine whether species differences in the occurrence of splitting were evident.

Food sharing: In order to assess social tolerance, an average score of close/touching group members (see table 4.3) to the owner during the trial was determined by adding each conspecific's total frequency within proximity and dividing by the number of instantaneous data points recorded for each trial. For example, if one conspecific was within proximity for all 10 data points and another was in proximity for only 2 data points, the average proximity score would be 1.2. Since this does not account for the total number of individuals within each group, the number of individuals present for each trial was used as a covariate for the ANOVA test with species as the fixed factor. A chi-square test of independence was used to examine species differences in the propensity to share food. The proportion of individuals within the group who received shared and unattended carrots was calculated by dividing the number of individuals by the number of group members for each trial. These data were analyzed using an ANOVA for species and sex differences. For all analyses, we report p-values as two-tailed tests and we set alpha to 0.05. All analyses were conducted in IBM SPSS (version 22).

Results

Social foraging: Significant species differences were found for proximity measures during the social foraging task, $F(3, 28) = 3.09$, $p = 0.043$. Univariate F-tests showed that chimpanzees spent more time within 0.5 m (i.e., very close) to their partner than bonobos did, $F(1, 31) = 8.97$, $p = 0.005$ (fig. 4.3a). Additionally, univariate tests indicated a species difference in the orientation of the pairs (fig. 4.3b) such that chimpanzees were less often in the mixed orientation than bonobos ($F(1, 31) = 6.73$, $p = 0.015$) but more often face-face than bonobos ($F(1, 31) = 6.19$, $p = 0.019$). Finally, there were significant species differences in the frequency of splitting the pile, $\chi^2 = 4.571$, $df = 1, 32$, $p = 0.033$, with bonobos splitting the pile more frequently than chimpanzees (fig. 4.3c).

Food sharing: Chimpanzee subjects (mean ± s.e.m. adjusted by group members = 0.941 ± 0.116) were within proximity to conspecifics in their

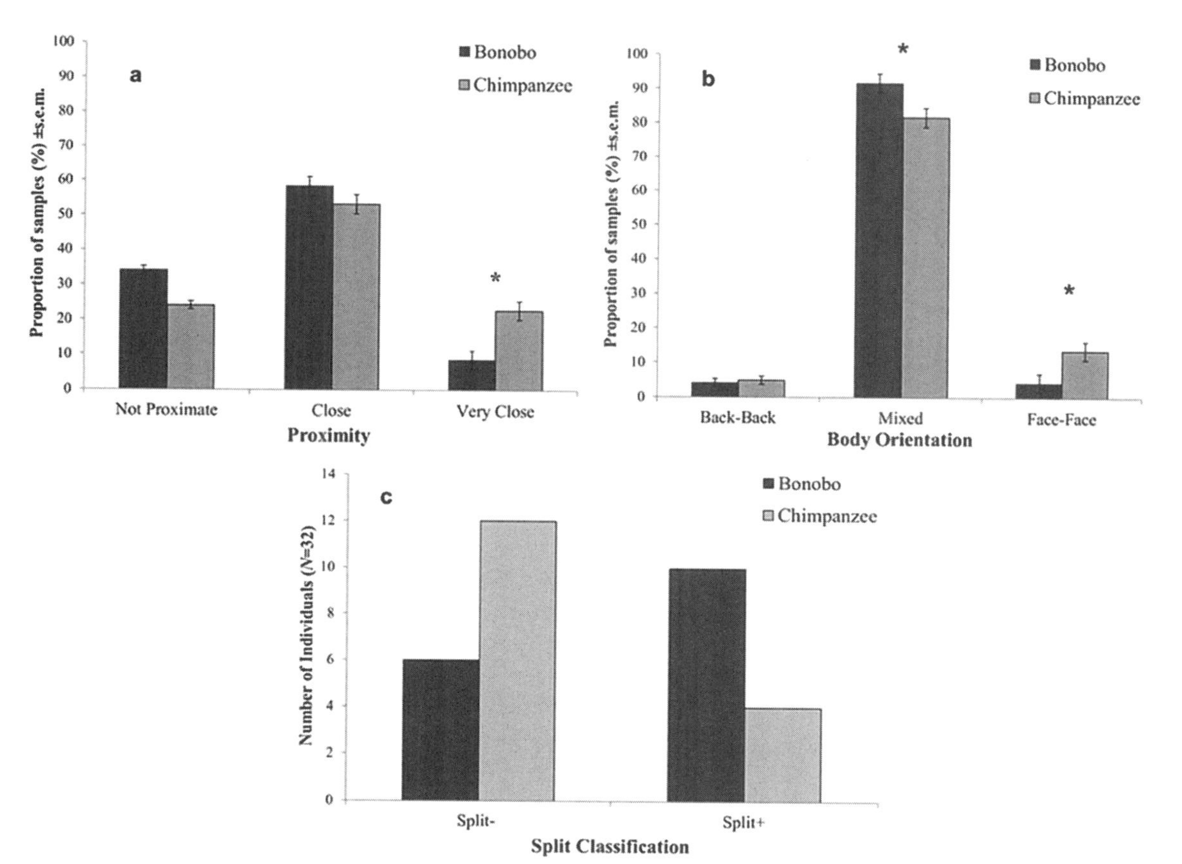

FIGURE 4.3. Prosocial analyses of the social foraging task regarding a) proximity measures from two trials in both species, b) body orientation category in which pairs from both bonobos and chimpanzees were positioned, c) how often one of the paired subjects took possession of a portion of the foraging pile in bonobos and chimpanzees with a total number of 32 individuals. (Significance at $P < 0.05$ is indicated by asterisk.)

group during food sharing sessions significantly more often than bonobos (mean ± s.e.m. adjusted by group members = 0.463 ± 0.116), $F(1, 31) = 7.29$, $p = 0.011$. Since no significant difference was found for the number of group mates between those owners who shared and those who did not, $t(15) = 1.51$, $p = 0.153$, we performed a chi-square test on the categorical scores of whether an owner shared food. The chi-square test revealed a significant species difference in the number of focal subjects who were observed to share food with group members, $\chi^2 = 7.40$, $N = 1, 34$, $p = 0.007$.

Following our definition, only one bonobo (5.9%) shared with a conspecific, while approximately half (47%) of the chimpanzees shared with at least one group member during their respective trials. Significant species differences were found for the percentage of individuals within each group who received a shared carrot (excluding mother-offspring), $F(1, 31) = 6.53$, $p = 0.016$, with chimpanzees sharing with a significantly greater percentage of group mates (26%) than bonobos (2.5%). The opposite was found when we examined the proportion of individuals within a group who were able to acquire an unattended carrot—a higher percentage of bonobo group members (80%) picked up unattended carrots than did chimpanzee conspecifics (9.3%), $F(1, 33) = 89.64$, $p < 0.001$. No sex differences were found for all analyses.

Discussion

Chimpanzees' and bonobos' close phylogenetic proximity to humans, and even closer genetic similarity to one another, make them ideal for investigations into the evolutionary origins of human sociality. The data presented here indicate that, despite their biological proximity, chimpanzees and bonobos exhibit different patterns of sociality. As indicated above (fig. 4.1a), bonobos spend more time in close proximity to conspecifics than chimpanzees do, and bonobos spend a greater proportion of time engaged in grooming behaviors compared to chimpanzees (fig. 4.1b). In addition, we found significant species differences in prosocial behavior in feeding contexts. Chimpanzees were more socially tolerant during both a social foraging and a food sharing task than their bonobo counterparts—tolerating closer proximity of group members significantly more often than bonobos. Furthermore, chimpanzees shared food significantly more often than bonobos during the food sharing task. Consistent with this finding, chimpanzees also co-fed at the forage pile during the social foraging task, whereas bonobos monopolized a portion of the forage pile by moving it away from their social partner more frequently than chimpanzees. These findings suggest that chimpanzees are in fact more tolerant of conspecifics and share food more often than bonobos—consistent with at least one previous report that similarly found that bonobos, on average, demonstrated lower social tolerance than chimpanzees (Cronin, de Groot, and Stevens 2015).

In their natural habitat, a large portion of the bonobo diet consists of THV, which is abundant and widely dispersed (Badrian and Malenky 1984; Chapman, White, and Wrangham 1994; Malenky and Stiles 1991; Malenky and Wrangham 1994; White and Wrangham 1988; Wrangham 1986). In contrast, wild chimpanzees forage primarily on clumped resources that require individuals to be in close proximity while feeding. These differences in feeding ecology may explain why bonobos exhibit higher inter-individual distances while foraging, as compared to chimpanzees (Stumpf 2007). The data presented in this chapter suggest captive chimpanzees are more socially tolerant during feeding contexts than are captive bonobos. It may be that feeding on high-value, clumped food resources (e.g., fruiting trees) has led to the selection for social tolerance during feeding and foraging in chimpanzees. In contrast, wide distribution and availability of relatively low nutritional value THV has not led to the same adaptations in bonobos. Furthermore, bonobos may seek closer social proximity outside of feeding contexts to compensate for the high inter-individual distances during foraging. Therefore, differences in feeding ecology may have led to differences in socio-communicative behavioral adaptations between the two species since their phylogenetic split some 1.5 million years ago. This hypothesis is supported by the data presented above indicating that bonobos spend more time in close proximity to conspecifics and spend more time grooming with social partners than their chimpanzee counterparts.

Two major findings emerge from the data presented here. First, when comparing measures of general sociality among bonobos and chimpanzees in similar captive settings, bonobos spend more time in close proximity to conspecifics as well as more time grooming with social partners. However, in experimental feeding situations in these same captive environments, chimpanzees are more likely to be in close proximity of conspecifics, and more likely to share food with social partners. Considered collectively, these data suggest that whereas bonobos may be the more social of the two *Pan* species, chimpanzees are actually more socially tolerant. While somewhat contradictory, reconciling these two findings is not as difficult as it may seem. In the time since diverging from their most recent common ancestor, chimpanzees, likely at least in part as a result of competitive exclusion with sympatric gorillas, have become ripe fruit specialists. Feeding on widely dispersed, but clumped, high-quality food resources likely led to selection for many of the territorial and sometimes violent behaviors exhibited by chimpanzees (see, for example, Goodall 1986). However, these same selection pressures have also likely led to social tolerance during feeding and foraging, as individuals were feeding on high quality food resources while in close proximity to conspecifics. In contrast, bonobos, given their increased reliance on dispersed, relatively low-quality THV, did not face such selection for social tolerance during feeding. Therefore, whereas both species of great apes exhibit high levels of sociality, chimpanzees may be the only species where social tolerance during feeding and foraging contexts was directly selected for.

All told, this study provides the first direct comparison of chimpanzee and bonobo social behavior in similar captive environments. However, much more work into the putative differences in socio-communicative behavior between the two species most closely related to humans are needed (see also Clay, chapter 12 this volume). These studies not only will elucidate the differences between bonobos and chimpanzees, but also will provide unprecedented insight into the evolutionary origins of human sociality and language by shedding light on how ecological factors can influence socio-communicative behavioral adaptations.

References

Altmann, J. 1974. "Observational study of behavior: Sampling methods." *Behaviour* 49 (3): 227–67.

Badrian, N., and R. K. Malenky. 1984. "Feeding ecology of *Pan paniscus* in the Lomako Forest, Zaire." In *The Pygmy Chimpanzee: Evolutionary Biology and Behavior*, edited by R. L. Susman, 275–99. Boston: Springer.

Becquet, C., N. Patterson, A. C. Stone, M. Przeworski, and D. Reich. 2007. "Genetic structure of chimpanzee populations." *PLoS Genetics* 3 (4): e66.

Boesch, C., and H. Boesch. 1989. "Hunting behavior of wild chimpanzees in the Taï National Park." *American Journal of Physical Anthropology* 78: 547–73.

Bulger, J., and W. J. Hamilton. 1987. "Rank and density correlates of inclusive fitness measures in a natural chacma baboon (*Papio ursinus*) troop." *International Journal of Primatology* 8 (6): 635–50.

Chapman, C. A., F. J. White, and R. W. Wrangham. 1994. "Party size in chimpanzees and bonobos: A reevaluation of theory based on two similarly forested sites." In *Chimpanzee Cultures*, edited by R. W. Wrangham, W. C. McGrew, F. B. M. de Waal, and P. G. Heltne, 41–58. Cambridge, MA: Harvard University Press.

Cronin, K. A., E. de Groot, and J. Stevens. 2015. "Bonobos show limited social tolerance in a group setting: A comparison with chimpanzees and a test of the relational model." *Folia Primatologica* 86 (3): 164–77.

de Waal, F. B. M. 1988. "The communicative repertoire of captive bonobos (*Pan paniscus*), compared to that of chimpanzees." *Behaviour* 106 (3): 183.

de Waal, F. B. M. 1989. "Behavioral contrasts between bonobo and chimpanzee." In *Understanding Chimpanzees*, edited by P. G. Heltne and L. A. Marquardt, 154–75. Cambridge, MA: Harvard University Press.

de Waal, F. B. M. 1992. "Appeasement, celebration, and food sharing in the two *Pan* species." In *Topics in Primatology, Volume 1: Human Origins*, edited by T. Nishida, W. C. McGrew, P. Marler, M. Pickford, and F. B. M. de Waal, 37–50. Tokyo: University of Tokyo Press.

de Waal, F. B. M. 2008. *Chimpanzee Politics: Power and Sex among Apes*. Portland: Ringgold Inc.

Doran, D. M., W. L. Jungers, Y. Sugiyama, J. G. Fleagle, and C. P. Heesy. 2002. "Multivariate and phylogenetic approaches to understanding chimpanzee and bonobo behavioral diversity." In *Behavioural Diversity in Chimpanzees and Bonobos*, edited by C. Boesch, G. Hohmann, and L. F. Marchant, 14–35. Cambridge: Cambridge University Press.

Engh, A. L., K. Esch, L. Smale, and K. E. Holekamp. 2000. "Mechanisms of maternal rank 'inheritance' in the spotted hyaena, *Crocuta crocuta*." *Animal Behaviour* 60 (3): 323–32.

Feistner, A. T. C., and W. C. McGrew. 1989. "Food-sharing in primates: A critical review." In *Perspectives in Primate Biology*, edited by P. K. Seth and S. Seth, 3:21–36. New Delhi, India: Today and Tomorrow's Printers and Publishers.

Freeberg, T. M., and E. M. Harvey. 2008. "Group size and social interactions are associated with calling behavior in Carolina chickadees (*Poecile carolinensis*)." *Journal of Comparative Psychology* 122 (3): 312–18.

Frère, C. H., M. Krützen, J. Mann, R. C. Connor, L. Bejder, W. B. Sherwin, and G. H. Orians. 2010. "Social and genetic interactions drive fitness variation in a free-living dolphin population." *Proceedings of the National Academy of Sciences* 107 (46): 19949–54.

Goodall, J. 1986. *The Chimpanzees of Gombe: Patterns of Behavior*. Cambridge, MA: Belknap Press of Harvard University Press.

Gruber, T., and Z. Clay. 2016. "A comparison between bonobos and chimpanzees: A review and update." *Evolutionary Anthropology* 25 (5): 239–52.

Head, J. S., C. Boesch, L. Makaga, and M. M. Robbins. 2011. "Sympatric chimpanzees (*Pan troglodytes troglodytes*) and gorillas (*Gorilla gorilla gorilla*) in Loango National Park, Gabon: Dietary composition, seasonality, and intersite comparisons." *International Journal of Primatology* 32 (3): 755–75.

Henzi, P., and L. Barrett. 2003. "Evolutionary ecology, sexual conflict, and behavioral differentiation among baboon populations." *Evolutionary Anthropology* 12 (5): 217–30.

Hey, J. 2010. "The divergence of chimpanzee species and subspecies as revealed in multipopulation isolation with migration analyses." *Molecular Biology and Evolution* 27 (4): 921–33.

Horn, H. S. 1968. "The adaptive significance of colonial nesting in the brewer's blackbird (*Euphagus cyanocephalus*)." *Ecology* 49 (4): 682–94.

Jaeggi, A. V., J. M. G. Stevens, and C. P. van Schaik. 2010. "Tolerant food sharing and reciprocity is precluded by despotism among bonobos but not chimpanzees." *American Journal of Physical Anthropology* 143 (1): 41–51.

Krams, I., T. Krama, T. M. Freeberg, C. Kullberg, and J. R. Lucas. 2012. "Linking social complexity and vocal complexity: A parid perspective." *Philosophical Transactions of the Royal Society of London B* 367 (1597): 1879–91.

Liebal, K., A. Vaish, D. Haun, and M. Tomasello. 2014. "Does sympathy motivate prosocial behaviour in great apes?" *PLoS One* 9 (1): e84299.

Malenky, R. K., and E. W. Stiles. 1991. "Distribution of terrestrial herbaceous vegetation and its consumption by *Pan paniscus* in the Lomako Forest, Zaire." *American Journal of Primatology* 23 (3): 153–69.

Malenky, R. K., and R. W. Wrangham. 1994. "A quantitative comparison of terrestrial herbaceous food consumption by *Pan paniscus* in the Lomako Forest, Zaire, and *Pan troglodytes* in the Kibale Forest, Uganda." *American Journal of Primatology* 32 (1): 1–12.

Marino, L. 2002. "Convergence of complex cognitive abilities in cetaceans and primates." *Brain, Behavior and Evolution* 59 (1–2): 21–32.

McComb, K., and S. Semple. 2005. "Coevolution of vocal communication and sociality in primates." *Biology Letters* 1 (4): 381–85.

Muroyama, Y., and Y. Sugiyama. 1994. "Grooming relationships in two species of chimpanzees." In *Chimpanzee Cultures*, edited by R. W. Wrangham, 169–79. Cambridge, MA: Harvard University Press.

Parish, A. R., F. B. M. de Waal, and D. Haig. 2000. "The other 'closest living relative': How bonobos (*Pan paniscus*) challenge traditional assumptions about females, dominance, intra- and intersexual interactions, and hominid evolution." *Annals of the New York Academy of Sciences* 907 (1): 97–113.

Prüfer, K., K. Munch, I. Hellmann, K. Akagi, J. R. Miller, B. Walenz, S. Koren, G. Sutton, C. Kodira, R. Winer, J. R. Knight, J. C. Mullikin, S. J. Meader, C. P. Ponting, G. Lunter, S. Higashino, A. Hobolth, J. Dutheil, E. Karakoç, C. Alkan, S. Sajjadian, C. R. Catacchio, M. Ventura, T. Marques-Bonet, E. E. Eichler, C. André, R. Atencia, L. Mugisha, J. Junhold, N. Patterson, M. Siebauer, J. M. Good, A. Fischer, S. E. Ptak, M. Lachmann, D. E. Symer, T. Mailund, M. H. Schierup, A. M. Andrés, J. Kelso, and S. Pääbo. 2012. "The bonobo genome compared with the chimpanzee and human genomes." *Nature* 486 (7404): 527.

Rubenstein, D. I., and R. W. Wrangham. 1986. *Ecological Aspects of Social Evolution: Birds and Mammals*. Princeton, NJ: Princeton University Press.

Sibbald, A. M., D. A. Elston, D. J. F. Smith, and H. W. Erhard. 2005. "A method for assessing the relative sociability of individuals within groups: An example with grazing sheep." *Applied Animal Behaviour Science* 91 (1): 57–73.

Silk, J. B. 2007a. "Social components of fitness in primate groups." *Science* 317 (5843): 1347–51.

Silk, J. B. 2007b. "The adaptive value of sociality in mammalian groups." *Philosophical Transactions of the Royal Society of London B* 362 (1480): 539–59.

Silk, J. B., S. C. Alberts, and J. Altmann. 2003. "Social bonds of female baboons enhance infant survival." *Science* 302 (5648): 1231–34.

Stanford, C. B. 1998. "The social behavior of chimpanzees and bonobos: Empirical evidence and shifting assumptions." *Current Anthropology* 39 (4): 399–420.

Stanford, C. B., and J. B. Nkurunungi. 2003. "Behavioral ecology of sympatric chimpanzees and gorillas in Bwindi Impenetrable National Park, Uganda: Diet." *International Journal of Primatology* 24 (4): 901–18.

Stevens, J., H. Vervaecke, H. de Vries, and L. Van Elsacker. 2005. "The influence of the steepness of dominance hierarchies on reciprocity and interchange in captive groups of bonobos (*Pan paniscus*)." *Behaviour* 142 (7): 941–60.

Stumpf, R. 2007. "Chimpanzees and bonobos: Diversity within and between species." In *Primates in Perspective*, edited by C. J. Campbell, A. Fuentes, K. C. MacKinnon, S. Bearder, and R. M. Stumpf, 321–344. New York: Oxford University Press.

Tutin, C. E. G., M. Fernandez, M. E. Rogers, E. A. Williamson, and W. C. McGrew, 1991. "Foraging profiles of sympatric lowland gorillas and chimpanzees in the Lope Reserve, Gabon [and Discussion]." *Philosophical Transactions of the Royal Society of London B* 334 (1270): 179–86.

Volterra, V. 1928. "Variations and fluctuations of the number of individuals in animal species living together." *ICES Journal of Marine Science* 3 (1): 3–51.

Watts, D. P. 2000. "Grooming between male chimpanzees at Ngogo, Kibale National Park. I. Partner number and diversity and grooming reciprocity." *International Journal of Primatology* 21 (2): 189–210.

White, F. J., and R. W. Wrangham. 1988. "Feeding competition and patch size in the chimpanzee species *Pan paniscus* and *Pan troglodytes*." *Behaviour* 105 (1–2): 148–64.

Wrangham, R. W. 1986. "Ecology and social relationships in two species of chimpanzee." In *Ecological Aspects of Social Evolution: Birds and Mammals*, edited by D. I. Rubenstein and R. W. Wrangham, 352–78. Princeton, NJ: Princeton University Press.

5

Endurance and Flexibility of Close Social Relationships: Comparing Chimpanzees (*Pan troglodytes verus*) and Sooty Mangabeys (*Cercocebus atys atys*)

ROMAN M. WITTIG, ALEXANDER MIELKE, JACK LESTER, AND CATHERINE CROCKFORD

Introduction

Many social animals form close and enduring social relationships, or social bonds, that have positive effects on health, fertility, and longevity (Brent et al. 2014; Seyfarth and Cheney 2012). In humans, the quality and the quantity of social relationships individuals maintain have been reported to affect mental health and mortality (Cohen and Wills 1985; House, Landis, and Umberson 1988). Perceived social support, provided by a close social bond partner, decreases the risk of mortality in humans, with an effect size comparable to well-established mortality-risk factors, such as alcoholism and smoking (Holt-Lunstad, Smith, and Layton 2010). In nonhuman primates and other social mammals, individuals that maintain close social bonds live longer (Archie et al. 2014; Silk et al. 2010) or have more offspring (Cameron, Setsaas, and Linklater 2009; Schülke et al. 2010; Silk 2007), which are more likely to survive (Silk, Alberts, and Altmann 2003; Silk et al. 2009; see also Mann, Stanton, and Murray, chapter 3 this volume).

Social bonds are close relationships that are based on reliable mutual affiliation patterns, predictable support, and low frequencies of escalated conflicts, and are stable over time (Crockford et al. 2013; Fraser, Schino, and Aureli 2008; Seyfarth and Cheney 2012). Bond partners groom each other at high frequencies (e.g., Silk, Seyfarth, and Cheney 1999) and can show high reciprocity of grooming (e.g., Frank and Silk 2009; Gomes, Mundry, and Boesch 2009). Grooming events between bond partners are linked to higher urinary oxytocin levels than control events (Crockford et al. 2013), oxytocin being a neuropeptide playing a major role in bond formation in mammals (Carter et al. 1992; Romero et al. 2014). The oxytocinergic system with its link to reward centers in the brain likely facilitates the maintenance of social bonds (Keverne and Curley 2004).

Grooming is also closely linked to several other behaviors that are informative about the quality of the relationship between two individuals. In many primate species, grooming is exchanged for agonistic support, food sharing, or feeding tolerance (Carne, Wiper, and Semple 2011; Gomes and Boesch 2011; Schino 2007; Ventura et al. 2006). Agonistic support, if predictable, is a powerful social tool that can be linked to fitness outcomes (e.g., reproductive success or offspring survival). Male bond partners in Assamese macaques (*Macaca assamensis*) form coalitions that predict future dominance ranks and reproductive success (Schülke et al. 2010). In chimpanzees (*Pan troglodytes*), agonistic support by bond partners is an effective tool that works to repel even dominant competitors (Wittig and Boesch 2003a; Wittig et al. 2014b), gaining coalition partners more copulations (Duffy, Wrangham, and Silk 2007). Food sharing also involves the oxytocinergic system, with food sharing events in chimpanzees showing higher urinary oxytocin levels than control events (Wittig et al. 2014a). Since higher oxytocin levels are independent from prior relationship quality, food sharing in chimpanzee might provide a mechanism for the formation of social bonds between non-kin adults (Mitani and Watts 2001; Wittig et al. 2014a; but see Gilby 2006).

In many monkey species that live in multi-male multi-female groups with female philopatry, social bonds are formed along matrilines. This appears to be especially true for monkey species with more despotic dominance structures like in *M. mulatta* (de Waal and Luttrell 1986), *Papio cynocephalus* (Silk, Alberts, and Altmann 2006a, 2006b), or *P. ursinus* (Silk, Seyfarth, and Cheney 1999), where social bonds are mainly formed and maintained between close maternal kin. Where dominance structures are less despotic, bonds are formed between rank neighbors, for example in *M. nigra* (Duboscq et al. 2017), *M. thibetana* (Xia et al. 2012), or *P. papio* (Kopp et al. 2015; Patzelt et al. 2011). Since dominance relationships in matrilinear societies follow matrilines, bonds formed with rank neighbors are also likely bonds with close maternal kin (Chapais and Berman 2004).

Substantial differences in the social structure of primate species should create different potential benefits from maintaining social bonds. Although social bonds are supposed to be predictable and enduring to provide best benefits, the duration of social bonds is most likely variable within and among species (Duboscq et al. 2017). Under the condition that bond partners are close maternal kin, bonds should be extremely stable and long lasting. If bond partners are chosen from the pool of close rank neighbors of a stable rank-order (e.g., based on matrilines), bonds should also be rather stable and long lasting. In case the decision for bond partners is, however, unrelated to kin and rank relationships, the pool of possible bond partners is growing, which

FIGURE 5.1. Grooming is important for the maintenance of bonds in (top) Western chimpanzee (*Pan troglodytes verus*) and (bottom) sooty mangabey (*Cercocebus atys atys*) living sympatrically in the Taï National Park, Côte d'Ivoire.

should make bonds more flexible but also less enduring (Duboscq et al. 2017). Therefore, here we investigate the quality of the relationships between individuals in two sympatric groups of Western chimpanzees (*P. t. verus*, fig. 5.1) and sooty mangabeys (*Cercocebus atys atys*, fig. 5.1) in the Taï National Park, Côte d'Ivoire. We investigate the effect of close maternal kin and rank on the formation of bonds in both species and examine the duration and stability of these close relationships.

Sooty mangabeys in the wild show female philopatry (Mielke et al. 2017; Range 2006; Range and Noë 2002). Females exhibit linear dominance hierarchies and form differentiated close social relationships with preferences for females of adjacent ranks (Range and Noë 2002). The ranks of juvenile females are close to their mothers and stable over time, indicating a matrilinear dominance system (Range 2006). Aggression among mangabey females in a feeding context increases in food patches and is usually directed down the hierarchy (Range and Noë 2002).

In contrast, wild chimpanzees show male philopatry (Inoue et al. 2008). Males form linear dominance relationships that can change over time (Muller and Wrangham 2004) while females' dominance hierarchy is linear in some populations but not in others (Wittig and Boesch 2003b). Maternal kin relationships in chimpanzees play only a minor role in predicting social bonds (Crockford et al. 2013; Langergraber, Mitani, and Vigilant 2007, 2009). Social bonds are formed between kin and non-kin (Crockford et al. 2013; Langergraber, Mitani, and Vigilant 2007, 2009), as well as between males and females (Lehmann and Boesch 2009; Mitani 2009; Wittig and Boesch 2010), with a preference for same sex partners (Surbeck et al. 2017).

Approach

STUDY ANIMALS AND SITE

The study site of the Taï Chimpanzee Project (TCP) is located in the Taï National Park, Côte d'Ivoire, where researchers have studied the behavior of four groups of Western chimpanzees since 1979 (Boesch and Boesch-Achermann 2000; Boesch and Wittig 2019). Since 2012, TCP researchers have also been following a group of sympatric-living sooty mangabeys in the chimpanzee territory. Groups are followed on a daily basis by qualified local field assistants and student researchers who conduct focal animal sampling of target individuals (Altmann 1974).

For these analyses, we used data for the south group of chimpanzees, collected from October 2012 to January 2016. During the study period, the group increased from 24 to 39 individuals due to extensive female immigration. For comparison we used data from the TCP mangabey group, collected from January 2014 to December 2016, during which time the group consisted of 58–67 individuals, with demographic changes due to deaths and births. For both species, we extracted data from the TCP long-term database, specifically grooming durations and aggressive events, for the focal targets with sufficient observation hours (>100 h chimpanzees, > 30 h mangabeys), which resulted in data for 11 adult chimpanzees (age > 12 years: 5♂, 6♀) and 24 adult mangabeys (age > 5 years: 4♂, 20♀). The mangabeys also faced some demographic changes among the adults during this time, with eleven focal animals dying or emigrating before the end of the observation time.

Behavioral data were collected from morning (~6 a. m.) to the evening (~6 p. m.) using full-day focal animal sampling for the chimpanzees and half-day or one hour focal animal sampling for the mangabeys. During focal observations all affiliative (e.g., grooming) and aggressive behaviors involving the focal animal were recorded including the names of the interaction partners. In addition, individuals resting or feeding in proximity of 1 m to the focal animal were recorded. Inter-observer reliability between data collectors exceeded 80% agreement.

SOCIAL RELATIONSHIPS AND BONDS

To quantify the change in relationships over time, we calculated dyadic affiliation strength using the Dynamic Dyadic Sociality Index (DDSI) (Kulik 2015) for the two communities. The DDSI represents relationships between all pairs of individuals, calculated for each day based on past positive and negative social interactions within each pair (for methods see Mielke et al. 2017). In short, for the chimpanzees, we used the duration of grooming interactions as positive, and directed aggression bouts as negative, interactions. For the mangabeys, we additionally included resting or feeding within 1 m proximity as socio-positive behavior (indicating tolerance between individuals). Dyads all are initially assigned with a value of 0.5, either from the beginning of the time period or when joining the group. Positive interactions increase the dyad's value, while negative interactions reduce it. Changes in the dyadic value of one dyad create changes in the dyads including the two interactors with all other community members in the opposite direction. For dyads to have consistently high values, they need to continually invest in each other. We calculated daily DDSI values for all dyads in each community, and then calculated a monthly average to reduce the impact of unstable short-term changes. As the DDSI needs some time to represent the social relationship of a dyad appropriately, we removed the first 6 months of data for each community as a burn-in phase and focused on the remaining 34 and 30 months of data for the chimpanzee and mangabey communities, respectively.

We calculated three measures from the DDSI: (1) We measured the distribution of the DDSI values within each species and defined the dyads with values above the mean plus two standard deviations (+2SD) as having especially strong positive relationships. Following Crockford et al. (2013), we called them enduring strong relationships when they had over six months continuously positive DDSI values above the threshold of mean +2SD; (2) We calculated the top three (TOP3) partners for each individual, the partners with whom an individual had the three highest average DDSI values over the entire observation time (Silk, Alberts, and Altmann 2003); (3) We identified the preferred (TOP1) partner for each individual, every month, which is the partner with whom an individual had the highest DDSI value out of all dyadic DDSI values that included the individual.

DOMINANCE RANKS

We cacluated hierarchy ranks of both communities using a modification of the Elo rank index (Foerster et al. 2016), as described in Mielke et al. (2017). We used unidirectional pant grunt vocalizations (given by the lower-ranking of two individuals: Noë, de Waal, and van Hooff 1980; Wittig and Boesch 2003b) in chimpanzees, using all available data for individuals above the age of nine years or after they were orphaned, from 1999 to 2017 (8,391 pant grunts between males, 846 between females). Staring with data from before our study period provides us with a well-known hierarchy at the starting point of our study. For the chimpanzee community, we calculated ranks within males and females separately and afterward combined the results, assuming that all males are higher-ranking than all females. In sooty mangabeys, females regularly supplant younger males. Therefore, one common hierarchy was established for both sexes. We used non-aggressive supplants in sooty mangabeys, executed by the higher-ranking of two individuals (Mielke et al. 2017; Range and Noë 2002) to establish hierarchies (2,909 supplants) between all individuals above three years of age.

While the original Elo index tracks the winning likelihood of one individual over the other using a fixed start value and change factor k, we used maximum likelihood estimation to optimize the k factor and allow individuals to enter the hierarchy with different start values. This reduces the need for a burn-in phase where ranks are relatively uncertain, does not assume rank changes where none exist, and reduces the need for a priori decisions by the researcher. Ordinal ranks were standardized between 0 and 1, with 1 being the highest-ranking individual on any given day.

KINSHIP

We defined close kin as dyads with close maternal kin relationships (e.g., mother–offspring or maternal siblings). We excluded paternal relationships from the analysis, since the effects of paternal relationships on the bond formation in chimpanzees are minimal (Langergraber, Mitani, and Vigilant 2007) and generally unclear in catarrhine monkeys (Widdig 2007; Widdig et al. 2001). Maternal relationship was determined by the known genealogy and confirmed using genetic analysis from noninvasively collected fecal samples (Arandjelovic et al. 2009), and regularly updated for the chimpanzees (Vigilant et al. 2001). For the mangabey group, however, our knowledge on the genealogy is incomplete, due to the fact that the study has been active for only a few years (Mielke et al. 2017). Therefore, to determine close maternal kin among the managbeys, we used microsatellite genotyping, the same method as for the chimpanzees, whereby DNA sequences located across different chromosomes are isolated and amplified through PCR and are used as genetic markers. We used microsatellites located in the noncoding regions of chromosomes that do not code for proteins and are therefore selectively neutral, making them ideal for population studies. Furthermore, we used autosomal microsatellites that are biparentally inherited, making them useful markers for parentage analyses. We typed between 8 and 10 microsatellite markers per DNA extract, the resulting combination of which created a unique DNA profile, known as a genotype, for each individual. In total we typed 69 individuals from the habituated group and an additional 7 individuals from neighboring groups. To assign parentage, we used a likelihood-based approach (Cervus 3.0.7; Kalinowski, Taper, and Marshall 2007) that compares the microsatellites of candidate parents and offspring and assigns parents with the highest statistical likelihood for each offspring. In order to ensure a high level of confidence in our analyses we used conservative input parameters for all simulations, which was especially important for the mangabeys, since we had just started to study this group in 2012 and most of the genealogy was still unknown. We took measures to account for all common sources of error in microsatellite analyses, such as the presence of null and false alleles, allelic dropout, and genotyping errors. We validated our data and methods by comparing the parentage assignments obtained from the genetic data to those assigned using behavioral data and found no errors in the assignments. We then followed up by looking for parent-offspring assignments among juveniles and adults, which yielded numerous mother-daughter dyads.

We were able to identify four adult mother–daughter dyads and one triad, where three adult females were assigned to each other as mother-offspring with significant confidence (table 5.1), but we were unable to detect maternal kin relationships where mothers died before we were able to collect fecal samples reliably. This means that for almost 50% of the adult females we were unable to identify close maternal kin amongst the adults, which, if true,

TABLE 5.1. The chimpanzees and mangabeys studied for these analyses

				TOP1 Partner		
Focal	*Sex*	*Species*	*# of Kin In Group**	*# Months$*	*# Months with Kin*	*Kin Ratio*
Ibrahim	♂	*P.t.v.*	1	34	1	0.029
Isha	♀	*P.t.v.*	1	34	6	0.176
Jacobo	♂	*P.t.v.*	1	34	6	0.176
Julia	♀	*P.t.v.*	1	34	3	0.088
Kinshasa	♀	*P.t.v.*	1	34	0	0
Kuba	♂	*P.t.v.*	1	34	0	0
Mbeli	♀	*P.t.v.*	0	34	0	0
Pemba	♀	*P.t.v.*	0	34	0	0
Shogun	♂	*P.t.v.*	1	34	22	0.647
Sumatra	♀	*P.t.v.*	1	34	34	1.000
Woodstock	♂	*P.t.v.*	0	34	0	0
Amboseli	♀	*C.a.a.*	1	30	13	0.433
Balancan	♀	*C.a.a.*	0	30	0	0
Budongo	♀	*C.a.a.*	1	30	0	0
Cayo	♀	*C.a.a.*	0	30	0	0
Curu	♀	*C.a.a.*	2	17	0	0
Fantom	♀	*C.a.a.*	1	11	4	0.364
Fongoli	♀	*C.a.a.*	1	30	9	0.300
Gombe	♀	*C.a.a.*	1	30	20	0.667
Kakamega	♂	*C.a.a.*	0	25	0	0
Kala	♂	*C.a.a.*	0	27	0	0
Kibale	♂	*C.a.a.*	0	21	0	0
Langtang	♀	*C.a.a.*	1	30	30	1.000
Lomas	♀	*C.a.a.*	0	30	0	0
Lope	♀	*C.a.a.*	0	23	0	0
Mahale	♀	*C.a.a.*	2	20	6	0.300
Makoku	♀	*C.a.a.*	1	30	28	0.933
Moremi	♀	*C.a.a.*	0	30	0	0
Ndoki	♀	*C.a.a.*	0	30	0	0
Odzala	♀	*C.a.a.*	2	8	8	1.000
Sonso	♀	*C.a.a.*	0	30	0	0
Tanjung	♀	*C.a.a.*	0	18	0	0
Tiningua	♀	*C.a.a.*	0	30	0	0
Wamba	♂	*C.a.a.*	0	29	0	0
Yakushima	♀	*C.a.a.*	1	28	23	0.821

Note: *P.t.v.* is *Pan troglodytes verus*; *C.a.a.* is *Cercocebus atys atys*; and TOP1 is the focal's preferred partner.

* And being a focal.

$ Variation due to demographic changes (death, migration).

would be a rather unusual situation for a matrilineal primate species. Therefore, identified close kin dyads are certainly close kin dyads, but non-kin dyads in our mangabey group have a possibility of being close kin. The data set for chimpanzees included three mother–son and one brother–sister dyad out of 55 dyads. In chimpanzees, no maternal brothers were detected, which are important and stable bond partners in other sites (Langergraber, Mitani, and Vigilant 2007).

STATISTICAL ANALYSIS

To test whether individuals preferred close maternal kin as bond partners over non-kin partners, we calculated the percentage of focal individuals whose top partner was close maternal kin and subtracted the percentage of kin when they were randomly assigned as top partners. To test the effect of rank on being top partners we calculated the rank difference between top partner and subjects and compared it with the average rank difference over all possible partners, indicating chance level for rank differences. The rank-related analysis was done only for mangabey females and chimpanzee females, since rank between the male chimpanzees was very unstable during the sampling period, creating several changes in ranks among the males. We used a bootstrap test (Wittig and Boesch 2010) and resampled the difference between observed and expected values for (a) kin bond partners and (b) rank with 200 repetitions and checked whether the 95% and 99% confidence interval included (CI of bootstrap distribution included 0) or excluded (CI of bootstrap distribution excluded 0) chance level. We employed bootstrap statistics in relation to the median and used the bootstrap plot program of a free statistic software programmed in R (Wessa 2017). In some cases, we did not apply statistical tests, but provide the average and the standard deviation for the measures of interest. We accept a meaningful difference between the two averages under the condition that the difference between them is larger than the sum of the two standard deviations.

Results

SOCIAL RELATIONSHIPS, STRONG POSITIVE RELATIONSHIPS, AND SOCIAL BONDS

The average DDSI of all dyads within a species is by definition 0.5. Chimpanzees showed a wider range of DDSI values compared to mangabeys (table 5.2). Chimpanzee same-sex dyads showed above average relationship scores,

TABLE 5.2. The chimpanzees' and mangabeys' average Dynamic Dyadic Sociality Index (DDSI) scores.

	Chimpanzees				*Mangabeys*			
	DDSI				*DDSI*			
	mean	*SD*	*range*	*N*	*mean*	*SD*	*range*	*N*
All	0.5	0.105	0.235–0.865	55	0.5	0.065	0.311–0.797	276
♂-♂	0.578	0.120	0.335–0.865	10	0.441	0.067	0.311–0.525	6
♂-♀	0.455	0.089	0.235–0.806	30	0.487	0.048	0.320–0.658	80
♀-♀	0.539	0.074	0.363–0.795	15	0.507	0.069	0.343–0.797	190

while the mean mixed-sex dyad DDSI score was below average (table 5.2). In contrast to chimpanzees, all mangabey dyads showed similar relationships scores, with female-female dyads having the highest average DDSI, male-male dyads having the lowest average DDSI, and mixed-sex dyads being in between (table 5.2).

Following the rule that strong relationships need to exceed the average DDSI plus two standard deviations, chimpanzee dyads needed to exceed a DDSI = 0.71, and mangabey dyads to exceed a DDSI = 0.61, to be considered as strong positive partners. We investigated how many dyads had monthly DDSI values above these thresholds for both species. For chimpanzees, 95 (5.1%) monthly DDSI values out of 1,850 total DDSI values were larger than 0.71. We found the majority of all monthly dyadic values showing a DDSI > 0.71 per month were amongst the male-male dyads (N = 58; representing 17.1% of all male-male DDSI values), and fewer among the mixed-sex dyads (N = 22; representing 2.2% of all mixed-sex DDSI values) and female-female dyads (N = 15; representing 3.0% of all female-female DDSI values). In contrast for mangabeys, we found 424 (6.9%) monthly DDSI values out of 6,108 total dyadic values exceeded a DDSI = 0.61 (average DDSI + 2 SD). We found the majority of monthly DDSIs > 0.61 among female-female dyads (N = 403; 9.5% of all female-female DDSI values), and only few among the mixed-sex dyads (N = 21; 1.2% of all mixed sex DDSI values), and none in male-male dyads (results summarized in table 5.3).

Reliable social bonds show high relationship strength over long periods of time. We compared the average DDSI over the entire study period and identified the TOP3 partners with the highest DDSI averages for every individual. In chimpanzees, we found that males had mostly male TOP3 partners ($N_{TOP3\ MALE}$ = 13, $N_{TOP3\ FEMALE}$ = 2), while female chimpanzees had mostly female TOP3 partners ($N_{TOP3\ MALE}$ = 4, $N_{TOP3\ FEMALE}$ = 14). Among the 33 chimpanzee TOP3 partners, five TOP3 partners were identified as close maternal kin (all mixed-sex dyads). Male and female chimpanzees had similar average DDSI values with their TOP3 partners ($DDSI_{TOP3\ Male} \pm SD = 0.605 \pm 0.024$; $DDSI_{TOP3\ Female} \pm SD = 0.588 \pm 0.040$). In contrast, in mangabeys all TOP3 partners were females, no matter whether they were partners of females ($N_{TOP3\ MALE} = 0$, $N_{TOP3\ FEMALE} = 60$) or of males ($N_{TOP3\ MALE} = 0$, $N_{TOP3\ FEMALE} = 12$). Among the 72 female TOP3 partners, nine were identified as close maternal kin. Female mangabeys showed higher average DDSI values with their TOP3 partners ($DDSI_{TOP3\ Female} \pm SD = 0.605 \pm 0.024$) compared to males ($DDSI_{TOP3\ Male} \pm SD = 0.540 \pm 0.022$).

KIN PARTNERS

Comparing the percentage of TOP3 partners that were close maternal kin, it appears that there was no striking difference between the two species. We found that in chimpanzees 15.2% of TOP3 partners were close maternal kin (5 out of 33), while in mangabeys 12.5% of TOP3 partners were close maternal kin (9 out of 72). However, the number of known maternal kin in mangabeys was most likely underestimated, while this was unlikely to be the case for the chimpanzees. To account for the variation in known genetic relationships across the two species, we ran an analysis considering only dyads with genetically established kin relationships. In mangabeys, we had identified four dyads and one triad with close kin relationships (table 5.1), creating 14 possible TOP3 kin partners in 11 focal females (8 females with one kin and 3 females with 2 kin each). In contrast, in chimpanzees there were four dyads with close kin relationships (table 5.1), creating eight possible TOP3 kin partners for four focal males and four focal females. In chimpanzees, only four individuals with close maternal kin had kin among their TOP3 partners (50%), while in mangabeys, we found nine kin partners among the TOP3 (64%).

TABLE 5.3 The relative strength of male-male, male-female, and female-female bonds among chimpanzees and mangabeys.

	Chimpanzees						*Mangabeys*					
	dyads	*bonds*		*study*	*bonds*		*dyads*	*bonds*		*study*	*bonds*	
	#	#	%	*duration*	*duration (months) ± SD*	%$	#	#	%	*duration*	*duration (months) ± SD*	%$
♂-♂	10	5	50%	34	8.8 ± 2.4	25.9%	6	0	0%	30 (25.5*)	0	0%
♂-♀	30	1	3.3%	34	14	41.2%	80	2	2.5%	30 (25.7*)	7 ± 1.4	27.2%
♀-♀	15	2	13.3%	34	7.5 ± 2.1	22.1%	190	21	11.1%	30 (25.8*)	17.2 ± 7.6	66.7%
total	55						276					

* Average presence in the group (due to demographic changes).

$ Of study duration.

The species difference became more obvious when we considered only the TOP1 partner (table 5.1). Close kin were the TOP1 partner for only two of eight chimpanzees (25%), while mangabeys had their close kin as their TOP1 partner in eight of 11 individuals (73%). Comparing how many months close maternal kin were the TOP1 partners compared to random expectations, we found that mangabeys' TOP1 partner was kin more often than expected (bootstrap test: median = 0.382, $CI_{2.5}$ = 0.244, $CI_{97.5}$ = 0.835, N = 11). This was not the case for chimpanzees (boostrap test: median = 0.032, $CI_{2.5}$ = −0.085, $CI_{97.5}$ = 0.547, N = 8).

RANK NEIGHBORS

Mangabey females chose individuals closer in rank than expected by random choice when forming relationships with the TOP1 partner (bootstrap test: median = −4.15, $CI_{2.5}$ = −4.9, $CI_{97.5}$ = −2.1, N = 20). Since maternal kin occupy neighboring ranks, we removed kin from the analysis (table 5.4). Mangabey females still chose individuals in closer rank than expected (bootstrap test: median = −1.8, $CI_{2.5}$ = −3.9, $CI_{97.5}$ = −0.05, N = 12), but the choice did not reach significance on a 1% level ($CI_{99.5}$ = 0.1). The average rank difference with the TOP1 partner was $median_{including\ kin}$ = 3 and $median_{excluding\ kin}$ = 6 respectively, indicating that the TOP1 partners were not rank neighbors. In contrast female chimpanzees (N = 6) chose TOP1 partners on average of a rank difference of $median_{including\ kin}$ = 3.5 and $median_{excluding\ kin}$ = 4 respectively.

DURATION OF BONDS

Social support should have the greatest effects when provided by reliable partners. Being a reliable partner implies that the relationship is stable and does not vary frequently. Therefore, we checked the duration of the enduring relationships (>6 months) with high DDSI values (>mean + 2SD). Bonds among male chimpanzees lasted on average 8.8 ± 2.4 months (mean ± SD) and among females for 7.5 ± 2.1 months (mean ± SD). The bonds among female mangabeys, the philopatric sex, lasted longer (average duration of bonds [months] ± SD = 17.2 ± 7.6) than bonds among chimpanzee females, and likely longer than among male chimpanzees, even though chimpanzees were observed for longer than mangabeys (results summarized in table 5.3).

We also investigated how often the TOP1 partner (per month) changed for each of the focal individuals, when the TOP1 partner stayed constant for at least three months. In chimpanzees, we found 22 changes of the TOP1 partner amongst the 11 individuals over 34 months, one top partner change on average

TABLE 5.4. Rank and affiliation in chimpanzees and mangabeys.

				Rank difference[$]	
			Rank	*Expected*	*TOP 1*
Isha	♀	*P.t.v.*	3	1.8	2
Julia	♀	*P.t.v.*	2	2.2	4
Kinshasa	♀	*P.t.v.*	4	1.8	3
Mbeli	♀	*P.t.v.*	5	2.2	4
Pemba	♀	*P.t.v.*	6	3	4
Sumatra	♀	*P.t.v.*	1	3	1* (male)
Amboseli	♀	*C.a.a.*	10	5.3	1*
Balancan	♀	*C.a.a.*	8	5.6	2
Budongo	♀	*C.a.a.*	19	9.1	7
Cayo	♀	*C.a.a.*	7	5.9	6
Curu	♀	*C.a.a.*	16	6.8	3
Fantom	♀	*C.a.a.*	20	10.0	1*
Fongoli	♀	*C.a.a.*	6	6.3	5
Gombe	♀	*C.a.a.*	3	8.2	3*
Langtang	♀	*C.a.a.*	9	5.4	1*
Lomas	♀	*C.a.a.*	18	8.2	8
Lope	♀	*C.a.a.*	12	5.4	7
Mahale	♀	*C.a.a.*	14	5.9	1*
Makoku	♀	*C.a.a.*	5	6.8	1*
Moremi	♀	*C.a.a.*	11	5.3	6
Ndoki	♀	*C.a.a.*	17	7.5	6
Odzala	♀	*C.a.a.*	15	6.3	1*
Sonso	♀	*C.a.a.*	1	10.0	6
Tanjung	♀	*C.a.a.*	2	9.1	3
Tiningua	♀	*C.a.a.*	13	5.6	1
Yakushima	♀	*C.a.a.*	4	7.5	1*
P.t.v. Mean (excluding kin TOP1*)					3 (3,4*)
P.t.v. Median (excluding kin TOP1*)					3,5 (4*)
C.a.a. Mean (excluding kin TOP1*)					3,5 (5*)
C.a.a. Median (excluding kin TOP1*)					3 (6*)

Note: *P.t.v.* is *Pan troglodytes verus*; *C.a.a.* is *Cercocebus atys atys*; and TOP1 is the focal's preferred partner.

* Kin partners are rank neighbors due to matriline constraints, so average in brackets is calculated without kin.

[$] According to female ranks.

every 17 months per individual. Focusing on males, we found 15 changes of the TOP1 partner among five males in 34 months; one TOP1 partner change every 11.3 months per individual. In comparison, mangabeys changed their TOP1 partner in 36 cases among 24 individuals over an average 25.7 months; four changes were due to the death of the TOP1 partner (11 mangabeys were observed for less than 30 months). Not counting the changes that happened due to the death of the TOP1 partner, for mangabeys, one TOP1 partner changed every 19.3 months on average.

Discussion

Using a dynamic and dyadic sociality index (DDSI), we investigated and compared the strength of social relationships among chimpanzees and sooty mangabeys, which partners they chose, how long they lasted, and how often they changed. Chimpanzees chose same-sex partners, while mangabeys chose only females for their strongest social partners, while adult mangabeys males do not groom each other nor tolerate each other in close proximity. Mangabey females, if not choosing close maternal kin, chose partners that were closer in rank than expected by random choice, but these were usually not rank neighbors. In contrast, rank difference did not influence the top partner choice of female chimpanzees. Although mangabeys showed a significant preference for maintaining top relationships with close maternal kin, chimpanzees did not show a kin preference for their top partner. These features suggest that both species have a preference for reliable partners, although for chimpanzees these are mainly same-sex (kin or non-kin), while for mangabeys their preference is for females, and preferably close maternal kin, if available, otherwise females with low rank disparity. In sum, though both species show flexibility in their bond partner choice, mangabeys' bonds are more enduring but less flexible in comparison to chimpanzees' bonds.

KIN BIAS OF BOND PARTNERS

The preference of mangabey females to choose close social partners who are kin-related fits the described matrilinear social system with despotic dominance relationships and female philopatry (Mielke et al. 2017; Range 2006; Range and Noë 2002, 2005). In the chimpanzee community, we did not find a strong impact of close maternal kin bonding partners, confirming previous results for chimpanzees living in Ngogo, Uganda (Langergraber, Mitani, and Vigilant 2007, 2009). However, our data set contained no maternal *brothers*. Given the chimpanzees' preference for same-sex partners, the available kin dyads in our chimpanzee group (mother–son and maternal brother-sister) did not fit their preference, which might explain why kin did not impact strongly on the observed bonds. Therefore, similar to other study sites (Budongo, Uganda: Crockford et al. 2013; Ngogo, Uganda: Mitani, Merriweather, and Zhang 2000), chimpanzees in Taï form and maintain close social relationships with both kin and non-kin partners. Two points remain to be investigated.

First, we considered only maternal kin relationships. There is a possibility that considering paternal kin relationships would change the picture. This would require, however, individuals to be able to recognize paternal relationships. There is evidence for fathers biasing their behavior toward their offspring in chimpanzees (Lehmann, Fickenscher, and Boesch 2006; Murray et al. 2016), capuchins (Sargeant et al. 2016), macaques (Ostner et al. 2013), and baboons (Moscovice et al. 2010), but not found in gorillas (Rosenbaum et al. 2015). This indicates that fathers in several species might have some understanding about their reproductive outcome (Widdig 2007). Whether or not this translates into understanding of paternal-kin relationships that goes beyond father–offspring relationships remains unclear. Close paternal relationships in chimpanzees seem not to play a role in bond partner selection (Langergraber, Mitani, and Vigilant 2007). Whether paternal kin discrimination plays a role in mangabeys, as it does in macaques (Widdig et al. 2001), needs to be investigated.

Second, up to twelve mangabey females and all four of the males maintained TOP1 partnerships with non-kin individuals. Whether these were indeed non-kin is uncertain since our method was unable to detect siblings without the mother present. It is, however, unlikely that all of these preferred partners were close maternal kin, related on a secondary level. Therefore, it seems reasonable to believe that mangabeys can, and do, form bonds with non-kin, especially if there is no maternal kin available in the group. A pattern like this is known from chacma baboons (*Papio ursinus*), where females without mothers or sisters present are more likely to form bonds with non-kin (Silk et al. 2010). Bonds between unrelated individuals, and the circumstances under which they are formed in mangabeys, need to be investigated further. In chimpanzees, it is less clear if non-kin bonds would be selected over kin bonds, if both were available. Any such preferences might also depend on the relationship between individuals. While mother-son relationships are common, and likely groom and support each other at high rates for male-female dyads, these relationships may not occur in the TOP3 for males, where grooming and association times are usually higher amongst males. However, if males have a male maternal sibling, it might be that large age gaps between siblings may preclude them from being the most beneficial partners.

RANK BIAS OF BOND PARTNERS

The influence of rank difference on the top partner identity in mangabey females remained significant even after removing top partners that were connected by a close maternal kin relationship. This suggests that in sooty mangabeys both qualities of partners—close kin and rank relationships—play a role when bonds are formed. It seems that mangabey females choose close ranking individuals as top partners in case there is no close kin available in the group. This would imply that sooty mangabeys bridge the structure of more despotic (de Waal and Luttrell 1986; Silk, Alberts, and Altmann 2006a; Silk, Seyfarth, and Cheney 1999) and more egalitarian (Duboscq et al. 2017; Xia et al. 2012) macaques and baboons. At the same time this conclusion needs caution, since our knowledge of maternal kinship in our group of mangabeys is not complete. Some of the closer-ranking females might indeed be from the same matriline, possibly being sisters.

ENDURANCE AND FLEXIBILITY OF BONDS

Chimpanzees maintained their closest partners on average for shorter periods, and changed them more frequently, than did mangabeys. Mangabeys had longer-lasting strong relationships, but chimpanzees were more flexible in their choice of partners. Several reasons might have driven this pattern. One reason might be that mangabeys' preference for kin over non-kin bonding partners could reduce their flexibility in partner choice compared to chimpanzees. Chimpanzees, not showing a preference along the lines of maternal kin, have a wider range of choices for potential partners, allowing them to change more opportunistically. Exhibiting such choice and forming new strong relationships might also be easier in fission-fusion societies, like in chimpanzees. Under normal group structures, where individuals can see each other most of the time, interventions can be observed by everybody and individuals can intervene against new bonds (Mielke et al. 2017).

Another reason might be that mangabey coalitions are conservative and are usually formed between rank neighbors against lower-ranking individuals (Range 2006), or the lowest-ranking individual in a triad would avoid the presence of two higher ranking ones (Mielke et al. 2017). They gain from forming coalitions with dominants, a strategy that maintains stable ranks over years similar to many despotic primate species (e.g., yellow baboons, *P. cynocephalus*, Silk, Alberts, and Altmann 2004; bonnet macaques, *M. radiata*, Silk 1982; see Thierry 1990). Chimpanzees' coalitions, in contrast, can be revolutionary. Bond partner coalitions are able to repel others (Wittig et al. 2014b), even turning the outcome of a conflict to benefit subordinate individuals (Nishida and Hosaka 1996). Chimpanzees therefore may gain more from flexible bonds than mangabeys, who might gain benefits related to long-term exchanges rather than coalition formation. This makes flexible bonds in chimpanzees likely adaptive, while they should not be in mangabeys.

Finally, another reason might be that the philopatric sex in chimpanzees (male) faces more social challenges than the philopatric sex (female) in mangabeys. At least during our observation time, dominance ranks amongst the chimpanzee males changed several times, including a period of a few months without a clear alpha male. Such social turmoil could lead to opportunistic changes in social relationships (chimpanzees: Newton-Fisher 2004, but not in chacma baboons, *P. ursinus*: Wittig et al. 2008). This period of social instability might have precipitated the short bond tenures found here: social bonds in chimpanzees under stable social conditions can also last for years (Mitani 2009). At the same time dominance relationships amongst the female mangabeys were stable, and only changed based on demographic dynamics. This reflects findings by Range and Noë (2002) showing similar stability of dominance relationships in another group of sooty mangabeys. If the social structure is stable, like in mangabeys, the advantage of strategic changes in bond partners is low. However, if the social structure is dynamic, like in chimpanzees, changing bonds can be used for strategic considerations.

Conclusion

In both species, chimpanzees and sooty mangabeys, individuals form social bonds with group members. Bonds in mangabeys, however, are more stable and longer lasting compared to those in chimpanzees. This seems to result from constraints that arise from fewer degrees of freedom in partner choice in mangabeys compared to chimpanzees. There seem to be several social consequences that arise from the different levels of flexibility and predictability of the bonds.

The flexibility of bonds allows chimpanzees to change bonding partners strategically. Chimpanzees have been observed to use partners with good relationships to reach political goals (Suchak et al. 2016), for example to overthrow the alpha male (Newton-Fisher 2004; Nishida and Hosaka 1996) or to gain mating access (Watts 1998). Strategic changes like that seem to be hampered in mangabeys, since neither kin nor ranks are flexible—both are inherited (Range and Noë 2002). It is likely that chimpanzees gain more direct benefits from maintaining social bonds, through agonistic support and coalitions, whilst mangabeys gain more indirect benefits, through inclusive fitness.

Furthermore, flexibility of bonds forces individuals to monitor others' relationships more frequently in order to keep track of transient relationships in the group (Crockford et al. 2007). Knowing whether third individuals are friends or foes of one's competitor is important in order to out-compete them

(Cheney and Seyfarth 2008). Using third party knowledge in competition has been postulated as the main reason for evolving complex social cognition (Jolly 1966; Zuberbühler and Byrne 2006) and brain size (Dunbar and Shultz 2007). The flexibility or stability of bonds and the underlying social structure that allows change, may impact, at least partially, on the social intelligence.

The capacity to form flexible bonds in chimpanzees is likely driven by the lack of opportunity to form social bonds with kin, since they have long interbirth intervals of 5–6 years (Boesch and Boesch-Achermann 2000). In sooty mangabeys females sometimes reproduce every year, or at least every other year, and they reach reproductive age at five years old. Such life history opens many options for available kin to bond with, especially if the known kin remain in one's group. In species where bond maintenance between adults is as likely to be between non-kin as kin, and flexible formation carries benefits, we hypothesize that the underlying mechanisms that sustain these relationships will be more complex than in species where bonds are predominantly between kin and based on inheritance.

Acknowledgments

We are grateful to the Ministry of Higher Education and Scientific Research of Côte d'Ivoire, the Office Ivorien des Parcs et Ressources, and the Park authorities for the permission to conduct this research. Special thanks for data collection are due to the staff of the Taï Chimpanzee Project and also to Anna Preis and Liran Samuni for the data on the chimpanzees, to Bomey Clement Gba and Jan Gogarten for data on the mangabeys, and to Linda Vigilant for supporting and supervising the genetic analysis. We thank the Centre Suisse de Recherches Scientifiques en Côte d'Ivoire for their support.

References

Altmann, J. 1974. "Observational study of behavior: Sampling methods." *Behaviour* 49 (3–4): 227–67.

Arandjelovic, M., K. Guschanski, G. Schubert, T. R. Harris, O. Thalmann, H. Siedel, and L. Vigilant. 2009. "Two-step multiplex polymerase chain reaction improves the speed and accuracy of genotyping using DNA from noninvasive and museum samples." *Molecular Ecology Resources* 9 (1): 28–36.

Archie, E. A., J. Tung, M. Clark, J. Altmann, and S. C. Alberts. 2014. "Social affiliation matters: Both same-sex and opposite-sex relationships predict survival in wild female baboons." *Proceedings of the Royal Society of London B* 281 (1793): 20141261.

Boesch, C., and H. Boesch-Achermann. 2000. *The Chimpanzees of the Taï Forest: Behavioural Ecology and Evolution*. Oxford: Oxford University Press.

Boesch, C., and R. Wittig. 2019. *The Chimpanzees of the Taï Forest: 40 Years of Research*. Cambridge: Cambridge University Press.

Brent, L. J. N., S. W. C. Chang, J.-F. Gariépy, and M. L. Platt. 2014. "The neuroethology of friendship." *Annals of the New York Academy of Sciences* 1316 (1): 1–17.

Cameron, E. Z., T. H. Setsaas, and W. L. Linklater. 2009. "Social bonds between unrelated females increase reproductive success in feral horses." *Proceedings of the National Academy of Sciences* 106 (33): 13850–53.

Carne, C., S. Wiper, and S. Semple. 2011. "Reciprocation and interchange of grooming, agonistic support, feeding tolerance, and aggression in semi-free-ranging Barbary macaques." *American Journal of Primatology* 73 (11): 1127–33.

Carter, C. S., J. R. Williams, D. M. Witt, and T. R. Insel. 1992. "Oxytocin and social bonding." *Annals of the New York Academy of Sciences* 652 (1): 204–11.

Chapais, B., and C. M. Berman. 2004. *Kinship and Behavior in Primates*. New York: Oxford University Press.

Cheney, D. L., and R. M. Seyfarth. 2008. *Baboon Metaphysics: The Evolution of a Social Mind*. Chicago: University of Chicago Press.

Cohen, S., and T. A. Wills. 1985. "Stress, social support, and the buffering hypothesis." *Psychological Bulletin* 98 (2): 310–57.

Crockford, C., R. M. Wittig, K. Langergraber, T. E. Ziegler, K. Zuberbühler, and T. Deschner. 2013. "Urinary oxytocin and social bonding in related and unrelated wild chimpanzees." *Proceedings of the Royal Society of London B* 280 (1755): 20122765.

Crockford, C., R. M. Wittig, R. M. Seyfarth, and D. L. Cheney. 2007. "Baboons eavesdrop to deduce mating opportunities." *Animal Behaviour* 73 (5): 885–90.

de Waal, F. B. M., and L. M. Luttrell. 1986. "The similarity principle underlying social bonding among female rhesus monkeys." *Folia Primatologica* 46 (4): 215–34.

Duboscq, J., C. Neumann, M. Agil, D. Perwitasari-Farajallah, B. Thierry, and A. Engelhardt. 2017. "Degrees of freedom in social bonds of crested macaque females." *Animal Behaviour* 123 (Supplement C): 411–26.

Duffy, K. G., R. W. Wrangham, and J. B. Silk. 2007. "Male chimpanzees exchange political support for mating opportunities." *Current Biology* 17 (15): R586–87.

Dunbar, R. I. M., and S. Shultz. 2007. "Evolution in the social brain." *Science* 317 (5843): 1344–47.

Foerster, S., M. Franz, C. M. Murray, I. C. Gilby, J. T. Feldblum, K. K. Walker, and A. E. Pusey. 2016. "Chimpanzee females queue but males compete for social status." *Scientific Reports* 6: 35404.

Frank, R. E., and J. B. Silk. 2009. "Impatient traders or contingent reciprocators? Evidence for the extended time-course of grooming exchanges in baboons." *Behaviour* 146 (8): 1123–35.

Fraser, O. N., G. Schino, and F. Aureli. 2008. "Components of relationship quality in chimpanzees." *Ethology* 114 (9): 834–43.

Gilby, I. C. 2006. "Meat sharing among the Gombe chimpanzees: Harassment and reciprocal exchange." *Animal Behaviour* 71 (4): 953–63.

Gomes, C. M., and C. Boesch. 2011. "Reciprocity and trades in wild West African chimpanzees." *Behavioral Ecology and Sociobiology* 65 (11): 2183.

Gomes, C. M., R. Mundry, and C. Boesch. 2009. "Long-term reciprocation of grooming in wild West African chimpanzees." *Proceedings of the Royal Society of London B* 276 (1657): 699–706.

Holt-Lunstad, J., T. B. Smith, and J. B. Layton. 2010. "Social relationships and mortality risk: A meta-analytic review." *PLoS Medicine* 7 (7): e1000316.

House, J. S., K. R. Landis, and D. Umberson. 1988. "Social relationships and health." *Science* 241 (4865): 540.

Inoue, E., M. Inoue-Murayama, L. Vigilant, O. Takenaka, and T. Nishida. 2008. "Relatedness in wild chimpanzees: Influence of paternity, male philopatry, and demographic factors." *American Journal of Physical Anthropology* 137 (3): 256–62.

Jolly, A. 1966. "Lemur social behavior and primate intelligence." *Science* 153 (3735): 501–6.

Kalinowski, S. T., M. L. Taper, and T. C. Marshall. 2007. "Revising how the computer program CERVUS accommodates genotyping error increases success in paternity assignment." *Molecular Ecology* 16 (5): 1099–1106.

Keverne, E. B., and J. P Curley. 2004. "Vasopressin, oxytocin and social behaviour." *Current Opinion in Neurobiology* 14 (6): 777–83.

Kopp, G. H., J. Fischer, A. Patzelt, C. Roos, and D. Zinner. 2015. "Population genetic insights into the social organization of Guinea baboons (*Papio papio*): Evidence for female-biased dispersal." *American Journal of Primatology* 77 (8): 878–89.

Kulik, L. 2015. "Development and consequences of social behavior in rhesus macaques (*Macaca Mulatta*)." PhD diss., University of Leipzig.

Langergraber, K. E., J. C. Mitani, and L. Vigilant. 2007. "The limited impact of kinship on cooperation in wild chimpanzees." *Proceedings of the National Academy of Sciences* 104 (19): 7786–90.

Langergraber, K. E., J. C. Mitani, and L. Vigilant. 2009. "Kinship and social bonds in female chimpanzees (*Pan Troglodytes*)." *American Journal of Primatology* 71 (10): 840–51.

Lehmann, J., and C. Boesch. 2009. "Sociality of the dispersing sex: The nature of social bonds in West African female chimpanzees, *Pan Troglodytes*." *Animal Behaviour* 77 (2): 377–87.

Lehmann, J., G. Fickenscher, and C. Boesch. 2006. "Kin biased investment in wild chimpanzees." *Behaviour* 143 (8): 931–55.

Mielke, A., L. Samuni, A. Preis, J. F. Gogarten, C. Crockford, and R. M. Wittig. 2017. "Bystanders intervene to impede grooming in western chimpanzees and sooty mangabeys." *Open Science* 4 (11): 171296.

Mitani, J. C. 2009. "Male chimpanzees form enduring and equitable social bonds." *Animal Behaviour* 77 (3): 633–40.

Mitani, J. C., D. A. Merriwether, and C. Zhang. 2000. "Male affiliation, cooperation and kinship in wild chimpanzees." *Animal Behaviour* 59 (4): 885–93.

Mitani, J. C., and D. P. Watts. 2001. "Why do chimpanzees hunt and share meat?" *Animal Behaviour* 61 (5): 915–24.

Moscovice, L. R., A. Di Fiore, C. Crockford, D. M. Kitchen, R. M. Wittig, R. M. Seyfarth, and D. L. Cheney. 2010. "Hedging their bets? Male and female chacma baboons form friendships based on likelihood of paternity." *Animal Behaviour* 79 (5): 1007–15.

Muller, M. N., and R. W. Wrangham. 2004. "Dominance, aggression and testosterone in wild chimpanzees: A test of the 'challenge hypothesis.'" *Animal Behaviour* 67 (1): 113–23.

Murray, C. M., M. A. Stanton, E. V. Lonsdorf, E. E. Wroblewski, and A. E. Pusey. 2016. "Chimpanzee fathers bias their behaviour towards their offspring." *Royal Society Open Science* 3 (11): 160441.

Newton-Fisher, N. E. 2004. "Hierarchy and social status in Budongo chimpanzees." *Primates* 45 (2): 81–87.

Nishida, T., and K. Hosaka. 1996. "Coalition strategies among adult male chimpanzees of the Mahale Mountains, Tanzania." In *Great Ape Societies*, edited by C. McGrew, L. F. Marchant, and T. Nishida, 114–34. New York: Cambridge University Press.

Noë, R., F. B. M. de Waal, and J. A. R. A. M. van Hooff. 1980. "Types of dominance in a chimpanzee colony." *Folia Primatologica* 34 (1–2): 90–110.

Ostner, J., L. Vigilant, J. Bhagavatula, M. Franz, and O. Schülke. 2013. "Stable heterosexual associations in a promiscuous primate." *Animal Behaviour* 86 (3): 623–31.

Patzelt, A., D. Zinner, G. Fickenscher, S. Diedhiou, B. Camara, D. Stahl, and J. Fischer. 2011. "Group composition of Guinea baboons (*Papio papio*) at a water place suggests a fluid social organization." *International Journal of Primatology* 32 (3): 652–68.

Range, F. 2006. "Social behavior of free-ranging juvenile sooty mangabeys (*Cercocebus torquatus atys*)." *Behavioral Ecology and Sociobiology* 59 (4): 511–20.

Range, F., and R. Noë. 2002. "Familiarity and dominance relations among female sooty mangabeys in the Taï National Park." *American Journal of Primatology* 56 (3): 137–53.

Range, F., and R. Noë. 2005. "Can simple rules account for the pattern of triadic interactions in juvenile and adult female sooty mangabeys?" *Animal Behaviour* 69 (2): 445–52.

Romero, T., M. Nagasawa, K. Mogi, T. Hasegawa, and T. Kikusui. 2014. "Oxytocin promotes social bonding in dogs." *Proceedings of the National Academy of Sciences* 111 (25): 9085–90.

Rosenbaum, S., J. P. Hirwa, J. B. Silk, L. Vigilant, and T. S. Stoinski. 2015. "Male rank, not paternity, predicts male–immature relationships in mountain gorillas, *Gorilla beringei beringei*." *Animal Behaviour* 104: 13–24.

Sargeant, E. J., E. C. Wikberg, S. Kawamura, K. M. Jack, and L. M. Fedigan. 2016. "Paternal kin recognition and infant care in white-faced capuchins (*Cebus Capucinus*)." *American Journal of Primatology* 78 (6): 659–68.

Schino, G. 2007. "Grooming and agonistic support: A meta-analysis of primate reciprocal altruism." *Behavioral Ecology* 18 (1): 115–20.

Schülke, O., J. Bhagavatula, L. Vigilant, and J. Ostner. 2010. "Social bonds enhance reproductive success in male macaques." *Current Biology* 20 (24): 2207–10.

Seyfarth, R. M., and D. L. Cheney. 2012. "The evolutionary origins of friendship." *Annual Review of Psychology* 63 (1): 153–77.

Silk, J. B. 1982. "Altruism among female Macaca radiata: Explanations and analysis of patterns of grooming and coalition formation." *Behaviour* 79 (2): 162–88.

Silk, J. B. 2007. "Social components of fitness in primate groups." *Science* 317 (5843): 1347–51.

Silk, J. B., S. C. Alberts, and J. Altmann. 2003. "Social bonds of female baboons enhance infant survival." *Science* 302 (5648): 1231–34.

Silk, J. B., S. C. Alberts, and J. Altmann. 2004. "Patterns of coalition formation by adult female baboons in Amboseli, Kenya." *Animal Behaviour* 67 (3): 573–82.

Silk, J. B., S. C. Alberts, and J. Altmann. 2006a. "Social relationships among adult female baboons (*Papio cynocephalus*) I. Variation in the strength of social bonds." *Behavioral Ecology and Sociobiology* 61 (2): 183–95.

Silk, J. B., S. C. Alberts, and J. Altmann. 2006b. "Social relationships among adult female baboons (*Papio cynocephalus*) II. Variation in the quality and stability of social bonds." *Behavioral Ecology and Sociobiology* 61 (2): 197–204.

Silk, J. B., J. C. Beehner, T. J. Bergman, C. Crockford, A. L. Engh, L. R. Moscovice, R. M. Wittig, R. M. Seyfarth, and D. L. Cheney. 2009. "The benefits of social capital: Close social bonds among female baboons enhance offspring survival." *Proceedings of the Royal Society of London B* 276 (1670): 3099–3104.

Silk, J. B., J. C. Beehner, T. J. Bergman, C. Crockford, A. L. Engh, L. R. Moscovice, R. M. Wittig, R. M. Seyfarth, and D. L. Cheney. 2010. "Female chacma baboons form strong, equitable, and enduring social bonds." *Behavioral Ecology and Sociobiology* 64 (11): 1733–47.

Silk, J. B., R. M. Seyfarth, and D. L. Cheney. 1999. "The structure of social relationships among female savanna baboons in Moremi Reserve, Botswana." *Behaviour* 136 (6): 679–703.

Suchak, M., T. M. Eppley, M. W. Campbell, R. A. Feldman, L. F. Quarles, and F. B. M. de Waal. 2016. "How chimpanzees cooperate in a competitive world." *Proceedings of the National Academy of Sciences* 113 (36): 10215–20.

Surbeck, M., C. Girard-Buttoz, C. Boesch, C. Crockford, B. Fruth, G. Hohmann, K. E. Langergraber, K. Zuberbühler, R. M. Wittig, and R. Mundry. 2017. "Sex-specific association patterns in bonobos and chimpanzees reflect species differences in cooperation." *Royal Society Open Science* 4 (5): 161081.

Thierry, B. 1990. "Feedback loop between kinship and dominance: The macaque model." *Journal of Theoretical Biology* 145 (4): 511–22.

Ventura, R., B. Majolo, N. F. Koyama, S. Hardie, and G. Schino. 2006. "Reciprocation and interchange in wild Japanese macaques: Grooming, cofeeding, and agonistic support." *American Journal of Primatology* 68 (12): 1138–49.

Vigilant, L., M. Hofreiter, H. Siedel, and C. Boesch. 2001. "Paternity and relatedness in wild chimpanzee communities." *Proceedings of the National Academy of Sciences* 98 (23): 12890–95.

Watts, D. P. 1998. "Coalitionary mate guarding by male chimpanzees at Ngogo, Kibale National Park, Uganda." *Behavioral Ecology and Sociobiology* 44 (1): 43–55.

Wessa, P. 2017. *Free Statistics Software, Office for Research Development and Education* (version 1.2.1). https://www.wessa.net/.

Widdig, A. 2007. "Paternal kin discrimination: The evidence and likely mechanisms." *Biological Reviews* 82 (2): 319–34.

Widdig, A., P. Nürnberg, M. Krawczak, W. J. Streich, and F. B. Bercovitch. 2001. "Paternal relatedness and age proximity regulate social relationships among adult female rhesus macaques." *Proceedings of the National Academy of Sciences* 98 (24): 13769–73.

Wittig, R. M., and C. Boesch. 2003a. "'Decision-making' in conflicts of wild chimpanzees (*Pan troglodytes*): An extension of the relational model." *Behavioral Ecology and Sociobiology* 54 (5): 491–504.

Wittig, R. M., and C. Boesch. 2003b. "Food competition and linear dominance hierarchy among female chimpanzees of the Taï National Park." *International Journal of Primatology* 24 (4): 847–67.

Wittig, R. M., and C. Boesch. 2010. "Receiving post-conflict affiliation from the enemy's friend reconciles former opponents." *PLoS One* 5 (11): e13995.

Wittig, R. M., C. Crockford, T. Deschner, K. E. Langergraber, T. E. Ziegler, and K. Zuberbühler. 2014a. "Food sharing is linked to urinary oxytocin levels and bonding in related and unrelated wild chimpanzees." *Proceedings of the Royal Society of London B* 281(1778): 20133096.

Wittig, R. M., C. Crockford, K. E. Langergraber, and K. Zuberbühler. 2014b. "Triadic social interactions operate across time: A field experiment with wild chimpanzees." *Proceedings of the Royal Society of London B* 281 (1779): 20133155.

Wittig, R. M., C. Crockford, J. Lehmann, P. L. Whitten, R. M. Seyfarth, and D. L. Cheney. 2008. "Focused grooming networks and stress alleviation in wild female baboons." *Hormones and Behavior* 54 (1): 170–77.

Xia, D., J. Li, P. A. Garber, L. Sun, Y. Zhu, and B. Sun. 2012. "Grooming reciprocity in female Tibetan macaques *Macaca thibetana*." *American Journal of Primatology* 74 (6): 569–79.

Zuberbühler, K., and R. W. Byrne. 2006. "Social cognition." *Current Biology* 16 (18): R786–90

6

Urinary Androgens, Dominance Hierarchies, and Social Group Structure among Wild Male Mountain Gorillas

STACY ROSENBAUM, RACHEL SANTYMIRE, AND TARA S. STOINSKI

Introduction

Competitive relationships among conspecifics have a host of health and reproductive correlates that have important evolutionary consequences. There is substantial variation between species in the type and degree to which competitive relationships express themselves (e.g., Mitchell, Boinski, and van Schaik 1991; Plavcan and van Schaik 1997). However, across species, competitive ability is related to outcomes such as lifetime fitness, longevity, and physiological stress, and is thus an important selective force (e.g., Fedigan 1983; Sapolsky 2005).

Competitive relationships occur among animals within the same social group (i.e., dominance hierarches), as well as between animals who live in different groups. Competition occurs when there are meaningful benefits associated with monopolizing a valuable resource, which generally fall into one of two categories: food and mates (Ellis 1995; Koenig 2002, reviewed in Kappeler and van Schaik 2002). The relative benefits of fighting for, or defending access to, one or the other (or both) depends on the specifics of the ecology and social system in which a species lives.

While there are many aspects of competitive ability that are likely under-appreciated and poorly understood (e.g., strategic ability, in brown-headed cowbirds, *Molothrus ater*: Gersick, Snyder-Mackler, and White 2012; personality, in water striders, *Gerridae* sp.: Sih, Chang, and Wey 2014), physical dominance (i.e., strength and size) undeniably plays an important role. Since testosterone (an androgenic steroid hormone) has significant anabolic effects, it is unsurprising that a considerable body of empirical work has shown that higher testosterone levels are correlated with willingness to engage in competition, as well as competitive outcomes, in a large number of species (e.g., California mice, *Peromyscus californicus*: Oyegbile and Marler 2005; ring-tailed lemurs, *Lemur catta*: Cavigelli and Pereira 2000; humans, *Homo sapiens*: Archer 2006; chimpanzees, *Pan troglodytes*: Anestis 2006; Gouldian finches, *Erythrura gouldiae*: Cain and Pryke 2017).

Testosterone is of interest to behavioral ecologists because it is part of the suite of costs males pay to compete with other males (e.g., Gesquiere et al. 2011; Muehlenbein and Watts 2010; Muller 2017) and confers both direct and indirect deleterious effects. There is some evidence that testosterone may act as an immunosuppressant or decrease immunocompetence (multi-taxa meta-analysis: Foo et al. 2017, but see also Gettler et al. 2014; Muller 2017; Trumble et al. 2016 for contradictory data from humans), as well as compromise the body's ability to fight oxidative stress (e.g., zebra finches, *Taeniopygia guttata*: Alonso-Alvarez et al. 2007, but see also Speakman and Garratt 2014). Crucially, testosterone also may lead to higher morbidity and mortality rates via behavioral pathways. In addition to potentially increasing the risk of physical conflict with other conspecifics, it also suppresses fear responses and acts as an analgesic, both of which promote risk-taking behavior (e.g., de Almeida, Cabral, and Narvaes 2015; Muller 2017). Understanding the relationship between testosterone and competition is key to understanding the tradeoffs animals experience, which drive the type and amount of reproductive effort in which they engage (e.g., Emery Thompson and Georgiev 2014; Muehlenbein and Bribiescas 2005).

TESTOSTERONE AND MALE-MALE COMPETITION IN GREAT APES

Across the three extant nonhuman great ape genera, there is considerable variation in social organization and ecology. Frugivorous orangutans (genus *Pongo*: Nater et al. 2017) live in extended social systems in peat swamp forests (van Schaik and van Hooff 1996). Their lack of cohesive, stable social groupings means that males functionally experience no intragroup sexual competition. Chimpanzees and bonobos (genus *Pan*), which occupy a wide variety of landscapes (and capitalize on a wide variety of foods) in central Africa (Hockings, Anderson, and Matsuzawa 2010; Hohmann et al. 2010; Malenky and Wrangham 1994; Newton-Fisher 1999; Pruetz 2006), live in multi-male, multi-female groups and form within-group dominance hierarchies. This means that in most cases, group mates are their most immediate source of intrasexual competition (e.g., Muller 2002; Surbeck, Mundry, and Hohmann 2011), even if the dangers of extra-group competitors are very real (Watts and Mitani 2001; Wilson and Wrangham 2003). For the genus *Gorilla*, male-male competition can come primarily either from group mates, or from extra-group competitors. Western gorillas (*Gorilla gorilla*) alternate between frugivory and herbivory, and live almost exclusively in one-male, multi-female groups, while mountain gorillas, their eastern cousins (*Gorilla beringei*), are high-elevation dwelling herbivores that live in a variety of social configurations (single-male, multi-female; multi-male, multi-female; all-male) (Caillaud et al. 2014; Gray et al. 2010; Parnell 2002; Remis 1997; Robbins 1995; Rogers et al. 2004; Schaller 1963). While competition over resources occurs among all the apes, variation in social structure and ecological niche is associated with differences in the strength and expression of their competitive relationships.

For male great apes, their primary limiting resource is usually mates. Male chimpanzees and bonobos who are higher in the dominance hierarchy generally are more reproductively successful than their lower-ranking counterparts (Bray, Pusey, and Gilby 2016; Newton-Fisher et al. 2010; Surbeck, Mundry, and Hohmann 2011; Surbeck et al. 2017; Wroblewski et al. 2009). Among western lowland gorillas (who live almost solely in single-male groups) and orangutans, intrasexual competition comes from extra-group competitors (i.e., animals that they do not directly interact with on a day-to-day basis). Similar to the dynamics in dominance hierarchies, winners gain priority-of-access, which also translates into reproductive success for males (western lowland gorillas: Inoue et al. 2013, orangutans: Banes, Galdikas, and Vigilant 2015).

Among the great apes, the relationship between testosterone and male-male competition has been best studied in chimpanzees and bonobos. For both species, a considerable portion of the male-male competition they experience takes the form of within-group dominance hierarches. There are many studies on the relationship between testosterone and dominance in both wild and captive chimpanzees (e.g., Anestis 2006; Klinkova, Heistermann, and Hodges 2004; Muehlenbein and Watts 2010; Muehlenbein, Watts, and Whitten 2004; Muller and Wrangham 2004). The overwhelming majority of studies have found that higher-ranked male chimpanzees have higher testosterone than their lower-ranked counterparts, though there is some evidence that there may be important nuances between basal-state testosterone levels versus testosterone effects associated with specific dominance-related behaviors such as sex and aggression (Klinkova, Heistermann, and Hodges 2004).

Data from closely related bonobos are more equivocal. Though some studies have found that higher-ranking male bonobos have higher testosterone than lower-ranking males (e.g., Marshall and Hohmann 2005), others have found no relationship or even a negative relationship between rank and testosterone (e.g., Sannen et al. 2004; Surbeck et al. 2012). Authors have generally speculated that the differences in the physiological mediation of dominance hierarchies in the two *Pan* species are due to their marked differences in social dynamics and male reproductive strategies. In bonobos, females are dominant, and overt male aggression is less commonly a part of males' mating strategies than it is for chimpanzees (e.g., Behringer et al. 2014; Surbeck and Hohmann 2013, 2017; Walker and Hare 2017; Wobber et al. 2013, see also Behringer et al., chapter 2 this volume).

Testosterone may also play an important role in species where males' primary competitors are not their group mates. Male orangutans' transitions among various morphs (Maggioncalda 1995), from smaller, unflanged varieties lacking most or all adult secondary sex characteristics, to the largest flanged morph exhibiting full secondary sex characteristics, is associated with higher testosterone (Emery Thompson, Zhou, and Knott 2012; Marty et al. 2015). To our knowledge, no data on the relationship between competitive outcomes and testosterone have been published for western lowland gorillas, though in captivity, males housed with females, males housed in all-male groups, and males housed alone were found to have similar urinary androgen metabolite levels (Stoinski et al. 2002).

Mountain gorillas are unusual because they are the only great ape species in which males occur in all the possible social configurations that other apes can live in (e.g., Harcourt and Stewart 2007; Yamagiwa 1987). In the wild, the modal group type is a single male with multiple females and their offspring, but ~40% of groups contain multiple males and multiple females (Gray et al. 2010). Males can also live in all-male groups or can be solitary for years at a time (Harcourt and Stewart 2007; Nsubuga et al. 2008; Robbins 1995, 1996). Their wide variety of potential social configurations makes them a particularly interesting species in which to evaluate the relationship between hormones and intrasexual competition. Males who live with other males experience constant intragroup competition, which males in single male groups (and solitary males) do not. However, during intergroup encounters, which are important events for males because females use them as opportunities to transfer between social groups, males in multi-male groups have the advantage of male allies that can help drive off rivals (e.g., Mirville et al. 2018a; Rosenbaum, Vecellio, and Stoinski 2016; Sicotte 1993). This may lower the risk that any one male incurs during such encounters, and put them at a competitive advantage over either solitary males, or males in single-male groups. Such differences could potentially be reflected in their levels of testosterone.

Approach

TERMINOLOGY: TESTOSTERONE VERSUS ANDROGENS

One of the challenges of measuring hormone levels in wild animals is that, instead of measuring the concentration of native hormone present in serum or whole blood, ethical and practical considerations require the use of urine or feces instead. This means we are evaluating hormone metabolites that the body excretes after the hormone of interest has been metabolized by the liver and (in the case of urine) kidneys. While structurally similar, these metabolites are not identical to the native hormone (e.g., Möstl, Rettenbacher, and Palme 2005). Available testosterone radio and enzyme immunoassays are known to cross-react with metabolites that share similar chemical structures (Hagey and Czekala 2003). Thus far, we have discussed testosterone specifically, because with a few exceptions (e.g., Stoinski et al. 2002), the literature under consideration primarily focuses on testosterone. In this chapter we use urine to explore the relationship between intrasexual competition and androgenic steroids, of which testosterone is one. Because we cannot eliminate the possibility of cross-reactivity (Stoinski et al. 2002), we refer to the substance we measure as urinary androgen metabolites (UAM). When discussing other published work, we follow the terminology used by the authors of a given paper.

REEVALUATING URINARY ANDROGEN METABOLITES (UAM) IN MOUNTAIN GORILLAS

Despite 50 years of research on the mountain gorillas monitored by the Dian Fossey Gorilla Fund's Karisoke Research Center in Rwanda, very little has been published on the relationship between dominance and testosterone in mountain gorillas. The one published preliminary study found that dominant males trended toward having higher testosterone levels than subordinate males, in two multi-male, multi-female groups and one all-male group (Robbins and Czekala 1997, see also Czekala and Robbins 2001). Robbins and Czekala did not find a difference in the testosterone levels of males living in the all-male group versus those living in the two groups that contained females. At the time the study was conducted, there were no samples available from either males in single-male groups or from solitary males.

This chapter reevaluates the relationship between UAM and dominance rank, and between UAM and social group configuration, in wild mountain gorillas. As well as using a larger sample size of both males and of individual urine samples than was available for the one previous study, this study expands the prior analyses to include males living in single-male groups, and solitary males.

We hypothesize that male-male competition is androgen mediated in this species. This is based both on the available preliminary data (Robbins and Czekala 1997), and on the fact that male gorillas have extreme adaptations for fighting for access to females (Harcourt et al. 1981; Leigh and Shea 1995). This strongly suggests that direct physical competition is an important component of their reproductive strategy, so androgen-mediated competition seems likely.

This hypothesis generates three predictions. First, we predict that within groups, male mountain gorillas of higher rank will have higher UAM levels than males that occupy lower ranks. Second, we predict that males living in multi-male groups will have higher UAM than males living in single-male, multi-female groups, due to constant (if usually not violent: Robbins 1996; Stoinski et al. 2009) within-group intrasexual competition. Third, we predict that solitary males will have higher UAM levels than males that live in any type of social groups with females. These males have little to lose by engaging in direct competition with other males, and face no social tradeoffs when acting aggressively that group-dwelling males might. Additionally, encounters between solitary males and groups are known to be more violent than encounters between mixed-sex groups (Robbins and Sawyer 2007; Rosenbaum, Vecellio, and Stoinski 2016; Watts 1994), and solitary males are more likely to initiate intergroup interactions than most categories of their group-living peers (Mirville et al. 2018b).

SUBJECTS

We collected behavior and hormone data on the mountain gorillas monitored by the Dian Fossey Gorilla Fund International's Karisoke Research Center in Rwanda. These animals have been monitored nearly continually since 1967, so extensive demographic and life history information are available for all monitored groups and individuals. Male mountain gorillas do not reach full size until they are ~15 years old (Galbany et al. 2017; Watts 1990). However, they are capable of siring offspring when they are as young as eight years old (Rosenbaum et al. 2015; Vigilant et al. 2015). Between eight and 12 years of age, they begin actively participating in intergroup encounters, and appear to be treated as low-ranking members of the adult males' dominance hierarches (Karisoke long-term records; Robbins 1995, 1996). To facilitate comparison to the only previous publication that examined the relationship between male rank and UAM in these animals (Robbins and Czekala 1997), we include all males who were at least 10 years old at the time urine sample(s) were collected.

SOCIAL GROUP STRUCTURE

Data were collected at two different points in time: 2003–2005 and 2011–2012. In 2003–2005, the animals lived in three multi-male, multi-female groups while in 2011–2012, they lived in three multi-male, multi-female groups, plus three single-male, multi-female groups (table 6.1). One of the 2011–2012

TABLE 6.1. Group composition and urine sample collection statistics.

Group Name	*BEE*	*SHI*	*PAB*		*ISA*	*KUY*	*BWE*	*INS*	*URU*
Years data collected	2003–05	2003–05	2003–04	2011–12	2011–12	2011–12	2011–12	2011–12	2011–12
Total group size*	26	27	52	45	12	14	9	6	5
# males ≥ 10 years old	8	8	6**	5	2	3/2***	1	1	1
# urine samples per male	Min: 3	Min: 11	Min: 9	Min: 4	Min: 24	Min: 8	46	14	2
	Max: 63	Max: 41	Max: 22	Max: 8	Max: 36	Max: 54			
	$\bar{x}$: 24	$\bar{x}$: 25.63	$\bar{x}$: 13.67	$\bar{x}$: 5.8	$\bar{x}$: 30	$\bar{x}$: 34			
Males' natal group					PAB	BEE	BEE	SHI	SHI

Note:

* As of midpoint of data collection.

** Group had a seventh male that turned 10 during data collection in 2003–2005. He was eliminated from the data presented here due to insufficient urine samples.

*** The beta male dispersed from the group 4 months into data collection. The formerly gamma-ranked male therefore holds two ranks in the data set, since he moved from gamma to beta during the study period.

groups was the same as one of the 2003–2005 groups. The others were groups that had formed in the intervening seven years and were composed of males who had dispersed from two of the three groups that were monitored in 2003–2005, along with female transfers and resultant offspring. This means there are data from two different points in time for five of the animals.

In addition to the males living in social groups, in 2003–2005 we collected eight urine samples from one solitary silverback when he was opportunistically encountered in the forest. In 2011–2012, we collected three urine samples from two solitary silverbacks. These are small sample sizes, but biomarker data from solitary silverbacks is logistically very difficult to gather and no information about their UAM values has been published. Therefore, we include their results here, limited to descriptive statistics where necessary.

RANK DETERMINATION

We determined the males' ranks using displacement patterns, described in Stoinski et al. (2009). Due to insufficient interactions between some males, it was difficult to determine dominance rank below the gamma position. Therefore, we classified males as alpha (the dominant male), beta, gamma, and subordinate. The subordinate category subsumes all males ranked lower than gamma. In general, older males are dominant over younger ones.

URINE COLLECTION, PRESERVATION, AND ANALYSIS

The first and third authors, along with other staff and scientists who had been tested on animal identification ability, collected urine samples opportunistically. After a subject was observed urinating, we waited for them to move away, then pipetted urine off the vegetation or ground and into polypropylene collection tubes. We did not collect samples if they were contaminated with rainwater, feces, or urine from other animals. After collection, we labeled samples with the relevant demographic, time, and date information, then placed the samples in a small insulated cooler bag containing an ice pack until returning to the Karisoke Research Center.

In 2003–2005, we placed the samples in a −20 C° freezer immediately upon returning from the field. We noted sample information in a logbook and, for a subset of the samples, the time that they were frozen. The samples were held for storage in the freezer until transport on ice to the Center for the Reproduction of Endangered Species at the San Diego Zoo, USA. The samples were then analyzed for UAM concentrations by radioimmunoassay (RIA), using ether extracted hydrolyzed urine. All extraction and analysis protocols for the samples collected in 2003–2005 may be found in Robbins and Czekala (1997) and Maggioncalda (1995).

In 2011–2012, the field collection protocol we used was identical to the one described above. However, we used a different preservation protocol to help mitigate problems associated with inconsistent electricity supply to the freezers. After returning from the field, we transferred samples to a flat surface, and particulate was allowed to settle for ~30 minutes. We then pipetted a 200uL aliquot off the top of the sample, being careful not to disturb any particulate pellet that had settled on the bottom. The sample was pipetted onto a filter paper (Whatman Grade 5, 55m diameter) placed on top of a sheet of aluminum foil, to ensure that the filter paper absorbed the entire volume. Once the urine was completely absorbed, we labeled the paper with the animal's name, and the date and time of collection, using a no. 2 pencil. We placed the filter papers inside a large dry box that contained reusable desiccants and noted the sample information and time of processing in a logbook. The samples were kept in the dry box, and desiccants replenished, for as long as necessary for the paper to dry completely (in the dry season, as short as a single day, and in the rainy season, as long as a week). Once the papers were completely dry, we stored them in individual plastic bags and placed them in a separate dry box with reusable desiccants (Shideler et al. 1995).

At the end of data collection in 2012, we transferred the dried samples to the Davee Center for Epidemiology and Endocrinology at Lincoln Park Zoo, USA. There, we stored the samples in a dark, dry, room-temperature cupboard until analysis. To extract the urine from the paper, we soaked the entire filter paper overnight in 5 mL of 100% methanol in a tightly capped tube, inside a refrigerator. The next day, we removed the filter paper from the methanol with forceps, squeezing it along the side of the tube to remove excess methanol. The filter paper was discarded, and the methanol was dried down from each sample using blown air. Once dry, we reconstituted the samples in 1 mL of dilution buffer with 3–5 small glass beads. Each one was then briefly vortexed, and sonicated for 20 minutes. This produced a neat dilution of 1:5. We used a Jaffe reaction (Taussky and Kurzmann 1954) to determine creatinine concentrations, to correct for differences in the amount of water in each sample.

The UAM concentrations in most samples were too high at the neat dilution to fall within the acceptable binding range (20–80%) of the testosterone enzyme immunoassay (EIA) we used to assess UAM concentrations (R156/7; C Munro, UC-Davis). Therefore, we diluted the samples before analysis; most samples were run at 1:30 (range = 1:10–1:33.33). Serial dilutions of the urine produced a parallel dose-response curve to the testosterone standard curve ($r = 0.998$). The inter-assay coefficient of variation was <15%; intra-assay coefficient of variation was <10%. Testosterone cross-reactivities for this assay have been previously published (Santymire and Armstrong 2010).

DATA SUMMARY AND ANALYSIS

Changes in sample preservation protocol, and the use of two different analysis methods (EIA versus RIA), could have resulted in different UAM values due to methodological rather than biological differences. We started by visualizing the two data sets separately, as well as checking basic descriptive statistics on each (table 6.2). The mean UAM value was 8.6% higher in 2011–2012 than in 2003–2005, while the standard deviation was 9.6% lower. Subjects were slightly older in 2011–2012. Both of these differences were statistically significant in a mixed effects regression model that controlled for repeat samples from the same individual (for UAM: $p = 0.001$; for age: $p < 0.0001$).

Inspection of the residuals revealed one extreme outlier among the UAM values in 2003–2005, which was double the next nearest value. We eliminated this sample from all analyses, including the descriptive statistics in table 6.2. There was insufficient statistical or theoretical justification for removing any other data points. Visual inspection indicated that the relationship between dominance rank and UAM was different among the 2003–2005 samples than in the 2011–2012 samples. Therefore, instead of including all the samples in the same model and controlling for the time period in which they were collected, we analyzed the two time periods separately.

The outcome variable for all of our linear mixed-effects models was UAM values expressed in ng/mg creatinine, based on the aforementioned immunoassay results. The models also included the following control variables, which were set as fixed effects: i) the age of the animal at the time the sample was collected, as age may affect UAM levels; ii) the date the sample was collected, to control for potential seasonal effects; and iii) the time of day the sample was collected, to control for known circadian variation in hormone levels. Initially, we also evaluated whether the lag time between when the sample was collected and when it was frozen was a significant predictor of UAM values. It was not ($p = 0.479$), so we removed this from the model results presented here.

In addition to the variables mentioned above, there are myriad social and ecological factors which may theoretically affect males' androgen production. These include, but are not limited to, female reproductive states, males' social dynamics, and group composition (e.g., variation in the presence or number of infants in the group). To help control for these, we used group identification as a fixed effect in our statistical models. While the preferred approach would be to nest animals within groups and treat both as random effects, due

TABLE 6.2. Descriptive statistics for urine samples, by data collection period.

Year	Protocol	*Subjects' age ± SD (Min, Max)*	*Mean T ± SD (Min, Max)*
2003–2005 n=487	Freeze urine until radioimmunoassay (RIA) analysis	16.49 ± 5.25 (10.13, 30.30)	123.91 ± 82.03 (4.53, 630.61)
2011–2012 n=256	Dry urine on filter paper until reconstitution for enzyme immunoassay (EIA) analysis	19.38 ± 3.95 (11.91, 33.29)	133.45 ± 72.61 (18.69, 729.34)

to the small number of groups in each of the two time periods, use of group identification as a random effect was statistically inappropriate.

The data were subset as appropriate for each analysis. For the model assessing the relationship between dominance rank and UAM, we limited the sample to males residing in multi-male groups at the time of sample collection (n = 670 samples, 27 males, 5 social groups). Rank was coded as a categorical variable where 0 = alpha, 1 = beta, 2 = gamma, and 3 = subordinate. For the model assessing differences between males living in different types of social group configurations (i.e., solitary males, males in single-male groups, and alpha males in multi-male groups), we did not include rank, but instead included a categorical variable coded as 0 = males in multi-male groups, 1 = males in single-male groups, and 2 = solitary males.

It was unclear whether the more biologically relevant comparison in this case was between males who did not live with other males and all males who lived in multi-male groups, or between males who did not live with other males and alpha males specifically in multi-male groups. We therefore chose to compare the males who did not live with other males to both categories. When comparing either males in single-male groups, or solitary males, to alpha males specifically, we removed group identification as a predictor since only one animal from each group was represented. We also changed animal identification from a random effect to a fixed-effect predictor, since the number of individual animals included in the sample was too small (<10) to set as a random effect.

Throughout, we used robust standard errors to correct for minor violation of underlying assumptions. Analyses were run in Stata 13 (StataCorp, College Station, TX).

Results

RELATIONSHIP BETWEEN DOMINANCE RANK AND UAM

Visual inspection of the raw, unadjusted UAM values did not indicate a clear relationship between rank and UAM in either 2003–2005 or 2011–2012 (fig. 6.1).

In 2003–2005 there was a positive, but not statistically significant, relationship between UAM and rank (i.e., lower-ranking males had higher values than higher-ranking males; table 6.3). In 2011–2012, there was a negative, but not statistically significant, relationship between UAM and dominance rank (i.e., higher-ranking males had higher values than lower-ranking males). There was, however, a trend for subordinate males to have lower UAM values than alpha males (table 6.3).

Comparing the two data collection periods, the signs of the coefficients associated with rank ran in opposite directions. In 2003–2005 it was positive and 2011–2012 it was negative (table 6.3). We therefore graphed the predictive margins of the rank variable for each of the two time periods, to visually assess any differences between them that were associated with rank specifically. In 2003–2005, males of lower ranks generally had higher UAM levels than males of higher rank, while the opposite was true in 2011–2012 (fig. 6.2). We stress that after controlling for collection time, collection date, and animal age and social group identity, there were no statistically significant differences in the UAM values of males of different ranks in either time period. However, the ability to detect significantly lower levels of UAM among lower-ranking males in 2011–2012 may have been hampered by the small number of samples available for gamma and subordinate ranked males during this period.

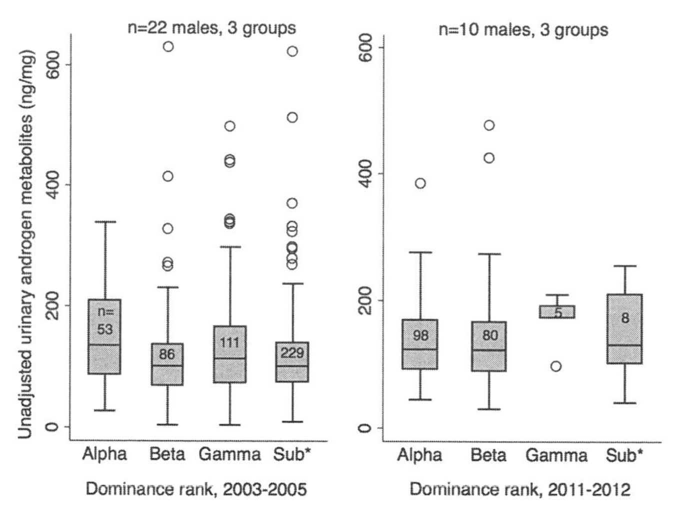

FIGURE 6.1. Unadjusted urinary androgen metabolite (UAM) values, by male dominance rank. *Sub = subordinate, which subsumes all males who held a rank of delta or below. Visually, there was no clear relationship between dominance rank and UAM in either 2003–2005 or 2011–2012, for males who lived in multi-male, multi-female groups. Box plots are unadjusted UAM values, and do not account for males' ages, time and date of sample collection, or expected intra-individual and/or intra-group UAM correlations. Numbers in box plots indicate the number of individual urine samples used in the analysis.

TABLE 6.3. Relationship between dominance rank and urinary androgen metabolite concentrations, for mountain gorilla males living in multi-male, multi-female groups.

	2003–2005 (n=479 samples, 22 males)			
	Coef	*Std Err*	*P*	*95% CI*
Rank (reference=alpha)				
Beta	3.31	27.63	0.905	−50.84, 57.46
Gamma	34.51	36.28	0.341	−36.59, 105.62
Subordinate	26.19	37.59	0.486	−47.48, 99.86
Age	3.88	2.26	0.086	−0.55, 8.31
Collection time	−2.70	7.06	0.000	−4.09, −1.32
Collection date	−0.07	0.03	0.040	−0.14, −0.003
Group (reference=PAB)				
BEE	7.85	14.97	0.600	−37.19, 21.50
SHI	−22.82	16.19	0.159	−54.54, 8.91
Constant	1315.27	549.02	0.017	239.20, 2391.34
	2011–2012 (n=191 samples, 10 males)			
Rank (reference=alpha)				
Beta	−3.28	10.10	0.746	−23.08, 16.52
Gamma	−16.55	20.46	0.418	−56.65, 23.54
Subordinate	−38.27	22.13	0.084	−81.65, 5.12
Age	−0.30	1.26	0.809	−2.77, 2.17
Collection time	9.85	6.99	0.159	−3.84, 2.36
Collection date	−0.10	0.03	0.001	−0.16, −0.04
Group (reference=ISA)				
KUY	11.38	6.97	0.102	−2.27, 25.04
PAB	57.72	13.12	0.000	−31.99, 83.45
Constant	2058.11	562.38	0.000	955.86, 3160.36

RELATIONSHIP BETWEEN URINARY ANDROGEN METABOLITES AND SOCIAL GROUP TYPE

In 2003–2005, the unadjusted UAM values of alpha males in multi-male groups were more than double the UAM values of the one solitary male for whom data were available (alpha males: mean = 151.21, SD = 83.27, n = 53 samples from three males; solitary male: mean = 70.03, SD = 43.89, n = 8 samples from one male). However, after controlling for age, collection date, and collection time, social configuration was not a significant predictor of UAM values (ß = 76.65, SE = 47.69, p = 0.108). The same was true when samples from the solitary male were compared to all group-living males (mean = 124.81, SD = 82.24, n = 479 samples), not only alphas (ß = −21.67, SE = 17.35, p = 0.211). There were no one-male, multi-female groups during this time period for comparison.

In 2011–2012, the unadjusted UAM values of alpha males in multi-male groups, and males in single-male groups, were very similar (alpha males:

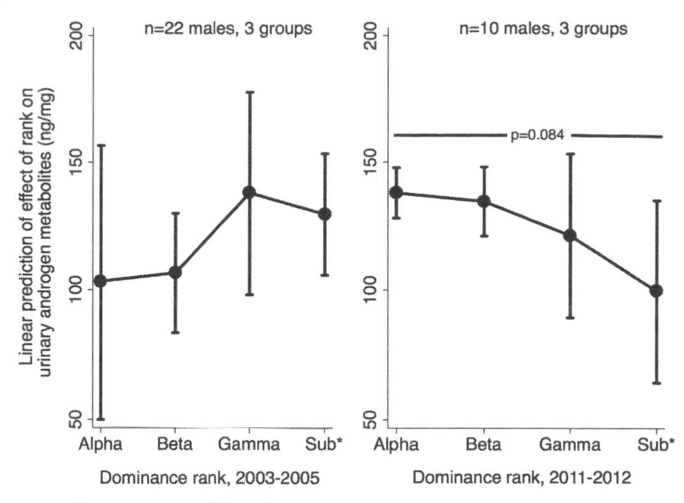

FIGURE 6.2. Predictive margins of the relationship between urinary androgen metabolites (UAM) and dominance rank. *Sub = subordinate, which subsumes all males who held a rank of delta or below. After controlling for males' ages, the time and date of sample collection, and the identification of social group and animal, there was little evidence of a relationship between dominance rank and UAM in multi-male, multi-female groups. In 2011–2012, there was a trend for alpha males to have significantly higher UAM concentrations than did subordinate males. The small number of samples available from subordinate males in this time period (n = 8; see fig. 6.1) may have prevented detection of a statistically significant relationship between dominance rank and UAM.

mean = 132.70, SD = 56.06, n = 98 samples from three males; males in single-male groups: mean = 131.18, SD = 97.08, n = 62 samples from three males). After controlling for age, collection date, and collection time, the difference was not significant (ß = −9.39, SE = 12.35, p = 0.447). The same was true when all males in multi-male groups (mean = 131.18, SD = 97.08, n = 191 samples) were compared to males in single-male groups (ß = −10.51, SE = 13.23, p = 0.427). Only three samples were available for solitary males in this time period. The mean was slightly lower (112.94, SD = 16.88), but the small sample size precludes statistical comparison.

Discussion

We found remarkably little evidence for androgen mediation of male mountain gorillas' intrasexual competition. There was no significant relationship between males' overall UAM levels and their within-group dominance hierarchies. This is in stark contrast to the known link between the two in chimpanzees, the other great ape in which males form strict dominance hierarchies. There was also no indication that gorilla males whose intrasexual competition came exclusively from extra-group males have different UAM levels than males who live with other males. Both results are surprising, given that gorillas' morphological characteristics indicate strong selection on males to engage in contest competition, where presumably androgens' anabolic effects would confer an advantage.

In the later of our two study periods (2011–2012), subordinate males in multi-male groups trended toward having lower UAM levels than alpha males. This finding is similar to previously reported data for this gorilla population (Robbins and Czekala 1997). However, data from the earlier time period (2003–2005), which included roughly double the number of both males and urine samples than either our 2011–2012 sample period or that of the Robbins and Czekala study, revealed no significant relationship between UAMs and rank. If anything, lower-ranking males had higher UAM values than higher-ranking males. The striking visual differences between the direction of the effect during our two points in time (fig. 6.2) suggest that there could be important nuances in the relationship between UAM and rank.

We speculate that the apparent (non-significant) differences in findings across our two study periods may be due to differences in social dynamics among males, and between social groups, at those two times. The associated androgen patterns are potentially consistent with the "challenge hypothesis," which predicts that males' androgen levels will up-regulate when needed for mating or competition, and will down-regulate at less critical times to minimize the long-term costs of androgen production (Hirschenhauser and Oliveira 2006; Wingfield et al. 1990). There is evidence for some version of the challenge hypothesis in a range of primate species (chimpanzees: Muller and Wrangham 2004; humans: Archer 2006; ring-tailed lemurs: Cavigelli and Pereira 2000; howler monkeys, *Alouatta* sp.: Cristóbal-Azkarate et al. 2006).

Many tests of the challenge hypothesis focus on males' androgen responses to the presence of cycling females, or other discrete competitive events. For example, in wild chimpanzees, males become more aggressive, and their testosterone levels rise, when parous females are in estrous (Muller and Wrangham 2004; Sobolewski, Brown, and Mitani 2013). These are presumably particularly high-stakes competitive events for male chimps (Muller and Wrangham 2004). An important limitation of our study is that opportunistic urine collection meant we could only address general (sometimes referred to as "baseline") UAM levels, rather than androgen responses to specific events, which may follow a very different pattern. Currently, there are no such data published for gorillas, though we will address males' androgen responses to competitive events in future publications (Rosenbaum et al. in prep). One methodological difficulty is that female gorillas' estrus periods are quite well-concealed compared to female chimpanzees, whose prominent sexual swellings advertise their reproductive status (Sillen-Tullberg and Moller 1993). While estrus events can be back-calculated from the timing of births, this does not capture cycles that do not end in births (Habumuremyi et al. 2016), and makes focused data collection around the time of the event difficult.

Rank (in)stability could also have generated the hormone pattern we observed. In 2003–2005, the dominant males in the three study groups were very well-established; Karisoke Research Center records show that each had already been dominant for between eight and 13 years, and each held their position for several years after the data collection period. Since their places at the top were firmly established, these high-ranking males may no longer have needed to maintain high androgen levels, while the lower-ranking males continued to jostle amongst themselves for more advantageous positions in their large groups' hierarchies. Dynamics between alpha males and lower-ranking males were different in 2011–2012. One of the three dominant males was elderly (34, which is near the upper range of wild male gorillas' lifespans: Bronikowski et al. 2011), and he was in the process of being actively challenged for the alpha position, albeit not generally aggressively, by his son. Another alpha male obtained his dominant position simultaneously with the start of our data collection, and so was not yet well-established in the role. The third alpha male had been the dominant animal in his group for four years, but he periodically had physically aggressive interactions with the beta male in his group (Karisoke Research Center long-term records).

Across species, there are fewer data available on the relationship between androgens and hierarchy stability than there are on the relationship between androgens and immediate mating opportunities. The data that do exist are somewhat equivocal. In chimpanzees, consensus suggests the relationship between high rank and high testosterone persists when ranks are stable, which is inconsistent with the most common interpretation of the challenge hypothesis (Muehlenbein, Watts, and Whitten 2004; Muller and Wrangham 2004, but see also Sobolewski, Brown, and Mitani 2013). Gesquiere and colleagues (2011) concluded the same was true for male baboons. Higher-ranking males had higher testosterone, regardless of the stability of the hierarchy they presided over. The same is true in mandrills, though their testosterone also increases during periods of rank instability (Setchell et al. 2008). Other studies have found that males' testosterone goes up during times when ranks are in flux, independent of a more generalized association between rank and testosterone (baboons, *Papio* sp.: Beehner et al. 2006; Sapolsky 1983). Notably, in bonobos, whose dominance hierarchies are less aggression-mediated than in chimpanzees, baboons, or mandrills (e.g., Surbeck and Hohmann 2017), there is no evidence of a relationship between rank stability and urinary testosterone metabolites (Sannen et al. 2004).

One possibility is that, in species where challenges may be hard to predict, males may need to maintain higher testosterone levels continuously (Muller and Wrangham 2004; Muehlenbein, Watts, and Whitten 2004). Chimpanzees have two characteristics that suggest this may explain why more dominant males have consistently higher testosterone, even when presiding over a stable dominance hierarchy. The first is their fission-fusion social system, which means that male chimps may never be entirely sure which within-group competitors they might encounter on a given day. The second is their lack of seasonal breeding, which means that breeding opportunities could be hard to predict (but see Emery Thompson 2013 for a thorough review of the relationship between seasonality and reproduction in female chimps). Mountain gorillas share the second characteristic (non-seasonal breeding: Watts 1998), but not the first (fission-fusion). Their groups are highly cohesive, so males do not face the within-group uncertainty that male chimpanzees do. Their continual (relative to chimpanzees) interactions may mean that higher-ranking males do not need to "re-litigate" their dominance status on an unpredictable schedule.

However, while male mountain gorillas may have more predictable within-group interactions, their interactions with extra-group males are arguably more unpredictable than is true for chimpanzees. Since chimpanzees are territorial (e.g., Mitani and Watts 2005; Williams et al. 2004) and frequently make use of predictably available seasonal resources (e.g., Hohmann et al. 2010), they may be able to make better predictions about their chances of encountering extra-group competitors in both space and time. Non-territorial mountain gorillas, whose food is widely distributed in both space and time, may encounter either lone silverbacks, or other mixed-sex groups, at any point (Fossey 1982; Mirville et al. 2018a; Sicotte 1993).

Just as there were differences in intragroup dynamics, there were also important differences in our two study periods in intergroup interactions. In 2011–2012, due to the larger number of groups and the rapidly growing gorilla population, males were involved in far more intergroup conflicts than they were in 2003–2005 (Caillaud et al. 2014). From 2000 to 2011, the frequency of intergroup encounters increased sixfold for the study population (Caillaud et al. 2014). This means the animals in our 2011–2012 sample were likely at much higher risk of intergroup-encounter associated injuries, and/or losing females to transfers or infants to infanticide, than the animals in our 2003–2005 sample.

Males of all ranks are involved in group defense (Mirville et al. 2018a; Sicotte 1993), so it is unclear whether continuous or unpredictable extra-group threat would result in a relationship between dominance hierarchy and androgens. Though we would not necessarily make an a priori prediction they would, since higher-ranking males have more to lose from inter-group competition than lower-ranking males, higher UAM levels in higher-ranking males in 2011–2012 could reflect a more competitive environment for these males specifically. Intergroup conflict causes strong stress hormone responses in adult male mountain gorillas (Eckardt et al. 2016), so it is reasonable to suppose they would also have a strong androgen response. In chimpanzees, males have higher testosterone when they go on territorial boundary patrols (Sobolewski, Brown, and Mitani 2012), and when they do more pant-hooting, which may function as a long-distance (i.e., either within or across group) form of male-male competition (Fedurek et al. 2016). Due to their similar functions for male reproductive investment, these are equivalent to non-territorial mountain gorillas' intergroup encounters (Sicotte 1993), and to their long-distance communication via chest-beats (Schaller 1963). We therefore predict that mountain gorillas will demonstrate a similar pattern, a topic that will need to be explored in future analyses.

Our study is the first to include any UAM data for male mountain gorillas in single male groups, or from solitary males. Contrary to our predictions, but consistent with the lack of a relationship between UAM and social group type in captive western lowland gorillas (Stoinski et al. 2002), these males had UAM concentrations similar to those of males living in multi-male groups. If anything, our preliminary data from solitary males suggest that their UAM values are lower than those of males in either single or multi-male groups. Due to our small sample sizes, this finding should be reevaluated with more data before firm conclusions are drawn. There should be few social costs associated with high androgen levels when living alone, and the selection for any adaptation that might make obtaining females more likely should be high, so this pattern is the opposite of what we predicted.

One interesting, though highly speculative, possibility worthy of future investigation is that lower-androgen males are more likely to be solitary than males with higher androgen levels. Perhaps females are less inclined to join these males, or perhaps they are less willing or able to compete (either physically or in more subtle ways) with males in their natal group. Evidence from rhesus macaques suggests that females prefer males whose facial coloration reflects their high testosterone levels (Waitt et al. 2003), while in humans, men with multiple concurrent sexual partners have higher testosterone than either single or monogamous men (van Anders, Hamilton, and Watson 2007).

Among primates whose social groups regularly contain more than one adult male, our results bear more resemblance to those published for male bonobos, whose intrasexual competition is less reliant on overt physical competition, than to results reported for, e.g., chimpanzees or baboons, where physical conflict is more common. Despite their morphology, gorillas' rates of contact aggression can be surprisingly low, even when many males live together (Robbins 1996; Stoinski et al. 2009). Their weaponry may raise the cost of aggression enough that they take particular pains to avoid it, primarily relying instead on lower-cost assessments of rivals (reviewed in Clutton-Brock 2016) and the relative peace that dominance hierarchies can afford (de Waal 1986). Although physical competitive ability is clearly important, our data suggest that there may be other underappreciated predictors of successful rank acquisition in this species. In humans, "dominance" is generally acknowledged to be the result of a complex suite of interacting traits (Cheng and Tracy 2014). Predictors such as personality, skills, experience, or intelligence have received far less attention in nonhuman animals than physical characteristics have, but likely play an important role for them as well (Anestis 2005; Briffa, Sneddon, and Wilson 2015; Eckardt et al. 2015; Fernald 2014; Sih et al. 2015).

Future Directions

Future research on gorillas specifically, and primates more generally, will need to carefully explore the relationship between androgens and both intra- and inter-social group dynamics, which may be an important predictor of basal-state androgens and/or of androgenic responses to specific competitive events (Klinkova, Heistermann, and Hodges 2004). The extensive work on the relationship between dominance rank and stress hormones in primates has demonstrated that context and ecology are very important determinants of the direction and magnitude of any association (Sapolsky 2005). It is certainly plausible that the same is true for other steroid hormones, especially for a species like mountain gorillas in which social dynamics can shift rapidly (Caillaud et al. 2014).

The relationship between male competition and androgens in great apes is clearly complex. These data are an interesting, if puzzling, addition to our understanding of the physiological mediators of intrasexual competition in these animals. To have a more complete picture of the extant apes, we would like to see additional data not only on solitary male gorillas (both western lowland and mountain gorillas), but also on the relationship between androgens and male competition within the various morphs of adult male orangutans. Since morphs with and without secondary sex characteristics use different strategies to compete for mating opportunities (Atmoko and van Hooff 2004), data on within-morph variation may help clarify under what circumstances we can expect to see a correlation between androgen levels and male intrasex competition. Such comparative data will help generate more nuanced hypotheses about the role that androgens play in mediating various levels and types of competition in great apes.

Acknowledgments

The authors wish to thank L.M. Hopper and S.R. Ross for inviting us to participate in this edited volume, K. Fowler for her assistance with laboratory analyses, and the many Karisoke Research Center employees and researchers, especially J. P. Hirwa and W. Eckardt, who assisted with urine collection over the years. The Karisoke Research Center is a project of the Dian Fossey Gorilla Fund International (DFGFI). The authors and DFGFI thank the Rwanda government and national park authorities for their long-term commitment to gorilla conservation and their support of the Karisoke Research Center. DFGFI is greatly indebted to the many Karisoke field assistants and researchers for their work on demographic and behavioral data over the last 48 years. We wish to acknowledge the many staff members who have shown extraordinary commitment under dangerous conditions while protecting and studying the gorillas. This work was funded by the Leakey Foundation, the Davee Center for Epidemiology and Endocrinology at the Lincoln Park Zoo, the National Science Foundation (Graduate Research Fellowship Program, BCS Directorate Doctoral Dissertation Improvement Grant #1122321), and the donors who support the DFGFI.

References

Alonso-Alvarez, C., S. Bertrand, B. Faivre, O. Chastel, and G. Sorci. 2007. "Testosterone and oxidative stress: The oxidation handicap hypothesis." *Proceedings of the Royal Society of London B* 274 (1611): 819–25.

Anestis, S. F. 2005. "Behavioral style, dominance rank, and urinary cortisol in young chimpanzees." *Behaviour* 142 (9–10): 1245–68.

Anestis, S. F. 2006. "Testosterone in juvenile and adolescent male chimpanzees (*Pan troglodytes*): Effects of dominance rank, aggression, and behavioral style." *American Journal of Physical Anthropology* 130 (4): 536–45.

Archer, J. 2006. "Testosterone and human aggression: An evaluation of the challenge hypothesis." *Neuroscience & Biobehavioral Reviews* 30 (3): 319–45.

Atmoko, S. U., and J. A. van Hooff. 2004. "Alternative male reproductive strategies: Male bimaturism in orangutans." In *Sexual Selection in Primates: New and Comparative Perspectives*, edited by P. M. Kappeler and C. P. van Schaik, 196. Cambridge: Cambridge University Press.

Banes, G. L., B. M. F. Galdikas, and L. Vigilant. 2015. "Male orang-utan bimaturism and reproductive success at Camp Leakey in Tanjung Puting National Park, Indonesia." *Behavioral Ecology and Sociobiology* 69 (11): 1785–94.

Beehner, J. C., T. J. Bergman, D. L. Cheney, R. M. Seyfarth, and P. L. Whitten. 2006. "Testosterone predicts future dominance rank and mating activity among male chacma baboons." *Behavioral Ecology and Sociobiology* 59 (4): 469–79.

Behringer, V., T. Deschner, C. Deimel, J. M. G. Stevens, and G. Hohmann. 2014. "Age-related changes in urinary testosterone levels suggest differences in puberty onset and divergent life history strategies in bonobos and chimpanzees." *Hormones and Behavior* 66 (3): 525–33.

Bray, J., A. E. Pusey, and I. C. Gilby. 2016. "Incomplete control and concessions explain mating skew in male chimpanzees." *Proceedings of the Royal Society of London B* 283 (1842): 20162071.

Briffa, M., L. U. Sneddon, and A. J. Wilson. 2015. "Animal personality as a cause and consequence of contest behaviour." *Biology Letters* 11 (3): 20141007.

Bronikowski, A. M., J. Altmann, D. K. Brockman, M. Cords, L. M. Fedigan, A. Pusey, T. Stoinski, W. F. Morris, K. B. Strier, and S. C. Alberts. 2011. "Aging in the natural world: Comparative data reveal similar mortality patterns across primates." *Science* 331 (6022): 1325–28.

Caillaud, D., F. Ndagijimana, A. J. Giarrusso, V. Vecellio, and T. S. Stoinski. 2014. "Mountain gorilla ranging patterns: Influence of group size and group dynamics." *American Journal of Primatology* 76 (8): 730–46.

Cain, K. E., and S. R. Pryke. 2017. "Testosterone production ability predicts breeding success and tracks breeding stage in male finches." *Journal of Evolutionary Biology* 30 (2): 430–36.

Cavigelli, S. A., and M. E. Pereira. 2000. "Mating season aggression and fecal testosterone levels in male ring-tailed lemurs (*Lemur catta*)." *Hormones and Behavior* 37 (3): 246–55.

Cheng, J. T., and J. L. Tracy. 2014. "Toward a unified science of hierarchy: Dominance and prestige are two fundamental pathways to human social rank." In *The Psychology of Social Status*, edited by J. T. Cheng, J. L. Tracy and C. Anderson, 3–27. New York: Springer.

Clutton-Brock, T. 2016. *Mammal Societies*. Oxford: John Wiley & Sons.

Cristóbal-Azkarate, J., R. Chavira, L. Boeck, E. Rodríguez-Luna, and J. J. Veàl. 2006. "Testosterone levels of free-ranging resident mantled howler monkey males in relation to the number and density of solitary males: A test of the challenge hypothesis." *Hormones and Behavior* 49 (2): 261–67.

Czekala, N., and M. M. Robbins. 2001. "Assessment of reproduction and stress through hormone analysis in gorillas." In *Mountain Gorillas: Three Decades of Research at Karisoke*, edited by M. M. Robbins, S. Pascale, and K. J. Stewart, 317. New York: Cambridge University Press.

de Almeida, R. M. M., J. C. C. Cabral, and R. Narvaes. 2015. "Behavioural, hormonal and neurobiological mechanisms of aggressive behaviour in human and nonhuman primates." *Physiology & Behavior* 143 (Supplement C): 121–35.

de Waal, F. B. M. 1986. "The integration of dominance and social bonding in primates." *Quarterly Review of Biology* 61 (4): 459–79.

Eckardt, W., H. D. Steklis, N. G. Steklis, A. W. Fletcher, T. S. Stoinski, and A. Weiss. 2015. "Personality dimensions and their behavioral correlates in wild Virunga mountain gorillas (*Gorilla beringei beringei*)." *Journal of Comparative Psychology* 129 (1): 26.

Eckardt, W., T. S. Stoinski, S. Rosenbaum, M. R. Umuhoza, and R. Santymire. 2016. "Validating faecal glucocorticoid metabolite analysis in the Virunga mountain gorilla using a natural biological stressor." *Conservation Physiology* 4 (1): cow029.

Ellis, L. 1995. "Dominance and reproductive success among nonhuman animals: A cross-species comparison." *Ethology and Sociobiology* 16 (4): 257–333.

Emery Thompson, M. 2013. "Reproductive ecology of female chimpanzees." *American Journal of Primatology* 75 (3): 222–37.

Emery Thompson, M., and A. V. Georgiev. 2014. "The high price of success: Costs of mating effort in male primates." *International Journal of Primatology* 35 (3): 609–27.

Emery Thompson, M., A. Zhou, and C. D. Knott. 2012. "Low testosterone correlates with delayed development in male orangutans." *PLoS One* 7 (10): e47282.

Fedigan, L. M. 1983. "Dominance and reproductive success in primates." *American Journal of Physical Anthropology* 26 (S1): 91–129.

Fedurek, P., K. E. Slocombe, D. K. Enigk, M. Emery Thompson, R. W. Wrangham, and M. N. Muller. 2016. "The relationship between testosterone and long-distance calling in wild male chimpanzees." *Behavioral Ecology and Sociobiology* 70 (5): 659–72.

Fernald, R. D. 2014. "Cognitive skills needed for social hierarchies." *Cold Spring Harbor Symposia on Quantitative Biology* 79: 229–36.

Foo, Y. Z., S. Nakagawa, G. Rhodes, and L. W. Simmons. 2017. "The effects of sex hormones on immune function: A meta-analysis." *Biological Reviews* 92 (1): 551–71.

Fossey, D. 1982. "Reproduction among free-living mountain gorillas." *American Journal of Primatology* 3 (S1): 97–104.

Galbany, J., D. Abavandimwe, M. Vakiener, W. Eckardt, A. Mudakikwa, F. Ndagijimana, T. S. Stoinski, and S. C. McFarlin. 2017. "Body growth and life history in wild mountain gorillas (*Gorilla beringei beringei*) from Volcanoes National Park, Rwanda." *American Journal of Physical Anthropology* 163 (3): 570–90.

Gersick, A. S., N. Snyder-Mackler, and D. J. White. 2012. "Ontogeny of social skills: social complexity improves mating and competitive strategies in male brown-headed cowbirds." *Animal Behaviour* 83 (5): 1171–77.

Gesquiere, L. R., N. H. Learn, M. C. M. Simao, P. O. Onyango, S. C. Alberts, and J. Altmann. 2011. "Life at the top: Rank and stress in wild male baboons." *Science* 333 (6040): 357–60.

Gettler, L. T., T. W. McDade, S. S. Agustin, A. B. Feranil, and C. W. Kuzawa. 2014. "Testosterone, immune function, and life history transitions in Filipino males (*Homo sapiens*)." *International Journal of Primatology* 35 (3): 787–804.

Gray, M., A. McNeilage, K. Fawcett, M. M. Robbins, B. Ssebide, D. Mbula, and P. Uwingeli. 2010. "Censusing the mountain gorillas in the Virunga Volcanoes: Complete sweep method versus monitoring." *African Journal of Ecology* 48 (3): 588–99.

Habumuremyi, S., C. Stephens, K. A. Fawcett, T. Deschner, and M. M. Robbins. 2016. "Endocrine assessment of ovarian cycle activity in wild female mountain gorillas (*Gorilla beringei beringei*)." *Physiology & Behavior* 157: 185–95.

Hagey, L. R., and N. M. Czekala. 2003. "Comparative urinary androstanes in the great apes." *General and Comparative Endocrinology* 130 (1): 64–69.

Harcourt, A. H., P. H. Harvey, S. G. Larson, and R. V. Short. 1981. "Testis weight, body weight and breeding system in primates." *Nature* 293 (5827): 55–57.

Harcourt, A. H., and K. J. Stewart. 2007. *Gorilla Society: Conflict, Compromise, and Cooperation Between the Sexes*. Chicago: University of Chicago Press.

Hirschenhauser, K., and R. F. Oliveira. 2006. "Social modulation of androgens in male vertebrates: Meta-analyses of the challenge hypothesis." *Animal Behaviour* 71 (2): 265–77.

Hockings, K. J., J. R. Anderson, and T. Matsuzawa. 2010. "Flexible feeding on cultivated underground storage organs by rainforest-dwelling chimpanzees at Bossou, West Africa." *Journal of Human Evolution* 58 (3): 227–33.

Hohmann, G., K. Potts, A. N'Guessan, A. Fowler, R. Mundry, J. U. Ganzhorn, and S. Ortmann. 2010. "Plant foods consumed by *Pan*: Exploring the variation of nutritional ecology across Africa." *American Journal of Physical Anthropology* 141 (3): 476–85.

Inoue, E., E. F. Akomo-Okoue, C. Ando, Y. Iwata, M. Judai, S. Fujita, S. Hongo, C. Nze-Nkogue, M. Inoue-Murayama, and J. Yamagiwa. 2013. "Male genetic structure and paternity in western lowland gorillas (*Gorilla gorilla gorilla*)." *American Journal of Physical Anthropology* 151 (4): 583–88.

Kappeler, P. M., and C. P. van Schaik. 2002. "Evolution of primate social systems." *International Journal of Primatology* 23 (4): 707–40.

Klinkova, E., M. Heistermann, and J. K. Hodges. 2004. "Social parameters and urinary testosterone level in male chimpanzees (*Pan troglodytes*)." *Hormones and Behavior* 46 (4): 474–81.

Koenig, A. 2002. "Competition for resources and its behavioral consequences among female primates." *International Journal of Primatology* 23 (4): 759–83

Leigh, S. R., and B. T. Shea. 1995. "Ontogeny and the evolution of adult body size dimorphism in apes." *American Journal of Primatology* 36 (1): 37–60.

Maggioncalda, A. N. 1995. "The socioendocrinology of orangutan growth, development and reproduction—An analysis of endocrine profiles of juvenile, developing adolescent, developmentally arrested adolescent, adult, and aged captive male orangutans." PhD diss., Duke University.

Malenky, R. K., and R. W. Wrangham. 1994. "A quantitative comparison of terrestrial herbaceous food consumption by *Pan paniscus* in the Lomako Forest, Zaire, and *Pan troglodytes* in the Kibale Forest, Uganda." *American Journal of Primatology* 32 (1): 1–12.

Marshall, A. J., and G. Hohmann. 2005. "Urinary testosterone levels of wild male bonobos (*Pan paniscus*) in the Lomako Forest, Democratic Republic of Congo." *American Journal of Primatology* 65 (1): 87–92.

Marty, P. R., M. A. van Noordwijk, M. Heistermann, E. P. Willems, L. P. Dunkel, M. Cadilek, M. Agil, and T. Weingrill. 2015. "Endocrinological correlates of male bimaturism in wild Bornean orangutans." *American Journal of Primatology* 77 (11): 1170–78.

Mirville, M. O., A. R. Ridley, J. P. M. Samedi, V. Vecellio, F. Ndagijimana, T. S. Stoinski, and C. C. Grueter. 2018a. "Factors influencing individual participation during intergroup interactions in mountain gorillas." *Animal Behaviour* 144: 75–86.

Mirville, M. O., A. R. Ridley, J. P. M. Samedi, V. Vecellio, F. Ndagijimana, T. S. Stoinski, and C. C. Grueter. 2018b. "Low familiarity and similar 'group strength' between opponents increase the intensity of intergroup interactions in mountain gorillas (*Gorilla beringei beringei*)." *Behavioral Ecology and Sociobiology* 72 (11): 178.

Mitani, J. C., and D. P. Watts. 2005. "Correlates of territorial boundary patrol behaviour in wild chimpanzees." *Animal Behaviour* 70 (5): 1079–86.

Mitchell, C. L., S. Boinski, and C. P. van Schaik. 1991. "Competitive regimes and female bonding in two species of squirrel monkeys (*Saimiri oerstedi* and *S. sciureus*)." *Behavioral Ecology and Sociobiology* 28 (1): 55–60.

Möstl, E., S. Rettenbacher, and R. Palme. 2005. "Measurement of corticosterone metabolites in birds' droppings: An analytical approach." *Annals of the New York Academy of Sciences* 1046 (1): 17–34.

Muehlenbein, M. P., and R. G. Bribiescas. 2005. "Testosterone-mediated immune functions and male life histories." *American Journal of Human Biology* 17: 527–58.

Muehlenbein, M. P., and D. P. Watts. 2010. "The costs of dominance: Testosterone, cortisol and intestinal parasites in wild male chimpanzees." *BioPsychoSocial Medicine* 4 (1): 21.

Muehlenbein, M. P., D. P. Watts, and P. L. Whitten. 2004. "Dominance rank and fecal testosterone levels in adult male Chimpanzees (*Pan troglodytes schweinfurthii*) at Ngogo, Kibale National Park, Uganda." *American Journal of Primatology* 64 (1):71–82.

Muller, M. N. 2002. "Agonistic relations among Kanyawara chimpanzees." In *Behavioural Diversity in Chimpanzees and Bonobos*, edited by C. Boesch, G. Hohmann, and L. Marchant, 112–24. Cambridge: Cambridge University Press.

Muller, M. N. 2017. "Testosterone and reproductive effort in male primates." *Hormones and Behavior* 91 (Supplement C): 36–51.

Muller, M. N., and R. W. Wrangham. 2004. "Dominance, aggression and testosterone in wild chimpanzees: A test of the 'challenge hypothesis.'" *Animal Behaviour* 67 (1): 113–23.

Nater, A., M. P. Mattle-Greminger, A. Nurcahyo, M. G. Nowak, M. de Manuel, T. Desai, C. Groves, M. Pybus, T. Bilgin Sonay, C. Roos, A. R. Lameira, S. A. Wich, J. Askew, M. Davila-Ross, G. Fredriksson, G. de Valles, F. Casals, J. Prado-Martinez, B. Goossens, E. J. Verschoor, K. S. Warren, I. Singleton, D. A. Marques, J. Pamungkas, D. Perwitasari-Farajallah, P. Rianti, A. Tuuga, I. G. Gut, M. Gut, P. Orozco-terWengel, C. P. van Schaik, J. Bertranpetit, M. Anisimova, A. Scally, T. Marques-Bonet, E. Meijaard, and M. Krützen. 2017. "Morphometric, behavioral, and genomic evidence for a new orangutan species." *Current Biology* 27 (22): 3487–98.e10.

Newton-Fisher, N. E. 1999. "The diet of chimpanzees in the Budongo Forest Reserve, Uganda." *African Journal of Ecology* 37 (3): 344–54.

Newton-Fisher, N. E., M. Emery Thompson, V. Reynolds, C. Boesch, and L. Vigilant. 2010. "Paternity and social rank in wild chimpanzees (*Pan troglodytes*) from the Budongo Forest, Uganda." *American Journal of Physical Anthropology* 142 (3): 417–28.

Nsubuga, A. M., M. M. Robbins, C. Boesch, and L. Vigilant. 2008. "Patterns of paternity and group fission in wild multimale mountain gorilla groups." *American Journal of Physical Anthropology* 135 (3): 263–74.

Oyegbile, T. O., and C. A. Marler. 2005. "Winning fights elevates testosterone levels in California mice and enhances future ability to win fights." *Hormones and Behavior* 48 (3): 259–67.

Parnell, R. J. 2002. "Group size and structure in western lowland gorillas (*Gorilla gorilla gorilla*) at Mbeli Bai, Republic of Congo." *American Journal of Primatology* 56 (4): 193–206.

Plavcan, J. M., and C. P. van Schaik. 1997. "Intrasexual competition and body weight dimorphism in anthropoid primates." *American Journal of Physical Anthropology* 103 (1): 37–68.

Pruetz, J. D. 2006. "Feeding ecology of savanna chimpanzees (*Pan troglodytes verus*) at Fongoli, Senegal." In *Feeding Ecology in Apes and Other Primates*, edited by G. Hohmann, M. M. Robbins, and C. Boesch, 161–82. Cambridge: Cambridge University Press.

Remis, M. J. 1997. "Ranging and grouping patterns of a western lowland gorilla group at Bai Hokou, Central African Republic." *American Journal of Primatology* 43 (2): 111–33.

Robbins, M. M. 1995. "A demographic analysis of male life history and social structure of mountain gorillas." *Behaviour* 132 (1): 21–47.

Robbins, M. M. 1996. "Male-male interactions in heterosexual and all-male wild mountain gorilla groups." *Ethology* 102 (7): 942–65.

Robbins, M. M., and N. Czekala. 1997. "A preliminary investigation of urinary testosterone and cortisol levels in wild male mountain gorillas." *American Journal of Primatology* 43 (1): 51–64.

Robbins, M., and S. Sawyer. 2007. "Intergroup encounters in mountain gorillas of Bwindi Impenetrable National Park, Uganda." *Behaviour* 144 (12): 1497–1519.

Rogers, E., K. Abernethy, M. Bermejo, C. Cipolletta, D. Doran, K. McFarland, T. Nishihara, M. Remis, and C. E. G. Tutin. 2004. "Western gorilla diet: A synthesis from six sites." *American Journal of Primatology* 64 (2): 173–92.

Rosenbaum, S., W. Eckardt, T. S. Stoinski, C. Kuzawa, and R. Santymire. In review. "Validation of an androgen enzyme immunoassay as a measure of fecal testosterone and 5alpha-dihydrotestosterone metabolites in wild male mountain gorillas (*Gorilla beringei beringei*)."

Rosenbaum, S., J. P. Hirwa, J. B. Silk, L. Vigilant, and T. S. Stoinski. 2015. "Male rank, not paternity, predicts male–immature relationships in mountain gorillas, *Gorilla beringei beringei*." *Animal Behaviour* 104: 13–24.

Rosenbaum, S., V. Vecellio, and T. Stoinski. 2016. "Observations of severe and lethal coalitionary attacks in wild mountain gorillas." *Scientific Reports* 6: 37018.

Sannen, A., L. Van Elsacker, M. Heistermann, and M. Eens. 2004. "Urinary testosterone-metabolite levels and dominance rank in male and female bonobos (*Pan paniscus*)." *Primates* 45 (2): 89–96.

Santymire, R. M., and D. M. Armstrong. 2010. "Development of a field-friendly technique for fecal steroid extraction and storage using the African wild dog (*Lycaon pictus*)." *Zoo Biology* 29 (3): 289–302.

Sapolsky, R. M. 1983. "Endocrine aspects of social instability in the olive baboon (*Papio anubis*)." *American Journal of Primatology* 5: 365–79.

Sapolsky, R. M. 2005. "The influence of social hierarchy on primate health." *Science* 308 (5722): 648–52.

Schaller, G. E. 1963. *The Mountain Gorilla: Ecology and Behavior*. Chicago: University of Chicago Press.

Setchell, J. M., T. Smith, E. J. Wickings, and L. A. Knapp. 2008. "Social correlates of testosterone and ornamentation in male mandrills." *Hormones and Behavior* 54 (3): 365–72.

Shideler, S. E., C. J. Munro, H. K. Johl, H. W. Taylor, and B. L. Lasley. 1995. "Urine and fecal sample collection on filter paper for ovarian hormone evaluations." *American Journal of Primatology* 37 (4): 305–15.

Sicotte, P. 1993. "Inter-group encounters and female transfer in mountain gorillas: Influence of group composition on male behavior." *American Journal of Primatology* 30 (1): 21–36.

Sih, A., A. T. Chang, and T. W. Wey. 2014. "Effects of behavioural type, social skill and the social environment on male mating success in water striders." *Animal Behaviour* 94 (Supplement C): 9–17.

Sih, A., K. J. Mathot, M. Moirón, P.-O. Montiglio, M. Wolf, and N. J. Dingemanse. 2015. "Animal personality and state–behaviour feedbacks: A review and guide for empiricists." *Trends in Ecology & Evolution* 30 (1): 50–60.

Sillen-Tullberg, B., and A. P. Moller. 1993. "The relationship between concealed ovulation and mating systems in anthropoid primates: A phylogenetic analysis." *American Naturalist* 141 (1): 1–25.

Sobolewski, M. E., J. L. Brown, and J. C. Mitani. 2012. "Territoriality, tolerance and testosterone in wild chimpanzees." *Animal Behaviour* 84 (6): 1469–74.

Sobolewski, M. E., J. L. Brown, and J. C. Mitani. 2013. "Female parity, male aggression, and the Challenge Hypothesis in wild chimpanzees." *Primates* 54 (1): 81–88.

Speakman, J. R., and M. Garratt. 2014. "Oxidative stress as a cost of reproduction: Beyond the simplistic trade-off model." *BioEssays* 36 (1): 93–106.

Stoinski, T. S., N. Czekala, K. E. Lukas, and T. L. Maple. 2002. "Urinary androgen and corticoid levels in captive, male Western lowland gorillas (*Gorilla g. gorilla*): Age- and social group-related differences." *American Journal of Primatology* 56 (2): 73–87.

Stoinski, T. S., V. Vecellio, T. Ngaboyamahina, F. Ndagijimana, S. Rosenbaum, and K. A. Fawcett. 2009. "Proximate factors influencing dispersal decisions in male mountain gorillas, *Gorilla beringei beringei*." *Animal Behaviour* 77 (5): 1155–64.

Surbeck, M., T. Deschner, G. Schubert, A. Weltring, and G. Hohmann. 2012. "Mate competition, testosterone and intersexual relationships in bonobos, *Pan paniscus*." *Animal Behaviour* 83 (3): 659–69.

Surbeck, M., and G. Hohmann. 2013. "Intersexual dominance relationships and the influence of leverage on the outcome of conflicts in wild bonobos (*Pan paniscus*)." *Behavioral Ecology and Sociobiology* 67 (11): 1767–80.

Surbeck, M., and G. Hohmann. 2017. "Affiliations, aggressions and an adoption: Male–male relationships in wild bonobos." In *Bonobos: Unique in Mind, Brain, and Behavior*, edited by B. Hare and S. Yamamoto, 35–46. Oxford: Oxford University Press.

Surbeck, M., K. E. Langergraber, B. Fruth, L. Vigilant, and G. Hohmann. 2017. "Male reproductive skew is higher in bonobos than chimpanzees." *Current Biology* 27 (13): R640–41.

Surbeck, M., R. Mundry, and G. Hohmann. 2011. "Mothers matter! Maternal support, dominance status and mating success in male bonobos (*Pan paniscus*)." *Proceedings of the Royal Society of London B* 278 (1705): 590–98.

Taussky, H. H., with the technical assistance of G. Kurzmann. 1954. "A microcolorimetric determination of creatine in urine by the Jaffe Reaction." *Journal of Biological Chemistry* 208 (2): 853–62.

Trumble, B. C., A. D. Blackwell, J. Stieglitz, M. Emery Thompson, I. Maldonado Suarez, H. Kaplan, and M. Gurven. 2016. "Associations between male testosterone and immune function in a pathogenically stressed forager-horticultural population." *American Journal of Physical Anthropology* 161 (3): 494–505.

van Anders, S. M., L. D. Hamilton, and N. V. Watson. 2007. "Multiple partners are associated with higher testosterone in North American men and women." *Hormones and Behavior* 51 (3): 454–59.

van Schaik, C. P., and J. A. R. A. M. van Hooff. 1996. "Toward an understanding of the orangutan's social system." In *Great Ape Societies*, edited by W. C. McGrew, L. F. Marchant, T. Nishida, 3–15. Cambridge: Cambridge University Press.

Vigilant, L., J. Roy, B. J. Bradley, C. J. Stoneking, M. M. Robbins, and T. S. Stoinski. 2015. "Reproductive competition and inbreeding avoidance in a primate species with habitual female dispersal." *Behavioral Ecology and Sociobiology* 69 (7): 1163–72.

Waitt, C., A. C. Little, S. Wolfensohn, P. Honess, A. P. Brown, H. M. Buchanan-Smith, and D. I. Perrett. 2003. "Evidence from rhesus macaques suggests that male coloration plays a role in female primate mate choice." *Proceedings of the Royal Society of London B* 270 (Supplement 2): S144–46.

Walker, K., and B. Hare. 2017. "Bonobo baby dominance: Did female defense of offspring lead to reduced male aggression?" In *Bonobos: Unique in Mind, Brain, and Behavior*, edited by B. Hare and S. Yamamoto, 49–64. Oxford: Oxford University Press.

Watts, D. P. 1990. "Mountain gorilla life histories, reproductive competition, and sociosexual behavior and some implications for captive husbandry." *Zoo Biology* 9 (3): 185–200.

Watts, D. P. 1994. "The influence of male mating tactics on habitat use in mountain gorillas (*Gorilla gorilla beringei*)." *Primates* 35 (1): 35–47.

Watts, D. P. 1998. "Seasonality in the ecology and life histories of mountain gorillas (*Gorilla gorilla beringei*)." *International Journal of Primatology* 19 (6): 929–48.

Watts, D. P., and J. C. Mitani. 2001. "Boundary patrols and intergroup encounters in wild chimpanzees." *Behaviour* 138 (3): 299–328.

Williams, J. M., G. W. Oehlert, J. V. Carlis, and A. E. Pusey. 2004. "Why do male chimpanzees defend a group range?" *Animal Behaviour* 68 (3): 523–32.

Wilson, M. L., and R. W. Wrangham. 2003. "Intergroup relations in chimpanzees." *Annual Review of Anthropology* 32 (1): 363–92.

Wingfield, J. C., R. E. Hegner Jr., A. M. Dufty, and G. F. Ball. 1990. "The 'Challenge Hypothesis': Theoretical implications for patterns of testosterone secretion, mating systems, and breeding strategies." *American Naturalist* 136 (6): 829–46.

Wobber, V., B. Hare, S. Lipson, R. Wrangham, and P. Ellison. 2013. "Different ontogenetic patterns of testosterone production reflect divergent male reproductive strategies in chimpanzees and bonobos." *Physiology & Behavior* 116 (Supplement C): 44–53.

Wroblewski, E. E., C. M. Murray, B. F. Keele, J. C. Schumacher-Stankey, B. H. Hahn, and A. E. Pusey. 2009. "Male dominance rank and reproductive success in chimpanzees, *Pan troglodytes schweinfurthii*." *Animal Behaviour* 77 (4): 873–85.

Yamagiwa, J. 1987. "Intra- and inter-group interactions of an all-male group of virunga mountain gorillas (*Gorilla gorilla beringei*)." *Primates* 28 (1): 1–30.

PART THREE

Studying Chimpanzees

7

Methods to Study Chimpanzee Social Learning from a Comparative Perspective

LYDIA M. HOPPER AND ALECIA J. CARTER

Introduction

Social learning describes how one individual can gain information or learn a new skill from observing the actions of another, or the outcomes of their actions (Zentall and Galef 2013). Social learning can potentially enable individuals to avoid time-consuming or even potentially lethal individual learning as they can learn from the success and failures of others. However, the efficacy and usefulness of socially provided information varies, and so *when* an individual should copy others, and *whom* they should copy, is a delicate balance. Interest in primate social learning has gained momentum in recent decades, most likely predicated by reports of cultural variation between wild groups of chimpanzees and other primates that have been suggested to rely on social learning. Here, we describe the approaches taken to study primate social learning, as well as the advantages and limitations of each, with a particular aim to review both observational and experimental approaches, and also studies with captive and wild populations. We aim to provide an overview of the sometimes seemingly disparate techniques used to study primate social learning in order to recommend practices typically restricted for use in one setting or species that could be translated to other settings to provide insights about chimpanzee social learning.

Approach

OBSERVATIONAL STUDIES OF SOCIAL LEARNING

Although others had previously identified potentially cultural chimpanzee behaviors (see Kroeber's 1928 discussion of Köhler 1925), it was McGrew and Tutin (1978), from their observations of wild chimpanzees in Tanzania, who proposed that certain behaviors could be considered cultural. Cultural behaviors, they argued, were not innate, but were acquired through the observation of others. Specifically, McGrew and Tutin reported the first observations of the grooming handclasp behavior—a specific form of reciprocal grooming—among Kasoge chimpanzees living in the Mahale mountains, Tanzania, that had not been reported for the neighboring chimpanzee troop living in Gombe National Park, Tanzania. The authors noted that the two groups were members of the same subspecies (*Pan troglodytes schweinfurthii*); they were likely historically a connected population; and both groups lived in comparable ecological environments. Thus, any differences observed in their behavior were likely due to learning, not genetics or ecology. Their approach represents the underpinnings of the "exclusion method" (Whiten et al. 1999; also known as the "ethnographic method": Laland and Janik 2006; or the "method of elimination": Koops et al. 2015; van Schaik 2003), which aims to exclude genetic and ecological explanations for variations in behavior across groups of animals when identifying potential cultures. For example, Whiten and colleagues (1999) compared the behavioral repertoires of chimpanzees living in seven different communities across Africa (in Guinea, Côte d'Ivoire, Tanzania, and Uganda) and by studying long-term behavioral records, identified 39 behaviors that had been reported for certain communities but not others. Whiten et al. (1999) argued that this variance represented socially learned behaviors. Subsequent observational studies, both of wild chimpanzees (e.g., Luncz and Boesch 2014; Luncz and van de Waal, chapter 18 this volume) and of those in captivity (e.g., Hook et al. 2002; van Leeuwen, Cronin, and Haun 2014), have provided more detailed insights into the emergence and transmission of novel behaviors, and the role of social learning specifically. Such observational data has also helped us to tease apart the mechanisms that underlie social learning in primates, providing more nuanced information about who learns from whom (e.g., Hobaiter et al. 2014; Kendal et al. 2015) and under what circumstances primates use social information (Barrett, McElreath, and Perry 2017; Vale, Flynn, et al. 2017) (for a review of social learning in non-primate mammals, see Thornton and Clutton-Brock 2011).

Although the exclusion method has been most extensively used to provide evidence of social learning (and by extension, culture) in chimpanzees (Crockford et al. 2004; Koops et al. 2015; McGrew 1992; McGrew and Tutin 1978; Schöning et al. 2008; Whiten et al. 1999), it has also provided evidence of social learning in other primates, including orangutans (Krützen, Willems, and van Schaik 2011; van Schaik 2003), gorillas (Robbins et al. 2016), white-faced capuchins (Perry et al. 2003), brown capuchins (Ottoni and Izar 2008), Japanese macaques (summarized in Huffman 1996), and spider monkeys

(Santorelli et al. 2011). In a similar manner, longitudinal observations of wild communities of primates have also revealed changes in groups' cultural behaviors over time. For example, Sapolsky and Share (2004) documented the spread and maintenance of a pacific culture in wild olive baboons after the selective death of the most aggressive and competitive individuals. In this regard, methods used to study wild chimpanzee social learning have been easily adapted for research in other primates in both a field and captive setting.

Despite the wide adoption of the exclusion method, the approach has several limitations and criticisms. First, it requires long-term studies of several groups and/or populations of a particular primate species, which can be understandably difficult to achieve. This may lead to a bias toward primate species that are abundant, relatively "easy" to work with, or of interest due to "closeness" to humans (through phylogeny or behavior, such as tool use). Second, although the exclusion method certainly reveals variation in behavior, it does not provide direct evidence that the behaviors were generated by social learning (Fragaszy and Perry 2003; Langergraber et al. 2010; Tennie, Hopper, van Schaik, chapter 19 this volume), nor does it reveal the interplay of the local environment on the emergence of certain behavioral variants (Schöning et al. 2008). Fragaszy and Perry (2003) further argue that the method is prone to false negatives, as one can a priori expect a correlation between ecology and behavior (see also Laland and Janik 2006; Perry and Manson 2003).

There are, however, two other ways long-term multi-group observational data sets can be used to identify social learning in wild primate populations: (1) "natural" translocations or diffusions (e.g., van de Waal, Borgeaud, and Whiten 2013) and (2) developmental studies (e.g., Lonsdorf and Bonnie 2010). We discuss each in turn next but wish to highlight that such methods could be applied to chimpanzees and other primates in a captive setting although, to date, such methods have typically been restricted to the study of abnormal behaviors (e.g., Hook et al. 2002; Hopper, Freeman, and Ross 2016) or experimental variants of the translocation method (Vale, Davis, et al. 2017; Whiten et al. 2007).

In studies of natural translocations among wild primate groups, researchers use opportunistic observations of natural diffusions of behavior to provide evidence of horizontal (i.e., peer to peer) social learning, either by following the behavior of migrant individuals to determine whether immigrants learn the behavioral variant of their new group or by documenting the spread of a novel behavior through the group (Luncz and van de Waal, chapter 18 this volume, discuss this technique in detail, see also Townsend, Watson, and Slocombe, chapter 11 this volume, for an example of vocal social learning among groups of captive chimpanzees that were relocated and merged).

The first approach is usually difficult to adopt in the wild, as it also requires documenting behavioral variation among multiple known groups of individuals. Despite this, such a case occurred when a female chimpanzee in Taï National Park, Côte d'Ivoire, progressively changed her tool-selection behavior to match the local group after immigration, suggesting that she socially learned to use a different tool type from that of her natal group (Luncz and Boesch 2014). In a more substantial sample of ten immigrant male vervet monkeys, van de Waal et al. (2013) showed that seven of these monkeys who had been trained to eat a particular color of corn in their natal group changed to eat a different color of corn after they immigrated to a new group. Although the monkeys' experience eating the dyed corn was the result of an experimental manipulation, the natural (non-experimental) immigration of these males to differently trained groups provided opportunistic evidence of horizontal social learning.

Natural diffusions can provide further evidence of social learning in the wild, particularly when combined with powerful statistical tools such as network-based diffusion analysis (NBDA) (Franz and Nunn 2009). NBDA and its variants test whether information is more likely to pass among individuals who frequently associate with one another according to a particular social network, thus providing evidence of social information transmission (Hoppitt and Laland 2013). Such an approach has been used to provide evidence of social learning of a novel foraging technique in humpback whales, for example (Allen et al. 2013). As with natural immigrations, observational studies of diffusions of behavior are difficult to complete as the spread of an innovated behavior must be observed and recorded. This approach been completed once in chimpanzees, where the innovation of "moss sponging" was observed (using moss, instead of leaves, to sponge water from a tree hole) and its spread across eight chimpanzees in a community in Budongo Forest, Uganda, was recorded over several days. A novel variant of NBDA was used to provide evidence that individuals who had observed another individual sponging with moss were more likely to adopt the behavior than those who had not observed another individual sponging with moss, thus providing evidence that the behavior was socially learned (Hobaiter et al. 2014).

Finally, developmental studies suggest that social learning can be inferred from tracking the development of behavior through time (Lonsdorf and Bonnie 2010). While the exclusion method and naturally arising diffusions can provide evidence of *whether* animals socially learn, the developmental approach can provide information about *what*, from *whom*, and *when* animals learn from others. However, by their nature, these studies are usually restricted to documenting vertical (from parent to offspring) or oblique (from adult to juvenile) social learning. For example, careful studies of juveniles' attention to particular group members have revealed from whom individuals are likely to learn. While chimpanzee, orangutan, and white-faced capuchin juveniles most frequently copy their mothers' food preferences and foraging techniques (chimpanzees: Humle, Snowdon, and Matsuzawa 2009; Lonsdorf 2005; orangutans: Jaeggi et al. 2010; white-faced capuchins: Perry 2009), semi-free-ranging brown capuchins have been reported to preferentially attend to, and learn from, more proficient individuals' use of tools (Ottoni, de Resende, and Izar 2005) and rarely use their mothers as models (Ottoni and Izar 2008). Developmental studies have also shed light on *who* learns, with female chimpanzees and white-faced capuchins tending to adopt the behavior of "experts" more closely than males, who can take longer to learn particular skills (Lonsdorf 2005; Perry 2009). Perhaps due to historical interest in primate parenting styles (summarized in Altmann 1980; Smith 2005), developmental studies of chimpanzees and other primates have progressed in tandem.

EXPERIMENTAL APPROACHES TO TESTING FOR SOCIAL LEARNING

As insightful as observational studies have been to our understanding of primate social learning, particularly in the wild, such studies suffer from reduced control over the numerous factors that may ultimately promote (group-specific) behaviors. Although, as discussed above, recent statistical advances have helped to overcome certain limitations in field studies (Hobaiter et al. 2014; Kendal et al. 2010), studying social learning experimentally has allowed researchers to more definitively identify the existence of social learning and to tease apart the mechanisms that underlie it in chimpanzees and other primate species. While observational research with captive chimpanzees allows the researcher to eliminate genetic and environmental influences on spontaneously emerging cultural behavior candidates (Bonnie and de Waal 2007; de Waal and Seres 1997; van Leeuwen, Cronin, and Haun 2014; Watson et al. 2015), experimental research allows for a more detailed understanding of the propagation and maintenance of socially learned traditions, both in captivity and in the field.

Most experimental research investigating chimpanzee social learning has been run in captive settings, although Matsuzawa and colleagues pioneered the use of experimental research with wild chimpanzees through the development of a "field laboratory" in Bossou, Guinea (Biro 2011 provides an overview). More recently, field experiments of social learning have been run with other populations of wild chimpanzees (Gruber et al. 2011) and with other primate species living in the wild (e.g., baboons: Carter et al. 2014; marmosets: Gunhold, Whiten, and Bugnyar 2014; vervet monkeys: van de Waal, Borgeaud, and Whiten 2013; see Reader and Biro 2010 for a review of field experiments of social learning).

In typical experiments testing social learning, the naive subjects are presented with a novel task to solve either asocially via trial-and-error learning (control condition) or after having observed a social model perform the solution (test condition) (e.g., Dean et al. 2012; Kendal et al. 2015; Tennie et al. 2010). To create the test condition in which subjects receive socially provided information, researchers can either train an individual to "demonstrate" the novel behavior to their group-mates (e.g., Whiten, Horner, and de Waal 2005) or the task can be presented "blind" to allow spontaneous innovation of the solution by an individual within the group, from which point researchers can study the potential transmission of the newly discovered skill via social learning (e.g., Kendal et al. 2015). The transmission of information can be studied within dyads, down experimental "generations," or within entire groups (Hopper 2017b; Whiten and Mesoudi 2008). Furthermore, in addition to asocial controls, in which no information about the novel task is provided, researchers have also used various "emulation" conditions to test whether a social model is necessary for the observer to learn the required actions or whether simply mechanical information about the task would suffice (as described by Tomasello 2009), and, in turn, whether social learning mediates the transmission of observed behaviors (e.g., Call, Carpenter, and Tomasello 2005; Hopper et al. 2007; Tennie, Call, and Tomasello 2010; see Hopper 2010 for a review).

One critique of such experimental paradigms is the limited context in which social learning has been studied. The test apparatuses used are typically puzzle boxes in which defenses have to be removed in a specific manner to obtain a food reward hidden inside, many building from the "artificial fruit" tasks originally developed by Whiten and colleagues (Caldwell and Whiten 2004; Custance, Whiten, and Fredman 1999; Whiten et al. 1996). More simply, researchers have presented primates with novel foods (e.g., Carter et al. 2014), novel colors of foods (e.g., van de Waal, Borgeaud, and Whiten 2013), or novel ways to obtain the food (e.g., by exchanging specific token types or in specific locations: Bonnie et al. 2007; Hopper et al. 2011) to determine whether subjects' success or solution type is mediated by the socially provided information (fig. 7.1). Although there is variation in the types of tasks presented to primates in such tests, most are food-centric, which some have argued may bias our understanding of when primates use social information and whom they may attend to preferentially (Watson and Caldwell 2009).

FIGURE 7.1. Chimpanzees exchanging one of two token types provided to them. In this case, they exchange the same color token as that exchanged by a dominant female within the group, even though doing so garners them only a piece of carrot, a less-preferred reward than they could obtain by exchanging the other token form (a grape). For details of this study, see Hopper et al. (2011). Photo: Gill Vale.

Accordingly, some researchers have employed experimental designs not reliant on testing social learning of foraging techniques (e.g., Tennie, Call, and Tomasello 2012; Watson, Buchanan-Smith, and Caldwell 2014). For example, and reflecting early studies in which conspecific vocalizations were played to groups of wild monkeys (e.g., Seyfarth, Cheney, and Marler 1980), Watson, Buchanan-Smith, and Caldwell (2014) played conspecific affiliative calls to captive marmosets generating short-lived "pacifistic" cultures in the test groups. However, more experimental testing of primates' ability to socially learn non-food-related behaviors are needed, especially given the range of purported socially mediated behaviors reported for chimpanzees and other primates that do not involve food acquisition, such as grooming techniques (e.g., McGrew and Tutin 1978; Tanaka 1998; van Leeuwen, Cronin, and Haun 2014) or even seemingly arbitrary self-directed behaviors (e.g., "grass-in-ear" behavior: van Leeuwen, Cronin, and Haun 2014).

A second criticism of an experimental approach is that, although experimental tests allow for tight control over the type and amount of social information provided, when researchers select which individual acts as a model it may bias for (or even against) the emergence of social learning (similar critiques have been raised against studies of other aspects of chimpanzee social cognition, such as cooperation; see, e.g., Suchak et al. 2014). This is of particular concern when considering dyadic or transmission-chain style studies, in which the researcher not only determines who the model is but also who has the opportunity to observe that model. An "open diffusion" study, in which a single trained animal is reintroduced to their social group (sensu Whiten, Horner, and de Waal 2005), is a more naturalistic approach, but it also can make it more difficult to accurately record who has watched whom, and how much social information each individual has received. Indeed, studies on primate social learning in the wild (Carter, Torrents Ticó, and Cowlishaw 2016; Carter et al. 2014) and captivity (Drea and Wallen 1999; Cronin et al. 2014) suggest that social and demographic constraints may prevent individuals from using social information they have acquired, such that the "amount" of social information they receive does not predict whether they express a (learned) behavior (Cronin et al. 2017). Thus, the absence of behavior by observers may reflect not a lack of learning, but rather a lack of production.

It has been proposed that humans' complex material culture, and our ability to learn detailed behaviors, rely on our ability to very closely copy the actions of others, even if those actions may appear to be redundant (Dean et al. 2014; Tennie, Hopper, van Schaik, chapter 19 this volume). To address this in chimpanzees, and test whether they, like humans, copy actions faithfully even when redundant, researchers have expanded on the basic social learning tests outlined above to tease apart how faithfully they copy the actions of others. Novel puzzle tasks have been designed such that they have more than one solution. For example, there might be a door that can be moved in two directions, either of which reveals a food reward (Hopper et al. 2008), or there might be two unique solutions for the same puzzle (Whiten, Horner, and de Waal 2005). Researchers have also developed methods to evaluate how closely chimpanzees replicate bodily actions of a human model (e.g., Custance, Whiten, and Bard 1995; Tennie, Call, and Tomasello 2012); whether they replicate demonstrated techniques even when they may be redundant (e.g., Horner and Whiten 2005); and whether they can develop their own solution to a task if the model's technique is not available to them (e.g., Tennie, Call, and Tomasello 2010). These studies have shown that chimpanzees are less likely to copy arbitrary or redundant actions than humans are, but rather will "emulate" and develop their own solution to a task (sensu Wood 1989). Comparable work with capuchin monkeys has shown that they can copy arbitrary actions associated with objects, but not simply body movements alone (Fragaszy et al. 2011); whether relating an action to an object would improve chimpanzees' likelihood to copy body movements is yet to be tested, although given their ability to learn how to solve puzzles and tool-use tasks via social learning it is probable that it would.

More recently, technological advances have augmented the experimental study of social learning. Video stimuli have been used to provide highly

FIGURE 7.2. A silverback gorilla participating in an eye-tracking study that investigated the interplay between memory and social learning in both gorillas and chimpanzees (Howard et al. 2017). Photo: Lydia Hopper.

controlled social demonstrations to naive chimpanzees (Hopper, Lambeth, and Schapiro 2012; Price et al. 2009; Ross et al. 2010), and other primates (e.g., Price and Caldwell 2007), providing the benefit that stimuli can be edited to show specific and repeatable information to observers (D'eath 1998). Video demonstrations of conspecifics solving novel tasks have also been presented to wild primate species (common marmosets: Gunhold, Whiten, and Bugnyar 2014), demonstrating the potential ability to transfer this traditionally captive method to a field setting. Eye-tracking devices have also been implemented to provide detailed data on what chimpanzees and other primate species attend to and for how long when observing such video demonstrations (e.g., Kano and Call 2014; Krupenye et al. 2017), with recent studies applying this to understand the mechanisms that might underpin social learning (Howard et al. 2017; Myowa-Yamakoshi, Yoshida, and Hirata 2015; fig. 7.2). Despite the implementation of video and eye-tracking techniques to study social learning, studies employing these technologies have still relied on footage of physical tasks as the stimuli. Although physical tasks, such as the puzzle boxes described above, have been the typical apparatus choice to test social learning, touchscreen computers have, more recently, been used to test

FIGURE 7.3. In a typical open diffusion experiment, the task to be learned is presented to naturally occurring groups (e.g., Carter et al. 2014). Here, a wild juvenile female baboon eats a novel food (dried apricot) after having watched an adult female eat one. Photo: Alecia Carter.

for social learning in chimpanzees (Martin, Biro, and Matsuzawa 2011) and other primate species (e.g., baboons: Claidière et al. 2014; macaques: Subiaul et al. 2004). To date, however, such paradigms are infrequently used despite the breadth of research using them to test other aspects of chimpanzee cognition (e.g., Inoue and Matsuzawa 2007; Leavens et al. 2001; Matsuzawa 1985, Martin and Idachi, chapter 8 this volume provide a review).

Experimental field studies, where the aim has been explicitly to induce a novel behavior and document its spread to other individuals in a wild population to test the role of social learning for the emergence of cultures, have been only recently adopted to study social learning in the wild. Experimental field studies of social learning are generally variations of captive diffusion experiments in which a novel stimulus, such as novel food (e.g., Biro 2011; Carter et al. 2014) or a novel task (e.g., Kendal et al. 2010; van de Waal et al. 2010), is provided to groups and the transmission of a novel behavior through the group is recorded (fig. 7.3). The original field observations of social learning in primates—those of the Japanese macaque Imo's innovation of sweet potato washing and its spread through her group (summarized in Nishida 1987)—is an example of such a diffusion experiment, though it was not the intention of the researchers to elicit such behavior. Despite their scarcity, experimental studies in the wild complement the findings of the observational studies, providing evidence of social learning in chimpanzees (Biro 2011) and other primates (e.g., baboons: Carter et al. 2014; vervet monkeys: van de Waal and Bshary 2011; lemurs: Kendal et al. 2010; Schnoell, Dittmann, and Fichtel 2014; Japanese macaques: Nishida 1987). In addition, experimental field studies have shown similar transmission biases in social learning to observational studies. For example, by presenting a novel food (grapes) covered in sand to groups of vervet monkeys, van de Waal et al. (2012) showed that infant vervet monkeys are more likely to adopt the grape-cleaning behavior of their mothers, a finding in line with observational studies of chimpanzees (see above). Similarly, vervet monkeys were more likely watch and copy the philopatric sex—females in vervet monkeys—who were manipulating an artificial fruit task (van de Waal et al. 2010), which is similar to observational findings in white-faced capuchins (Perry 2009) and chimpanzees (Lonsdorf 2005, though we note that juvenile chimpanzees have far fewer opportunities to watch demonstrators other than their mothers). More recently, a field experiment with vervet monkeys revealed that they did not show a bias to copy dominant individuals (Botting et al. 2018), which contrasts with data from experiments with captive chimpanzees who have been shown to copy dominant and "expert" individuals (Kendal et al. 2015).

One potential limitation of experimental field studies is that the researcher often cannot choose the identity or numbers of the individuals providing the social information if testing in a group setting (Cronin et al. 2017), as preferred resources are often monopolized by dominant individuals (e.g., van de Waal et al. 2010). However, these types of experiments can provide information not usually available in observational studies about the characteristics of individual learners and those who are more likely to acquire social information. For example, sex, age class, and personality can influence whether individuals learn to solve new problems individually or socially. In chimpanzees, individual learning about new species of nuts to crack was more common in juveniles, and the first individuals to watch and socially learn about the new nuts were also juveniles (Biro 2011). This pattern of juvenile individual and social learning is similar in other primates: juvenile chacma baboons were both more likely to initially eat and watch others eating a novel food (Carter et al. 2014), juvenile red-fronted lemurs were more likely to be the first to solve a two-action task (Schnoell, Dittmann, and Fichtel 2014), and juvenile Japanese macaques were more likely to adopt potato washing and wheat cleaning behaviors (Nishida 1987), for example. In addition to age-class differences in individual and social learning, sex and personality play a role, with female red-fronted lemurs more likely to learn to solve a two-action task than males (Schnoell, Dittmann, and Fichtel 2014) and bolder individuals more likely to socially learn about a novel food than shyer individuals (Carter et al. 2014). To date, experimental field studies of social learning have been more commonly adopted in other primates (see above), though field experiments are becoming more common in chimpanzees (Biro 2011; Gruber et al. 2011).

Discussion

Observational and experimental research, with both captive and wild populations of chimpanzees, has begun to provide a detailed picture of how, when, and whom chimpanzees copy, and yet many questions still remain. Furthermore, while our understanding of chimpanzee social learning is growing, chimpanzees are over-represented in studies of social cognition, and other aspects of primate cognition more widely (Cronin et al. 2017; Hopper 2017a). More work with other primate species is required, and, in particular, research using comparable methods across species and settings will provide us with a richer understanding of comparative primate cognition.

Research with chimpanzees has revealed that they can learn a number of skills from observing others, including how to manufacture and use tools (Price et al. 2009) and how to solve novel puzzle boxes containing food (Watson et al. 2017; Whiten et al. 1996), and that these socially learned skills are remembered and maintained over long periods of time (Vale et al. 2016; Whiten, Horner, and de Waal 2005). Indeed, social information can be so potent for chimpanzees that they persevere in using a socially learned behavior even when it is not necessary (Price et al. 2009) and they will copy others' potentially suboptimal techniques if the model is prestigious (Hopper et al. 2011). The strong experimental evidence that chimpanzees are able to closely observe and replicate the behaviors of their group mates, has relegated the question *can chimpanzees copy?* and promoted the questions *how faithfully do chimpanzees copy?* and *when do chimpanzees copy?*.

It appears that chimpanzees can copy specific aspects of a demonstrated behavior (e.g., whether to slide a door to either the left or to the right, Hopper et al. 2008, or which of two colored tokens to exchange for food, Hopper et al. 2011), but that they appear less able to copy the bodily actions of others (Tennie, Call, and Tomasello 2012; although see Fuhrmann et al. 2014). It has also been proposed that while certain cultural behaviors may be simple enough for chimpanzees to independently innovate without the need for social demonstration (Tennie, Hopper, van Schaik, chapter 19 this volume), more complex behaviors may necessitate socially provided information to be learned (Hopper et al. 2015). Indeed, the importance of social information for chimpanzees to learn behaviors has been proposed to be related to the fact that it enhances their memory for the demonstration they observe (Howard et al. 2017).

Controlled experimental protocols have also facilitated cross-species comparisons of animals' social learning mechanisms. The most common species to be compared to chimpanzees in tests of social learning is humans (or vice versa), and typically children (e.g., Hopper et al. 2008; Horner and Whiten 2005; Tennie, Call, and Tomasello 2006; Tomasello, Savage-Rumbaugh, and Kruger 1993; Whiten et al. 1996; see Hopper, Marshall-Pescini, and Whiten 2012 for a review). Unlike chimpanzees, humans appear better able to copy from "ghost" conditions testing for emulative learning (Caldwell et al. 2012; Hopper et al. 2010) and are more likely to replicate fine-grained, and even irrelevant, actions demonstrated by a social model (Horner and Whiten 2005; Lyons, Young, and Keil 2007; but see Keupp et al. 2015). It has been argued that it is this difference in the way in which we copy others, as compared to chimpanzees, that underlies our material culture, because even complex or opaque behaviors can be replicated, often mediated by active teaching (Burdett, Dean, and Ronfard 2017).

Although early studies assigned specific learning mechanisms to species (e.g., chimpanzees emulate not imitate: Horner and Whiten 2005; or that monkeys do not imitate: Fragaszy and Visalberghi 2004), more recent work suggests that species have an array of social learning mechanisms, and that the interesting question is to determine when, and how optimally, individuals copy others and in what ways (Acerbi, Tennie, and Nunn 2011; Hopper et al. 2015; Whiten et al. 2009). Although only a few apparatuses have been used both with chimpanzees and another nonhuman species to facilitate direct comparisons (e.g., Hopper et al. 2013), a number of test paradigms that have been used with chimpanzees have been used to test other species. For example, ghost conditions have been used with animals including insects (Loukola et al. 2017), reptiles (Kis, Huber, and Wilkinson 2015), birds (Auersperg et al. 2014), dogs (Miller, Rayburn-Reeves, and Zentall 2009), and other primates (Subiaul et al. 2004), as have tests using video demonstrations to show specific information (Burkart et al. 2012; Gunhold, Whiten, and Bugnyar 2014; Price and Caldwell 2007). These studies have revealed that, like chimpanzees, other species evidence a range of social learning mechanisms that are not restricted by species, and research into the social learning strategies of other species have revealed interesting patterns. For example, nine-spined stickleback fish show greater copying as the number of demonstrators increases (i.e., "conformity": Pike and Laland 2010; as has been shown for a chimpanzee: Luncz and Boesch 2014), and that they selectively copy those who are gaining more rewards (Kendal et al. 2009), in contrast to chimpanzees who appear to focus more on the relative value of their own rewards, but not in comparison to others' (Hopper et al. 2013; Vale, Flynn, et al. 2017).

More than *what* is copied, researchers have also studied *who* chimpanzees copy; so-called transmission biases, or social learning strategies (Laland 2004; Rendell et al. 2011). As detailed above, observations from the wild suggest that chimpanzees are more likely to copy those that are the same age or older (Biro et al. 2003) and that young females are more likely than young males to copy their mother's behavior (Lonsdorf 2005). Mirroring this, most experiments with captive chimpanzees that have seeded groups with trained models have specifically chosen dominant or mid-ranking older females for the models (Call, Carpenter, and Tomasello 2005; Hopper et al. 2007; Whiten, Horner, and de Waal 2005). Although many of these studies have demonstrated transmission of seeded behaviors within these captive groups from these female models, fewer studies have been run to explicitly test transmission biases, but this is changing. Recent studies have begun to tease apart which individuals within a group chimpanzees are more likely to copy (Horner et al. 2010), and which individuals are more likely to use socially provided information (Kendal et al. 2015). Although currently data are mixed (Vale, Davis, et al. 2017; Watson et al. 2017), it appears that chimpanzees are more likely to copy dominant, prestigious, and/or knowledgeable individuals, and that those who are low ranking are more likely to use social information. It is likely that other, yet-to-be-identified, social relationships also mediate these patterns, based on what has been described in other primates, and so using comparable methods across species will be the key for identifying these.

Future Directions

As for any field, the more we learn about how chimpanzees and other primates socially learn, the more we realize we do not yet know. The results from the studies we review above highlight a curious conflict in findings from captivity and the wild: juvenile primates are more likely to innovate new behaviors yet adults rarely learn from them. This raises the question about the adaptive value of chimpanzees' social learning strategies. To address this, we recommend that researchers investigate the biases that dictate and predict chimpanzees' use of social information. Furthermore, traditionally, there has been a divide between field research and studies of captive chimpanzees. Through this chapter, we have demonstrated some of the overlaps between these disciplines and the similarities in their findings. We have highlighted the methods that can be adapted for both settings, and specific areas for greater exploration. In addition, while field experiments are more common with a number of primate taxa (e.g., vervet monkeys, baboons, and capuchins) they are not typically run with great apes, including chimpanzees, and we have suggested methods that could be adopted by chimpanzee researchers. Similarly, while the long-term field sites studying wild chimpanzee communities have produced detailed longitudinal data sets, such consistency in the observation of captive and semi-free-ranging chimpanzee populations is far less common, but could provide novel insights into social learning strategies (e.g., Hannah and McGrew 1987; de Waal and Seres 1997).

References

Acerbi, A., C. Tennie, and C. L. Nunn. 2011. "Modeling imitation and emulation in constrained search spaces." *Learning & Behavior* 39 (2): 104–14.

Allen, J., M. Weinrich, W. Hoppitt, and L. Rendell. 2013. "Network-based diffusion analysis reveals cultural transmission of lobtail feeding in humpback whales." *Science* 340 (6131): 485.

Altmann, J. 1980. *Baboon Mothers and Infants*. Cambridge, MA: Harvard University Press.

Auersperg, A. M. I., A. M. I. von Bayern, S. Weber, A. Szabadvari, T. Bugnyar, and A. Kacelnik. 2014. "Social transmission of tool use and tool manufacture in Goffin cockatoos (*Cacatua goffini*)." *Proceedings of the Royal Society of London B* 281 (1793): 20140972.

Barrett, B. J., R. L. McElreath, and S. E. Perry. 2017. "Pay-off-biased social learning underlies the diffusion of novel extractive foraging traditions in a wild primate." *Proceedings of the Royal Society of London B* 284 (1856): 20170358.

Biro, D. 2011. "Clues to culture? The coula- and panda-nut experiments." In *The Chimpanzees of Bossou and Nimba*, edited by T. Matsuzawa, T. Humle, and Y. Sugiyama, 165–73. Tokyo: Springer.

Biro, D., N. Inoue-Nakamura, R. Tonooka, G. Yamakoshi, C. Sousa, and T. Matsuzawa. 2003. "Cultural innovation and transmission of tool use in wild chimpanzees: Evidence from field experiments." *Animal Cognition* 6 (4): 213–23.

Bonnie, K. E., and F. B. M. de Waal. 2007. "Copying without rewards: Socially influenced foraging decisions among brown capuchin monkeys." *Animal Cognition* 10 (3): 283–92.

Bonnie, K. E., V. Horner, A. Whiten, and F. B. M. de Waal. 2007. "Spread of arbitrary conventions among chimpanzees: A controlled experiment." *Proceedings of the Royal Society of London B* 274 (1608): 367–72.

Botting, J., A. Whiten, M. Grampp, and E. van de Waal. 2018. "Field experiments with wild primates reveal no consistent dominance-based bias in social learning." *Animal Behaviour* 136: 1–12.

Burdett, E. R. R., L. G. Dean, and S. Ronfard. 2017. "A diverse and flexible teaching toolkit facilitates the human capacity for cumulative culture." *Review of Philosophy and Psychology* 9 (4): 807–18.

Burkart, J., A. Kupferberg, S. Glasauer, and C. van Schaik. 2012. "Even simple forms of social learning rely on intention attribution in marmoset monkeys (*Callithrix jacchus*)." *Journal of Comparative Psychology* 126 (2): 129.

Caldwell, C. A., K. Schillinger, C. L. Evans, and L. M. Hopper. 2012. "End state copying by humans (*Homo sapiens*): Implications for a comparative perspective on cumulative culture." *Journal of Comparative Psychology* 126 (2): 161.

Caldwell, C. A., and A. Whiten. 2004. "Testing for social learning and imitation in common marmosets, Callithrix jacchus, using an artificial fruit." *Animal Cognition* 7 (2): 77–85.

Call, J., M. Carpenter and M. Tomasello. 2005. "Copying results and copying actions in the process of social learning: Chimpanzees (*Pan troglodytes*) and human children (*Homo sapiens*)." *Animal Cognition* 8: 151–63.

Carter, A. J., H. H. Marshall, R. Heinsohn, and G. Cowlishaw. 2014. "Personality predicts the propensity for social learning in a wild primate." *PeerJ* 2: e283.

Carter, A. J., M. Torrents Ticó, and G. Cowlishaw. 2016. "Sequential phenotypic constraints on social information use in wild baboons." *eLife* 5: e13125.

Claidière, N., K. Smith, S. Kirby, and J. Fagot. 2014. "Cultural evolution of systematically structured behaviour in a non-human primate." *Proceedings of the Royal Society of London B* 281 (1797): 20141541.

Crockford, C., I. Herbinger, L. Vigilant, and C. Boesch. 2004. "Wild chimpanzees produce group-specific calls: A case for vocal learning?" *Ethology* 110 (3): 221–43.

Cronin, K. A., S. L. Jacobson, K. E. Bonnie, and L. M. Hopper. 2017. "Studying primate cognition in a social setting to improve validity and welfare: A literature review highlighting successful approaches." *PeerJ* 5: e3649.

Cronin, K. A., B. A. Pieper, E. J. C. Van Leeuwen, R. Mundry, and D. B. M. Haun. 2014. "Problem solving in the presence of others: How rank and relationship quality impact resource acquisition in chimpanzees (*Pan troglodytes*)." *PLoS One* 9 (4): e93204.

Custance, D. M., A. Whiten, and K. A. Bard. 1995. "Can young chimpanzees (*Pan troglodytes*) imitate arbitrary actions? Hayes & Hayes (1952) revisited." *Behaviour* 132 (11): 837–59.

Custance, D. M., A. Whiten, and T. Fredman. 1999. "Social learning of an artificial fruit task in capuchin monkeys (*Cebus apella*)." *Journal of Comparative Psychology* 113 (1): 13.

Dean, L. G., R. L. Kendal, S. J. Schapiro, B. Thierry, and K. N. Laland. 2012. "Identification of the social and cognitive processes underlying human cumulative culture." *Science* 335 (6072): 1114–18.

Dean, L. G., G. L. Vale, K. N. Laland, E. Flynn, and R. L. Kendal. 2014. "Human cumulative culture: A comparative perspective." *Biological Reviews* 89 (2): 284–301.

D'eath, R. B. 1998. "Can video images imitate real stimuli in animal behaviour experiments?" *Biological Reviews* 73 (3): 267–92.

de Waal, F. B. M., and M. Seres. 1997. "Propagation of handclasp grooming among captive chimpanzees." *American Journal of Primatology* 43 (4): 339–46.

Drea, C. M., and K. Wallen. 1999. "Low-status monkeys 'play dumb' when learning in mixed social groups." *Proceedings of the National Academy of Sciences* 96 (22): 12965–69.

Fragaszy, D., B. Deputte, E. Johnson Cooper, E. N. Colbert-White, and C. Hémery. 2011. "When and how well can human-socialized capuchins match actions demonstrated by a familiar human?" *American Journal of Primatology* 73 (7): 643–54.

Fragaszy, D., and S. E. Perry. 2003. "Towards a biology of traditions." In *The Biology of Traditions: Models and Evidence*, edited by D. Fragaszy and S. E. Perry. Cambridge: Cambridge University Press.

Fragaszy, D., and E. Visalberghi. 2004. "Socially biased learning in monkeys." *Learning & Behavior* 32 (1): 24–35.

Franz, M., and C. L. Nunn. 2009. "Network-based diffusion analysis: A new method for detecting social learning." *Proceedings of the Royal Society of London B* 276: 1829–36.

Fuhrmann, D., A. Ravignani, S. Marshall-Pescini, and A. Whiten. 2014. "Synchrony and motor mimicking in chimpanzee observational learning." *Scientific Reports* 4: 5283.

Gruber, T., M. N. Muller, V. Reynolds, R. Wrangham, and K. Zuberbühler. 2011. "Community-specific evaluation of tool affordances in wild chimpanzees." *Scientific Reports* 1: 128.

Gunhold, T., A. Whiten, and T. Bugnyar. 2014. "Video demonstrations seed alternative problem-solving techniques in wild common marmosets." *Biology Letters* 10 (9): 20140439.

Hannah, A. C., and W. C. McGrew. 1987. "Chimpanzees using stones to crack open oil palm nuts in Liberia." *Primates* 28 (1): 31–46.

Hobaiter, C., T. Poisot, K. Zuberbühler, W. Hoppitt, and T. Gruber. 2014. "Social network analysis shows direct evidence for social transmission of tool use in wild chimpanzees." *PLoS Biology* 12 (9): e1001960.

Hook, M. A., S. P. Lambeth, J. E. Perlman, R. Stavisky, M. A. Bloomsmith, and S. J. Schapiro. 2002. "Inter-group variation in abnormal behavior in chimpanzees (*Pan troglodytes*) and rhesus macaques (*Macaca mulatta*)." *Applied Animal Behaviour Science* 76 (2): 165–76.

Hopper, L. M. 2010. "'Ghost' experiments and the dissection of social learning in humans and animals." *Biological Reviews* 85 (4): 685–701.

Hopper, L. M. 2017a. "Cognitive research in zoos." *Current Opinion in Behavioral Sciences* 16: 100–110.

Hopper, L. M. 2017b. "Social learning and decision making." In *Handbook of Primate Behavior Management*, edited by S. J. Schapiro, 225–42. Boca Raton: CRC Press.

Hopper, L. M., E. G. Flynn, L. A. N. Wood, and A. Whiten. 2010. "Observational learning of tool use in children: Investigating cultural spread through diffusion chains and learning mechanisms through ghost displays." *Journal of Experimental Child Psychology* 106 (1): 82–97.

Hopper, L. M., H. D. Freeman, and S. R. Ross. 2016. "Reconsidering coprophagy as an indicator of negative welfare for captive chimpanzees." *Applied Animal Behaviour Science* 176: 112–19.

Hopper, L. M., S. P. Lambeth, and S. J. Schapiro. 2012. "An evaluation of the efficacy of video displays for use with chimpanzees (*Pan troglodytes*)." *American Journal of Primatology* 74 (5): 442–49.

Hopper, L. M., S. P. Lambeth, S. J. Schapiro, and S. F. Brosnan. 2013. "When given the opportunity, chimpanzees maximize personal gain rather than 'level the playing field.'" *PeerJ* 1: e165.

Hopper, L. M., S. P. Lambeth, S. J. Schapiro, and A. Whiten. 2008. "Observational learning in chimpanzees and children studied through 'ghost' conditions." *Proceedings of the Royal Society of London B* 275 (1636): 835–40.

Hopper, L. M., S. P. Lambeth, S. J. Schapiro, and A. Whiten. 2015. "The importance of witnessed agency in chimpanzee social learning of tool use." *Behavioural Processes* 112: 120–29.

Hopper, L. M., S. Marshall-Pescini, and A. Whiten. 2012. "Social learning and culture in child and chimpanzee." In *The Primate Mind: Built to Connect with Other Minds*, edited by F. B. M. de Waal and P. F. Ferrari, 99–118. Cambridge, MA: Harvard University Press.

Hopper, L. M., S. J. Schapiro, S. P. Lambeth, and S. F. Brosnan. 2011. "Chimpanzees' socially maintained food preferences indicate both conservatism and conformity." *Animal Behaviour* 81 (6): 1195–1202.

Hopper, L. M., A. Spiteri, S. P. Lambeth, S. J. Schapiro, V. Horner, and A. Whiten. 2007. "Experimental studies of traditions and underlying transmission processes in chimpanzees." *Animal Behaviour* 73 (6): 1021–32.

Hoppitt, W., and K. N. Laland. 2013. *Social Learning: An Introduction to Mechanisms, Methods, and Models*. Princeton, NJ: Princeton University Press.

Horner, V., D. Proctor, K. E. Bonnie, A. Whiten, and F. B. M. de Waal. 2010. "Prestige affects cultural learning in chimpanzees." *PLoS One* 5 (5): e10625.

Horner, V., and A. Whiten. 2005. "Causal knowledge and imitation/emulation switching in chimpanzees (*Pan troglodytes*) and children (*Homo sapiens*)." *Animal Cognition* 8 (3): 164–81.

Howard, L. H., K. E. Wagner, A. L. Woodward, S. R. Ross, and L. M. Hopper. 2017. "Social models enhance apes' memory for novel events." *Scientific Reports* 7: 40926.

Huffman, M. A. 1996. "Acquisition of innovative cultural behaviors in nonhuman primates: A case study of stone handling, a socially transmitted behavior in Japanese macaques." In *Social Learning in Animals: The Roots of Culture*, edited by C. M. Heyes and B. G. Galef. San Diego: Academic Press.

Humle, T., C. T. Snowdon, and T. Matsuzawa. 2009. "Social influences on ant-dipping acquisition in the wild chimpanzees (*Pan troglodytes verus*) of Bossou, Guinea, West Africa." *Animal Cognition* 12 (1): 37–48.

Inoue, S., and T. Matsuzawa. 2007. "Working memory of numerals in chimpanzees." *Current Biology* 17 (23): R1004–5.

Jaeggi, A. V., L. P. Dunkel, M. A. Van Noordwijk, S. A. Wich, A. A. L. Sura, and C. P. Van Schaik. 2010. "Social learning of diet and foraging skills by wild immature Bornean orangutans: Implications for culture." *American Journal of Primatology* 72 (1): 62–71.

Kano, F., and J. Call. 2014. "Great apes generate goal-based action predictions: An eye-tracking study." *Psychological Science* 25 (9): 1691–98.

Kendal, R. L., D. M. Custance, J. R. Kendal, G. L. Vale, T. S. Stoinski, N. L. Rakotomalala, and H. Rasamimanana. 2010. "Evidence for social learning in wild lemurs (*Lemur catta*)." *Learning & Behavior* 38 (3): 220–34.

Kendal, R. L., L. M. Hopper, A. Whiten, S. F. Brosnan, S. P. Lambeth, S. J. Schapiro, and W. Hoppitt. 2015. "Chimpanzees copy dominant and knowledgeable individuals: Implications for cultural diversity." *Evolution and Human Behavior* 36 (1): 65–72.

Kendal, J. R., L. Rendell, T. W. Pike, and K. N. Laland. 2009. "Nine-spined sticklebacks deploy a hill-climbing social learning strategy." *Behavioral Ecology* 20 (2): 238–44.

Keupp, S., T. Behne, J. Zachow, A. Kasbohm, and H. Rakoczy. 2015. "Over-imitation is not automatic: Context sensitivity in children's overimitation and action interpretation of causally irrelevant actions." *Journal of Experimental Child Psychology* 130: 163–75.

Kis, A., L. Huber, and A. Wilkinson. 2015. "Social learning by imitation in a reptile (*Pogona vitticeps*)." *Animal Cognition* 18 (1): 325–31.

Köhler, W. 1925. *The Mentality of Apes*. London: Kegan Paul, Trench, Trubner.

Koops, K., C. Schöning, M. Isaji, and C. Hashimoto. 2015. "Cultural differences in ant-dipping tool length between neighbouring chimpanzee communities at Kalinzu, Uganda." *Scientific Reports* 5: 12456.

Kroeber, A. L. 1928. "Sub-human culture beginnings." *Quarterly Review of Biology* 3 (3): 325–42.

Krupenye, C., F. Kano, S. Hirata, J. Call, and M. Tomasello. 2017. "A test of the submentalizing hypothesis: Apes' performance in a false belief task inanimate control." *Communicative & Integrative Biology* 10 (4): e1343771.

Krützen, M., E. P. Willems, and C. P. van Schaik. 2011. "Culture and geographic variation in orangutan behavior." *Current Biology* 21 (21): 1808–12.

Laland, K. N. 2004. "Social learning strategies." *Animal Learning & Behavior* 32 (1): 4–14.

Laland, K. N., and V. M. Janik. 2006. "The animal cultures debate." *Trends in Ecology & Evolution* 21 (10): 542–47.

Langergraber, K. E., C. Boesch, E. Inoue, M. Inoue-Murayama, J. C. Mitani, T. Nishida, A. Pusey, V. Reynolds, G. Schubert, R. W. Wrangham, E. Wroblewski, and L. Vigilant. 2010. "Genetic and 'cultural' similarity in wild chimpanzees." *Proceedings of the Royal Society of London B* 278 (1704): 408–16.

Leavens, D. A., F. Aureli, W. D. Hopkins, and C. W. Hyatt. 2001. "Effects of cognitive challenge on self-directed behaviors by chimpanzees (*Pan troglodytes*)." *American Journal of Primatology* 55 (1): 1–14.

Lonsdorf, E. V. 2005. "Sex differences in the development of termite-fishing skills in the wild chimpanzees, *Pan troglodytes schweinfurthii*, of Gombe National Park, Tanzania." *Animal Behaviour* 70 (3): 673–83.

Lonsdorf, E. V., and K. E. Bonnie. 2010. "Opportunities and constraints when studying social learning: Developmental approaches and social factors." *Learning & Behavior* 38 (3): 195–205.

Loukola, O. J., C. J. Perry, L. Coscos, and L. Chittka. 2017. "Bumblebees show cognitive flexibility by improving on an observed complex behavior." *Science* 355 (6327): 833–36.

Luncz, L. V., and C. Boesch. 2014. "Tradition over trend: Neighboring chimpanzee communities maintain differences in cultural behavior despite frequent immigration of adult females." *American Journal of Primatology* 76 (7): 649–57.

Lyons, D. E., A. G. Young, and F. C. Keil. 2007. "The hidden structure of overimitation." *Proceedings of the National Academy of Sciences* 104 (50): 19751–56.

Martin, C. F., D. Biro, and T. Matsuzawa. 2011. "Chimpanzees' use of conspecific cues in matching-to-sample tasks: Public information use in a fully automated testing environment." *Animal Cognition* 14 (6): 893–902.

Matsuzawa, T. 1985. "Use of numbers by a chimpanzee." *Nature* 315 (6014): 57–59.

McGrew, W. C. 1992. *Chimpanzee Material Culture: Implications for Human Evolution.* Cambridge: Cambridge University Press.

McGrew, W. C., and C. E. G. Tutin. 1978. "Evidence for a social custom in wild chimpanzees?" *Man* 13 (2): 234–51.

Miller, H. C., R. Rayburn-Reeves, and T. R. Zentall. 2009. "Imitation and emulation by dogs using a bidirectional control procedure." *Behavioural Processes* 80 (2): 109–14.

Myowa-Yamakoshi, M., C. Yoshida, and S. Hirata. 2015. "Humans but not chimpanzees vary face-scanning patterns depending on contexts during action observation." *PLoS One* 10 (11): e0139989.

Nishida, T. 1987. "Local traditions and cultural transmission." In *Primate Societies*, edited by B. B. Smuts, D. L. Cheney, R. M. Seyfarth, R. W. Wrangham, and T. T. Struhsaker, 462–74. Chicago: University of Chicago Press.

Ottoni, E. B., B. D. de Resende, and P. Izar. 2005. "Watching the best nutcrackers: What capuchin monkeys (*Cebus apella*) know about others' tool-using skills." *Animal Cognition* 8 (4): 215–19.

Ottoni, E. B., and P. Izar. 2008. "Capuchin monkey tool use: Overview and implications." *Evolutionary Anthropology* 17 (4): 171–78.

Perry, S. E. 2009. "Conformism in the food processing techniques of white-faced capuchin monkeys (*Cebus capucinus*)." *Animal Cognition* 12 (5): 705–16.

Perry, S. E., M. Baker, L. Fedigan, J. GrosLouis, K. Jack, K. C. MacKinnon, J. H. Manson, M. Panger, K. Pyle, and L. Rose. 2003. "Social conventions in wild white-faced capuchin monkeys." *Current Anthropology* 44 (2): 241–68.

Perry, S. E., and J. H. Manson. 2003. "Traditions in monkeys." *Evolutionary Anthropology* 12 (2): 71–81.

Pike, T. W., and K. N. Laland. 2010. "Conformist learning in nine-spined sticklebacks' foraging decisions." *Biology Letters* 6 (4): 466–68.

Price, E. E., and C. A. Caldwell. 2007. "Artificially generated cultural variation between two groups of captive monkeys, *Colobus guereza kikuyuensis*." *Behavioural Processes* 74 (1): 13–20.

Price, E. E., S. P. Lambeth, S. J. Schapiro, and A. Whiten. 2009. "A potent effect of observational learning on chimpanzee tool construction." *Proceedings of the Royal Society of London B* 276 (1671): 3377–83.

Reader, S. M., and D. Biro. 2010. "Experimental identification of social learning in wild animals." *Learning & Behavior* 38 (3): 265–83.

Rendell, L., L. Fogarty, W. J. E. Hoppitt, T. J. H. Morgan, M. M. Webster, and K. N. Laland. 2011. "Cognitive culture: Theoretical and empirical insights into social learning strategies." *Trends in Cognitive Sciences* 15 (2): 68–76.

Robbins, M. M., C. Ando, K. A. Fawcett, C. C. Grueter, D. Hedwig, Y. Iwata, J. L. Lodwick, S. Masi, R. Salmi, T. S. Stoinski, A. Todd, V. Vercellio, and J. Yamagiwa. 2016. "Behavioral variation in gorillas: Evidence of potential cultural traits." *PLoS One* 11 (9): e0160483.

Ross, S. R., M. S. Milstein, S. E. Calcultt, and E. V. Lonsdorf. 2010. "Preliminary assessment of methods used to demonstrate nut-cracking behavior to five captive chimpanzees (*Pan troglodytes*)." *Folia Primatologica* 81: 224–32.

Santorelli, C. J., C. M. Schaffner, C. J. Campbell, H. Notman, M. S. Pavelka, J. A. Weghorst, and F. Aureli. 2011. "Traditions in spider monkeys are biased towards the social domain." *PLoS One* 6 (2): e16863.

Sapolsky, R. M., and L. J. Share. 2004. "A pacific culture among wild baboons: Its emergence and transmission." *PLoS Biology* 2 (4): e106.

Schnoell, A. V., M. T. Dittmann, and C. Fichtel. 2014. "Human-introduced long-term traditions in wild redfronted lemurs?" *Animal Cognition* 17 (1): 45–54.

Schöning, C., T. Humle, Y. Möbius, and W. C. McGrew. 2008. "The nature of culture: Technological variation in chimpanzee predation on army ants revisited." *Journal of Human Evolution* 55 (1): 48–59.

Seyfarth, R. M., D. L. Cheney, and P. Marler. 1980. "Monkey responses to three different alarm calls: Evidence of predator classification and semantic communication." *Science* 210 (4471): 801–3.

Smith, H. J. 2005. *Parenting for Primates*. Cambridge, MA: Harvard University Press.

Subiaul, F., J. F. Cantlon, R. L. Holloway, and H. S. Terrace. 2004. "Cognitive imitation in rhesus macaques." *Science* 305 (5682): 407–10.

Suchak, M., T. M. Eppley, M. W. Campbell, and F. B. M. de Waal. 2014. "Ape duos and trios: Spontaneous cooperation with free partner choice in chimpanzees." *PeerJ* 2: e417.

Tanaka, I. 1998. "Social diffusion of modified louse egg-handling techniques during grooming in free-ranging Japanese macaques." *Animal Behaviour* 56 (5): 1229–36.

Tennie, C., J. Call, and M. Tomasello. 2006. "Push or pull: Imitation vs. emulation in great apes and human children." *Ethology* 112 (12): 1159–69.

Tennie, C., J. Call, and M. Tomasello. 2010. "Evidence for emulation in chimpanzees in social settings using the floating peanut task." *PLoS One* 5 (5): e10544.

Tennie, C., J. Call, and M. Tomasello. 2012. "Untrained chimpanzees (*Pan troglodytes schweinfurthii*) fail to imitate novel actions." *PLoS One* 7 (8): e41548.

Tennie, C., K. Greve, H. Gretscher, and J. Call. 2010. "Two-year-old children copy more reliably and more often than nonhuman great apes in multiple observational learning tasks." *Primates* 51 (4): 337–51.

Thornton, A., and T. Clutton-Brock. 2011. "Social learning and the development of individual and group behaviour in mammal societies." *Philosophical Transactions of the Royal Society of London B* 366: 978–87.

Tomasello, M. 2009. *The Cultural Origins of Human Cognition*. Cambridge, MA: Harvard University Press.

Tomasello, M., S. Savage-Rumbaugh, and A. C. Kruger. 1993. "Imitative learning of actions on objects by children, chimpanzees, and enculturated chimpanzees." *Child Development* 64 (6): 1688–1705.

Vale, G. L., S. J. Davis, E. van de Waal, S. J. Schapiro, S. P. Lambeth, and A. Whiten. 2017. "Lack of conformity to new local dietary preferences in migrating captive chimpanzees." *Animal Behaviour* 124: 135–44.

Vale, G. L., E. G. Flynn, J. R. Kendal, B. Rawlings, L. M. Hopper, S. J. Schapiro, S. P. Lambeth, and R. L. Kendal. 2017. "Testing differential use of payoff-biased social learning strategies in children and chimpanzees." *Proceedings of the Royal Society of London B* 284 (1868): 20171751.

Vale, G. L., E. G. Flynn, L. Pender, E. Price, A. Whiten, S. P. Lambeth, S. J. Schapiro, and R. L. Kendal. 2016. "Robust retention and transfer of tool construction techniques in chimpanzees (*Pan troglodytes*)." *Journal of Comparative Psychology* 130 (1): 24.

van de Waal, E., C. Borgeaud, and A. Whiten. 2013. "Potent social learning and conformity shape a wild primate's foraging decisions." *Science* 340 (6131): 483.

van de Waal, E., and R. Bshary. 2011. "Social-learning abilities of wild vervet monkeys in a two-step task artificial fruit experiment." *Animal Behaviour* 81 (2): 433–38.

van de Waal, E., M. Krützen, J. Hula, J. Goudet, and R. Bshary. 2012. "Similarity in food cleaning techniques within matrilines in wild vervet monkeys." *PLoS One* 7 (4): e35694.

van de Waal, E., N. Renevey, C. M. Favre, and R. Bshary. 2010. "Selective attention to philopatric models causes directed social learning in wild vervet monkeys." *Proceedings of the Royal Society of London B* 277 (1691): 2105–11.

van Leeuwen, E. J. C., K. A. Cronin, and D. B. M. Haun. 2014. "A group-specific arbitrary tradition in chimpanzees (*Pan troglodytes*)." *Animal Cognition* 17 (6): 1421–25.

van Schaik, C. P. 2003. "Local traditions in orangutans and chimpanzees: Social learning and social tolerance." In *The Biology of Traditions: Models and Evidence*, edited by D. Fragaszy and S. E. Perry. Cambridge: Cambridge University Press.

Watson, C. F. I., H. M. Buchanan-Smith, and C. A. Caldwell. 2014. "Call playback artificially generates a temporary cultural style of high affiliation in marmosets." *Animal Behaviour* 93: 163–71.

Watson, C. F. I., and C. A. Caldwell. 2009. "Understanding behavioral traditions in primates: Are current experimental approaches too focused on food?" *International Journal of Primatology* 30 (1): 143–67.

Watson, S. K., L. A. Reamer, M. C. Mareno, G. L. Vale, R. A. Harrison, S. P. Lambeth, S. J. Schapiro, and A. Whiten. 2017. "Socially transmitted diffusion of a novel behavior from subordinate chimpanzees." *American Journal of Primatology* 79 (6): e22642.

Watson, S. K., S. W. Townsend, A. M. Schel, C. Wilke, E. K. Wallace, L. Cheng, V. West, and K. E. Slocombe. 2015. "Vocal learning in the functionally referential food grunts of chimpanzees." *Current Biology* 25 (4): 495–99.

Whiten, A., D. M. Custance, J.-C. Gomez, P. Teixidor, and K. A. Bard. 1996. "Imitative learning of artificial fruit processing in children (*Homo sapiens*) and chimpanzees (*Pan troglodytes*)." *Journal of Comparative Psychology* 110 (1): 3.

Whiten, A., J. Goodall, W. C. McGrew, T. Nishida, V. Reynolds, Y. Sugiyama, C. E. G. Tutin, R. W. Wrangham, and C. Boesch. 1999. "Cultures in chimpanzees." *Nature* 399 (6737): 682–85.

Whiten, A., V. Horner, and F. B. M. de Waal. 2005. "Conformity to cultural norms of tool use in chimpanzees." *Nature* 437 (7059): 737.

Whiten, A., N. McGuigan, S. Marshall-Pescini, and L. M. Hopper. 2009. "Emulation, imitation, over-imitation and the scope of culture for child and chimpanzee." *Philosophical Transactions of the Royal Society of London B* 364 (1528): 2417–28.

Whiten, A., and A. Mesoudi. 2008. "Establishing an experimental science of culture: Animal social diffusion experiments." *Philosophical Transactions of the Royal Society of London B* 363 (1509): 3477.

Whiten, A., A. Spiteri, V. Horner, K. E. Bonnie, S. P. Lambeth, S. J. Schapiro, and F. B. M. de Waal. 2007. "Transmission of multiple traditions within and between chimpanzee groups." *Current Biology* 17 (12): 1038–43.

Wood, D. 1989. "Social interaction as tutoring." In *Interaction in Human Development*, edited by M. H. Bornstein and J. S. Bruner, 59–80. Hillsdale, NJ: Lawrence Erlbaum Associates.

Zentall, T. R., and B. G. Galef Jr. 2013. *Social Learning: Psychological and Biological Perspectives*. New York: Psychology Press.

8

Automated Methods and the Technological Context of Chimpanzee Research

CHRISTOPHER FLYNN MARTIN
AND IKUMA ADACHI

Introduction

"A monkey hitting keys at random on a typewriter keyboard for an infinite amount of time will almost surely type a given text, such as the complete works of William Shakespeare" ("Infinite Monkey Theorem"). This imaginary scenario is a popular entry point for mathematical discussions about probability and random sequence generation. But from an animal cognition standpoint, we'd like to know more about the imaginary monkey. What compels them to type in the first place? We suspect food rewards are involved. Perhaps a computerized system determines whether the monkey's keypresses are random enough to eventually write *Hamlet*, and automatically dispenses food accordingly. Imaginary monkeys aside, for the greater part of the past century, automated research methods like these have contributed to a deeper understanding of the chimpanzee (*Pan troglodytes*) mind by providing a robust empirical platform that highlights objectivity, precision, and detailed record keeping. As one of the animal species closest to humans, chimpanzees have historically been the target of unique interest from a diverse range of academic disciplines, including biology, psychology, behavior analysis, linguistics, ethology, and cognitive science. Automated methods are a common denominator for much research done by scientists from these varied fields, owing to the applicability and sustainability of the paradigm.

In broad terms, the technology of automated methods is built on two pillars: task facilitation and data collection. Task facilitation is characterized by devices that enable stimulus presentation by flexibly and precisely controlling the size, location, and order of stimuli presented to the subject. Task facilitation also includes manipulanda, which are devices for behavioral input such as buttons, levers, joysticks, or touch-sensitive surfaces, and a means of dispensing reinforcement automatically via a food hopper or feeder machine.

Data collection involves the automatic measuring and recording of the events of the apparatus and the behavioral responses of the subject. While these elements have remained constant fixtures in animal research since the advent of operant chambers and cumulative recorders in the 1930s (Lattal 2004), advances in electronics have given rise to modern equipment that is capable of combining many if not all of the disparate components of task facilitation and data collection into a single computerized task platform. Perhaps the best example of this trend was the introduction of the touch-sensitive monitor, which enabled one-to-one mapping of stimulus presentation and response onto the same two-dimensional surface (Elsmore, Parkinson, and Mellgren 1989), leading to an explosion of studies on chimpanzee cognition from the 1990s onward. Moreover, it can be argued that the methodological possibilities afforded by new technologies like the touch-panel have had a recursive effect on the subject matter under investigation. As the technology has advanced, scientists have been compelled to maximize the potential of new machinery by addressing research topics in novel ways, by integrating ethological considerations into experiment designs, and by elaborating on existing findings gathered from non-automated methods like sliding trays, puzzle boxes, and face-to-face interactions with human experimenters, which potentially could have had unexpected or confounding impacts on subjects' performance.

Approach

This chapter gives a historical overview of the advancement of automated methods in chimpanzee research. We have divided the chapter into sections that are characterized by the technology of the era, the context in which it was used, and the major guiding principles behind a variety of research efforts. The first sections cover the era of Robert Yerkes and the origins of automated methods, as well as the application of operant conditioning practices to chimpanzee research under the framework of behaviorism and learning theory. Subsequent sections cover learning-set theory and the utilization of automated methods to examine language acquisition and use by chimpanzees. These projects were followed by the cognitive revolution and advancements in chimpanzee psychophysics research that were made possible by the advent of modern computer programming and the introduction of touchscreen monitors. A detailed overview of one project, the Ai Project at Kyoto University Japan, serves as an exemplar of the integration and continued development of automated methods over several decades as part of a unified research program. The final section of the chapter extends to the current era, where cognitive research continues to flourish under new automated paradigms, where the scope of the method is being expanded to encompass animal welfare and enrichment, and where there is a growing effort to reach beyond research in university and primate centers toward applications in zoo settings.

THE YERKES ERA

The first chimpanzee study using electronic automated technological methods consisted of a social learning task conducted by Meredith P. Crawford at the Yerkes Laboratories of Primate Biology in Orange Park, Florida (Crawford 1941). Driven by an interest in the communicative abilities of chimpanzees, and building on an earlier study on their joint problem-solving skills (Crawford 1937), Crawford designed a social task whereby a pre-trained chimpanzee might solicit and direct, through gestures and vocalizations, the help of a naive chimpanzee partner to jointly operate an apparatus by pushing buttons in a particular order (see also Duguid et al., chapter 13 this volume). Recognizing the value of objectivity that could be attained through automated means, Crawford sought "to provide an account of the development of cooperative behavior between two chimpanzees without human tuition" (Crawford 1941, p. 260). The apparatus consisted of four standalone boxes, each with a differently colored button on its face. The boxes and two automated food vendors were connected to an electrical relay board and were spread evenly across the exterior of two rooms, with one chimpanzee situated in each room. Food was released when the buttons were pressed in the correct serial order through the combined actions of the subjects. Results indicated that while subjects were keen to observe each other's actions and some solicitation behavior did occur, none of it was referentially directed by the knowledgeable individual toward the correct button a partner should push. The absence of active teaching in chimpanzee social learning is a topic that was later echoed by future studies in the wild (Matsuzawa et al. 2001). Crawford's early study built on, and contributed to, language-related topics in developmental psychology at the time, and while the task facilitation element was automated, the data collection aspect relied on the experimenters' observation of the behavior of the chimpanzees. Subsequent studies involving chimpanzees' language abilities using more-fully-automated systems were introduced decades later (Matsuzawa 2003; Rumbaugh and Washburn 2003) yet in the intervening decades automation studies with chimpanzees shifted to topics in behaviorism and learning theory.

THE BEHAVIORIST PARADIGM

In America in the 1930s, B. F. Skinner introduced the first fully automated methods for animal research, consisting of operant conditioning chambers for task facilitation with rats and pigeons, and a cumulative recorder for data collection (Lattal 2004). Two decades later, a team led by Charles Ferster began an effort at Orange Park, USA, to apply Skinner's methods to chimpanzees (Dewsbury 2003). Since the era of its founding by Robert Yerkes, the Orange Park laboratory had primarily involved face-to-face research tasks on discrimination learning, tool use, and patterned string problems. These studies emphasized the mean performance and behavioral variances of groups of subjects. In contrast, the behaviorists utilized single-subject designs with repeated measures, and their focus was almost solely directed at the topic of conditioned responses to reinforcement schedules. The operant research was carried out with new fully automated electronic equipment consisting of telegraph switches, lights of different colors, automated food hoppers, and cumulative recorders (Ferster 1957; Ferster and Skinner 1957; Kelleher 1958). There was also some effort made to automate the methods for what traditionally had been experimenter-facilitated tasks, such as conditional discrimination tasks, which led to an increase in chimpanzee performance (Dewsbury 2003).

In the early 1960s, Ferster continued to develop automated methods as part of a study examining the arithmetic abilities of chimpanzees at the Institute for Behavioral Research in Maryland, USA (Ferster and Hammer 1966). Two young chimpanzees were given ad libitum access to a fully automated apparatus consisting of a small in-line display panel, light arrays, switches, and a food hopper. Among other tasks, subjects were required to match the number of shapes presented on the display panel to one of two light arrays showing the sample number's binary equivalent. Responses were made by pressing a switch located directly under each light array, and food was delivered via an automated hopper. Following in the tradition of strict behaviorism, the study involved a closed economy of food such that the chimpanzees received their daily diet solely through the usage of the apparatus, averaging between four and seven thousand trials per day over the span of four to five hours. Remarkably, the subjects achieved near-perfect performance after prolonged exposure to the task, with as little as one or two errors per one thousand trials. Referring to the automated setup as a "suitably organized environment," Ferster noted that "chimpanzees can acquire many of the elements of a symbolic repertory such as arithmetic" (Ferster 1964, p. 106).

Ferster's methods were adopted by the United States' spaceflight program, which included tests with chimpanzees at the Holloman Air Force Base in New Mexico (Belleville et al. 1963; Rohles 1961). The project necessitated a means of examining the effects of spaceflight on chimpanzee physiological and cognitive functioning without a human present to facilitate the tasks, and the fully automated operant paradigm provided a solution. Two young chimpanzees, Ham and Enos, were given multiple schedule reinforcement and matching-to-sample tasks to perform during spaceflight, with Ham doing so for an eighteen-minute trip to space and Enos for an hour in low Earth orbit. In the years following these events, psychology research with chimpanzees continued at the air base, including a study by Farrer (1967) that examined the pictographic memory skills of thirty chimpanzees using an automated apparatus.

The strict behaviorist paradigm for chimpanzee studies was short-lived, with behaviorism research at the Yerkes Institute lasting only a few years, but it served a practical purpose for the spaceflight program, and it contributed methodological advancements to the future direction of chimpanzee research. Automated machinery consisting of circuit relays, single-subject experimental designs and statistics, intermittent reinforcement patterns, and the introduction of a self-start key to initiate a trial, were all features of the behaviorist paradigm that were adopted in later chimpanzee cognitive research.

WGTA AND VIDEO-SCREEN METHOD

Concurrent to the mid-twentieth-century behaviorism studies using fully automated electronic apparatuses, the general field of primate psychology was increasingly reliant on another kind of partially automated (though not electronic) device: the Wisconsin General Testing Apparatus (WGTA). Developed for use with primates by Harry Harlow (1949), the WGTA consisted of a tray with food wells above which stimulus objects were manually placed by experimenters, and a moving partition that gave the subjects variable access to the tray. The WGTA was used to examine discrimination learning, where certain objects would be positively reinforced, and *learning-sets*, which involved the transference of reinforcement value to novel stimuli combinations (see Levine 1959 for discussion). Comparative research with the WGTA included multiple species including macaques (*Macaca* sp.), gibbons (Hylobatidae), gorillas (*Gorilla gorilla*), chimpanzees, and various other primates (for reviews, see Fobes and King 1982; Meador et al. 1987).

The success and widespread use of the WGTA led to efforts to electronically automate different aspects of the device, mainly for use with macaques. Schrier (1961) motorized the moving partition between the subject and the tray, and Davenport, Chamove, and Harlow (1970) utilized an illuminated switch for the subject to initiate trials, as well as micro-switches in the food wells to detect the latency of subject responses. In 1989, Washburn, Hopkins, and Rumbaugh (1989) proposed a complete overhaul of the WGTA, to be replaced by a fully automated video-task paradigm that could accommodate learning-set research. Their system consisted of a computer screen for stimulus presentation, and a joystick for subject input. The video-task system they created was inspired by reports from Savage-Rumbaugh (1986) that two chimpanzees, Sherwin and Austin, were adept at using joysticks to move an on-screen cursor toward a target for a food reward. Subsequent research using Washburn's video-task with those two chimpanzees, as well as others at the Georgia State University's Language Research Center, USA, was undertaken to examine counting numerals (Beran, Rumbaugh, and Savage-Rumbaugh 1998), long-term memory for lexigrams (Beran et al. 2000), and, more recently, visuospatial memory in a three-dimensional environment (Dolins et al. 2014) (for other examples of these joystick studies, see Beran, Perdue, and Parris, chapter 20 this volume; Vale and Brosnan, chapter 15 this volume). The stream of research utilizing joysticks as input devices also expanded to include additional comparative psychology research groups, including the work with baboons (Deruelle and Fagot 1998), and work comparing capuchins and chimpanzees (Fragaszy et al. 2009).

LANA PROJECT

For much of the 1960s and '70s language acquisition was a primary focus of chimpanzee research. Ape-language studies typically involved face-to-face interactions with humans and utilized American Sign Language (Gardner and Gardner 1969; Terrace et al. 1979) as well as plastic lexigram tokens (Premack 1971). The first attempt at removing the face-to-face element was made by Duane Rumbaugh in the 1970s with the LANA Project (Rumbaugh and Washburn 2003), which sought to train a chimpanzee named Lana to use a computerized lexigram board that consisted of 100 distinct buttons, each of which could be independently programmed to illuminate. Pressing certain combinations of lexigram buttons activated food delivery from multiple Universal Feeders (automated dispensers capable of dispensing any kind of appropriately sized food), sounds from a music player, photos from a slide projector, or film from a video player. These discrete components of the system were controlled by a PDP-8 minicomputer that evaluated Lana's button presses. A row of in-line projectors (each capable of displaying up to nine distinct images) presented lexigram-based commands from a remotely located human, thus enabling a non-face-to-face channel of communication. The goal of completely automating the language learning process was not achieved, however, due to Lana's neglect of the machine without a human demonstrator to personally train her. Nonetheless, many language-related topics were investigated, and the project was a milestone in incorporating a fully automated communication device with a wide variety of features and capabilities. Modified versions of Lana's keyboard apparatus were later used by Sue Savage-Rumbaugh in subsequent language research projects with chimpanzees Sherman and Austin, as well as bonobos (*Pan paniscus*) Kanzi and Panbanisha (Savage-Rumbaugh and Lewin 1994). These projects leveraged the keyboard technology to facilitate vocabulary acquisition and usage, and in a notable experiment Sherman and Austin successfully used the lexigram apparatus to make requests of one another in order to exchange tools needed to acquire out-of-reach food (Savage-Rumbaugh and Lewin 1994).

AI PROJECT

The modern field of comparative cognitive science emerged as a synthesis of the above-mentioned approaches, including the traditional psychology tasks of the Yerkes era and WGTA paradigm, as well as the technological approaches of the LANA ape language project and the behaviorism paradigm. The Ai Project of Kyoto University (Matsuzawa 2003) serves as a useful exemplar of the progression of the methods, technologies, and scientific scopes for comparative cognitive science. The project began in 1977 under the leadership of Kiyoko Murofushi in collaboration with Toshio Asano and Tetsuro Matsuzawa. Named after the principal subject, a female chimpanzee named Ai, the project was an integrated research effort that continues to the present day, now with a multi-generational group of chimpanzees (Matsuzawa, Tomonaga, and Tanaka 2006; see also Hirata et al., chapter 9 this volume). The forty-year history of the Ai project has included several generations of scientists, chimpanzee participants, research topics, and technological methods. It has also emphasized the importance of maintaining a naturalistic physical and social environment for chimpanzee subjects, including a semi-natural enriched living space with fifteen-meter-high climbing towers, and a multi-generational group of conspecifics (Ochai and Matsuzawa 1999). A key feature that has remained constant since the beginning has been the simple scenario of a chimpanzee facing to an automated device, with little or no direct human involvement. The requisite technology to facilitate such a scenario has played a central role in the history of the Ai Project.

Project Development Using an Integrated Approach

The Ai Project initially replicated many of the technological aspects of the LANA project's computer-controlled apparatus, including a lexigram board, in-line projection displays, universal feeder, and PDP-8 minicomputer. Drawing on methods from prior monkey research on discrimination learning tasks (Matsuzawa 2003) and incorporating ethological considerations from the foundational field work of Japanese primatology (Langlitz 2017), the Ai project's goal was to examine the perceptual world of chimpanzees using objectivity, precision, and detailed record keeping (Matsuzawa 2003).

The project began with a focus on language acquisition using the Kyoto University Lexigram System, which was monochromatic and inspired by Kanji characters. Ai performed discrete trial matching-to-sample tasks in which physical objects were manually presented in a small window, requiring a response to the corresponding lexigram button on the keyboard (Asano et al. 1982). In contrast to other ape-language studies that focused on communicative interactions with humans, the incorporation of the matching-to-sample paradigm leveraged an existing empirical framework from prior studies on monkeys, rats, and pigeons to elucidate aspects of chimpanzee perception. Data were gathered on color perception (Matsuzawa 1985a), shape perception (Matsuzawa 1990), the representation of symbols (Itakura 1992; Kojima 1984), and the concept of number (Matsuzawa 1985b).

Expansion of Cognitive Testing

Over time the Ai project widened its scope of inquiry beyond language-related topics to include topics of short-term memory (Fujita and Matsuzawa 1990; Kawai and Matsuzawa 2000), serial learning (Biro and Matsuzawa 1999), and psychophysics (Fagot and Tomonaga 1999). These advancements were made possible by increasing the number of subjects, and by adding new technology to the experimental procedures. The lexigram keyboard system was replaced by touch-sensitive screens, and new computers allowed for tasks to be developed using BASIC, a general purpose, high-level programming language, to facilitate experimental procedures.

The introduction of the touch-sensitive screen was a landmark event in primate research because it enabled one-to-one mapping of stimulus presentation and subject response (Elsmore, Parkinson, and Mellgren 1989), and it drastically increased the flexibility for presenting two-dimensional stimuli, and for varying their size and location. Moreover, the combination of touchscreens and modern computer programming led to new kinds of tasks. Traditional conditional discrimination tasks such as matching-to-sample had always been constrained by a limited number of stationary physical buttons available to chimpanzee subjects. With the touchscreen, many choice alternatives could be presented simultaneously and in randomized locations from trial to trial, thereby reducing the effects of potential artifacts like location-bias.

With touch-sensitive screens it also became possible to have multiple of the same stimuli presented in conjunction with a single odd stimulus. Such kinds of tasks, known as oddity tasks or multiple-alternative match-to-sample, required the subject to detect a target stimulus from among distractor stimuli. This paradigm, often referred to as Visual Search, was introduced to chimpanzee research for the first time by Masaki Tomonaga (1993) to investigate a variety of psychophysical phenomenon, including perception of depth and shadow (Imura and Tomonaga 2003), complex geometric forms (Tomonaga and Matsuzawa 1992), biological motion (Tomonaga 2001), faces (Tomonaga 2007), and gaze direction (Tomonaga and Imura 2010). Crucially, the progression in methodological procedures from Match-to-Sample to Visual Search involved a corresponding decrease in the amount of training required by subjects to comprehend the task requirements. Whereas for standard match-to-sample (MTS) tasks the subject had to learn to touch a sample stimulus followed by the matching stimulus, with Visual Search the subject had only to make a single touch to the screen toward the odd stimulus. Thus, the progression in technological methods corresponded to a simplified task design that decreased the workload of the chimpanzee while maintaining efficacy and experimental rigor.

In another stream of development of the paradigm, researchers aimed to analyze subjects' spontaneous and "natural" cognition with automated tasks. For example, Tanaka (2007) developed a task to measure subjects' preference using a touch-sensitive screen. Each subject was presented with digitized color photographs of different species of primates, and the order of the subjects' touches to the stimuli were used as an index of their preference toward various primate species. The subjects were trained only to touch all the stimulus pictures rather than being selectively reinforced to choose any particular order of the pictures. Thus, the order of the touch could reflect the subjects' spontaneous sorting of the stimuli. More recently, Adachi and colleagues modified a traditional MTS paradigm and successfully revealed chimpanzee natural bias to match certain audio-visual characteristics. They first trained subjects with MTS paradigm for black and white discrimination, and then presented in test trials an auditory stimulus of either high or low pitch preceding to the sample stimulus onset (Ludwig, Adachi, and Matsuzawa 2011). The sound stimulus had no contingency with reinforcement, and thus any systematic impact on subjects' performance was reflective of their natural cognitive style. With this procedure, the authors found that chimpanzees spontaneously map the high-pitch sound to a bright color and low-pitch sound to a dark color, as has been shown in humans. These procedures were different from previous paradigms in that they did not analyze performance in the light of a generalization process based on prior training in order to explore the roles of stimuli aspects. Rather, the setup and test conditions enabled a spontaneous exploration of subjects' natural cognitive styles (also see Adachi 2014; Dahl and Adachi 2013).

Social tasks have migrated into the touch-panel paradigm with apparatuses that accommodate multiple chimpanzee participants using interconnected (Martin, Biro, and Matsuzawa 2011) or shared (Martin, Biro, and Matsuzawa 2014) touchscreens. Social tasks that incorporate multiple subjects and touchscreens have been used to investigate cognitive imitation in macaques (Subiaul et al. 2004), as well as public information use (Martin, Biro, and Matsuzawa 2011), competitive game-theoretic behavior (Martin, Bhui, et al. 2014), and temporal coordination (Martin et al. 2017) in chimpanzees.

By providing a universal platform, the touchscreen reduced the technological requirements for facilitating chimpanzee cognitive research. Discrete components like button boards, relay switches, and display panels were combined into a single integrated computerized device that could holistically handle all the experimental events. Modern applications of the method have branched out to re-incorporate additional peripheral technologies that may function alongside the touchscreen, including optional trackball input for studying self-agency (Kaneko and Tomonaga 2012), audio stimuli for examining cross-modal comprehension (Martinez and Matsuzawa 2009), and the addition of token vending machines to examine food preference (Sousa and Matsuzawa 2001).

Eye-tracking is another example of a recent developments in automated methods pioneered by Ai Project research. Developed by Fumihiro Kano for use with chimpanzees, these methods involve devices that passively track the subjects' eye movements as they view photos or videos while a juice dispenser maintains a stationary head posture. This method has been used to investigate attention toward conspecific bodily features (Kano and Tomonaga 2009), as well as anticipatory looking patterns during false-belief tasks (Krupenye et al. 2016). In the same way that the Visual Search task reduced the workload of the chimpanzee compared to Match-to-Sample, the eye-tracking method eliminates altogether the need for training prior to the commencement of data collection.

Future Directions

The Ai project synthesized existing technology and multidisciplinary research. We foresee this trend continuing with the three core aspects of objectivity, precision, and detailed record keeping. In the past, developments in the overall paradigm were largely confined to cognitive research, but in the future, the expansion will likely involve applications of new technology to welfare and enrichment, as well as dynamic changes in the overall scheme of comparative cognitive studies.

HYBRID APPROACHES

While the focus of this chapter has been on technological methods that are fully automated and thus rely entirely on electronic equipment to facilitate a task and record the behavior of a chimpanzee subject, there are also studies that have applied a hybrid approach of technological methods in combination with physical object tasks, or in combination with facilitation by a human experimenter. For example, Boysen utilized a television screen to present numerical stimuli for a task that also involved counting of physical objects, manipulation of paper placards, and manual delivery of food items by an experimenter (Boysen and Berntson 1989b). Other research undertaken by Boysen introduced novel technological means of data collection, including an electrocardiographic recording of chimpanzees' heart rates while viewing conspecific faces (Boysen and Berntson 1989a), and monitoring of brain waves (Berntson, Boysen, and Torello 1993). More recently, Satoshi Hirata and colleagues have examined the brain activity of chimpanzees while listening to and viewing stimuli of conspecifics (Hirata et al. 2013; Ueno et al. 2010). Automatic recording of brain activity to determine neural correlates of handedness has been undertaken by Hopkins and colleagues using PET brain scanning methods (Hopkins et al. 2010).

COMPARISONS WITH OTHER METHODOLOGIES

Every research methodology has its own advantages and disadvantages. Much prior research on chimpanzee mental abilities has employed tasks that require subjects to manipulate physical objects, which can include puzzle boxes that require manual or tool-guided manipulation to unlock baited food rewards or sliding trays that must be manipulated individually or in tandem to attain food. Such kinds of physical tasks have the advantage of presenting chimpanzees with a clear problem-solving scenario involving visible but out-of-reach food, offering the advantages of being motivational and comprehensible to naive subjects with little or no prior training. While computerized tasks retain the same basic scenario of solving problems to acquire food, the means of attaining the food is more abstract and less intuitive, requiring subjects to be trained on the association between touching stimuli on a screen and being rewarded accordingly. However, once chimpanzee participants are familiar with this process, computerized tasks have some key advantages from a methodological standpoint. On the one hand, they offer fully automated experimental events with rapid trial progressions, and on the other, a data-capturing method that measures responses on the millisecond time scale. Such timing precision enables an in-depth analysis of subjects' response latencies toward different kinds of stimuli, shedding light on the cognitive processes underlying recognition and comprehension. Moreover, when experimental events are controlled by a computer that is connected to automatic food dispensers, there is no need for a human experimenter to handle or place food on a trial-to-trial basis, which carries the potential confounds of shifting the subjects' attention away from the task and of introducing unintentional biases (i.e., cues that animals can take from the experimenter's behavior). Finally, the use of a computer at the core of the procedure can increase the sophistication and flexibility of the tasks administered, which may include pictorial stimuli, moving stimuli, and many more possibilities for task-design elements that are not or are less readily available in physical object manipulation studies.

EXPANSION TO NEW SPECIES AND LOCATIONS

In zoo settings, automated touch-panel tasks can provide a platform for cognitive research, enrichment, and education (Egelkamp and Ross 2019). Major efforts to provide zoo-housed chimpanzees with touch panel tasks include the Lincoln Park Zoo (Egelkamp et al. 2016) and Kyoto City Zoo, Japan. Other zoos, including the Smithsonian National Zoo, Zoo Atlanta, and Indianapolis Zoo have developed similar cognitive research programs for orangutans (*Pongo pygmaeus*) and gorillas. With a goal of expanding future adoption to more institutions and multiple primate species, this chapters' authors have developed a portable mesh-mounted apparatus that consists of an integrated touch-panel, PC, battery, and automatic feeder. The device, which is currently in use at several zoos in the United States, emphasizes simplicity and ease of use and has gone through several iterations of design and construction. A key benefit of widespread adoption of automated methods at more zoos will be the creation of opportunities for cognitive studies with a diverse range of species and relatively large sample sizes of subjects compared to studies drawn from a single institution.

POTENTIAL EXPANSION OF PARADIGM

The future of automated methods with chimpanzees will likely include increasing integration with new technologies and advances in computer science. For example, the social touch-panel paradigm developed by Martin, Biro, and Matsuzawa (2014) may be expanded to include tasks carried out by multiple individuals across a computer network or internet connection. Such a scenario could also enable game theory tasks that are remotely played by individuals of different species, thereby creating a direct between-species comparison.

Another possibility for applying new computer methods is the application of machine learning and artificial intelligence to cognitive research. For example, Lee et al. (2004) conducted a macaque study whereby a computer algorithm modeled the subjects' decision-making heuristics during a binary-choice task in real-time and adjusted the stimuli and reward parameters accordingly from trial to trial. Such kinds of dynamic optimization algorithms offer finely tuned targeting of specific cognitive traits while requiring fewer trials to make robust analytical determinations, while also avoiding overtraining toward developing undesirable response bias. Recent advances in artificial intelligence aimed at making adaptive use of big data, such as the deep learning algorithms behind the board-game playing AlphaGo program (Silver et al. 2016), could also provide a new kind of tool for finding and elucidating previously overlooked trends in detailed records that have been accumulated through automated computer tasks with animals.

QUALITY OF LIFE, WELLNESS, AND ENRICHMENT

Quality of life and environmental enrichment are fundamental priorities for research laboratories, zoos, and sanctuaries, and technological methods are increasingly being used in these settings. A key feature of this trend is the utilization of technology to create opportunities for captive primates to forage for food in ways that match the behavioral patterns of their wild counterparts in natural habitats. In a parallel way to wild foraging, cognitive touch-panel tasks offer an opportunity for chimpanzees to feed intermittently over an extended period based on their own mental and physical efforts. Along these lines, Yamanashi and Hayashi (2011) compared the activity budgets of the Kyoto University chimpanzee group to those of a wild chimpanzee group in Bossou, West Africa. It was shown that access to cognitive research sessions with touchscreens resulted in a comparable amount of time spent feeding between the groups. Moreover, the daily feeding schedule for the Ai Project's chimpanzee group, who are fed three meals and given four additional opportunities to earn food during research tasks, was informed by observations by Matsuzawa that wild chimpanzees in Bossou engage in about seven feeding bouts per day (Matsuzawa, pers. comm.). Creating more naturalistic foraging behavior in captivity can also be accomplished by providing constant access to opportunities to gain food resources though computer tasks. For example, Fagot and Paleressompoulle (2009) developed touch-panel systems situated inside of the outdoor living space of baboons, thereby enabling ad libitum access to the machines (see also Paxton Gazes et al. 2019 for comparable work with macaques). A similar system is under development for chimpanzees at Kyoto University Primate Research Institute (Matsuzawa, pers. comm.).

Automated computer task methods have been used successfully to aid in physical rehabilitation processes. In 2006, a 24-year-old male chimpanzee named Reo at the Kyoto University Primate Research Institute developed acute transverse myelitis, resulting in impaired leg movement. The illness necessitated a move to a recovery room and an extended rehabilitation process lasting several years (Hayashi et al. 2013). To encourage locomotion during his recovery, a touchscreen was installed on one wall of Reo's room connected to a universal feeder machine two meters away on the opposite wall. Sakuraba and colleagues (2016) found that the automated setup, which encouraged the chimpanzee to move between the screen and the feeder during a four-hour window of cognitive testing each day, led to a significant increase in locomotion. Reo's summative travel distance between the touchscreen and feeder averaged 500 meters a day compared to the baseline mean of 136 meters without the automated task turned on (Sakuraba, Tomonaga, and Hayashi 2016).

An additional technological effort to improve the well-being of chimpanzee Reo during his recovery included the utilization of video-conferencing software to connect him with former group members, including his mother, Reiko, whom he had not seen for several years. This endeavor was launched up based on a prior "ChimpSkype" pilot study to connect chimpanzees in the Primate Research Institute and Lincoln Park Zoo, USA, using Skype (Martin, Ross, pers. comm.). This chapter's authors and Yoko Sakuraba placed laptops running Skype in front of Reo in his isolated recovery room, and his mother, Reiko, in a separate location of the Primate Research Institute. The project was carried out for enrichment purpose for a single isolated chimpanzee, Reo, yet opens the door for future implementation of this technology for scientific studies with a quantitative focus. For example, the lively interactions suggested the possible comprehension of the real-time aspect of the videolink, and this topic is deserving of further empirical attention.

Conclusions

In sum, the history of automated methods in chimpanzee research has involved not only the development of new technologies but also the synthesis of a variety of devices, task designs, and research paradigms, leading to a deeper and more broadened understanding of chimpanzee cognition. Looking forward to the future, this trend will likely continue with technologies like eye tracking, internet networking, and artificial intelligence potentially being integrated in ways that capitalize on the core strengths of automated methods: objectivity, precision, and detailed data collection. Moreover, the refinement of existing technologies, such as the effort to make portable and easy-to-operate touch-panel apparatuses and software, may help to pave the way for more widespread adoption across species, individuals, and institutions, thereby enhancing the scope and applicability of comparative cognitive science.

References

Adachi, I. 2014. "Spontaneous spatial mapping of learned sequence in chimpanzees: Evidence for a SNARC-like effect." *PLoS One* 9 (3): e90373.

Asano, T., T. Kojima, T. Matsuzawa, K. Kubota, and K. Murofushi. 1982. "Object and color naming in chimpanzees (*Pan troglodytes*)." *Proceedings of the Japan Academy* 58: 118–22.

Belleville, R. E., F. H. Rohles, M. E. Grunzke, and F. C. Clark. 1963. "Development of a complex multiple schedule in the chimpanzee." *Journal of the Experimental Analysis of Behavior* 6: 549–56.

Beran, M. J., J. L. Pate, W. K. Richardson, and D. M. Rumbaugh. 2000. "A chimpanzee's (*Pan troglodytes*) long-term retention of lexigrams." *Animal Learning & Behavior* 28: 201–7.

Beran, M. J., D. M. Rumbaugh, and E. S. Savage-Rumbaugh. 1998. "Chimpanzee (*Pan troglo dytes*) counting in a computerized testing paradigm." *Psychological Record* 48: 3–19.

Berntson, G. G., S. T. Boysen, and M. W. Torello. 1993. "Vocal perception: Brain event-related potentials in a chimpanzee." *Developmental Psychobiology* 26 (6): 305–19.

Biro, D., and T. Matsuzawa. 1999. "Numerical ordering in a chimpanzee (*Pan troglodytes*): Planning, executing, and monitoring." *Journal of Comparative Psychology* 113: 178–85.

Boysen, S. T., and G. G. Berntson. 1989a. "Conspecific recognition in the chimpanzee (*Pan troglodytes*): Cardiac responses to significant others." *Journal of Comparative Psychology* 103 (3): 215.

Boysen, S. T., and G. G. Berntson. 1989b. "Numerical competence in a chimpanzee (*Pan troglodytes*)." *Journal of Comparative Psychology* 103 (1): 23.

Crawford, M. P. 1937. "The cooperative solving of problems by young chimpanzees." *Comparative Psychology Monographs* 14: 1–88.

Crawford, M. P. 1941. "The cooperative solving by chimpanzees of problems requiring serial responses to color cues." *Journal of Social Psychology* 13: 259–80.

Dahl, C. D., and I. Adachi. 2013. "Conceptual metaphorical mapping in chimpanzees (*Pan troglodytes*)." *eLife* 2013 (2): e00932.

Davenport, J. W., A. S. Chamove, and H. F. Harlow. 1970. "The semiautomatic Wisconsin general test apparatus." *Behavior Research Methods and Instrumentation* 2: 135–38.

Deruelle, C., and J. Fagot. 1998. "Visual search for global/local stimulus features in humans and baboons." *Psychonomic Bulletin and Review* 5: 476–81.

Dewsbury, D. 2003. "Conflicting approaches: Operant psychology arrives at a primate laboratory." *Behavioral Analysis* 26: 253–65.

Dolins, F. L., C. Klimowicz, J. Kelley, and C. R. Menzel. 2014. "Using virtual reality to investigate comparative spatial cognitive abilities in chimpanzees and humans."*American Journal of Primatology* 76: 496–513.

Egelkamp, C. E., L. M. Hopper, K. A. Cronin, S. L. Jacobson, and S. R. Ross. 2016. "Using touch-screens to explore the welfare and cognition of zoo-housed primates." *PeerJ Preprints* 4: e2312v1

Egelkamp, C. E., and S. R. Ross. 2019. "A review of zoo-based cognitive research using touch-screen interfaces." *Zoo Biology* 38 (2): 220–35.

Elsmore, T. F., J. K. Parkinson, and R. L. Mellgren. 1989. "Video touch-screen stimulus-response surface for use with primates." *Bulletin of the Psychonomic Society* 27: 60–63.

Fagot, J., and D. Paleressompoulle. 2009. "Automatic testing of cognitive performance in baboons maintained in social groups." *Behavior Research Methods* 41: 396–404.

Fagot, J., and M. Tomonaga. 1999. "Global and local processing in humans (*Homo sapiens*) and chimpanzees (*Pan troglodytes*): Use of a visual search task with compound stimuli." *Journal of Comparative Psychology* 113: 3–12.

Farrer, D. N. 1967. "Picture memory in the Chimpanzee." *Perception and Motor Skills* 25: 305–15.

Ferster, C. B. 1957. "Concurrent schedules of reinforcement in the chimpanzee." *Science* 125 (3257): 1090–91.

Ferster, C. B. 1964. "Arithmetic behavior in chimpanzees." *Scientific American* 210: 98–106.

Ferster, C. B., and C. E. J. Hammer. 1966. "Synthesizing the components of arithmetic behavior." In *Operant Behavior: Areas of Research and Application*, edited by W. K. Honig, 634–76. New York: Appleton-Century-Crofts.

Ferster, C. B., and B. F. Skinner. 1957. *Schedules of Reinforcement*. New York: Appleton-Century-Crofts.

Fobes, J. L., and J. E. King. 1982. "Measuring primate learning abilities." In *Primate Behavior*, edited by J. L. Fobes and J. E. King, 289–326. New York: Academic Press.

Fragaszy, D. M., E. Kennedy, A. Murnane, C. Menzel, G. Brewer, J. Johnson-Pynn, and W. Hopkins. 2009. "Navigating two-dimensional mazes: Chimpanzees (*Pan troglodytes*) and capuchins (*Cebus apella sp.*) profit from experience differently." *Animal Cognition* 12: 491–504.

Fujita, K., and T. Matsuzawa. 1990. "Delayed figure reconstruction by a chimpanzee (*Pan troglodytes*) and humans (*Homo sapiens*)." *Journal of Comparative Psychology* 104: 345–51.

Gardner, R. A., and B. T. Gardner. 1969. "Teaching sign language to a chimpanzee." *Science* 165 (3894): 664–72.

Harlow, H. F. 1949. "The formation of learning sets." *Psychological Review* 56: 51–65.

Hayashi, M., Y. Sakuraba, S. Watanabe, A. Kaneko, and T. Matsuzawa. 2013. "Behavioral recovery from tetraparesis in a captive chimpanzee." *Primates* 54: 237–43.

Hirata, S., G. Matsuda, A. Ueno, H. Fukushima, K. Fuwa, K. Sugama, and T. Hasegawa. 2013. "Brain response to affective pictures in the chimpanzee." *Scientific Reports* 3: 1342.

Hopkins, W. D., J. P. Taglialatela, J. Russell, T. M. Nir, and J. Schaeffer. 2010. "Cortical representation of lateralized grasping in chimpanzees (*Pan troglodytes*): A combined MRI and PET study." *PLoS One* 5 (10): e13383.

Imura, T., and M. Tomonaga. 2003. "Perception of depth from shading in infant chimpanzees (*Pan troglodytes*)." *Animal Cognition* 6: 253–58.

"Infinite Monkey Theorem." Wikipedia. Accessed October 18, 2018. https://en.wikipedia.org/wiki/InfiniteMonkeyTheorem.

Itakura, S. 1992. "A chimpanzee with the ability to learn the use of personal pronouns." *Psychological Record* 42: 157–72.

Kaneko, T., M. Tomonaga. 2012. "Relative contributions of goal representation and kinematic information to self-monitoring by chimpanzees and humans." *Cognition* 125: 168–78.

Kano, F., and M. Tomonaga. 2009. "How chimpanzees look at pictures: A comparative eye-tracking study." *Proceedings of the Royal Society of London B* 276: 1949–55.

Kawai, N., and T. Matsuzawa. 2000. "Numerical memory span in a chimpanzee." *Nature* 403: 39–40.

Kelleher, R. T. 1958. "Fixed-ratio schedules of conditioned reinforcement with chimpanzees." *Journal of the Experimental Analysis of Behavior* 1: 281–89.

Kojima, T. 1984. "Generalization between productive use and receptive discrimination of names in an artificial visual language by a chimpanzee." *International Journal of Primatology* 5: 161–82.

Krupenye, C., F. Kano, S. Hirata, J. Call, and M. Tomasello. 2016. "Great apes anticipate that other individuals will act according to false beliefs." *Science* 354 (6308): 110–14.

Langlitz, N. 2017. "Synthetic primatology: What humans and chimpanzees do in a Japanese laboratory and the African field." *BJHS Themes* 2: 101–25.

Lattal, K. 2004. "Steps and pips in the history of the cumulative recorder." *Journal of the Experimental Analysis of Behavior* 82: 329–55.

Lee, D., M. L. Conroy, B. P. McGreevy, and D. J. Barraclough. 2004. "Reinforcement learning and decision making in monkeys during a competitive game." *Cognitive Brain Research* 22: 45–58.

Levine, M. 1959. "A model of hypothesis behavior in discrimination learning set." *Psychological Review* 66: 353–66.

Ludwig, V. U., I. Adachi, and T. Matsuzawa. 2011. "Visuoauditory mappings between high luminance and high pitch are shared by chimpanzees (*Pan troglodytes*) and humans." *Proceedings of the National Academy of Sciences* 108: 20661–65.

Martin, C. F., R. Bhui, P. Bossaerts, M. Tetsuro, and C. Camerer. 2014. "Chimpanzee choice rates in competitive games match equilibrium game theory predictions." *Scientific Reports* 4: 5182.

Martin, C. F., D. Biro, and T. Matsuzawa. 2011. "Chimpanzees' use of conspecific cues in matching-to-sample tasks: Public information use in a fully automated testing environment." *Animal Cognition* 14: 893–902.

Martin, C. F., D. Biro, and T. Matsuzawa. 2014. "The arena system: A novel shared touch-panel apparatus for the study of chimpanzee social interaction and cognition." *Behavior Research Methods* 46: 611–18.

Martin, C. F., D. Biro, and T. Matsuzawa. 2017. "Chimpanzees spontaneously take turns in a shared serial ordering task." *Scientific Reports* 7: 14307.

Martinez, L., and T. Matsuzawa. 2009. "Visual and auditory conditional position discrimination in chimpanzees (*Pan troglodytes*)." *Behavioural Processes* 82: 90–94.

Matsuzawa, T. 1985a. "Colour naming and classification in a chimpanzee (*Pan troglodytes*)." *Journal of Human Evolution* 14: 283–91.

Matsuzawa, T. 1985b. "Use of numbers by a chimpanzee." *Nature* 315: 57–59.

Matsuzawa, T. 1990. "Form perception and visual acuity in a chimpanzee." *Folia Primatologica* 55 (1): 24–32.

Matsuzawa, T. 2003. "The Ai project: Historical and ecological contexts." *Animal Cognition* 6: 199–211.

Matsuzawa, T., D. Biro, T. Humle, N. Inoue-Nakamura, R. Tonooka, and G. Yamakoshi. 2001. "Emergence of culture in wild chimpanzees: Education by master-apprenticeship." In

Primate Origins of Human Cognition and Behavior, edited by T. Matsuzawa, 557–74. Tokyo: Springer.

Matsuzawa, T., M. Tomonaga, and M. Tanaka. 2006. *Cognitive Development in Chimpanzees*. Tokyo: Springer.

Meador, D. M., D. M. Rumbaugh, J. L. Pate, and K. A. Bard. 1987. "Learning, problem solving, cognition, and intelligence." In *Comparative Primate Biology*, edited by G. Mitchell and J. Erwin, 17–83. New York: Liss.

Ochai, T., and T. Matsuzawa. 1999. "Environmental enrichment for captive chimpanzees (*Pan troglodytes*): Introduction of cimbing frames 15m high." *Primate Research* 15: 289–96.

Paxton Gazes, R., M. C. Lutz, M. J. Meyer, T. C. Hassett, and R. R. Hampton. 2019. "Influences of demographic, seasonal, and social factors on automated touchscreen computer use by rhesus monkeys (*Macaca mulatta*) in a large naturalistic group." *PLoS One* 14 (4): e0215060.

Premack, D. 1971. "On the assessment of language competence in the chimpanzee." In *Behavior of Non-Human Primates*, edited by A. M. Schrier and F. Stolnitz, 4:185–228. New York: Academic Press.

Rohles, J. F. H. 1961. "The development of an instrumental skill sequence in the chimpanzee." *Journal of the Experimental Analysis of Behavior* 4: 323–25.

Rumbaugh, D. M., and D. A. Washburn. 2003. *Intelligence of Apes and Other Rational Beings*. New Haven: Yale University Press.

Sakuraba, Y., M. Tomonaga, and M. Hayashi. 2016. "A new method of walking rehabilitation using cognitive tasks in an adult chimpanzee (*Pan troglodytes*) with a disability: A case study." *Primates* 57: 403–12.

Savage-Rumbaugh, E. S. 1986. *Ape Language: From Conditioned Response to Symbol*. New York: Columbia University Press.

Savage-Rumbaugh, E. S., and R. Lewin. 1994. *Kanzi: The Ape at the Brink of the Human Mind*. New York: Wiley.

Schrier, A. M. 1961. "A modified version of the Wisconsin general test apparatus." *Journal of Psychology* 52: 193–200.

Silver, D., A. Huang, C. J. Maddison, A. Guez, L. Sifre, G. van den Driessche, J. Schrittwieser, I. Antonoglou, V. Panneershelvam, M. Lancot, S. Dieleman, D. Grewe, J. Nham, N. Kalchbrenner, I. Sutskever, T. Lillicrap, M. Leach, K. Kavukcuoglu, T. Graepel, and D. Hassabis. 2016. "Mastering the game of go with deep neural networks and tree search." *Nature* 529: 484–89.

Sousa, C., and T. Matsuzawa. 2001. "The use of tokens as rewards and tools by chimpanzees (*Pan troglodytes*)." *Animal Cognition* 4: 213–21.

Subiaul, F., J. F. Cantlon, R. L. Holloway, and H. S. Terrace. 2004. "Cognitive imitation in rhesus macaques." *Science* 305 (5682): 407–10.

Tanaka, M. 2007. "Development of the visual preference of chimpanzees (*Pan troglodytes*) for photographs of primates: Effect of social experience." *Primates* 48: 303–9.

Terrace, H. S., L.-A. Petitto, R. J. Sanders, and T. G. Bever. 1979. "Can an ape create a sentence?" *Science* 206 (4421): 891–902.

Tomonaga, M. 1993. "Use of multiple-alternative matching-to-sample in the study of visual search in a chimpanzee (*Pan troglodytes*)." *Journal of Comparative Psychology* 107: 75–83.

Tomonaga, M. 2001. "Visual search for biological motion patterns in chimpanzees (*Pan troglodytes*)." *Psychologia* 44: 46–59.

Tomonaga, M. 2007. "Visual search for orientation of faces by a chimpanzee (*Pan troglodytes*): Face-specific upright superiority and the role of facial configural properties." *Primates* 48: 1–12.

Tomonaga, M., and T. Imura. 2010. "Visual search for human gaze direction by a chimpanzee (*Pan troglodytes*)." *PLoS One* 5 (2): e9131.

Tomonaga, M., and T. Matsuzawa. 1992. "Perception of complex geometric figures in chimpanzees (*Pan troglodytes*) and humans (*Homo sapiens*): Analyses of visual similarity on the basis of choice reaction time." *Journal of Comparative Psychology* 106: 43–52.

Ueno, A., S. Hirata, K. Fuwa, K. Sugama, K. Kusunoki, G. Matsuda, and T. Hasegawa. 2010. "Brain activity in an awake chimpanzee in response to the sound of her own name." *Biology Letters* 6 (3): 311–13.

Washburn, D. A., W. D. Hopkins, and D. M. Rumbaugh. 1989. "Automation of learning-set testing: The video-task paradigm." *Behavior Research Methods, Instruments, and Computers* 21: 281–84.

Yamanashi, Y., and M. Hayashi. 2011. "Assessing the effects of cognitive experiments on the welfare of captive chimpanzees (*Pan troglodytes*) by direct comparison of activity budget between wild and captive chimpanzees." *American Journal of Primatology* 73: 1231–38.

9

The Establishment of Sanctuaries for Former Laboratory Chimpanzees: Challenges, Successes, and Cross-Cultural Context

SATOSHI HIRATA, NARUKI MORIMURA, KOSHIRO WATANUKI, AND STEPHEN R. ROSS

Introduction

Chimpanzees (*Pan troglodytes*) have been managed by humans in captive settings around the world for well over 100 years and the context of that care has ranged widely over that period. From the initial captures of chimpanzees in equatorial Africa for import into zoological parks, and use in circuses and menageries, to the adoption of captive breeding programs, and their use as subjects of behavioral, cognitive and medical research, our relationship with our closest living relatives has been complex and often controversial. Given that the United States and Japan are the two countries holding the highest number of captive chimpanzees in the world, and most involved in past biomedical experimentation on this species leading to their subsequent retirement to sanctuaries, the value of exploring the context of captive chimpanzee care across and between these two very different cultures is considerable. Therefore, in this chapter, we describe the history of captive chimpanzees in these two countries, providing a parallel perspective on the origination of two major chimpanzee sanctuaries: Kumamoto Sanctuary (formerly, Chimpanzee Sanctuary Uto), in Uto peninsula, Kumamoto prefecture, Japan, and Chimp Haven, the National Chimpanzee Sanctuary located in Keithville, Louisiana, United States. We aim to compare and contrast the paths that led to the establishment of these organizations and highlight the challenges faced in the maintenance and sustainability of these sanctuaries. While some of the characteristics we discuss are culturally specific, others are shared across the two countries and organizations. Together, we hope to demonstrate the overall value of the chimpanzee sanctuary paradigm and reveal key facets that will aid in its success and ultimately evolve our complex relationship with our closest living relatives. So we begin with a short history lesson detailing how all these chimpanzees, over 2,300 at the time of writing, came to be in these two countries situated thousands of miles apart, and even further from the natural range of this species in equatorial Africa.

The Historical Context of Chimpanzees in Biomedical Research and Retirement to Sanctuaries

In Japan, the United States, Africa, and Europe, invasive biomedical studies have been conducted with chimpanzees for the development of vaccines to combat infectious diseases such as hepatitis B and C, HIV, and malaria (Cohen 2007; Bennett and Panicker 2016; Fukui 1987 in Japanese). The recent increase in ethical and welfare considerations in these countries, however, has contributed to the widespread termination of invasive biomedical studies in those areas (Kaiser 2015; Knight 2008; Morimura, Idani, and Matsuzawa 2011; Vermij 2003). Because the euthanasia of chimpanzees has not been an option in Japan and the United States (National Research Council 1997, see also Hua and Ahuja 2013), the need for sanctuaries to house chimpanzees no longer needed for biomedical research programs developed quickly. The result is a relatively young global community of organizations with shared missions, but set in vastly different cultural contexts. While issues related to the discontinuation of invasive studies and the value of animal sanctuaries have been widely discussed in North America and Europe (Bennett and Panicker 2016; Hua and Ahuja 2013; Ross 2014; Vermij 2003; Wadman 2011), there are very few reports and comments published in English on the situation in Japan, and those that exist are sometimes inaccurate (Goodman and Check 2002). Therefore, here, we provide an overview of the history of captive chimpanzees in Japan, the current scenario regarding Japanese sanctuary-housed chimpanzees, and how this compares with the historic and contemporary captive chimpanzee environments in America.

CHIMPANZEES IN JAPAN

The number of captive chimpanzees living in Japan has varied widely over the last 90 years (fig. 9.1) and the history of this population has been well-described by both Watanuki et al. (2014) and Ochiai et al. (2015). Chimpanzees have been housed primarily in zoos, but also as subjects of medical and behavioral/cognitive research. The first reported case of a chimpanzee being brought to Japan was accompanying an Italian circus during an exposition in 1921 (Ochiai et al. 2015), but it was not until 1926, when two chimpanzees were brought into the country by a collector from New York, that this species permanently resided in Japan (Ochiai et al. 2015; Takashima 1955) (table 9.1).

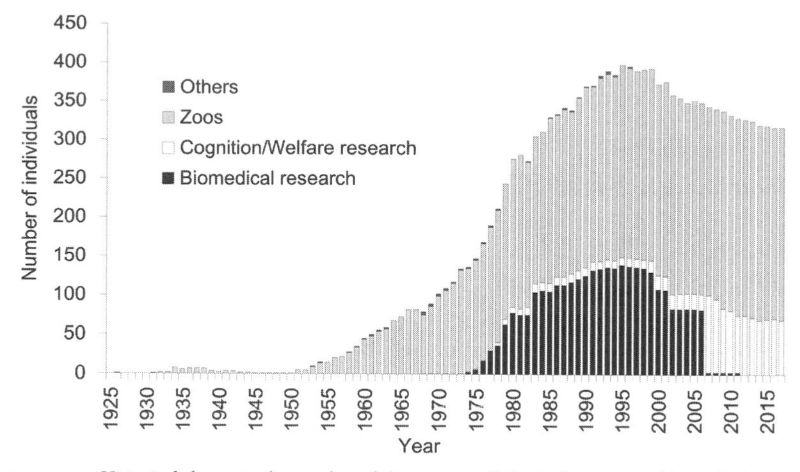

FIGURE 9.1. Historical change in the number of chimpanzees living in Japan across biomedical research facilities, cognition/welfare research facilities, zoos, and in other institutions.

One of the two chimpanzees was obtained by a Japanese animal dealer in an exchange for endemic Japanese animals (such as monkeys, raccoon dogs, and pheasants) and was kept in the dealer's house for 10 months. The chimpanzee was later taken to a zoo in Japan for exhibition, but she died one month later. In 1931, the same zoo obtained a young female chimpanzee through a different animal dealer. That chimpanzee had undergone training to perform actions such as brushing its teeth and smoking tobacco for public entertainment. The chimpanzee's performances in the zoo aroused considerable public interest and increased visitor attendance. Subsequently, many other Japanese zoos began obtaining chimpanzees from the wild. By the end of World War II, in 1945, the number of chimpanzees in Japan had decreased to just one individual, but the number began to increase again in 1951. The number of chimpanzees kept in zoos has constituted the majority of all chimpanzees in Japan from 1932 to the present day, peaking in 2002 when there were 252 chimpanzees in Japanese zoos.

In 1968, the Primate Research Institute (PRI) of Kyoto University housed a single chimpanzee (Matsuzawa 2006). This was the start of behavioral and cognitive studies of chimpanzees in Japan—half a century after the pioneering work on captive chimpanzees by Yerkes (1925) in the United States and Köhler (1925) on the Canary Islands. PRI obtained seven additional chimpanzees between 1977 and 1985, and those individuals produced seven offspring. The institute's research program, named the Ai Project after a female chimpanzee who arrived at the institute in 1977, began as an investigation

TABLE 9.1. A timeline of chimpanzees in Japan and the United States of America.

Japan	*Year*	*USA*
	1902	First chimpanzee arrives in USA
First chimpanzee arrives in Japan	1921	
	1961	Ham, the chimpanzee, launched into space as part of USA's space program
Start of invasive biomedical studies in Japan	1974	
	1975	End of importation of wild chimpanzees to USA
End of importation of wild chimpanzees to Japan	1983	
	1995	Breeding moratorium on federally owned and supported chimpanzees
	2005	First chimpanzees arrive at Chimp Haven
Sanwa facility becomes Chimpanzee Sanctuary Uto (later, Kumamoto Sanctuary)	2007	
Last three chimpanzees in biomedical facility retire to Kumamoto Sanctuary	2012	
	2018	For the first time in USA, more chimpanzees housed in sanctuaries than research centers

into the development of language by chimpanzees, but later expanded to cover broader areas of cognition and behavior (Matsuzawa 2003). To characterize his approach, Matsuzawa coined a Japanese term that can be translated as "comparative cognitive science" (Matsuzawa 2003). The project illustrates various aspects of chimpanzee cognition, such as working memory, behavioral development, social learning, and social intelligence (Matsuzawa 2001; Matsuzawa, Tomonaga, and Tanaka 2006; Tomonaga, Matsuzawa, and Tanaka 2003; see also Martin and Adachi, chapter 8 this volume). Matsuzawa also investigated behavior and cognition among wild chimpanzees in Bossou and Nimba, Guinea, and developed a unique method of field experimentation characterized by a combination of field and laboratory observations (Matsuzawa 1994, 2011; see also Yamamoto, chapter 14 this volume).

While invasive biomedical research with chimpanzees was never conducted at PRI, the use of chimpanzees for this purpose started in 1974 at the University of Tokyo (Sugiyama 1985; Watanuki et al. 2014). In 1975, Japan's Ministry of Health and Welfare set up a committee to develop vaccines for hepatitis B. The University of Tokyo worked closely with the ministry and imported additional chimpanzees from an American dealer (Koshimizu et al. 1977). The ministry began providing grants for studying hepatitis using chimpanzees and the number of chimpanzees used for biomedical research increased in the late 1970s. Most chimpanzees were imported directly from Africa, which was legal until 1980 when Japan ratified the Convention on International Trade in Endangered Species of Wild Fauna and Flora (CITES). Even after Japan's compliance with CITES in 1980, importing wild captured chimpanzees into the country for research was possible for a period of time if both the exporting and importing countries reached an agreement.

In total, more than 150 chimpanzees were imported to Japan for invasive biomedical research conducted at eight different institutions (Matsubayashi 1993). With the decline in the study of hepatitis B, the focus shifted to investigating hepatitis C, embryonic stem cells, and malaria. A total of 205 chimpanzee have been subjects of biomedical research in Japan (including both wild-caught and captive-born individuals), peaking in 1995 when there were 139 individuals available for such use.

CHIMPANZEES IN THE UNITED STATES

Compared to that of Japan, the nature of the captive chimpanzee population in the United States is considerably more convoluted (see table 9.1 for a summary timeline for both countries). The timeframe is longer, the number of chimpanzees is higher, and the categories of housing types and sites are more varied. Furthermore, there was no centralized database to monitor their numbers or where they lived, and so estimates for chimpanzee numbers are much more ambiguous historically. This gap in knowledge led to the formation of Lincoln Park Zoo's Project ChimpCARE (www. chimpcare. org) in 2009, which is the closest approximation of a comprehensive database tracking the US captive chimpanzee population. By stitching together records from various information sets, including the North American Regional studbook for chimpanzees (tracking zoo-housed chimpanzees) and governmental records of laboratory with Lincoln Park Zoo's Project ChimpCARE database, we can provide an overview of the historical trajectory of chimpanzee numbers in the US (fig. 9.2).

The first chimpanzees to arrive in the United States were likely those brought into zoos. The earliest record in the North American Regional studbook for chimpanzees lists a young chimpanzee named Polly arriving at the Bronx Zoo in New York in January 1902 via the Hamburg Zoo in Germany. Like many of those early arrivals, Polly did not live a long life, dying in 1907, but dozens of other chimpanzees began to stream in over the next several

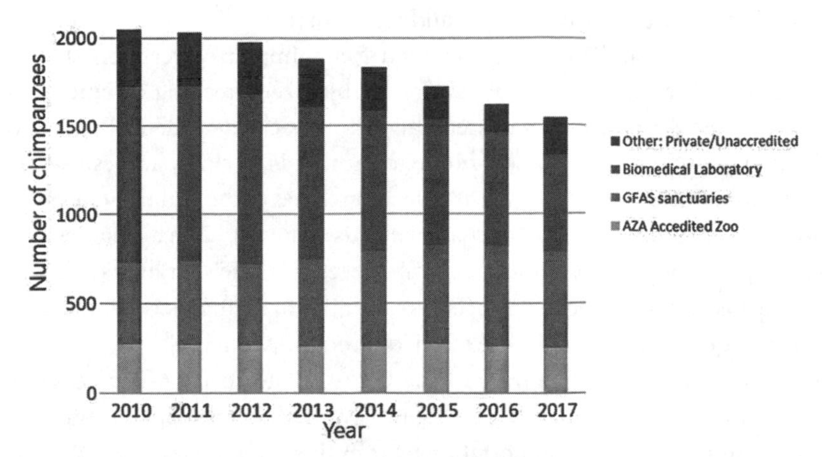

FIGURE 9.2. Recent history of changes to the number of chimpanzee living in the USA across zoos accredited by the Association of Zoos and Aquariums (AZA), sanctuaries accredited by the Global Federation of Animal Sanctuaries (GFAS), biomedical research facilities, and in other facilities including unaccredited zoos and private ownership.

decades. Later, chimpanzees began to be imported for research. The website http://first100chimps.wesleyan.edu/ carefully details their numbers and describes the first two chimpanzees to arrive for research: a pair named Bill and Dwina, who were originally captured in Cameroon, and then purchased by Robert Yerkes in May of 1925 from an importer. Dwina died in 1930 and Bill in 1933, though both apparently produced offspring despite their young age. Yerkes continued to acquire chimpanzees for his research, initially at Yale University's primate laboratory, which was later relocated to Orange Park, Florida. By 1936, this colony had grown to 42 individuals through a combination of captive breeding and continued importation from Africa.

By the middle of the twentieth century, there were likely several hundred chimpanzees living in the United States, not only in laboratory settings and in zoological parks, but also in more public and less regulated settings, being used as pets and performers. However, the real growth of the chimpanzee population was for use in research. First by Yerkes and later by many more, interest in chimpanzee physiology and reproduction grew throughout the 1930s and 1940s. This was followed by a growth in biomedical studies of infectious diseases and immunology as scientists used chimpanzees as models for diseases such as hepatitis and polio. Around this time, there was also increased attention to the cognitive and communicative abilities of chimpanzees. Several chimpanzees were broadly exposed to human environments and reared as if human children in an attempt to uncover potential language capabilities (Gardner and Gardner 1975; Hayes and Hayes 1951).

In the 1950s and 1960s, the demand for chimpanzee research subjects resulted in the development of several chimpanzee breeding colonies. One of those, the Aeromedical Research Laboratory located at the Holloman Air Force base in New Mexico, was home to hundreds of chimpanzees used for both biomedical research as well as research related to the United States space program (Martin and Adachi, chapter 8 this volume). Other colonies were more specifically used to produce chimpanzees as models for human disease. While it is difficult to obtain an accurate number of chimpanzees at this time, the total populations may have exceeded 1,000 individuals.

Importation of chimpanzees to the United States from Africa ceased in 1975 with the adoption of CITES, about five years before Japan ratified the accord and began to end importations. From this point forward, the demand for chimpanzees relied on captive breeding programs. There was widespread breeding success throughout the 1970s and 1980s, though many chimpanzee infants were removed from their mothers at birth to be raised by human caretakers, which has been shown to have a range of negative developmental outcomes for these individuals (see Ross, chapter 24 this volume). Among those outcomes were reduced breeding and maternal rearing capabilities, which in turn led to growing concerns about the long-term sustainability of the captive chimpanzee population. In 1981, a governmental Task Force developed plans for a national breeding program to boost the number of chimpanzees to be used in research, and just a few years later the first chimpanzees were experimentally infected with the HIV virus to combat the shockingly rapid rise of AIDS epidemic. A group of five chimpanzee facilities convened in 1986 to create the National Chimpanzee Breeding and Research Plan in an attempt to consolidate the chimpanzee population and maintain a sustainable population for research needs (Graham 1981; Hobson, Graham, and Rowell 1991).

In 2015, the United States Fish and Wildlife Service updated the status of chimpanzees to Endangered, thereby providing them the extra protections afforded by the Endangered Species Act, and essentially ending the commercial exchanges of the species that drove their trade as pets. Zoo populations remain relatively steady through a limited cooperative breeding program managed by the Chimpanzee Species Survival Plan and are the only population that are expected to persist into the next century. At the time of writing, 247 chimpanzees live in zoos accredited by the Association of Zoos and Aquariums in the United States.

THE ORIGINS OF CHIMPANZEE SANCTUARIES IN JAPAN

Despite the need for chimpanzee sanctuary housing in Japan, the Ministry of Health and Welfare did not appear to have any specific policy to guide that process. A ministry official went on record saying it was assumed that no chimpanzees would remain after the conclusion of research (Yamamori 1992), and some have interpreted this to mean that chimpanzees were to be administered as one might do with research equipment and expendables that need to be cleared away at the conclusion of a study. Regardless, there was no financial support available from the government for retiring chimpanzees from biomedical studies.

In 1978, one of the eight biomedical research institutions, a private pharmaceutical company called Sanwa Kagaku Kenkyusho Co., Ltd. (hereafter, Sanwa), began to receive chimpanzees no longer used in research. Initially, Sanwa's purpose was to rehabilitate chimpanzees after use for biomedical research at other institutions, but in 1982 the company built a new, larger enclosure and altered its focus to breeding chimpanzees for biomedical research. In the 1990s Sanwa began to conduct funded biomedical research and allowed researchers to use its chimpanzees for experiments on hepatitis C, malaria, and embryonic stem cells. The number of chimpanzees at Sanwa continued to grow and reached 117 individuals in 2000, at which time most other biomedical institutions had discontinued the research and housing of chimpanzees.

In 1998, Tetsuro Matsuzawa, along with Juichi Yamagiwa and other colleagues at Kyoto University, organized the voluntary organization Support for African/Asian Great Apes (SAGA) (Matsuzawa 2016). The primary motivation in establishing the organization was to halt invasive biomedical research on chimpanzees in Japan by promoting awareness of the need for better care of chimpanzees and other great apes. SAGA set out three main principles: (1) to assist in the conservation of great apes and their natural habitats; (2) to enhance the quality of life of great apes in captivity; and (3) to prevent the use of great apes as subjects in invasive studies and promote scientific understanding through noninvasive techniques. Some years later, in 2002, Matsuzawa and colleagues established the Great Ape Information Network (GAIN) (Matsuzawa 2016) to develop an up-to-date database on captive great apes in Japan in order to improve communication among institutions housing great apes and to help promote the postmortem utilization of chimpanzees. The use of chimpanzee biomaterials following their natural deaths served as a compromise for researchers who desired biomedical samples from chimpanzees.

In 2000, Sanwa halted chimpanzee breeding for financial reasons. However, in the same year, Sanwa increased its invasive biomedical experiments at a new facility. Led by Matsuzawa, SAGA continued to raise concerns about invasive studies using great apes by holding annual symposiums and appealing to Sanwa to discontinue the practice (Matsuzawa 2016). In 2006, after approximately 10 years of discussions, Sanwa decided to halt biomedical studies with chimpanzees and an arrangement was made for Kyoto University to collaborate in managing Sanwa's chimpanzees and facility. This was a result of the efforts of SAGA and GAIN, which led to increased awareness among researchers and the general public about the welfare and conservation needs of great apes. This era-defining group of Japanese ape researchers, all from Kyoto University, included not only Matsuzawa, but also Yamagiwa, who led gorilla (*Gorilla* sp.) research in the field (Yamagiwa 1983), Sugiyama, who initiated the study of wild chimpanzees in Bossou, Guinea (Sugiyama and Koman 1979), Nishida, who undertook longitudinal observations of wild chimpanzees in Mahale, Tanzania (Nishida 1968), and Kano, who pioneered the study of wild bonobos in the Democratic Republic of Congo (*Pan paniscus*, Kano 1982). Each of these researchers wrote books in Japanese to promote understanding about great apes in Japan. For example, Matsuzawa's many books have been cited in formal textbooks for elementary, junior high, and high schools in Japan (e.g., Matsuzawa 2011), which has led to increased public awareness about apes in captivity and in the wild (Matsuzawa 1998).

In 2007, the Sanwa facility was reformed and renamed as the Chimpanzee Sanctuary Uto (Morimura, Hirata, and Matsuzawa in press; Morimura, Idani, and Matsuzawa 2011). It is the only chimpanzee sanctuary to have ever been established in Japan. When the sanctuary was founded, it housed 79 individuals (40 males, 39 females). It was decided that Sanwa would continue to maintain the facility and the chimpanzees, but that experts at Kyoto University would assume responsibility for management. Sanwa provided financial assistance in hiring professional academic staff to manage the sanctuary. The main purpose for Chimpanzee Sanctuary Uto was to establish care for the physical and psychological well-being of chimpanzees and to promote scientific studies with chimpanzees using noninvasive techniques toward understanding the origin of human nature.

Initial efforts focused on socialization as 22 of the 79 chimpanzees were individually housed in indoor spaces. Ultimately, all the females (except for one blind chimpanzee) were integrated into mixed-sex social groups, and the remaining 14 males were formed into all-male social groups. In 2008, PRI, led by Matsuzawa and colleagues, established a new independent research center called the Wildlife Research Center of Kyoto University (Matsuzawa 2009). In 2008, Chimpanzee

FIGURE 9.3. Aerial view of Kumamoto Sanctuary, Japan.

Sanctuary Uto came under the management of the Wildlife Research Center of Kyoto University, and in 2011 Chimpanzee Sanctuary Uto was renamed Kumamoto Sanctuary (fig. 9.3). Management of the facility, land, and chimpanzees was completely transferred to the Wildlife Research Center of Kyoto University, while some additional financial support was received from Sanwa for the care of the chimpanzees (Tomonaga 2012). In 2012, the final three chimpanzees living in another biomedical research facility were transferred to Kumamoto Sanctuary (Hirata et al. 2012). Thus, there have been no chimpanzees in biomedical research institutions in Japan since 2012. Halting biomedical research with chimpanzees in Japan was not the result of a governmental act, and Japan has no laws specifying the types of research that can be conducted using chimpanzees or other great apes (Bennett and Panicker 2016). Instead, as described above, it was a result of efforts made by researchers and other individuals to sway public opinion and understanding about captive chimpanzees' welfare needs.

THE ORIGINS OF CHIMPANZEE SANCTUARIES IN THE UNITED STATES

By 1997, the laboratory population of chimpanzees in the United States leveled off at about 1,500 individuals. There were an additional 300 in accredited

TABLE 9.2. List of large (>20 chimpanzees) sanctuaries in Japan and the United States of America.

Year established	*Sanctuary*	*Total population (as of January 2020)*
2007	Kumamoto Sanctuary (Japan)	53
1995	Chimp Haven (USA)	300
1997	Save the Chimps (USA)	239
2014	Project Chimps (USA)	59
1993	Center for Great Apes (USA)	31

zoos and several hundred more in unaccredited facilities and under private management. At this time, the US federal government acknowledged that fewer chimpanzees were needed for HIV research than was initially anticipated (National Research Council 1997). This, along with growing ethical concerns about the use of this species in invasive research and the substantial financial burden of maintaining large colonies of chimpanzees, resulted in a careful reevaluation of the population. The National Research Council recommended a cessation of breeding and provided an endorsement against the use of euthanasia as a method of population control, thus the laboratory population was often referred to as a "surplus population."

With these changes in characterizing the laboratory-housed population came the recognition that alternatives may be needed to house chimpanzees no longer needed in research. Several sanctuaries were established in the last decade of the twentieth century, including those that were started with chimpanzees from laboratory settings and from the pet and entertainment industry (table 9.2); this is in contrast to Japan in which there is only one sanctuary for chimpanzees. The two largest efforts to create sanctuary homes for ex-research chimpanzees in the United States were Save the Chimps and Chimp Haven. Save the Chimps was the result of work by Dr. Carole Noon, who successfully negotiated the retirement of hundreds of chimpanzees from the Coulston Foundation in Alamagordo, New Mexico, a biomedical laboratory with a long history of animal cruelty violations. Though Save the Chimps started initially by renovating that laboratory and resocializing the chimpanzees there (as had been the process at Kumamoto Sanctuary in Japan), hundreds of chimpanzees were eventually transferred 1,800 miles away to a new facility in Florida composed of twelve 3-acre islands (Noon 1999).

The other major chimpanzee retirement initiative, and the one we will focus primarily on here, was the result of work by Dr. Linda Brent and Amy Fultz, the cofounders of Chimp Haven in Louisiana (fig. 9.4). With the growing recognition of the "surplus" chimpanzee issue facing laboratories, the

FIGURE 9.4. Aerial view of Chimp Haven, USA.

United States government passed the CHIMP Act (Chimpanzee Health Improvement Maintenance Protection Act) into law in 2000, giving the means for the federal government to provide funding to establish the National Chimpanzee Sanctuary System. Government funds were appropriated to build the chimpanzee sanctuary facilities and to provide 75% of the lifetime care costs for the government-owned chimpanzees retired from research. The remaining 25% of care costs, approximately $4,500 per chimpanzee annually, were to be raised by the organization. At that time, approximately 1,000 government-owned chimpanzees were housed in laboratories, though it was not determined at that time how many would be retired.

Years before the establishment of the CHIMP Act, Brent and Fultz, then chimpanzee researchers at the Southwest Biomedical Foundation in Texas, had recognized the need for a chimpanzee sanctuary and established the nonprofit organization Chimp Haven in 1995. Five years later, the Parish of Caddo donated 200 acres of forested land near the small town of Keithville, Louisiana, on which to build the sanctuary. In 2003, the National Institutes of Health (NIH) selected Chimp Haven, among several candidate organizations, as the group to administer the national chimpanzee sanctuary system (Brent 2004). Two years later, the first retired chimpanzees arrived. Since that time, over 400 chimpanzees have been transferred from laboratories across the United States to retirement at Chimp Haven.

Although hundreds of chimpanzees were retired to Chimp Haven in their first eight years, there remained uncertainty about how many individuals would remain subjects of research in laboratories. In a landmark decision on June 22, 2013, the NIH announced that all but 50 federally owned chimpanzees (intended as a reserve colony) were officially retired, paving the way for more than 300 additional chimpanzees to make their way to Chimp Haven in the future. Several years later, the NIH conceded that no chimpanzees were necessary for further medical research and the idea of maintaining a small reserve colony was eliminated as well. As of the writing of this chapter, Chimp Haven is the largest chimpanzee sanctuary in the world, housing over 300 chimpanzees, although over 200 others remain eligible for future retirement.

Challenges of Maintenance and Sustainability in Japan and the United States

Chimpanzees are incredibly complex animals. They have diverse psychological and behavioral needs to be met, and they are physically powerful animals that require specialized housing to keep them safely contained. All of these factors contribute to the fact that chimpanzees are incredibly expensive to maintain and because of their relatively long life span, the financial burden of their lifetime care is considerable.

In the past, chimpanzees have primarily been maintained either in research or in zoological settings. Those facilities associated with income-generating practices, such as pharmaceutical companies or zoos, had access to a flow of funds that could help offset the care and housing of resident chimpanzees. In contrast, research centers conducting behavioral and cognitive research have typically relied on private or governmental grants to relieve the financial burden. However, because sanctuaries are typically nonprofit organizations and not open to a potentially paying audience of visitors, the means by which sanctuaries maintain sustainable funding for their chimpanzee colonies is a complex issue. Here we describe the differential challenges faced by sanctuaries in Japan and in the United States.

PRIVATE SECTOR SUPPORT

In North America (Canada and United States) and Europe (Austria, Netherlands, Spain, and the United Kingdom), there are several sanctuaries for retired chimpanzees that rely broadly on financial support from the private sector (Ross 2014; Vermij 2003). This is done through a variety of means, including direct mailings, public appeals, and targeted fundraising to private foundations that may have specific interests in chimpanzees or in animal welfare issues. While these approaches have been successful in these countries, relying on the private sector for financial support has proven to be much more difficult in Japan, where philanthropy is less common. The total annual giving in the United States (approximately $400 billion) is considerably more than in Japan (about $7 billion), equating to $1,100 per person in the United States and $22 per person in Japan (Giving USA 2017; Japan Fundraising Association 2017). Put differently, the proportion of the charitable donations to gross domestic product (GDP) in the United States is approximately 1.85% but only 0.22% in Japan (Johns Hopkins Comparative Nonprofit Sector Project 2004). The sources of these philanthropic funds are very different as well. About three-quarters of the giving in the United States derives from individuals whereas only one-third of giving in Japan comes from private donations. Conversely, about two-thirds of giving in Japan is from corporations, whereas this makes up only around 5% of philanthropic gifts in the United States. A culture of donating is not well rooted in Japan, and the nonprofit sector is not yet well developed; and they do not have intermediary support or strong fundraising functions. In contrast, sanctuaries in the United States benefit from the fact that environmental and animal organizations received 3% of all charitable donations, totaling 10.7 billion USD (Giving USA 2017). Stark differences in the income-tax benefits derived from charitable giving in the two countries likely only serve to amplify the contrast in philanthropic climate for chimpanzee sanctuaries. Fundraising for chimpanzee sanctuaries is a difficult proposition for many reasons, including public confidence in sustainability and the lack of public access, but those challenges are especially foreboding in Japan.

One other notable programmatic difference between Kumamoto Sanctuary and Chimp Haven relates to public support; they differ in their ability to accommodate public access to the sanctuary. Zoos in both countries are well-known entities in the local community and some are even recognized nationally or internationally. Ideally, presence in the public mind serves to engender public support and drive philanthropic activities. Traditionally, sanctuaries do not engage in direct public access for a number of reasons, including some philosophical opposition to displaying the animals for fear of negative outcomes. Additionally, most sanctuaries are simply not built with public access in mind, and as such there is no safe and worthwhile means to accommodate visitors to view the animals. Kumamoto Sanctuary is one such example of this, as it originated from a private research facility with no intention of

FIGURE 9.5. Chimpanzees at Chimp Haven extracting food from an artificial termite mound in view of visitors to the sanctuary. During Chimpanzee Discovery Days at Chimp Haven the general public is invited to come to the sanctuary to view the chimpanzees in their forested habitats and learn from staff about the mission and activities of the sanctuary.

serving visiting guests. It is located on a steep incline, without comfortable pathways or areas for most visitors to view the chimpanzees. Chimp Haven is like most other sanctuaries in that it is generally not open to the public, but since its inception the sanctuary has accommodated visitors to view the chimpanzees in their forested habitats on a small number of days each year. These "Chimpanzee Discovery Days" occur four times per year and allow interested people to see the chimpanzees and learn from staff about the sanctuary's mission and activities (fig. 9.5). Hansen et al. (2020) found that, in general, such public programs have little effect on the welfare of the chimpanzees and that public access, though limited, has the potential to result in greater visibility and public support.

GOVERNMENTAL RELATIONS AND OTHER AFFILIATIONS

In the United States, chimpanzees used in biomedical research were split between those owned by the federal government (specifically NIH) and those owned by universities and research centers. In Japan, no such governmental ownership has existed, though governmental grants for chimpanzee management have been a significant source of financial support, as in the United States. Because of the difficulties in raising money from the private sector, and the fact that the Japanese government held no responsibility for the former research chimpanzees in that country, Kumamoto Sanctuary became affiliated with Kyoto University. Without governmental intervention, the sanctuary was free to pursue its goals of providing the chimpanzees with a comfortable retirement and the continuation of noninvasive behavioral and cognitive research (see below).

As noted above, Chimp Haven was selected by NIH as the organization that would receive the federal contract to operate the National Chimpanzee Sanctuary System. As such, the sanctuary is an independent nonprofit organization, but also is closely affiliated as a contractor of the federal government. Such affiliation comes with both benefits and challenges. While the terms of the CHIMP act define that 75% of all chimpanzee care costs are provided by NIH, the sanctuary must independently raise the remaining 25%, as well as those funds to support specific education, research programs, and any new housing that is needed for an expanded population. Additionally, as contractors of the federal government, certain restrictions and limitations are in place that may serve to limit activities in which an otherwise independent organization might prefer to engage.

Research in a Sanctuary Setting

Since the establishment of Kumamoto Sanctuary in 2011, researchers affiliated with Kyoto University have conducted scientific studies to investigate cognitive processes and also have applied science in evaluating welfare conditions (fig. 9.6). For example, scientific studies have revealed apes' long-term memory of single events (Kano and Hirata 2015a), implicit false-belief comprehension (Krupenye et al. 2016), attention to conspecific pictures (Kano and Hirata 2015b), and skin temperature change in response to emotional videos (Kano et al. 2016). Other scientific studies with welfare considerations have included the evaluation of hair cortisol as an indicator of stress (Yamanashi et al. 2016), the change in hair cortisol levels in different housing conditions and social situations (Yamanashi et al. 2016, 2018b), the relationship between the frequency of social play and social tension (Yamanashi et al. 2018a), and the diagnosis and care of a blind chimpanzee with Down syndrome (Hirata et al. 2017). Given the historical background and financial environment in Japan detailed in this chapter, conducting research is very important for the successful and sustained management of Kumamoto Sanctuary.

Research in American chimpanzee sanctuaries is less common than the

FIGURE 9.6. At Kumamoto Sanctuary, touch-screen tasks are offered to a group of chimpanzees living in an indoor-outdoor space.

level maintained at Kumamoto Sanctuary, but is not entirely absent (Ross and Leinwand 2020). Since its inception, and given that its founders were themselves chimpanzee researchers, Chimp Haven has always implemented a program of observational behavioral research with the resident chimpanzees. The primary focus has been on questions directly relating to their care and management in a sanctuary setting, such as studies of tool-use enrichment (Fultz and Case 2018), human-animal relationships (Case et al. 2015), and nest-building (Fultz et al. 2013). Other chimpanzee sanctuaries also participate in some noninvasive studies, though at relatively low frequencies. Though outside the scope of this chapter, it is worth noting that for many years, noninvasive cognitive research has been taking place in African sanctuaries as well (reviewed in Ross and Leinwand 2020).

Conclusions and Future Directions

The history of chimpanzee sanctuaries in Japan and their history in the United States share both distinct commonalities (a seismic shift in sentiment about the use of chimpanzees in invasive medical research followed by widespread retirement efforts) as well as defining differences (cultural context of how sanctuaries are supported and managed, with or without inherent governmental involvement). It is noteworthy that, together, these two countries made up the vast majority of active biomedical use of chimpanzees throughout the last quarter of the twentieth century, and though taking somewhat separate paths, they both exist today as countries that do not partake in such activities. Between the two facilities considered in this chapter, over 500 chimpanzees have transitioned from lives as biomedical research subjects (either active or in reserve) to those of retirees. As time continues to pass, the period of active chimpanzee use will continue to diminish in relative terms and one can imagine a period in the future when it will appear as only a tiny blip in our memory.

With no breeding occurring in Japanese or American sanctuaries (see also Cronin and Ross, chapter 22 this volume), the span of history including chimpanzee sanctuaries is likely to be even shorter than that in which biomedical chimpanzee research took place. Even with a few young chimpanzees (the result of failed birth control) living in today's sanctuaries, most population models predict that the need for such institutions will cease before 2060. So what might the priorities be for these next forty years in which we have this dwindling population of sanctuary-housed chimpanzees?

First and foremost, the culture of sanctuary life revolves around providing the best possible living conditions for the chimpanzees and meeting the behavioral, cognitive, and emotional needs. Sanctuaries are uniquely equipped to study chimpanzee welfare and behavior in ways that have not been possible in traditional zoo and laboratory settings (Brent 2007; Ross and Leinwand 2020). With social groups and physical spaces that much more closely approximate the conditions found for wild chimpanzees, sanctuaries provide a potentially intriguing venue as a form of "hybrid" setting in which the controlled aspects of captive work are melded with the naturalism of in situ conditions. Furthermore, applied noninvasive studies on how chimpanzees think and feel will be useful to guide the care and management of chimpanzee populations that persist into the future, such as those that are part of sustainable breeding programs in accredited zoos (Ross 2010).

Secondly, while the tenure of chimpanzee sanctuaries may extend only over several more decades, there is no reason to believe the need for animal sanctuaries will diminish that quickly for other species. Chimpanzee sanctuaries are among the largest and best organized sanctuaries in the world and may provide excellent models for future sanctuary organizations aimed at providing lifetime care for other species in need. As today's chimpanzee sanctuaries work to answer questions about the challenges and benefits of visitor (Hansen et al. 2020) and research programs (Hansen et al. 2018; Ross et al.

2019) in such settings, we anticipate that progressive programmatic growth in the sanctuary world as a whole will benefit.

Finally, because the sanctuary community is relatively young, there remain opportunities to pursue partnerships with other organizations that can provide specialized expertise. One such recent partnership is the collaboration between Chimp Haven and Lincoln Park Zoo, USA—the first multifaceted programmatic collaboration between an accredited zoo and an accredited chimpanzee sanctuary (Ross et al. 2019). The alliance serves to leverage each institution's programmatic strengths as means for growth in the other. Such programmatic growth is often outside the capabilities of sanctuaries, which are focused almost exclusively on chimpanzee care for large populations. By leveraging the established organizational strengths of Kyoto University and Lincoln Park Zoo's Lester Fisher Center for the Study and Conservation of Apes, both Kumamoto Sanctuary and Chimp Haven, respectively, are provided expanded opportunities for programmatic growth and outreach.

Collaborations such as that between Chimp Haven and Lincoln Park Zoo have proven to be valuable, but even more so might be partnerships between sanctuaries. To date, sanctuaries in Japan, the United States, and around the world have not communicated and collaborated with one another in any substantial way and as such, we propose a global collaboration of sanctuaries in different countries. Such collaboration would allow issues to be addressed that cannot be solved in one particular country and would also allow for the sharing of information, resources, and sustainable strategies that could be adopted between regions. Much like the social and cooperative nature of the apes themselves, such partnerships may prove essential to the fitness and sustainability of the sanctuaries that house them.

Acknowledgments

Work presented in this chapter was financially supported by Great Ape Information Network and JSPS (#26245069, #16H06283, #18H05524, and LGP-U4).

References

Bennett, A. J., and S. Panicker. 2016. "Broader impacts: International implications and integrative ethical consideration of policy decisions about US chimpanzee research." *American Journal of Primatology* 78: 1282–1303.

Brent, L. 2004. "Solutions for research chimpanzees." *Lab Animal* 33 (1): 37–43.

Brent, L. 2007. "Life-long wellbeing: Applying animal welfare science to nonhuman primates in sanctuaries." *Journal of Applied Animal Welfare Science* 10 (1): 55–61.

Case, L., A. Fultz, L. Cohen, and E. Loeser. 2015. "A comparison of proximity values between chimpanzees (*Pan troglodytes*) in a stable and unstable group." *American Journal of Primatology* 77: 91.

ChimpCARE. 2017."Chimpanzees in the U.S." http://www.chimpcare.org/map.

Cohen, J. 2007. "The endangered lab chimp." *Science* 315: 450–52.

The First 100 Chimpanzees. 2006. "The First 100 Chimpanzees." first100chimps.wesleyan.edu.

Fukui, M. 1987. "The present and future of chimpanzees for medical and biological research and experiments in and outside Japan." *Report of the Primate Group of the Hepatitis Study Committee of the Ministry of Health and Welfare, Japan*: 1–9.

Fultz, A., L. Brent, S. D. Breaux, and A. P. Grand. 2013. "An evaluation of nest-building behavior by sanctuary chimpanzees with access to forested habitats." *Folia Primatologica* 84 (6): 405–20.

Fultz, A., and C. Case. 2018. "The development of simulated termite mounds for sanctuary chimpanzees (*Pan troglodytes*): Construction methods and materials." *Animal Keepers Forum* 45 (1): 10–13.

Gardner, R. A., and B. T. Gardner. 1975. "Early signs of language in child and chimpanzee." *Science* 187: 752–53.

Giving USA. 2017. "Giving USA 2017." https://givingusa.org/.

Goodman, S., and E. Check. 2002. "Animal experiments: The great primate debate." *Nature* 417: 684–87.

Graham, C. E. 1981. "A national chimpanzee breeding plan." *American Journal of Primatology* 1: 99–101.

Hansen, B. K., A. L. Fultz, L. M. Hopper, and S. R. Ross. 2018. "An evaluation of video cameras for collecting observational data on sanctuary-housed chimpanzees (*Pan troglodytes*)." *Zoo Biology* 37: 156–61.

Hansen, B. K., L. M. Hopper, S. R. Ross, and A. L. Fultz. 2020. "Understanding the effects of public programs on sanctuary-housed chimpanzees." *Anthrozoös* 33 (4): 481–95.

Hayes, K. J., and C. Hayes. 1951. "The intellectual development of a home-raised chimpanzee." *Proceedings of the American Philosophical Society* 95 (2): 105–9.

Hirata, S., H. Hirai, E. Nogami, N. Morimura, and T. Udono. 2017. "Chimpanzee down syndrome: A case study of trisomy 22 in a captive chimpanzee." *Primates* 58: 267–73.

Hirata, S., T. Udono, M. Tomonaga, and T. Matsuzawa. 2012. "First sky after 30 years: The number of chimpanzees in invasive biomedical institutions has become zero." *Kagaku* 82: 866–67.

Hobson, W. C., C. E. Graham, and T. J. Rowell. 1991. "National chimpanzee breeding program: Primate Research Institute." *American Journal of Primatology* 24: 257–63.

Hua, J., and N. Ahuja. 2013. "Chimpanzee sanctuary: 'Surplus' life and the politics of transspecies care." *American Quarterly* 65: 619–37.

Japan Fundraising Association. 2017. "Giving Japan 2015." http://jfra.jp/en.

Johns Hopkins Comparative Nonprofit Sector Project. 2004. Posted September 1. "Comparative data tables (2004)." http://ccss.jhu.edu/research-projects/comparative-nonprofit-sector-project/.

Kaiser, J. 2015. "An end to US chimp research." *Science* 350: 1013.

Kano, F., and S. Hirata. 2015a. "Great apes make anticipatory looks based on long-term memory of single events." *Current Biology* 25: 2513–17.

Kano, F., and S. Hirata. 2015b. "Social attention in the two species of Pan: Bonobos make more eye contact than chimpanzees." *PLoS One* 10 (6): e0129684

Kano, F., S. Hirata, T. Deschner, V. Behringer, and J. Call. 2016. "Nasal temperature drop in response to a playback of conspecific fights in chimpanzees: A thermo-imaging study." *Physiology & Behavior* 155: 83–94.

Kano, T. 1982. "The social group of pygmy chimpanzees (*Pan paniscus*) of Wamba." *Primates* 23: 171–88.

Knight, A. 2008. "The beginning of the end for chimpanzee experiments?" *Philosophy, Ethics, and Humanities in Medicine* 3: 16.

Köhler, W. 1925. *The Mentality of Apes*. New York: Harcourt Brace.

Koshimizu, K., T. Magaribuchi, M. Ito, K. Uchizono, and T. Shikata. 1977. "Rearing and management of chimpanzees for experimental infection with hepatitis B virus." *Experimental Animal* 26: 51–64.

Krupenye, C., F. Kano, S. Hirata, J. Call, M. Tomasello. 2016. "Great apes anticipate that other individuals will act according to false beliefs." *Science* 354: 110–14.

Matsubayashi, K. 1993. "Problems in the experimental use, breeding and care of chimpanzees." *Primate Research* 9: 145–49.

Matsuzawa, T. 1994. "Field experiments on use of stone tools by chimpanzees in the wild." In *Chimpanzee Cultures*, edited by R. W. Wrangham, W. C. McGrew, F. B. M. de Waal, and P. Heltne, 351–70. Cambridge, MA: Harvard University Press.

Matsuzawa, T. 1998. "Chimpanzees in Japan need help." *Pan Africa News* 5: 20–21.

Matsuzawa, T., ed. 2001. *Primate Origins of Human Cognition and Behavior*. Tokyo: Springer.

Matsuzawa, T. 2003. "The Ai project: Historical and ecological contexts." *Animal Cognition* 6: 199–211.

Matsuzawa, T. 2006. "Sociocognitive development in chimpanzees: A synthesis of laboratory work and fieldwork." In *Cognitive Development in Chimpanzees*, edited by T. Matsuzawa, M. Tomonaga, and M. Tanaka, 3–33. Tokyo: Springer.

Matsuzawa, T. 2009. "Tetsuro Matsuzawa." *Current Biology* 19: R310–12.

Matsuzawa, T. 2011. *Power of Imagination: What Chimpanzees Tell Us about Human Mind*. Tokyo: Iwanami Shoten Publishers.

Matsuzawa, T. 2016. "SAGA and GAIN for great apes." *Primates* 57: 1–2.

Matsuzawa, T., M. Tomonaga, and M. Tanaka., eds. 2006. *Cognitive Development in Chimpanzees*. Tokyo: Springer.

Morimura, N., S. Hirata, and T. Matsuzawa. In press. "Challenging cognitive enrichment." In *Primate Welfare*, edited by A. Weiss. Oxford: Oxford University Press.

Morimura, N., G. I. Idani, and T. Matsuzawa. 2011. "The first chimpanzee sanctuary in Japan: An attempt to care for the 'surplus' of biomedical research." *American Journal of Primatology* 73: 226–32.

National Research Council. 1997. *Chimpanzees in Research: Strategies for Their Ethical Care, Management, and Use*. Washington, DC: National Academies Press.

Nishida, T. 1968. "The social group of wild chimpanzees in the Mahali Mountains." *Primates* 9: 167–224.

Noon, C., 1999. "Chimpanzees and retirement." *Journal of Applied Animal Welfare Science* 2 (2): 141–46.

Ochiai, T., K. Watanuki, T. Udono, N. Morimura, S. Hirata, M. Tomonaga, G. Idani, and T. Matsuzawa. 2015. "The history of captive chimpanzees (*Pan troglodytes*) in Japan 1920–1950." *Primate Research* 31: 19–29.

Ross, S. R. 2010. "How cognitive studies help shape our obligation for the ethical care of chimpanzees." In *The Mind of the Chimpanzee: Ecological and Experimental Perspectives*, edited by E. V. Lonsdorf, S. R. Ross, and T. Mastsuzawa, 309–19. Chicago: University of Chicago Press.

Ross, S. R. 2014. "Captive chimpanzees." In *The Ethics of Captivity*, edited by L. Gruen, 57–76. New York: Oxford University Press.

Ross, S. R., B. K. Hansen, L. M. Hopper, and A. Fultz. 2019. "A unique zoo-sanctuary collaboration for chimpanzees." *American Journal of Primatology* 81 (5): e22941.

Ross, S. R., and J. Leinwand. 2020. "A review of research conducted in primate sanctuaries." *Biology Letters* 16 (4): 20200033.

Sugiyama, Y. 1985. "Importation and experimental use of chimpanzees." *Kagaku* 55: 127–30.

Sugiyama, Y., and J. Koman. 1979. "Tool-using and -making behavior in wild chimpanzees at Bossou, Guinea." *Primates* 20: 513–24.

Takashima, H. 1955. *A Story about Importation of Animals*. Tokyo: Gakufu-shoin.

Tomonaga, M. 2012. "Welcome to Kumamoto Sanctuary." *Kagaku* 82: 38–39.

Tomonaga, M., T. Matsuzawa, and M. Tanaka, eds. 2003. *Development of Cognition and Behavior in Chimpanzees*. Kyoto: Kyoto University Press.

Vermij, P. 2003. "Europe's last research chimps to retire." *Nature Medicine* 9: 981.

Wadman, M. 2011. "Animal rights: Chimpanzee research on trial." *Nature* 474: 268–71.

Watanuki, K., T. Ochiai, S. Hirata, N. Morimura, M. Tomonaga, G. Idani, and T. Matsuzawa. 2014. "Review and long-term survey of the status of captive chimpanzees in Japan in 1926–2013." *Primate Research* 30: 147–56.

Yamagiwa, J. 1983. "Diachronic changes in two eastern lowland gorilla groups (*Gorilla gorilla graueri*) in the Mt. Kahuzi Region, Zaire." *Primates* 24: 174–83.

Yamamori, E. 1992. "Dirge of laboratory chimpanzees: After-care of chimpanzees in a breeding colony with deficit balance." *AERA* (February 25): 36.

Yamanashi, Y., E. Nogami, M. Teramoto, N. Morimura, and S. Hirata. 2018a. "Adult-adult social play in captive chimpanzees: Is it indicative of positive animal welfare?" *Applied Animal Behaviour Science* 199: 75–83.

Yamanashi, Y., M. Teramoto, N. Morimura, S. Hirata, J. Suzuki, M. Hayashi, K. Kinoshita, M. Murayama, and G. Idani. 2016. "Analysis of hair cortisol levels in captive chimpanzees: Effect of various methods on cortisol stability and variability." *MethodsX* 3: 110–17.

Yamanashi, Y., M. Teramoto, N. Morimura, E. Nogami, and S. Hirata. 2018b. "Social relationship and hair cortisol level in captive male chimpanzees (*Pan troglodytes*)." *Primates* 59 (2): 145–52.

Yerkes, R. M. 1925. *Almost Human*. New York: Century Company.

PART FOUR

Communication

10

Gestural Communication in the Great Apes: Tracing the Origins of Language

CATHERINE HOBAITER

Introduction

Whether spoken, signed, or written, language is at the heart of human behavior. The accumulation of cultural change that has taken us from using knapped stone tools to space travel in an evolutionary heartbeat appears almost impossible without it. As we learn and develop our language skills, they in turn shape us: acquiring a new word changes our perception of the world around us (Winawer et al. 2007). So while philosophers and scientists continue to argue about what language is and how it evolved (e.g., Everaert et al. 2017; Fitch 2017; Lieberman 2015), there is general agreement that language is the cognitive capacity that speaks directly to what makes us human. Nevertheless, in exploring the evolutionary origins of language, much has been made of the claim that language is the most complex system of animal communication (e.g., Chomsky 2017; Everaert et al. 2017; Hauser, Chomsky, and Fitch 2002; Tomasello and Call 1997).

But what do we mean by complex? A species' system of communication is as "complex" as it needs to be to meet the adaptive demands of the species' niche. Many species across the animal kingdom employ rich repertoires of vocalizations, facial expressions, body postures, olfactory cues, gestures, and more to communicate nuanced information. And, like us, other species use the structure of signals—for example the order in which they're produced—to encode additional information. The phonological-syntax of many bird species allows recombination of a fixed repertoire of notes into different songs (e.g., Engesser, Ridley, and Townsend 2016; Kakishita et al. 2009; Nelson, Hallberg, and Soha 2004; Suzuki, Wheatcroft, and Griesser 2016; Wohlgemuth, Sober, and Brainard 2010). Similarly, the combinatorial information encoded in some monkey species' alarm calls is based on the order in which the calls are produced (e.g., Arnold and Zuberbühler 2006, 2008; Candiotti, Zuberbühler, and

Lemasson 2012; Ouattara, Lemasson, and Zuberbühler 2009a, 2009b; Schlenker et al. 2017). Even linguistic cultures are not unique. Like humans, other species acquire the regional dialects of their species' songs by learning socially from others around them (for example, orcas, *Orcinus orca*: Deecke, Ford, and Spong 2000; Ford 1991; sperm whales, *Physeter macrocephalus*: Rendell and Whitehead 2003; saddlebacks, *Philesturnus carunculatus*: Jenkins 1978; and white-crowned sparrows, *Zonotrichia leucophrys*: Baker 1975; Nelson, Hallberg, and Soha 2004).

Yet, while a rich source of information with a social function, many species' signals exist as fixed responses of the signaler to cues from stimuli and context, and are produced irrespective of a recipient's attention, knowledge, or even presence (e.g., Cheney and Seyfarth 1990; Marler 1961; Schel et al. 2013; Seyfarth and Cheney 2003). Humans produce these types of signals too (Bates, Camaioni, and Volterra 1975); our sighs, oohs, ahhs, and groans encode a rich array of information (Cowen et al. 2018). We give an involuntary yelp picking up a too-hot pan from the stove, and any potential recipients in the room receive useful information from this signal: the pan is hot! But we did not yelp with the goal of communicating—we would have yelped whether or not someone else was there. Language is different. We choose whether or not to tell someone who was out of the room that the pan is hot.

With language we do more than broadcast information, we intend to communicate a goal to a partner who we recognize as having their own behavior, goals, and knowledge. The intentional goal-directed nature of human language takes it beyond information: it has meaning (Dennett 1987; Grice 1957). Rather than fixed information encoded in the physical form of the signal, the relationship between spoken or signed words and their meaning is a function of the signaler's intention. This fundamental property of language has very rarely been observed in other species (Rendall, Owren, and Ryan 2009; Seyfarth and Cheney 2003). There are exceptions. One of the alarm-calls given by chimpanzees (*Pan troglodytes*) is produced taking into account the presence and response of the recipient (Crockford, Wittig, and Zuberbühler 2015, 2017; Schel et al. 2013), and may be adjusted to their knowledge of the danger (Crockford, Wittig, and Zuberbühler 2017; Crockford et al. 2012, see also Townsend, Watson, and Slocombe, chapter 11 this volume). Grouper fish (*Plectropomus pessuliferus marisrubri*) employ a referential body gesture in inter-species hunting behavior with moray eels (Vail, Manica, and Bshary 2013). Wild ravens (*Corvus corax*) use a "showing" gesture with their head when offering food to their partners (Pika and Bugnyar 2011). With improving methods we are able to detect intentional communication more widely. Nevertheless, these cases typically represent individual signals used by a spe-

FIGURE 10.1. Chimpanzees, and other great apes, employ their large repertoires of 60–80 gestures to communicate everyday goals. These gestures represent an intentional *system* of communication used by individuals of all age-sex groups, across the full range of behavioral contexts.

cies in a highly specified way, often associated with an evolutionary urgent need (food, danger).

Great ape gestures are different. Great apes (the genera *Gorilla*, *Pan*, and *Pongo*) employ their large repertoires of 60–80 gestures intentionally to communicate everyday goals (fig. 10.1; e.g., Bard and Vauclair 1984; Fröhlich, Wittig, and Pika 2016; Genty et al. 2009; Graham, Furuichi, and Byrne 2017; Hobaiter and Byrne 2011a, 2011b, 2014; Knox et al. 2019; Liebal, Pika, and Tomasello 2006; Perlman, Tanner, and King 2012; Pika, Liebal, and Tomasello 2003, 2005; Plooij 1978; Pollick and de Waal 2007; Roberts, Vick, and Buchanan-Smith 2012a; Tanner and Byrne 1999; Tomasello, Gust, and Frost 1989; Tomasello et al. 1985). If this is true, great ape gestures represent the only nonhuman *system* of communication described, to date, in which we can explore not just information exchange but meaning in the linguistic sense. It is a significant claim, and it rests on our ability to detect goals or intentions in another species.

The Evidence for Intentionality

Wild apes' use of gestures and vocalizations was noted in the first field studies of chimpanzees and gorillas (Goodall 1968; Plooij 1978; Schaller 1963), but was described in a captive chimpanzee as early as 1935 (Ladygina-Kohts [1935]

2002). Detailed case studies of individual gesture types in wild chimpanzees, such as leaf-clipping (Nishida 1980), the grooming hand-clasp (McGrew et al. 2001), and the directed scratch (Pika and Mitani 2006), sparked wider interest in the field; but it was the study of gesture as a system of communication in captive apes that allowed substantial progress in our understanding of the intentional nature of ape gesture (Bard 1992; Bard et al. 2014; Cartmill and Byrne 2007, 2010; Genty et al. 2009; Leavens and Hopkins 1998; Leavens, Hopkins, and Bard 1996; Liebal, Pika, and Tomasello 2006; McCarthy, Jensvold, and Fouts 2012; Pika, Liebal, and Tomasello 2003, 2005; Pollick and de Waal 2007; Tanner and Byrne 1993, 1999; Tomasello, Gust, and Frost 1989; Tomasello et al. 1985).

This body of work was initiated by Tomasello and Call in the 1980s, inspired by early field work with chimpanzees (Plooij 1978) and studies of communication in pre-verbal human infants (Bates, Camaioni, and Volterra 1975). At around the same time, the philosopher Dennett (1983, 1987) was operationalizing communicative intent in nonhuman animals using a Gricean framework. Grice argued that within linguistic communication both the signaler and the recipient take into account each other's mental states of mind (Grice 1957). Early human developmental research explored similar distinctions between "perlocutory" acts, in which a signal may have an effect on a recipient but without any evidence that this effect is intended by the signaler, and intentional "illocutory" acts, in which an infant employed a conventionalized signal to order a socially recognized goal (Bates, Camaioni, and Volterra 1975). As with other animals, we have no way to interrogate a one-year-old girl about the meaning of her signals; instead we must infer her goal from cues in her observable behavior. Does she check whether or not her audience can see her? Or wait for their response, and then persist if she doesn't receive one? Take another human example: I walk into a busy bar and see you already being served. I would like you to get me a beer, but I can't wave to you as you're facing the wrong direction. Others in the pub would be able to receive a wave signal, but I have a specific recipient in mind: you. So instead I should call out or tap you on the shoulder. The stimulus leading to my signal production (my desire for a pint) is a steady state, but when signaling I don't opt for a continuous string of "beer-beer-beer-beer." Instead I ask, and then wait to see what you do. If you change your behavior in a way that indicates you understand me, I stop signaling (despite the fact that my desire for a beer will not be satisfied until I get to drink it). If your behavior suggests you haven't understood, I would persist, or even perhaps elaborate—perhaps pointing to the beer tap. All of these are behavioral cues that indicate that my signal is an intentional one, directed to a specific audience with a particular goal in mind. We see these same behavioral cues in the intentional gestural communication of other apes.

A system of communication that shares an evolutionary foundation with language must employ *intentional* signals. In the search for the evolutionary scaffold on which human language was built, this represented a key missing piece of the puzzle that, despite decades of research, had remained absent in studies of nonhuman communication (Rendall, Owren, and Ryan 2009; Seyfarth and Cheney 2003). Tomasello and colleagues employed ape behavior, such as response-waiting and gaze-alternation, as a means to distinguish intentional gestural communication (Tomasello et al. 1985). From their work, and the subsequent studies of ape gesture it inspired, researchers have found abundant evidence for intentional signal use in gestural communication, irrespective of the study site, the species, or the specific methodology used. Even with a strict definition in which each individual gestural token is required to have evidence of intentional use in order to be analyzed, individual studies are left with thousands of cases to analyze (Genty et al. 2009; Graham, Furuichi, and Byrne 2017; Hobaiter and Byrne 2011a). As in the busy bar example above, one line of evidence showing that gestural signalers take their recipient into account is in apes' selection of gestures based on their recipient's visual attention. Gesture is unusual (as compared to vocalization, or facial expression) in that it contains a variety of different modalities in which information can be communicated—all gestures are visible, but signalers can choose to employ gestures that also incorporate audible and/or tactile information. Signalers exploit this flexibility, using visual-silent gestures toward recipients who are looking in their direction, and using audible or contact gestures to those who are not (Genty et al. 2009; Hobaiter and Byrne 2011a; McCarthy, Jensvold, and Fouts 2012; Pika and Mitani 2006; Pika, Liebal, and Tomasello 2003; Tanner and Byrne 1996; Tomasello and Call 2006, 2007), or, alternatively, moving into the recipient's line of sight (Liebal, Call, and Tomasello 2004; Liebal et al. 2004).

It is this widespread intentional use across a large repertoire that makes gesture, to date, unique in nonhuman animal communication. While the same operational definition of intentional signal use has been applied successfully to the nonhuman signals described above (raven gesture: Pika and Bugnyar 2011; grouper gesture: Vail, Manica, and Bshary 2013; chimpanzee vocalization: Schel et al. 2013), there is no similar evidence of a large repertoire of intentional signals, employed every day, across contexts, outside of great ape gesture and human language. Today, ape gesture is, by definition, intentional. A typical definition might read: a "discrete, mechanically ineffective physical movement of the body *observed during periods of intentional communication*" (Hobaiter and Byrne 2011a). Specific studies vary on whether gestures are limited only to movements of the hands (Leavens and Hopkins 1998; Leavens, Russell, and Hopkins 2010; Pollick and de Waal 2007; Roberts, Vick, and Buchanan-Smith 2012a), or encompass body postures and some facial actions (Cartmill and Byrne 2007; Genty et al. 2009). However, all definitions require that the signals be used within intentional communication to be coded as gesture.

A brief aside here. While I argue that our understanding of great ape gesture is important for our understanding of the evolutionary trajectory of language, and that our hominin ancestors likely communicated a set of language-like meanings using gesture, it does not *necessarily* follow, as some have argued (see Condillac 2001; Corballis 2002, 2017; Hewes 1973, 1999), that language emerged from a gestural origin (i.e., the "gesture-first" hypothesis). In many ways, the distinction between gestural and spoken modalities is an unhelpful one. Where do we categorize an orang-utan vocalization when its sound is modified by a manual gesture (Hardus et al. 2009)? Voiceless-calls, such as blown raspberries, are under voluntary control (Leavens et al. 2004), and may serve to promote social bonding in chimpanzees (Watts 2016). Lip-smacking in geladas (*Theropithecus gelada*) shows similarities to human speech in terms of rhythm and development (Ghazanfar et al. 2012). So where does gesture end and speech begin? The unhelpful nature of making a distinction on the basis of modality is made more apparent if we compare paralinguistic gesture to sign-language. Sign language is language in its complete form (see, for example: Bellugi and Klima 1979; Stokoe 1960); it is not the modality that provides a distinction between gesture and language, but the way in which the two systems are used—for example, in the intentionality of human language (whether spoken or signed) as opposed to human gesture.

While the weight of current evidence for intentional use in great ape communication is on the gestural end of the spectrum, as we have seen (Crockford, Wittig, and Zuberbühler 2015, 2017; Schel et al. 2013), there is no clear boundary at which intentional use stops. One way to address this issue, is to acknowledge that in addition to there being no clear boundary, modeling the evolutionary origins of language is not a zero-sum game. As Kendon (2017) points out, "You can build a language in any modality." Positive evidence for language-like capacities in gesture, does not—on its own—indicate that vocalization was not a component of the system of communication from which language evolved. Indeed, there is an increasing body of work focusing on the combined use of different channels of ape communication, such as gesture and vocalization (Genty and Zuberbühler 2014; Hobaiter, Byrne, and Zuberbühler 2017; Leavens and Hopkins 2005; Leavens, Russell, and Hopkins 2010;

Liebal, Waller, and Slocombe 2011, Liebal et al. 2013; Pollick, Jeneson, and de Waal 2008; Wilke et al. 2017). With that in mind, I focus here on the evidence for what gesture *can* tell us about language-like capacities in chimpanzee and bonobo (*Pan paniscus*) communication, but without making assumptions about what other modalities did or did not contribute to language evolution.

After all, the definition for intentional communication is not without problems. Gricean communication appears to require the ability to make complex inferences about others' mental states for even simple requests (e.g., "I want you to recognize that I intend that you pass me the cake"; see for example: Scott-Phillips 2014, 2015). If true, we are faced with the conundrum that a two- or three-year-old child, capable of using language, may not be—explicitly—capable of attributing these mental-states (Liddle and Nettle 2006; Wimmer and Perner 1983). How do children learn to use language before they have the cognitive capacity to understand it? It may be that the standard interpretation of Gricean intent has overstated the cognitive skills needed to acquire intentional language (Gómez 1994; Moore 2016, 2017; Townsend et al. 2017). Alternatively, infants and great apes may employ non-Gricean forms of intentional communication (e.g., Bar-On 2013; Leavens, Bard, and Hopkins 2017). To date, alternative definitions for intentional signal use tend to suffer from the same issue as the original Gricean definition: a bias toward recording visual attention to detect directedness. For example, Townsend et al. (2017) argued that we can move beyond Gricean frameworks, but still required that a signal be "directed to a recipient," and interpreting the cognitive intention to "direct" from observable behavior is not straightforward. If the information in the signal is primarily visual there are reliable behavioral indications: for example, the signaler monitors the recipient's viewpoint and adjusts their signaling to match it. But if the information in the signal is primarily audible, as is the case in vocalizations and long-distance gestures (such as buttress-drumming), for example, how do we measure this? The signaler has no need to check the recipient's auditory state of attention; as long as they are aware of the recipient's location they are aware of whether the signal can be received. As a result we probably underestimate the frequency with which nonhuman animals employ intentional communication.

The absence of any strong evidence to date in areas such as cetacean or elephant communication, or in a wider range of great ape vocalizations, likely has more to do with our method of determining intentional use, than with the true absence of intentional communication in these species. Faced with the increasing evidence for the intentional use of at least some signals across a diverse range of species and orders (primates, fish, birds, etc.), we have two options. We can shift the goal posts, arguing that our current behavioral measures of cognitive intent are insufficient—adjusting them to provide a new definition of what intentional communication is. This approach may work for a little while, but I suspect that over time more and more cases, in more and more species, will pass whatever new criteria are set, and we will have devoted a substantial amount of effort to something that doesn't necessarily allow us to answer the question we are often using these criteria to address: whether and how human language is different from other species' communication. Alternatively we can accept that the capacity for this form of Gricean intentional communication is likely more widespread than previously considered and move away from tick-lists of language features, instead asking what language and other nonhuman systems of communication are for. Language, like other systems of communication, is adapted to the socio-ecological niche in which it is used. Describing the tool kit of properties in language or ape gesture may never be enough to tell us what they are used for: to explore that, we have to look at how they are used.

What Does It All Mean?

Given the evidence that great ape gestures are employed intentionally with language-like meaning, perhaps the most obvious question is: what do they mean? But this is not a straightforward question to answer. It can be challenging enough to consider what someone else *means* when we are from the same species and culture, and speak the same language! It requires us to assess not only the words used, but also the speaker's goal in using them (Smith 1965). While the intentional goal-directed nature of ape gestures was described in the 1980s (Call and Tomasello 2007; Tomasello, Gust, and Frost 1989; Tomasello et al. 1985), knowing that gestures are meaningful is not the same as knowing what a gesture means. Over the next three decades a gesture's goal was typically classified by recording the behavioral context in which it was used: feeding, travel, play, etc. (e.g., Liebal, Pika, and Tomasello 2006; Pika, Liebal, and Tomasello 2003, 2005; Plooij 1978; Pollick and de Waal 2007; Tomasello, Gust, and Frost 1989; Tomasello et al. 1985). This method allows clear categorization on the basis of observable behavioral data; but it is problematic in terms of assessing the signaler's intended goal.

One feature of gestural communication that was highlighted in early descriptions was its openness or flexible use: one gesture appeared to be used to achieve several goals, and several gestures appeared to achieve the same goal (Call and Tomasello 2007). In his studies of gestural ontogeny Plooij (1978) described this openness as "one of the most characteristic design features of human language" (p. 127). However, measuring the context in which a gesture occurs does not measure the signaler's intended goal, and risks exaggerating the flexibility of individual gestures. Imagine a gesture with a single fixed meaning: "Stop!" A signaler could use this when playing, fighting, or feeding, and the gesture's meaning would appear highly flexible, whereas the signaler's intention is very specific.

Another approach to investigating meaning is to assess the recipient's response to the gesture. Used to explore the flexibility of gestures, this method showed that while the outcome appeared to be independent of the context in which the gestures were used, it was sensitive to differences in social rank between the signaler and recipient (Roberts, Vick, and Buchanan-Smith 2012a). Given that the signaler intends to change the recipient's behavior, their behavior should indicate the signaler's goal. Here, we almost approach meaning in the linguistic-sense, but not quite. The recipient's response could be a refusal, a misunderstanding, or the start of a negotiation. We need one more step in the process—we can define a signaler's goal by looking at the recipient's behavior that *stops the signaler from continuing to signal* (the "apparently satisfactory outcome"; Cartmill and Byrne 2010; Genty et al. 2009; Graham et al. 2018; Hobaiter and Byrne 2014; Knox et al. 2019). In the busy bar described above, what finally stops me asking for a beer is when your behavior indicates that you understand my goal. A chimpanzee signaler's goal is to use their gestures to change the behavior, or perhaps even the understanding (Cartmill and Byrne 2010), of their recipient. Of course, in any one communication there is still the possibility that there has been miscommunication, or misunderstanding, and that the signaler gives up, not because they achieved their goal, but because they decided not to pursue it. However, over hundreds of cases from dozens of dyads, across behavioral contexts, and days, and partners, we can examine the data for *consistent patterns of behavior* that indicate what a gesture means.

Data from across ape species (Cartmill and Byrne 2010; Genty et al. 2009; Genty, Neumann, and Zuberbühler 2015a; Graham et al. 2018; Hobaiter and Byrne 2014) have shown that gesture types are associated with specific meanings, and that these meanings do not vary between signalers. In chimpanzees, the 70–80 gestures identified so far have been shown to be used to achieve 19 goals, four of which are play related. However, the redundancy within the system is not evenly distributed, with some goals achieved by a single gesture type, and others achieved by several (Hobaiter and Byrne 2014). It may be that, like some words, some gestures can have a range of meanings that are disambiguated by the social or environmental context in which they are used. Whether or not the word "bark" makes you think of a dog vocalizing, the skin of a branch or tree, or a sailing ship (barque), depends on when you use it, with whom you're talking, and which other words and signals you combine it with.

Across ape species, the repertoires of gestures an ape employs over its lifetime appear to be largely species-typical, with significant overlap between all species and the largest overlap between those most closely related (fig. 10.2; Byrne et al. 2017; Graham, Furuichi, and Byrne 2017; Hobaiter and Byrne 2011a). Within the *Pan* genus, chimpanzees and bonobos share 88–96% of gesture types (Graham, Furuichi, and Byrne 2017; Graham et al. 2018; Schneider, Call, and Liebal 2012). In fact, we see many of the same shared *Pan* gestures in the communication of one- to two-year-old children (Kersken et al. 2018), who are on the cusp of language use, but who don't yet have a well-developed repertoire of words with which to communicate their goals. Within chimpanzees and bonobos, both the range of meanings they use gestures to convey, and the specific meaning of each gesture type, also show large overlap. Variation between them tends to be in the frequency with which a meaning is associated with a gesture rather than the type of meaning. For example: the "Directed push" gesture is used by both chimpanzees and bonobos to request that an infant "Climb on me." While this was the primary goal for bonobo signalers, it was only a secondary goal in chimpanzees, who employed this gesture more commonly to request a "Reposition" during grooming (Graham et al. 2018).

While valuable, the use of apparently satisfactory outcomes to discriminate gesture meanings has restrictions. It limits us to describing meanings that require a change in the behavior of the recipient. We are able to recognize an imperative demand to "Stop that!" or "Come here!" but not a declarative: "What lovely fig tree." Nevertheless, because we are able to assign a meaning to the majority of the gestures we see used, we are reasonably sure that we are not missing a large set of alternative types of meaning in great ape gesture (Graham et al. 2018; Hobaiter and Byrne 2014). Similarly, the regular spontaneous use of pointing in captive apes (Bohn, Call, and Tomasello 2015; Leavens and Hopkins 1998; Leavens, Hopkins, and Bard 1996; Moore, Call, and Tomasello 2015), and the few examples of potentially referential gestures in wild populations (chimpanzees: Hobaiter, Leavens, and Byrne 2013; Pika and Mitani 2006; bonobos: Genty et al. 2014), are typically used in imperative rather than declarative communication (Tomasello 2007; although cf. Lyn, Russell, and Hopkins 2010 for comprehension of declarative communication in apes). However, the method also overlooks imperative meanings that represent a request *not* to do something. "Don't hit me" or "Stay where you are" are indistinguishable from the recipient refusing to respond. The specific exploration of failed signals, and the integration of information from the social

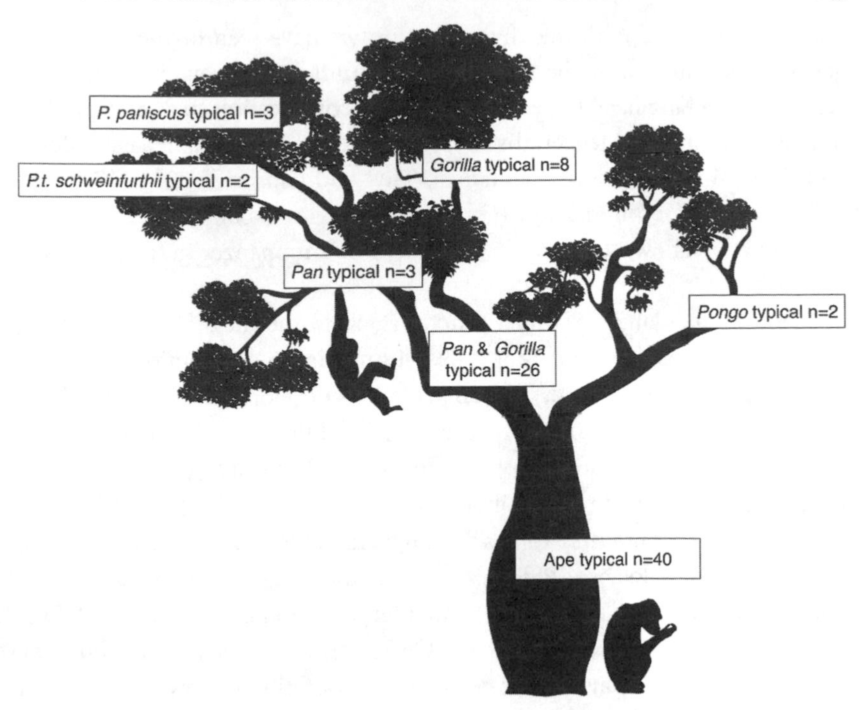

FIGURE 10.2. The overlap of gesture types across the great ape family. Large proportions of the great ape repertoire are shared across species and families, with the greatest overlap between the most closely related species (Graham et al. 2017; Byrne et al. 2017). Some species remain un- or under-studied (e.g. orangutans, mountain gorillas); but further research and improving methods are providing more detailed resolution (Kersken et al. 2018; Knox et al. 2019).

context with other cues and signals such as vocalization or facial expression, may allow more nuanced discrimination of the different meanings that chimpanzees use their gestures to communicate.

Beyond Repertoires and Meaning

At present, support for the hypothesis that language emerged from a system that included a repertoire of gestures for the day-to-day communication of goals, rests heavily on the evidence for language-like meaning in a large and diverse repertoire of signals. What other evidence exists to support the role of gesture within language evolution? Over the past five years, we have seen an increasing diversity of studies across, not only species, but also populations and communities of chimpanzees and bonobos; and increasing data from wild populations, under which communication is deployed in the natural conditions that require its full expression (Hobaiter and Byrne 2011a; Seyfarth and Cheney 2017). With this diversification we have seen further hints that gestural communication shared some of the fundamental organizational principles seen in language. For example: laws of compression such as Zipf's law of abbreviation and Menzerath's law, found across human languages (Bentz and Ferrer-i-Cancho 2016; Piantadosi, Tily, and Gibson 2011), and in nonhuman vocal communication (Gustison et al. 2016; Luo et al. 2013), have recently been shown to apply to sequences of chimpanzee gesture (Heesen et al. 2019).

Gesture, like language, must function within the messy "real world," in which individuals have different states of knowledge or attention, social relationships, and goals. And, like language, the development of the available species-typical system, and its expression by different individuals in their day-to-day communication, appears to be highly flexible (e.g., Bard 1992; Bard et al. 2013, 2014, 2019; Fröhlich, Wittig, and Pika 2016; Fröhlich et al. 2016; Halina, Rossano, and Tomasello 2013; Hobaiter and Byrne 2011b; Liebal, Pika, and Tomasello 2006; Perlman, Tanner, and King 2012; Pika, Liebal, and Tomasello 2003, 2005; Plooij 1978; Schneider, Call, and Liebal 2012, 2017; Tanner and Byrne 1993, 1999; Tomasello, Gust, and Frost 1989; Tomasello et al. 1985). Across individuals, we see considerable variation in how the system of gestures is deployed (Bard et al. 2013). Similarly, with substantial redundancy available in the system, the selection of specific gesture types from within the large available repertoire varies between individuals (Fröhlich, Wittig, and Pika 2016; Genty, Neumann, and Zuberbühler 2015a; Halina, Rossano, and Tomasello 2013; Roberts and Roberts 2016), at different periods within an individual's lifetime (Hobaiter and Byrne 2011b; Pika and Fröhlich 2018), and depends on both the current recipient and the social history of previous partners (Bard et al. 2013, 2019; Fröhlich et al. 2016, 2017; Genty, Neumann, and Zuberbühler 2015b; Roberts and Roberts 2016; Schneider, Call, and Liebal 2012). These studies highlight gesture's flexibility, and in particular the role of the social environment in how gestures are used. Recent studies of contact gestures showed striking variation in form and the location of contact to the recipient (e.g., Touch; Bard et al. 2019), as well as tuning to the physical and social context (e.g., Push; Perlman, Tanner, and King 2012), highlighting the scope of even single gesture types for nuanced communication.

Here, chimpanzees and bonobos are of particular interest. All apes use large repertoires of gestures across a range of contexts to express a range of meanings; but the social world in which chimpanzees and bonobos communicate every day is different from that of the other great apes. Their large communities can contain over a hundred individuals, and there is a complex web of distinct biological and social relationships. While bonobos and

West African chimpanzees are more cohesive than East African chimpanzees (Chapman, White, and Wrangham 1994; Lehmann and Boesch 2004), both species employ fission-fusion social behavior, so while they may see and interact with some individuals almost every day, others may be absent for weeks, months, or even years. In this social world, where nuanced social relationships and ranks impact both male and female fitness (Ishizuka et al. 2018; Newton-Fisher et al. 2010; Pusey, Williams, and Goodall 1997), who knows what about whom, and when they find it out, may be crucial information. Research on the "victim screams" that chimpanzees produce when they are threatened or attacked by another chimpanzee showed that the victim regulated, and perhaps even "exaggerated," the information in their scream about the severity of the attack depending on the relative rank of individuals in their audience to the attacker (which impacts his or her ability to intervene in the fight; Slocombe and Zuberbühler 2007). Fission-fusion social behavior within a dense rainforest means that visual signals can be limited to a specific nearby audience, but audible ones, whether vocal or gestural, may be received by individuals that are out-of-sight to the signaler. In addition, the acoustic information in a vocal signal includes the signaler's identity (e.g., Crockford et al. 2004; Herbinger et al. 2009; Kojima, Izumi, and Ceugniet 2003), whereas only chimpanzee drumming gestures have been suggested to do the same (Arcadi, Robert, and Mugurusi 2004). Revealing your identity, as well as the information in your signal, to an out-of-sight audience could be both a benefit and a cost. Chimpanzees appear to exploit these different types of information, taking into account different audiences, including third-party "eavesdropping," and select gesture types in such a way that they communicate publicly or limit their gestures to a "private" audience (Hobaiter and Byrne 2012; Hobaiter, Byrne, and Zuberbühler 2017).

Chimpanzees and bonobos diverge less than half of one percent in their genotype (de Manuel et al. 2016), but the variation between the two species in their social phenotype is profound (see Behringer et al., chapter 2 this volume; Taglialatela et al., chapter 4 this volume; Yamamoto, chapter 14 this volume). They show striking differences in social hierarchy (Furuichi 2011; Surbeck, Mundry, and Hohmann 2011), aggressive interactions (Tokuyama and Furuichi 2016), xenophobia (Ryu, Hill, and Furuichi 2015; Sakamaki et al. 2015), and socio-sexual behavior (Furuichi 1987; Woods and Hare 2011), and more subtle differences in social behavior such as grooming (Sakamaki 2013) or in their aversion to risk taking (Heilbronner et al. 2008). As a result, it is perhaps unsurprising that we see substantial overlap between the two species in the aspects of gestural communication that appear to be biologically inherited (Byrne et al. 2017; Graham, Furuichi, and Byrne 2017; Graham et al.

2018), together with differences in the way in which they use their gestures in social interactions. Bonobo mothers gaze more before gesturing, and infants respond more quickly to requests; whereas chimpanzee mothers wait longer for a response, and needed to persist more often—resulting in more frequent use of gesture sequences (Fröhlich et al. 2016). Across ape species it was found that chimpanzee and bonobo, but particularly bonobo, mothers were more proactive in their gesturing toward their infants (Schneider, Call, and Liebal 2017). One point of interest going forward is that while East and West African chimpanzees are more genotypically similar to each other than to bonobos, bonobos and West African chimpanzees show interesting phenotypic similarities in social group size and cohesion (Chapman, White, and Wrangham 1994; Lehmann and Boesch 2004), which appear quite different to East African chimpanzees. With increasing data from a wider variety of bonobo communities, we may see more phenotypic variation in their behavior, allowing for greater overlap (or distinction) from the range of behavior seen in the different subspecies and populations of chimpanzees.

Future Directions

As discussed in the introduction, two key features of human language—syntax and dialect—appear to be based on mechanisms widely shared in other species' communication: the use of structure to vary the information encoded in a signal, and social learning. In human communication these are biologically inherited universals seen across languages. We all use syntactic structures to reorder words and express a vast range of meaning. We all learn signals and how to use them by imitating others around us, resulting in the immense cultural variation of dialects and accents we see around the world. But, surprisingly, these core features of human language, based on mechanisms regularly found in other species, appear absent in great ape gesture. The development of these complex, human-universal language features in the human lineage through a single, recent genetic leap (macro-mutation) is extremely implausible (Lieberman 2015). If true, this would represent a significant evolutionary puzzle. Instead, their precursors were likely present in the communication of our evolutionary ancestors, and should be shared by our modern great ape relatives. But where is the evidence?

Even more puzzling, the underlying cognitive capacities for both syntax and dialect appear present in the wider repertoire of great ape behavior. Tool using in chimpanzees and the manual food processing in mountain gorillas require hierarchical "programs" of manual actions: an "action-grammar" (Byrne 2003; Hayashi and Inoue-Nakamura 2011). It has been argued that the cognitive capacity for this behavioral syntax may be exapted to other behavior, and so could be employed in communication (Barton and Venditti 2014; Byrne 2003). Indeed, captive bonobos show strong evidence for the capacity to encode information within symbol sequences (Savage-Rumbaugh and Lewin 1994; Savage-Rumbaugh and Rumbaugh 1978) and food calls (Clay and Zuberbühler 2011). Great apes regularly employ sequences of gestures (Genty et al. 2009; Hobaiter and Byrne 2011b; Liebal, Call, and Tomasello 2004; McCarthy, Jensvold, and Fouts 2012; Roberts, Vick, and Buchanan-Smith 2012b); however, studies across species find no evidence for meaning encoded within the *structure* of these sequences (Genty et al. 2009; Hobaiter and Byrne 2011b; Liebal, Call, and Tomasello 2004).

One approach to exploring syntax in great ape gesture may be to reexamine what we consider a gesture to be. This may seem an odd suggestion after several decades of study—but it could be key to understanding whether apes use sequences to construct meaning. At present there is widespread variation in definitions of what makes a distinct gesture "type"—is raising one arm a different gesture to raising two arms? Does the orientation of your palm, or even individual fingers, matter when you reach out toward someone? How do we parse out variation that is the result of the environment (whether an ape sits or stands) from variation that results from the ape's choice to encode different information in her movement? Understanding whether or not, from the ape's perspective, she is repeating the same gesture twice, or producing two different gestures in combination, is fundamental to understanding whether or not the structure of signal combinations impacts meaning. Imagine not knowing whether or not a change in tone leads to a change in emphasis (as it might in English or German) or a change in meaning (as it can in Thai or Vietnamese). Even in one language, being able to classify two words as being the same "word" despite differences in accent, tone, and emphasis (whether spoken by an elderly man from Bath or a young girl from Glasgow) is fundamental to being able to parse out the meaning of the sentence. To date, the approach has been to make an arbitrary decision based on our (anthropocentric) understanding of ape movement and behavior, for example: we have coded an arm swing as being a gesture type distinct from a leg swing (e.g., Byrne et al. 2017; Hobaiter and Byrne 2011a), something that has led to different "repertoires" being described by different researchers. But there may be a way forward—with increasingly large data sets on gestural meanings we may be able to reverse engineer what a gesture is by asking what is relevant to the apes who are using them. Does an ape signaler consistently achieve a different goal when they raise one arm rather than two? Swing an arm rather than a leg?

We recently made a first attempt at this—defining the potential repertoire of chimpanzee gestures, for which we have the largest data sets, on the basis of features we have used to discriminate different gesture types: movement, body part, single/double limb, rhythmic repetition, detached object use, etc. The hypothetical maximum repertoire (every possible combination of these features) was over 6,000 gesture types! But this could be rapidly reduced by eliminating those that were physically impossible (e.g., movement = spin + body part = head); leaving a repertoire of just over 1,000 possible gesture types. From a data set of over 4,500 instances of chimpanzee gestures, we showed that just 124 of the possible gestures were used (fig. 10.3), showing that chimpanzees use only a small proportion of the ways in which they can move their body to produce their gestures. Within this data set we examined the pattern of meanings associated with each of the features to explore whether meanings changed depending on, for example, body part for a given movement. The use of one limb or two within a gesture did not appear to change its core meaning for some movements (e.g., raise, shake object) but did for others (e.g., hit, grab-pull); a signaler achieves the same goal by hitting with either the hand (Slap) or fist (Punch), but the same movement and body part combination achieved a different goal when it was produced with or without a detached object (Hobaiter and Byrne 2017). A first attempt, in which we describe a "repertoire" of 81 gesture types, this study showed how we can use the apes' own behavior to better inform our understanding of their communication. To investigate sequence use and syntax we now need to explore the single and combined use of individual gestures. The catch is that the data sets required to do so are truly massive. The key to achieving this, then, is likely to be in collaboration across the increasing number of studies. In doing so, we may also be able to address the intriguing question of culture in communication.

The highly successful studies of great ape material and social cultural behavior were possible only through widespread collaboration across sites (e.g., Kühl et al. 2016; van Schaik et al. 2003; Whiten et al. 1999) and the field has used similar systematic collaboration to understand rarely observed behavior, such as conspecific killing (Wilson et al. 2014). In doing the same for communication we may be able to answer the question of whether or not there is social learning or culture in great ape gesture. Outside of communication, we have evidence from both captive experiments (e.g., Hopper et al. 2007) and field observation (e.g., Hobaiter et al. 2014) that great apes are capable of learning socially from other individuals (e.g., in their tool use and food processing), and that this leads to cultural variation in their behavior (Hopper and Carter, chapter 7 this volume; Luncz and van de Waal, chapter 18 this volume). However, we have almost no evidence for the expression of this social

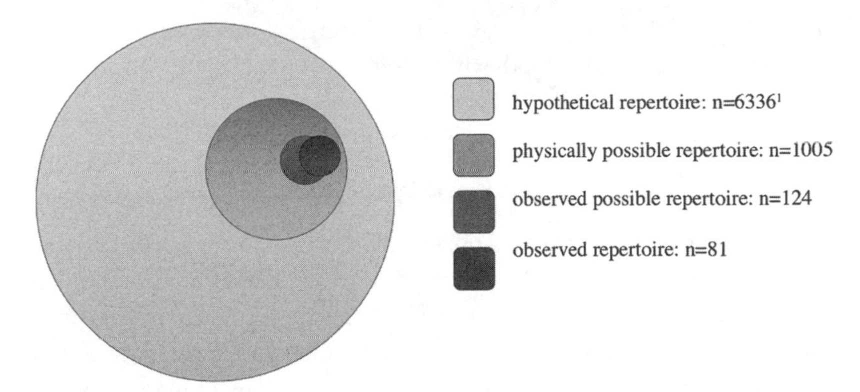

FIGURE 10.3. Number of gesture types observed in wild East African chimpanzees (*Pan troglodytes schweinfurthii*) expressed as a proportion of those that are possible. Many of the physical actions that, if accompanied by indications of intentional use, could be described as a gesture are not expressed in the chimpanzee repertoire ([1]data: Hobaiter and Byrne 2017).

learning capacity in their communication (for discussion of possible cases see: Crockford et al. 2004; Taglialatela et al. 2012; Townsend, Watson, and Slocombe, chapter 11 this volume; Watson et al. 2015). Captive apes have even shown a remarkable ability to learn complex artificial systems of communication, enhancing their communicative repertoires through the acquisition of hundreds of signs in languages such as ASL or Yerkish (Gardner and Gardner 1969; Savage-Rumbaugh and Lewin 1994; Savage-Rumbaugh and Rumbaugh 1978), but their available repertoires of natural gestures appear fixed (Byrne et al. 2017). However, there are tantalizing hints of the potential for culture in ape communication. Orang-utans show group-level variation in some call types (van Schaik et al. 2003) and in the modification of their natural calling (Hardus et al. 2009). An individual ape's use of gestures from within their available species repertoire appears to be sensitive to their social environment (see for example: Hobaiter and Byrne 2011b, Pika and Fröhlich 2018). Chimpanzee hand-clasp behavior, which while not a gesture may signal the desire to groom, varies in presence and form between groups (Whiten et al. 1999; Wrangham et al. 2016), and a number of behaviors, including possible gestures, have been suggested as candidates for cultural behavior in bonobos (Hohmann and Fruth 2003). At present, the majority of studies have taken place in just a handful of sites, and there remains only one longitudinal field study of gesture. We could not describe the features of human language by studying people from only one city; it would be impossible to know whether their speech pattern is a human universal, a socially learned local dialect, or an artifact of their city environment. Within the nine great ape subspecies there are hundreds of groups in habitats as diverse as rainforest and savannah.

To reveal the mechanisms behind how ape gestures are acquired and used, we must compare communities both within and across populations, including humans, across the full diversity of their natural range.

Acknowledgments

I thank Lydia Hopper and Steve Ross for inviting me to take part in this book and their work in putting this collection together, and Brittany Fallon and Kirsty Graham for thoughtful discussions and very helpful comments on the manuscript. I thank the collaborators and funders who have made our research possible, especially the past and present members of the St. Andrews Gesture group: Erica Cartmill, Mawa Dafreville, Brittany Fallon, Emilie Genty, Kirsty Graham, Simone Pika, Adrian Soldati, and Joanne Tanner, and in particular the continued mentorship and support of Dick Byrne, as well as all the staff of the Budongo Conservation Field Station in Uganda. Without all of these people the past decade of work on great ape gestural communication would have been neither possible nor anywhere near as much fun.

References

Arcadi, A. C., D. Robert, and F. Mugurusi. 2004. "A comparison of buttress drumming by male chimpanzees from two populations." *Primates* 45: 135–39.

Arnold, K., and K. Zuberbühler. 2006. "Language evolution: Sematic combinations in primate calls." *Nature* 441 (7091): 303.

Arnold, K., and K. Zuberbühler. 2008. "Meaningful call combinations in a non-human primate." *Current Biology* 18: R202–3.

Baker, M. C. 1975. "Song dialects and genetic differences in white-crowned sparrow." *Evolution* 29: 226–41.

Bard, K. A. 1992. "Intentional behavior and intentional communication in young free-ranging orang-utans." *Child Development* 63: 1186–97.

Bard, K. A., S. Dunbar, V. Maguire-Herring, Y. Veira, K. G. Hayes, and K. MacDonald. 2014. "Gestures and social-emotional communicative development in chimpanzee infants." *American Journal of Primatology* 76: 14–29.

Bard, K. A., and J. Vauclair. 1984. "The communicative context of object manipulation in ape and human adult-infant pairs." *Journal of Human Evolution* 13: 181–90.

Bard, K. A., B. Hewlett, K. Ross, B. Wallauer, H. Keller, S. Boysen, and T. Matsuzawa. 2013. "The effects of lived experiences on primate social cognition." *Folia Primatologica* 84 (3–5): 246.

Bard, K. A., V. Maguire-Herring, M. Tomonaga, and T. Matsuzawa, T. 2019. "The gesture 'Touch': Does meaning-making develop in chimpanzees' use of a very flexible gesture?" *Animal Cognition* 22 (4): 535–50.

Bar-On, D. 2013. "Origins of meaning: Must we 'go Gricean'?" *Mind and Language* 28: 342–75.

Barton, R., and C. Venditti. 2014. "Rapid evolution of the cerebellum in humans and other great apes." *Current Biology* 24: 2440–44.

Bates, E., L. Camaioni, and V. Volterra. 1975. "The acquisition of performatives prior to speech." *Merrill-Palmer Quarterly of Behavior and Development* 21: 205–26.

Bellugi, U., and E. S. Klima. 1979. *The Signs of Language.* Cambridge, MA: Harvard University Press.

Bentz, C., and R. Ferrer-i-Cancho. 2016. "Zipf's law of abbreviation as a language universal." *Capturing Phylogenetic Algorithms for Linguistics.* Lorentz Centre Workshop, Leiden, October 2015.

Bohn, M., J. Call, and M. Tomasello. 2015. "Communication about absent entities in great apes and human infants." *Cognition* 145: 63–72.

Byrne, R. W. 2003. "Imitation as behaviour parsing." *Philosophical Transactions of the Royal Society of London B* 358: 529–36.

Byrne, R. W., E. Cartmill, E. Genty, K. E. Graham, C. Hobaiter, and J. Tanner. 2017. "Great ape gestures: Intentional communication with a rich set of innate signals." *Animal Communication* 4: 755–69.

Call, J., and M. Tomasello. 2007. "The gestural repertoire of chimpanzees (*Pan troglodytes*)." In *The Gestural Communication of Apes and Monkeys*, edited by J. Call and M. Tomasello, 17–40. London: Lawrence Erlbaum Associates.

Call, J., and M. Tomasello, eds. 2007. *The Gestural Communication of Apes and Monkeys.* London: Lawrence Erlbaum Associates.

Candiotti, A., K. Zuberbühler, and A. Lemasson. 2012. "Context-related call combinations in female Diana monkeys." *Animal Cognition* 15: 327–39.

Cartmill, E., and R. W. Byrne. 2007. "Orangutans modify their gestural signaling according to their audience's comprehension." *Current Biology* 17: 1345–48.

Cartmill, E., and R. W. Byrne. 2010. "Semantics of primate gestures: Intentional meanings of orangutan gestures." *Animal Cognition* 13: 793–804.

Chapman, C. A., F. J. White, and R. W. Wrangham. 1994. "Party size in chimpanzees and bonobos." In *Chimpanzee Cultures*, edited by R. W. Wrangham, 41–58. Cambridge, MA: Harvard University Press.

Cheney, D. L., and R. M. Seyfarth. 1990. "Attending to behaviour versus attending to knowledge: Examining monkeys' attribution of mental states." *Animal Behaviour* 40: 742–53.

Chomsky, N. 2017. "Language architecture and its import for evolution." *Neuroscience & Biobehavioral Reviews* 81: 295–300.

Clay, Z., and K. Zuberbühler. 2011. "Bonobos extract meaning from call sequences." *PLoS One* 6: e18786.

Condillac, É. B. 2001. *Essay on the Origin of Human Knowledge.* Translated by H. Aasleff. Cambridge: Cambridge University Press.

Corballis, M. C. 2002. *From Hand to Mouth: The Origins of Language.* Princeton, NJ: Princeton University Press.

Corballis, M. C. 2017. "A word in the hand: The gestural origins of language." In *Neural Mechanisms of Language, Innovations in Cognitive Neuroscience*, edited by M. Mody, 199–218. New York: Springer.

Cowen, A. S., H. A. Elfenbein, P. Laukka, and D. Keltner. 2018. "Mapping 24 emotions conveyed by brief human vocalization." *American Psychologist* 74 (6): 698–712.

Crockford, C., I. Herbinger, L. Vigilant, and C. Boesch. 2004. "Wild chimpanzees produce group-specific calls: A case for vocal learning?" *Ethology* 110: 221–43.

Crockford, C., R. M. Wittig, R. Mudry, and K. Zuberbühler. 2012. "Wild chimpanzees inform ignorant group members of danger." *Current Biology* 22: 142–46.

Crockford, C., R. M. Wittig, and K. Zuberbühler. 2015. "An intentional vocalization draws others' attention: A playback experiment with wild chimpanzees." *Animal Cognition* 18: 581–91.

Crockford, C., R. M. Wittig, and K. Zuberbühler. 2017. "Vocalizing in chimpanzees is influence by social-cognitive processes." *Science Advances* 3: e1701742.

Deecke, V. B., J. K. B. Ford, and P. Spong. 2000. "Dialect change in resident killer whales: Implications for vocal learning and cultural transmission." *Animal Behaviour* 60: 629–38.

de Manuel, M., M. Kuhlwilm, P. Frandsen, V. C. Sousa, T. Desai, J. Prado-Martinex, J. Hernandez-Rodriguez, et al. 2016. "Chimpanzee genomic diversity reveals ancient admixture with bonobos." *Science* 354: 477–81.

Dennett, D. C. 1983. "Intentional systems in cognitive ethology." *The Brain and Behavioural Sciences* 6: 343–90.

Dennett, D. C. 1987. *The Intentional Stance*. Cambridge, MA: MIT Press.

Engesser, S., A. P. Ridley, and S. W. Townsend. 2016. "Meaningful call combinations and compositional processing in the southern pied babbler." *Proceedings of the National Academy of Sciences* 113: 5976–81.

Everaert, M. B. H., M. A. C. Huybregts, R. C. Berwick, N. Chomsky, I. Tattersall, A. Moro, and J. J. Bolhuis. 2017. "What is language and how could it have evolved?" *Trends in Cognitive Sciences* 21: 569–71.

Fitch, W. T. 2017. "Preface to the special issue on the biology and evolution of language." *Psychonomic Bulletin and Review* 24: 1–2.

Ford, J. K. B. 1991. "Vocal traditions among resident killer whales (*Orcinus orcas*) in coastal waters of British Columbia." *Canadian Journal of Zoology* 69: 1454–83.

Fröhlich, M., P. Kuchenbuch, G. Müller, B. Fruth, T. Furuichi, R. M. Wittig, and S. Pika. 2016. "Unpeeling the layers of language: Bonobos and chimpanzees engage in cooperative turn-taking sequences." *Scientific Reports* 6: 25887.

Fröhlich, M., G. Müller, C. Zeiträg, R. M. Wittig, and S. Pika. 2017. "Gestural development of chimpanzees in the wild: The impact of interactional experience." *Animal Behaviour* 134: 271–82.

Fröhlich, M., R. M. Wittig, and S. Pika. 2016. "Play-solicitation gestures in chimpanzees in the wild: Flexible adjustment to social circumstances and individual matrices." *Royal Society Open Science* 3: 160278.

Furuichi, T. 1987. "Sexual swelling, receptivity, and grouping of wild pygmy chimpanzee females at Wamba, Zaïre." *Primates* 28: 309–18.

Furuichi, T. 2011. "Female contributions to the peaceful nature of bonobo society." *Evolutionary Anthropology* 20: 131–42.

Gardner, R. A., and B. T. Gardner. 1969. "Teaching sign language to a chimpanzee." *Science* 165: 664–72.

Genty, E., T. Breuer, C. Hobaiter, and R. W. Byrne. 2009. "Gestural communication of the gorilla (*Gorilla gorilla*): Repertoire, intentionality and possible origins." *Animal Cognition* 12: 527–46.

Genty, E., Z. Clay, C. Hobaiter, and K. Zuberbühler. 2014. "Multi-modal use of a socially directed call in bonobos." *PLoS One* 9: e8473.

Genty, E., C. Neumann, and K. Zuberbühler. 2015a. "Complex patterns of signalling to convey different social goals of sex in bonobos, *Pan paniscus*." *Scientific Reports* 5: 16135.

Genty, E., C. Neumann, and K. Zuberbühler. 2015b. "Bonobos modify communication signals according to recipient familiarity." *Scientific Reports* 5: 16442.

Genty, E., and K. Zuberbühler. 2014. "Spatial reference in a bonobo gesture." *Current Biology* 24: 1601–5.

Ghazanfar, A. A., D. Y. Takahashi, N. Mathur, and W. T. Fitch. 2012. "Cineradiography of monkey lipsmacking reveals putative precursors of speech dynamics." *Current Biology* 22: 1176–82.

Gómez, J. C. 1994. "Mutual awareness in primate communication: A Gricean approach." In *Self-Awareness in Animals and Humans*, edited by S. T. Parker, R. W. Mitchell, and M. L. Boccia, 61–80. Cambridge: Cambridge University Press.

Goodall, J. 1968. "The behaviour of free-living chimpanzees in the Gombe Stream Reserve." *Animal Behaviour Monographs* 1: 161–311.

Graham, K. E., T. Furuichi, and R. W. Byrne. 2017. "The gestural repertoire of the wild bonobo (*Pan paniscus*): A mutually understood communication system." *Animal Cognition* 20: 171–77.

Graham, K. E., C. Hobaiter, J. Ounsley, T. Furuichi, and R. W. Byrne. 2018. "Bonobo and chimpanzee gestures overlap extensively in meaning." *PLoS Biology* 16: e2004825.

Grice, H. P. 1957. "Meaning." *Philosophical Review* 64: 377–88.

Gustison, M. L., S. Semple, R. Ferrer-i-Cancho, and T. J. Bergman. 2016. "Gelada vocal sequences follow Menzerath's linguistic law." *Proceedings of the National Academy of Sciences* 113: E2750–58.

Halina, M., F. Rossano, and M. Tomasello. 2013. "The ontogenetic ritualization of bonobo gestures." *Animal Cognition* 16: 653–66.

Hardus, M. E., A. R. Lameira, C. P. Van Schaik, and S. A. Wich. 2009. "Tool use in wild orangutans modifies sound production: A functionally deceptive innovation?" *Proceedings of the Royal Society of London B* 276 (1673): 3689–94.

Hauser, M. D., N. Chomsky, and W. T. Fitch. 2002. "The faculty of language: What is it, who has it, and how did it evolve?" *Science* 298: 1569–79.

Hayashi, M., and N. Inoue-Nakamura. 2011. "From handling stones and nuts to tool-use." In *The Chimpanzees of Bossou and Nimba*, edited by T. Matsuzawa, T. Humle, and Y. Sugiyama, 175–183. Tokyo: Springer.

Heesen, R., C. Hobaiter, R. Ferrer-i-Cancho, and S. Semple. 2019. "Linguistic laws in chimpanzee gestural communication." *Proceedings of the Royal Society of London B* 286: 20182900

Heilbronner, S. R., A. G. Rosati, J. R. Stevens, B. Hare, and M. D. Hauser. 2008. "A fruit in the hand or two in the bush? Divergent risk preferences in chimpanzees and bonobos." *Biology Letters* 4 (3): 246–49.

Herbinger, I., S. Papworth, C. Boesch, and K. Zuberbühler. 2009. "Vocal, gestural and locomotor responses of wild chimpanzees to familiar and unfamiliar intruders: A playback study." *Animal Behaviour* 78: 1389–96.

Hewes, G. W. 1973. "Primate communication and the gestural origins of language." *Current Anthropology* 14: 5–24.Hewes, G. W. 1999. "A history of the study of language origins and the gestural primacy hypothesis." In *Handbook of Human Symbolic Evolution*, edited by A. Lock and C. R. Peters, 571–95. Oxford: Oxford University Press, Clarendon Press.

Hobaiter, C., and R. W. Byrne. 2011a. "The gestural repertoire of the wild chimpanzee." *Animal Cognition* 14: 745–67.

Hobaiter, C., and R. W. Byrne. 2011b. "Serial gesturing by wild chimpanzees: Its nature and function for communication." *Animal Cognition* 14: 827–38.

Hobaiter, C., and R. W. Byrne. 2012. "Gesture use in consortship." In *Developments in Primate Gesture Research*, edited by S. Pika and K. Liebal, 129–46. Amsterdam: John Benjamins Publishing Company.

Hobaiter, C., and R. W. Byrne. 2013. "Laterality in the gestural communication of wild chimpanzees." *Annals of the New York Academy of Sciences* 1288: 9–16.

Hobaiter, C., and R. W. Byrne. 2014. "The meanings of chimpanzee gestures." *Current Biology* 24: 1596–1600.

Hobaiter, C., and R. W. Byrne. 2017. "What is a gesture? A meaning based approach to defining gestural repertoires." *Neuroscience & Biobehavioral Reviews* 82: 3–12.

Hobaiter, C., R. W. Byrne, and K. Zuberbühler. 2017. "Wild chimpanzees use of single and combined vocal and gestural signals." *Behavioral Ecology and Sociobiology* 71: 96.

Hobaiter, C., D. A. Leavens, and R. W. Byrne. 2013. "Deictic gesturing in wild chimpanzees (*Pan troglodytes*)? Some possible cases." *Journal of Comparative Psychology* 128 (1): 82–87.

Hobaiter, C., T. Poisot, K. Zuberbühler, W. Hoppitt, and T. Gruber. 2014. "Social network analysis shows direct evidence for social transmission of tool use in wild chimpanzees." *PLoS Biology* 12: e1001960.

Hohmann, G., and B. Fruth. 2003. "Culture in bonobos? Between species and within-species variation in behaviour." *Current Anthropology* 44: 563–71.

Hopper, L. M., A. Spiteri, S. P. Lambeth, S. J. Schapiro, V. Horner, and A. Whiten. 2007. "Experimental studies of traditions and underlying transmission processes in chimpanzees." *Animal Behaviour* 73: 1021–32.

Ishizuka, S., Y. Kawamoto, T. Sakamaki, N. Tokuyama, K. Toda, H. Okamura, and T. Furuichi. 2018. "Paternity and kin structure among neighbouring groups in wild bonobos at Wamba." *Royal Society Open Science* 5: 171006.

Jenkins, P. F. 1978. "Cultural transmission of song patterns and dialect development in a free-living bird population." *Animal Behaviour* 26: 50–78.

Kakishita, Y., K. Sasahara, T. Nishino, M. Takahashi, and K. Okanoya. 2009. "Ethological data mining: An automata-based approach to extract behavioural units and rules." *Data Mining and Knowledge Discovery* 18: 446–71.

Kendon, A. 2017. "Reflections on the 'gesture-first' hypothesis of language origins." *Psychonomic Bulletin & Review* 24 (1): 163–70.

Kersken, V., J.-C. Goméz, U. Liszkowski, A. Soldati, and C. Hobaiter. 2018. "A gestural repertoire of 1–2 year old children: In search of the ape gestures." *Animal Cognition* 22 (4): 577–95.

Knox, A., J. Markx, E. How, A. Azis, C. Hobaiter, F. J. F. van Veen, and H. Morrogh-Bernard. 2019. "Gesture use in communication between mothers and offspring in wild orang-utans (*Pongo pygmaeus wurmbii*) from the Sabangau peat-swamp forest, Borneo." *International Journal of Primatology* 40 (3): 393–416.

Kojima, S., A. Izumi, and M. Ceugniet. 2003. "Identification of vocalizers by pant hoots, pant grunts and screams in a chimpanzee." *Primates* 44: 225–30.

Kühl, H. S., A. K. Kalan, M. Arandjelovic, F. Aubert, L. D'Auvergne, A. Goedmakers, S. Jones, et al. 2016. "Chimpanzee accumulative stone throwing." *Scientific Reports* 6: 22219.

Ladygina-Kohts, N. N. (1935) 2002. *Infant Chimpanzee and Human Child: A Classic 1935 Comparative Study of Ape Emotions and Intelligence*. Translated by B. Wekker. Oxford: Oxford University Press.

Leavens, D. A., K. A. Bard, and W. D. Hopkins. 2017. "The mismeasure of ape social cognition." *Animal Cognition* 22 (4): 487–504.

Leavens, D. A., and W. D. Hopkins. 1998. "Intentional communication by chimpanzees: A cross-sectional study of the use of referential gestures." *Developmental Psychology* 34: 813–22.

Leavens, D. A., and W. D. Hopkins. 2005. "Multimodal concomitants of manual gesture by chimpanzees (*Pan troglodytes*): Influence of food size and distance." *Gesture* 5: 1–2.

Leavens, D. A., W. D. Hopkins, and K. A. Bard. 1996. "Indexical and referential pointing in chimpanzees (*Pan troglodytes*)." *Journal of Comparative Psychology* 110: 346–53.

Leavens, D. A., A. B. Hostetter, M. J. Wesley, and W. D. Hopkins. 2004. "Tactical use of unimodal and bimodal communication in chimpanzees, *Pan troglodytes*." *Animal Behaviour* 67: 467–76.

Leavens, D. A., J. L. Russell, and W. D. Hopkins. 2010. "Multimodal communication by captive chimpanzees (*Pan troglodytes*)." *Animal Cognition* 13: 33–44.

Lehmann, J., and C. Boesch. 2004. "To fission or to fusion: Effects of community size on wild chimpanzee (*Pan troglodytes verus*) social organization." *Behavioral Ecology and Sociobiology* 56: 207–16.

Liddle, B., and D. Nettle. 2006. "Higher-order theory of mind and social competence in school-age children." *Journal of Cultural and Evolutionary Psychology* 4: 231–44.

Liebal, K., J. Call, and M. Tomasello. 2004. "The use of gesture sequences in chimpanzees." *American Journal of Primatology* 64: 377–96.

Liebal, K., J. Call, M. Tomasello, and S. Pika. 2004. "To move or not to move; How apes adjust to the attentional state of others." *Interactional Studies* 5: 199–219.

Liebal, K., S. Pika, and M. Tomasello. 2006. "Gestural communication of orang-utans (*Pongo pygmaeus*)." *Gesture* 6: 1–36.

Liebal, K., B. M. Waller, and K. E. Slocombe. 2011. "The language void: The need for multimodality in primate communication research." *Animal Behaviour* 81: 919–24.

Liebal, K., B. M. Waller, K. E. Slocombe, A. M. Burrows. 2013. *Primate Communication: A Multimodal Approach.* Cambridge: Cambridge University Press.

Lieberman, P. 2015. "Language did not spring forth 100,000 years ago." *PLoS Biology* 13 (2): e1002064.

Luo, B., T. Jiang, Y. Liu, J. Wang, A. Lin, X. Wei, and J. Feng. 2013. "Brevity is prevalent in bat short-range communication." *Journal of Comparative Physiology A* 199 (4): 325–33.

Lyn, H., J. L. Russell, and W. D. Hopkins. 2010. "The impact of environment on the comprehension of declarative communication in apes." *Psychological Science* 21: 360–65.

Marler, P. 1961. "The logical analysis of animal communication." *Journal of Theoretical Biology* 1: 295–317.

McCarthy, M. S., M. L. A. Jensvold, and D. H. Fouts. 2012. "Use of gesture sequences in captive chimpanzee (*Pan troglodytes*) play." *Animal Cognition* 16 (3): 471–81.

McGrew, W. C., L. F. Marchant, S. E. Scott, and C. E. G. Tutin. 2001. "Intergroup differences in a social custom of wild chimpanzees: The grooming hand-clasp of the Mahale Mountains, Tanzania." *Current Anthropology* 42: 148–53.

Moore, R. 2016. "Meaning and ostension in great-ape gestural communication." *Animal Cognition* 19: 223–31.

Moore, R. 2017. "Social cognition, Stag Hunts, and the evolution of language." *Biology and Philosophy* 32 (6): 797–818.

Moore, R., J. Call, and M. Tomasello. 2015. "Production and comprehension of gestures between orang-utans (*Pongo pygmaeus*) in a referential communication game." *PLoS One* 10: e0129726

Nelson, D. A., K. I. Hallberg, and J. A. Soha. 2004. "Cultural evolution of Puget sound white-crowned sparrow song dialects." *Ethology* 110 (11): 879–908.

Newton-Fisher, N. E., M. Emery Thompson, V. Reynolds, C. Boesch, and L. Vigilant. 2010. "Paternity and social rank in Budongo Forest chimpanzees." *American Journal of Physical Anthropology* 142: 417–28.

Nishida, T. 1980. "The leaf-clipping display: A newly-discovered expressive gesture in wild chimpanzees." *Journal of Human Evolution* 9: 117–28.

Ouattara, K., A. Lemasson, and K. Zuberbühler. 2009a. "Campbell's monkeys concatenate vocalizations into context-specific call sequences." *Proceedings of the National Academy of Sciences* 106 (51): 22026–31.

Ouattara, K., A. Lemasson, and K. Zuberbühler. 2009b. "Campbell's monkeys use affixation to alter call meaning." *PLoS One* 4: e7808.

Perlman, M., J. E. Tanner, and B. J. King. 2012. "A mother gorilla's variable use of touch to guide her infant." In *Developments in Primate Gesture Research*, edited by S. Pika and K. Liebal, 55. Amsterdam: John Benjamins Publishing.

Piantadosi, S. T., H. Tily, and E. Gibson. 2011. "Word lengths are optimized for efficient communication." *Proceedings of the National Academy of Sciences* 108 (9): 3526–29.

Pika, S., and T. Bugnyar. 2011. "The use of referential gestures in ravens (*Corvus corax*) in the wild." *Nature Communications* 2 (560).

Pika, S., and M. Fröhlich. 2018. "Gestural acquisition in great apes: The social negotiation hypothesis." *Animal Cognition* 22 (4): 551–65.

Pika, S., K. Liebal, and M. Tomasello. 2003. "Gestural communication in young gorillas (*Gorilla gorilla*): Gestural repertoire, learning and use." *American Journal of Primatology* 60: 95–111.

Pika, S., K. Liebal, and M. Tomasello. 2005. "Gestural communication in subadult bonobos (*Pan paniscus*): Repertoire and use." *American Journal of Primatology* 65: 39–61.

Pika, S., and J. C. Mitani. 2006. "Referential gestural communication in wild chimpanzees (*Pan troglodytes*)." *Current Biology* 16: 191–92.

Plooij, F. X. 1978. "Some basic traits of human language in wild chimpanzees." In *Action, Gesture and Symbol: The Emergence of Language*, edited A. Locke. London: Academic Press.

Pollick, A. S., and F. B. M. de Waal. 2007. "Ape gestures and language evolution." *Proceedings of the National Academy of Sciences* 104: 8184–89.

Pollick, A. S., A. Jeneson, and F. B. M. de Waal. 2008. "Gestures and multimodal signalling in bonobos." In *The Bonobos*, edited by T. Furuichi and J. Thompson, 75–94. New York: Springer.

Pusey, A., J. Williams, and J. Goodall. 1997. "The influence of dominance rank on the reproductive success of female chimpanzees." *Science* 277 (5327): 828–31.

Rendall, D., M. J. Owren, and M. J. Ryan. 2009. "What do animal signals mean?" *Animal Behaviour* 78 (2): 233–40.

Rendell, L. E., and H. Whitehead. 2003. "Comparing repertoires of sperm whale codas: a multiple methods approach." *Bioacoustics* 14 (1): 61–81.Roberts, A. I., and S. G. B. Roberts. 2016. "Wild chimpanzees modify modality of gestures according to the strength of social bonds and personal network size." *Scientific Reports* 6: 33864.

Roberts, A. I., S.-J. Vick, and H. Buchanan-Smith. 2012a. "Usage and comprehension of manual gestures in wild chimpanzees." *Animal Behaviour* 84: 459–70.

Roberts, A. I., S.-J. Vick, and H. Buchanan-Smith. 2012b. "Communicative intentions in wild chimpanzees: Persistence and elaboration in gestural signalling." *Animal Cognition* 16 (2): 187–96.

Ryu, H., D. A. Hill, and T. Furuichi. 2015. "Prolonged maximal sexual swelling in wild bonobos facilitates affiliative interactions between females." *Behaviour* 152 (3–4): 285–311.

Sakamaki, T. 2013. "Social grooming among wild bonobos (*Pan paniscus*) at Wamba in the Luo Scientific Reserve, DR Congo, with special reference to the formation of grooming gatherings." *Primates* 4: 349–59.

Sakamaki, T., I. Behncke, M. Laporte, M. Mulavwa, H. Ryu, H. Takemoto, N. Tokuyama, S. Yamamoto, and T. Furuichi. 2015. "Intergroup transfer of females and social relationships between immigrants and residents in bonobo (*Pan paniscus*) societies." In *Dispersing Primate Females*, edited by T. Furuichi, J. Yamagiwa, and F. Aureli, 127–64. Tokyo: Springer.

Savage-Rumbaugh, E. S., and R. Lewin. 1994. *Kanzi: The Ape at the Brink of the Human Mind.* New York: Wiley.

Savage-Rumbaugh, E. S., and D. M. Rumbaugh. 1978. "Symbolization, language, and chimpanzees: A theoretical reevaluation based on initial language acquisition processes in four young Pan troglodytes." *Brain and Language* 6: 265–300.

Schaller, G. E. 1963. *The Mountain Gorilla: Ecology and Behavior.* Chicago: University of Chicago Press.

Schel, A. M., S. Townsend, Z. Machanda, K. Zuberbühler, and K. Slocombe. 2013. "Chimpanzee alarm call production meets key criteria for intentionality." *PLoS One* 8: e76674.

Schlenker, P., E. Chemla, C. Cäsar, R. Ryder, and K. Zuberbühler. 2017. "Titi semantics: Context and meaning in Titi monkey call sequences." *Natural Language and Linguistic Theory* 35: 271–98.

Schneider, C., J. Call, and K. Liebal. 2012. "What role do mothers play in the gestural acquisition of bonobos (*Pan paniscus*) and chimpanzees (*Pan troglodytes*)?" *International Journal of Primatology* 33: 246–62.

Schneider, C., J. Call, and K. Liebal. 2017. "'Giving' and 'responding' difference in gestural communication between nonhuman great ape mothers and infants." *Developmental Psychobiology* 59: 303–13.

Scott-Phillips, T. 2014. *Speaking Our Minds: Why Human Communication Is Different, and How Language Evolved to Make It Special.* London: Palgrave MacMillan.

Scott-Phillips, T. C. 2015. "Meaning in animal and human communication." *Animal Communication* 18 (3): 801–5.

Seyfarth, R. M., and D. L. Cheney. 2003. "Signalers and receivers in animal communication." *Annual Review of Psychology* 54 (1): 145–73.

Seyfarth, R. M., and D. L. Cheney. 2017. "The origin of meaning in animal signals." *Animal Behaviour* 124: 339–46.

Slocombe, K. E., and K. Zuberbühler. 2007. "Chimpanzees modify recruitment screams as a function of audience composition." *Proceedings of the National Academy of Sciences* 104 (43): 17228–33.

Smith, W. J. 1965. "Message, meaning and context in ethology." *American Naturalist* 99: 405–9.

Stokoe, W. C., Jr. 1960. "Sign language structure: An outline of the visual communication systems of the American deaf." *Studies in Linguistics Occasional Papers* 8: 1–78.

Surbeck, M., R. Mundry, and G. Hohmann. 2011. "Mother's matter! Maternal support, dominance status and mating success in male bonobos (*Pan paniscus*)." *Proceedings of the Royal Society of London B* 278: 590.

Suzuki, T. N., D. Wheatcroft, and M. Griesser. 2016. "Experimental evidence for compositional syntax in bird calls." *Nature Communications* 7: 10986.

Taglialatela, J. P., L. Reamer, S. J. Schapiro, and W. D. Hopkins. 2012. "Social learning of a communicative signal in captive chimpanzees." *Biology Letters* 8 (4): 498–501.

Tanner, J. E., and R. W. Byrne. 1993. "Concealing facial evidence of mood: Perspective-taking in a captive gorilla?" *Primates* 34: 451–57.

Tanner, J. E., and R. W. Byrne. 1996. "Representation of action through iconic gesture in a captive lowland gorilla." *Current Anthropology* 37: 162–73.

Tanner, J. E., and R. W. Byrne. 1999. "Spontaneous gestural communication in captive lowland gorillas." In *The Mentalities of Gorillas and Orang-utans in Comparative Perspective*, edited by S. T. Parker, R. W. Mitchell, and H. L. Miles, 211–40. Cambridge: Cambridge University Press.

Tokuyama, N., and T. Furuichi. 2016. "Do friends help each other? Patterns of female coalition formation in wild bonobos at Wamba." *Animal Behaviour* 119: 27–35.

Tomasello, M. 2007. "If they're so good at grammar, then why don't they talk? Hints from apes' and humans' use of gestures." *Language Learning and Development* 3: 133–56.

Tomasello, M., and J. Call. 1997. *Primate Cognition*. New York: Oxford University Press.

Tomasello, M. and J. Call. 2006. "Do chimpanzees know what others see—or only what they are looking at?" In *Rational Animals?*, edited by S. Hurley and M. Nudds, 371–84. New York: Oxford University Press.

Tomasello, M., B. L. George, A. C. Kruger, M. Jeffrey, and A. Evans. 1985. "The development of gestural communication in young chimpanzees." *Journal of Human Evolution* 14: 175–86.

Tomasello, M., D. Gust, and G. T. Frost. 1989. "A longitudinal investigation of gestural communication in young chimpanzees." *Primates* 30: 35–50.

Townsend, S. W., S. E. Koski, R. W. Byrne, K. E. Slocombe, B. Bickel, M. Boeckle, I. Braga Goncalves, et al. 2017. "Exorcising Grice's ghost: An empirical approach to studying intentional communication in animals." *Biological Reviews* 92: 1427–33.

Vail, A. L., A. Manica, and R. Bshary. 2013. "Referential gestures in fish collaborative hunting." *Nature Communications* 4 (1): 1–7.

van Schaik, C. P., M. Ancrenaz, G. Borgen, B. Galdikas, C. D. Knott, I. Singleton, A. Suzuki, S. S. Utami, and M. Merrill. 2003. "Orangutan cultures and the evolution of material culture." *Science* 299: 102–5.

Watson, S. K., S. W. Townsend, A. M. Schel, C. Wilka, E. K. Wallace, L. Cheng, V. West, and K. E. Slocombe. 2015. "Vocal learning in the functionally referential food grunts of chimpanzees." *Current Biology* 25: 495–99.

Watts, D. P. 2016. "Production of grooming-associated sounds by chimpanzees (*Pan troglodytes*) at Ngogo: Variation, social learning, and possible functions." *Primates* 57: 61–72.

Whiten, A., J. Goodall, W. C. McGrew, T. Nishida, V. Reynolds, Y. Sugiyama, C. E. G. Tutin, R. W. Wrangham, and C. Boesch. 1999. "Culture in chimpanzees." *Nature* 399: 682–85.

Wilke, C., E. Kavanagh, E. Donnellan, B. M. Waller, Z. P. Machanda, and K. E. Slocombe. 2017. "Production of and responses to unimodal and multimodal signals in wild chimpanzees, *Pan troglodytes schweinfurthii*." *Animal Behaviour* 123: 305–16.

Wilson, M. L., C. Boesch, B. Fruth, T. Furuichi, I. C. Gilby, C. Hashimoto, C. L. Hobaiter, G. Hohmann, N. Itoh, K. Koops, J. N. Lloyd, T. Matsuzawa, J. C. Mitani, D. C. Mjungu, D. Morgan, M. N. Muller, R. Mundry, M. Nakamura, J. Pruetz, A. E. Pusey, J. Riedel, C. Sanz, A. M. Schel, N. Simmons, M. Waller, D. P. Watts, F. White, R. M. Wittig, K. Zuberbühler, and R. W. Wrangham. 2014. "Lethal aggression in Pan in better explained by adaptive strategies than human impacts." *Nature* 513: 414–17.

Wimmer, H., and J. Perner. 1983. "Beliefs about beliefs: Representation and constraining function of wrong beliefs in young children's understanding of deception." *Cognition* 13: 103–28.

Winawer, J., N. Witthoft, M. C. Frank, L. Wu, A. R. Wade, and L. Boroditsky. 2007. "Russian blues reveal effects of language on color discrimination." *Proceedings of the National Academy of Sciences* 104: 7780–85.

Wohlgemuth, M. J., S. J. Sober, and M. S. Brainard. 2010. "Linked control of syllable sequence and phonology in birdsong." *Journal of Neuroscience* 29: 12936–49.

Woods, V., and B. Hare. 2011. "Bonobo but not chimpanzee infants use socio-sexual contact with peers." *Primates* 52: 111–16.

Wrangham, R. W., K. Koops, Z. P. Machanda, S. Worthington, A. B. Bernard, N. F. Brazeau, R. Donovan, J. Rosen, C. Wilke, E. Otali, and M. N. Muller. 2016. "Distribution of a chimpanzee social custom is explained by matrilineal relationship rather than conformity." *Current Biology* 26: 3033–37.

11

Flexibility in Great Ape Vocal Production

SIMON W. TOWNSEND*, STUART K. WATSON*, AND KATIE E. SLOCOMBE

Introduction

Traits that are often cited to distinguish spoken language from other, nonhuman vocal communication systems range from the way that language combines a finite number of sounds into an infinite number of meaningful expressions (Collier et al. 2014; Hurford 2011), to its rich semantic layer (Hurford 2007). However, a large body of comparative data on the vocal behavior of primates has emerged over the last five decades indicating these differences could be considered continuous or quantitative in nature (Watson et al. 2015a; Zuberbühler 2005; Zuberbühler, Cheney, and Seyfarth 1999). For example, a number of monkey and ape species (and, indeed, non-primate mammals and birds, see Gill and Bierema 2013; Townsend and Manser 2013) use vocalizations to refer to external objects and events in the environment in a way that appears comparable to language's semanticity or reference (Clay, Smith, and Blumstein 2012; Seyfarth, Cheney, and Marler 1980; Slocombe and Zuberbühler 2005, although see Townsend and Manser 2013; Wheeler and Fischer 2012). Moreover, primates are also capable of combining sounds and calls together in different ways to encode meaning (Cäsar et al. 2013; Clarke, Reichard, and Zuberbühler 2006; Clay and Zuberbühler 2009; Spillmann et al. 2010) and even, in some instances, to increase communicative output (Arnold and Zuberbühler 2006; Collier et al. 2014; Coye, Zuberbühler, and Lemasson 2016; Coye et al. 2015; Ouattara, Lemasson, and Zuberbühler 2009). These rudimentary semantic and syntactic abilities in our close relatives have been argued to support a gradual evolutionary scenario for human language abilities (Arnold and Zuberbühler 2006; Slocombe and Zuberbühler 2005; Zuberbühler 2005), building on preexisting cognitive and communicative capacities in our primate ancestors.

* Joint first author

While semantics, phonology, and syntax are central to facilitating the communication of an unbounded array of thoughts and ideas in language, an essential facet of human linguistic systems is that they comprise arbitrary conventions, acquired culturally through social learning (Deacon and Poeppel 1997; Ghazanfar and Rendall 2008; Wheeler and Fischer 2012; Zuberbühler 2006). In contrast, primate vocal communication is generally considered to be largely innate and tightly tied to the emotional state of the signaler, akin to human emotional vocalizations, such as laughter and screams, which are universal in all human cultures (Wheeler and Fischer 2012; Tomasello 2010). If true, this might suggest that the evolutionary transition that led to flexible control and a decoupling of emotion/arousal and call production (a fundamental step, giving rise to arbitrary form-meaning pairings and later, conventional communication systems) occurred independently within the hominid lineage and did not build upon specific, preexisting abilities within the primate lineage. Here, we review existing literature on this topic and propose, first, that monkeys may actually demonstrate greater vocal flexibility than hitherto generally recognized. Second, we argue that apes, particularly chimpanzees (*Pan troglodytes*), demonstrate a degree of control beyond monkeys, pointing toward a potential continuum in vocal flexibility across the primate order.

Vocal Flexibility

Vocal flexibility is traditionally separated into flexibility at the production level and flexibility at the receiver level (Egnor and Hauser 2004; Seyfarth and Cheney 2010). A further distinction is often made between production flexibility in terms of usage and flexibility of the articulators (larynx and supralaryngeal vocal tract; Egnor and Hauser 2004; Hopkins, Russell, and Cantalupo 2007; Owren, Amoss, and Rendall 2011; Seyfarth and Cheney 2010). Given this, we aim to examine the proximate psychological mechanisms underlying call production in primates. We focus solely on flexibility at the producer level and consider both articulatory and usage flexibility, as both have been argued to represent key evolutionary innovations central to the acquisition of arbitrary, shared lexicons (Hauser, Chomsky, and Fitch 2002).

ARTICULATORY AND USAGE FLEXIBILITY IN MONKEYS

Monkeys, like many other groups of mammals and birds, possess limited vocal repertoires (Seyfarth and Cheney 2010). For example, vervet monkeys (*Chlorocebus pygerythrus*) and baboons (*Papio* spp.) utilize only a small number of different call types to negotiate the full extent of events imposed by the social and ecological environment (Cheney and Seyfarth 1990, 1992, 2008, see Bouchet, Blois-Heulin, and Lemasson 2013 for a review). In this sense, monkey call systems already seem to differ qualitatively to the open-ended, productive lexicons that characterize human language (Hurford 2011). This view has been compounded by seminal studies suggesting, in stark contrast to human language (Tomasello 2010), that the social environment has little influence on the ontogeny of monkey vocalizations. For example, cross-fostering experiments with Japanese (*Macaca fuscata*) and rhesus (*M. mulatta*) macaques has been shown to have a limited effect on the development of species-specific vocalizations (Owren et al. 1992, 1993). Furthermore, squirrel monkeys (*Saimiri sciureus*) raised in acoustic isolation from other individuals, one of whom underwent a deafening operation, were found to develop a vocal repertoire that was not acoustically different from infants raised under social conditions (Winter et al. 1973). A more recent study, examining the development of squirrel monkey vocalizations from infancy to maturity, also found that a socially isolated individual and a congenitally deaf individual developed repertoires that were acoustically within the variability range of socially raised monkeys (Hammerschmidt, Freudenstein, and Jürgens 2001). Taken together, these findings seem to suggest that the acoustic structure of monkey vocalizations is developmentally fixed rather than socially learned. However, not all evidence supports this view and recent work employing state of the art recording and analysis techniques suggests social experience can influence vocal development in monkeys.

Using high density vocal sampling and detailed quantitative acoustic tracking, Takahashi et al. (2015) demonstrated that the transition from "cry" production to adult-like "phee" calls in infant marmosets (*Callithrix jacchus*) is contingent upon parental vocal feedback. This pattern has subsequently been likened to preverbal vocal development in humans where verbal cues from parents sensitize infants to important acoustic features of speech (Goldstein, King, and West 2003). In order to rule out the alternative explanation that these individuals may have vocally matured at a faster rate regardless of feedback, this study was followed up by a twin study in which each sibling received different levels of vocal feedback (Takahashi, Liao, and Ghazanfar 2017). Supporting the argument that vocal development is indeed contingent upon parental feedback, it was found that siblings who received a greater amount of feedback had a faster rate of vocal development. The comparison of twins in this study also allowed the authors to rule out genetic differences and perinatal experiences as alternative explanations.

In adulthood, there is also evidence for articulatory flexibility in monkeys, albeit often to a limited extent and generally involving a fixed call template. For example, the coos of rhesus macaques and the recruitment screams of pig-tailed macaques (*M. nemestrina*) vary systematically across matrilines (Gouzoules and Gouzoules 1990; Hauser 1992). Furthermore, the acoustic structure of the shrill barks of barbary macaques (*M. sylvanus*), the phee calls of Wied's black-tufted ear marmosets (*C. kuhlii*), and the grunts of olive baboons (*P. anubis*) are subject to the influence of the social and physical environment respectively (Ey et al. 2009; Fischer, Hammerschmidt, and Todt 1998; Rukstalis, Fite, and French 2003). Interestingly, Hihara et al. (2003) reported evidence for a spontaneous example of vocal plasticity in a Japanese macaque. A single individual was trained to produce two "coo" calls: first to request food to be provisioned (out of reach) and second to request a raking tool to access the food. Initially, the monkey produced acoustically identical calls for both purposes, but after five days, while the tasks remained the same, the structure of the calls had drifted and become acoustically distinct (i.e., it produced one call to request food and a different one to request the tool). One interpretation forwarded by Hihara et al. (2003) was that their study provides empirical evidence for non-emotionally-driven vocal plasticity, because the emotional valence of the context eliciting the calls remained the same while the acoustic structure changed. However, again, when compared with the highly plastic, experience-mediated acquisition of human language systems, any similarity in monkey articulatory flexibility has been argued to remain modest at best, constrained to a few acoustic parameters, and largely expressed in contact calls that lack clear referential meaning.

In contrast to call production, current evidence suggests that monkeys are comparatively less constrained in call usage. For example, a number of studies have shown that, similar to humans, monkeys produce species-specific vocalizations flexibly and selectively, suggesting a degree of voluntary control (Egnor and Hauser 2004; Seyfarth and Cheney 2010). For example, wild vervet monkeys do not produce alarm calls when alone and are more likely to vocalize when in the presence of kin than non-kin (Cheney and Seyfarth 1985, 1990). The presence of an audience (so-called audience effects, McGregor 1993; Zuberbühler 2008) also influences food-calling behavior in red-bellied tamarins (*Saguinus labiatus*) and spider monkeys (*Ateles geoffroyi*) (Chapman and Lefebvre 1990; Caine, Addington, Windfelder 1995). While the effect of the presence, or absence, of an audience upon vocal production may be reducible to basic arousal processes (Zajonc and Sales 1966), more complex effects where dynamically changing stimuli, such as the composition or

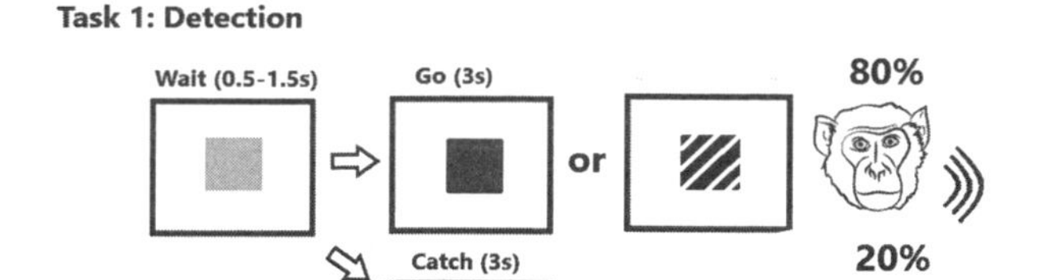
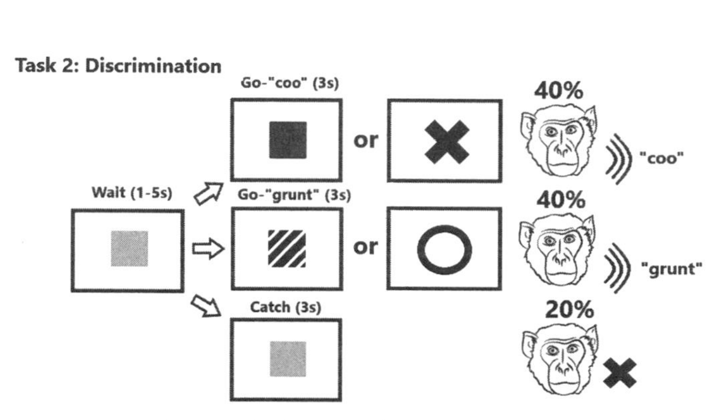

FIGURE 11.1. Experimental design used by Hage, Gavrilov, and Nieder (2013). Task 1: Two monkeys were trained in a go/no-go protocol to vocalize whenever a visual cue appeared. Task 2: One monkey was further trained to utter distinct vocalizations in response to specific visual cues.

behavior of the audience, affect call production (e.g., Papworth et al. 2008; Wich and deVries 2006) are arguably more likely to reflect voluntary control (Schel et al. 2013b). Moreover, a small number of captive-based studies have highlighted that monkeys can learn to produce vocalizations on command (fig. 11.1). Utilizing operant conditioning paradigms, researchers have shown that captive rhesus macaques can be successfully trained to produce social contact calls in response to food rewards (Coudé et al. 2011) or even arbitrary visual stimuli (Hage, Gavrilov, and Nieder 2013). While this required intensive shaping by the experimenters, these findings do at least suggest that it is within the capabilities of monkeys to decouple certain species-specific calls from the contexts or arousal states that typically elicit them.

A review of the behavioral literature on monkey vocal articulatory and usage flexibility therefore seems to indicate that, despite some intriguing parallels, a considerable gap still separates human and monkey vocal plasticity (Wheeler and Fischer 2012) and that these differences are often attributed to variation in the accompanying neurobiology facilitating vocal control in humans and monkeys. Much of what is known about the neurobiology of vocal behavior in primates originates from seminal work in squirrel monkeys and brain-injured aphasic human patients (Ackermann, Hage, and Ziegler 2014; Jürgens 1998; Owren, Amoss, and Rendall 2011; Simonyan and Horwitz 2011). Comparative neurobiologists often distinguish between three key mid-brain and brainstem based sub-systems when analyzing the vocal pathway, the first of which is shared among nonhuman primates and humans and the other two being potentially derived, or particularly well developed, in humans (see Petkov and Jarvis 2012; Pisanski et al. 2016 for detailed reviews). First, in response to an external or internal stimulus, the anterior cingulate cortex or limbic brain regions (via the periaqueductal gray) initiate innate vocal behavior, such as laughter or screams (common to both nonhuman primates and humans, Jürgens 2009). Motor-neurons directly linking the motor cortex with laryngeal muscles represents the second system, while the formatio reticularis, housed in the brain stem, and the motor-neurons connecting the cortex with the supra laryngeal articulators (jaw, lips) constitute the third system that, in conjunction with the second, facilitates voluntary, yet precise, motor control of vocal production apparatus (Pisanksi et al. 2016; Wheeler and Fischer 2012). In humans, stimulation or damage to the laryngeal motor cortex (LMC) respectively produces or eliminates production of learned vocalizations, yet lesion and stimulation studies in monkeys produce limited effects (Petkov and Jarvis 2012), suggesting vocal production in monkeys may not engage functionally analogous brain areas. However, again, the argument regarding whether monkey vocal production is underpinned by similar brain areas is not yet firmly closed.

Single-cell recordings of the ventrolateral prefrontal cortex in macaques during vocal production revealed that the activity of call-related neurons predicted the production of calls that were previously trained to be given in response to visual stimuli, but did not predict the onset of spontaneous calls (Hage and Nieder 2013). Once calls are brought under volitional control, in this case through extensive conditioning training, in a highly artificial, socially isolated setting, the prefrontal cortex does seem to be involved in the production of these calls, showing interesting parallels with human language. However, whether the rich socio-ecological context that influences naturalistic call production is sufficient to engage this circuitry remains ambiguous. Furthermore, a recent study using probabilistic diffusion tractography in humans and rhesus macaques has identified that the LMC structural network is largely comparable in these species, with the key difference being the strength of connections between LMC and the inferior parietal and somatosensory cortices (Kumar, Croxson, and Simonyan 2016). These connections were found to be seven times stronger in humans, and this has been argued to support a continuum theory of vocal learning (Petkov and Jarvis 2012) suggesting that pathway differences may be better characterized as differences of degree rather than kind (Kumar, Croxson, and Simonyan 2016). Together, this body of neurobiological data suggests that a key differentiating feature between monkeys and humans may be the relative poverty of cortical control and connections governing the larynx and articulators in monkeys compared to humans.

Given the behavioral and neurobiological differences between monkey and human vocal flexibility, arguing for a gradual evolution of vocal control, building on neurobiological and behavioral capabilities present in monkeys, is challenging (Arbib et al. 2008, Tomasello 2010). However, as we have highlighted, there is an emerging body of data suggesting monkey call production may be under greater volitional control than hitherto appreciated. Furthermore, these conclusions are derived mostly from a small number of monkey species, are often generalized to all nonhuman primates, and rarely consider evidence of great ape vocal flexibility and neurobiology. If we wish to paint a complete evolutionary picture, incorporating ape data is critical (Wheeler and Fischer 2012) and doing so might help us better understand how and when human vocal production skills evolved.

Great Ape Vocal Production

Over the last 15 years, there has been a concerted effort to systematically probe the vocal communication skills of great apes, both in the wild and captivity. Until recently, the majority of research has focused on vocalizations produced by chimpanzees (*Pan troglodytes*), presumably due to phylogenetic proximity to humans. However, in the last five years the vocal systems of other members of the great apes—bonobos (*P. paniscus*, who are equally as related to humans as chimpanzees), orangutans (*Pongo pygmaeus*), and gorillas (*Gorilla gorilla*)—have been revisited and subject to empirical investigation. One consequence of this is that we are now in a better position to compare chimpanzee and other ape vocal production not only with humans, but also directly with monkeys.

GREAT APE ARTICULATORY FLEXIBILITY

As with monkeys, species-typical ape vocalizations have also been shown to be modifiable at the structural level. The long distance social contact-calls, or pant hoots, of chimpanzees diverge structurally amongst communities (Crockford et al. 2004). Critically, this finding cannot be purely explained as an artifact of genetic differences, as the pant hoots of communities that are closer to each other (and hence more likely to be genetically similar) are acoustically more different than those from more distant communities (Crockford et al. 2004). Pant hoots can also converge acoustically over short time scales. Mitani and Gros-Louis (1998) demonstrated that the pant hoots of males who called together were structurally more similar than solitary pant hoots. Given the small sample sizes of these earlier studies, the exact proximate mechanism driving convergence and divergence in the pant hoots of chimpanzees can only be speculated on, but at the very least it suggests a latent capacity to modify the fine acoustic structure of existing vocalizations based on auditory input.

In a recent study, articulatory flexibility has also been demonstrated in the functionally referential food calls of chimpanzees (Watson et al. 2015a). Historically, functionally referential calls have been differentiated from other call types in a species' repertoire due to their surface-level similarities to human referential words (Macdeonia and Evans 1993; Townsend and Manser 2013; Seyfarth, Cheney, and Marler 1980; Wheeler and Fischer 2012). However, their relevance to understanding the evolutionary origins of human referential abilities has been called into question given that, among other things, these vocalizations are not learned (Wheeler and Fischer 2012). It has been argued that, as with other non-referential call types, their acoustic structures are tightly tied to the underlying arousal state of the signaler (Fischer, Wheeler, and Higham 2015; Wheeler and Fischer 2012). By taking advantage of a unique integration event, in which an entire group of captive chimpanzees was translocated from Beekse Bergen Safari Park in the Netherlands, and integrated with a resident group living at Edinburgh Zoo, Scotland (also described in Herrelko et al., chapter 23 this volume), Watson et al. (2015a) provided intriguing data suggesting that the acoustic structure of chimpanzee's functionally referential food calls *can* be decoupled from the underlying arousal state experienced by the signaler. Specifically, those authors demonstrated that before being integrated, the two groups of chimpanzees had structurally distinct calls for the same referent: apples. This intergroup difference was, however, entirely explicable through variance in preference. The "Scottish" chimpanzees at Edinburgh Zoo had a low preference for apples, while the "Dutch" chimpanzees had a high preference for them. Not only did the acoustic differences accurately reflect the divergence in preference, they also supported previous work demonstrating that chimpanzee food-call structure and preference are highly correlated (Slocombe and Zuberbühler

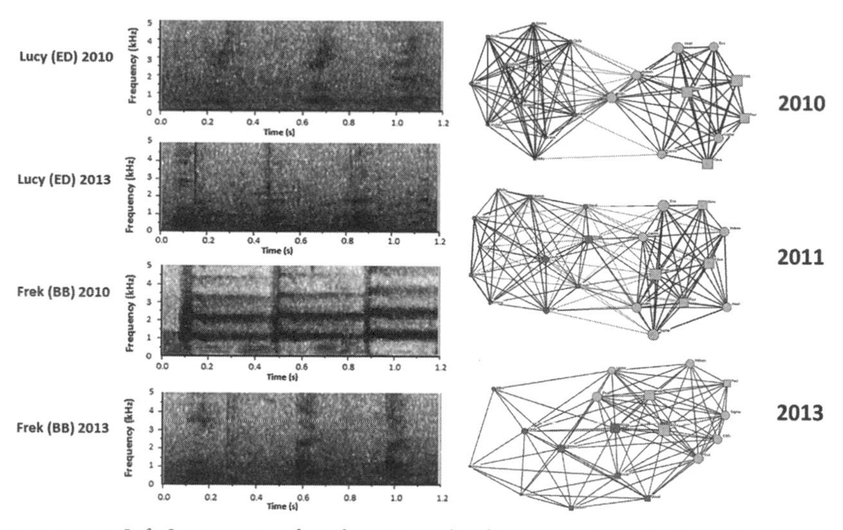

FIGURE 11.2. Left: Spectrograms of rough grunts produced in response to apples from a Scottish and Dutch chimpanzee in 2010 and 2013. BB, Beekse Bergen Safari Park chimpanzee (Dutch); ED, Edinburgh Zoo chimpanzee (Scottish). Right: Sociograms illustrating social convergence between the two subgroups across three years. Squares = males. Circles = females. Dark gray = Scottish. Light gray = Dutch. Sociograms reproduced with permission from Watson et al. (2015a).

2006). Following three years of living together, data indicated that although the Dutch chimpanzees had maintained a high preference for apples, they had modified the acoustic structure of their calls to converge on those produced by the Edinburgh chimpanzees in response to apples (fig. 11.2). Thus, the Dutch chimpanzees demonstrated a decoupling of call structure and preference, which had previously been shown to drive variation in the structure of these calls. Critically, these differences mapped onto changes in social cohesion: social network analyses showed that only after the Dutch and Scottish chimpanzees were integrated and strong between-group relationships had developed, to the extent that they could be considered a single bonded group (i.e., after three years), did their calls converge (Watson et al. 2015a).

While vocal convergence is certainly not a new phenomenon in chimpanzees, or indeed other primate species, to our knowledge Watson et al. (2015a) present the first data suggesting that referential yet seemingly emotionally fixed vocalizations are open to basic social learning processes. From a vocal learning perspective, it is unclear whether these data are best characterized as a case of usage learning or modification of an existing call type. Regardless, the key finding is that call structure can be decoupled from the preference for the referent of the call (a behavioral proxy for the arousal state the referent elicits in the caller). Although direct physiological measures of arousal would have allowed a stronger test of the decoupling of arousal and call structure than behavioral measures alone, it was not possible to collect such measures in this zoo-housed population. Fischer, Wheeler, and Higham (2015) argued that preference was not a suitable behavioral indicator of arousal, and that changes in general arousal levels after integration may have been responsible for the change in call structure. Watson et al. (2015b) countered these arguments and showed that changes in general arousal levels were unlikely to explain the results of Watson et al. (2015a) and that preference was the most pertinent behavioral proxy for arousal elicited by different food types. Although these debates continue, minimally, these data suggest an extra, subtle layer of complexity in the production of functionally referential calls in apes, with comparable results currently absent in monkeys. Hence, these findings represent a potentially important and relevant difference between the two sub-orders.

What Watson et al.'s (2015a) data do not provide evidence for is social learning of an acoustically novel vocalization. All acoustic changes documented occurred within the range known to characterize chimpanzee food calls (Fischer, Wheeler, and Higham 2015; Watson et al. 2015b). In contrast, there is tentative evidence that great apes can learn to produce novel unvoiced sounds. A noteworthy example here comes from the pant hoots of captive chimpanzees in which a translocated male chimpanzee arrived at Lion Country Safari Park, Florida, USA, with a novel pant hoot containing an unvoiced sound, the "Bronx cheer" (Marshall, Wrangham, and Arcadi 1999). Prior to the arrival of the new male, none of the "resident" individuals had been observed producing this syllable within their pant hoots, but all resident males subsequently incorporated this element into their individually-specific pant hoot vocalizations, suggesting they had socially acquired it from the incoming male (Marshall, Wrangham, and Arcadi 1999).

Captive chimpanzees have also been reported to use novel "raspberry" sounds in an intentional manner when begging for food from human caretakers (Hopkins, Taglialatela, and Leavens 2007) and these sounds seem to be socially learned by offspring from their mothers (Taglialatela et al. 2012). Similar comparative data has been forwarded from orangutans where different wild populations have been argued to use distinct unvoiced signals, indicating some form of local cultural innovation (Hardus et al. 2010; see also Lameira 2017). This flexibility is also evident in captive populations, where certain individuals have learned new unvoiced consonant-like vocalizations and even imitated human whistling (Lameira et al. 2013; Wich et al. 2009). Again, these kinds of data suggest a potential degree of articulatory control in apes that is yet to be shown in monkeys.

Until recently, data demonstrating fine laryngeal control and learning of novel voiced vocalizations in apes was largely absent, representing a key discontinuity with human speech (Fitch 2010). However, more recently it has been found that the novel "extended grunts" produced by captive chimpanzees (Hopkins, Taglialatela, and Leavens 2007) are, like their unvoiced "raspberry" counterparts, socially learned (Russell et al. 2013; Taglialatela et al. 2012). A captive orangutan has also been found to produce novel "wookie" calls, which exist outside of the species-typical repertoire (Lameira 2017; Lameira et al. 2016). As the study of vocal flexibility in apes progresses it seems plausible that further examples will accumulate.

USAGE FLEXIBILITY IN GREAT APES

Similar to monkeys, great apes show a more sophisticated level of control and flexibility in how they use their existing vocal repertoire, compared to how they modify the structure of their vocalizations. Great apes can be conditioned to produce species-typical calls and novel attention-getting sounds in response to arbitrary cues (Pierce 1985; Russell et al. 2013). Behavioral data also indicate that great apes have voluntary control over whether to call or not call.

As with monkeys, chimpanzees have also been shown to be sensitive to audience composition in their call production. Chimpanzees have the ability to suppress calling in contexts that are normally associated with call production. For example, females living in highly competitive environments show inhibition of copulatory vocalizations when other high-ranking females are in the vicinity (Townsend, Deschner, and Zuberbühler 2008). Similarly, individuals who had seen a model snake in the forest, selectively increased alarm call production in response to the arrival of close social partners into the area, but not to the arrival of less-preferred partners (Schel et al. 2013a). Equally, in the wild, the production of food calls in males is associated with the presence of preferred social partners in the caller's vicinity (Fedurek and Slocombe 2013; Slocombe et al. 2010) or nearby estrus females (Kalan and Boesch 2015). In line with this, a field experiment showed that lone male chimpanzees, feeding silently, selectively started to produce food calls in response to playbacks that simulated the arrival of other particular males under their feeding tree (Schel et al. 2013a). The probability of the feeding male directing calls at the simulated arrival was greatest when the "arriving" male was higher ranking and a preferred social partner, showing that calls are produced tactically to recruit the most socially valuable individuals (Schel et al. 2013a). Travel "hoo" calls are most likely to be produced by males during travel initiation, when allies are close by (Gruber and Zuberbühler 2013), and victims of aggression are more likely to scream when there is an adult or late adolescent male in the audience, who may be able to support them effectively in the fight (Fedurek, Slocombe, and Zuberbühler 2015). In agonistic contexts, the composition of the audience affects not only the likelihood of victims producing screams, but also the structure of those screams. When facing severe rather than mild aggression, chimpanzee victims increase the pitch and duration of their screams if there is a bystander in the immediate vicinity who outranks their aggressor and therefore could effectively stop the attack (Slocombe and Zuberbühler 2007).

Bonobo and gorilla vocal production is also influenced by the composition and nature of the audience. Bonobos selectively direct contest hoots at equal or higher-ranking individuals (Genty et al. 2014) and during female-female genital contact events, copulation calls were most likely to be produced if the alpha female of the group was present (Clay and Zuberbühler 2012). Lastly, the production rate of gorilla close calls has been shown to be mediated by the visibility and distance of their nearest neighbor. High rates of calling were found to be produced when their nearest neighbor was visible but relatively far away, and when their nearest neighbor was close but not visible (Hedwig et al. 2015).

Taken together, there is an impressive array of evidence across call types and contexts indicating that apes use their calls selectively and flexibly. Furthermore, recent studies with chimpanzees have also revealed that certain alarm calls seem to be produced intentionally in a goal-directed manner. Schel et al. (2013b) tested the production of three types of graded alarm calls against criteria that were originally developed to identify intentional production of gestures in pre-linguistic human infants (Bates et al. 1979) and later adapted for great ape gestures (as reviewed in Liebal et al. 2013). Through field experiments, where model pythons were revealed to certain individuals, Schel et al. (2013b) found that two specific types of alarm calls, alarm "huus" and "waa" barks, were associated with audience checking prior to signaling and were given in response to potentially ignorant "friends" arriving into the area. Alarm calls were also associated with gaze alternation between the python and group members and, crucially, calls seem to be produced in a goal-directed manner, as calls stopped only when all group members were safe (i.e., had approached and seen the ambush predator or moved a safe distance from it).

While this study indicates that alarm calls are given intentionally (see Townsend et al. 2017), with the goal of changing the behavior of others, Crockford et al. (2012) suggest that chimpanzee alarm call production may be influenced not only by the behavior of recipients, but also by their mental states. Crockford and colleagues found that alarm call production was more likely as the audience's potential knowledge regarding the snake decreased, indicating that the goal of alarm calling may be to change the knowledge state of recipients. A follow-up study also found that individuals who heard group member alarm calls immediately before they themselves saw a model snake were less likely to call and show marking behavior (gaze alternation between snake and audience) than individuals who heard non-alarm calls from group members (Crockford, Wittig, and Zuberbühler 2017). The authors suggest that chimpanzees use vocalizations to influence the mental states of others, and target ignorant individuals with important information. This claim contrasts sharply with evidence for macaque monkey mothers who fail to modulate their calling according to the knowledge of their offspring about a predator (Cheney and Seyfarth 1990), but fits with the more sophisticated theory of mind skills shown for chimpanzees compared to monkeys (Call and Tomasello 2008). However, there are a number of lower-level explanations for the pattern of results observed (Liebal et al. 2013, Schel et al. 2013b) that need to be ruled out with further data before mental state attribution can be confidently concluded in chimpanzees.

Although the evidence to date has shown great ape vocalizations can be used flexibly in relation to the presence, identity, behavior, and even mental states of recipients, great apes may not have much flexibility over which call type they produce in any given context, as many great ape vocalizations seem to be highly context bound. However, this perception might be partly fueled by the historical focus vocal researchers have put on highly context-specific calls and their ability to function referentially. This has likely created a corresponding bias in the literature and interpretation of results in favor of context specificity at the expense of reporting vocal signals used flexibly over several contexts (Liebal et al. 2013; Wheeler and Fischer 2012). Although Pollick and de Waal (2007) found that captive bonobo and chimpanzee vocalizations and facial expressions were tied to predictable contexts to a higher degree than gestural signals, recent research has identified functional flexibility in bonobo peep calls (Clay, Archbold, and Zuberbühler 2015). Following a study showing that human infants produced acoustically similar vocalizations in positive, neutral, and negative contexts (Oller et al. 2013), demonstrating a decoupling of affective state and vocal production, Clay, Archbold, and Zuberbühler (2015) examined the peeps of wild bonobos in different affective contexts. Although peeps produced in negative contexts were acoustically distinct, peeps produced across neutral and positive affective contexts were acoustically indiscriminate, indicating some degree of functional flexibility. It seems likely as less context-bound, more flexible, vocal signals are being increasingly identified as sharing important features with human language (e.g., Wheeler and Fischer 2012), more empirical studies will focus on these types of signals and further evidence for functional flexibility is likely to be found.

Neurobiological Foundations of Great Ape Vocal Flexibility

In light of the emerging data demonstrating further articulatory and usage flexibility in great ape calls, a sensible question to pose at this point is: can this variation be accounted for by any neurobiological differences? Understandably, due to ethical and legal constraints on invasive research with great apes, very little research probing the exact neurological substrates involved in great-ape vocal production has been conducted. Thus, in comparison to the macaque and squirrel monkey work, we know very little about the presence or strength of different neural pathways involved in great ape vocal production and the few available data come from single individuals that were studied 50–60 years ago (e.g., Bailey, Bonin, and McCulloch 1950; Kuypers 1958).

Modern techniques are starting to challenge our previous understanding about vocal production neural pathways (e.g., Kumar, Croxson, and Simonyan 2016), so currently we just do not have the relevant comparable data to make inferences about the presence, or nature, of neural pathways underpinning great ape, as opposed to monkey, vocal production. A different approach has been successfully taken, however, by Taglialatela and colleagues (2008), who examined whether the chimpanzee homologue to Broca's area, which seems to play an important role in human language production (Lieberman et al. 2007), is active during chimpanzee vocal production. Through employing positron emission tomography, Taglialatela et al. (2008) elegantly demonstrated that chimpanzee vocal production engages subcortical and neocortical structures including the homologue of Broca's area. While it should be pointed out the vocalizations under investigation were human-directed attention-getting calls, some of which were unvoiced (Hopkins, Taglialatela, and Leavens 2007; Russell et al. 2013; Taglialatela et al. 2012), these data nevertheless indicate important similarities between the cortical structures involved in vocal production in humans and chimpanzees. Further work probing the extent to which cortical structures also mediate the production of voiced and species-typical vocalizations in chimpanzees, as has been demonstrated in monkeys (Hage and Nieder 2013), is now key to unraveling how similar and how different the neural structures underpinning ape, monkey, and human vocal production are.

Great Ape Vocal Flexibility: Implications for Language Evolution?

Here, we have reviewed recent behavioral and neurobiological developments in chimpanzee and other ape vocal flexibility at both the articulatory and usage level. These data are important for two reasons. First, and more generally, they provide a window into the minds of our closest living relatives and support behavioral data collected in parallel over the last 50 years, from the field and captivity, suggesting apes excel in terms of socio-cognitive abilities (Call and Tomasello 2008; de Waal and Ferrari 2012). Second, and most critically, these data can help shed light on the evolutionary progression of human vocal flexibility.

The general consensus among researchers in the field is that the human ability to modulate and acquire vocalizations independently of arousal state, which is so central to the instantiation of a culturally inherited, conventional communication system, is a derived trait and not built upon capacities inherited from our primate ancestors (Arbib et al. 2008; Tomasello 2010; Wheeler and Fischer 2012). This conclusion has, however, been predominantly based on comparative data from a very limited number of monkey species and rarely have our closest living relatives, chimpanzees, or indeed other apes been considered. Here, we have reviewed more recent studies that update this view, demonstrating vocal flexibility in both monkeys and apes. By bringing to light the most recent advances in understanding vocal flexibility and its neurobiological underpinnings, we argue that previous claims that the vocal production abilities of monkeys are inexorably tied to emotional states may be premature (Coudé et al. 2011; Hage and Nieder 2013; Hihara et al. 2003). Furthermore, it is inappropriate to extrapolate from these capabilities to those of the great apes, who appear to demonstrate even greater voluntary control in naturalistic contexts. In the past, this was partially justified by a lack of systematic data on ape vocal flexibility, but as we demonstrate here, this argument no longer holds. A number of studies now show that not only are apes capable of flexibly using vocalizations in response to subtle social influences, in some circumstances independently from associated arousal states, they also demonstrate a finer degree of control over their articulators than monkeys (Hopkins, Taglialatela, and Leavens 2007; Lameira et al. 2013; Wich et al. 2009), leading to greater acoustic plasticity. The fact that, on a behavioral level, ape vocal production skills appear more sophisticated than those of monkeys is suggestive of a gradual evolutionary, rather than a derived, de novo scenario (Petkov and Jarvis 2012).

Acknowledgments

We thank the organizers and participants of the Chimpanzees in Context meeting in Chicago, 2016, as well as Adriano Lameira and Jared Taglialetela for stimulating discussion and input. S. W. T. and S. K. W. were funded by the University of Zurich and Swiss National Science Foundation (PP00P3_1163850), and K. E. S. was funded by University of York.

References

Ackermann, H., S. R. Hage, and W. Ziegler. 2014. "Brain mechanisms of acoustic communication in humans and nonhuman primates: An evolutionary perspective." *Behavioral and Brain Sciences* 37 (6): 529.

Arbib, M. A., K. Liebal, S. Pika, M. C. Corballis, C. Knight, D. A. Leavens, D. Maestripieri, J. E. Tanner, M. A. Arbib, and K. Liebal. 2008. "Primate vocalization, gesture, and the evolution of human language." *Current Anthropology* 49 (6): 1053–76.

Arnold, K., and K. Zuberbühler. 2006. "Language evolution: Semantic combinations in primate calls." *Nature* 441 (7091): 303.

Bailey, P., G. Von Bonin, and W. S. McCulloch. 1950. *The Isocortex of the Chimpanzee*. Urbana: University of Illinois Press.

Bates, E., L. Benigni, I. Bretherton, L. Camaioni, and V. Volterra. 1979. *The Emergence of Symbols: Communication and Cognition in Infancy*. New York: Academic Press.

Bouchet, H., C. Blois-Heulin, and A. Lemasson. 2013. "Social complexity parallels vocal complexity: A comparison of three non-human primate species." *Frontiers in Psychology* 4: 390.

Caine, N. G., R. L. Addington, and T. L. Windfelder. 1995. "Factors affecting the rates of food calls given by red-bellied tamarins." *Animal Behaviour* 50 (1): 53–60.

Call, J., and M. Tomasello. 2008. "Does the chimpanzee have a theory of mind? 30 years later." *Trends in Cognitive Sciences* 12 (5): 187–92.

Cäsar, C., K. Zuberbühler, R. J. Young, and R. W. Byrne. 2013. "Titi monkey call sequences vary with predator location and type." *Biology Letters* 9 (5): 20130535.

Chapman, C. A., and L. Lefebvre. 1990. "Manipulating foraging group size: Spider monkey food calls at fruiting trees." *Animal Behaviour* 39 (5): 891–96.

Cheney, D. L., and R. M. Seyfarth. 1985. "Vervet monkey alarm calls: Manipulation through shared information?" *Behaviour* 94 (1): 150–66.

Cheney, D. L., and R. M. Seyfarth. 1990. "Attending to behaviour versus attending to knowledge: Examining monkeys' attribution of mental states." *Animal Behaviour* 40 (4): 742–53.

Cheney, D. L., and R. M. Seyfarth. 1992. *How Monkeys See the World: Inside the Mind of Another Species*. Chicago: University of Chicago Press.

Cheney, D. L., and R. M. Seyfarth. 2008. *Baboon Metaphysics: The Evolution of a Social Mind*. Chicago: University of Chicago Press.

Clarke, E., U. H. Reichard, and K. Zuberbühler. 2006. "The syntax and meaning of wild gibbon songs." *PLoS One* 1 (1): e73.

Clay, Z., J. Archbold, and K. Zuberbühler. 2015. "Functional flexibility in wild bonobo vocal behaviour." *PeerJ* 3: e1124.

Clay, Z., C. L. Smith, and D. T. Blumstein. 2012. "Food-associated vocalizations in mammals and birds: What do these calls really mean?" *Animal Behaviour* 83 (2): 323–30.

Clay, Z., and K. Zuberbühler. 2009. "Food-associated calling sequences in bonobos." *Animal Behaviour* 77 (6): 1387–96.

Clay, Z., and K. Zuberbühler. 2012. "Communication during sex among female bonobos: Effects of dominance, solicitation and audience." *Scientific Reports* 2: 291.

Collier, K., B. Bickel, C. P. van Schaik, M. B. Manser, and S. W. Townsend. 2014. "Language evolution: Syntax before phonology?" *Proceedings of the Royal Society of London B* 281 (1788): 20140263.

Coudé, G., P. F. Ferrari, F. Rodà, M. Maranesi, E. Borelli, V. Veroni, F. Monti, S. Rozzi, and L. Fogassi. 2011. "Neurons controlling voluntary vocalization in the macaque ventral premotor cortex." *PLoS One* 6 (11): e26822.

Coye, C., K. Ouattara, K. Zuberbühler, and A. Lemasson. 2015. "Suffixation influences receivers' behaviour in non-human primates." *Proceedings of the Royal Society of London B* 282 (1807): 20150265.

Coye, C., K. Zuberbühler, and A. Lemasson. 2016. "Morphologically structured vocalizations in female Diana monkeys." *Animal Behaviour* 115: 97–105.

Crockford, C., I. Herbinger, L. Vigilant, and C. Boesch. 2004. "Wild chimpanzees produce group-specific calls: A case for vocal learning?" *Ethology* 110 (3): 221–43.

Crockford, C., R. M. Wittig, R. Mundry, and K. Zuberbühler. 2012. "Wild chimpanzees inform ignorant group members of danger." *Current Biology* 22 (2): 142–46.

Crockford, C., R. M. Wittig, and K. Zuberbühler. 2017. "Vocalizing in chimpanzees is influenced by social-cognitive processes." *Science Advances* 3 (11): e1701742.

Deacon, T., and D. Poeppel. 1997. "The symbolic species: The co-evolution of language and the brain." *Nature* 388 (6644): 734.

de Waal, F. B. M., and P. F. Ferrari, eds. 2012. *The Primate Mind*. Cambridge, MA: Harvard University Press.

Egnor, S. E. R., and M. D. Hauser. 2004. "A paradox in the evolution of primate vocal learning." *Trends in Neurosciences* 27 (11): 649–54.

Ey, E., C. Rahn, K. Hammerschmidt, and J. Fischer. 2009. "Wild female olive baboons adapt their grunt vocalizations to environmental conditions." *Ethology* 115 (5): 493–503.

Fedurek, P., and K. E. Slocombe. 2013. "The social function of food-associated calls in male chimpanzees." *American Journal of Primatology* 75 (7): 726–39.

Fedurek, P., K. E. Slocombe, and K. Zuberbühler. 2015. "Chimpanzees communicate to two different audiences during aggressive interactions." *Animal Behaviour* 110: 21–28.

Fischer, J., K. Hammerschmidt, and D. Todt. 1998. "Local variation in Barbary macaque shrill barks." *Animal Behaviour* 56 (3): 623–29.

Fischer, J., B. C. Wheeler, and J. P. Higham. 2015. "Is there any evidence for vocal learning in chimpanzee food calls?" *Current Biology* 25 (21): R1028–29.

Fitch, W. T. 2010. *The Evolution of Language*. Cambridge: Cambridge University Press.

Genty, E., Z. Clay, C. Hobaiter, and K. Zuberbühler. 2014. "Multi-modal use of a socially directed call in bonobos." *PLoS One* 9 (1): e84738.

Ghazanfar, A. A., and D. Rendall. 2008. "Evolution of human vocal production." *Current Biology* 18 (11): 457–60.

Gill, S. A., and A. M.-K. Bierema. 2013. "On the meaning of alarm calls: A review of functional reference in avian alarm calling." *Ethology* 119 (6): 449–61.

Goldstein, M. H., A. P. King, and M. J. West. 2003. "Social interaction shapes babbling: Testing parallels between birdsong and speech." *Proceedings of the National Academy of Sciences* 100 (13): 8030–35.

Gouzoules, H., and S. Gouzoules. 1990. "Matrilineal signatures in the recruitment screams of pigtail macaques, *Macaca nemestrina*." *Behaviour* 115 (3): 327–47.

Gruber, T., and K. Zuberbühler. 2013. "Vocal recruitment for joint travel in wild chimpanzees." *PLoS One* 8 (9): e76073.

Hage, S. R., N. Gavrilov, and A. Nieder. 2013. "Cognitive control of distinct vocalizations in rhesus monkeys." *Journal of Cognitive Neuroscience* 25 (10): 1692–1701.

Hage, S. R., and A. Nieder. 2013. "Single neurons in monkey prefrontal cortex encode volitional initiation of vocalizations." *Nature Communications* 4 (2409): 1–11.

Hammerschmidt, K., T. Freudenstein, and U. Jürgens. 2001. "Vocal development in squirrel monkeys." *Behaviour* 138 (9): 1179–1204.

Hardus, M. E., A. R. Lameira, I. Singleton, H. C. Morrogh-Bernard, C. D. Knott, M. Ancrenaz, S. S. Utami Atmoko, and S. A. Wich. 2010. "A description of the orangutan's vocal and sound repertoire, with a focus on geographic variation." In *Orangutans: Geographic Variation in Behavioral Ecology and Conservation*, edited by S. A. Wich, S. S. Utami Atmoko, T. Mitra Setia, and C. P. Van Schaik, 49–64. Oxford: Oxford University Press.

Hauser, M. D. 1992. "Articulatory and social factors influence the acoustic structure of rhesus monkey vocalizations: A learned mode of production?" *Journal of the Acoustical Society of America* 91 (4): 2175–79.

Hauser, M. D., N. Chomsky, and W. T. Fitch. 2002. "The faculty of language: What is it, who has it, and how did it evolve?" *Science* 298 (5598): 1569–79.

Hedwig, D., R. Mundry, M. M. Robbins, and C. Boesch. 2015. "Audience effects, but not environmental influences, explain variation in gorilla close distance vocalizations—A test of the acoustic adaptation hypothesis." *American Journal of Primatology* 77 (12): 1239–52.

Hihara, S., H. Yamada, A. Iriki, and K. Okanoya. 2003. "Spontaneous vocal differentiation of coo-calls for tools and food in Japanese monkeys." *Neuroscience Research* 45 (4): 383–89.

Hopkins, W. D., J. L. Russell, and C. Cantalupo. 2007. "Neuroanatomical correlates of handedness for tool use in chimpanzees (*Pan troglodytes*) implication for theories on the evolution of language." *Psychological Science* 18 (11): 971–77.

Hopkins, W. D., J. P. Taglialatela, and D. A. Leavens. 2007. "Chimpanzees differentially produce novel vocalizations to capture the attention of a human." *Animal Behaviour* 73 (2): 281–86.

Hurford, J. R. 2007. *The Origin of Meaning*. Oxford: Oxford University Press.

Hurford, J. R. 2011. *The Origins of Grammar: Language in the Light of Evolution II*. Oxford: Oxford University Press.

Jürgens, U. 1998. "Neuronal control of mammalian vocalization, with special reference to the squirrel monkey." *Naturwissenschaften* 85 (8): 376–88.

Jürgens, U. 2009. "The neural control of vocalization in mammals: A review." *Journal of Voice* 23 (1): 1–10.

Kalan, A. K., and C. Boesch. 2015. "Audience effects in chimpanzee food calls and their potential for recruiting others." *Behavioral Ecology and Sociobiology* 69 (10): 1701–12.

Kumar, V., P. L. Croxson, and K. Simonyan. 2016. "Structural organization of the laryngeal motor cortical network and its implication for evolution of speech production." *Journal of Neuroscience* 36 (15): 4170–81.

Kuypers, H. G. J. M. 1958. "Some projections from the peri-central cortex to the pons and lower brain stem in monkey and chimpanzee." *Journal of Comparative Neurology* 110 (2): 221–55.

Lameira, A. R. 2017. "Bidding evidence for primate vocal learning and the cultural substrates for speech evolution." *Neuroscience & Biobehavioral Reviews* 83: 429–39.

Lameira, A. R., M. E. Hardus, B. Kowalsky, H. de Vries, B. M. Spruijt, E. H. M. Sterck, R. W. Shumaker, and S. A. Wich. 2013. "Orangutan (*Pongo spp.*) whistling and implications for the emergence of an open-ended call repertoire: A replication and extension." *Journal of the Acoustical Society of America* 134 (3): 2326–35.

Lameira, A. R., M. E. Hardus, A. Mielke, S. A. Wich, and R. W. Shumaker. 2016. "Vocal fold control beyond the species-specific repertoire in an orang-utan." *Scientific Reports* 6: 30315.

Liebal, K., B. M. Waller, K. E. Slocombe, and A. M. Burrows. 2013. *Primate Communication: A Multimodal Approach*. Cambridge: Cambridge University Press.

Lieberman, P., S. Fecteau, H. Théoret, R. R. Garcia, F. Aboitiz, A. MacLarnon, R. Melrose, T. Riede, I. Tattersall, and P. Lieberman. 2007. "The evolution of human speech: Its anatomical and neural bases." *Current Anthropology* 48 (1): 39–66.

Macedonia, J. M., and C. S. Evans. 1993. "Essay on contemporary issues in ethology: Variation among mammalian alarm call systems and the problem of meaning in animal signals." *Ethology* 93 (3): 177–97.

Marshall, A. J., R. W. Wrangham, and A. C. Arcadi. 1999. "Does learning affect the structure of vocalizations in chimpanzees?" *Animal Behaviour* 58 (4): 825–30.

McGregor, P. K. 1993. "Signalling in territorial systems: A context for individual identification, ranging and eavesdropping." *Philosophical Transactions of the Royal Society of London B* 340 (1292): 237–44.

Mitani, J. C., and J. Gros-Louis. 1998. "Chorusing and call convergence in chimpanzees: Tests of three hypotheses." *Behaviour* 135 (8): 1041–64.

Oller, D. K., E. H. Buder, H. L. Ramsdell, A. S. Warlaumont, L. Chorna, and R. Bakeman. 2013. "Functional flexibility of infant vocalization and the emergence of language." *Proceedings of the National Academy of Sciences* 110 (16): 6318–23.

Ouattara, K., A. Lemasson, and K. Zuberbühler. 2009. "Campbell's monkeys use affixation to alter call meaning." *PLoS One* 4 (11): e7808.

Owren, M. J., R. T. Amoss, and D. Rendall. 2011. "Two organizing principles of vocal production: Implications for nonhuman and human primates." *American Journal of Primatology* 73 (6): 530–44.

Owren, M. J., J. A. Dieter, R. M. Seyfarth, and D. L. Cheney. 1992. "Evidence of limited modification in the vocalizations of cross-fostered rhesus (*Macaca mulatta*) and Japanese (*M. fuscata*) macaques." In *Topics in Primatology: Human Origins*, edited by T. Nishida, W. C. McGrew, P. Marler, M. Pickford, and F. B. M. de Waal, 257–70. Tokyo: University of Tokyo Press.

Owren, M. J., J. A. Dieter, R. M. Seyfarth, and D. L. Cheney. 1993. "Vocalizations of rhesus (*Macaca mulatta*) and Japanese (*M. Fuscata*) macaques cross-fostered between species show evidence of only limited modification." *Developmental Psychobiology* 26 (7): 389–406.

Papworth, S., A.-S. Böse, J. Barker, A. M. Schel, and K. Zuberbühler. 2008. "Male blue monkeys alarm call in response to danger experienced by others." *Biology Letters* 4 (5): 472–75.

Petkov, C. I., and E. D. Jarvis. 2012. "Birds, primates, and spoken language origins: Behavioral phenotypes and neurobiological substrates." *Frontiers in Evolutionary Neuroscience* 4 (12).

Pierce, J. D. 1985. "A review of attempts to condition operantly alloprimate vocalizations." *Primates* 26 (2): 202–13.

Pisanski, K., V. Cartei, C. McGettigan, J. Raine, and D. Reby. 2016. "Voice modulation: A window into the origins of human vocal control?" *Trends in Cognitive Sciences* 20 (4): 304–18.

Pollick, A. S., and F. B. M. de Waal. 2007. "Ape gestures and language evolution." *Proceedings of the National Academy of Sciences* 104 (19): 8184–89.

Rukstalis, M., J. E. Fite, and J. A. French. 2003. "Social change affects vocal structure in a Callitrichid primate (*Callithrix kuhlii*)." *Ethology* 109 (4): 327–40.

Russell, J. L., J. M. McIntyre, W. D. Hopkins, and J. P. Taglialatela. 2013. "Vocal learning of a communicative signal in captive chimpanzees, *Pan troglodytes*." *Brain and Language* 127 (3): 520–25.

Schel, A. M., Z. Machanda, S. W. Townsend, K. Zuberbühler, and K. E. Slocombe. 2013a. "Chimpanzee food calls are directed at specific individuals." *Animal Behaviour* 86 (5): 955–65.

Schel, A. M., S. W. Townsend, Z. Machanda, K. Zuberbühler, and K. E. Slocombe. 2013b. "Chimpanzee alarm call production meets key criteria for intentionality." *PLoS One* 8 (10): e76674.

Seyfarth, R. M., and D. L. Cheney. 2010. "Production, usage, and comprehension in animal vocalizations." *Brain and Language* 115 (1): 92–100.

Seyfarth, R. M., D. L. Cheney, and P. Marler. 1980. "Monkey responses to three different alarm calls: Evidence of predator classification and semantic communication." *Science* 210 (4471): 801–3.

Simonyan, K., and B. Horwitz. 2011. "Laryngeal motor cortex and control of speech in humans." *Neuroscientist* 17 (2): 197–208.

Slocombe, K. E., T. Kaller, L. Turman, S. W. Townsend, S. Papworth, P. Squibbs, and K. Zuberbühler. 2010. "Production of food-associated calls in wild male chimpanzees is dependent on the composition of the audience." *Behavioral Ecology and Sociobiology* 64 (12): 1959–66.

Slocombe, K. E., and K. Zuberbühler. 2005. "Functionally referential communication in a chimpanzee." *Current Biology* 15 (19): 1779–84.

Slocombe, K. E., and K. Zuberbühler. 2006. "Food-associated calls in chimpanzees: Responses to food types or food preferences?" *Animal Behaviour* 72 (5): 989–99.

Slocombe, K. E., and K. Zuberbühler. 2007. "Chimpanzees modify recruitment screams as a function of audience composition." *Proceedings of the National Academy of Sciences* 104 (43): 17228–33.

Spillmann, B., L. P. Dunkel, M. A. Van Noordwijk, R. N. A. Amda, A. R. Lameira, S. A. Wich, and C. P. Van Schaik. 2010. "Acoustic properties of long calls given by flanged male orangutans (*Pongo pygmaeus wurmbii*) reflect both individual identity and context." *Ethology* 116 (5): 385–95.

Taglialatela, J. P., L. Reamer, S. J. Schapiro, and W. D. Hopkins. 2012. "Social learning of a communicative signal in captive chimpanzees." *Biology Letters* 8 (4): 498–501.

Taglialatela, J. P., J. L. Russell, J. A. Schaeffer, and W. D. Hopkins. 2008. "Communicative signaling activates 'Broca's' homolog in chimpanzees." *Current Biology* 18 (5): 343–48.

Takahashi, D. Y., A. R. Fenley, Y. Teramoto, D. Z. Narayanan, J. I. Borjon, P. Holmes, and A. A. Ghazanfar. 2015. "The developmental dynamics of marmoset monkey vocal production." *Science* 349 (6249): 734–38.

Takahashi, D. Y., D. A. Liao, and A. A. Ghazanfar. 2017. "Vocal learning via social reinforcement by infant marmoset monkeys." *Current Biology* 27 (12): 1844–52.

Tomasello, M. 2010. *Origins of Human Communication*. Cambridge, MA: MIT Press.

Townsend, S. W., T. Deschner, and K. Zuberbühler. 2008. "Female chimpanzees use copulation calls flexibly to prevent social competition." *PLoS One* 3 (6): e2431.

Townsend, S. W., S. E. Koski, R. W. Byrne, K. E. Slocombe, B. Bickel, M. Boeckle, I. B. Goncalves, J. M. Burkart, T. Flower, and F. Gaunet. 2017. "Exorcising Grice's ghost: An empirical approach to studying intentional communication in animals." *Biological Reviews* 92 (3): 1427–33.

Townsend, S. W., and M. B. Manser. 2013. "Functionally referential communication in mammals: The past, present and the future." *Ethology* 119 (1): 1–11.

Watson, S. K., S. W. Townsend, A. M. Schel, C. Wilke, E. K. Wallace, L. Cheng, V. West, and K. E. Slocombe. 2015a. "Reply to Fischer et al." *Current Biology* 25 (21): R1030–31.

Watson, S. K., S. W. Townsend, A. M. Schel, C. Wilke, E. K. Wallace, L. Cheng, V. West, and K. E. Slocombe. 2015b. "Vocal learning in the functionally referential food grunts of chimpanzees." *Current Biology* 25 (4): 495–99.

Wheeler, B. C., and J. Fischer. 2012. "Functionally referential signals: A promising paradigm whose time has passed." *Evolutionary Anthropology* 21 (5): 195–205.

Wich, S. A., and H. de Vries. 2006. "Male monkeys remember which group members have given alarm calls." *Proceedings of the Royal Society of London B* 273 (1587): 735–40.

Wich, S. A., K. B. Swartz, M. E. Hardus, A. R. Lameira, E. Stromberg, and R. W. Shumaker. 2009. "A case of spontaneous acquisition of a human sound by an orangutan." *Primates* 50 (1): 56–64.

Winter, P., P. Handley, D. Ploog, and D. Schott. 1973. "Ontogeny of squirrel monkey calls under normal conditions and under acoustic isolation." *Behaviour* 47 (3): 230–39.

Zajonc, R. B, and S. M. Sales. 1966. "Social facilitation of dominant and subordinate responses." *Journal of Experimental Social Psychology* 2 (2): 160–68.

Zuberbühler, K. 2005. "The phylogenetic roots of language: Evidence from primate communication and cognition." *Current Directions in Psychological Science* 14 (3): 126–30.

Zuberbühler, K. 2006. "Language evolution: The origin of meaning in primates." *Current Biology* 16 (4): 123–25.

Zuberbühler, K. 2008. "Audience effects." *Current Biology* 18 (5): 189–90.

Zuberbühler, K., D. L. Cheney, and R. M. Seyfarth. 1999. "Conceptual semantics in a nonhuman primate." *Journal of Comparative Psychology* 113 (1): 33.

12

Vocal Communication in Chimpanzees and Bonobos: A Window into the Social World

ZANNA CLAY

Introduction

Chimpanzees and bonobos are humans' closest living relatives, with our lineage diverging from theirs only around five to seven million years ago (MYA). The two species are themselves very closely related, with recent estimates suggesting a divergence occurring between 1.5 to 2.5 MYA (Langergraber et al. 2012; Prüfer et al. 2012). As the contents of this book beautifully attest, substantial progress has been made in the last several decades in our understanding of chimpanzees. A particular acceleration has occurred in the past 15 years, with data now available from over ten habituated field sites and numerous semi-habituated communities and a wealth of data from captivity (McGrew 2017). Comparatively less is known about bonobos, which is perhaps not surprising, given their later discovery, more remote and restricted geographic range, and smaller distribution in captivity. Nevertheless, a balanced comparison of the two species is necessary in order to understand the context in which chimpanzees have evolved, as well as to understand the evolutionary basis for our own capacities, as humans (Gruber and Clay 2016; Taglialatela et al., chapter 4 this volume; Yamamoto, chapter 14 this volume). Fortunately, we are now in a position to make more informed comparisons. The publication of an edited book by Hare and Yamamoto (2017), documenting recent progress in bonobo research, demonstrates the considerable advances that have been made in understanding bonobos and how they compare to chimpanzees.

Vocalizations present a useful window through which we can explore the minds and social worlds of our great ape cousins. This follows the logic that in any social animal, communication and cognition are intimately related (Seyfarth and Cheney 2003). This approach has already proved fruitful in studying other nonhuman primates, with research into monkey vocal communication revealing intricate information about how monkeys see the world and navigate their social landscapes (Cheney and Seyfarth 1990; Seyfarth et al. 2010). In this chapter, I investigate what vocalizations can tell us about the social awareness of chimpanzees and bonobos and what clues their vocalizations can provide to their underlying social organization.

Chimpanzees in Context

At the time of writing this chapter, four sub-species of chimpanzees (*P. t. schweinfurthii*; *P. t. verus*; *P. t. troglodytes*; *P. t. eliotii*) collectively occupy a geographic range covering 21 countries in Africa. By comparison, bonobos are restricted to the Democratic Republic of the Congo and are geographically separated from chimpanzees by the Congo River. Although there is still no consensus about what might have led to *Pan* speciation, evidence from riverine sediments and paleotopography of the region indicate that the formation of the Congo River predates the chimpanzee-bonobo speciation by over 30 million years, thus cannot have been the cause. Using an accumulation of biogeographical evidence, Takemoto, Kawamoto, and Furuichi (2015) propose that the divergence may have occurred during a rare dry period in the Pleistocene, during which there was a pronounced and extended reduction in water flow. During this time, it is plausible that a founder population could have crossed the river on foot to reach its left bank. This "corridor hypothesis" is further supported by similar divergence times and lower diversity of other primate species found on the river's left bank. Genetic analyses also support a monophyletic origin of bonobos (Fischer et al. 2011; Zsurka et al. 2010), which suggests that the ancestral population of bonobos was probably small (Hey 2010) and may have experienced a genetic bottleneck. Certain mutations could thus have become fixed relatively quickly, which could explain some of the apparently rapid divergences in bonobo morphology, genetics, and behavior from those of chimpanzees.

One of the most obvious communicative differences between bonobos and chimpanzees is their vocal pitch (Grawunder et al. 2018). Compared to chimpanzees and most other mammals of equivalent size, bonobos produce exceptionally high-pitched vocalizations, with bonobo vocalizations being around an octave higher than those produced by chimpanzees (Grawunder et al. 2018). Given that lower frequencies travel more efficiently in dense environments than higher frequencies, it is unclear what adaptive pressures might underlie this this pitch shift or whether it was a result of genetic drift. Alternatively, it has been suggested that higher pitch might have accompanied a suite of other paedomorphic traits that were selected as a result of female choice for selection against aggression (Hare, Wobber, and Wrangham 2012). This is supported by evidence that bonobos display a number of paedomorphic traits as compared to chimpanzees (see also Behringer et al., chapter 2 this volume). As well as being physically smaller, they show reduced cranial, mandible, and tooth size (Cramer 1977; Latimer et al. 1981; Pilbrow 2006), feature reduced sexual dimorphism, and retain the juvenile "white tail-tuft" and "pink lips" into adulthood (Kano 1992). Behaviorally, bonobos are less aggressive than chimpanzees, and show enhanced playfulness and socio-sexual behavior as well as high levels of social tolerance (reviewed in Hare, Wobber, and Wrangham 2012). Some aspects of bonobo development also appear to be delayed as compared to those of chimpanzees. In this chapter, I will examine how certain features of bonobo vocal communication might relate to this hypothesis.

Pan Vocal Repertoires

Although chimpanzees and bonobos experience considerable ecological diversity (Boesch, Hohmann, and Marchandt. 2002), both species *typically* dwell in dense tropical rainforest habitats. In this environment, individuals are regularly out of sight of each other and thus must frequently depend on vocalizations to coordinate their movements and activities. The fission-fusion dynamics that characterize both species place further demands on long- distance communication, where vocalizations enable foraging sub-parties to maintain contact and coordinate decision making.

Given overlaps in their socio-ecologies and phylogeny, similarities in their vocal communication should be expected. Both species have highly graded vocal repertoires (de Waal 1988; Marler and Tenaza 1977), where gradation refers to fuzzy scaling of acoustic similarity between call types (Marler 1976). Compared to discrete vocal systems, graded vocal systems are notoriously difficult to analyze, which might explain why there have been few attempts to quantitatively describe the chimpanzee and bonobo vocal repertoires. Nevertheless, acoustic variation present in graded repertoires creates potential for considerable communicative complexity, provided that receivers can perceive acoustic differences among them, which they can link to signal meaning (Hauser 1998).

Early descriptions of the chimpanzee vocal repertoire were made by Reynolds (1965) and by Goodall (1968), although both were primarily based on contextual and acoustic descriptions. A more quantitative approach was undertaken by Marler and Tenaza (1977), who identified more than a dozen call types. This repertoire, still widely accepted today, includes multiple variants of screams, hoots, barks, grunts, pants, and squeaks.

The only major description of the bonobo vocal repertoire has been made by de Waal (1988), who compared the vocal repertoire of captive bonobos with what has been described for chimpanzees (Marler and Tenaza 1977; van Hooff 1973). According to de Waal (1988), the bonobo vocal repertoire consists of a similar number of call types, which includes three hoots, three peeps, two barks, as well as grunts, pant laughs, pout moans, and screams. Since this study, one preliminary analysis was made in the wild (Bermejo and Omedes 1999), which largely confirmed these findings. In addition, the authors stressed the role of call combinations in bonobo vocal communication. The flexible use of heterogeneous vocal sequences highlights a potential for calls to be combined in different ways to provide different meanings, something which has received support from analysis of their food-associated call sequences (Clay and Zuberbühler 2009, 2011a). Although promising, the repertoire study by Bermejo and Omedes (1999) was limited in that the study community was not habituated and no acoustic analyses were made. A detailed quantitative analysis of the wild bonobo vocal repertoire is required, and remains outstanding.

As noted above, the most striking difference between the vocal repertoires of the two species is the pitch: bonobos have remarkably high-pitched vocalizations compared to chimpanzees, despite overlapping in physical size (de Waal 1988; Grawunder et al. 2018; Mitani and Gros-Louis 1995). In a direct comparison, Mitani and Gros-Louis (1995) found that chimpanzee screams had a mean frequency of 1275 Hz compared to 2846 Hz for bonobos. Building upon this, we recently compared the highest and lowest pitched species-typical calls of individuals in two communities of wild chimpanzees and two communities of wild bonobos (Grawunder et al. 2018). Our results confirmed that bonobo vocalizations were around twice the acoustic frequency of those of chimpanzees; moreover, they achieve a much greater frequency range, being able to produce low frequency vocalizations despite generally vocalizing in higher ranges.

In terms of proximate causes, our recent morphometric analyses revealed that higher fundamental frequencies can be explained by laryngeal morphology: bonobos have significantly shorter vocal folds than chimpanzees (Grawunder et al. 2018). The ultimate causes underlying these species differences are not yet well understood. It has been suggested that the climactic peaks in chimpanzee pant hoots may be an indicator of male fitness, whereby the ability to maintain vocal control at the edge of the frequency range can be an honest indicator of male quality (Riede, Arcadi, and Owren 2007). Although comparisons of long-distance vocalizations remain outstanding, the male display vocalization in bonobos—the "contest hoot"—is among the highest frequency vocalizations in the adult vocal repertoire (Clay et al. in prep). Higher-pitched vocalizations might also reflect shifts away from both between- and within-group aggression in bonobos. In chimpanzees, the lowest frequency calls—roars and pant hoot build ups—are associated with between- and within-group conflicts (Marler and Tenaza 1977). Given the reduced levels of aggression in bonobos (Wilson et al. 2014), pressure to maintain calls with a low fundamental frequency in contest contexts may have been reduced.

It is possible that higher vocal pitch in bonobos could also be related to selection against aggression and to retention of juvenile traits, a process which some have likened to "self-domestication" (Hare, Wobber, and Wrangham 2012). This would be consistent with similar patterns observed in artificially domesticated Siberian red foxes, who, unlike their control counterparts, retained a juvenile communicative repertoire into adulthood (Gogoleva et al. 2008). Domestic cat vocalizations are higher in mean frequency as compared to their wild counterparts (Nicastro 2004). Intriguingly, bonobo calls analyzed were also shorter in duration than those of chimpanzees (Grawunder et al. 2018), a pattern that also resembles domesticated cats as compared to wild cat species (Nicastro 2004). Anatomical retention of juvenilized features could plausibly gi ve rise to related variations in laryngeal mechanisms and shorter vocal tract length (Mitani and Gros-Louis 1995, see also Behringer et al., chapter 2 this volume). Although a quantitative comparison of the vocal repertoires of immature and mature chimpanzees and bonobos is outstanding, our recent analyses of fundamental frequency and vocal tract morphology lend support to this hypothesis.

Despite striking differences in vocal pitch, we nevertheless find numerous parallels in both the acoustic form and contextual usage between the two *Panins*. For instance, the bonobo pant laugh, pout moan, low hoots, and wieew-bark appear homologous to those produced by chimpanzees. Chimpanzees use the "pant hoot" vocalization for long-distance vocalization; whereas bonobos produce the "high hoot" (Marler and Tenaza 1977; de Waal 1988; Schamberg et al. 2016, 2017). Structurally, the chimpanzee "pant hoot" is a composite vocalization, composed of four distinct phases (introduction, build up, climax, and downward phase), while the bonobo "high hoot" is a "whooping" call that can be composed of rapid sequences of both staccato and legato elements, which do not possess the phrase-like form of chimpanzee pant hoots (Hohmann and Fruth 1994) (see fig. 12.1).

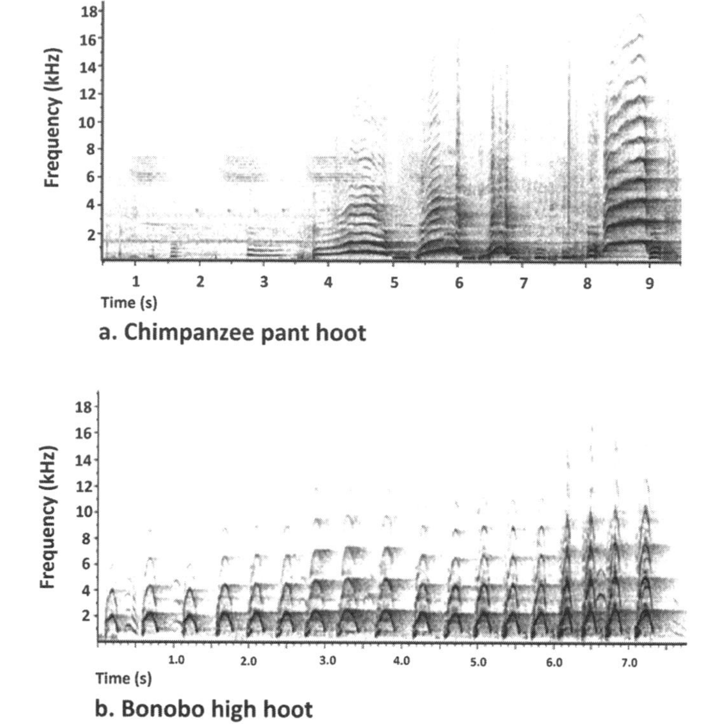

FIGURE 12.1. Time-frequency spectrograms showing representative examples of long-distance vocalizations in (a) chimpanzees (pant hoots) and (b) bonobos (high hoots). Recordings were made by Zanna Clay from an adult male chimpanzee in the Sonso community in Budongo Forest (Uganda) and from an adult male bonobo in the Bompusa community in Lui Kotale, Salonga National Park (Democratic Republic of Congo).

In both species, there is extensive variation in gradation within and between call types in their repertoire. For instance, pant laughing in both species is fairly discrete (de Waal 1988; Marler 1976) whereas there is notable gradation in their screams and barks, which increases the richness of information to be conveyed. For instance, the "waa-bark" of chimpanzees and the homologue "wieew bark" of bonobos are both highly graded and produced in similar alarm contexts (de Waal 1988; Marler and Tenaza 1977). In both species, the acoustic structure of victim screams vary as a function of caller identity and the intensity of attack (Clay, Furuichi, and de Waal 2016; Slocombe, Townsend, and Zuberbühler 2009; Slocombe and Zuberbühler 2005, 2007). For chimpanzees, screams have been shown to acoustically vary according to the caller's social role (Slocombe, Townsend, and Zuberbühler 2009; Slocombe and Zuberbühler 2007) and in bonobos, by the degree of social expectation (Clay et al. 2016). In sum, despite some differences in the vocal repertoires of bonobos and chimpanzees, we find many similarities. Understanding the social and ecological pressures driving these similarities and divergences are important for understanding how communication systems evolve.

Pan Social Structure and Vocal Communication

Although chimpanzees and bonobos share core similarities in their social organization, they also show some striking differences, particularly in regard to patterns in social bonding and social dominance. Vocal behavior can provide a useful window to explore this variation.

Both chimpanzees and bonobos live in large, mixed-sex, and fission-fusion societies, where individuals travel in fluid sub-groups that may vary in composition depending on social, reproductive, and ecological factors (Goodall 1986; Kano 1992). In contrast to most catarrhine primates (Sterck, Watts, and van Schaik 1997), both species show male philopatry and female migration (Goodall 1986; Kano 1992). Being the philopatric sex—and thus more closely related—chimpanzee males form strong social bonds and cooperate to share food, hunt, and support each other in social conflicts and territorial protection (Boesch 2009; Nishida 2011). Males are also highly dominant; they maintain a strict linear hierarchy and are dominant over all females in their communities (Boesch 2009; Nishida 2011).

In species such as chimpanzees, which form linear dominance hierarchies that are structured through aggression, subordinates commonly produce formalized signals to express their subordinate status (Preuschoft and van Schaik 2000). This can serve to reduce social tension and mitigate the risks associated with aggression received from dominants. In chimpanzees, subordinates produce formalized vocal greetings, known as "pant-grunts" (Bygott 1979; Laporte and Zuberbühler 2010; Newton-Fisher 2004), toward higher-ranking individuals. Pant-grunts are an acoustically heterogeneous signal, consisting of a series of grunts, typically combined with pant inhalations that may grade into barks or screams (Crockford and Boesch 2005; Goodall 1986; Laporte and Zuberbühler 2010). Although there is notable variation across individuals, sites, and contexts (Boesch and Boesch-Achermann 2000; Laporte and Zuberbühler 2010; Nishida and Hosaka 1996), pant-grunts are produced reliably by both sexes upwards in the social hierarchy. As a result, they are considered strong indicators of social dominance (Laporte and Zuberbühler 2010; Mitani et al. 2002; Newton-Fisher 2004; Sakamaki 2011).

Pant-grunting among female chimpanzees is rarer, presumably due to their lower dominance status and reduced sociability as compared to males (Muller 2002; Williams et al. 2002). Nevertheless, pant-grunting among Western chimpanzee females in the Taï forest of the Côte d'Ivoire (Wittig and Boesch 2003) is more frequent than that of Eastern chimpanzee females, which is probably due to their enhanced sociability (Budongo, Uganda: Newton-Fisher 2006; Kibale, Uganda: Muller 2002; Gombe, Tanzania: Pusey, Williams, and Goodall 1997). Females chimpanzees in Taï are more gregarious than females in most chimpanzee communities (Williams et al. 2002), and more involved in community activities, such as hunting, patrols, and meat sharing (Boesch 2009).

Recent research contributes relevant insights into the expression of social dominance in chimpanzees and bonobos. Among wild male bonobos, data indicate consistently steep and linear dominance hierarchies and persistent dyadic rank relationships (Furuichi 1997; Surbeck and Hohmann 2013; Surbeck et al. 2012). Males aggressively compete in the presence of fertile females and rates of male-male aggression appear to correlate positively with mating success (Surbeck et al. 2012). Evidence from two different sites revealed high reproductive skew among males, which was significantly higher than that observed in chimpanzees (Ishizuka et al. 2018; Surbeck et al. 2017). Although these patterns would predict that male bonobos should invest in expressing their dominance status, it has long been assumed that bonobos generally *lack* a formalized vocal signal of subordinance (Furuichi and Ihobe 1994; Paoli, Palagi, and Tarli 2006; Vervaecke, De Vries, and Van Elsacker 2000). This has been interpreted as a reflection of their generally reduced aggressiveness and increased social tolerance as compared to chimpanzees. However, our recent analysis of their greeting behavior (Clay et al. in prep) refutes this conclusion. Contrary to current assumptions, we found that subordinate wild male bonobos in Lui Kotale regularly produced pant-grunts, and did so reliably upwards in a linear dominance hierarchy of males. Rates of receiving pant-grunts were positively correlated with rates of aggressiveness, a pattern that is consistent with that shown in chimpanzees (Newton-Fisher 2004). In this regard, patterns of pant-grunting highlight closer similarities in the dominance relationships among male chimpanzees and bonobos than might have been previously assumed.

Despite an absence of genetic ties with other females, female bonobos have high social status in their communities and form strong social bonds with one another, as well as with males (Furuichi 2009, 2011; Surbeck and Hohmann 2013; Tokuyama and Furuichi 2016). The alpha position in bonobo societies is typically occupied by a female and some females can dominate even high-ranking males (Furuichi 1989, 1997, 2011; Parish 1994; Stevens et al. 2006; Surbeck, Roger, and Hohmann 2011; Surbeck et al. 2012; Surbeck and Hohmann 2013). Given their high status, we might expect males to submissively greet females, as they do for males. However, this is not what we found (Clay et al. in prep). Even though females in this community showed high dominance status, vocal greeting toward females by either sex was extremely rare.

The apparent absence of formal greetings to female bonobos may reflect differences in the structure and expression of female dominance relationships as compared to those of males. Unlike males, female bonobo dominance relationships are characterized by low levels of linearity and multiple tied relationships (Douglas, Ngonga, and Hohmann 2017; Moscovice et al. 2017). Their dominance relationships also depend on context and can regularly fluctuate due to rank reversals and switching between decided and undecided relationships (Douglas, Ngonga, and Hohmann 2017; Moscovice et al. 2017; Stevens et al. 2007). Inter-sexually, female dominance tends to be tied to coalitionary contexts rather than being absolute. Moreover, dominance is generally not enforced through physical aggression with the outcome of conflicts often independent of dominance status (Surbeck and Hohmann 2013). In this regard, non-linear dominance relationships and reduced aggression may have released vocal greetings from their dominance signaling function for females.

Lack of female-directed vocal greetings may also reflect enhanced inter- and intra-sexual social bonding with females in bonobo society (Furuichi 2011). Males are rarely aggressive to females (Hohmann and Fruth 2003; Surbeck et al. 2012), and instead invest in developing affiliative relationships with them (Surbeck and Hohmann 2013; Surbeck et al. 2012), something which appears to confer reproductive advantages (Surbeck et al. 2012). Males maintain prolonged social relationship with their mothers (Furuichi 1997; Surbeck, Roger, and Hohmann 2011) but also form stable bonds with non-related females as well (Surbeck and Hohmann 2013). In this regard, reduced pant-grunting may highlight shifts toward friendlier intra-sexual relations in this species.

Ritualized Vocal Displays in Bonobos: A Case of Vocal Paedomorphism?

As immature members of their community, young male chimpanzees in both the wild (Nishida 2003) and captivity (Adang 1984) regularly "pester" adult group-members, such as by stamping or throwing sand/sticks but quickly running away afterward. This behavior appears to function for social provocation and has thus been interpreted to be a form of social exploration that enables young males to "learn and expand their social limits" before ascending their

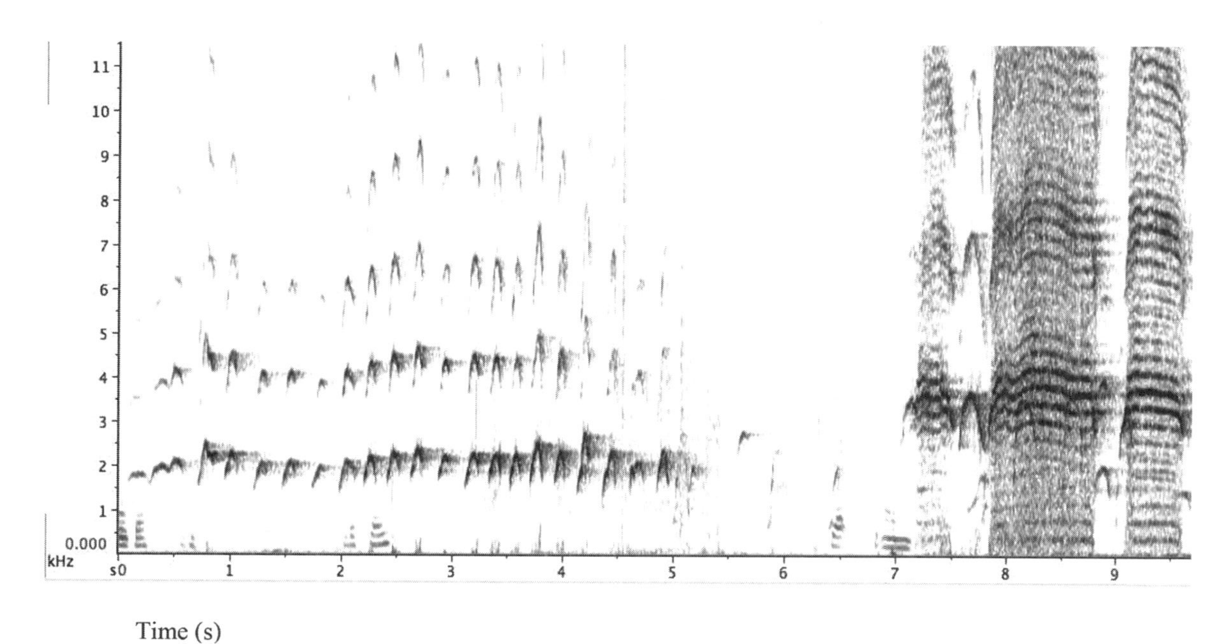

FIGURE 12.2. Time-frequency spectrogram showing contest hoot series produced by a wild male bonobo in the Bompusa community in Lui Kotale, Salonga National Park (Democratic Republic of Congo). Recordings were made by Zanna Clay. From 7 s, the spectrogram shows the submissive scream response of the signaler's social target, a lower-ranked male.

social hierarchy (Adang 1984). This function reflects evidence that males gradually abandon pestering as they approach adulthood and come to attain their own social rank (Adang 1984).

Interestingly, starting in early adolescence and continuing into adulthood, male bonobos produce a conspicuous vocal display—the contest hoot—that appears to resemble the pestering behavior of immature male chimpanzees. The contest hoot is a male-specific vocalization consisting of a series of high-pitched, arched hoot vocalizations produced in a rapid sequence (de Waal 1988; Genty et al. 2014) (see fig. 12.2). Typically, contest hoots consist of an introductory phase, an escalation phase composed of a series of stereotyped units, and a let-down phase. Previously, we showed that contest hoots are typically directed at specific social targets, of either sex, whereby the vocalizer orients toward or approaches the target while giving some form of display, such as lunging, swaying, running past, or throwing objects (Genty et al. 2014). As with chimpanzee pestering behavior (Adang 1984; Nishida 2003), our analyses suggest that contest hoots primarily function to initiate social interactions and provoke behavioral responses, usually in the form of an aggressive chase or submissive flee by their target. In most cases, the hooting male runs away or ceases the provocation once they have been effective in eliciting a response, indicative of intentionality. Recently, we showed that in a wild community at LuiKotale, male bonobos preferentially target individuals that are close to or higher in rank to them, suggestive of a dominance testing function (Clay et al. in prep). We found that the target's response type (passive, aggressive, or submissive) reliably related to the dominance relationship between signaler and target. In this regard, contest hoots may enable males to test and express their vocalizations using a conspicuous and ritualized display that can be observed by group members. This reflects evidence of audience effects, with male contest hoot more likely to occur following fusion events, where parties change in composition and increase in size.

Contest hooting suggests that, in the absence of pronounced male aggression, vocalizations may provide males with an alternative outlet to express and explore their social dominance relationships. The apparent similarity between bonobo contest hoots and the immature chimpanzee pestering behavior appears consistent with the hypothesis that bonobos have retained certain paedomorphic traits into adulthood (Hare, Wobber, and Wrangham 2012).

Copulation Calls in *Pan*: A Reproductive or Social Signal?

Like humans and some monkey species, female chimpanzees and bonobos produce acoustically distinct vocalizations during sexual interactions, known as "copulation calls" (Clay et al. 2010; Townsend, Deschner, and Zuberbühler 2008). Female copulation calling is particularly prevalent amongst species such as chimpanzees and bonobos that show multi-male and multi-female groups in which females mate promiscuously and advertise receptivity with pronounced sexual swellings (Dixson 1998; Pradhan et al. 2006). Considered as sexually selected signals, copulation calls are thought to function to promote the caller's reproductive success through inciting male-male completion and/or confusing paternity (see Pradhan et al. 2006). In Barbary macaques (*Macaca sylvanus*) and yellow baboons (*Papio cynocephalus*), for example, copulation calls convey information about caller identity (Semple 2001), sexual swelling size (an approximate cue to fertility status, e.g., Nunn 1999; Semple and McComb 2000; Semple et al. 2002), and, for yellow baboons, male partner rank (Semple et al. 2002). By advertising sexual activity, females can attract potential mates and therefore incite mating competition, which can increase the quality or number of sexual partners. In barbary macaques, the acoustic structure of female copulation calls co-varies with the occurrence of ejaculation (Pfefferle et al. 2008a), which may incite male-male competition by enabling males to discriminate ejaculatory from non-ejaculatory copulations (Pfefferle et al. 2008b).

Copulation calls by female chimpanzees and bonobos share a number of similarities. In both species, these calls carry reliable acoustic cues to caller identity as well as partner rank (Clay and Zuberbühler 2011a, 2011b; Townsend, Deschner, and Zuberbühler 2008). Females are also more likely to call with high-ranking partners as compared to low-ranking partners (Townsend, Deschner, and Zuberbühler 2008). There is also evidence of audience effects in both species, which suggests that females have some control over call production. Wild chimpanzee females are more likely to call in the presence of high-ranking males but less likely to call in the presence of high-ranking females; the former being thought to incite male-male competition (as high-ranking males tend to co-associate) and the latter being thought to reduce female mating competition (Townsend, Deschner, and Zuberbühler 2008). Female reproductive competition can be severe in chimpanzees, such as through female-led infanticides and even lethal aggression (Townsend et al. 2007). By inhibiting copulation calls in the presence of high-ranking females—the most likely perpetrators of infanticide—lower-ranking females may reduce risks of aggression from other females toward their own offspring.

Although we find similarities in bonobo copulation calling, we find some relevant differences. Most pertinent is how copulation calling apparently reflects the shift of sexual behavior in bonobos to being a social behavior. Compared

FIGURE 12.3. Genito-genital contact between a lower-ranking female bonobo (on top) with a higher-ranking female partner. Lower-ranking females are more likely to call when engaging in sexual contact with high-ranking partners, regardless of their sex, and calling can be sensitive to audience effects. Photo by Zanna Clay, taken at Lola ya Bonobo Sanctuary (Democratic Republic of Congo).

to chimpanzees, bonobos exhibit a notably heightened socio-sexuality, with females engaging in sexual interactions in all age and sex combinations (Hohmann and Fruth 2000; Kano 1992, see fig. 12.3). Sex serves as a social tool for bonobos, such as for reducing social tension, for promoting social tolerance, and for expressing, establishing, and repairing social relationships (Clay and de Waal 2013). Socio-sexual contacts among females, known as genito-genital contacts, appear to be especially important in facilitating the social integration of immigrants into their new community and their subsequent coexistence with non-related group members (de Waal 1987; Furuichi 1989; Hohmann and Fruth 2000; Kano 1992, see fig. 12.3).

During same-sex interactions, female bonobos sometimes produce "copulation calls," as when mating with males (Clay et al. 2010). Despite evident differences in the physical nature of the interaction, copulation calls produced with female partners share the same acoustic structure of those produced with males and the calls contain strong cues of caller identity in both contexts (Clay and Zuberbühler 2011a, 2011b). Within female-female pairings, we found that calling was not explained by physical stimulation alone, as genital contact duration and spatial position had no effects (Clay and Zuberbühler 2012).

Although females were more likely to produce copulation calls with males than female partners, partner rank has a strong effect such that females are more likely to call with high-ranking partners, regardless of their sex (Clay et al. 2010). For female partners, this interacts with initiation, with low-ranking females more likely to vocalize when *invited* to have sex with high-ranking females. One interpretation is that low-ranking females may call to express or advertise their socio-sexual interactions with socially important group members. Developing affiliations with dominant females is critical to a female's social position, thus having been chosen by a higher-ranking partner may enhance a female's general social standing. This reflects recent data from wild bonobos highlighting the important role that high-ranking females play in supporting low-ranking females in coalitions against males (Tokuyama and Furuichi 2016). Future research using playback experiments to test receiver knowledge are required to test this hypothesis.

In terms of audience effects, unlike female chimpanzees (Townsend, Deschner, and Zuberbühler 2008), bonobos are more likely to call in the presence of the highest-ranking female (Clay and Zuberbühler 2012). This species difference may relate to the central position that females play in bonobo society as compared to chimpanzees (Furuichi 2011). In bonobo society, there may be benefits to signaling your association with established females, a sign of social recognition.

In conclusion, research suggests overlaps in chimpanzees' and bonobos' copulation calling, but also relevant differences, which appear to reflect different levels of reproductive competition as well as intra-sexual bonding. In both species, female copulation calls appear to convey a considerable amount of information to listeners, and females appear to be sensitive to their audience composition, which highlights an underlying social awareness. Compared to chimpanzees, bonobo copulation calls appear to have become partly ritualized away from their original reproductive function to be used as social signals.

Vocal Coordination: Decision Making in the Forest

As well as for expressing social relationships and promoting reproductive and social strategies, vocalizations play a critical role in coordinating decisions and group movements across distances. In chimpanzees, pant hoots give individuals some control over their social environment by communicating with group members that are out of sight (Arcadi 1996; Goodall 1968; Marshall, Wrangham, and Arcadi 1999; Mitani and Gros-Louis 1998; Mitani and Nishida 1993). Pant hoots appear to have a number of functions, which include coordinating party movements, attracting conspecifics to food sources, signaling caller location (Ghiglieri 1984; Reynolds and Reynolds 1965; Wrangham 1977), as well as signaling social status (Clark and Wrangham 1994; Mitani and Nishida 1993) and promoting social bonding (Fedurek, Donnellan, and Slocombe 2014, Fedurek et al. 2013; Mitani and Brandt 1994). Pant hoots seem to be used in socially structured ways (Mitani and Nishida 1993), with males more likely to call in the presence of socially bonded partners (Mitani and Nishida 1993; Wilson, Hauser, and Wrangham 2007). Fedurek, Donnellan, and Slocombe (2014) recently showed that pant hooting among male chimpanzees in Kibale, Uganda, was positively correlated with caller rank, the presence of estrus females, as well as the discovery of highly preferred food. Together these findings highlight the important social, ecological, and reproductive functions that pant hoots appear to play, including for signaling male social status and bonding, especially during periods of high male-male competition.

Evidence of acoustic convergence among socially bonded males further highlights the social functions of pant hooting (Marshall, Wrangham, and Arcadi 1999; Mitani and Brandt 1994; Mitani and Gros-Louis 1998). This process is probably also responsible for evidence that chimpanzees have group-specific pant hoot "dialects" (Crockford et al. 2004). Evidence that vocalizations are shaped by a history of social interactions challenges the hypothesis that primate vocalizations are genetically fixed and inflexible. Although the mechanisms underlying this acoustic flexibility are still poorly understood, they highlight the importance of social variables for shaping vocal communication in this species.

Comparatively less is known about long-distance communication in bonobos. Nevertheless, Hohmann and Fruth (1994) found notable levels of acoustic convergence and behavioral synchronization in wild bonobo high hoot choruses, consistent with a social bonding function. These convergent findings suggest that both chimpanzees and bonobos may be able to control and modify their long-distance vocalizations in flexible ways.

Recent work by Schamberg and colleagues (2016, 2017) highlighted the combinatorial nature of bonobo long-distance communication. As well as producing high hoots in a single series, bonobos regularly combine them with whistles and low-hoots to make "whistle-high hoot' and "low hoot-high hoot" combinations. These two call combinations appear to have specific functions for coordinating group movements (Schamberg et al. 2016, 2017). Whistles are high-pitched, tonally flat vocalizations that occur in a wide variety of contexts, mostly preceding other calls (Bermejo and Omedes 1999). While neither whistles nor high hoots appear to exhibit a high degree of context specificity, we found that the combination of these two signals together into the whistle-high hoot sequence reliably signals a vocalizer's imminent travel to join a different party (Schamberg et al. 2016). Combining high hoots with low hoots appears to fulfil another function (Schamberg et al. 2017). Low hoots are an acoustically noisy, low-pitched vocalization in which the caller produces sound through both inspirations and expirations (Bermejo and Omedes 1999; de Waal 1988). In contrast to whistle-high hoots, the low hoot-high hoot combination is more likely to lead to individuals from other parties approaching the caller (i.e., functioning in inter-party recruitment; Schamberg et al. 2017). The production of distinct call combinations by bonobos in the context of inter-party decision further suggests a sophisticated degree of vocal flexibility in this species. Although research into chimpanzee call combinations requires more attention, a study by Crockford and Boesch (2005) demonstrated that chimpanzees also combine signals together in context-specific ways to create new meanings. Future research is needed to understand more about the context in which call combinations are given in these species and how these signal combinations might map onto social relationships.

Conclusion

In sum, vocalizations provide a relevant opportunity to investigate underlying social awareness of nonhuman animals. Moreover, for closely related species, such as chimpanzees and bonobos, they can provide important insights into similarities and differences in underlying social organization. This chapter highlights notable overlaps as well as divergences in the vocal communication behavior of chimpanzees and bonobos. Given large overlaps in their socio-ecologies and underlying social systems, we find many similarities both in the form and usage of their vocalizations. Nevertheless, we find important differences, which appear to map onto broader differences in their social tendencies, including in patterns of social dominance, reduced aggressiveness, and enhanced socio-sexuality in bonobos. Certain features of the bonobo vocal system, including their higher vocal pitch and retention of juvenile vocal traits, such as male vocal pestering, are consistent with the hypothesis of behavioral paedomorphism or selection against aggression in this species. Nevertheless, there are also many other areas of their vocal communication, some of which were touched on here, which cannot be explained within the paedomorphic framework. Understanding the ultimate and proximate mechanisms that might drive these apparent species differences, as well as their core similarities, is the task for future research.

References

Adang, O. M. J. 1984. "Teasing in young chimpanzees." *Behaviour* 88 (1): 98–121.

Arcadi, A. C. 1996. "Phrase structure of wild chimpanzee pant hoots: Patterns of production and interpopulation variability." *American Journal of Primatology* 39 (3): 159–78.

Bermejo, M., and A. Omedes. 1999. "Preliminary vocal repertoire and vocal communication of wild bonobos (*Pan paniscus*) at Lilungu (Democratic Republic of Congo)." *Folia Primatologica* 70 (6): 328–57.

Boesch, C. 2009. *The Real Chimpanzee: Sex Strategies in the Forest.* Cambridge: Cambridge University Press.

Boesch, C., and H. Boesch-Achermann. 2000. *The Chimpanzees of the Taï Forest: Behavioural Ecology and Evolution.* Oxford: Oxford University Press.

Boesch, C., G. Hohmann, and L. Marchandt. 2002. *Behavioural Diversity in Chimpanzees and Bonobos.* Cambridge: Cambridge University Press.

Bygott, D. 1979. "Agonistic behavior and dominance among wild chimpanzees." In *The Great Apes*, edited by D. Hamburg and E. McCown, 405–27. Menlo Park, CA: Benjamin/Cummings.

Cheney, D. L., and R. M. Seyfarth. 1990. *How Monkeys See the World: Inside the Mind of Another Species.* Chicago: University of Chicago Press.

Clark, A. P., and R. W. Wrangham. 1994. "Chimpanzee arrival pant-hoots: Do they signify food or status?" *International Journal of Primatology* 15 (2): 185–205.

Clay, Z., and F. B. M. de Waal. 2013. "Development of socio-emotional competence in bonobos." *Proceedings of the National Academy of Sciences* 110 (45): 18121–26.

Clay, Z., T. Furuichi, and F. B. M. de Waal. 2016. "Obstacles and catalysts to peaceful coexistence in chimpanzees and bonobos." *Behaviour* 153: 1293–1300.

Clay, Z., S. Pika, T. Gruber, and K. Zuberbühler. 2010. "Female bonobos use copulation calls as social signals." *Biology Letters* 7: 513–16.

Clay, Z., L. Ravaux, F. B. M. de Waal, and K. Zuberbühler. 2016. "Bonobos (*Pan paniscus*) vocally protest against violations of social expectations." *Journal of Comparative Psychology* 130 (1): 44.

Clay, Z., and K. Zuberbühler. 2009. "Food-associated calling sequences in bonobos." *Animal Behaviour* 77 (6): 1387–96.

Clay, Z., and K. Zuberbühler. 2011a. "Bonobos extract meaning from call sequences." *PLoS One* 6 (4): e18786.

Clay, Z., and K. Zuberbühler. 2011b. "The structure of bonobo copulation calls during reproductive and non-reproductive sex." *Ethology* 117 (12): 1158–69.

Clay, Z., and K. Zuberbühler. 2012. "Communication during sex among female bonobos: Effects of dominance, solicitation, and audience." *Scientific Reports* 2: 291.

Cramer, D. L. 1977. "Craniofacial morphology of *Pan paniscus*: A morphometric and evolutionary appraisal." *Contributions to Primatology* 10: 1.

Crockford, C., and C. Boesch. 2005. "Call combinations in wild chimpanzees." *Behaviour* 142: 397–421.

Crockford, C., I. Herbinger, L. Vigilant, and C. Boesch. 2004. "Wild chimpanzees produce group-specific calls: A case for vocal learning?" *Ethology* 110 (3): 221–43.

de Waal, F. B. M. 1987. "Tension regulation and nonreproductive functions of sex in captive bonobos." *National Geographic Research* 3 (3): 318–35.

de Waal, F. B.M. 1988. "The communication repertoire of captive bonobos (*Pan paniscus*), compared to that of chimpanzees." *Behaviour* 106: 183–251.

Dixson, A. 1998. *Primate Sexuality*. Oxford: Oxford University Press.

Douglas, H., A. C. Ngonga, and G. Hohmann. 2017. "A novel approach for dominance assessment in gregarious species: ADAGIO." *Animal Behaviour* 123: 21–32.

Fedurek, P., E. Donnellan, and K. E. Slocombe. 2014. "Social and ecological correlates of long-distance pant hoot calls in male chimpanzees." *Behavioral Ecology and Sociobiology* 68 (8): 1345–55.

Fedurek, P., Z. P. Machanda, A. M. Schel, and K. E. Slocombe. 2013. "Pant hoot chorusing and social bonds in male chimpanzees." *Animal Behaviour* 86 (1): 189–96.

Fischer, A., K. Prüfer, J. M. Good, M. Halbwax, V. Wiebe, C. André, R. Atencia, L. Mugisha, S. E. Ptak, and S. Pääbo. 2011. "Bonobos fall within the genomic variation of chimpanzees." *PLoS One* 6 (6): e21605.

Furuichi, T. 1989. "Social interactions and the life history of female *Pan paniscus* in Wamba, Zaire." *International Journal of Primatology* 10 (3): 173–97.

Furuichi, T. 1997. "Agonistic interactions and matrifocal dominance rank of wild bonobos (*Pan paniscus*) at Wamba." *International Journal of Primatology* 18 (6): 855–75.

Furuichi, T. 2009. "Factors underlying party size differences between chimpanzees and bonobos: A review and hypotheses for future study." *Primates* 50 (3): 197–209.

Furuichi, T. 2011. "Female contributions to the peaceful nature of bonobo society." *Evolutionary Anthropology* 20 (4): 131–42.

Furuichi, T., and H. Ihobe. 1994. "Variation in male relationships in bonobos and chimpanzees." *Behaviour* 130: 212–28.

Genty, E., Z. Clay, C. Hobaiter, and K. Zuberbühler. 2014. "Multi-modal use of a socially directed call in bonobos." *PLoS One* 9 (1): e84738.

Ghiglieri, M. P. 1984. *The Chimpanzees of Kibale Forest: A Field Study of Ecology and Social Structure*. New York: Columbia University Press.

Gogoleva, S. S., J. A. Volodin, E. V. Volodina, and L. N. Trut. 2008. "To bark or not to bark: Vocalizations by red foxes selected for tameness or aggressiveness toward humans." *Bioacoustics* 18 (2): 99–132.

Goodall, J. 1968. "The behaviour of free-living chimpanzees in the Gombe Stream Reserve." *Animal Behaviour Monographs* 1: 161–311.

Goodall, J. 1986. *The Chimpanzees of Gombe, Patterns of Behaviour*. Cambridge, MA: Belknap Press of Harvard University Press.

Grawunder, S., C. Crockford, Z. Clay, A. K. Kalan, J. M. Stevens, A. Stoessel, and G. Hohmann. 2018. "Higher fundamental frequency in bonobos is explained by larynx morphology." *Current Biology* 28 (20): R1188–89.

Gruber, T., and Z. Clay. 2016. "A comparison between bonobos and chimpanzees: A review and update." *Evolutionary Anthropology* 25: 239–52.

Hare, B., V. Wobber, and R. Wrangham. 2012. "The self-domestication hypothesis: Evolution of bonobo psychology is due to selection against aggression." *Animal Behaviour* 83 (3): 573–85.

Hare, B., and S. Yamamoto. 2017. *Bonobos: Unique in Mind, Brain, and Behavior*. Oxford: Oxford University Press.

Hauser, M. D. 1998. "Functional referents and acoustic similarity: Field playback experiments with rhesus monkeys." *Animal Behaviour* 55: 1647–58.

Hey, J. 2010. "The divergence of chimpanzee species and subspecies as revealed in multipopulation isolation-with-migration analyses." *Molecular Biology and Evolution* 27 (4): 921–33.

Hohmann, G., and B. Fruth. 1994. "Structure and use of distance calls in wild bonobos (*Pan paniscus*)." *International Journal of Primatology* 15 (5): 767–82.

Hohmann, G., and B. Fruth. 2000. "Use and function of genital contacts among female bonobos." *Animal Behaviour* 60: 107–20.

Hohmann, G., and B. Fruth. 2003. "Intra- and inter-sexual aggression by bonobos in the context of mating." *Behaviour* 140: 1389–1413.

Ishizuka, S., Y. Kawamoto, T. Sakamaki, N. Tokuyama, K. Toda, H. Okamura, and T. Furuichi. 2018. "Paternity and kin structure among neighbouring groups in wild bonobos at Wamba." *Royal Society Open Science* 5 (1): 171006.

Kano, T. 1992. *The Last Ape: Pygmy Chimpanzee Behavior and Ecology*. Stanford: Stanford University Press.

Langergraber, K. E., K. Prüfer, C. Rowney, C. Boesch, C. Crockford, K. Fawcett, E. Inoue, M. Inoue-Muruyama, J. C. Mitani, and M. N. Muller. 2012. "Generation times in wild chimpanzees and gorillas suggest earlier divergence times in great ape and human evolution." *Proceedings of the National Academy of Sciences* 109 (39): 15716–21.

Laporte, M. N. C., and K. Zuberbühler. 2010. "Vocal greeting behaviour in wild chimpanzee females." *Animal Behaviour* 80 (3): 467–73.

Latimer, B. M., T. D. White, W. H. Kimbel, D. C. Johanson, and C. O. Lovejoy. 1981. "The pygmy chimpanzee is not a living missing link in human evolution." *Journal of Human Evolution* 10 (6): 475–88.

Marler, P. 1976. "Social organization, communication and graded signals: The chimpanzee and the gorilla." In *Growing Points in Ethology*, edited by P. G. Bateson and R. A. Hinde, 239–80. Cambridge: Cambridge University Press.

Marler, P., and R. Tenaza. 1977. "Signaling behavior of apes with special reference to vocalizations." In *How Animals Communicate*, edited by T. A. Sebeok, 965–1033. London: Indiana University Press.

Marshall, A. J., R. W. Wrangham, and A. C. Arcadi. 1999. "Does learning affect the structure of vocalizations in chimpanzees?" *Animal Behaviour* 58 (4): 825–30.

McGrew, W. C. 2017. "Field studies of *Pan troglodytes* reviewed and comprehensively mapped, focussing on Japan's contribution to cultural primatology." *Primates* 58 (1): 237–58.

Mitani, J. C., and K. L. Brandt. 1994. "Social factors influence the acoustic variability in the long-distance calls of male chimpanzees." *Ethology* 96 (3): 233–52.

Mitani, J. C., and J. Gros-Louis. 1995. "Species and sex differences in the screams of chimpanzees and bonobos." *International Journal of Primatology* 16 (3): 393–411.

Mitani, J. C., and J. Gros-Louis. 1998. "Chorusing and call convergence in chimpanzees: Tests of three hypotheses." *Behaviour* 135: 1041–64.

Mitani, J. C., and T. Nishida. 1993. "Contexts and social correlates of long-distance calling by male chimpanzees." *Animal Behaviour* 45: 735–46.

Mitani, J. C., D. P. Watts, J. W. Pepper, and A. D. Merriwether. 2002. "Demographic and social constraints on male chimpanzee behavior." *Animal Behaviour* 64: 727–37.

Moscovice, L. R., P. H. Douglas, L. Martinez-Iñigo, M. Surbeck, L. Vigilant, and G. Hohmann. 2017. "Stable and fluctuating social preferences and implications for cooperation among female bonobos at LuiKotale, Salonga National Park, DRC." *American Journal of Physical Anthropology* 163 (1): 158–72.

Muller, M. N. 2002. "Agonistic relations among Kanyawara chimpanzees." In *Behavioral Diversity in Chimpanzees and Bonobos*, edited by C. Boesch, G. Hohmann, and L. Marchant, 112–24. Cambridge: Cambridge University Press.

Newton-Fisher, N. E. 2004. "Hierarchy and social status in Budongo chimpanzees." *Primates* 45 (2): 81–87.

Newton-Fisher, N. E. 2006. "Female coalitions against male aggression in wild chimpanzees of the Budongo Forest." *International Journal of Primatology* 27 (6): 1589–99.

Nicastro, N. 2004. "Perceptual and acoustic evidence for species-level differences in meow vocalizations by domestic cats (*Felis catus*) and African wild cats (*Felis silvestris lybica*)." *Journal of Comparative Psychology* 118 (3): 287.

Nishida, T. 2003. "Harassment of mature female chimpanzees by young males in the Mahale Mountains." *International Journal of Primatology* 24 (3): 503–14.

Nishida, T. 2011. *Chimpanzees of the Lakeshore: Natural History and Culture at Mahale.* Cambridge: Cambridge University Press.

Nishida, T., and K. Hosaka. 1996. "Coalition strategies among adult male chimpanzees of the Mahale Mountains, Tanzania." In *Great Ape Societies*, edited by W. C. McGrew, T. Nishida, and L. F. Marchandt, 114–34. Cambridge: Cambridge University Press.

Nunn, C. L. 1999. "The evolution of exaggerated sexual swellings in primates and the graded-signal hypothesis." *Animal Behaviour* 58: 229–46.

Paoli, T., E. Palagi, and S. M. Tarli. 2006. "Re-evaluation of dominance hierarchy in bonobos (*Pan paniscus*)." *American Journal of Physical Anthropology* 130 (1): 116–22.

Parish, A. R. 1994. "Sex and food control in the 'uncommon chimpanzee': How bonobo females overcome a phylogenetic legacy of male dominance." *Ethology and Sociobiology* 15 (3): 157–79.

Pfefferle, D., K. Brauch, M. Heistermann, J. K. Hodges, and J. Fischer. 2008a. "Female barbary macaque (*Macaca sylvanus*) copulation calls do not reveal the fertile phase but influence mating outcome." *Proceedings of the Royal Society of London B* 275 (1634): 571–78.

Pfefferle, D., M. Heistermann, J. K. Hodges, and J. Fischer. 2008b. "Male barbary macaques eavesdrop on mating outcome: A playback study." *Animal Behaviour* 75: 1885–91.

Pilbrow, V. 2006. "Population systematics of chimpanzees using molar morphometrics." *Journal of Human Evolution* 51 (6): 646–62.

Pradhan, G. R., A. Engelhardt, C. P. van Schaik, and D. Maestripieri. 2006. "The evolution of female copulation calls in primates: A review and a new model." *Behavioral Ecology and Sociobiology* 59 (3): 333–43.

Preuschoft, S., and C. P. van Schaik. 2000. "Dominance and communication: Conflict management in various social settings." In *Natural Conflict Resolution*, edited by F. Aureli and F. B. M. de Waal, 77–105. Berkley: University of California Press.

Prüfer, K., K. Munch, I. Hellmann, K. Akagi, J. R. Miller, B. Walenz, S. Koren, G. Sutton, C. Kodira, and R. Winer. 2012. "The bonobo genome compared with the chimpanzee and human genomes." *Nature* 486 (7404): 527–31.

Pusey, A., J. Williams, and J. Goodall. 1997. "The influence of dominance rank on the reproductive success of female chimpanzees." *Science* 277 (5327): 828–31.

Reynolds, V. 1965. *Budongo: A Forest and Its Chimpanzees.* York: Methuen.

Reynolds, V., and F. Reynolds. 1965. "Chimpanzees of the Budongo Forest." In *Primate Behaviour: Field Studies of Monkeys and Apes*, edited by I. DeVore, 368–424. New York: Holt, Rinehart and Winston.

Riede, T., A. C. Arcadi, and M. J. Owren. 2007. "Nonlinear acoustics in the pant hoots of common chimpanzees (*Pan troglodytes*): Vocalizing at the edge." *Journal of the Acoustical Society of America* 121 (3): 1758–67.

Sakamaki, T. 2011. "Submissive pant–grunt greeting of female chimpanzees in Mahale Mountains National Park, Tanzania." *African Study Monographs* 32 (1): 25–41.

Schamberg, I., D. L. Cheney, Z. Clay, G. Hohmann, and R. M. Seyfarth. 2016. "Call combinations, vocal exchanges and interparty movement in wild bonobos." *Animal Behaviour* 122: 109–16.

Schamberg, I., D. L. Cheney, Z. Clay, G. Hohmann, and R. M. Seyfarth. 2017. "Bonobos use call combinations to facilitate inter-party travel recruitment." *Behavioral Ecology and Sociobiology* 71 (4): 75.

Semple, S. 2001. "Individuality and male discrimination of female copulation calls in the yellow baboon." *Animal Behaviour* 61 (5): 1023–28.

Semple, S., and K. McComb. 2000. "Perception of female reproductive state from vocal cues in a mammal species." *Proceedings of the Royal Society of London B* 267 (1444): 707–12.

Semple, S., K. McComb, S. Alberts, and J. Altmann. 2002. "Information content of female copulation calls in yellow baboons." *American Journal of Primatology* 56 (1): 43–56.

Seyfarth, R. M., and D. L. Cheney. 2003. "Signalers and receivers in animal communication." *Annual Review of Psychology* 54: 145–73.

Seyfarth, R. M., D. L. Cheney, T. Bergman, J. Fischer, K. Zuberbühler, and K. Hammerschmidt. 2010. "The central importance of information in studies of animal communication." *Animal Behaviour* 80 (1): 3–8.

Slocombe, K. E., S. W. Townsend, and K. Zuberbühler. 2009. "Wild chimpanzees (*Pan troglodytes schweinfurthii*) distinguish between different scream types: Evidence from a playback study." *Animal Cognition* 12 (3): 441–49.

Slocombe, K. E., and K. Zuberbühler. 2005. "Agonistic screams in wild chimpanzees (*Pan troglodytes schweinfurthii*) vary as a function of social role." *Journal of Comparative Psychology* 119 (1): 67–77.

Slocombe, K. E., and K. Zuberbühler. 2007. "Chimpanzees modify recruitment screams as a function of audience composition." *Proceedings of the National Academy of Sciences* 104: 17228–33.

Sterck, E. H., D. P. Watts, and C. P. van Schaik. 1997. "The evolution of female social relationships in nonhuman primates." *Behavioral Ecology and Sociobiology* 41 (5): 291–309.

Stevens, J. M. G., H. Vervaecke, H. de Vries, and L. van Elsacker. 2006. "Social structures in *Pan paniscus*: Testing the female bonding hypothesis." *Primates* 47 (3): 210–17.

Stevens, J. M. G., H. Vervaecke, H. de Vries, and L. van Elsacker. 2007. "Sex differences in the steepness of dominance hierarchies in captive bonobo groups." *International Journal of Primatology* 28 (6): 1417–30.

Surbeck, M., T. Deschner, G. Schubert, A. Weltring, and G. Hohmann. 2012. "Mate competition, testosterone and intersexual relationships in bonobos, *Pan paniscus*." *Animal Behaviour* 83 (3): 659–69.

Surbeck, M., and G. Hohmann. 2013. "Intersexual dominance relationships and the influence of leverage on the outcome of conflicts in wild bonobos (*Pan paniscus*)." *Behavioral Ecology and Sociobiology* 67: 1767–80.

Surbeck, M., K. E. Langergraber, B. Fruth, L. Vigilant, and G. Hohmann. 2017. "Male reproductive skew is higher in bonobos than chimpanzees." *Current Biology* 27 (13): R640–41.

Surbeck, M., M. Roger, and G. Hohmann. 2011. "Mothers matter! Maternal support, dominance status and mating success in male bonobos (*Pan paniscus*)." *Proceedings of the Royal Society of London B* 278 (1705): 590–98.

Takemoto, H., Y. Kawamoto, and T. Furuichi. 2015. "How did bonobos come to range south of the Congo River? Reconsideration of the divergence of *Pan paniscus* from other Pan populations." *Evolutionary Anthropology* 24 (5): 170–84.

Tokuyama, N., and T. Furuichi. 2016. "Do friends help each other? Patterns of female coalition formation in wild bonobos at Wamba." *Animal Behaviour* 119: 27–35.

Townsend, S. W., K. E. Slocombe, M. E. Thompson, and K. Zuberbühler. 2007. "Female-led infanticide in wild chimpanzees." *Current Biology* 17 (10): R355–56.

Townsend, S. W., T. Deschner, and K. Zuberbühler. 2008. "Female chimpanzees use copulation calls flexibly to prevent social competition." *PLoS One* 3 (6): e2431.

Townsend, S. W., and K. Zuberbühler. 2009. "Audience effects in chimpanzee copulation calls." *Communicative & Integrative Biology* 2 (3): 282–84.

Van Hooff, J. A. R. A. M. 1973. "A structural analysis of the social behaviour of a semi-captive group of chimpanzees." In *Social Communication and Movement, Studies of Interaction and Expression in Man and Chimpanzee*, edited by M. von Cranach and I. Vine, 75–162. London: Academic Press.

Vervaecke, H., H. De Vries, and L. Van Elsacker. 2000. "Dominance and its behavioral measures in a captive group of bonobos (*Pan paniscus*)." *International Journal of Primatology* 21 (1): 47–68.

Williams, J. M., A. E. Pusey, J. V. Carlis, B. P. Farm, and J. Goodall. 2002. "Female competition and male territorial behavior influence female chimpanzees' ranging patterns." *Animal Behaviour* 63 (2): 347–60.

Wilson, M. L., C. Boesch, B. Fruth, T. Furuichi, I. C. Gilby, C. Hashimoto, C. L. Hobaiter, G. Hohmann, N. Itoh, K. Koops, J. N. Lloyd, T. Matsuzawa, J. C. Mitani, D. C. Mjungu, D. Morgan, M. N. Muller, R. Mundry, M. Nakamura, J. Pruetz, A. E. Pusey, J. Riedel, C. Sanz, A. M. Schel, N. Simmons, M. Waller, D. P. Watts, F. White, R. M. Wittig, K. Zuberbühler, and R. W. Wrangham. 2014. "Lethal aggression in *Pan* is better explained by adaptive strategies than human impacts." *Nature* 513: 414–17.

Wilson, M. L., M. D. Hauser, and R. W. Wrangham. 2007. "Chimpanzees (*Pan troglodytes*) modify grouping and vocal behaviour in response to location-specific risk." *Behaviour* 144: 1621–53.

Wittig, R. M., and C. Boesch. 2003. "Food competition and linear dominance hierarchy among female chimpanzees of the Tai National Park." *International Journal of Primatology* 24 (4): 847–67.

Wrangham, R. W. 1977. "Feeding behaviour of chimpanzees in Gombe National Park, Tanzania." In *Primate Ecology*, edited by T. H. Clutton-Brock, 503–38. London: Academic Press.

Zsurka, G., K. Tatiana, P. Viktoriya, K. Hallmann, C. E. Elger, K. Khrapko, and W. S. Kunz. 2010. "Distinct patterns of mitochondrial genome diversity in bonobos (*Pan paniscus*) and humans." *BMC Evolutionary Biology* 10 (1): 270.

PART FIVE

Cooperation

13

Cooperation and Communication in Great Apes

SHONA DUGUID, MATTHIAS ALLRITZ, AFRICA DE LAS HERAS, SUSKA NOLTE, AND JOSEP CALL

Introduction

It is often said that a discovery is a two-step process: making a new observation and realizing its significance. In 1937, Crawford published a seminal study on chimpanzee (*Pan troglodytes*) cooperation. He trained pairs of chimpanzees to pull together on a rope to obtain food placed on top of a heavy box that no single individual could pull alone. Pairs of chimpanzees coordinated their efforts to solve this task and more importantly, they occasionally communicated with each other to jump start cooperation.

In retrospect, it is difficult to understand why such an important finding did not catch on, particularly given the central role that communication plays in human cooperation (Tomasello 2008) and the influence that Crawford's paradigm has recently played in the expansion of experimental studies of primate cooperation. It is not that Crawford's work was completely ignored. On the contrary, several studies over the years sporadically investigated cooperation using his paradigm (e.g., Chalmeau and Gallo 1996; Mendres and de Waal 2000; Povinelli and O'Neill 2000). However, during a substantial portion of that time, the research focus was on whether apes are motivated to help others and not how they work together toward common goals. In the studies that did look at cooperation—by which we mean here how two or more individuals act together for mutual gains—very few subjects were tested, only a minority of subjects cooperated, and when cooperation broke down, or did not occur in the first place, communication did not come to the rescue. Communication did not convincingly emerge as the grease lubricating the wheels of coordination. Such a fragmentary body of evidence on experimental cooperation paired with the prevailing emphasis on social competition, not cooperation, coalesced to stifle the development of this area (Schmelz and Call 2016).

Things have changed more recently, and multiple studies have produced accumulating evidence of coordination in experimental settings, in chimpanzees and in non-primate species (e.g., Heaney, Gray, and Taylor 2017; Marshall-Pescini et al. 2017; Melis, Hare, and Tomasello 2006a; Plotnik et al. 2011). Several studies have illustrated how cooperation can spontaneously develop over time and they have shown an expansion on subjects' temporal horizon when making social decisions. Furthermore, many cooperation studies have embraced the tenets of experimental economics, and consequently, they have become more sophisticated from a theoretical and methodological perspective. Instead of investigating whether two individuals will pull two ropes simultaneously, studies are asking questions about whether individuals will make choices that take into account the options available to their partners (e.g., Jensen, Call, and Tomasello 2007; Sánchez-Amaro et al. 2017).

Despite these advances, one thing that is still puzzling in these studies is the minor role that communication seems to play in facilitating coordination, especially because apes use communication routinely to mediate in their social activities (Call and Tomasello 2000; Cartmill and Byrne 2010; Graham et al. 2018; Hobaiter and Byrne 2014). Even when communication appears between test partners, it is often ignored. Is this lack of communication a methodological artifact or is there a deeper theoretical reason that underlies this apparent disconnection? The aim of this chapter is to find an answer to this question by reviewing the evidence available and presenting some new data.

This chapter is organized as follows: we start with a more detailed description of the early studies by Crawford and how his observations are still acutely relevant to the discussion today, followed by a review of the experimental work since Crawford that has contributed to our knowledge of how apes use communication to facilitate cooperation. In the second part of this chapter, we summarize three recent experiments that have focused on how apes (*Pan troglodytes*, *P. paniscus*, *Pongo pygmaeus*) coordinate their actions for successful cooperation and the role that communication plays in coordination success. Finally, we identify open questions regarding how apes use communication for cooperation and suggest ways to address these in future studies.

Crawford's Observations of Cooperation in Chimpanzees

The foundations of the experimental study of ape cooperation can be traced back to Meredith Crawford's work with captive chimpanzees in the 1930s and 1940s (Crawford 1937, 1941). Not only did his experiments inspire one of the main current experimental paradigms in comparative studies of cooperation (the loose-string task; Hirata 2003; Massen, Schaake, and Bugnyar, chapter 16 this volume), they also provide quintessential examples of how chimpanzees use communication to coordinate with one another.

The box-pulling task has been particularly influential (Crawford 1937). In this task, chimpanzees faced out-of-reach food placed on top of a box outside their testing enclosure. They learned to move the food within reach individually by pulling a rope attached to the box, after which the weight of the box was increased until it was too heavy for single individuals to move it. Eventually two ropes were attached to the heavy box. Pairs of chimpanzees, tested together, could access the food on top of the box only by pulling the ropes simultaneously. The four chimpanzees (about five years old at the start of the study) were in adjacent cages but the separating bars were wide enough to reach through with an entire arm. When first presented with the cooperative problem-solving task, Crawford noted "how completely lacking any ability for cooperation, of the type required by the experiment" these apes were (1937, p. 49). Nevertheless, after the initial failure to coordinate, and the introduction of various training sessions, they eventually achieved success, working together to pull the box toward them.

It was not until about 90 sessions of this task that communication between partners appeared. Crawford observed that when one chimpanzee was not engaging with the task, her partner might solicit action from them in the form of begging gestures (an outstretched arm and hand in the direction of the recipient), a touch on the shoulder or back of the head, and in some cases, a gentle push toward the box. This behavior was displayed mainly by two chimpanzees, Bula and Bimba, particularly when they were paired together. When their partner did not respond, or responded incorrectly, they persisted and often succeeded (except when partnered with chimpanzee Alpha, with whom the behavior quickly ceased). However, this behavior was not ubiquitous; there were many occasions in which a partner did not take part but no attempts were made to solicit help.

In a later experiment Crawford presented the same chimpanzees with a different (more complex) serial ordering coordination task (Crawford 1941). In this experiment they learned individually to press four colored boxes in a specific order to release food. Pairs of chimpanzees were then tested in adjacent cages, with each individual having access to two of the four boxes (either the first and third, or second and fourth boxes). To solve the task the participants would need to take turns to press the boxes in the correct order. Crawford noted that the same pair (Bimba and Bula) showed similar solicitation behavior (touch, gentle pushing, and begging gestures), always after attempting, unsuccessfully, to reach through and press the appropriate box themselves. As with the box-pulling task, the participants were not very successful at solving the task, nor did they show any of the solicitation behavior seen in the previous study until they received extra training and the separator between the cages was changed so that they could reach their arm through to their partner.

Crawford was rather conservative with his overall interpretation of this communicative behavior. He considered how long the soliciting behavior took to appear and the low frequency and lack of consistency in its usage, and provided a cautious interpretation of its function. He noted that the solicitation gestures did not seem to include any specific information about what the recipient should do, but rather, elicited *some* activity from the partner more generally and this activity would include pulling the rope because there were limited possibilities in the testing area for the expression of other behaviors. He also identified several possible proximate factors that might influence the use of communication (or the lack thereof) and reasons for why he observed this behavior mainly in one pair. Interestingly, the relevance of these factors for explaining the variation has been rediscovered by contemporary authors using new methods. One factor was the general motivation of the participants to solve the task and their perseverance in trying to solve it. A second factor was the responsiveness of the partner: Crawford observed that when attempts to solicit were unsuccessful, these behaviors were quickly extinguished. Crawford also suggested that those individuals who used communication to solicit their partner were generally considered the more intelligent individuals. These individuals might be more likely to communicate to coordinate with a partner because communicating is a more intelligent way to solve the task, but it may also indicate that they simply understood the task better. This is consistent with a lack of communication until the chimpanzees had extensive experience with the tasks. Finally, the most successful pair in Crawford's study were also known to have a close social bond, suggesting that the dominance or tolerance relationship within pairs may be important to the likelihood to communicate and cooperate.

Crawford's Continuing Legacy

Since Crawford's experiments, there has been considerable effort to study how and when apes cooperate, supporting observations of the complexity and variation in social relationships and cooperation from wild communities of chimpanzees (Mitani 2009; Surbeck, Boesch, et al. 2017; Surbeck, Girard-Buttoz, et al. 2017). Seminal experiments by Melis and colleagues found that, not only can chimpanzees coordinate their actions for mutual gain in a cooperative rope-pulling task, but they understand the need for a partner and recruit the better partner (Melis, Hare, and Tomasello 2006a). Crucially, these experiments, as well as many others, also identified some of the constraints on cooperation: in particular the preference chimpanzees have for working independently when possible and the role of increased tolerance in promoting cooperation (e.g., Bullinger, Melis, and Tomasello 2011; Cronin et al. 2014; Hare et al. 2007; Melis, Hare, and Tomasello 2006b; Suchak et al. 2014). These studies did not explicitly consider the role of communication (or at least, did not report communication). The following review will focus on those experiments that explicitly coded and reported inter-subject communication.

FROM TWO ROPES TO ONE ROPE

The legacy of Crawford's work is clear in studies that have used methods based on Crawford's box-pulling task and, to a certain extent, these studies have made similar observations regarding communication. Hirata and Fuwa (2007) made a key change to the original Crawford box-pulling task: one single rope was threaded through the apparatus such that if only one individual pulls on one end of the rope the rope moves out of reach for the partner and the apparatus remains stationary (see fig. 16.1 in Massen, Schaake, and Bugnyar, chapter 16 this volume). This means that trying to pull the rope individually is a costlier action and accidental simultaneous pulling much less likely. In addition, the authors exchanged the box for a longer plank that reduced the necessity for very close proximity between the two test subjects and thus reduced the tolerance demands on the subjects. Hirata and Fuwa (2007) tested one pair of chimpanzees and, as with the earlier studies, they did achieve some cooperation success after significant amounts of training. There were no signs of communication between the two chimpanzees, however. One of the two chimpanzees did solicit help when partnered with a human experimenter (by taking the hand of the experimenter and pulling them toward the task) but they did not continue this behavior when retested with their original conspecific partner, pointing to the importance of the identity of the partner. Similar solicitation behavior has been reported for both chimpanzees and orangutans tested with modified versions of the original paradigm (Chalmeau 1994; Chalmeau and Gallo 1996; Chalmeau et al. 1997; Hirata, Morimura, and Fuwa 2010). Conversely, Povinelli and O'Neill (2000) paired knowledgeable and naive chimpanzees on the original box-pulling task and found no evidence of solicitation or other communication between conspecifics. Similarly, Warneken, Chen, and Tomasello (2006) found that when a human partner disengaged during a cooperative task, chimpanzees showed no signs of trying to re-engage them to complete the task. These studies tested a very small sample of young chimpanzees (or orangutans), so it is difficult to make systematic comparisons that explain the observed differences in communication. However, they do provide further evidence that apes *can* use communication appropriately to recruit partners when needed.

FROM SYMMETRICAL TO COMPLEMENTARY ROLES

Tasks based on the box-pulling paradigm involve symmetrical roles and simultaneous actions, but newer paradigms have included sequential and complementary actions (e.g., Fletcher, Warneken, and Tomasello 2012; Martin, Biro, and Matsuzawa 2017). Several studies also incorporate "social tool" use. Völter, Rossano, and Call (2015) found that orangutan mothers will enlist the cooperation of their infants, not only by giving them the tool to complete the task that they themselves cannot reach, but also by physically manipulating their infant's actions toward completing the required actions. While this does not constitute communication, and may be specific to the mother-infant relationship, it does suggest that these orangutans understood the necessary actions of a cooperative partner and were capable of manipulating them toward achieving a common goal. This is functionally very similar to the role communication would have in the same context, albeit through instrumentally ineffective gestures. More recently, Schweinfurth et al. (2018) observed an adult male chimpanzee using begging gestures toward youngsters that he used as "social tools" to obtain fruit juice from a fountain. However, just like the orangutan mothers tested by Völter, Rossano, and Call (2015), this male did not share the juice with the youngsters. Chimpanzees have also been shown to facilitate cooperation by providing tools to a partner and by adjusting the visibility of their actions (Grueneisen et al. 2017; Melis and Tomasello 2013). In two experiments with chimpanzees, which required a single tool transfer from one individual to another who could then release mutual rewards, pairs of chimpanzees could not physically manipulate a partner but potential recipients did occasionally request the tool, thus soliciting action from a partner in a similar way to the symmetrical tasks (Bullinger, Melis, and Tomasello 2014; Melis and Tomasello 2013).

Tasks in which there is asymmetry in the knowledge required for successful coordination, as opposed to asymmetry of actions, offer a particularly interesting context to study communication. This is because when one individual has privileged knowledge (e.g., of the location of food or a tool), to solve a task efficiently, communication is necessary to succeed and it needs to be more specific than simply soliciting action. In an experiment by Bullinger,

Melis, and Tomasello (2014), knowledgeable chimpanzees could indicate the location of a hidden tool or hidden food for an ignorant conspecific partner to release food for both of them. The rates of overt communication were still low in both cases (tool: 8% trials; food: 13%). In the tool experiment all of the communication was done by only three of 12 subjects and in the food experimental condition, communication rates were not higher than the non-social baseline. It is perhaps surprising that these chimpanzees failed to inform a conspecific even when they incurred a cost to themselves: the lack of communication led to a breakdown in coordination success as partners were effectively guessing the location. However, these results are consistent with previous findings that captive apes reliably point to request tools or food from human experimenters, including pointing to a specific location from several potential ones, but they rarely point merely to inform others, or to request from conspecifics (see e.g., Bohn, Call, and Tomasello 2015; Bullinger et al. 2011b; Leavens, Russell, and Hopkins 2005; Roberts et al. 2014, but see Pelé et al. 2009; Moore, Call, and Tomasello 2015, for examples of one orangutan pointing for conspecifics).

An interesting exception is the "alert hoo" by which chimpanzees warn other group members to the presence of a snake. Findings from several field experiments found that callers were sensitive to the apparent knowledge state of potential recipients (see also Townsend, Watson, and Slocombe, chapter 11 this volume). In particular, chimpanzees were more likely to inform recipients that were apparently ignorant to the snake. In addition to the call, chimpanzees have also been observed to perform "marking behavior"—the caller places themselves such that they can see the snake and the recipient and then alternating gaze between them—which was similarly modulated by receiver knowledge (Crockford et al. 2012; Crockford, Wittig, and Zuberbühler 2017). There are no reports of these calls being used for coordinating cooperative actions, but they do suggest that there may be forms of informative communication that are yet to be observed in captive experimental settings.

FROM PULLING BOXES TO EXPERIMENTAL ECONOMICS

The final set of studies in this review draws heavily from interdependent games in experimental economics: they look at decision making when the optimal actions depend on the decisions of others (and vice versa; Brosnan et al. 2011; Jensen, Call, and Tomasello 2007; Proctor et al. 2013; Sánchez-Amaro et al. 2016, 2017). Using classic pay-off matrices from game theory also provides the opportunity to relate cooperative decision making in nonhuman primates to the extensive literature on human cooperation, and to extend these comparisons to a range of species (see Vale and Brosnan, chapter 15 this volume).

One major difference from the human literature is that the tasks developed for nonhuman primates generally allow communication, providing further insights into how apes use communication to mitigate the risk of coordination failure and to negotiate conflicts of interest. Melis, Hare, and Tomasello (2009) created one such situation by adapting the loose-string paradigm by presenting two boards. Just as in the original studies, food could be accessed only by two chimpanzees simultaneously pulling the ends of the rope for one board, but in this case there were two boards for the chimpanzees to choose from. An additional challenge was that one of the two boards had an unequal distribution of food. Thus, the more dominant partner (the one that would get access to the higher value reward) had an incentive to choose the board with the single highest reward, while the subordinate partner would gain more from the board with the equal food distribution. If the chimpanzees did not coordinate their actions towards the same board, then neither would get anything (i.e., the dominant needed the subordinate just as much as the other way around). Yet, there was a noticeable lack of overt communication between the chimpanzees. Perhaps this was due to the competitive aspect of the coordination task (see also Sánchez-Amaro et al. 2017).

The stag-hunt coordination problems presented to pairs of chimpanzees (Brosnan et al. 2011; Bullinger et al. 2011a; Duguid et al. 2014; Vale and Brosnan, chapter 15 this volume) are drawn almost directly from behavioral economics (Skyrms 2004). This coordination game provides an interesting comparison to the negotiation tasks because both individuals have to choose between two options of differing value but there is no conflict of interest. In the version of the task used by Bullinger, Melis, and Tomasello (2011) and Duguid et al. (2014) with chimpanzees the game is played as follows: two individuals each start collecting a lower value reward individually. While they collect this, a higher value reward appears for a short time (20 seconds) in a second location. This higher reward can be retrieved only through cooperation (again, via simultaneous pulling of ropes). During this time, they have the choice to stay with the lower value reward or try to cooperate with a partner for the higher value reward. Further access to the low value reward is denied as soon as a subject chooses to cooperate. In this non-competitive context communication is more common than in the previously described negotiation tasks, in the form of attention-getting gestures by individuals waiting for a partner to join them at the cooperative option. However, communication rates were low, with communication occurring in 7.9% and 11.5% of trials in the two versions of the task. As the waiting time increased, participants were more likely to communicate with a partner, perhaps responding to the urgency of the need to coordinate (although this interpretation would need a non-social control to rule out other, non-social factors such as frustration). In the first version of the task, the chimpanzees were very successful at coordinating despite the lack of communication, but in the second (harder) version the rate of coordination failure was much higher (Duguid et al. 2014). So, as in Bullinger, Melis, and Tomasello (2014), not using communication to coordinate their decisions potentially had serious consequences for their ability to cooperate.

Communication, Comprehension, and Cooperation

Thus far, we have focused on the *production* of communication, but the *success* of communication is dependent upon the responses of the recipient. For a comprehensive account of ape communication during cooperation, it is important to consider both production and comprehension (Hobaiter, chapter 10 this volume). If the responses do not help the communicator to achieve their goals, then they are unlikely to show this behavior in the first place, or at least it is unlikely to persist. The scarcity of communication and experimental design often makes it difficult to interpret responses to communication by the receiver without a suitable control condition in which communication is not possible. Situations with asymmetrical knowledge between partners are a good source of evidence about what recipients interpret from the communication because one can measure whether this information improves success rates. For example, Bullinger, Melis, and Tomasello (2014) found little evidence that chimpanzees made use of indications from a partner about the location of a hidden tool or food. Despite not knowing the location of the tool, the chimpanzees were unlikely to even wait for information from a partner (see also Herrmann and Tomasello 2006; Moore, Call, and Tomasello 2015; Templemann, Kaminski, and Liebal 2013), though there are some exceptions (e.g., Mulcahy and Call 2009; Povinelli, Nelson, and Boysen 1992).

Some evidence for responding to communication comes from a different cooperative context: instrumental helping tasks. In these tasks one individual can provide low-cost help to a conspecific, with no immediate benefit for themselves (for example, by providing a tool or opening a door). Several studies find that requests for action from the potential recipient, in the form of begging gestures or attention-getters, increase helping rates (Melis, Engelmann, and Warneken 2018; Melis, Schneider, and Tomasello 2011; Yamamoto, Humle, and Tanaka 2009, 2012). However, there is some debate about whether increased rates of helping are a response to successfully communicating a request or are a way to minimize harassment from the requestor, or a result of stimulus enhancement (Melis, Engelmann, and Warneken 2018; Stevens 2004; Tennie, Jensen, and Call 2016).

Communication for Cooperation in the Context of Other Species

How does the emerging picture of how apes use communication for cooperation compare to other species? Adaptations of Crawford's box-pulling task have been used to test cooperation in a substantial number of different species (see Massen, Schaake, and Bugnyar, chapter 16 this volume, for a review). Many species achieve a reasonable level of cooperation success but few species have demonstrated an understanding of the role of the partner and an ability to coordinate actions with them. Thus far, only three species have successfully passed the "delay" version of the task that aims to test whether individuals understand that they need a partner to complete the task by waiting for the partner to arrive before pulling the ropes: Asian elephants (*Elephas maximus*, Plotnik et al. 2011), wolves (*Canis lupus*, Marshall-Pescini et al. 2017), and kea (*Nestor notabilis*, Heaney, Gray, and Taylor 2017, though in a previous study with a larger sample they did not pass, see Schwing et al. 2016). Perhaps surprisingly, there is very little communication reported in any of these experiments. There is an increase in gaze alternation between the task and the partner in dogs and wolves (Marshall-Pescini et al. 2017), hyenas (*Crocuta crocuta*, Drea and Carter 2009), and also capuchins in a slightly different two-action task (*Cebus apella*, Hattori, Kuroshima, and Fujita 2005).

Before we can draw any conclusions about how apes (mainly chimpanzees) compare to other species we should remember that chimpanzees also showed very little communication in the loose-string task. Even in Crawford's study they started to show solicitation behavior only after 90 sessions, which are many more sessions than are included in most recent experiments. Unfortunately, with the exception of some tasks, such as a sequential two-action box with capuchins (Hattori, Kuroshima, and Fujita 2005) or a four-choice coordination task with rhesus macaques (*Macaca mulatta*, Mason and Hollis 1962; see also Povinelli, Parks, and Novak 1992), there is very little variation in the cooperation tasks with other species. Moreover, the species tested in these paradigms represent only a small fraction of those we would need for an informative phylogenetic comparison. It is clear that there is still a lot to learn about how other species use communication to coordinate their actions and how this compares with the great apes.

There is one species for which there is a substantial literature on cooperation and communication: humans. As in the nonhuman literature, the focus on coordinating decisions and actions toward common goals is relatively new but evidence of how communication supports cooperation and efficient coordination in humans comes from a number of experimental disciplines. Many forms of coordination problems have been formalized by game theory, in the same way that costly cooperation has in the form of games such as the prisoner's dilemma (see Vale and Brosnan, chapter 15 this volume). Traditionally, the experimental versions of these games prevent any form of communication between players, but some experiments have manipulated the ability to communicate. They have revealed that communication is particularly helpful for solving coordination problems with common goals, such as the stag hunt (Cooper et al. 1992; Duffy and Feltovich 2002). Communication can be so efficient in these games that it can make some coordination problems trivial. However, the communication exhibited by human participants in these experiments usually relies heavily on written language and restricted forms of communication, such as predetermined messages sent via the computer, making it difficult to compare the results to other species.

More challenging coordination problems, which are more like real-life interactions, come in the form of experiments devised by behavioral ecologists and psychologists. Participants are presented with challenges, often similar in structure to one of the economic games, embedded in a story or an interactive foraging game (King et al. 2010; McClung et al. 2017; Thomas et al. 2014). These studies largely support the conclusions from behavioral economics that communication facilitates coordination. In addition, there is often greater communicative freedom so researchers can analyze non-linguistic communication, or the content and features of the language that people use to coordinate with others. For example, when teams of people were given the task of "foraging" for tokens in a small arena, the rate of gestural communication was a positive predictor of group success even though people could speak as much as they wished (King et al. 2010). In another foraging task, analysis of the speech content indicated that direct reference to common goals predicted cooperation (McClung et al. 2017).

Another way to find out which aspects of human communication are most important for coordinating with others is to study the development of cooperation and communication. Children can coordinate cooperative actions with adult partners in the second year of life (e.g., Warneken, Chen, and Tomasello 2006). These are highly scaffolded by the adults and children begin to coordinate with peers from around two years of age in simple tasks (e.g., Brownell and Carriger 1990). Similarly, at this age children help others and point informatively, but they do not follow informative points of peers when looking for a hidden object (Hepach, Kante, and Tomasello 2017; Kachel,

Moore, and Tomasello 2018). This is an interesting parallel with apes (Bullinger, Melis, and Tomasello 2014; Herrmann and Tomasello 2006; Moore, Call, and Tomasello 2015). From three years of age onward, children demonstrate the ability to coordinate in more complex tasks similar to the foraging-like tasks used with adults (e.g., Ashley and Tomasello 1998; Fletcher, Warneken, and Tomasello 2012; Goldvicht-Bacon and Diesendruck 2016; Grueneisen, Wyman, and Tomasello 2015). Communication also plays an important role in coordinating by this age through a variety of task-specific solutions: imperative directions (e.g., Ashley and Tomasello 1998), establishing common ground with a partner (e.g., Duguid et al. 2014), as well as non-linguistic cues, such as eye contact, can all facilitate coordination (Wyman, Rakoczy, and Tomasello 2013).

So how does ape communication for coordination compare to humans? Communication clearly plays a central role in human cooperation: it increases coordination efficiency and when given the opportunity, adults and children will use both language and non-linguistic forms of communication to solve coordination problems. Often, communication makes the task of coordinating so trivial it can hardly be called a coordination problem—so much so that Tomasello (2008) argues that the fundamental difference between human and nonhuman ape communication is that human communication is cooperative. According to this view, it is perfectly suited to cooperative problem solving in a way that ape communication is not. In a slightly different interpretation of the findings, Moore (2016, 2017) argues that chimpanzees do have the communicative skills required but they do not use them for coordination because 1) they are unable to adapt communication from other contexts to solve coordination problems and 2) the pressure to do so in their everyday lives is low. Both accounts emphasize the importance of studying how apes apply their communication skills to different cooperative contexts. They also highlight the importance of considering how the limitations in motivation to cooperate may hinder communication as well as how communication may limit cooperation.

Moving beyond Crawford: Examples of Changing Paradigms

Next, we turn our attention to studies that require higher communicative specificity beyond simply producing a general solicitation to "do something," and which allow participants to communicate in ways other than distal pointing. This includes tasks in which participants are able to move between and to touch different options, or in which they can make relevant information visible to a partner (e.g., Grueneisen et al. 2017; Karg et al. 2015). In the following section, we briefly report three recent experiments conducted at the Wolfgang Köhler Primate Research Center in Leipzig, Germany, that incorporate such methodological elements.

TOOL EXCHANGE IN COLLABORATIVE AND HELPING CONTEXTS

Our two closest living relatives, bonobos and chimpanzees, seem to differ on some types of behaviors that are thought to influence cooperation (Behringer et al. 2012; Hare et al. 2007; Hare, Wobber, and Wrangham 2012; Kano 1992; Wobber et al. 2010; Wobber, Wrangham, and Hare 2010; see also Taglialatela et al., chapter 4 this volume; Yamamoto, chapter 14 this volume). Moreover, recent research showed that bonobos, in comparison to chimpanzees, focus more on the face and eyes of conspecifics (Kano, Hirata, and Call 2015) and have more gray matter in brain regions associated with processing emotionally relevant stimuli (Rilling et al. 2012). Yet, while these differences suggest that bonobos are more socially adept, there are only a handful of studies directly comparing the two species during cooperative tasks (e.g., Hare et al. 2007), and we know very little about how bonobos use communication for coordination. Therefore, directly comparing both species in experimental studies of cooperation is crucial to draw definite conclusions as to which factors influence cooperation and how this is mediated by communication.

In a recent study, Nolte and Call (in review) directly compared both species in a tool-transfer task and investigated the relationship between altruistic helping and mutual cooperation (see also Nolte 2019). This combination offers a unique opportunity to understand the role of communication in these two contexts while controlling for task difficulty, individual differences, and tolerance between participants. Nolte and Call presented six chimpanzee and six bonobo dyads with a helping and a collaborative version of a task. The individuals in each dyad were in neighboring rooms that allowed communicative and object-sharing interactions between the apes. The researchers found a stark difference between species in their tool-sharing behavior. Chimpanzees did not share tools, with the exception of a mother-daughter pair, in which the mother shared a tool on two occasions in the helping experiment. In contrast, all the female-female bonobo dyads transferred a tool on multiple trials. Even though bonobos frequently shared tools altruistically, they did so more consistently when benefiting from it. Importantly, the use of the reaching gesture increased the likelihood of a transfer in bonobos, supporting very similar findings with chimpanzees in previous instrumental helping tasks (Melis et al. 2010; Yamamoto, Humle, and Tanaka 2009, 2012). However, there was no significant difference in reaching behavior between species (nor whether they stayed in close proximity to the helper). Although the small sample size and the infrequency of reaching behavior prescribe caution, these results suggest that behavioral differences in the production of communicative behavior do not explain the species differences in cooperation success in this task. Instead, part of the difference may be in the response to the reaching behavior. The bonobos may have been better able to interpret the needs of the recipient or were more motivated to help once their attention had been drawn to their needs. An alternative explanation for increased helping responses to requests is that requests are perceived as harassment and the helping responses are a means to alleviate this (Stevens 2004). In this case bonobos could have been more sensitive, not to the needs of others, but to harassment.

THE BOUNCING BALL TASK

A recent study by Voinov et al. (2020) was designed to study communication for coordination in pairs of captive chimpanzees in a simple game in which they bounced a virtual ball back and forth between two touch screens located in adjacent testing rooms. Either subject could start a trial by touching an initiation symbol, upon which the ball appeared in the center of their screen. Once a subject touched the ball, it moved toward the other screen until it was either returned by their partner (by touching it multiple times) or crossed their partner's screen completely and left it. Subjects were able to see their partner and part of the second screen on most trials. Playing the ball back and forth for multiple *turns* resulted in a fruit reward for both subjects, whereas allowing the ball to leave the screen counted as a failure, with no reward given to either subject. Unlike previous studies, which were designed to challenge the ability to coordinate or motivation to cooperate (e.g., Duguid et al. 2014; Sánchez-Amaro et al. 2016), this study did not provide subjects with tempting alternatives to coordinating, or other motivations to defect.

Five pairs of chimpanzees (nine individuals) participated in the study. All pairs completed individual and social training. After completing training, they participated in two test conditions. In the social test condition, the dyads had to coordinate to move the ball back and forth between two rooms (fig. 13.1). In the individual test condition, both monitors were also set up in two adjacent rooms, but the chimpanzees played by themselves, without a partner, controlling the ball's movement across both screens.

In the social test condition, of Voinov et al.'s (2020) study, successful coordination could be achieved in several mutually non-exclusive ways. Chimpanzees could focus solely on their part of the task, moving the ball toward

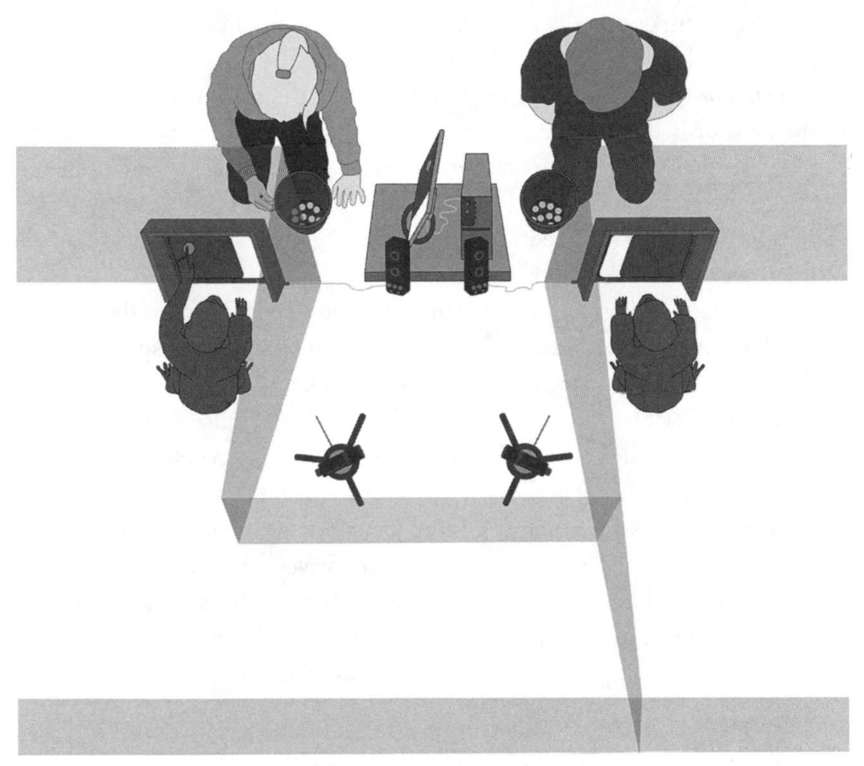

FIGURE 13.1. Setup of the bouncing ball game by Voinov et al. (in prep.). Illustration by Laura Därr, Max Planck Institute for Evolutionary Anthropology, Leipzig, Germany.

the other screen. However, the authors reasoned that the chimpanzees might further increase their success if they were sensitive to their partner's role, and to the possibility of encouraging action from the partner by means of communication. Voinov et al. recorded all instances of behaviors that qualified as potentially communicative events. As in previous studies (e.g., Bullinger, Melis, and Tomasello 2011; Crawford 1937; Duguid et al. 2014; Hirata and Fuwa 2007), audible and non-audible gestures were very rare and were primarily shown by a few specific individuals. Out of nine subjects, four occasionally showed potentially communicative behaviors, ranging between 1.5 and 24.3% of game turns with communicative behaviors among social training trials for these subjects.

In addition to regular ball playing trials, the social test condition also included *probe turns* in which the flow of the game was interrupted by the experimental program: for example, the ball would get "stuck" (it would rest and be unresponsive to further touches). Voinov et al. (2020) hypothesized that if the gestures described above served to promote cooperation by the partner, they should be produced more often when the game was interrupted. For the majority of subjects, this was indeed what the authors found: the percentage of turns in which a subject produced gestures was higher for these probe trials than for regular trials. Five of eight chimpanzees who completed both the individual and social test conditions produced more gestures in social probe turns, and for two individuals this difference was significant, while only two of them showed a reverse pattern.

While these findings await replication, a tentative conclusion is that at least some chimpanzees used forms of communication to solve this specific coordination problem (compare with Leavens, Russell, and Hopkins 2010). Their gestures served as *communication for coordination* in the sense that (a) gestures were produced at a higher rate when coordination was interrupted experimentally than when it was not and (b) at least some chimpanzees engaged in them only when there was a partner in front of the other screen to communicate *to*. As such, the findings of Voinov et al. (2020) lend even stronger support to the view that such gestures are not mere expressions of frustration or impatience in the face of failure to obtain a reward, but may in fact be produced to request from a partner that they coordinate their actions with one's own.

INFORMING A PARTNER TASK

In many studies, the lack of communication did not prevent high rates of successful coordination. When the task demanded similar and simultaneous actions, subjects performed the correct actions without direction from their partners (sometimes spontaneously and others after intensive training) (Bullinger, Melis, and Tomasello 2011; Duguid et al. 2014; Hirata and Fuwa 2007; Melis, Hare, and Tomasello 2006a). Tasks with asymmetrical roles present a more challenging coordination problem (Bullinger, Melis, and Tomasello 2014), but even then, communication remains infrequent. One reason for this may have been that the need for communication was not crucial because success by chance was still 50 percent.

De las Heras, Sperber, and Call (in review) designed a study to elicit communication in the context of cooperation: a mutualistic task in which pairs of conspecifics needed to collaborate to retrieve food from an apparatus by transferring a tool and, crucially, with a high risk of failure if subjects did not inform their partners of the location of the hidden food. Ten dyads of chimpanzees and eight dyads of orangutans participated in the test. Participants played either a communicator or an operator role from adjacent rooms. A box with four possible food locations was attached to the mesh separating these rooms. From the communicator side, subjects could see which one of the four locations contained food through a window in the box. The insertion of a tool in the baited location from the operator side delivered food for both participants. Subjects received individual training in both conditions prior to the test to learn how to operate the apparatus and experience with the visibility of the food from both rooms. In order for both to get a reward, the communicator had to transfer a tool to the operator through the mesh. In addition, to retrieve the reward reliably, the communicator needed to inform the operator where to insert the tool. In the hidden condition, only the communicator could see which one of four tubes was baited whereas in the visible condition operators also had visual access to the food's location because the apparatus' lid was transparent.

A key difference in this study compared to previous studies of communication in cooperative contexts is that the tool transfer action can be a communicative act in itself. Although communicators could pass the tool through the mesh at any position, they preferentially transferred the tool above the baited location (56% of trials), which explains the operators' success as they inserted the tool at the location in which they received it in 41.5% of trials. This resulted in a success rate (34.5%) above chance levels (25%). Although previous experiments showed that apes have an understanding of what their conspecifics can or cannot see (Bräuer, Call, and Tomasello 2007; Hare et al. 2000), the comparison of the hidden and visible conditions revealed that communicators did not adapt their tool-transfer strategy according to the information available to operators. This suggests that the motivation for tool transfers was imperative. They may have been trying to influence the operator's behavior but may not have been trying to share information declaratively, in the way infants do from an early age (Liszkowski et al. 2006).

While increasing the need for specific communication, this study did not require the participants to use gestures beyond their natural repertoire. Instead, communication was grounded on the tool transfer because a purely instrumental action could be used as an indicator of the food's location. Just as in Voinov et al. (2020), this study marks a subtle but important change in how communication is examined in these experiments. By considering a broader range of potentially communicative behaviors these findings highlight potential forms of communication not previously considered.

Conclusion

We can say with confidence that Crawford's (1937, 1941) seminal findings have now received the full attention that they deserve, 80 years after their publication. Much contemporary work on the experimental study of primate cooperation is based on his procedure (or variations thereof) and many of the results obtained recently corroborate his findings: 1) chimpanzees can coordinate with others in novel tasks, 2) they can also communicate to support cooperation, although at a much lower rate than it is observed in humans, and 3) there are large individual differences (determined in part by factors like tolerance) in the likelihood that two individuals will work together to solve a task. But the data that have accumulated in the last two decades have also extended Crawford's original findings in significant ways. First, several studies have shown that cooperation (and communication) can occur spontaneously, without human scaffolding. Second, there are now tasks testing for coordination in which individuals play different roles, not just the same role. Third, coordination has been reported in multiple species, not just primates but also other mammals and birds. However, so far, not much information has been reported in terms of the role that communication may play in supporting coordination in those other species. Finally, the kinds of behaviors that may be used as communicative devices may include instrumental actions that have acquired a communicative function and, crucially, they are not merely an unspecific request to respond in some unspecified way. We anticipate that interest in this area will continue to grow in the coming years. Endowed with the necessary methodological sophistication and ingenuity, researchers may uncover new ways in which communication does lubricate the wheels of cooperation in great apes and other nonhuman animals.

Future Directions

Our aim with this chapter was to discuss whether and when chimpanzees, orangutans, and bonobos communicate to solve coordination problems. In those studies that have explicitly sought to study communication there is a small but consistent use of communication to solicit action from a conspecific or a human experimenter in the form of attention-getters and imperative pointing. So far, with the exception of the study of de las Heras et al. (in review), there is little evidence of apes informing each other of, for example, the location of a hidden food or tool.

We have also highlighted that this is a developing field and we do not yet fully understand how chimpanzees coordinate their decisions and the role communication plays in achieving this. One indicator of this is that the communication reported in these experiments is limited in comparison to the range of documented communication, particularly in gestures, used in the wild and in captivity (Call and Tomasello 2000; Graham, Furuichi, and Byrne 2017; Hobaiter and Byrne 2011; Hobaiter, chapter 10 this volume). Several studies, especially those looking at production and comprehension of informative communication, have relied heavily on the use or potential use of pointing gestures. Future studies could add to the breadth of experimental paradigms by creating coordination problems designed to tap into some of the potential coordination mechanisms observed in wild chimpanzees (see Yamamoto, chapter 14 this volume). For example, communication for coordinating travel in wild chimpanzees often involves touching the receiver (e.g., solicitations to follow or travel together, cf. Hobaiter and Byrne 2014) but many of the experiments reviewed in this chapter did not allow, or allowed only minimal, physical contact between partners. Additionally, the types of tasks subjected to experimental scrutiny are not necessarily representative of the complex cooperative activities such as hunting, patrolling, or coalitionary support observed in wild chimpanzees. Nevertheless, we argue that they represent an important stepping stone for developing more sophisticated tasks that can be used to investigate the psychological underpinnings of cooperation and its relation to communication. This is not to say that previous studies, including those where apes did not communicate, are invalid, but that greater variation in experimental design incorporating additional dimensions of complex cooperative interactions will advance our knowledge in this area and help us to better interpret the existing pattern of results.

As already noted by Crawford (1937), there is considerable individual variation within each study and there are indications that this individual variation is consistent. For example, the same chimpanzees that were most communicative in the study by Voinov et al. (2020), were the most communicative when tested by Duguid et al. (2014). Similiarly, the same orangutan pointed for conspecifics in studies by Moore, Call, and Tomasello (2015) and Pelé et al. (2009). It would require a large-scale collation and analysis of data but the source of this variation can tell us about features of the communication. There are several, non-mutually exclusive, explanations for variation in communication. First, it is possible that individual differences in sociocognitive abilities enable certain subjects to understand more readily that the task is a cooperative one with multiple roles and a mutual goal. This may in turn result in a higher motivation for these more socially adept individuals to communicate when necessary and appropriate. If true, one may expect that the same subjects who produce gestures frequently in a coordination task should also excel in other tasks that tap into theory of mind and joint action understanding. Second, it is possible that the specific gestures that were produced very often by some subjects are those that have a history of being successfully followed by food delivery in interactions with human caretakers and researchers (Leavens, Hopkins, and Bard 2005). If true, one may expect that the same subjects who produced gestures frequently in this coordination task should also produce the same or very similar gestures frequently when interacting with humans who may provide food. Finally, it is possible that it is the combination of individuals working together that predicts the level of communication. The level of tolerance and dominance relationship between two individuals has already been shown to be an important predictor in co-ordination success (Hare et al. 2007; Melis et al. 2006b).

A word must be said about comparing ape species. There is a heavy bias toward chimpanzees in the number of different experiments and only a handful have compared two species. In some cases, several species of great ape are included but the sample size is generally too small to detect systematic species differences. The recent study by Nolte and Call (in review) found a noticeable difference in the behavior of bonobos and chimpanzees (see also Hare et al. 2007; Taglialatela et al., chapter 4 this volume; Yamamoto, chapter 14 this volume). However, it is possible that this species difference is a product of individual differences, or the relationships between the individuals tested. Although the overall gestural repertoires of chimpanzees and bonobos have been found to be similar (Graham et al. 2018), there is also evidence that the way they use these gestures differs between the two species (Fröhlich et al. 2016). Thus, treating them as one group may mask important species differences.

References

Ashley, J., and M. Tomasello. 1998. "Cooperative problem-solving and teaching in preschoolers." *Social Development* 7 (2): 143–63.

Behringer, V., T. Deschner, E. Möstl, D. Selzer, and G. Hohmann. 2012. "Stress affects salivary alpha-amylase activity in bonobos." *Physiology & Behavior* 105 (2): 476–82.

Bohn, M., J. Call, and M. Tomasello. 2015. "Communication about absent entities in great apes and human infants." *Cognition* 145: 63–72.

Bräuer, J., J. Call, and M. Tomasello. 2007. "Chimpanzees really know what others can see in a competitive situation." *Animal Cognition* 10 (4): 439–48.

Brosnan, S. F., A. Parrish, M. J. Beran, T. Flemming, L. Heimbauer, C. F. Talbot, S. P. Lambeth, S. J. Schapiro, and B. J. Wilson. 2011. "Responses to the assurance game in monkeys, apes, and humans using equivalent procedures." *Proceedings of the National Academy of Sciences* 108 (8): 3442–47.

Brownell, C. A., and M. S. Carriger. 1990. "Changes in cooperation and self-other differentiation during the second year." *Child Development* 61 (4): 1164–74.

Bullinger, A. F., A. P. Melis, and M. Tomasello. 2011. "Chimpanzees, *Pan troglodytes*, prefer individual over collaborative strategies towards goals." *Animal Behaviour* 82 (5): 1135–41.

Bullinger, A. F., A. P. Melis, and M. Tomasello. 2014. "Chimpanzees (*Pan troglodytes*) instrumentally help but do not communicate in a mutualistic cooperative task." *Journal of Comparative Psychology* 128 (3): 251–S60.

Bullinger, A. F., E. Wyman, A. P. Melis, and M. Tomasello. 2011. "Coordination of chimpanzees (*Pan troglodytes*) in a stag hunt game." *International Journal of Primatology* 32 (6): 1296–1310.

Bullinger, A. F., F. Zimmermann, J. Kaminski, and M. Tomasello. 2011. "Different social motives in the gestural communication of chimpanzees and human children." *Developmental Science* 14 (1): 58–68.

Call, J., and M. Tomasello. 2000. "Use and comprehension of gestures in chimpanzees and orangutans." *American Journal of Primatology* 51 (S1): 29–30.

Cartmill, E. A., and R. W. Byrne. 2010. "Semantics of primate gestures: Intentional meanings of orangutan gestures." *Animal Cognition* 13 (6): 793–804.

Chalmeau, R. 1994. "Do chimpanzees cooperate in a learning task?" *Primates* 35 (3): 385–92.

Chalmeau, R., and A. Gallo. 1996. "What chimpanzees (*Pan troglodytes*) learn in a cooperative task." *Primates* 37 (1): 39–47.

Chalmeau, R., K. Lardeaux, P. Brandibas, and A. Gallo. 1997. "Cooperative problem solving by orangutans (*Pongo pygmaeus*)." *International Journal of Primatology* 18 (1): 23–32.

Cooper, R., D. V. DeJong, R. Forsythe, and T. W. Ross. 1992. "Communication in coordination games." *Quarterly Journal of Economics* 107 (2): 739–71.

Crawford, M. P. 1937. "The cooperative solving of problems by young chimpanzees." *Comparative Psychology Monographs* 14: 1–88.

Crawford, M. P. 1941. "The cooperative solving by chimpanzees of problems requiring serial responses to color cues." *Journal of Social Psychology* 13 (2): 259–80.

Crockford, C., R. M. Wittig, R. Mundry, and K. Zuberbühler. 2012. "Wild chimpanzees inform ignorant group members of danger. *Current Biology* 22 (2): 142–46.

Crockford, C., R. M. Wittig, and K. Zuberbühler. 2017. "Vocalizing in chimpanzees is influenced by social-cognitive processes." *Science Advances* 3 (11): e1701742.

Cronin, K. A., B. A. Pieper, E. J. Van Leeuwen, R. Mundry, and D. B. Haun. 2014. "Problem solving in the presence of others: How rank and relationship quality impact resource acquisition in chimpanzees (*Pan troglodytes*)." *PLoS One* 9 (4): e93204.

De las Heras, A., D. Sperber, and J. Call. In review. "Chimpanzees and orangutans intentionally communicate to solve a cooperative problem, but fail to adapt their communication to the knowledge state of their partners."

Drea, C. M., and A. N. Carter. 2009. "Cooperative problem solving in a social carnivore." *Animal Behaviour* 78 (4): 967–77.

Duffy, J., and N. Feltovich. 2002. "Do actions speak louder than words? An experimental comparison of observation and cheap talk." *Games and Economic Behavior* 39 (1): 1–27.

Duguid, S., E. Wyman, A. F. Bullinger, K. Herfurth-Majstorovic, and M. Tomasello. 2014. "Coordination strategies of chimpanzees and human children in a stag hunt game." *Proceedings of the Royal Society of London B* 281 (1796): 20141973.

Fletcher, G. E., F. Warneken, and M. Tomasello. 2012. "Differences in cognitive processes underlying the collaborative activities of children and chimpanzees." *Cognitive Development* 27 (2): 136–53.

Fröhlich, M., P. Kuchenbuch, G. Müller, B. Fruth, T. Furuichi, R. M. Wittig, and S. Pika. 2016. "Unpeeling the layers of language: Bonobos and chimpanzees engage in cooperative turn-taking sequences." *Scientific Reports* 6: 25887.

Goldvicht-Bacon, E., and G. Diesendruck. 2016. "Children's capacity to use cultural focal points in coordination problems." *Cognition* 149: 95–103.

Graham, K. E., T. Furuichi, and R. W. Byrne. 2017. "The gestural repertoire of the wild bonobo (*Pan paniscus*): A mutually understood communication system." *Animal Cognition* 20 (2): 171–77.

Graham, K. E., C. Hobaiter, J. Ounsley, T. Furuichi, and R. W. Byrne. 2018. "Bonobo and chimpanzee gestures overlap extensively in meaning." *PLoS Biology* 16 (2): e2004825.

Grueneisen, S., S. Duguid, H. Saur, and M. Tomasello. 2017. "Children, chimpanzees, and bonobos adjust the visibility of their actions for cooperators and competitors." *Scientific Reports* 7 (1): 8504.

Grueneisen, S., E. Wyman, and M. Tomasello. 2015. "Conforming to coordinate: Children use majority information for peer coordination." *Developmental Psychology* 33 (1): 136–47.

Hare, B., J. Call, B. Agnetta, and M. Tomasello. 2000. "Chimpanzees know what conspecifics do and do not see." *Animal Behaviour* 59 (4): 771–85.

Hare, B., A. P. Melis, V. Woods, S. Hastings, and R. W. Wrangham. 2007. "Tolerance allows bonobos to outperform chimpanzees on a cooperative task." *Current Biology* 17 (7): 619–23.

Hare, B., V. Wobber, and R. W. Wrangham. 2012. "The self-domestication hypothesis: Evolution of bonobo psychology is due to selection against aggression." *Animal Behaviour* 83 (3): 573–85.

Hattori, Y., H. Kuroshima, and K. Fujita. 2005. "Cooperative problem solving by tufted capuchin monkeys (*Cebus apella*): Spontaneous division of labor, communication, and reciprocal altruism." *Journal of Comparative Psychology* 119 (3): 335–42.

Heaney, M., R. D. Gray, and A. H. Taylor. 2017. "Keas perform similarly to chimpanzees and elephants when solving collaborative tasks." *PLoS One* 12 (2): e0169799.

Hepach, R., N. Kante, and M. Tomasello. 2017. "Toddlers help a peer." *Child Development* 88 (5): 1642–52.

Herrmann, E., and M. Tomasello. 2006. "Apes' and children's understanding of cooperative and competitive motives in a communicative situation." *Developmental Science* 9 (5): 518–29.

Hirata, S. 2003. "Cooperation in chimpanzees." *Hattatsu* 95: 103–11.

Hirata, S., and K. Fuwa. 2007. "Chimpanzees (*Pan troglodytes*) learn to act with other individuals in a cooperative task." *Primates* 48 (1): 13–21.

Hirata, S., N. Morimura, and K. Fuwa. 2010. "Intentional communication and comprehension of the partner's role in experimental cooperative tasks." In *The Mind of the Chimpanzee: Ecological and Experimental Perspectives*, edited by E. V. Lonsdorf, S. R. Ross, and T. Matsuzawa, 251–63. Chicago: University of Chicago Press.

Hobaiter, C., and R. W. Byrne. 2011. "The gestural repertoire of the wild chimpanzee." *Animal Cognition* 14 (5): 745–67.

Hobaiter, C., and R. W. Byrne. 2014. "The meanings of chimpanzee gestures." *Current Biology* 24 (14): 1596–1600.

Jensen, K., J. Call, and M. Tomasello. 2007. "Chimpanzees are rational maximizers in an ultimatum game." *Science* 318 (5847): 107–9.

Kachel, G., R. Moore, and M. Tomasello. 2018. "Two-year-olds use adults' but not peers' points." *Developmental Science* 21 (5): e12660.

Kano, F., S. Hirata, and J. Call. 2015. "Social attention in the two species of *Pan*: Bonobos make more eye contact than chimpanzees." *PLoS One* 10 (6): e0129684.

Kano, T. 1992. *The Last Ape: Pygmy Chimpanzee Behavior and Ecology*. Stanford, CA: Stanford University Press.

Karg, K., M. Schmelz, J. Call, and M. Tomasello. 2015. "Chimpanzees strategically manipulate what others can see." *Animal Cognition* 18 (5): 1069–76.

King, A. J., C. Narraway, L. Hodgson, A. Weatherill, V. Sommer, and S. Sumner. 2010. "Performance of human groups in social foraging: The role of communication in consensus decision making." *Biology Letters* 7 (2): 237–40.

Leavens, D. A., W. D. Hopkins, and K. A. Bard. 2005. "Understanding the point of chimpanzee pointing: Epigenesis and ecological validity." *Current Directions in Psychological Science* 14 (4): 185–89.

Leavens, D. A., J. L. Russell, and W. D. Hopkins. 2005. "Intentionality as measured in the persistence and elaboration of communication by chimpanzees (*Pan troglodytes*)." *Child Development* 76 (1): 291–306.

Leavens, D. A., J. L. Russell, and W. D. Hopkins. 2010. "Multimodal communication by captive chimpanzees (*Pan troglodytes*)." *Animal Cognition* 13 (1): 33–40.

Liszkowski, U., M. Carpenter, T. Striano, and M. Tomasello. 2006. "12- and 18-month-olds point to provide information for others." *Journal of Cognition and Development* 7 (2): 173–87.

Marshall-Pescini, S., J. F. Schwarz, I. Kostelnik, Z. Virányi, and F. Range. 2017. "Importance of a species' socioecology: Wolves outperform dogs in a conspecific cooperation task." *Proceedings of the National Academy of Sciences* 114 (44): 11793–98.

Martin, C. F., D. Biro, and T. Matsuzawa. 2017. "Chimpanzees spontaneously take turns in a shared serial ordering task." *Scientific Reports* 7: 14307.

Mason, W. A., and J. H. Hollis. 1962. "Communication between young rhesus monkeys." *Animal Behaviour* 10 (3–4): 211–21.

McClung, J. S., S. Placì, A. Bangerter, F. Clément, and R. Bshary. 2017. "The language of cooperation: Shared intentionality drives variation in helping as a function of group membership." *Proceedings of the Royal Society of London B* 284 (1863): 20171682.

Melis, A. P., J. M. Engelmann, and F. Warneken. 2018. "Correspondence: Chimpanzee helping is real, not a byproduct." *Nature Communications* 9: 615.

Melis, A. P., B. Hare, and M. Tomasello. 2009. "Chimpanzees coordinate in a negotiation game." *Evolution and Human Behavior* 30 (6): 381–92.

Melis, A. P., B. Hare, and M. Tomasello. 2006a. "Chimpanzees recruit the best collaborators." *Science* 311 (5765): 1297–1300.

Melis, A. P., B. Hare, and M. Tomasello. 2006b. "Engineering cooperation in chimpanzees: Tolerance constraints on cooperation." *Animal Behaviour* 72 (2): 275–86.

Melis, A. P., A. C. Schneider, and M. Tomasello. 2011. "Chimpanzees, *Pan troglodytes*, share food in the same way after collaborative and individual food acquisition." *Animal Behaviour* 82 (3): 485–93.

Melis, A. P., and M. Tomasello. 2013. "Chimpanzees' (*Pan troglodytes*) strategic helping in a collaborative task." *Biology Letters* 9 (2): 20130009.

Melis, A. P., F. Warneken, K. Jensen, A. C. Schneider, J. Call, and M. Tomasello. 2010. "Chimpanzees help conspecifics obtain food and non-food items." *Proceedings of the Royal Society of London B* 278 (1710): 1405–13.

Mendres, K. A., and F. B. M. de Waal. 2000. "Capuchins do cooperate: The advantage of an intuitive task." *Animal Behaviour* 60 (4): 523–29.

Mitani, J. C. 2009. "Cooperation and competition in chimpanzees: Current understanding and future challenges." *Evolutionary Anthropology* 18 (5): 215–27.

Moore, R. 2016. "Gricean communication, joint action, and the evolution of cooperation." *Topoi* 37 (2): 329–41.

Moore, R. 2017. "Social cognition, stag hunts, and the evolution of language." *Biology and Philosophy* 32 (6): 797–818.

Moore, R., J. Call, and M. Tomasello. 2015. "Production and comprehension of gestures between orang-utans (*Pongo pygmaeus*) in a referential communication game." *PLoS One* 10 (6): e0129726.

Mulcahy, N. J., and J. Call. 2009. "The performance of bonobos (*Pan paniscus*), chimpanzees (*Pan troglodytes*), and orangutans (*Pongo pygmaeus*) in two versions of an object-choice task." *Journal of Comparative Psychology* 123 (3): 304–9.

Nolte, S. 2019. "Cooperation and teaching in the context of cumulative culture." PhD diss., University of St. Andrews.

Nolte, S., and J. Call. In review. "Targeted helping and cooperation in zoo-living chimpanzees and bonobos."

Pelé, M., V. Dufour, B. Thierry, and J. Call. 2009. "Token transfers among great apes (*Gorilla gorilla*, *Pongo pygmaeus*, *Pan paniscus*, and *Pan troglodytes*): Species differences, gestural requests, and reciprocal exchange." *Journal of Comparative Psychology* 123 (4): 375–84.

Plotnik, J. M., R. Lair, W. Suphachoksahakun, and F. B. M. de Waal. 2011. "Elephants know when they need a helping trunk in a cooperative task." *Proceedings of the National Academy of Sciences* 108 (12): 5116–21.

Povinelli, D. J., K. E. Nelson, and S. T. Boysen. 1992. "Comprehension of role reversal in chimpanzees: Evidence of empathy?" *Animal Behaviour* 43 (4): 633–40.

Povinelli, D. J., and D. K. O'Neill. 2000. "Do chimpanzees use their gestures to instruct each other?" In *Understanding Other Minds: Perspectives from Autism*, edited by S. Baron-Cohen, H. Tager-Flusberg, and D. J. Cohen, 459–87. New York: Oxford University Press.

Povinelli, D. J., K. A. Parks, and M. A. Novak. 1992. "Role reversal by rhesus monkeys, but no evidence of empathy." *Animal Behaviour* 44 (2): 269–81.

Proctor, D., R. A. Williamson, F. B. M. de Waal, and S. F. Brosnan. 2013. "Chimpanzees play the ultimatum game." *Proceedings of the National Academy of Sciences* 110 (6): 2070–75.

Rilling, J. K., J. Scholz, T. M. Preuss, M. F. Glasser, B. K. Errangi, and T. E. Behrens. 2012. "Differences between chimpanzees and bonobos in neural systems supporting social cognition." *Social Cognitive and Affective Neuroscience* 7 (4): 369–79.

Roberts, A. I., S. J. Vick, S. G. B. Roberts, and C. R. Menzel. 2014. "Chimpanzees modify intentional gestures to coordinate a search for hidden food." *Nature Communications* 5: 3088.

Sánchez-Amaro, A., S. Duguid, J. Call, and M. Tomasello. 2016. "Chimpanzees coordinate in a snowdrift game." *Animal Behaviour* 116: 61–74.

Sánchez-Amaro, A., S. Duguid, J. Call, and M. Tomasello. 2017. "Chimpanzees, bonobos and children successfully coordinate in conflict situations." *Proceedings of the Royal Society of London B* 284 (1856): 20170259.

Schmelz, M., and J. Call. 2016. "The psychology of primate cooperation and competition: A call for realigning research agendas." *Philosophical Transactions of the Royal Society of London B* 371 (1686): 20150067.

Schweinfurth, M. K., S. E. DeTroy, E. J. Van Leeuwen, J. Call, and D. Haun. 2018. "Spontaneous social tool use in chimpanzees (*Pan troglodytes*)." *Journal of Comparative Psychology* 132 (4): 455.

Schwing, R., E. Jocteur, A. Wein, R. Noe, and J. J. Massen. 2016. "Kea cooperate better with sharing affiliates." *Animal Cognition* 19 (6): 1093–1102.

Skyrms, B. 2004. *The Stag Hunt and the Evolution of Social Structure*. Cambridge: Cambridge University Press.

Stevens, J. R. 2004. "The selfish nature of generosity: Harassment and food sharing in primates." *Proceedings of the Royal Society of London B* 271 (1538): 451–56.

Suchak, M., T. M. Eppley, M. W. Campbell, and F. B. M. de Waal. 2014. "Ape duos and trios: Spontaneous cooperation with free partner choice in chimpanzees." *PeerJ* 2: e417.

Surbeck, M., C. Boesch, C. Girard-Buttoz, C. Crockford, G. Hohmann, and R. M. Wittig. 2017. "Comparison of male conflict behavior in chimpanzees (*Pan troglodytes*) and bonobos (*Pan paniscus*), with specific regard to coalition and post-conflict behavior." *American Journal of Primatology* 79 (6): e22641.

Surbeck, M., C. Girard-Buttoz, C. Boesch, C. Crockford, B. Fruth, G. Hohmann, K. E. Langergraber, K. Zuberbühler, R. M. Wittig, and R. Mundry. 2017. "Sex-specific association patterns in bonobos and chimpanzees reflect species differences in cooperation." *Royal Society Open Science* 4 (5): 161081.

Templemann, S., J. Kaminski, and K. Liebal. 2013. "When apes point the finger." *Interaction Studies* 14 (1): 7–23.

Tennie, C., K. Jensen, and J. Call. 2016. "The nature of prosociality in chimpanzees." *Nature Communications* 7: 13915.

Thomas, K. A., P. DeScioli, O. S. Haque, and S. Pinker. 2014. "The psychology of coordination and common knowledge." *Journal of Personality and Social Psychology* 107 (4): 657.

Tomasello, M. 2008. *Origins of Human Communication*. Cambridge, MA: MIT Press.

Voinov, P. V., J. Call, G. Knoblich, M. Oshkina, and M. Allritz. 2020. "Chimpanzee coordination and potential communication in a two-touchscreen turn-taking game." *Scientific Reports* 10: 3400.

Völter, C. J., F. Rossano, and J. Call. 2015. "From exploitation to cooperation: Social tool use in orang-utan mother–offspring dyads." *Animal Behaviour* 100: 126–34.

Warneken, F., F. Chen, and M. Tomasello. 2006. "Cooperative activities in young children and chimpanzees." *Child Development* 77 (3): 640–63.

Wobber, V., B. Hare, J. Maboto, S. Lipson, R. W. Wrangham, and P. T. Ellison. 2010. "Differential changes in steroid hormones before competition in bonobos and chimpanzees." *Proceedings of the National Academy of Sciences* 107 (28): 12457–62.

Wobber, V., R. W. Wrangham, and B. Hare. 2010. "Bonobos exhibit delayed development of social behavior and cognition relative to chimpanzees." *Current Biology* 20 (3): 226–30.

Wyman, E., H. Rakoczy, and M. Tomasello. 2013. "Non-verbal communication enables children's coordination in a 'Stag Hunt' game." *European Journal of Developmental Psychology* 10 (5): 597–610.

Yamamoto, S., T. Humle, and M. Tanaka. 2009. "Chimpanzees help each other upon request." *PLoS One* 4 (10): e7416.

Yamamoto, S., T. Humle, and M. Tanaka. 2012. "Chimpanzees' flexible targeted helping based on an understanding of conspecifics' goals." *Proceedings of the National Academy of Sciences* 109 (9): 3588–92.

14

The Evolution of Cooperation in Dyads and in Groups: Comparing Chimpanzees and Bonobos in the Wild and in the Laboratory

SHINYA YAMAMOTO

Introduction

Humans demonstrate cooperation in a variety of contexts: in pairs, in groups, and sometimes even with strangers. Some nonhuman animals demonstrate similar traits, including at least fundamental forms of cooperation, and studying these animals can help to deepen our knowledge of how human cooperation evolved. Chimpanzees (*Pan troglodytes*) and bonobos (*Pan paniscus*), our closest living evolutionary relatives, are ideal subjects for this research. The *Pan* and human lineages diverged approximately 7 million years ago, while *Pan troglodytes* and *Pan paniscus* separated approximately 1 million years ago. Comparing these three species thus helps disentangle the evolutionary history of each species as they evolved their modern day characteristics.

It is now conspicuous that there are significant differences between chimpanzees and bonobos despite their phylogenetically close relationship (see also Behringer et al., chapter 2 this volume; Clay, chapter 12 this volume; Taglialatela et al., chapter 4 this volume). Table 14.1 lists some of the traits that were thought to be central to explaining the evolution of our own species and shows that they are shared with both *Pan* species. Interestingly, chimpanzees and bonobos demonstrate traits more akin to those observed in humans than in each other (Hare and Yamamoto 2015, 2017). This suggests that it may be misleading to consider only one of these two *Pan* species when determining which behavioral traits are shared with humans. Instead, careful comparison and understanding of how chimpanzees and bonobos diverged from one another can allow for inferences about cognitive evolution in related traits in our own species (Hare 2011).

Chimpanzees and bonobos differ significantly from one another in their social behavior. Chimpanzees are characterized by clear dominance within group hierarchies and hostile between-group relationships. Males are domi-

TABLE 14.1. Comparison among chimpanzees, bonobos, and humans

Trait	*Bonobo*	*Chimpanzee*	*Human Foragers*
Extractive foraging	Only captivity	Frequent	Frequent
Non-conceptive sexual behavior	Frequent	Absent	Frequent
Lethal aggression between groups	Absent	Present	Present
Mother's importance to adult offspring	High	Low	High
Infanticide/ female coercion	Absent	Present	Present
Levels of adult play	High	Low	High
Cooperative hunting	Absent	Present	Present
Sharing between strangers	Present	Absent	Present
Male-male alliances	Absent	Frequent	Frequent
Female gregariousness	High	Low	High

Source: Hare and Yamamoto 2017.

nant, coercing females, committing infanticide, and striving for alpha status. They are also highly territorial and xenophobic, displaying coalitionary aggression that can be lethal (Wilson et al. 2014). In contrast, bonobos show much less intense forms of aggression both within and between groups. These findings contribute to the broadly held view that chimpanzees are more competitive and aggressive whereas bonobos are more tolerant and peaceful.

Does this then mean that chimpanzees are less cooperative than bonobos or that bonobos are a more suitable target to investigate the evolution of cooperation? The answer should be "no." Instead, recent comparative research emphasizes the importance of studying both species (e.g., Hare and Yamamoto 2015, 2017; see also Taglialatela et al., chapter 4 this volume). As chimpanzees and bonobos live in divergent environments with different social structures, cooperative behavior likely evolved with adaptive modifications in accordance with these societies and environments. It is necessary to investigate cooperation among both chimpanzees and bonobos not only from the cognitive perspectives at the individual level but also from the viewpoints of broader social and environmental scales.

Approach

This chapter discusses the evolution of cooperation based on empirical studies with a two-by-two research paradigm, studying chimpanzees and bonobos both in the wild and captivity (table 14.2). Bossou, in Guinea, is one of the longest-running research sites studying wild chimpanzees, with researchers collecting behavioral data on a daily basis since 1976 (Matsuzawa, Humle, and Sugiyama 2011). Wamba, in the Democratic Republic of Congo, is the longest-running wild bonobo research site, founded in 1973 and still ongoing (Kano

TABLE 14.2. Two-by-two research paradigm, studying chimpanzees and bonobos both in the wild and captivity.

	Chimpanzee	*Bonobo*
In the wild	Bossou (Guinea)	Wamba & Mbali (Democratic Republic of Congo)
Controlled experiment	Kumamoto Sanctuary & Primate Research Institute, Kyoto University, Japan	Kumamoto Sanctuary & Primate Research Institute, Kyoto University, Japan

1992). The demographic records at Wamba can be traced back for almost half a century, even with the interruption of the civil war (Furuichi et al. 2012), providing invaluable behavioral data on wild bonobos. Conversely, in captivity, an experimental environment has been set up at the Primate Research Institute and Kumamoto Sanctuary of Kyoto University, both in Japan. Chimpanzees in the Primate Research Institute have been a part of cognitive and behavioral research since 1978 (Matsuzawa, Tomonaga, and Tanaka 2006). Kumamoto Sanctuary is home to the biggest community of chimpanzees in Japan and is the only facility in Japan that houses bonobos (Hirata et al., chapter 9 this volume). Here, I focus on these four sites.

Running studies in the wild and in captivity is a powerful way to generate a comprehensive understanding of the evolution of cooperation. Studies in the wild allow us to observe animals' natural behaviors and take a broader perspective in pursuing the socio-ecological and environmental factors that affect the animals' behavioral repertoire. In captive settings, the animals can be tested under controlled conditions, which enables investigation of the animals' cooperative traits from a cognitive viewpoint. With this two-by-two research paradigm, it is possible to investigate the evolution of cooperation from a variety of perspectives—from individual cognition to broader social and environmental factors.

CHIMPANZEE COOPERATION IN THE WILD

Wild chimpanzees share food not only with their kin, but also with non-kin individuals. This food sharing is a typical cooperative behavior observed in the wild, which benefits the recipient at the expense of the owner. Typically, male chimpanzees share meat with other group members after hunting a prey animal (Boesch 1994; Mitani and Watts 2001). In Bossou, where the chimpanzees do not eat meat, they share large cultivated fruits, such as papayas and pineapples (Hockings et al. 2007). In this case, as with meat sharing at other sites, the food is desirable, but the distribution of food items is generally limited to specific individuals, primarily adult males. Non-food-possessors surround the owner and beg for a share. The owner could disregard the beggars because the owner is normally dominant to them, but instead the owner often shares some portion with others or is tolerant of the recipients' scrounging.

Although food sharing appears to be a cooperative behavior, the underlying psychology is ambiguous. Food sharing is typically initiated by a beggar, while an owner rarely proactively shares food. Some have argued that chimpanzees share food because they want to avoid harassment from beggars (Gilby 2006). If this is true, chimpanzees' food sharing does not come from their altruism or prosociality, even though it may appear so. Other researchers have revealed a reciprocal relationship between food sharers and recipients. Chimpanzees, at least at certain field sites, seem to barter food for grooming, support during agonistic interaction, and/or sex (Gomes and Boesch 2009; Hockings et al. 2007; Mitani and Watts 2001). However, this is primarily a correlative observation, and thus the underlying psychological mechanism remains unclear even though the form apparently fits the reciprocal hypothesis.

The above examples are cooperative interactions between dyadic pairs of individuals, but other forms of cooperation in chimpanzees can be seen in larger groups. In Bossou, village roads go through the chimpanzee habitat, and the chimpanzees must cross the roads to go back and forth between forests. Altruistic helping has been observed during this risky endeavor, such as when a subadult male helped a non-kin adult female to escort her two infants across the road (Matsuzawa 2018). The male carried one of the female's infants, and after crossing the road, he gave the infant back to its mother. Importantly, during road-crossing events, chimpanzees demonstrate well-coordinated group movement (fig. 14.1; Hockings, Anderson, and Matsuzawa 2006; Yamamoto et al. in prep). Before crossing the road, if the group has spread out, members wait for others to join them. An adult male may then start to cross and scan the road. Interestingly, as depicted in the photo of a typical crossing (fig. 14.1), the leading male often stops in the middle of crossing the road instead of entering the safer forest and looks after the others, not only his kin but also non-kin group members. Under this male's supervision, females, elders, and youngsters cross the road quickly. The one who takes the rear position is also normally another adult male. Thus, two adult males seemingly take the role of guardians for the group. This behavioral coordination during road crossing, especially the males' division of roles to guard the other group members, may reflect concern for the group rather than simply concern for oneself or a specific individual.

FIGURE 14.1. Chimpanzees' group cooperation when crossing a risky road. (Photo by Shinya Yamamoto at Bossou, Guinea.)

Another form of in-group chimpanzee cooperation can be seen in cooperative hunting (Boesch 1994). Chimpanzees, especially adult males, chase a prey animal as a group and sometimes demonstrate a kind of division of labor: some chase the prey from behind, and others go forward to wait in a circling maneuver. This type of cooperation is different from the cooperation as part of road crossings (described above) in that the hunting individuals may pursue their own benefit (i.e., meat) but may or may not be altruistically motivated. Thus, it might be better to describe this as *coordination* rather than *cooperation*. Regardless, this example shows chimpanzees' ability to deftly coordinate their behavior in group settings.

CHIMPANZEE COOPERATION TESTED EXPERIMENTALLY

Observations of wild animals can reveal their natural behavior in their natural environment, but it is often difficult to determine the underlying mechanisms and psychological states that drive such behavior. In this regard, controlled experiments in captive environments can confer an advantage.

In 2005, Silk and colleagues published a paper that surprised many chimpanzee researchers. The title of their paper was "Chimpanzees are indifferent to the welfare of unrelated group members" (Silk et al. 2005). In their study, they tested chimpanzees at two different captive sites, providing them with a so-called prosocial choice test. The acting chimpanzee could choose a "1/0" option that delivered a food reward only to the actor or a "1/1" option that delivered the same reward both to the actor and to his/her unrelated partner. Despite the fact that the acting chimpanzee could benefit their partner at no material cost, they seemed to not take advantage of this opportunity, suggesting that chimpanzee behavior is not motivated by other-regarding preferences. This study has been replicated by other research teams (e.g., Jensen et al. 2006; Yamamoto and Tanaka 2010) with some modifications, though the results were virtually the same. Even between mother and infant, or under turn-taking conditions in which actors could build reciprocal cooperation, the chimpanzees showed no proactive other-regarding preferences (Yamamoto and Tanaka 2010).

Sharing experiments have also failed to identify proactive altruism or prosociality in chimpanzees. Ueno and Matsuzawa (2004) observed in detail food sharing between mother-infant pairs in an experiment in which initially only mothers were provided with fruits while the infants had nothing. This experiment demonstrated that most food transfers were initiated by the recipient infants, and that proactive sharing by the mothers was infrequent and involved only the inedible parts of the fruits, such as seeds and peel. Tanaka and Yamamoto (2009) also found the same pattern when they provided chimpanzees with tokens that could be exchanged for food. The chimpanzees shared tokens only when they were subject to begging by recipients (see also Horner et al. 2011). These studies raise questions about chimpanzees' prosociality even when they demonstrated an apparently prosocial behavior such as sharing.

Despite these results, it seems unwise to conclude that chimpanzees are not equipped with the psychological abilities needed to show prosocial behavior. They help others without any direct benefit to themselves in other contexts; and in experiments involving helping tasks, chimpanzees help not only human experimenters (Warneken and Tomasello 2006) but also their conspecific partners (Warneken et al. 2007; Yamamoto, Humle, and Tanaka 2009, Yamamoto, Humle, and Tanaka 2012). In these experiments, unlike in situations in the wild or in the above-described sharing experiments, the chimpanzees were safe from attacks or harassment by their partner, a potential recipient. For example, in one set of experiments (Yamamoto, Humle, and Tanaka 2009), two chimpanzees were kept separate in two adjacent booths, but could interact with their partner through a small hole opened in a separating panel. They could still avoid any physical interaction with their partner, and the duration of the trials were the same whether they helped or not.

In this situation, the chimpanzees helped their partners by giving them a tool that was useful only for the partner.

Although chimpanzees seemingly help others without any direct benefit to themselves, researchers have also found limitations to their helping behavior. First, chimpanzees appear to rarely help proactively, doing so only when their recipient obviously makes a request (Warneken and Tomasello 2006; Yamamoto, Humle, and Tanaka 2009). Helping behavior was not prompted by simple observation of another's efforts to attain a goal (Yamamoto, Humle, and Tanaka 2009), demonstrating that chimpanzees lack proactivity in their helping. Second, their helping can be observed only in very restricted contexts. Warneken and Tomasello (2006) tested human infants and chimpanzees in several different situations in which the experimenter was faced with an instrumental problem and was unable to reach a goal. The children, as young as 18 months of age, readily helped the experimenter in most situations involving out-of-reach objects, access thwarted by a physical obstacle, achievement of a wrong (correctable) result, and use of a wrong (correctable) means. However, chimpanzees reliably helped only in situations where the experimenter struggled to get out-of-reach objects (Warneken and Tomasello 2006).

Why do chimpanzees help others only when help is explicitly requested and seemingly only in limited circumstances? The lack of proactivity and the limited circumstances in which chimpanzees help have led some researchers to argue that this is because they lack the necessary cognitive ability to interpret the other's need for help. However, recent studies have suggested that it is their lack of motivation, rather than their cognitive ability, that explains the differences with human helping behaviors. Yamamoto, Humle, and Tanaka (2012) examined whether chimpanzees chose the appropriate tool from a set of seven objects for a partner who needed the tool to achieve his/her goal. Their results showed that, when they were able to observe their partner's situation, chimpanzees could select the appropriate tool according to their partner's situation and provide it to them (fig. 14.2). In the control condition, chimpanzees were unaware of their partner's situation, and so could not select the appropriate tool even when their partner demonstrated an explicit request by reaching out his/her arm. This suggests an understanding of their partner's need by visually inspecting the situation. These results have two important implications. First, chimpanzees demonstrate flexible targeted helping according to their partner's situation. Second, chimpanzees can understand their partner's need, but nonetheless seldom help proactively. This experiment, as well as others with larger sample sizes (Melis and Tomasello 2013), suggests that chimpanzees seem to be equipped with the essential

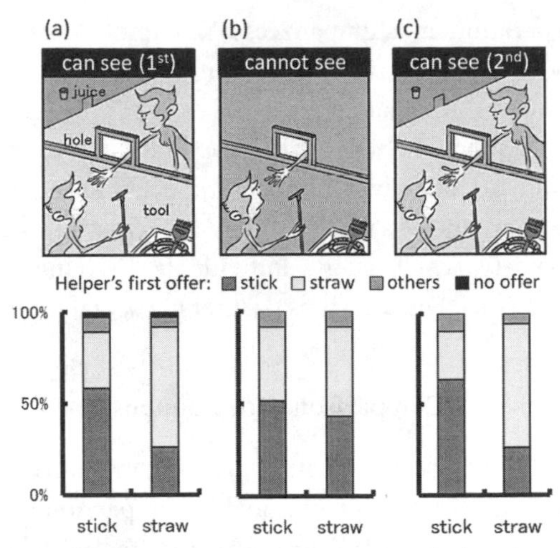

FIGURE 14.2. Chimpanzees can understand what a partner needed when they can see the situation, though they seldom help proactively. (Source: Yamamoto et al. 2012.)

cognitive abilities, such as perspective taking, to exhibit helping behavior, but a lack of motivation may result in the absence of proactive helping.

What about reciprocal cooperation? Yamamoto and Tanaka (2009a, 2009b) tested chimpanzees' reciprocity in an "other-rewarding token insertion task." In this task, two chimpanzees were let into two adjacent experimental booths separated by transparent panels. Each booth was equipped with a "vending machine" that, when a token was inserted, delivered an apple piece to the adjacent booth but not to its own booth. When experimenters facilitated turn-taking such that the chimpanzees had to take turns inserting tokens, the chimpanzees continued inserting tokens and both obtained rewards. However, in a free situation in which both chimpanzees could insert tokens whenever they wanted, they stopped such reciprocal cooperation. Specifically, those chimpanzees who had inserted more tokens than their partners, subsequently declined to insert more tokens, thereby disrupting what had been reciprocal cooperation. Such disadvantageous inequity aversion has been well-documented in chimpanzees (Brosnan, Schiff, and de Waal 2005; Brosnan et al. 2010), in which individuals demonstrate a sensitivity to inequality when receiving a lesser reward than a conspecific for equal effort. As such, while chimpanzees seem to have the cognitive ability to respond negatively to inequity, an ability considered to be essential for the evolution of reciprocal cooperation (Brosnan 2011; Vale and Brosnan, chapter 15 this volume), this might often result in the corruption of reciprocal cooperation, rather than promoting cooperation in chimpanzees (Yamamoto and Takimoto 2012). Melis, Hare, and Tomasello (2008) also posit that reciprocity may not play a large role in guiding the social decisions of chimpanzees, although their experimental results support chimpanzees' capability for contingent reciprocity. Reciprocal interaction, which has been reported in the wild (Gomes and Boesch 2009; Mitani and Watts 2001), may be achieved as a result of their fission-fusion society, which allows individuals to change their partner(s) flexibly and avoid non-cooperators as partners in social interactions.

Comparison with Bonobos

Studies on bonobos are scarcer than those on chimpanzees but have been accumulating recently, enabling a reasonable comparison and new perspectives (Hare and Yamamoto 2015; Yamamoto 2017; Taglialatela et al., chapter 4 this volume). This section describes some empirical studies on cooperation among bonobos, both in the wild and in the laboratory, and compares both *Pan* species.

Like chimpanzees, bonobos are one of the nonhuman animals demonstrating the most frequent food sharing (Kano 1980; Kuroda 1984). However, the characteristics of sharing between the two species differ in several respects (Yamamoto 2015; Yamamoto and Furuichi 2017). Wild chimpanzees share meat, but fruit sharing is rare between non-kin individuals; whereas wild bonobos share fruits, including abundantly available fruits, as well as meat. In typical cases of chimpanzee food sharing, dominant males horde the food and control the sharing of it with other males and females. With bonobos, food transfers are normally observed from dominant females to subordinate females. As mentioned above, chimpanzee sharing is often explained by reciprocal contexts; however, there seems to be no evidence supporting the existence of reciprocity in bonobos' sharing interactions. These differences allow us to consider the different functions of food sharing in chimpanzees and bonobos. Yamamoto and Furuichi (2017) proposed a hypothesis that shared food serves as valuable "currency" among chimpanzees, whereas in bonobo sharing, the transferred food has little inherent value itself, serving instead as a "catalyst" for a positive social interaction that potentially strengthens social bonds. Sharing in bonobos thus goes beyond the nutritional function or value (Goldstone et al. 2016; Yamamoto 2015; Yamamoto and Furuichi 2017), and shows similarities to human sharing. This disparity may be due to differences in inter-individual tolerance formed by the different environments in which chimpanzees and bonobos evolved. This environmental influence on the evolution of cooperation will be further discussed later.

Experimental studies on captive bonobos support the idea that they are more socially tolerant and more cooperative in dyadic interactions than chimpanzees (Hare et al. 2007). Bonobos prefer sharing monopolizable foods with others rather than eating them alone (Hare and Kwetuenda 2010) and will help and give food to conspecifics (Krupenye, Tan, and Hare 2018), including members of other groups (Tan and Hare 2013; Tan, Ariely, and Hare 2017). Bonobos have shown these cooperative behaviors more proactively than chimpanzees and although it is too early to define the underlying psychological mechanisms that drive these cooperative behaviors, bonobos seem to outperform chimpanzees in these dyadic cooperation tasks.

What, then, is group cooperation like in wild bonobos? While data are currently limited, what we do know about group cooperation in bonobos shows a marked difference to that shown by chimpanzees. In contrast to their high levels of dyadic cooperation, wild bonobos seem to show well-organized group coordination only on rare occasions. Bonobos have never been observed cooperating as a group to hunt prey animals. Rather, their hunting is more individualistic and opportunistic (Hohmann and Fruth 2008; Ihobe 1992). This difference in group coordination between chimpanzees and bonobos may also be found in road crossings, which have been observed among wild bonobos in Wamba, similar to those demonstrated by chimpanzees in Bossou, described earlier. At Wamba, the bonobos, unlike chimpanzees, cross the road quickly and do not wait for other group members. This suggests a much less conspicuous division of roles to protect other group members in bonobos than in chimpanzees (Yamamoto et al. in prep).

Discussion and Future Directions

Observations from the field as well as experimental research in captive environments suggest that chimpanzees are less cooperative in dyadic interactions than bonobos, with chimpanzees' dyadic cooperation limited in both its proactivity and contexts. Chimpanzees have sophisticated perspective-taking abilities (Hare, Call, and Tomasello 2001; Hare et al. 2000) and can understand what others need and desire in problem-solving situations (Yamamoto, Humle, and Tanaka 2012), but they tend not help others proactively. Bonobos, by contrast, exhibit more social tolerance and help others proactively. Food sharing in bonobos is highly socialized, while chimpanzees' sharing remains within the bounds of its nutritional function. In this sense, chimpanzees are more dissimilar to humans than bonobos are.

However, when we shift our focus to cooperation in a group, chimpanzees may outperform bonobos, more closely resembling human cooperation.

Some wild chimpanzee populations are known to be good cooperative hunters, while bonobos are not. The same applies for territorial boundary patrol behavior as chimpanzees show strong cohesiveness in this group activity. Our preliminary observations and analyses of chimpanzee and bonobo road crossings suggest that chimpanzees are better organized and more protective of other group members than bonobos. Although more research, especially experimental studies, is required to definitively compare chimpanzees and bonobos in this respect, this difference between the two *Pan* species deserves substantial attention since it seemingly contradicts the conventional view that chimpanzees are more competitive and less cooperative than bonobos.

Based on these empirical studies, hypotheses of different evolutionary paths for dyadic cooperation and group cooperation could be proposed (fig. 14.3). Dyadic cooperation may have evolved in accordance with the development of social tolerance and empathy—mechanisms underpinning cooperative interaction between individuals. Yamamoto (2017) defined empathy with three elements: (1) understanding of others (with a self-other distinction), (2) matching with others, and (3) prosociality. Combination of these three elements is important for emergence of sophisticated cooperation. Chimpanzees are good at exhibiting cognitive ability for the first element, that is, they can understand and predict what others are doing and/or will do. However, they may be worse than bonobos and humans with regard to the latter two elements, possibly leading to chimpanzees' lack of proactivity in their helping behavior. Moreover, Hare, Wobber, and Wrangham (2012) argue that bonobos and humans may have developed their inter-individual tolerance through a self-domestication process. As a result of their high social tolerance, they may have increased chances of being in proximity to other individuals during some activity and hence strengthened their ability and tendency for preferentially state matching with others and prosociality. This increased empathetic and prosocial ability may promote proactive dyadic cooperation in both bonobos and humans, but much less so in chimpanzees.

Another evolutionary path may be applied to cooperation in groups. One hypothesis is that group cooperation is not simply an expanded version of dyadic cooperation but is instead closely associated with the existence of an outgroup threat acting as a strong driving factor. The idea that intergroup competition leads to greater intragroup tolerance has been developed in theoretical models (e.g., Choi and Bowles 2007) as well as in primatology (e.g., Majolo, de Bortoli Vizioli, and Lehmann 2016; Sterck, Watts, and van Schaik 1997). Investigation of the specific conspicuous differences between two *Pan* species regarding their intergroup competition and intragroup tolerance should most directly benefit our understanding of the evolution of human cooperative

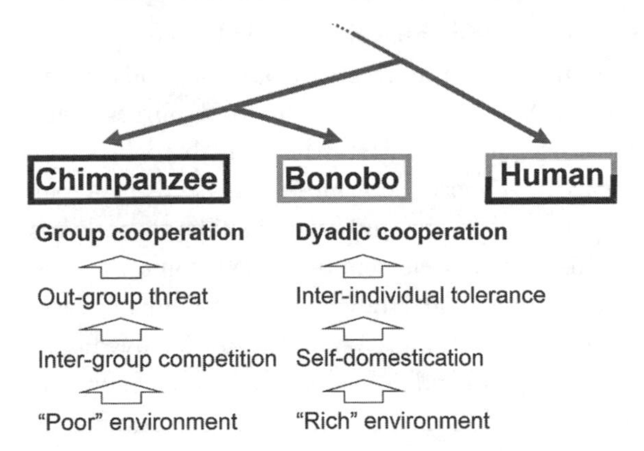

FIGURE 14.3. Hypothesis on different evolutionary paths of group cooperation and dyadic cooperation.

(and also warlike) societies. A scarcity of resources and the resulting hostile intergroup relationships are assumed to be the primary out-group threat, promoting intragroup cooperation in both *Pan* species and in humans. When intergroup competition is high, greater group solidarity should benefit animals in combating other groups. Chimpanzees, who inhabit not only deep tropical rainforests but also much drier forest and savanna landscapes with less abundant, more fluctuating and dispersed food resources, have hostile intergroup relationships and sometimes demonstrate lethal violence against out-group individuals. In this species, in-group cooperation should be essential to survive. Their sophisticated cooperative hunting and well-organized road crossing behavior may reflect this tendency for group cooperation. By contrast, bonobos, who mostly inhabit largely stable tropical rainforests, have largely peaceful intergroup relationships. They may therefore have had many fewer opportunities to cooperate as a group. In humans, who are considered to have evolved in a drier environment with far fewer food resources, group cooperation may have been essential not only for competing with rival conspecific groups but also for protection from predators, another major out-group threat.

Thus, the continued investigation of environmental influences on the evolution of cooperation is of great interest. Which type of environment enhances cooperation? Here, again, there may be two different ways for dyadic cooperation and group cooperation to evolve. One possible hypothesis is that a stable environment with large food patches (i.e., a "rich" environment) promotes dyadic cooperation while a fluctuating environment with dispersed resources (i.e., a "poor" environment) solicits group cooperation. The "rich" environment can accommodate larger parties with less competition over food resources, and thus animals do not need to compete with each other. This enables animals to be more gregarious and leads to higher tolerance among individuals. With a higher density of food sources, females can be gregarious and form coalitions against males, which enables females to obtain high social status and choose less aggressive males as mate partners, leading to selection against aggression and favoring tolerant individuals (Furuichi 2011). Indeed, this is the key element of the self-domestication hypothesis in bonobos (Hare, Wobber, and Wrangham 2012). This increased social tolerance provides animals with more opportunities for dyadic cooperation. As the Japanese proverb says, *only when basic needs for living are met, can animals spare the effort to be polite*. By contrast, environments may have influenced the evolution of group cooperation in the opposite way. Animals may have developed group cooperation to survive in poor environments. It will be more beneficial to have strong intragroup solidarity to outperform other groups when competing for food resources, hunting prey animals that are difficult to obtain without collective efforts, and protecting group members from predation risks. All of these are more plausible in the harsher conditions of open-land environments. Thus, group cooperation may have evolved in humans and chimpanzees who inhabit drier and seasonally more fluctuating environments than do bonobos.

To test these hypotheses on the evolution of cooperation, especially group cooperation, which has received less attention than has dyadic cooperation, it is essential to develop and conduct further empirical studies, both in the wild and captivity. Although controlling conditions in the wild is difficult, the road crossings of Bossou chimpanzees and Wamba bonobos have the potential to provide important insights through direct comparisons between those populations. Detailed, comparative analyses of road crossing behavior could reveal group organization in a specific, natural, and replicable situation. This would inform us about how the two *Pan* species react to out-group threats, and what and whom they are concerned about. This kind of chimpanzee-bonobo comparison is one of the most important and challenging research themes for the future. To this end, captive studies are, of course, also helpful. Research environments that enable us to compare the two *Pan* species directly in similar setups have become well established around the world, (e.g., from African sanctuaries: Hare et al. 2007; from European zoos: Jaeggi, Stevens, and van Schaik 2010). In Kumamoto Sanctuary, Japan, which houses the largest community of chimpanzees and the only bonobos in Japan (Hirata et al., chapter 9 this volume), researchers can develop the same experimental setups for both species. High-tech devices, such as eye-trackers and thermal imaging cameras, are also very powerful tools for investigating the animals' cognition and underlying physiological mechanisms (e.g., Kano, Hirata, and Call 2015). There is little doubt that comparing different species across different settings, in both dyadic and group conditions, will yield the most insight in understanding cooperation in our closest living relatives.

References

Boesch, C. 1994. "Cooperative hunting in wild chimpanzees." *Animal Behaviour* 48: 653–67.

Brosnan, S. F. 2011. "A hypothesis of the co-evolution of cooperation and responses to inequity." *Frontiers in Neuroscience* 5: 43.

Brosnan, S. F., H. C. Schiff, and F. B. M. de Waal. 2005. "Tolerance for inequity may increase with social closeness in chimpanzees." *Proceedings of the Royal Society of London B* 1560: 253–58.

Brosnan, S. F., C. Talbot, M. Ahlgren, S. P. Lambeth, and S. J. Schapiro. 2010. "Mechanisms underlying responses to inequitable outcomes in chimpanzees, *Pan troglodytes*." *Animal Behaviour* 79: 1229–37.

Choi, J.-K., and S. Bowles. 2007. "The coevolution of parochial altruism and war." *Science* 318: 636–40.

Furuichi, T. 2011. "Female contributions to the peaceful nature of bonobo society." *Evolutionary Anthropology* 20: 131–42.

Furuichi, T., G. Idani, H. Ihobe, C. Hashimoto, Y. Tashiro, T. Sakamaki, M. N. Mulavwa, K. Yangozene, and S. Kuroda. 2012. "Long-term studies on wild bonobos at Wamba, Luo Scientific Reserve, D.R. Congo: Towards the understanding of female life history in a male-philopatric species." In *Long-Term Field Studies of Primates*, edited by P. M. Kappeler and D. P. Watts, 143–433. Berlin: Springer.

Gilby, I. C. 2006. "Meat sharing among the Gombe chimpanzees: Harassment and reciprocal exchange." *Animal Behaviour* 71: 953–63.

Goldstone, L. G., V. Sommer, N. Nurmi, C. Stephens, and B. Fruth. 2016. "Food begging and sharing in wild bonobos (*Pan paniscus*): Assessing relationship quality?" *Primates* 57 (3): 367–76.

Gomes, C. M., and C. Boesch. 2009. "Wild chimpanzees exchange meat for sex on a long-term basis." *PLoS One* 4 (4): e5116.

Goodall, J. 1986. *The Chimpanzees of Gombe: Patterns of Behavior*. Cambridge, MA: Belknap Press of Harvard University Press.

Hare, B. 2011. "From hominoid to hominid mind: What changed and why?" *Annual Review of Anthropology* 40: 293–309.

Hare, B., J. Call, B. Agnetta, and M. Tomasello. 2000. "Chimpanzees know what conspecifics do and do not see." *Animal Behaviour* 59 (4): 771–85.

Hare, B., J. Call, and M. Tomasello. 2001. "Do chimpanzees know what conspecifics know?" *Animal Behaviour* 61: 139–51.

Hare, B., and S. Kwetuenda. 2010. "Bonobos voluntarily share their own food with others." *Current Biology* 20: R230–31.

Hare, B., A. Melis, V. Woods, S. Hastings, and R. W. Wrangham. 2007. "Tolerance allows bonobos to outperform chimpanzees in a cooperative task." *Current Biology* 17: 619–23.

Hare, B., V. Wobber, and R. W. Wrangham. 2012. "The self-domestication hypothesis: Evolution of bonobo psychology is due to selection against aggression." *Animal Behaviour* 83: 573–85.

Hare, B., and S. Yamamoto. 2015. "Moving bonobos off the scientifically endangered list." *Behaviour* 152: 247–58.

Hare, B., and S. Yamamoto. 2017. *Bonobos: Unique in Mind, Brain, and Behavior.* Oxford: Oxford University Press.

Hockings, K. J., J. R. Anderson, and T. Matsuzawa. 2006. "Road-crossing in chimpanzees: A risky business." *Current Biology* 16: 668–70.

Hockings, K. J., T. Humle, J. R. Anderson, D. Biro, C. Sousa, G. Ohashi, and T. Matsuzawa. 2007. "Chimpanzees share forbidden fruit." *PLoS One* 2 (9): e886.

Hohmann, G., and B. Fruth. 2008. "New records on prey capture and meat eating by bonobos at Lui Kotale, Salonga National Park, Democratic Republic of Congo." *Folia Primatologica* 79: 103–10.

Horner, V., J. D. Carter, M. Suchak, and F. B. M. de Waal. 2011. "Spontaneous prosocial choice by chimpanzees." *Proceedings of the National Academy of Sciences* 108 (33): 13847–51.

Ihobe, H. 1992. "Observations on the meat-eating behavior of wild bonobos (*Pan paniscus*) at Wamba, Republic of Zaire." *Primates* 33: 247–50.

Jaeggi, A. V., J. M. G. Stevens, and C. P. van Schaik. 2010. "Tolerant food sharing and reciprocity is precluded by despotism in bonobos but not chimpanzees." *American Journal of Physical Anthropology* 143: 41–51.

Jensen, K., B. Hare, J. Call, and M. Tomasello. 2006. "What's in it for me? Self-regard precludes altruism and spite in chimpanzees." *Proceedings of the Royal Society of London B* 273: 1013–21.

Kano, F., S. Hirata, and J. Call. 2015. "Social attention in the two species of *Pan*: Bonobos make more eye contact than chimpanzees." *PLoS One* 10: e0133573

Kano, T. 1980. "Social behavior of wild pygmy chimpanzees (*Pan paniscus*) of Wamba: A preliminary report." *Journal of Human Evolution* 9: 243–60.

Kano, T. 1992. *The Last Ape: Pygmy Chimpanzee Behavior and Ecology.* Stanford: Stanford University Press.

Krupenye, C., J. Tan, and B. Hare. 2018. "Bonobos voluntarily hand food to others but not toys or tools." *Proceedings of the Royal Society of London B* 285: 20181536

Kuroda, S. 1984. "Interaction over food among pygmy chimpanzees." In *The Pygmy Chimpanzee: Evolutionary Biology and Behavior*, edited by R. L. Susman, 301–24. New York: Plenum Press.

Majolo, B., A. de Bortoli Vizioli, and J. Lehmann. 2016. "The effect of intergroup competition on intragroup affiliation in primates." *Animal Behaviour* 114: 13–19.

Matsuzawa, T. 2018. "Chimpanzee Velu: The wild chimpanzee who passed away at the estimated age of 58." *Primates* 59: 107–11.

Matsuzawa, T., T. Humle, and Y. Sugiyama. 2011. *The Chimpanzees in Bossou and Nimba*. Tokyo: Springer.

Matsuzawa, T., M. Tomonaga, and M. Tanaka. 2006. *Cognitive Development in Chimpanzees.* Tokyo: Springer.

Melis, A. P., B. Hare, and M. Tomasello. 2008. "Do chimpanzees reciprocate received favours?" *Animal Behaviour* 76: 951–62.

Melis, A. P., and M. Tomasello. 2013. "Chimpanzees' (*Pan troglodytes*) strategic helping in a collaborative task." *Biology Letters* 9: 20130009.

Mitani, J. C., and D. P. Watts. 2001. "Why do chimpanzees hunt and share meat?" *Animal Behaviour* 61: 915–24.

Silk, J. B., S. F. Brosnan, J. Vonk, J. Henrich, D. J. Povinelli, A. S. Richardson, S. P. Lambeth, J. Mascaro, and S. J. Schapiro. 2005. "Chimpanzees are indifferent to the welfare of unrelated group members." *Nature* 437: 1357–59.

Sterck, E. H. M., D. P. Watts, and C. P. van Schaik. 1997. "The evolution of female social relationships in nonhuman primates." *Behavioral Ecology and Sociobiology* 41: 291–309.

Tan, J., D. Ariely, and B. Hare. 2017. "Bonobos respond prosocially toward members of other groups." *Scientific Reports* 7: 14733.

Tan, J., and B. Hare. 2013. "Bonobos share with strangers." *PLoS One* 8 (1): e51922.

Tanaka, M., and S. Yamamoto. 2009. "Token transfer between mother and offspring chimpanzees (*Pan troglodytes*): Mother–offspring interaction in a competitive situation." *Animal Cognition* 12: S19–26.

Ueno, A., and T. Matsuzawa. 2004. "Food transfer between chimpanzee mothers and their infants." *Primates* 45: 231–39.

Warneken, F., B. Hare, A. P. Melis, D. Hanus, and M. Tomasello. 2007. "Spontaneous altruism by chimpanzees and young children." *PLoS Biology* 5 (7): e184.

Warneken, F., and M. Tomasello. 2006. "Altruistic helping in human infants and young chimpanzees." *Science* 311: 1301–3.

Wilson, M., C. Boesch, B. Fruth, T. Furuichi, I. C. Gilby, C. Hashimoto, C. L. Hobaiter, G. Hohmann, N. Itoh, and K. Koops. 2014. "Lethal aggression better explained by adaptive strategies than by human impact." *Nature* 513: 414–17.

Yamamoto, S. 2017. "Primate empathy: Three factors and their combinations for empathy-related phenomena." *WIREs Cognitive Science* 8: e1431..

Yamamoto, S. 2015. "Non-reciprocal but peaceful fruit sharing in wild bonobos in Wamba." *Behaviour* 152 (3-4): 335-357.

Yamamoto, S., and T. Furuichi. 2017. "Courtesy food sharing characterized by begging for social bonds in wild bonobos." In *Bonobos: Unique in Mind, Brain, and Behavior*, edited by B. Hare and S. Yamamoto, 125–39. Oxford: Oxford University Press.

Yamamoto, S., T. Humle, and M. Tanaka. 2009. "Chimpanzees help each other upon request." *PLoS One* 4 (10): e7416.

Yamamoto, S., T. Humle, and M. Tanaka. 2012. "Chimpanzees' flexible targeted helping based on an understanding of conspecifics' goals." *Proceedings of the National Academy of Sciences* 109: 3588–92.

Yamamoto, S., and A. Takimoto. 2012. "Empathy and fairness: Psychological mechanisms for eliciting and maintaining prosociality and cooperation in primates." *Social Justice Research* 25: 233–55.

Yamamoto, S., and M. Tanaka. 2009a. "Do chimpanzees (*Pan troglodytes*) spontaneously take turns in a reciprocal cooperation task?" *Journal of Comparative Psychology* 123: 242–49.

Yamamoto, S., and M. Tanaka. 2009b. "Selfish strategies develop in social problem situations in chimpanzee (*Pan troglodytes*) mother–infant pairs." *Animal Cognition* 12: S27–36.

Yamamoto, S., and M. Tanaka. 2010. "The influence of kin relationship and a reciprocal context on chimpanzees' (*Pan troglodytes*) sensitivity to a partner's payoff." *Animal Behaviour* 79: 595–602.

15

Putting Chimpanzee Cooperation in Context

GILLIAN L. VALE AND SARAH F. BROSNAN

Introduction

Cooperation plays a fundamental role in the success of humans, yielding collective action that pervades almost every aspect of modern life. Humans cooperate more frequently than other mammalian species and more extensively with unrelated individuals (Boyd and Richerson 2009). This ability to rely on collective action and knowledge (i.e., culture) has enabled *Homo sapiens* to invade many of the most uninhabitable climates of Earth, a feat that remains unchallenged by most other species. Of course, we are not alone in our capability to work with others, so what is it that makes human cooperation so distinctive? One way to answer this question is to study other animals, including other primates, to look for evolutionary explanations for this behavior. Before we do so, however, it is necessary to define what we mean by "cooperation." Definitions vary substantially across disciplines (see Noe 2006). In some cases, cooperation assumes a cost to the individuals involved (Dugatkin 1997; Trivers 1971), whereas in others, the focus is purely on whether the action helps another (Bshary and Bergmuller 2008; Melis and Semmann 2010). We do not wade into this debate, but instead focus on what is key for our approach. First, we assume that an act is cooperative if one or more individuals can achieve a better outcome by working with others than by working alone, although we specifically do not assume that they either acted intentionally or understood the outcome of their action (see also Bshary and Bergmuller 2008). Second, we include failed attempts to cooperate as a cooperative effort; that is, one individual can behave cooperatively but the other individuals involved may not reciprocate this act of helping (see Brosnan and de Waal 2002).

It is noteworthy that primates are not the most cooperative taxon; that honorific arguably goes to eusocial insects (i.e., termites, ants, and some bees and wasps) that live in very large colonies and are obligately cooperative. In these societies, division of labor occurs in which, for example, some individuals forage, some defend the nest, some tend to eggs, and a queen (or a very few queens) lay all the eggs. This level of interdependency has led to these societies being dubbed "superorganisms" (Hölldobler and Wilson 2009). However, while the cooperation seen in these species is impressive, it is also largely inflexible. In most cases, individuals are predisposed through either their genes, their early diet or another aspect of their environment, or their age for a certain task (although see Loukola et al. 2017). So while the society or hive functions as a single cooperative organism, no individual ever has much autonomy in deciding whether, when, or with whom they will cooperate. In contrast, primates, including humans, must decide for any interaction whether or not to cooperate and, if they do, with whom and to what degree. How do individuals make these decisions and what strategies and control mechanisms do they employ? Moreover, how closely related are the types of cooperation we see in humans and nonhuman primates? These are the key questions that interest us.

Although most studies of cooperation in the laboratory focus on individuals working together to acquire food, in wild primates diets consist mainly of fruits, vegetation, nuts, and sometimes a small number of animal foods, so the need to work together to acquire food is slight (Hirata, Morimura, and Fuwa 2010). Instead, most documented cooperation in wild primates happens in the social realm, including the formation of coalitions (Harcourt and de Waal 1992), grooming (Henazi and Barrett 1999; Russell and Phelps 2013; Schino and Aureli 2008), territory defense (Boesch 2003), and cooperative breeding (Burkart and van Schaik 2010). However, some of these activities are related to food (i.e., group defense of quality food sources), and more direct food-related cooperation does occur, specifically in the contexts of food sharing (Jaeggi and Gurven 2013; Taglialatela et al., chapter 4 this volume; Yamamoto, chapter 14 this volume) and acquisition of foods that are difficult to obtain (Hockings et al. 2007), in particular, hunting (Boesch 2003).

Cooperative hunting may be the best known example of primate cooperation, although it has been documented primarily in chimpanzees (*Pan troglodytes*). One of the most complex variants appears in the chimpanzees in the Taï forest, Côte d'Ivoire. When these chimpanzees hunt, individuals adopt specific, complementary roles (described as driver, chaser, blocker, and ambusher: Boesch 2003) to surround monkey prey and drive them to one individual, who captures them (see also Watts and Mitani 2002 for similar hunting behavior documented in the Ngogo chimpanzees at the Kibale National Park, Uganda). Moreover, the resulting prey tend to be shared among most members of the group, with access biased toward individuals who participated in the hunt. This is unlike other sites, where food is often confiscated by the alpha male and shared in response to scrounging or, possibly, as a means to avoid conflict (Gilby et al. 2010, 2015) or to affirm coalitionary bonds (Nishida et al. 1992). One debate in the literature has revolved around why, among chimpanzees, those in the Taï forest adopt such complex and complementary hunting roles. Boesch (2002) argues that the continuous canopy at Taï makes hunting very difficult, requiring the coordination of multiple individuals. At sites where the canopy is discontinuous a single individual can successfully hunt on his own (most hunters are male, although see Pruetz and Bertolani 1997). Unfortunately, opportunities to study wild chimpanzees are vanishing due to anthropogenic pressures (see Chapman et al., chapter 25 this volume; Hartel et al., chapter 26 this volume; Morgan et al., chapter 27 this volume), but it is critical to further study this ecologically driven hypothesis in order to better understand how chimpanzees decide when and with whom, if anyone, to hunt.

Interestingly, given that chimpanzees cooperate in several domains (i.e., hunting, coalitions and alliances, group defense), there are far fewer documented cases of cooperation among the other great apes, although to what degree this reflects differential research effort is unclear. The other *Pan* species, bonobos (*P. paniscus*), rarely cooperatively hunt (although see Surbeck and Hohmann 2008) or engage in group patrols; however, they do show extensive coalitions and alliances, particularly among females (Gruber and Clay 2016; Clay, chapter 12 this volume). Male bonobos also form alliances; however, they are formed predominantly with the opposite sex. Males, for example, form an important bond with their mothers, who provide support during agonistic encounters with other males and during competition over potential mates (Surbeck, Mundry, and Hohmann 2010). In addition to the sex reversals we see in the alliances formed across *Pan* species, there is a sex difference in their hunting behavior. While chimpanzees predominantly hunt in male parties, the opposite is true for bonobos: when there are hunts, females bonobos take a more active role and even lead mammalian hunts, and when it comes to the distribution of food, females appear to lay claim to a higher proportion than we often see in their chimpanzee relatives (Gruber and Clay 2016).

Some monkeys also cooperate in the wild. A full discussion of monkey cooperation is beyond the scope of this chapter, but given our focus on capuchin monkeys, it is relevant to consider them in more detail. Wild capuchins cooperate under similar circumstances as chimpanzees, with documented cases of group territorial behavior (e.g., Scarry 2017), reciprocated grooming (Manson et al. 2004), (infrequent) food transfer (Rose 1997), and group hunting (Rose 1997). Tufted capuchins (*Sapajus [Cebus] apella*), for example, engage in collective intergroup aggression toward neighboring animals that involves both male and female monkeys of subordinate and dominant rank (Di Bitetti 2001). By engaging in intergroup conflict, successful groups protect potentially high-quality resources within their home ranges, securing future access to valuable sustenance (Scarry 2017). This suggests that capuchins participate in collective action in response to between-group competition, although the exact nature of resources that are defended (e.g., potential mates, food, home range) remains a topic of some debate.

Studies of cooperation in the wild are essential if we are to understand the pressures that enable cooperation to evolve; however, field studies are also extremely time consuming and often provide only indirect evidence of the cooperative processes involved. We cannot control the interactions and therefore are left to hypothesize causation based primarily on observations (although the ability to study genetics, endocrinology, and so forth in the field is making it possible to study mechanisms that were heretofore invisible). Thus, in order to untangle the underlying mechanisms, particularly cognitive ones, we turn to experimental studies in captivity. In more tightly controlled settings, where we can control both who participates and key features of the interaction, as well as introduce control conditions, we can more rigorously uncover the *potential* processes that underpin cooperation in primates, including exploring the proximate causes and strategies suggested by field studies (Janson and Brosnan 2013).

Approach

FOOD RETRIEVAL TASKS

Cooperation studies tend to rely on one of a few procedures that are widespread across studies and species (see also Duguid et al., chapter 13 this volume; Massen, Schaake, and Bugnyar, chapter 16 this volume). This improves our ability to compare across species and contexts (although it does limit the contexts in which cooperation has been studied). One widely used test has been the "cooperative food retrieval task" (Hirata and Fuwa 2007; see Melis and Semmann 2010). In this task, two individuals are presented with out-of-reach rewards that require synchronized action for reward retrieval (e.g., simultaneous pulling of a rope draws the food toward the participants). Early investigations with chimpanzees found that subjects were capable of solving the task, but only in juvenile chimpanzees and only after they were trained extensively (Crawford 1937, 1941). Later studies with adults also found limited evidence for chimpanzee cooperation (Chalmeau 1994; Chalmeau and Gallo 1996; Povinelli and O'Neil 2000).

These results were surprising for several reasons. First, as discussed above, chimpanzees coordinate actions for a common outcome in the wild, in at least some populations (Boesch 2003), and this task seems, at least to us, to be similar (although how we interpret a study may be very different from another species' interpretation, Brosnan 2017; de Waal 2016; Schmelz et al. 2017). Second, chimpanzees do very well individually at raking in out-of-reach rewards by using a tool to pull food toward them (Price et al. 2009), suggesting that they should at least understand the mechanism of cooperative food retrieval tasks. Third, even if only one individual tested understands the task, the actions of that individual should draw the other to the apparatus (i.e., to pull on the rope), promoting success. This is a basic social facilitation process that is found in a broad range of animal species, including primates (see Hoppitt and Laland 2008 for a review of social learning mechanisms).

It is perhaps unsurprising then, that chimpanzees in more recent investigations have shown greater success with this task, with pairs of chimpanzees pulling in the food trays at very high rates. Moreover, these studies have revealed something of the conditions in which cooperation can occur and when it breaks down (see also Duguid et al., chapter 13 this volume; Massen, Schaake, and Bugnyar, chapter 16 this volume). For example, chimpanzees preferentially collaborate with individuals that are less inclined to monopolize resources, and, when given a choice, actively choose partners with whom they have a previous track record of successful collaboration (Melis, Hare, and Tomasello 2006a, 2006b). Similarly, in a completely free-choice situation in which the entire group was present, chimpanzees skillfully collaborated with others, frequently selecting partners of close dominance rank or kin that should be socially tolerant and unlikely to exploit the resources gained from any cooperative acts (Suchak et al. 2014, 2016). Thus, more recent evidence indicates the importance of partner choice, which might be expected based on studies from the field.

Similar tasks have also been run with monkeys, primarily (but not exclusively) capuchin monkeys and callithrichids. The first capuchin study indicated that capuchins lacked awareness that synchronized behavior with their partner was necessary to solve the task (Chalmeau, Visalberghi, and Gallo 1997). Nonetheless, evidence that they were aware of and understood at least something about the need for their partner was later reported by Mendres and de Waal (2000). They found that capuchins were more likely to pull in a counterweighted tray (which brought in food) when a partner was present than when absent. Because we cannot tell whether any individual attempt was contingent on their partner's behavior, we cannot rule out a social facilitation effect (i.e., the mere presence of a partner could account for this finding). However, capuchins' success decreased when visual access to their partner was blocked, despite the capuchins continuing to pull on the apparatus (Mendres and de Waal 2000). This suggests that capuchins were engaging in a degree of behavioral coordination when they could see their partner, which facilitated cooperation. Further work indicates that the capuchins do poorly on tasks where they jointly use a lever to access a distant food reward, even those who have previously done well on analogous cooperative food retrieval tasks, possibly because of the lack of kinesthetic feedback when the lever doesn't itself bring the food toward the subjects (Brosnan and de Waal 2002).

Callitrichids have been tested on comparable tasks and show a similar level of understanding of their partner's role. In one such test, partnered cotton-top tamarins (*Sanguinus oedipus*) successfully solved the joint action task and, much like capuchins (Mendres and de Waal 2000), preferred to pull on the apparatus when a partner was present rather than absent (Cronin, Kurian, and Snowdon 2005). As cooperative breeders, callitrichids display cognitive and social attributes that likely facilitate their success in these cooperative settings, including their routine coordination with conspecifics, high social tolerance, and prosocial tendencies (Burkart et al. 2007; Cronin and Snowdon 2008). Indeed, it is important to compare subjects across a variety of different social systems to determine the degree to which shared behaviors are or are not due to shared mechanisms. Although capuchins show some behaviors related to cooperative breeding, such as allonursing, they are not cooperative breeders, nor are chimpanzees, indicating that there may be multiple mechanisms for cooperation in primates.

In addition to behavioral coordination, a growing line of evidence indicates that capuchins pay attention to their potential benefits, that is, the distribution of food rewards, in these tasks. For example, when capuchins require assistance to pull in the tray (i.e., it is weighted so heavily that they cannot do it themselves), they are more likely to share the food they obtain with the partner than they are when they are able to pull in the tray by themselves (i.e., the tray is light enough to pull alone). This seems to work both ways, as partners are more likely to pull again after having had food shared with them (de Waal and Berger 2000). Moreover, when subjects must work together to pull in the tray, they are more successful when the food is separated, such that each individual can acquire their own food immediately after pulling in the tray, and without approaching the other monkey. Indeed, when the food rewards are positioned together in the middle of the tray, cooperation plummets, even though the foods are still separated. This is true for both kin (mother/offspring) and non-kin pairs, and occurs from the first trial, indicating that the subjects are extrapolating something from their interactions in the social group to this new situation (de Waal and Davis 2003).

This latter task indicates that subjects are unwilling to participate if their partner could monopolize the rewards, but one additional study adds nuance to this. In a study by Brosnan, Freeman, and de Waal (2006) subjects were given the option to pull together for spatially segregated food rewards that were either equal or unequal. Importantly, the subjects were free to decide who would pull for the lower or higher value food. Based on previous results, we anticipated that pulling rates would be the lowest in the unequal condition. To the contrary, subjects were equally likely to pull across conditions, but pairs in which the dominant routinely laid claim to the higher value reward were far less likely to cooperate in *any* condition. This contrasted with partnerships in which both partners received the lower-value outcome about half the time; these partnerships continued to cooperate at more than double the rates of the less equitable partnerships across all conditions. This indicates that it was not the outcome per se, but the partner's behavior that was key.

This relates to another line of work in our lab that has found that, in some contexts, a number of primates, including both capuchin monkeys and chimpanzees, respond negatively to getting a reward less preferred than the reward received by a social partner (reviewed in Brosnan and de Waal 2014; Price and Brosnan 2012). In these studies, subjects sequentially complete a task (typically exchanging a token with an experimenter) for rewards that are either the same as (equity condition) or different than (inequity condition) what their partner receives. If subjects are inequity averse, they should react more in the inequity condition, in which they receive a different reward than their partner, relative to the equity condition. To control for potential "frustration effects," in which the subjects' reaction is due to the expectation that they will receive a food simply because it is visible, we run a control for contrast effects. For this, both subjects and partners are shown a favored food before being given a less preferred food item after they complete the task. In this way, their attention is drawn to the more preferred food, but the situation is still equitable as both subjects are shown the more preferred but receive the less preferred food (in addition, both foods are always present in all conditions, providing more control). Inequity aversion is thus inferred only in subjects that also respond more in the inequity condition than in this control for individual contrast effects.

One of the main contributions of this avenue of research has been to identify which species do, and which do not, respond to inequity. There is now a substantial body of work indicating that, in at least some contexts, capuchin monkeys, chimpanzees, and macaques (*Macaca* spp.) respond negatively to inequity (see Brosnan and de Waal 2014 for a list of all studies to that date, including those not finding a response to inequity). In contrast, orangutans (*Pongo pygmaeus*), squirrel monkeys (*Saimiri* spp.), callitrichids (*Callithrix* spp. and *Sanguinus* spp.), and owl monkeys (*Aotus* spp.) do not react (reviewed in Brosnan and de Waal 2014, although see Mustoe et al. 2016 for evidence that marmosets respond in some contexts). Thus far, the data indicate that the key feature shared by primates that respond to inequity is their pronounced tendency to cooperate with non-kin (Brosnan 2011). This makes sense if, as has been hypothesized previously, inequity cues individuals to the value of their cooperative partner and helps them to determine whether they should continue cooperating, or find a new partner (Brosnan and de Waal 2014; Fehr and Schmidt 1999). Moreover, the potentially surprising lack of response to inequity in the callitrichids may be due to the fact that they are cooperative breeders, who therefore cannot easily find new partners. It may be that they are more sensitive to inequity in the early stages of a forming relationships or to particularly large-scale inequity, but not to the relatively small inequities present in these food distribution tasks (Freeman et al. 2013).

Key to discussion of both cooperation and inequity is the tendency to talk about what "the animals" do or what "a species" does, as if they were a monolith that all performed in the same way, despite the fact that we all know that our inferential statistics and graphs are hiding what may be both substantial and interesting individual differences. Indeed, individual differences are (or ought to be) of central importance when considering behaviors that *should* vary across individual and context, such as cooperation. For example, consider the above-cited work showing that in cooperative contexts, both chimpanzees and capuchin monkeys are sensitive to the behavior of their partner (Brosnan, Freeman, and de Waal 2006; de Waal and Berger 2000; Melis et al. 2006a, 2006b) and the distribution of foods (de Waal and Davis 2003); clearly subjects are not uniformly cooperative, nor should they be. Cooperating indiscriminately does not promote one's fitness; rather, the adaptive consequences of cooperation depend upon factors such as how frequently it occurs, whether it is reciprocated, and the payoffs to the individuals involved. Thus, animals should be selective in who they cooperate with and when they should cooperate. We see the same variability in response to inequity. Various studies show that subjects' responses are affected by their sex, rank, personality, or even the orientation of subjects to one another and the rewards used in the tasks (Brosnan et al. 2015; Engelmann et al. 2017; Talbot et al. 2018). In the future, in addition to group-level analyses that help us to be certain of a phenomenon's validity and magnitude, it is important to include deeper explorations into individual differences. Studying these may be the key to understanding which factors are the most important mechanisms underpinning behavior, as well as the characteristics of individuals that may be prone to cooperate or not.

Experimental Economic Games

Food-retrieval tasks have been immensely informative for understanding cooperation. They are intuitive and allow researchers to test numerous mechanisms related to cooperation. However, fully understanding a phenomenon requires a mixed approach, utilizing different procedures, to best understand the limits and possibilities. Recently, researchers have adopted new paradigms to study cooperation that are based on experimental economics (Brosnan et al. 2013; Watzek, Smith, and Brosnan 2017). Games in experimental economics simplify complex decisions into a number of restricted and simple choices—often a dichotomous choice between options—that correspond to different payoffs and risks, making it a useful framework for the study of the mechanics of primates' decisions (Brosnan 2013). Additionally, these comparable methodologies allow for a direct comparison across multiple taxa. This is aided by the fact that these tasks require very little training and make no assumptions about species' underlying cognitive abilities. Animals need only sufficient numerosity skills to prefer more food items to fewer and the capacity to make a choice (e.g., choose one of two physical tokens or icons on a computer screen).

Of course we cannot guarantee that even identical procedures will be perceived, interpreted, or acted upon in identical ways across species, or result in similar levels of motivation. However, being able to utilize the same task across species is the best starting point for a comparative program (Smith, Watzek, and Brosnan 2018). Following this, we can make predictions about species' play based on these simplified games and then test them using more species-typical approaches to ensure that we do not accidentally underestimate a species' abilities or miss a species-specific manifestation of the trait. Indeed, one key element of our approach is that we provide no instruction or pre-testing to any species, including humans. By taking this approach, we are able to directly compare across species and, we hope, therefore better understand the ways in which cooperation evolved across primates.

All of our tasks follow the same general procedure, in which subjects must choose one of two options, and outcomes are contingent upon partners' choices (in some cases, "risk free" options may pay the same regardless of the partner's choice; this includes the *Hare* choice in the Stag Hunt game,

FIGURE 15.1. Chimpanzees participating in an exchange-based manual task experimental economic game.

discussed below). In order to give subjects the best opportunity to succeed, we have utilized two different approaches: an exchange-based manual task (fig. 15.1) and a joystick-based computerized task. In the manual version, adjacent pairs of individuals with visual access to one another are given tokens that represent each of the options of the game being played (differentiated by color/pattern). Subjects select one of the two tokens and return it to the experimenter, who shows the subjects both individuals' choices and delivers rewards according to the payoff matrix. The computerized version is very similar, except that the options are represented by two differently colored/patterned icons presented on a shared computer screen, and joysticks are used to move a cursor to select one of the icons. Two versions of the computer task have been employed. In the synchronous version, individuals see their partner's choice only after both individuals have made their selection, and in the asynchronous version, subjects can visually see the choice of their partner before they make their choice. This allows us to determine the degree to which subjects are making their decision based on an internal rule (which could be a strategy, or based on a non-contingent preference for one outcome) or are making their choice contingent on their partner's choice (note that even this does not imply that they understand this contingency; they could be making their choices depending upon what other option they see without realizing that it was chosen by a partner).

THE STAG HUNT GAME

Our first study utilized the Stag Hunt, or Assurance game, which has been argued to be the best available model of low-cost human cooperation (Skyrms 2003). In this game, individuals choose between two options, *Stag* and *Hare* (fig. 15.2). If both subjects play *Stag*, they receive the highest payoff, whereas any partner that plays *Hare* receives a low payoff, irrespective of what their partner plays. The game has two Nash equilibria: players can coordinate on *Stag*, which is the payoff dominant equilibrium, or *Hare*, which is the risk dominant option because the subject receives a small, but guaranteed, reward. Aside from its hypothesized importance for humans, we chose to begin our research with this game for a practical reason: it is very straightforward, and thus subjects who understand the payoffs, and trust their partners, should always play *Stag*. Therefore, if pairs failed to find the *Stag-Stag* outcome, they likely had not learned the payoffs or did not understand the game structure. If, however, they did find the payoff dominant outcome, there was a degree of certainty that they understood the task and that it provides a fairly reliable method for studying primate social decision-making, which was important for future, more complex games for which it would be less easy to tell what they understood.

As might be expected, both chimpanzees and capuchins showed the capacity to coordinate on the payoff-dominant *Stag-Stag* outcome (Brosnan, Wilson, and Beran 2012; Brosnan et al. 2011). However, consistent with the secondary theme of individual differences, there was large variation within and across species and across contexts. Capuchin monkeys, at the Language Research Center, Georgia State University, USA (LRC), initially were tested on the manual task, and did very poorly; of the six capuchin pairs, only one consistently played *Stag-Stag*. When given the computerized version, overall performance increased, but only in the asynchronous condition, in which they could see their partner's choice prior to play. Thus, we hypothesized that the capuchins were solving the task via matching their partner, with a slight *Stag* bias. Considering the difference in performance between the computerized task and the manual task (where they could see their partner's choice), we additionally hypothesized that the enhanced play of *Stag-Stag* in the computer task may have been due to the shorter delays between choice and reward and the greater number of trials in a session, both of which promote

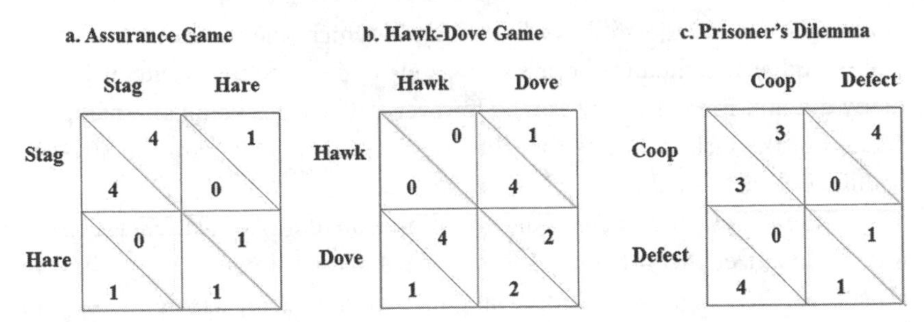

FIGURE 15.2. Payoff matrices for (a) the Assurance game (Stag hunt), (b) the Hawk-Dove game, and (c) the Prisoner's Dilemma.

learning, rather than any intrinsic aspect of the computerized format (Brosnan, Wilson, and Beran 2012).

Rhesus macaques, which cooperate in social contexts (Maestripieri 2007) but not food contexts, were highly successful. All pairs tested at the LRC found the payoff-dominant outcome, irrespective of whether they could see their partner's play (Brosnan, Wilson, and Beran 2012). To begin to explore the mechanism that the macaques were using to solve the task, they were then tested on a simulation version in which they played computerized "partners" with pre-set strategies. In this context, the macaques showed a very strong *Stag* bias, indicating that their success in the original coordination task could be due to a preference for the option that, in some contexts, paid them the largest number of rewards (Parrish et al. 2014). This suggests that rhesus macaques may have solved the Assurance game through reinforcement learning, rather than by developing a cooperative joint strategy.

Chimpanzees were the most variable in their performance. We initially tested chimpanzees from two different sites who, although housed in similarly enriched environments, differed substantially in their experience with cognitive training and testing. Ten chimpanzee pairs were housed at the National Center for Chimpanzee Care, USA. Although these chimpanzees were very accustomed to working with humans, they had relatively little, or no, previous experience with cognitive testing. None of the pairs played the payoff-dominant *Stag-Stag* strategy, but most did show consistent preferences. Six matched their partner's choice (which we note was a reasonable strategy, paying on average 2.5 rewards per trial as compared to the four rewards for the *Stag-Stag* choice and the single reward for playing *Hare*), and two pairs "anti-coordinated," selecting the token not selected by their partner (*Hare-Stag*). This strategy leads to the lowest possible average payoffs (0.5 rewards per trial), so it is unclear why the subjects would follow this strategy. Interestingly, relevant to the matching strategy, there was no strong evidence to suggest that partners were showing a "leader-follower" strategy, in which one partner played first and the other matched (although a few appeared to use this strategy, just as many did not; Brosnan et al. 2011). However, other work using an analogous version of the task indicates that this may be a relatively common strategy (Bullinger et al. 2011).

In other cases, chimpanzees did coordinate on *Stag*. At the LRC, we tested four chimpanzees who had extensive experience with cognitive studies, in most cases since infancy, including language training using a lexigram system (see also Beran, Perdue, and Parrish, chapter 20 this volume). Among these four chimpanzees, two of four pairs (consisting of three unique individuals) settled on the *Stag-Stag* outcome. To see whether this would generalize to a condition in which they could not see their partner's choice, we added a visual barrier, and their choice pattern persisted, even when exposed to new stimuli (Brosnan et al. 2011). These results suggest that there is a strong experience effect, with chimpanzees who are more familiar with cognitive testing outperforming those who are less experienced. There are, however, two caveats to add. First, a substantial amount of experience may be required; we recently repeated the study with chimpanzees who had an intermediate level of testing (frequent cognitive testing, but not since infancy and without any language training) and again found no evidence of a *Stag-Stag* strategy (Hall et al. 2019). Secondly, the form of the task may be key in whether chimpanzees can solve it; as mentioned previously, chimpanzees succeeded in an analogue task (Bullinger et al. 2011). Indeed, even among more experienced chimpanzees the ability to see their partner and the relative risk of playing *Stag* or *Hare* have been found to influence performance (Duguid et al. 2014). Not surprisingly, contextual factors matter and cooperative strategies are associated with the costs, risks, and payoffs involved in any given situation.

Finally, we tested humans, using a procedure that was identical to that used with other primates. In the manual task, humans exchanged tokens with experimenters to earn rewards, and in the computerized task, subjects played on the same program the primates had used, on a shared computer screen using a joystick to make decisions and receiving their rewards from a coin dispenser (Brosnan et al. 2011, 2012). To our surprise, in the manual task a substantial number of human participants—10 of 26 pairs—settled on the *Hare-Hare* strategy, and only five of the pairs played *Stag-Stag*. Intriguingly, none of the ten pairs playing *Hare-Hare* ever experienced *Stag-Stag* as they explored the payoff space. Thus, we hypothesized that humans may have thought they had found the best coordinated outcome despite not trying every option (post-game debriefings supported this hypothesis). In the computerized version, which was played using the synchronous condition, 22 of 27 pairs played

Stag-Stag, but the remaining participants again settled on *Hare-Hare*. Supporting our earlier hypothesis, the participants who played *Hare-Hare* never experienced *Stag-Stag*. Interestingly, participants in the computerized game talked to one another, and those who found the *Stag-Stag* outcome discussed the game with their partner. Conversely, participants who consistently chose *Hare-Hare* never spoke to one another about the game (they spoke, but about different topics). It appears that humans did better when they were able to use language to explore the parameter space (see also Duguid et al., chapter 13 this volume).

Overall, the results show very strong consistencies as well as very strong discontinuities across the primates. At least some members of all the species tested found the payoff-dominant coordinated solution. However, capuchins relied on matching their partner, which limits the contexts in which they can coordinate. That being said, of the species tested, capuchins live in the smallest groups and routinely maintain visual contact with one another (Fragaszy, Visalberghi, and Fedigan 2004). This suggests that there may have been very little selective pressure to develop strategies that work in visually isolated contexts. Despite having large brain-to-body ratios, they also have the smallest absolute brain size of all of these species (Rilling and Insel 1999), which may mean that they lack a key cognitive ability (Deaner et al. 2007). Humans and chimpanzees showed the most variability across pairs, but it is not clear whether the root causes of this variability are the same. In humans, failing to use language to explore the parameter space may have inhibited the participants' performance. In the case of chimpanzees, there seems to be an experience effect, with subjects who are more accustomed to routine and extensive cognitive testing outperforming those who are not. If this is the case, then we must be careful to recognize the effect of extensive experience, especially in working on cognitive tasks with humans, when making pronouncements about chimpanzees' abilities. It is likely that experience is one of the key factors influencing individual differences in cooperation, and that to some extent cooperation can be learned (e.g., Hirata and Fuwa 2007; Massen, Schaake, and Bugnyar, chapter 16 this volume).

THE HAWK-DOVE GAME

Although the focus of this chapter is on cooperation, we wish to briefly touch on a game of conflict, as conflict is the flip side of cooperation. One of the benefits of the experimental economics approach is that it is easy to change the question by simply adjusting the payoff matrix. Thus, to compare responses to cooperation and conflict, we have employed the Hawk-Dove, or

Chicken game. A player benefits when they play *Hawk* and the other plays *Dove*; however, if they both play *Hawk*, neither receives anything. Finally, if both players choose *Dove*, they receive an intermediate payoff (fig. 15.2). There are two asymmetric Nash equilibria in this game (i.e., one player plays *Hawk* and the other *Dove*). Although this a game of conflict, the best outcome for both partners is the "alternating Nash," in which participants take turns playing *Hawk* and *Dove*, which is, indeed, a cooperative outcome.

We recently compared the decision making of our four species in the Hawk-Dove game. Replicating the task formats used with the Stag Hunt game, we employed both the manual and computerized (asynchronous and synchronous) versions. As before, capuchins did very poorly in the manual version; however, in the computerized version, all pairs found the uncoordinated Nash equilibrium as long as they could see their partner's choice (e.g., the asynchronous version; Brosnan et al. 2017). Because we found an interesting pattern in which the subject who habitually played *Hawk* tended to be the dominant partner, we repeated the game with new stimuli (and the same payoffs). Again, subjects found the alternating Nash; however, in this case, the partners' choices switched (the subject who had played *Hawk* played *Dove*, and vice versa). This indicated that dominance was not influencing their choices in this particular context, but that which monkey more frequently played *Hawk* was determined stochastically. Rhesus monkeys also found the asymmetric Nash, although contrary to their performance in the Stag Hunt game, subjects showed a consistent strategy only in the asynchronous game, when they could see their partner's play (Brosnan et al. 2017). This may indicate that macaques were using a different mechanism to make choices in the Hawk-Dove than in the Stag Hunt game. As in the Stag Hunt game, chimpanzees with only moderate experience did not find a consistent strategy (these chimpanzees were at the National Center for Chimpanzee Care and the Yerkes National Primate Research Center; we were unable to test the LRC chimpanzees, who had extensive experience; Hall et al. 2019). In the case of humans, participants did surprisingly well; despite having no instruction, approximately a third of human pairs found the alternating Nash strategy in both the manual and computerized versions of the task (Brosnan et al. 2017).

Conclusions and Future Directions

Testing multiple species using methodologies that are very similar is key to comparative work that seeks to understand the evolutionary trajectory of social decision making. We can learn the evolutionary history of behaviors and identify the potential processes and traits that may be shared broadly among primates, as well as those that may be species unique. In particular, the study of multiple species with different socio-ecological niches helps to tease apart potential social and ecological pressures that influenced the evolution of social decision-making (Brosnan et al. 2017). Although there are certainly drawbacks to direct comparison (as mentioned above), we see this as the key first step to understanding how species' responses differ, or not.

Chimpanzees are an important species to study in this regard. Aside from being one of our two most closely related extant relatives (with bonobos), fieldwork indicates that chimpanzees are among the most cooperative primates, cooperating in a variety of different contexts and with a variety of different partners. This makes it important to study their behavior in captive experiments in order to better understand why, when, and how they cooperate (Janson and Brosnan 2013). Fortunately, we have an extensive set of data on chimpanzees and other primates in very similar conditions in order to begin to put chimpanzee cooperation in context (Cronin 2017).

One result that stands out as anomalous is how poorly the chimpanzees have done in experimental economic game–based studies relative to their behavior on cooperative food retrieval tasks. We see four possible reasons for this, none of which are mutually exclusive, and all of which are broadly relevant to cooperation studies in general. First, as we have emphasized throughout, there are individual differences in cooperative behavior, and it may be that experience with cognitive tasks is a key aspect of this. We note that the capuchin and rhesus results that we discussed come from individuals who participate in cognitive testing every day and have done so for many years. This may have given them an extra level of experience or motivation that allowed them to interpret the task in a different way.

Second, it may be that these tasks are not sufficiently motivating for the chimpanzees. These are relatively simple tasks that are repeated many times within a session, and it may be that the chimpanzees simply do not care. Similar problems have plagued other social decision-making studies; for example, early work indicated that chimpanzees did not make prosocial choices that benefitted another individual (Silk et al. 2005; Vonk et al. 2008). One explanation for this is the chimpanzees lacked any motivating reason to behave in a prosocial way (Brosnan 2017; de Waal 2016; Schmelz et al. 2017). Subsequent work supports this proposition, finding that in different contexts, particularly those in which subjects were given additional experience or motivation to make a prosocial choice, subjects' behavior changed markedly (Claidière et al. 2015; Horner et al. 2011; Schmelz et al. 2017). Aside from motivation issues, subjects may not even understand the experimenters' intended purpose for the original tasks. Supporting this, in other game theory–based studies, chimpanzees have shown behavior similar to that of other species (i.e., in the Stag Hunt game: Brosnan et al. 2011; Bullinger et al. 2011; Duguid et al. 2014). To promote task understanding, one of these tasks used extensive training to ensure that subjects (i) recognized that two individuals were required to gain a *Stag* payoff, (ii) monitored their partner's behavior, and (iii) chose the *Stag* option in pre-tests. This study found very high, almost ceiling, levels of *Stag-Stag* coordination (Bullinger et al. 2011).

Third, and related to the previous point, it may be that these games do not measure what we think they are measuring. We know that chimpanzees—and many other primates—attend to their social partners' actions and outcomes. This is the foundation of social learning, a skill at which chimpanzees do very well (Hopper and Carter, chapter 7 this volume). What we do not know is whether they understand that their payoffs are dependent upon their partner's choices. As our simulation study emphasizes (Parrish et al. 2014), all of these games can be solved in the absence of a partner through individuals focusing on their own average payoffs. This makes it extremely difficult to determine the degree to which the subjects recognize their partner's role.

Fourth and finally, despite the fact that these games, designed by humans and originally for humans, seem very intuitive to us, they may not be so for nonhuman primates. As we discussed above, this has precedent in other cooperation paradigms in which task understanding is lacking (Brosnan and de Waal 2002; Suchak et al. 2018). Although it appears thus far that the primates understand the tasks that we have discussed above, it is important to keep in mind that they may perceive them differently.

Future research, we believe, needs to focus on species-level and individual differences. Why are some species and individuals outperforming others? Is it their genetic make-up (i.e., hormone levels), their selective environment, their current environment, the cooperative context, the social context, their individual cognitive ability, their level of experience, or something else altogether? Of course, these are not mutually exclusive explanations and they may interact in ways that we do not understand. We also need insight into how these individuals see these tasks. Do chimpanzees, for instance, realize that they are working jointly, or are they "cooperating" individually, using associative learning or other cognitive mechanisms to obtain, on average, the highest payoff? Once we consider each of these, we will more fully understand the cooperative tendencies of one of our closest living relatives, and thereby better understand how and why we cooperate.

Acknowledgments

We thank the many, many researchers who have worked in our lab on these projects throughout the years. They are too numerous to name, but their contributions are greatly appreciated. In particular, we thank Bart Wilson, Steven Schapiro, Susan Lambeth, Larry Williams, and Michael Beran, who have been central to the experimental economics program from the start. We also thank Lydia Hopper and Steve Ross for bringing together such an invigorating group of speakers for the Chimpanzees in Context meeting, which resulted in this edited volume. We also gratefully acknowledge support that has funded the research described in this chapter, including NSF CAREER 0847351 to S. F. Brosnan; NSF HDS 0729244, NSF SES 1123897, and 1658867 to SF Brosnan, B. J. Wilson, S. J. Schapiro, and M. J. Beran; NSF SES 1425216 to S. F. Brosnan, M. J. Beran, S. J. Schapiro, and L. E. Williams; and support from the John Templeton Foundation and Georgia State University to S. F. Brosnan.

References

Boesch, C. 2002. "Cooperative hunting roles among Tai chimpanzees." *Human Nature* 13 (1): 27–46.

Boesch, C. 2003. "Complex cooperation among Taï chimpanzees." In *Animal Social Complexity Intelligence Culture, and Individualized Societies*, edited by F. B. M. de Waal and P. L. Tyack, 93–110. Cambridge, MA: Harvard University Press.

Boyd, R., and P. J. Richerson. 2009. "Culture and the evolution of human cooperation." *Philosophical Transactions of the Royal Society of London B* 364: 3281–88.

Brosnan, S. F. 2011. "A hypothesis of the co-evolution of cooperation and responses to inequity." *Frontiers in Neuroscience* 5: 43.

Brosnan, S. F. 2013. "Justice- and fairness-related behaviors in nonhuman primates." *Proceedings of the National Academy of Sciences* 110 (2): 10416–23.

Brosnan, S. F. 2017. "Understanding social decision-making from another species' perspective." *Learning & Behavior* 46 (2): 101–2.

Brosnan, S. F., M. J. Beran, A. E. Parrish, S. A. Price, and B. J. Wilson. 2013. "Comparative approaches to studying strategy: Towards an evolutionary account of primate decision making." *Evolutionary Psychology* 11 (3): 606–27.

Brosnan, S. F., and F. B. M. de Waal. 2002. "A proximate perspective on reciprocal altruism." *Human Nature* 13: 129.

Brosnan, S. F., and F. B. M. de Waal. 2014. "Evolution of responses to (un) fairness." *Science* 346 (6207): 1251776.

Brosnan, S. F., C. Freeman, and F. B. M. de Waal. 2006. "Partner's behavior, not reward distribution, determines success in an unequal cooperative task in capuchin monkeys." *American Journal of Primatology* 68 (7): 713–24.

Brosnan, S. F., L. M. Hopper, S. Richey, H. D. Freeman, C. F. Talbot, S. D. Gosling, S. P. Lambeth, and S. J. Schapiro. 2015. "Personality influences responses to inequity and contrast in chimpanzees." *Animal Behaviour* 101: 75–87.

Brosnan, S. F., A. Parrish, M. J. Beran, T. Flemming, L. Heimbauer, C. F. Talbot, S. P. Lambeth, S. J. Schapiro, and B. J. Wilson. 2011. "Responses to the assurance game in monkeys, apes, and humans using equivalent procedures." *Proceedings of the National Academy of Sciences* 108 (8): 3442–47.

Brosnan, S. F., S. A. Price, K. Leverett, L. Prétôt, M. Beran, and B. J. Wilson. 2017. "Human and monkey responses in a symmetric game of conflict with asymmetric equilibria." *Journal of Economic Behavior & Organization* 142: 293–306.

Brosnan, S. F., B. J. Wilson, and M. J. Beran. 2012. "Old World monkeys are more similar to humans than New World monkeys when playing a coordination game." *Proceedings of the Royal Society of London B* 279 (1733): 1522–30.

Bshary, R., and R. Bergmuller. 2008. "Distinguishing four fundamental approaches to the evolution of helping." *Journal of Evolutionary Biology* 21: 405–20.

Bullinger, A. F., E. Wyman, A. P. Melis, and M. Tomasello. 2011. "Coordination of chimpanzees (*Pan troglodytes*) in a stag hunt game." *International Journal of Primatology* 32 (6): 1296–1310.

Burkart, J. M., E. Fehr, C. Efferson, and C. P. van Schaik. 2007. "Other-regarding preferences in a non-human primate, the common marmoset (*Callithrix jacchus*)." *Proceedings of the National Academy of Sciences* 104 (50): 19762–66.

Burkart, J. M., and C. P. van Schaik. 2010. "Cognitive consequences of cooperative breeding in primates?" *Animal Cognition* 13 (1): 1–19.

Chalmeau, R. 1994. "Do chimpanzees cooperate in a learning task?" *Primates* 35: 385–92.

Chalmeau, R., and A. Gallo. 1996. "Cooperation in primates: Critical analysis of behavioural criteria." *Behavioural Processes* 35: 101–11.

Chalmeau, R., E. Visalberghi, and A. Gallo. 1997. "Capuchin monkeys (*Cebus apella*) fail to understand a cooperative task." *Animal Behaviour* 54: 1215–25.

Claidière, N., A. Whiten, M. C. Mareno, E. J. Messer, S. F. Brosnan, L. M. Hopper, S. P. Lambeth, S. J. Schapiro, and N. McGuigan. 2015. "Selective and contagious prosocial resource donation in capuchin monkeys, chimpanzees and humans." *Scientific Reports* 5: 7631.

Crawford, M. P. 1937. *The Cooperative Solving of Problems by Young Chimpanzees.* Baltimore: Johns Hopkins University Press.

Crawford, M. P. 1941. "The cooperative solving by chimpanzees of problems requiring serial responses to color cues." *Journal of Social Psychology* 13 (2): 259–80.

Cronin, K. A. 2017. "Comparative studies of cooperation: Collaboration and prosocial behavior in animals." In *APA Handbook of Comparative Psychology*, edited by J. Call, G. B. Burghardt, I. Pepperberg, C. T. Snowdon, and T. Zental, 915–29. Washington, DC: American Psychological Association.

Cronin, K. A., A. V. Kurian, and C. T. Snowdon. 2005. "Cooperative problem solving in a cooperatively breeding primate (*Saguinus oedipus*)." *Animal Behaviour* 69 (1):133–42.

Cronin, K. A., and C. T. Snowdon. 2008. "The effects of unequal reward distributions on cooperative problem solving by cottontop tamarins, *Saguinus oedipus*." *Animal Behaviour* 75: 245–57.

Deaner, R. O., K. Isler, J. Burkart, and C. P. van Schaik. 2007. "Overall brain size, and not encephalization quotient, best predicts cognitive ability across non-human primates." *Brain, Behavior and Evolution* 70: 115–24.

de Waal, F. B. M. 2016. *Are We Smart Enough to Know How Smart Animals Are?* New York: W. W. Norton.

de Waal, F. B. M., and M. L. Berger. 2000. "Payment for labour in monkeys." *Nature* 404 (6778): 563.

de Waal, F. B. M., and J. M. Davis. 2003. "Capuchin cognitive ecology: Cooperation based on projected returns." *Neuropsychologia* 41 (2): 221–28.

Di Bitetti, M. S. 2001. "Home-range use by the tufted capuchin monkey (*Cebus apella nigritus*) in a subtropical rainforest of Argentina." *Journal of Zoology* 253 (1): 33–45.

Dugatkin, L. A. 1997. *Cooperation among Animals: An Evolutionary Perspective*. Oxford: Oxford University Press.

Duguid, S., E. Wyman, A. F. Bullinger, K. Herfurth-Majstorovic, and M. Tomasello. 2014. "Coordination strategies of chimpanzees and human children in a stag hunt game." *Proceedings of the Royal Society of London B* 281 (1796): 20141973.

Engelmann, J. M., J. B. Clift, E. Herrmann, and M. Tomasello. 2017. "Social disappointment explains chimpanzees' behaviour in the inequity aversion task." *Proceedings of the Royal Society of London B* 284 (1861): 20171502.

Fehr, E., and K. M. Schmidt. 1999. "A theory of fairness, competition, and cooperation." *Quarterly Journal of Economics* 114 (3): 817–68.

Fragaszy, D. M., E. Visalberghi, and L. M. Fedigan. 2004. *The Complete Capuchin: The Biology of the Genus Cebus*. Cambridge: Cambridge University Press.

Freeman, H. D., J. Sullivan, L. M. Hopper, C. F. Talbot, A. N. Holmes, N. Schultz-Darken, L. E. Williams, and S. F. Brosnan. 2013. "Different responses to reward comparisons by three primate species." *PLoS One* 8 (10): e76297.

Gilby, I. C., M. Emery, J. D. Thompson, and R. Wrangham. 2010. "No evidence of short-term exchange of meat for sex among chimpanzees." *Journal of Human Evolution* 59: 44–53.

Gilby, I. C., Z. P. Machanda, D. C. Mjungu, J. Rosen, M. N. Muller, A. E. Pusey, and R. W. Wrangham. 2015. "Impact hunters' catalyse cooperative hunting in two wild chimpanzee communities." *Philosophical Transactions of the Royal Society of London B* 370 (1683): 20150005.

Gruber, T., and Z. Clay. 2016. "A comparison between bonobos and chimpanzees: A review and update." *Evolutionary Anthropology* 25 (5): 239–52.

Hall, K., M. Smith, J. L. Russell, S. P. Lambeth, S. J. Schapiro, and S. F. Brosnan. 2019. "Chimpanzees rarely settle on consistent patterns of play in the Hawk Dove, Assurance, and Prisoner's Dilemma games, in a token exchange task." *Animal Behavior and Cognition* 6 (1): 48–70.

Harcourt, A. H., and F. B. M. de Waal. 1992. *Coalitions and Alliances in Humans and Other Animals*. Oxford: Oxford University Press.

Henazi, P. S., and L. Barrett. 1999. "The value of grooming to female primates." *Primates* 40 (1): 47–59.

Hirata, S., and K. Fuwa. 2007. "Chimpanzees (*Pan troglodytes*) learn to act with other individuals in a cooperative task." *Primates* 48 (1): 13–21.

Hirata, S., N. Morimura, and K. Fuwa. 2010. "Intentional communication and comprehension of the partner's role in experimental cooperative tasks." In *The Mind of the Chimpanzee: Ecological and Experimental Perspectives*, edited by E. V. Lonsdorf, S. R. Ross, and T. Matsuzawa, 251–64. Chicago: University of Chicago Press.

Hockings, K. J., T. Humle, J. R. Anderson, D. Biro, C. Sousa, G. Ohashi, and T. Matsuzawa. 2007. "Chimpanzees share forbidden fruit." *PLoS One* 2 (9): e886.

Hölldobler, B., and E. O. Wilson. 2009. *The Superorganism: The Beauty, Elegance, and Strangeness of Insect Societies*. New York: W. W. Norton.

Hoppitt, W., and K. N. Laland. 2008. "Social processes influencing learning in animals: A review of the evidence." In *Advances in the Study of Behavior*, vol. 39, edited by J. Brockmann, T. Roper, M. Naguib, K. Wynne-Edwards, C. Barnard, and J. Mitani, 105–65. San Diego: Academic Press.

Horner, V. J., C. D. Carter, M. Suchak, and F. B. M. de Waal. 2011. "Spontaneous prosocial choice by chimpanzees." *Proceedings of the National Academy of Sciences* 108 (33): 13847–51.

Jaeggi, A. V., and M. Gurven. 2013. "Natural cooperators: Food sharing in humans and other primates." *Evolutionary Anthropology* 22 (4): 186–95.

Janson, C. H., and S. F. Brosnan. 2013. "Experiments in primatology: From the lab to the field and back again." In *Primate Ecology and Conservation: A Handbook of Techniques*, edited by M. Blair, E. Sterling, and N. Bynum, 177–94. Oxford: Oxford University Press.

Loukola, O. J., C. J. Perry, L. Coscos, and L. Chittka. 2017. "Bumblebees show cognitive flexibility by improving on an observed complex behavior." *Science* 355 (6327): 833–36.

Maestripieri, D. 2007. *Macachiavellian Intelligence*. Chicago: University of Chicago Press.

Manson, J. H., C. D. Navarrete, J. B. Silk, and S. Perry. 2004. "Time-matched grooming in female primates? New analyses from two species." *Animal Behaviour* 67 (3): 493–500.

Melis, A. P., B. Hare, and M. Tomasello. 2006a. "Engineering cooperation in chimpanzees: Tolerance constraints on cooperation." *Animal Behaviour* 72 (2): 275–86.

Melis, A. P., B. Hare, and M. Tomasello. 2006b. "Chimpanzees recruit the best collaborators." *Science* 311 (5765): 1297–1300.

Melis, A. P., and D. Semmann. 2010. "How is human cooperation different?" *Philosophical Transactions of the Royal Society of London B* 365: 2663–74.

Mendres, K. A., and F. B. M. de Waal. 2000. "Capuchins do cooperate: The advantage of an intuitive task." *Animal Behaviour* 609 (4): 523–29.

Mustoe, A. C., A. M. Harnisch, B. Hochfelder, J. Cavanaugh, and J. A. French. 2016. "Inequity aversion strategies between marmosets are influenced by partner familiarity and sex but not by oxytocin." *Animal Behaviour* 114: 69–79.

Nishida, T., T. Hasegawa, H. Hayaki, Y. Takahata, and S. Uehara. 1992. "Meat-sharing as a coalition strategy by an alpha male chimpanzee?" In *Topics in Primatology*, edited by T. Nishida, 159–74. Tokyo: University of Tokyo Press.

Noe, R. 2006. "Cooperation experiments: Coordination through communication versus acting apart together." *Animal Behaviour* 71: 1–18.

Parrish, A. E., S. F. Brosnan, B. J. Wilson, and M. J. Beran. 2014. "Differential responding by rhesus monkeys (*Macaca mulatta*) and humans (*Homo sapiens*) to variable outcomes in the assurance game." *Animal Behavior and Cognition* 1 (3): 215–29.

Povinelli, D. J., and D. K. O'Neil. 2000. "Do chimpanzees use their gestures to instruct each other?" In *Understanding Other Minds: Perspectives from Developmental Cognitive Neuroscience*, edited by S. Baron-Cohen, H. Tager-Flusberg, and D. J. Cohen, 459–87. Oxford: Oxford University Press.

Price, E. E., S. P. Lambeth, S. J. Schapiro, and A. Whiten. 2009. "A potent effect of observational learning on chimpanzee tool construction." *Proceedings of the Royal Society of London B* 276 (1671): 3377–83.

Price, S. A., and S. F. Brosnan. 2012. "To each according to his need? Variability in the responses to inequity in non-human primates." *Social Justice Research* 25 (2): 140–69.

Pruetz, J. D., and P. Bertolani. 1997. "Savanna chimpanzees, *Pan troglodytes verus*, hunt with tools." *Current Biology* 17 (5): 412–17.

Rilling, J. K., and T. R. Insel. 1999. "Differential expansion of neural projection systems in primate brain evolution." *Neuroreport* 10 (7): 1453–59.

Rose, L. M. 1997. "Vertebrate predation and food-sharing in *Cebus* and *Pan*." *International Journal of Primatology* 18 (5): 727–65.

Russell, Y. I., and S. Phelps. 2013. "How do you measure pleasure? A discussion about intrinsic costs and benefits in primate allogrooming." *Biology and Philosophy* 28 (6): 1005–20.

Scarry, C. J. 2017. "Male resource defence during intergroup aggression among tufted capuchin monkeys." *Animal Behaviour* 123: 169–78.

Schino, G., and F. Aureli. 2008. "Grooming reciprocation among female primates: A meta-analysis." *Biology Letters* 4 (1): 9–11.

Schmelz, M., S. Grueneisen, A. Kabalak, J. Jost, and M. Tomasello. 2017. "Chimpanzees return favors at a personal cost." *Proceedings of the National Academy of Sciences* 114 (28): 7462–67.

Silk, J. B., S. F. Brosnan, J. Vonk, J. Henrich, D. J. Povinelli, A. S. Richardson, S. P. Lambeth, J. Mascaro, and S. J. Schapiro. 2005. "Chimpanzees are indifferent to the welfare of unrelated group members." *Nature* 437 (7063): 1357.

Skyrms, B. 2003. *The Stag Hunt and the Evolution of Social Structure*. Cambridge: Cambridge University Press.

Smith, M., J. Watzek, and S. F. Brosnan. 2018. "The importance of a truly comparative methodology for comparative psychology." *International Journal of Comparative Psychology* 31.

Suchak, M., T. M. Eppley, M. W. Campbell, and F. B. M. de Waal. 2014. "Ape duos and trios: Spontaneous cooperation with free partner choice in chimpanzees." *PeerJ* 2: e417.

Suchak, M., T. M. Eppley, M. W. Campbell, R. A. Feldman, L. F. Quarles, and F. B. M de Waal. 2016. "How chimpanzees cooperate in a competitive world." *Proceedings of the National Academy of Sciences* 113 (36): 10215–20.

Suchak, M., J. Watzek, L. F. Quarles, and F. B. M. de Waal. 2018. "Novice chimpanzees cooperate successfully in the presence of experts, but may have limited understanding of the task." *Animal Cognition* 21: 87–98.

Surbeck, M., and G. Hohmann. 2008. "Primate hunting by bonobos at LuiKotale, Salonga National Park." *Current Biology* 18 (19): R906–7.

Surbeck, M., R. Mundry, and G. Hohmann. 2010. "Mothers matter! Maternal support, dominance status and mating success in male bonobos (*Pan paniscus*)." *Philosophical Transactions of the Royal Society of London B* 278 (1705): 590–98.

Talbot, C. F, A. E. Parrish, J. Watzek, J. L. Essler, K. L. Leverett, A. Paukner, and S. F. Brosnan. 2018. "The influence of reward quality and quantity and spatial proximity on the responses to inequity and contrast in capuchin monkeys (*Cebus [Sapajus] apella*)." *Journal of Comparative Psychology* 132 (1): 75.

Trivers, R. L. 1971. "The evolution of reciprocal altruism." *Quarterly Review of Biology* 46 (1): 35–57.

Vonk, J., S. F. Brosnan, J. B. Silk, J. Henrich, A. S. Richardson, S. P. Lambeth, S. J. Schapiro, and D. J. Povinelli. 2008. "Chimpanzees do not take advantage of very low cost opportunities to deliver food to unrelated group members." *Animal Behaviour* 75 (5): 1757–70.

Watts, D. P., and J. C. Mitani. 2002. "Hunting behavior of chimpanzees at Ngogo, Kibale National Park, Uganda." *International Journal of Primatology* 23 (1): 1–28.

Watzek, J., M. F. Smith, and S. F. Brosnan. 2017. "Comparative economics: Using experimental economics paradigms to understand primate social decision-making." In *The Evolution of Primate Social Cognition*, edited by L. Di Paolo, F. Di Vincenzo, and A. d'Almeida. Cham: Springer.

16

A Comparison of Cooperative Cognition in Corvids, Chimpanzees, and Other Animals

JORG J. M. MASSEN, WOUTER A. A. SCHAAKE, AND THOMAS BUGNYAR

Introduction

Through cooperation an individual can achieve goals that it cannot reach by itself. Cooperation can be observed in a wide variety of species with varying goals, and, consequently, in a wide variety of (social) contexts (see also Duguid et al., chapter 13 this volume; Vale and Brosnan, chapter 15 this volume; Yamamoto, chapter 14 this volume). Across taxonomic groups, species breed cooperatively (mammals: Lukas and Clutton-Brock 2013; birds: Riehl 2013; fish: Dey et al. 2017), hunt cooperatively (chimpanzees, *Pan troglodytes*: Boesch 2002; other mammals: e.g., Pitman and Durban 2012; birds: e.g., Bednarz 1988; fish: e.g., Vail, Manica, and Bshary 2013), and defend their territories cooperatively against predators (chimpanzees: Boesch 1991; other mammals: e.g., Bshary and Noë 1997; birds: e.g., Cornell, Marzluff, and Pecoraro 2012; fish: Milinski et al. 1997) and conspecifics (chimpanzees: Goodall 1986; other mammals: e.g., Gersick et al. 2015; birds: e.g., Bossema and Benus 1985; fish: Clifton 1990); and some species build alliances within their social group (primates: reviewed in Schino 2007; coatis, *Nasua nasua*: Romero and Aureli 2008; ravens, *Corvus corax*: Fraser and Bugnyar 2012). The evolution of cooperation among kin can be explained by the indirect fitness benefits through kin-selection (Hamilton 1964). Yet, many species cooperate with unrelated individuals (Clutton-Brock 2009), which entails the risk of exploitation; i.e., cooperative partners may try to minimize their own costs while reaping the benefits.

To study proximate and ultimate mechanisms that have nonetheless facilitated the evolution of cooperation, many researchers have adopted a comparative and experimental approach while testing their species. Experimental investigations of cooperation started, not surprisingly, with tests on our closest living relatives, chimpanzees (Crawford 1937). Crawford presented young

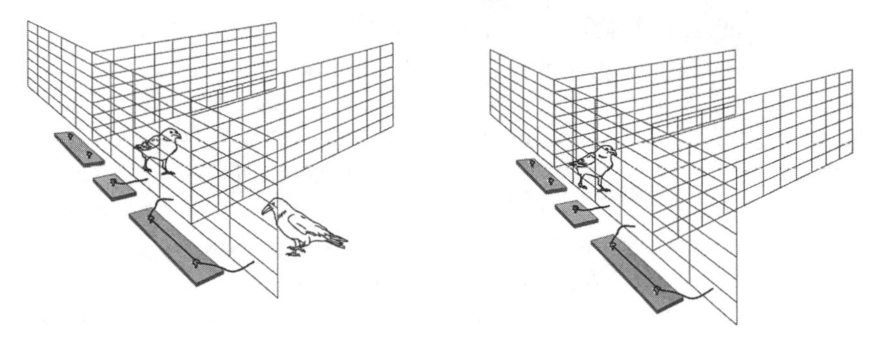

FIGURE 16.1. Experimental set-up for study 1. The bird in the middle compartment can choose to either pull in the platform in the middle on its own (single string-pulling apparatus), or to work on the loose-string string-pulling apparatus (here on its left), when the partners are present (left) or absent (right). Drawing by Nadja Kavcik-Graumann.

chimpanzees with two ropes attached to a box filled with fruit outside of their cage that was too heavy to pull in by one chimpanzee on its own. Only when two of the youngsters pulled the two ropes simultaneously were they able to pull the box into their reach see Duguid et al., chapter 13 this volume for a detailed review of Crawford's studies of cooperation). Since the 1930s this paradigm has been tested in several other primate species (Mendres and de Waal 2000; Molesti and Majolo 2016), and other paradigms have been developed (Chalmeau, Visalberghi, and Gallo 1997; Chalmeau et al. 1997; Cronin, Kurian, and Snowdon 2005; Drea and Carter 2009; Eskelinen et al. 2016; Fady 1972; Hirata 2003; Petit, Desportes, and Thierry 1992; Suchak et al. 2014, 2018; Tebbich, Taborsky, and Winkler 1996; Werdenich and Huber 2002), often inspired by the original paradigm of Crawford. The paradigm that has gained most traction in recent years is the "loose-string" paradigm, originally designed by Hirata (2003). In this task, food is placed on a platform that is out of reach for the subjects. The platform contains two loops through which a rope is threaded (fig. 16.1). The two ends of that rope are then placed in the cage of the subjects, far enough apart such that one individual cannot take both of them. If only one individual pulls one end of the rope, it slips through the loops and all the individual can achieve is simply pulling the rope into their cage. However, if two individuals pull on both ends simultaneously, the platform will move into reach. To solve the paradigm thus requires cooperation (Hirata 2003).

The advantage of the loose-string task is that it can be easily adapted to accommodate testing of different species. Consequently, it has been used to test chimpanzees (Hirata 2003; Hirata and Fuwa 2007; Melis, Hare, and Tomasello 2006a, 2006b), bonobos (*Pan paniscus*, Hare et al. 2007), rooks (*Corvus frugilegus*, Scheid and Noë 2010; Seed, Clayton, and Emery 2008), Asian elephants (*Elephas maximus*, Plotnik et al. 2011), African grey parrots (*Psittacus erithacus*, Péron et al. 2011), dogs (*Canis familiaris*, Marshall-Pescini et al. 2017; Ostojić and Clayton 2014), ravens (Asakawa-Haas et al. 2016; Massen, Ritter, and Bugnyar 2015), keas (*Nestor notabilis*, Heaney, Gray, and Taylor 2017; Schwing et al. 2016), giant- and Asian small-clawed otters (*Pteronura brasiliensis* and *Aonyx cinerea*, Schmelz et al. 2017a), wolves (*Canis lupus*, Marshall-Pescini et al. 2017), long-tailed macaques (*Macaca fascicularis*, Stocker et al. in prep.), and common marmosets (*Callithrix jacchus*, Martin et al. in prep). A common denominator for success in this paradigm is interindividual tolerance or affiliation between the two partners (Hare et al. 2007; Massen, Ritter, and Bugnyar 2015; Marshall-Pescini et al. 2017; Melis, Hare, and Tomasello 2006b; Schwing et al. 2016; Seed, Clayton, and Emery 2008). Also, the relative rank between the two individuals cooperating seems to have an influence on cooperation success, albeit one study has reported that individuals closer in rank cooperate better (Marshall-Pescini et al. 2017), whereas another reported the opposite pattern (Massen et al. 2015). Finally, it has been shown that individuals prefer to cooperate with more proficient partners (Melis, Hare, and Tomasello 2006a) and with their "friends" (Asakawa-Haas et al. 2016), and that they cease cooperating with those individuals that cheated on them (i.e., took more than their share of the rewards) in a previous trial (Massen, Ritter, and Bugnyar 2015; Schwing et al. 2016).

Apart from social constraints, a physical understanding of the task at hand seems paramount for success. To test whether their subjects understand the contingencies of the loose-string paradigm, researchers have reverted to different knowledge probes. Typically, researchers test their subjects in either an alone condition, where pulling is pointless and thus the ability of the subject to inhibit pulling is tested, or a delay condition, where the partner enters the testing room only after a given delay, and the subject thus has to wait with pulling until its partner has arrived. Results on these knowledge probes are mixed, with some species passing them (chimpanzees, Melis, Hare, and Tomasello 2006a; elephants, Plotnik et al. 2011; one African grey parrot, Péron et al. 2011; dogs, Ostojić and Clayton 2014; keas, Heaney, Gray, and Taylor 2017; wolves, Marshall-Pescini et al. 2017), while other studies have reported negative results for different, and also for the *same* (in italics), species in these knowledge probes (i.e., rooks, Seed, Clayton, and Emery 2008; ravens, Massen, Ritter, and Bugnyar. 2015; *keas*, Schwing et al. 2016; otters, Schmelz et al. 2017a; *dogs*, Marshall-Pescini et al. 2017).

Although all previous studies use the same paradigm, the procedures of testing, and especially the prior experience and/or training of the different animals, differ dramatically between the studies (table 16.1), making comparisons between the species difficult (Bugnyar 2008). This is especially relevant, as it has been shown that, for example, chimpanzees need to learn to act cooperatively in this (Hirata and Fuwa 2007) or other (Suchak et al. 2018) cooperation paradigms, and, for example, ravens start waiting for their partner only after enough experience with the task (Asakawa-Haas et al. 2016). Therefore, we retested ravens that originally failed both knowledge probes (Massen, Ritter, and Bugnyar 2015), but started to wait for their partner given more experience (Asakawa-Haas et al. 2016), in two new experiments that tested whether these ravens have some sort of understanding of the need for a partner in the loose-string paradigm and how well they can coordinate with that partner respectively.

Ravens are a species particularly well-suited to study the evolution of (social) cognition. In contrast to, for example, chimpanzees, any parallels in cognition with humans can hardly be attributed to a common ancestor, and thus have likely evolved convergently in both species. The study of such convergently evolved cognition can give us important insights into the (socio-) ecological selective pressures that led to the evolution of that parallel intelligence (Fitch, Huber, and Bugnyar 2010). And there are many such parallels between ravens, humans, and chimpanzees. Ravens show marked preferences for particular individuals (Fraser and Bugnyar 2010a), as do chimpanzees (Fraser, Schino, and Aureli 2008); they support preferred individuals in conflicts (Fraser and Bugnyar 2012), as do chimpanzees (Hemelrijk and Ek 1991); they reconcile conflicts with preferred partners (Fraser and Bugnyar 2011), as do chimpanzees (de Waal and van Roosmalen 1979); and they console their partners who have had a conflict with a third party (Fraser and Bugnyar 2010b), as do chimpanzees (Fraser et al. 2009). Moreover, ravens remember the value of such relationships over years (Boeckle and Bugnyar 2012), and also remember the cooperative willingness of humans in an experimental setting (Müller et al. 2017). Their social knowledge is not restricted to their own relationships, as, like chimpanzees (Slocombe and Zuberbühler 2007), they also seem to understand the relationship of others (Massen et al. 2014a), so called third-party understanding, and use that knowledge in an almost political way to sabotage the relationships of others (Massen et al. 2014b, cf. chimpanzees: de Waal 1982; Mielke et al. 2017). Furthermore, ravens are able to deceive conspecifics in competition for hidden food (Bugnyar and Kotrschal 2002, cf. chimpanzees: Hare, Call, and Tomasello 2006), and seem to use Theory-of-Mind-like abilities to do so (Bugnyar, Reber, and Buckner 2016).

TABLE 16.1. Studies using the loose-string paradigm and the differences in procedure. Species, study, sample size, whether the animals were trained prior to the experiment (and how), whether there was a knowledge probe conducted, and whether the animals passed this knowledge probe.

Species and Studies	*N*	*Training prior to Test*	*Knowledge Probe*	*Passed (Y/N)*	*Notes*
Chimpanzees					
Melis, Hare, and Tomasello 2006b[#]	51	Yes: Individuals could operate the apparatus by themselves as ropes were offered close to each other and the distance between the ropes was then incrementally increased.	No	-	Whereas there was no knowledge probe, the animals were actually tested in an alone condition and had to (and did) open a door for their partner to enter too, suggesting that they understood the need of a partner.
Hirata and Fuwa 2007	2	Yes: Individuals could operate the apparatus by themselves as ropes were offered close to each other.	No	-	The chimpanzees initially did not succeed in the cooperation condition, but did so after some experience.
Bonobos					
Hare et al. 2007	20	Yes: Individuals could operate the apparatus by themselves as ropes were offered close to each other.	No	-	
Rooks					
Seed, Clayton, and Emery 2008	8	Yes: Individuals could operate the apparatus by themselves as ropes were offered close to each other and the distance between the ropes was then incrementally increased.	Yes: Partners entered after a ± 10 sec delay.	No	Delay caused by opened door; neophobia likely played a role, at least in some trials.
Scheid and Noë 2011	9	Yes: Individual "simple" string pulling using shaping procedures.	No	-	Waiting behavior was scored, albeit that this occurred sporadically only.
Asian elephants					
Plotnik et al. 2011	12	Yes: Individual "simple" string pulling.	Yes: Trials in which the partner arrived with incrementally increasing delays & a condition where the partner could not pull.	Yes	The increasing delays (5s increments up to 25s) were shaped such that elephants proceeded to the next delay only after being successful in the previous delay length.

TABLE 16.1. *(continued)*

African grey parrot					
Péron et al. 2011	3	Yes: First, individual training with both ends connected to each other such that one individual could pull by itself, then while "cooperating" with an experimenter.	Yes: Partners entered after a ± 15 sec delay.	Yes	Note, only one individual passed.
Domestic dogs					
Ostojić and Clayton 2014	11	Yes: Individuals could operate the apparatus by themselves as ropes were offered close to each other and the distance between the ropes was then incrementally increased.	Yes: partners entered after a ± 2 sec delay.	Yes	Subjects were (trained) pet dogs.
Marshall-Pescini et al. 2017	14	Yes: Individual "simple" string pulling, and after the dogs initially did not succeed in the cooperation condition they received additional training in which individuals could operate the apparatus by themselves as ropes were offered close to each other and the distance between the ropes was then incrementally increased.	No	-	Subjects were hand-raised dogs kept in packs. Even after more training the dogs did only sporadically succeed in the cooperation condition and therefore did not proceed to further testing (see wolves).
Ravens					
Massen, Ritter, and Bugnyar 2015[!]	9	No: They solved it spontaneously. At the time of the knowledge probe, all animals (apart from two) had ample experience with the paradigm.	Yes: Partners entered after a delay of ± 10 sec & an alone condition.	No	
Keas					
Schwing et al. 2016	14	Yes: Individuals received training working on the apparatus with a human.	Yes: Two different conditions in which the human did not pull.	No	Whereas training and knowledge-probes were with humans, subsequent tests were with conspecifics.

(continues)

TABLE 16.1. *(continued)*

Species and Studies	*N*	*Training prior to Test*	*Knowledge Probe*	*Passed (Y/N)*	*Notes*
Heaney, Gray, and Taylor 2017	4	Yes: Individuals were trained to gather and pull both ends of the rope by themselves.	Yes: Trials in which the partner arrived with incrementally increasing delays.	Yes	Keas were first shaped to wait up to 25s (following Plotnik et al. 2011). Once trained, the keas were tested in three delay conditions: delays ranging 0–25s; delays ranging 26–45s; and delays ranging 46–65s.
Otters (giant and Asian small-clawed)					
Schmelz et al. 2017a	5/4	Yes: Individual "simple" string pulling.	Yes: Two delay conditions based on the position of the ropes and the place were the otters entered.	No	
Wolves					
Marshall-Pescini et al. 2017	12	Yes: Individual "simple" string pulling, and some (4/7 dyads) received additional training in which individuals could operate the apparatus by themselves as ropes were offered close to each other and the distance between the ropes was then incrementally increased.	Yes: Partners entered after a delay of ± 10 sec.	Yes	

Note: In all studies at least some of the animals solved the loose-string paradigm. Success rates differed depending on experience, inter-individual tolerance, and rank differences.

[#] Melis, Hare, and Tomasello 2006a is a follow-up to Melis, Hare, and Tomasello 2006b and therefore not included in this table.

[!] Asakawa-Haas et al. 2016 is a follow-up to Massen, Ritter, and Bugnyar 2015 and therefore not included in this table.

Consequently, along with other corvids, their intelligence is often compared to that of apes (Clayton and Emery 2007; Güntürkün and Bugnyar 2016), and a recent study suggested that there are also some interesting parallels regarding the neural substrate underlying such complex (social) cognition (Olkowicz et al. 2016). Particularly relevant for our study is that ravens cooperate in the wild in various contexts, including hunting (Hendricks and Schlang 1998) and coalitionary support (Braun and Bugnyar 2012), and have been shown to spontaneously cooperate in the loose-string paradigm (Massen, Ritter, and Bugnyar 2015).

Approach: Study 1

SUBJECTS AND HOUSING

We tested 11 ravens (7 males, 4 females; age 3–5; table 16.2) that were housed as one group in a large outdoor aviary (15 x 15 x 5 meter) at the Haidlhof Research Station, in Bad Vöslau, Austria. All birds had gained experience with the loose-string paradigm while participating in previous tests (Asakawa-Haas et al. 2016; Massen, Ritter, and Bugnyar 2015), albeit their experience

TABLE 16.2. Raven subjects: name, sex, year of birth (YoB), and the amount of experience with the paradigm they had before study 1 and 2, as well as to which dyad they belonged if they participated in experiment 2.

Name	*Sex*	*YoB*	*Experience prior to Exp. 1*#	*Dyad*	*Experience prior to Exp. 2*
Tom	♂	2012	956 trials		
Rufus	♂	2012	396 trials		
George	♂	2012	1016 trials		
Adele	♀	2012	1056 trials		
Nobel	♀	2012	1056 trials		
Paul	♂	2012	396 trials	1	456 trials
Louise	♀	2012	1016 trials	1	1096 trials
Rocky	♂	2011	140 trials	2	180 trials
Joey	♀	2010	160 trials	2	180 trials
Laggie	♂	2012	976 trials	3	1016 trials
Bobbie	♀	2013	-*	3	570 trials
Horst	♂	2012	1016 trials	4	1096 trials
Astrid	♀	2010	-*	4	0 trials

#Note that this is the maximum number of cooperation trials that they *could* have completed in previous experiments.

*These two ravens did not participate in experiment 1.

varied because not all of them participated in all previous experiments, and some birds acted as a partner in these experiments more often than others (table 16.2). Furthermore, all birds also had experience operating a single string-pulling apparatus (see below). Experiment 1 was conducted between December 2014 and January 2015.

PROCEDURE

The experiment was performed in the birds' home aviary. Prior to testing, all birds that were not in the test were moved to an adjacent aviary. The home aviary contained three adjacent compartments: the two outside ones, both 6 x 15 x 5 m, and the inner one, 3 x 4.8 x 5 m. In the experiment, the test subject was in the middle compartment. In that location, it could choose to operate either a single string-pulling apparatus that would bring a platform containing a reward into reach, or one of two loose-string string-pulling apparatuses, that, if their partner cooperated, would also bring a platform, containing the same reward for both partner and test subject, into the reach of both birds (cf. Seed, Clayton, and Emery 2008; see fig. 16.1). Note, for the test subject the reward for cooperating and for working alone was always the same.

The birds were tested in two sessions of 20 trials: one when two "neutral" (cf. Asakawa-Haas et al. 2016) partners were present, one on either side of the subject, and one in which partners were absent (fig. 16.1). In the latter condition, choosing to operate the loose-string string-pulling apparatus was futile and thus we used this probe condition to test whether the birds understood the need for a partner. The order of "alone" vs. "partners-present" sessions was counter-balanced across the subjects. Per trial, the ropes of only one of the two loose-string string-pulling apparatuses, as well as the rope of the single string-pulling apparatus, were offered to the subjects simultaneously by two experimenters (fig. 16.1). The side of the loose-string string-pulling apparatus that was activated, as well as the experimenter that operated either the loose-string or single string-pulling apparatuses, and the side of the partners (when present) were counterbalanced over the trials (i.e., activation of which loose-string string-pulling apparatus and function of experimenter changed after the fifth trial, and then again after the fifteenth trial, and in the partners-present condition, the two partners exchanged compartments after the tenth trial).

All experiments were filmed with two cameras (Canon Legria HF G10), and these videos were coded for behavior afterward. All analyses were conducted in SPSS 23 (IBM), all tests were two-tailed, and alpha was set to 0.05.

Results: Study 1

When the ravens had to choose between operating either an apparatus that required cooperation or an apparatus that they could operate alone, they chose the former significantly more often when partners were present in comparison to when no partner was present ($T^+ = 66$, $N = 11$, $p = 0.003$; fig. 16.2). In the partners-present condition, four ravens significantly preferred (chi-square tests: $p < 0.05$) to cooperate with a partner, three preferred to work alone, and the remaining birds did not show a preference to work either alone or together. Moreover, none of the subjects showed a difference in their preference to work alone or together based on the identity of the partner with whom they could cooperate (chi-square tests: $p > 0.16$ for all pairwise comparisons), corroborating our choice of neutral partners, while also controlling for the effect of possible preferences to be in either of the partners proximity (cf. Asakawa-Haas et al. 2016) on the subject's choice for cooperation or working alone.

Approach: Study 2

SUBJECTS AND HOUSING

We initiated our experiment with a total of four male/female raven pairs. By the time of experiment 2, three of these pairs were housed as breeding couples

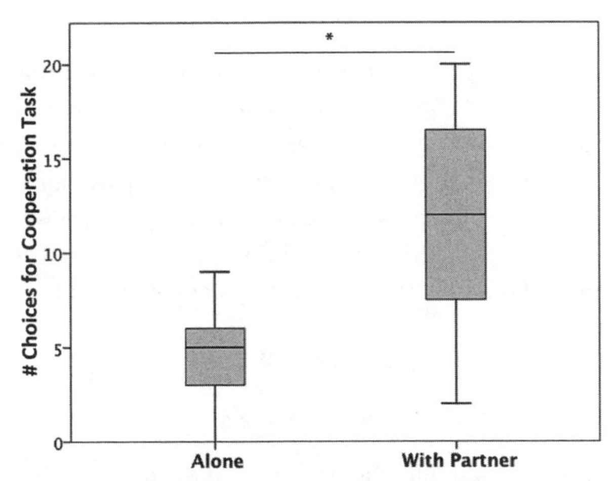

FIGURE 16.2. Median, inter-quartile range, and range of the number of trials in which subjects chose to operate the loose-string string-pulling task (rather than the single string-pulling task), when the subject was alone (20 trials) or when partners were present (20 trials). * $P < 0.05$

FIGURE 16.3. Experimental set-up for study 2. The pair of ravens had to coordinate their actions to simultaneously pull the strings of the same loose-string string-pulling apparatus. Drawing by Nadja Kavcik-Graumann.

in separate aviaries (all 8 x 10 x 5 m), while the remaining pair was still housed in the social group (see above) and separated from the other ravens of that group prior to testing in one of the large (6 x 15 x 5 m) compartments of their enclosure. Six of the eight birds in experiment 2 also participated in experiment 1; one of the additional two birds did have experience with the loose-string paradigm, whereas the other additional bird had no experience with the paradigm at all (table 16.2). Experiment 2 was conducted between August and October 2016.

PROCEDURE

Two loose-string string-pulling apparatuses, approximately 2 m apart, were installed outside the aviaries of the pairs (see fig. 16.3). First, the pairs received four sessions of 10 trials in which both apparatuses were baited and were simultaneously offered to the ravens. In order to solve either or both of the loose-string string-pulling tasks the ravens had to coordinate their actions with each other (cf. Marshall-Pescini et al. 2017). As the ravens were not very successful at first (fig. 16.4), we then tried to re-familiarize them with the paradigm by offering them the strings of only one apparatus at a time, again for four sessions of 10 trials. Finally, to see whether the birds' performance would increase with more experience, we retested the birds in another four sessions of 10 trials in which both loose-string string-pulling apparatuses were simultaneously offered to them. The sides of the experimenters, and in the one-apparatus condition also the side of the active apparatus, were counterbalanced across the trials.

Results: Study 2

Unfortunately, during the first sequence both birds of dyad 2 (see table 16.2) ceased working, and therefore were excluded from the rest of the study and analyses. The birds of dyads 1, 3, and 4 worked successfully on one of the two apparatuses in only 5, 0, and 2 trials (out of 40) respectively during the first sequence, and none of the dyads solved both loose-string string-pulling tasks in any trial. After the re-familiarization sequence, in which the dyads solved the one loose-string string-pulling task in 13, 10, and 14 trials respectively, the

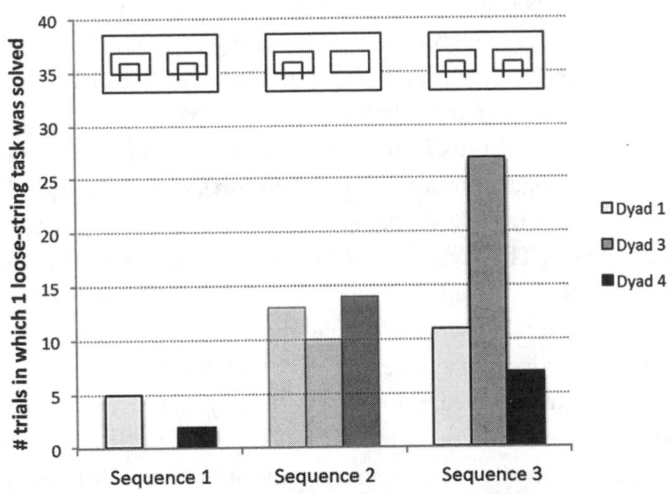

FIGURE 16.4. Performance (number of trials in which the dyad solved at least one loose-string string-pulling task) of the three dyads in the three different sequences of 40 trials. In sequence 1 and 3 the birds could operate two loose-string string-pulling tasks and thus had to coordinate more extensively, whereas in sequence 2 only one of the two loose-string string-pulling tasks was activated (represented by muted color scheme).

birds increased their performance significantly in the two loose-string string-pulling tasks paradigm (sequence 1 vs. 3: chi-square tests: $p < 0.05$).

In the third sequence, the three dyads solved one of both loose-string string-pulling tasks in 11, 27, and 7 trials respectively (fig. 16.4). If both birds pulled randomly on one end of a rope, there are six possible configurations in which two birds can do this on the four ends of the two loose-string string-pulling tasks, only two of which are successful. Thus, if we compare the birds' performance in the third sequence to random pulling, we find that the birds of dyad 3 perform significantly above chance level (chi-square test: $\chi^2 = 21.1$, $p < 0.001$). Notably, in six trials the birds of this dyad also managed to solve both loose-string string-pulling tasks, one after the other, emphasizing that they were able to coordinate their actions.

Discussion

Using two different experiments based on the loose-string paradigm (Hirata 2003), we tested whether ravens have an understanding of the need for a partner to solve this task, and whether they can coordinate their actions when two loose-string tasks are presented simultaneously. We showed that in a forced-choice set-up, the ravens chose the loose-string apparatus over a simple string-pulling apparatus more often when there was a potential cooperative partner present, indicating a basic understanding of the need for a partner for this task. Nevertheless, when tested as free-roaming pairs, they still had considerable problems coordinating their actions between two loose-string apparatuses, and only one pair succeeded in solving both tasks in the same trial, and then only six times out of 80 trials.

The results of the ravens in the choice task are comparable to those of chimpanzees (Bullinger, Melis, and Tomasello 2011), and African grey parrots (Péron et al. 2011), tested in a similar set-up, albeit that the results of the three parrots were mixed. The ravens' successful performance in this task is, however, in contrast to that of another corvid species—i.e., rooks (Seed, Clayton, and Emery 2008). It is important to note, though, that the ravens in our experiment had considerably more experience with the loose-string paradigm than the rooks tested by Seed, Clayton, and Emery (2008). Moreover, the results, and subsequent interpretation, of this experiment heavily depend on the individuals' motivation to cooperate; if the animals prefer to work alone, it is difficult to find a difference between the two conditions, yet this does not have to mean that they did not understand it, and such a floor effect may explain (some of) the results of the rooks. Considering this, it is interesting to see what the different species do in the partner-present condition only.

Although the chimpanzees in the experiment of Bullinger and colleagues (2011) seemingly understood the need of a partner, when given the choice, they did prefer to work alone (similar results have also been reported for keas, Heaney, Gray, and Taylor 2017). In contrast, Rekers, Haun, and Tomasello (2011) reported that human children prefer to work together, suggesting that this intrinsic preference to cooperate arose late in the evolution of our species. These results seem to corroborate findings that chimpanzees, in contrast to humans, appear to lack, or show decreased, other-regarding preferences (Jensen et al. 2006; Silk et al. 2005; Tennie, Jensen, and Call 2016; Vonk et al. 2008; Yamamoto and Tanaka 2010). As with cooperation studies, however, the paradigms and procedures used when studying prosociality, even in one species, vary drastically (Duguid et al., chapter 13 this volume; House et al. 2014), and in several experiments chimpanzees have shown spontaneous prosociality (Horner et al. 2011; Mendonça et al. 2018; Warneken and Tomasello 2006), or prosocial preferences contingent on their partner's identity or behavior (Claidière et al. 2015; House et al. 2014; Schmelz et al. 2017b; Yamamoto, Humle, and Tanaka 2009). Our results fall somewhere in the middle: while four of the 11 ravens we tested did show a preference to cooperate, three preferred to work alone; however, it stands in contrast to experimental studies that failed to show any prosocial concern in ravens (Di Lascio et al. 2013; Lambert et al. 2017; Massen et al. 2015). Understanding what drives these individual differences will be an important avenue for future work. For example, in the wild cooperation notably occurs between bonded pairs (Braun and Bugnyar 2012; Hendricks and Schlang 1998). Future studies should investigate whether and to what degree a preference to cooperate and prosociality align.

In study 2, after being given increased experience with a single loose-string task, one dyad of ravens was eventually able to coordinate their actions in such a fashion that they solved both loose-string tasks available, one after the other. However, even though the other pairs also increased their performance with experience, they had considerable problems coordinating their actions well enough to solve one of the two loose-string tasks offered, let alone both in succession. These results contrast with those from pairs of wolves tested in the same set-up and which succeeded in operating both loose-string tasks in 74% of the trials on average (Marshall-Pescini et al. 2017). Possibly these differences in the ability to coordinate across ravens and wolves reflect ecological differences; although ravens have been observed hunting together (Hendricks and Schlang 1998) they are typically scavengers, whereas wolves depend heavily on highly coordinated group hunts for their nourishment (Mech, Smith, and MacNulty 2015). To date, chimpanzees have not yet been tested in this set-up, but considering the highly coordinated group hunts of chimpanzees (Boesch 2002), and their ability to take turns (Martin, Biro, and Matsuzawa 2017), it will be particularly interesting to see how chimpanzees perform in this set-up.

Future Directions

The ravens in our experiment initially failed the knowledge probes that are typically conducted when testing the loose-string paradigm (Massen, Ritter, and Bugnyar 2015). After considerable experience, however, they learned to wait for a partner (Asakawa-Haas et al. 2016), and in the current study, we showed that they seem to understand the need for a partner in a forced-choice set-up. We also showed that they could learn to coordinate when more than one task was presented simultaneously. The critical point here is experience, and therefore, we stress that when comparisons are made between species, the difference in experience (or training) between these species needs to be considered too (table 16.1). Generally, we advise against comparisons about "intelligence" based on conclusions drawn from experiments using such a variety of experience in the procedure. If one chooses to do so, we suggest refraining from (operant) training prior to the task, and recommend investigating the physical cognition required for the task with much more detail (see also Albiach-Serrano 2015). For example, when animals (learn to) wait for a partner, what are they waiting for, the partner or tension in the rope they are holding? To study the cooperative abilities of different species, which inevitably needs to be done in rather artificial set-ups, we thus need to adapt our paradigms to their physical cognition and give them sufficient experience with the paradigm at hand.

Acknowledgments

We are very grateful to the editors of this volume, Lydia Hopper and Steve Ross, for including a chapter on "feathered apes" in this book about chimpanzees. Similarly, J. J. M. M. thanks Lydia and Steve for inviting him to speak at the 4th Understanding Chimpanzees: Chimpanzees in Context symposium. Finally, we thank Kenji Asakawa-Haas, Martina Schiestl, Diana Neuerer, Lisa Stadler, and Ricarda Resch for their help with conducting the experiments, all animal keepers for their care for our animals, Nadja Kavcik-Graumann for the drawings of our set-ups, and Sarah Marshall-Pescini for insightful discussions about test set-ups. These studies were funded by the Austrian Science Fund (FWF); projects M1351-B17 and P26806-B22 granted to J. J. M. M. and Y366-B17 granted to T. B.

Ethics Statement

Participation in our experiments was on a voluntary basis and birds were never forced to enter a test compartment. Moreover, the ravens in these experiments were never food deprived, and water was available ad libitum during and outside of the experiments. The design of our experiments meets the latest Association for the Study of Animal Behaviour/ Animal Behavior Society guidelines, complies with Austrian law, and was evaluated and approved by the ethical board of the behavioral research group at the Faculty of Life Science of the University of Vienna (case number: 2015-003).

References

Albiach-Serrano, A. 2015. "Cooperation in primates: A critical, methodological review." *Interaction Studies* 16: 361–82.

Asakawa-Haas, K., M. Schiestl, T. Bugnyar, and J. J. M. Massen. 2016. "Partner choice in raven (*Corvus corax*) cooperation. " *PLoS One* 11: e0156962.

Bednarz, J. C. 1988. "Cooperative hunting in Harris' Hawks (*parbuteo unicinctus*)." *Science* 239: 1525–27.

Boeckle, M., and T. Bugnyar. 2012. "Long-term memory for affiliates in ravens." *Current Biology* 22 (9): 801–6.

Boesch, C. 1991. "The effects of leopard predation on grouping patterns in forest chimpanzees." *Behaviour* 117: 220–41.

Boesch, C. 2002. "Cooperative hunting roles among Tai chimpanzees." *Human Nature* 13: 27–46.

Bossema, I., and R. F. Benus. 1985. "Territorial defence and intra-pair cooperation in the carrion crow (*Corvus corone*)." *Behavioral Ecology and Sociobiology* 16: 99–104.

Braun, A., and T. Bugnyar. 2012. "Social bonds and rank acquisition in raven nonbreeder aggregations." *Animal Behaviour* 84: 1507–15.

Bshary, R., and R. Noë. 1997. "Red colobus and Diana monkeys provide mutual protection against predators." *Animal Behaviour* 54 (6): 1461–74.

Bugnyar, T. 2008. "Animal cognition: Rooks team up to solve a problem." *Current Biology* 18: R530–32.

Bugnyar, T., and K. Kotrschal. 2002. "Observational learning and the raiding of food caches in ravens, *Corvus corax*: Is it 'tactical' deception?" *Animal Behaviour* 64: 185–95.

Bugnyar, T., S. Reber, and C. Buckner. 2016. "Ravens attribute visual access to unseen competitors." *Nature Communications* 7: 10506.

Bullinger, A. F., A. P. Melis, and M. Tomasello. 2011. "Chimpanzees, *Pan troglodytes*, prefer individual over collaborative strategies towards goals." *Animal Behaviour* 82: 1135–41.

Chalmeau, R., K. Lardeux, P. Brandibas, and A. Gallo. 1997. "Cooperative problem solving by orangutans (*Pongo pygmaeus*)." *International Journal of Primatology* 18: 23–32.

Chalmeau, R., E. Visalberghi, and A. Gallo. 1997. "Capuchin monkeys, *Cebus apella*, fail to understand a cooperative task." *Animal Behaviour* 54: 1215–25.

Claidière, N., A. Whiten, M. C. Mareno, E. J. E. Messer, S. F. Brosnan, L. M. Hopper, S. P. Lambeth, S. J. Schapiro, and N. McGuigan. 2015. "Selective and contagious prosocial resource donation in capuchin monkeys, chimpanzees and humans." *Scientific Reports* 5: 7631.

Clayton, N. S., and N. J. Emery. 2007. "The social life of corvids." *Current Biology* 17: R652–56.

Clifton, K. E. 1990. "The cost and benefits of territory sharing for the Caribbean coral reef fish, *Scarus iserti*." *Behavioral Ecology and Sociobiology* 26: 139–47.

Clutton-Brock, T. 2009. "Cooperation between non-kin in animal societies." *Nature* 462: 51–57.

Cornell, H. N., J. M. Marzluff, and S. Pecoraro. 2012. "Social learning spreads knowledge about dangerous humans among American crows." *Proceedings of the Royal Society of London B* 279: 499–508.

Crawford, M. P. 1937. "The cooperative solving of problems by young chimpanzees." *Comparative Psychology Monographs* 14: 1–88.

Cronin, K. A., A. V. Kurian, and C. T. Snowdon. 2005. "Cooperative problem solving in a cooperatively breeding primate (*Saguinus oedipus*)." *Animal Behaviour* 69: 133–42.

de Waal, F. B. M. 1982. *Chimpanzee Politics: Power and Sex among Apes*. Baltimore: Johns Hopkins University Press.

de Waal, F. B. M., and A. van Roosmalen. 1979. "Reconciliation and consolation among chimpanzees." *Behavioral Ecology and Sociobiology* 5: 55–66.

Dey, C. J., C. M. O'Conner, H. Wilkinson, S. Schultz, S. Balshine, and J. L. Fitzpatrick. 2017. "Direct benefits and evolutionary transitions to complex societies." *Nature Ecology & Evolution* 1: 0137.

Di Lascio, F., F. Nyffeler, R. Bshary, and T. Bugnyar. 2013. "Ravens (*Corvus corax*) are indifferent to the gains of conspecific recipients or human partners in experimental tasks." *Animal Cognition* 16: 35–43.

Drea, C. M., and A. N. Carter. 2009. "Cooperative problem solving in a social carnivore." *Animal Behaviour* 78: 967–77.

Eskelinen, H. C., K. A. Winship, B. L. Jones, A. E. M. Ames, and S. A. Kuczaj. 2016. "Acoustic behavior associated with cooperative success in bottlenose dolphins (*Tursiops truncatus*)." *Animal Cognition* 19: 789–97.

Fady, J. C. 1972. "Absence of cooperation of the instrumental type under natural conditions in *Papio papio*." *Behaviour* 43: 157–64.

Fitch, W. T., L. Huber, and T. Bugnyar. 2010. "Social cognition and the evolution of language: Constructing cognitive phylogenies." *Neuron* 65: 795–814.

Fraser, O. N., and T. Bugnyar. 2010a. "The quality of social relationships in ravens." *Animal Behaviour* 79: 927–33.

Fraser, O. N., and T. Bugnyar. 2010b. "Do ravens show consolation? Responses to distressed other." *PLoS One* 5: e10605.

Fraser, O. N., and T. Bugnyar. 2011. "Ravens reconcile after conflicts with valuable partners." *PLoS One* 6: e18118.

Fraser, O. N., and T. Bugnyar. 2012. "Reciprocity of agonistic support in ravens." *Animal Behaviour* 83: 171–77.

Fraser, O. N., S. E. Koski, R. M. Wittig, and F. Aureli. 2009. "Why are bystanders friendly to recipients of aggression?" *Communicative & Integrative Biology* 2: 285–91.

Fraser, O. N., G. Schino, and F. Aureli. 2008. "Components of relationship quality in chimpanzees." *Ethology* 114: 834–43.

Gersick, A. S., D. L. Cheney, J. M. Schneider, R. M. Seyfarth, and K. E. Holekamp. 2015. "Long-distance communication facilitates cooperation among wild spotted hyaenas, *Crocuta crocuta*." *Animal Behaviour* 103: 107–16.

Goodall, J. 1986. *The Chimpanzees of Gombe: Patterns of Behavior*. Cambridge, MA: Harvard University Press.

Güntürkün, O., and T. Bugnyar. 2016. "Cognition without cortex." *Trends in Cognitive Sciences* 20: 291–303.

Hamilton, W. D. 1964. "The genetical evolution of social behaviour I-II." *Journal of Theoretical Biology* 7: 1–52.

Hare, B., J. Call, and M. Tomasello. 2006. "Chimpanzees deceive a human competitor by hiding." *Cognition* 101: 495–514.

Hare, B., A. P. Melis, V. Woods, S. Hastings, and R. W. Wrangham. 2007. "Tolerance allows bonobos to outperform chimpanzees on a cooperative task." *Current Biology* 17: 619–23.

Heaney, M., R. Gray, and A. H. Taylor. 2017. "Keas perform similarly to chimpanzees and elephants when solving collaborative tasks." *PLoS One* 2: e0169799.

Hemelrijk, C. K., and A. Ek. 1991. "Reciprocity and interchange of grooming and 'support' in captive chimpanzees." *Animal Behaviour* 41: 923–35.

Hendricks, P., and S. Schlang. 1998. "Aerial attacks by common ravens, *Corvus corax*, on adult feral pigeons, *Columba livia*." *Canadian Field Naturalist* 112: 702–3.

Hirata, S. 2003. "Cooperation in chimpanzees." *Hattatsu* 95: 103–11.

Hirata, S., and K. Fuwa. 2007. "Chimpanzees (*Pan troglodytes*) learn to act with other individuals in a cooperative task." *Primates* 48: 13–21.

Horner, V., D. Carter, M. Suchak, and F. B. M. de Waal. 2011. "Spontaneous prosocial choice by chimpanzees." *Proceedings of the National Academy of Sciences* 108: 13847–51.

House, B. R., J. B. Silk, S. P. Lambeth, and S. J. Schapiro. 2014. "Task design influences prosociality in captive chimpanzees (*Pan troglodytes*)." *PLoS One* 9: e103422.

Jensen, K., B. Hare, J. Call, and M. Tomasello. 2006. "What's in it for me? Self-regard precludes altruism and spite in chimpanzees." *Proceedings of the Royal Society B: Biological Sciences* 273 (1589): 1013–21.

Lambert, M. L., J. J. M. Massen, A. M. Seed, T. Bugnyar, and K. E. Slocombe. 2017. "An 'unkindness' of ravens? Measuring prosocial preferences in *Corvus corax*." *Animal Behaviour* 123: 383–93.

Lukas, D., and T. Clutton-Brock. 2013. "Life histories and the evolution of cooperative breeding in mammals." *Proceedings of the Royal Society of London B* 279: 4065–70.

Marshall-Pescini, S., J. F. L. Schwarz, I. Kostelnik, Z. Virányi, and F. Range. 2017. "Importance of a species' socioecology: Wolves outperform dogs in a conspecific cooperation task." *Proceedings of the National Academy of Sciences* 114 (44): 11793–98.

Martin, C. F., D. Biro, and T. Matsuzawa. 2017. "Chimpanzees spontaneously take turns in a shared serial ordering task." *Scientific Reports* 7: 14307.

Martin, J. S., S. E. Koski, T. Bugnyar, A. V. Jaeggie, and J. J. M. Massen. In prep. "Prosociality enhances cooperation in common marmosets (*Callithrix jacchus*)."

Massen, J. J. M., M. Lambert, M. Schiestl, and T. Bugnyar. 2015. "Subadult ravens generally don't transfer valuable tokens to conspecifics when there is nothing to gain for themselves." *Frontiers in Comparative Psychology* 6: 885.

Massen, J. J. M., A. Pasukonis, J. Schmidt, and T. Bugnyar. 2014a. "Ravens notice dominance reversals among conspecifics within and outside their social group." *Nature Communications* 5: 3679.

Massen, J. J. M., C. Ritter, and T. Bugnyar. 2015. "Tolerance and reward equity predict cooperation in ravens (*Corvus corax*)." *Scientific Reports* 5: 1–11.

Massen, J. J. M., G. Szipl, M. Spreafico, and T. Bugnyar. 2014b. "Ravens intervene in others' bonding attempts." *Current Biology* 24: 1–4.

Mech, L. D., D. W. Smith, and D. R. MacNulty. 2015. *Wolves on the Hunt: The Behavior of Wolves Hunting Wild Prey*. Chicago: University of Chicago Press.

Melis, A. P., B. Hare, and M. Tomasello. 2006a. "Chimpanzees recruit the best collaborators." *Science* 311: 1297–1300.

Melis, A. P., B. Hare, and M. Tomasello. 2006b. "Engineering cooperation in chimpanzees: Tolerance constraints on cooperation." *Animal Behaviour* 72: 275–86.

Mendonça, R. S., C. D. Dahl, S. Carvalho, T. Matsuzawa, and I. Adachi. 2018. "Touch-screen-guided task reveals a prosocial choice tendency by chimpanzees (*Pan troglodytes*)." *PeerJ* 6: e5315

Mendres, K. A., and F. B. M. de Waal. 2000. "Capuchins do cooperate: The advantage of an intuitive task." *Animal Behaviour* 60 (4): 523–29.

Mielke, A., L. Samuni, A. Preis, J. F. Gogarten, C. Crockford, and R. Wittig. 2017. "Bystanders intervene to impede grooming in western chimpanzees and sooty mangabeys." *Royal Society Open Science* 4: 171296.

Milinski, M., J. H. Lüthi, R. Eggler, and G. A. Parker. 1997. "Cooperation under risk: Experiments on costs and benefits." *Proceedings of the Royal Society of London B* 264: 831–37.

Molesti, S., and B. Majolo. 2016. "Cooperation in wild barbary macaques: Factors affecting free partner choice." *Animal Cognition* 19: 133–46.

Müller, J. J. A., J. J. M. Massen, T. Bugnyar, and M. Osvath. 2017. "Ravens remember the nature of a single reciprocal interaction sequence over 2 days and even after a month." *Animal Behaviour* 128: 69–78.

Olkowicz, S., M. Kocourek, R. K. Lučan, M. Porteš, W. T. Fitch, S. Herculano-Houzel, and P. Němec. 2016. "Birds have primate-like numbers of neurons in the forebrain." *Proceedings of the National Academy of Sciences* 113: 7255–60.

Ostojić, L., and N. S. Clayton. 2014. "Behavioural coordination of dogs in a cooperative problem-solving task with a conspecific and a human partner." *Animal Cognition* 17: 445–59.

Péron, F., L. Rat-Fisher, M. Lalot, and D. Bovet. 2011. "Cooperative problem solving in African grey parrots (*Psittacus erithacus*)." *Animal Cognition* 14: 545–53.

Petit, O., C. Desportes, and B. Thierry. 1992. "Differential probability of coproduction in 2 species of macaque (*Macaca tonkeana, Macaca mulatta*)." *Ethology* 90: 107–20.

Pitman, R. L., and J. W. Durban. 2012. "Cooperative hunting behavior, prey selectivity and prey handling by pack ice killer whales (*Orcinus* orca), type B, in Antarctic Peninsula waters." *Marine Mammal Science* 28: 16–36.

Plotnik, J. M., R. Lair, W. Suphachoksahakun, and F. B. M. de Waal. 2011. "Elephants know when they need a helping trunk in a cooperative task." *Proceedings of the National Academy of Sciences* 108: 5116–21.

Rekers, Y., D. B. Haun, and M. Tomasello. 2011. "Children, but not chimpanzees, prefer to collaborate." *Current Biology* 21: 1756–58.

Riehl, C. 2013. "Evolutionary routes to non-kin cooperative breeding in birds." *Proceedings of the Royal Society of London B* 280: 20132245.

Romero, T., and F. Aureli. 2008. "Reciprocity of support in coatis (Nasua nasua)." *Journal of Comparative Psychology* 122: 19–25.

Scheid, C., and R. Noë. 2010. "The performance of rooks in a cooperative task depends on their temperament." *Animal Cognition* 13 (3): 545–53.

Schino, G. 2007. "Grooming and agonistic support: A meta-analysis of primate reciprocal altruism." *Behavioral Ecology* 18: 115–20.

Schmelz, M., S. Duguid, M. Bohn, and C. J. Völter. 2017a. “Cooperative problem solving in giant otters (*Pteronura brasiliensis*) and Asian small-clawed otters (*Aonyx cinerea*).” *Animal Cognition* 20: 1107–14.

Schmelz, M., S. Grueneisen, A. Kabalak, J. Jost, and M. Tomasello. 2017b. “Chimpanzees return favors at a personal cost.” *Proceedings of the National Academy of Sciences* 114: 7462–67.

Schwing, R., E. Jocteur, A. Wein, R. Noë, and J. J. M. Massen. 2016. “Kea cooperate better with sharing affiliates.” *Animal Cognition* 19: 1093–1102.

Seed, A. M., N. S. Clayton, and N. J. Emery. 2008. “Cooperative problem solving in rooks (*Corvus frugilegus*).” *Proceedings of the Royal Society of London B* 275: 1421–29.

Silk, J. B., S. F. Brosnan, J. Vonk, J. Henrich, D. J. Povinelli, A. S. Richardson, S. P. Lambeth, J. Mascaro, and S. J. Schapiro. 2005. “Chimpanzees are indifferent to the welfare of unrelated group members.” *Nature* 437: 1357–59.

Slocombe, K. E., and K. Zuberbühler. 2007. “Chimpanzees modify recruitment screams as a function of audience composition.” *Proceedings of the National Academy of Sciences* 104: 17228–33.

Stocker, M., E. H. M. Sterck, T. Bugnyar, and J. J. M. Massen. In prep. “Cooperation with closely bonded individuals reduces cortisol levels in long-tailed macaques.”

Suchak, M., T. M. Eppley, M. W. Campbell, and F. B. M. de Waal. 2014. “Ape duos and trios: Spontaneous cooperation with free partner choice in chimpanzees.” *PeerJ* 2: e417.

Suchak, M., J. Watzek, L. F. Quarles, and F. B. M. de Waal. 2018. “Novice chimpanzees cooperate successfully in the presence of experts, but may have limited understanding of the task.” *Animal Cognition* 21: 87–98.

Tebbich, S., M. Taborsky, and H. Winkler. 1996. “Social manipulation causes cooperation in keas.” *Animal Behaviour* 52: 1–10.

Tennie, C., K. Jensen, and J. Call. 2016. “The nature of prosociality in chimpanzees.” *Nature Communications* 7: 13915.

Vail, A. L., A. Manica, and R. Bshary. 2013. “Referential gestures in fish collaborative hunting.” *Nature Communications* 4 (1): 1–7.

Vonk, J., S. F. Brosnan, J. B. Silk, J. Henrich, A. S. Richardson, S. P. Lambeth, S. J. Schapiro, and D. J. Povinelli. 2008. “Chimpanzees do not take advantage of very low cost opportunities to deliver food to unrelated group members.” *Animal Behaviour* 75: 1757–70.

Warneken, F., and M. Tomasello. 2006. “Altruistic helping in human infants and young chimpanzees. *Science* 311: 1301–3.

Werdenich, D., and L. Huber. 2002. “Social factors determine cooperation in marmosets.” *Animal Behaviour* 64: 771–78.

Yamamoto, S., T. Humle, and M. Tanaka. 2009. “Chimpanzees help each other upon request.” *PLoS One* 4: e7416.

Yamamoto, S., and M. Tanaka. 2010. “The influence of kin relationship and reciprocal context on chimpanzees’ other-regarding preferences.” *Animal Behaviour* 79: 595–602.

PART SIX

Tool Use, Cognition, and Culture

17

Extractive Foraging in an Extreme Environment: Tool and Proto-tool Use by Chimpanzees at Fongoli, Senegal

JILL D. PRUETZ, STEPHANIE L. BOGART, AND STACY LINDSHIELD

Introduction

Chimpanzees (*Pan troglodytes*) are dispersed across sub-Saharan Africa (Humle et al. 2016) in a range of habitat types from tropical forests to grassland savannas (Russak and McGrew 2008). Most research sites across Africa have a mean annual rainfall greater than 1,500 mm and are considered forest-woodland habitat types, including Gombe and Mahale, Tanzania; Budongo and Kibale, Uganda; Kahuzi-Biega, Democratic Republic of Congo; Lope, Gabon; Goualougo Triangle, Republic of Congo; Gashaka, Nigeria; Bossou, Guinea; and Taï, Côte d'Ivoire (for reviews see Bogart 2009; Russak and McGrew 2008). From here on, these chimpanzees from these sites are referred to as "forest-dwelling." Thus far, four sites have a mean annual rainfall of less than 1,500 mm and are characterized as savanna-woodland habitat types: Fongoli and Mount Assirik, Senegal; Ugalla, Tanzania; and Semliki, Uganda (for reviews see Bogart 2009; Ogawa et al. 2007; Russak and McGrew 2008). Chimpanzees from these sites are labelled "savanna-dwelling." Only Fongoli currently has habituated chimpanzees with long-term data on their behavior. Fongoli chimpanzees face pressures in their savanna-woodland habitat in southeastern Senegal including intense heat, scattered foods, and water scarcity (Bertolani and Pruetz 2011; Bogart and Pruetz 2008, 2011; Pruetz 2006, 2007; Pruetz and Bertolani 2009; Pruetz and LaDuke 2010; Skinner and Pruetz 2012; Sponheimer et al. 2006; Wessling et al. 2018). These apes appear to rely on tool and proto-tool use during foraging to successfully combat such pressures.

In this chapter we focus on the roles of tool and proto-tool use in the chimpanzees' foraging behavior of *Macrotermes* termites, baobab fruit (*Adansonia digitate*), and bushbabies (*Galago senegalensis*). Termites and baobab fruit encompass a large portion of the Fongoli chimpanzees' diet, while *Galago* are hunted less frequently, and using tools. Fongoli chimpanzees are the

TABLE 17.1. Foraging hypotheses and associated predictions related to tool and proto-tool use as they relate to Fongoli chimpanzees.

Hypothesis	*Description*	*Study Predictions*
Opportunity[1,2]	Favorable environments facilitate tool use. Tool use occurs in environments where individuals routinely encounter problems that are solvable with tools.	1. Tool use frequency is positively correlated with the abundance of foods targeted with tools, and 2. Tool use frequency is unrelated to the availability of other foods.
Necessity[1,2]	Tool use is a strategy that remedies food scarcity. Following innovation, tool use continues to occur when it buffers against food scarcity.	Tool use increases as the availability of other foods decreases.
Relative profitability[3,4]	Tools are used to acquire foods that are relatively high in energy or nutrients.	Foods acquired with tools are more profitable than foods that are accessible without tools at that time.

[1]Fox, Sitompul, and van Schaik 1999, [2]Spagnoletti et al. 2012, [3]Rutz and St. Clair 2012, [4]Sanz and Morgan 2013.

only apes recorded to hunt systematically with tools, with almost 500 cases recorded. Specifically, in this chapter we examine hypotheses regarding three prominent foraging (proto)tool behaviors at Fongoli: tool-assisted hunting, termite fishing, and baobab cracking. We explore the necessity, opportunity, and relative profitability hypotheses to examine foraging behaviors, presented in table 17.1.

The *necessity hypothesis* maintains that foraging tool behaviors stem from ecological or environmental pressures on an animals' fitness, such that the lack of preferred resources instigates the need to use tools as a fallback strategy (Fox, Sitompul, and van Schaik 1999; Lambert 2007). The *opportunity hypothesis* posits that such foraging tool behaviors rely on encounter rates with tool-acquired resources and may contribute to an individual's fitness but is less constrained (Fox, Sitompul, and van Schaik 1999). Finally, the *relative profitability hypothesis* puts the target behavior in the context of other available resources and specifically suggests the energetic gains achieved by embedded foods exceed those of non-embedded foods (i.e., using tools allows individuals to gain better rewards than could be otherwise obtained: Rutz and St. Clair 2012; Sanz and Morgan 2013). As Gruber (2013) pointed out, rather than considering hypotheses in isolation, a dynamic model that considers two or more of them together likely wields better explanatory power, especially given the anthropogenic changes that characterize most wild primate habitats today.

Although tool and proto-tool use has been studied extensively in nonhuman primates, explanations for such behavior vary according to species as well as ecological context (Pruetz et al. in press). The major hypotheses that attempt to explain tool-assisted feeding and foraging in primates have been tested for the most prolific users at some sites (i.e., chimpanzees, *Pan troglodytes*, orangutans, *Pongo* sp., and capuchins, *Sapajus* sp. and *Cebus* sp.), with most evidence supporting the opportunity hypothesis (chimpanzees: Koops, Furuichi, and Hashimoto 2015; Koops, McGrew, and Matsuzawa 2013; Koops et al. 2015; Sanz and Morgan 2013, orangutans: Koops, Visalberghi, and van Schaik 2014, capuchins: Spagnoletti et al. 2012). However, the necessity hypothesis has been supported for tool-assisted feeding by chimpanzee behavior at Bossou, Guinea (Yamakoshi 1998). While the chimpanzees' tool kit at Fongoli is less varied than that of some forest-living chimpanzees (Pruetz et al. in press), tool and proto-tool behavior is frequent and could be instrumental for their survival in this harsh environment. Senegal's savanna environment poses many pressures that the Fongoli chimpanzees must cope with annually and, therefore, may exacerbate conditions supporting the ecological necessity hypothesis, and predictions based on fruit scarcity. However, previous findings have not supported this hypothesis at Fongoli in regard to termite fishing (Bogart 2009; Bogart and Pruetz 2008).

TOOL-ASSISTED HUNTING

Chimpanzees living in a savanna environment at Fongoli, Senegal, are the only known mammals besides humans that routinely hunt with tools (Pruetz and Bertolani 2007; Pruetz et al. 2015), with almost 500 cases now recorded. Chimpanzees at Fongoli use modified stick tools ("spears") to hunt nocturnal prosimians (*Galago senegalensis*) as they sleep in tree cavities during the day. Additionally, female chimpanzees at Fongoli exhibit this hunting behavior significantly more than expected (Pruetz and Bertolani 2007; Pruetz et al. 2015), which is unlike the male-dominated chimpanzee hunting behavior at all other long-term chimpanzee study sites (Boesch and Boesch-Achermann 2000; Fahy et al. 2013; Goodall 1986; McGrew 1979; Mitani and Watts 2001; Stanford 1998, 1999; Watts and Mitani 2002). Predictions stemming from this hypothesis would thus be that seasonal tool-assisted hunting is correlated negatively with Fongoli chimpanzees' fruit availability (table 17.1).

TERMITE FISHING

Termite fishing is the most common type of tool use at Fongoli, recorded in all months of the year and comprising a large portion of their activity and feeding budget (Bogart 2009; Bogart and Pruetz 2011; Pruetz et al. in press).

The chimpanzees rely heavily on insects in their diet and spend more time foraging and feeding on termites (*Macrotermes subhyalinus*) (23.7% of the adult male feeding and foraging budget) than any other habituated chimpanzee populations that termite fish (Bogart and Pruetz 2011). Annually, 7.4% of the Fongoli male's daily activity budget is spent termite fishing, with a peak during the dry season that can encompass almost 20% of their activity budget time (Bogart and Pruetz 2011). Fongoli chimpanzees concentrate on termites particularly in the late dry season (April) until early wet season (July) when they spend up to 57% of their monthly feeding time on termites, surpassing time spent consuming fruit (Bogart and Pruetz 2011).

Our previous research indicated that fruit scarcity does not correlate with termite fishing at Fongoli (Bogart and Pruetz 2008), and thus does not support the prediction of the necessity hypothesis. Following the opportunity hypothesis, an increase in the availability of termites should positively correlate with termite fishing. Previous data indicated that the *Macrotermes* nest density at Fongoli (23.6 nests per hectare) is greater than (generally double) other sites with comparable data (Bogart and Pruetz 2011), even considering more recent data (Nimba, Guinea: Koops, McGrew, and Matsuzawa 2013, Kasekela/Gombe, Tanzania: O'Malley 2011, Semliki, Uganda: Webster et al. 2014). Thus, we predict the opportunity hypothesis to support termite fishing at Fongoli. Finally, the relative profitability hypothesis is examined in regard to termite fishing, predicting termites living in subterranean mounds provide a nutritious resource. Data to inform this hypothesis will come from profitability estimates and feeding rates presented here.

BAOBAB CRACKING

Chimpanzees at Fongoli (Gašperšič 2008) and Assirik (Marchant and McGrew 2005) in southeastern Senegal routinely crack the hard shell of baobab fruits on stone or wooden anvils. Baobab fruit cracking is classified as proto-tool use because individuals primarily use non-portable anvils, including tree branches, boulders, and outcroppings of laterite, but not movable anvils that can be manipulated during food procurement (Shumaker, Walkup, and Beck 2011). Movable anvils are easily accessible on the savanna-woodland mosaic landscapes in southeastern Senegal, and it is possible for chimpanzees to manipulate such anvils while cracking fruits, but this potential tool-use behavior has not been extensively explored. Thus, baobab cracking resides in the proto-tool use domain. Similar forms of proto-tool use in chimpanzees, also called food pounding or smashing, have been reported at Gombe, Tanzania (Goodall 1986), Bili-Uéré, Democratic Republic of Congo (Hicks et al. 2019), and Taï, Côte d'Ivoire (Luncz and Boesch 2015; Luncz and de Waal, chapter 18 this volume). In this chapter, we report time allocation data for baobab procurement and ingestion, and apply the theoretical frameworks of tool-use ecology to anvil use at Fongoli. Chimpanzees at Fongoli primarily ingest baobab fruit during the fresh fruiting months of November to January, when they allocate most of their feeding time to this food (Lindshield et al. 2017; Pruetz 2006).

Approach

Here, we examine Fongoli chimpanzee tool-assisted hunting, termite fishing, and baobab-cracking behaviors separately in relation to possible explanations for tool and proto-tool use. We place our findings into the broader context of tool use in chimpanzees and other prolific tool users and discuss how the savanna environment may contribute to (proto)tool use at Fongoli. Finally, our analyses of savanna chimpanzee tool use can help inform hypotheses regarding the behavioral ecology of some of the earliest hominins living under similar environmental pressures during the Miocene-Pliocene transition (Pickering 2013; Pickering and Dominguez-Rodrigo 2010; White et al. 2009).

We provide information on Fongoli chimpanzees' tool and proto-tool behavior collected over the course of more than 10 years (2006–2016) in this hot, dry, and open habitat. Research was conducted at the Fongoli site in the Department of Kedougou in Senegal, where chimpanzees live in a Sudano-Guinean savanna-woodland habitat, the northernmost extent of the species' range. The Fongoli community has been studied since April 2001, and adult males were habituated for follows from night nest to night nest in early 2005. Females were calm when in the presence of males, but they were not used as focal subjects to reduce the risk of overhabituation due to the slight, but real, risk that they may be targeted by human hunters in order to acquire their infants for the pet trade (Pruetz and Kante 2010).

The community averaged 32 individuals in 2005–2016 (range 28–34) and at the time of writing numbers 34 chimpanzees (13 adult males, 7 adult females, 14 immature individuals). Adult males served as focal subjects for recording behavior instantaneously at five-minute intervals (Altmann 1974) during all-day focal follows.

TOOL-ASSISTED HUNTING

We used behavioral sampling (Altmann 1974) to record any tool-assisted hunting event during all-day follows of chimpanzees between March 2005 and September 2017. For analyses, however, we used data collected between January 2006, after all individual chimpanzees of the Fongoli community were finally identified, and September 2017 (n = 141 months), during which time we averaged 19.7 observation days per month (calculated from n = 96 months). The variables we collected during tool-assisted hunting included (1) date, (2) time, (3) habitat type (woodland, shrubland, gallery forest, bamboo woodland), (4) location (GPS coordinates), and (5) individual chimpanzee identity. The number of events recorded each month are presented as monthly averages across all years.

TERMITE FISHING

We summarized monthly rates of termite fishing by calculating the proportion of days during which a focal individual engaged in these behaviors, relative to the total number of sampling days per month. For feeding rate, we used video recording when possible in order to estimate the feeding rate (termites per minute). Details of methods can be found in Bogart (2009).

BAOBAB CRACKING

In addition to five-minute instantaneous sampling of focal subjects from 2005 to 2016, we used the continuous recording method (Altmann 1974) to sample the baobab feeding behavior of focal adult males during 537 sampling hours in November 2011–January 2012 and November 2012–January 2013. We recorded the time allocated by adult males to the procurement and ingestion of 106 baobab fruits. For a subsample of these fruits (n = 58), we estimated the relative contribution of baobab cracking to the time that an individual allocated to feeding on a single baobab fruit. Three feeding categories were used in this analysis: baobab fruit cracking, all other baobab procurement behaviors (traveling within a feeding tree, biting a fruit open), and ingesting baobab fruit pulp and/or seed. Our analysis excludes seed reingestion because it follows the complete gut passage of the original fruit (Bertolani and Pruetz 2011). We excluded associations between baobab cracking and fruit phenophase or size in this analysis due to the small number of fruit cracking cases in our sample. We summarized monthly rates of baobab cracking by calculating the proportion of days during which a focal individual engaged in these behaviors, relative to the total number of sampling days per month. Lindshield et al. (2017) provide additional details of the methods we used.

FOOD AVAILABILITY

Phenology data on fruit (ripe/unripe), leaf (immature/mature), and flower availability have been collected since 2003, with early observations published elsewhere (Bogart 2009; Pruetz 2006). Transect length ranged from 2 km and approximately 300 trees, shrubs, and liane monitored in early years of the project to 3.4 km and approximately 900 trees, shrubs, and liane monitored since 2012. We also systematically sample a number of tree species (e.g., *Cola cordifolia*, *Adansonia digitata*) that are underrepresented along the phenology transect. Phenology data is collected monthly, at mid-month.

FOOD PROFITABILITY

The profitability of termites and baobab fruit, relative to other available foods, was measured during continuously recorded feeding bouts of focal adult males during the 2011–2012 and 2012–2013 baobab fruiting seasons. Profitability data were not available for the *Galago* hunting season. Profitability is the amount of metabolizable energy ingested per unit of time (Conklin and Wrangham 1994), given here as kJ g·hr^{-1}. The inclusion of the feeding rate into this calculation accounts for differences in handling and processing times among foods, and provides a more accurate measure of energy intake than ingestion time or energy density per unit of weight alone. Thus, profitability facilitates comparative analyses of foods that highly vary in ingestion rates in addition to energy values. More information about measuring metabolizable energy and feeding rates are provided in Lindshield (2014).

Results

FOOD AVAILABILITY

Fruit is most abundant in the peak dry month of April, as well as May when the first rains of the season usually appear (table 17.2). Relatively low fruit availability occurs during the transition from wet to early dry season (October–January) and during peak wet season (June–July). Yearly variation in monthly fruit availability, given as the standard error of the mean, shows that fruit availability varies most in the late dry season (March–April) and in August, a wet season month.

TABLE 17.2. Monthly fruit availability, shown as the percentage of feeding trees bearing fruit. The number of years per month in this sample is reported in parentheses.

Month	Season*	*Average % feeding trees bearing fruit ± SE*	*Average rainfall mm ± SE*
January (5)	Dry	9.66 ± 1.09	0 ± 0
February (5)	Dry	11.84 ± 1.38	0 ± 0
March (4)	Dry	14.84 ± 3.73	0 ± 0
April (3)	Dry	18.83 ± 3.83	8.1 ± 1.98
May (5)	Transitional	19.38 ± 1.84	59.5 ± 4.73
June (3)	Wet	9.20 ± 2.55	164.4 ± 4.97
July (4)	Wet	9.54 ± 2.49	231.7 ± 9.84
August (4)	Wet	12.99 ± 3.77	214.2 ± 13.79
September (5)	Wet	12.02 ± 2.93	283.2 ± 20.52
October (5)	Transitional	8.85 ± 2.58	99.9 ± 10.72
November (5)	Dry	8.28 ± 1.95	1.3 ± 0.41
December (4)	Dry	10.61 ± 2.85	0 ± 0
Annual monthly average		12.17 ± 1.08	88.5 ± 30.94

*Monthly precipitation and temperature records from Pruetz and Bertolani 2009, Bogart and Pruetz 2011, and Pruetz (unpublished data).

TOOL-ASSISTED HUNTING

Fongoli chimpanzees were observed to use tools to hunt *Galago* prey a total of 490 times between March 2005 (when the first event was recorded) and September 30, 2017 (n = 151 months) and 485 times between January 2006 and September 2017 (n = 141 months; range 18–66 hunts annually), after all individuals in the community had been identified. Most tool-assisted hunting occurred during the early rainy season or the transitional period between the dry and rainy seasons, with the months of June (n = 243 total tool-assisted hunts) and July (n = 152 total tool-assisted hunts) being outliers in this regard (fig. 17.1a).

There was a significant correlation between mean number of days observing chimpanzees per month and the mean number of tool-assisted hunts recorded per month ($R^2 = 0.512$, n = 12, F = 10.487, p = 0.0089). Therefore, we transformed each monthly average into a daily rate based on the average number of days we observed chimpanzees each month and used this rate to calculate a new monthly average for analyses. The uncorrected, overall mean number of tool-assisted hunts per month was 3.38, while the corrected mean was 5.08 (SD = 7.867; upper 95% CI = 10.080, lower 95% CI = 0.084). These weighted values for tool-assisted hunting varied significantly by month (two-sided t-test, p = 0.047).

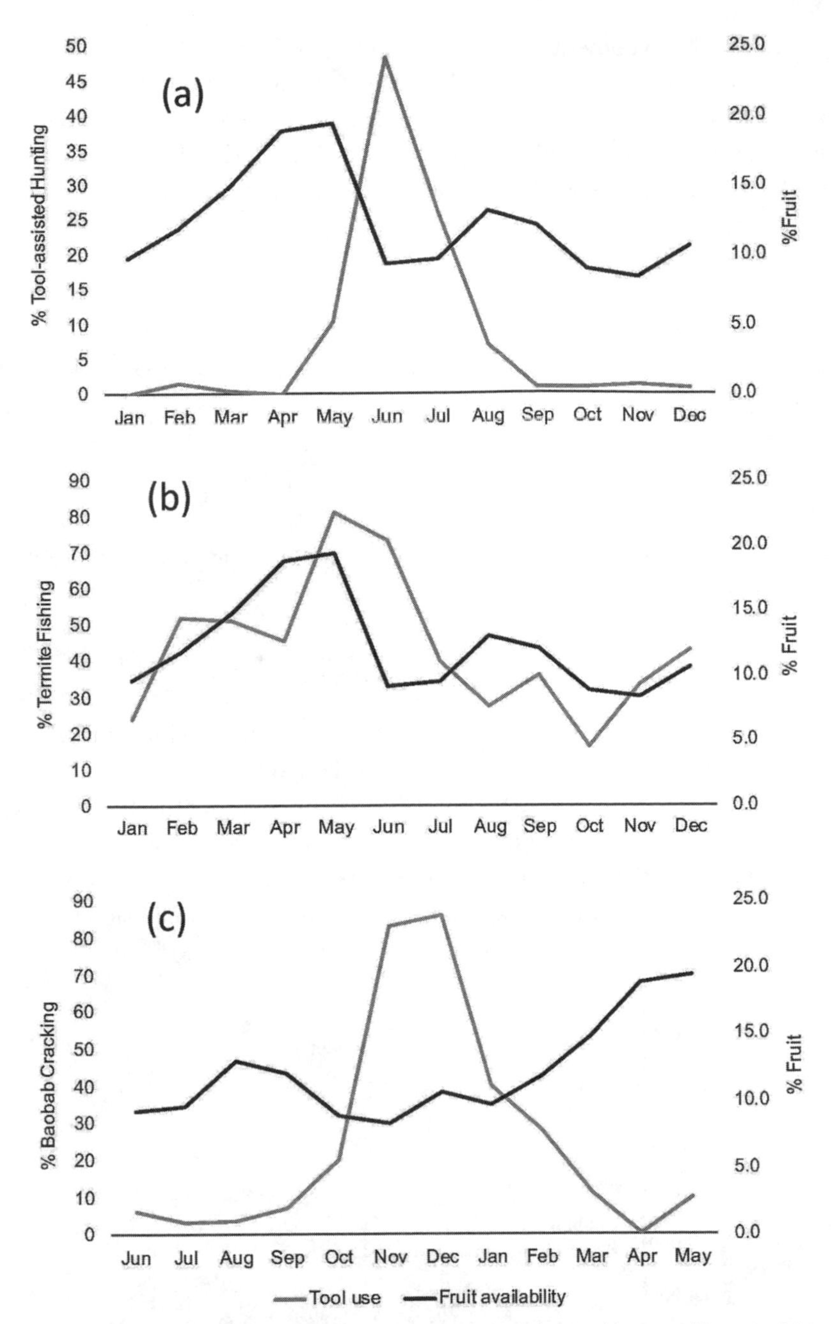

FIGURE 17.1. Monthly percentage of observation days with (a) tool-assisted hunting, (b) termite fishing, and (c) baobab cracking, relative to fruit availability. Peak activity in tool-use is centered on the x-axis of each graph, thus the x-axis of (c) differs from the other graphs.

To test the hypothesis that fruit availability influenced Fongoli chimpanzee tool-assisted hunting, we used these weighted monthly values for tool-assisted hunting in an analysis of variance (ANOVA) and average percentage of fruit available to Fongoli chimps (table 17.2). Results indicated no significant effect of fruit availability on tool-assisted hunting rates ($F = 0.308$, $df = 7$, $p = 0.917$, fig. 17.1a).

TERMITE FISHING

Termites have lower profitability than other resources at Fongoli (fig. 17.2) and the chimpanzees termite fished more frequently from May through July, which coincides with high fruit availability (fig. 17.1b). Furthermore, and as can be seen in fig. 17.1b, termite fishing rates follow the general trend of monthly fruit availability at Fongoli. A total of 33 bouts of termite fishing were analyzed for feeding rates on termites. The average rate of intake was 6.32 termites per minute (SD 2.03, range: 0.11–32.5).

BAOBAB CRACKING

Ripe baobab fruit was well above average in profitability, while unripe baobab fruit was around the average (fig. 17.2). Fongoli chimpanzees primarily ingested baobab fruit during the fresh fruiting season months of November to January (70% of observations), but they also ingested baobab fruit during most months of the year (fig. 17.1c). The peak baobab season coincided with below-average monthly fruit availability, and the onset of baobab cracking

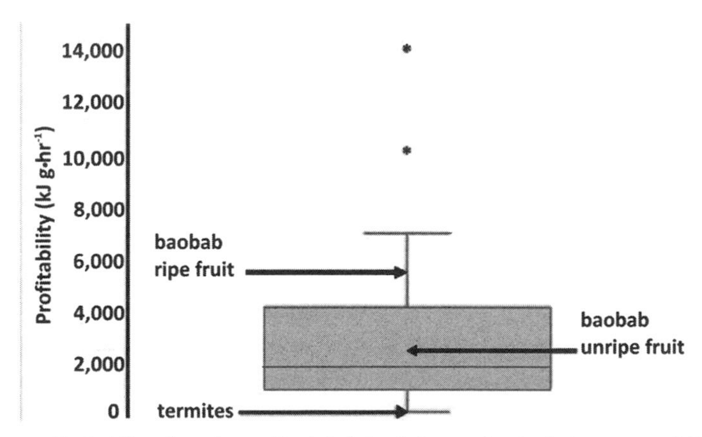

FIGURE 17.2. Profitability of termites and baobab fruit relative to other foods that were available during the 2011–2012 and 2012–2013 baobab seasons.

TABLE 17.3. Time allocated (decimal hours) to baobab fruit cracking relative to overall baobab feeding (%) in adult male chimpanzees at Fongoli. Feeding behaviors are partitioned into ingestive and non-ingestive components.

	Procurement time (in hours)		
	Fruit Cracking	*Other Non-ingesting*	*Fruit Ingesting*
Average (%)	0.005 (5.1)	0.053 (18.2)	0.153 (76.7)
SD	0.004	0.159	0.188
Range (%)	0.0008–0.0210 (0.0–25.0)	0.0003–0.9235 (0.4–73.3)	0.0131–0.8587 (26.1–98.0)
N	58	57	58

and ingestion of unripe fruit occurred in November, the month of the lowest fruit availability (table 17.2).

Fongoli chimpanzees opened baobab fruits with and without anvils, and routinely and concomitantly used biting and cracking actions to open a single fruit. Anvil use was not required for baobab fruit ingestion, as we observed 48 cases where a chimpanzee opened a fruit with biting actions only. Of these cases, 35% (n = 17) involved unripe fruit, 13% (n = 6) were semi-ripe, and the remaining 52% (n = 25) were ripe. About one-quarter of baobab feeding time was allocated to procurement, and non-cracking procurement comprised the majority of this fraction (table 17.3). Overall, baobab cracking comprised a small percentage (5%) of total feeding time.

Discussion

While the chimpanzees' tool kit at Fongoli is less varied than that of some forest-living chimpanzees, tool and proto-tool behavior is frequent and could be instrumental for their survival in this harsh environment. Our results indicate that no single hypothesis—necessity, opportunity, or relative profitability—adequately explains their behavior. We found that the opportunity hypothesis is best supported for tool-assisted hunting and termite fishing, even though the necessity hypothesis might be expected given the pressures of the savanna-mosaic habitat. In contrast, we found that the necessity (and possibly the relative profitability) hypothesis may better explain baobab cracking.

Based on phenological data on food availability and the monthly frequency of *Galago* hunting, the scarcity of ripe fruit at Fongoli does not appear to explain the chimpanzees' tool-assisted hunting. Tool-assisted hunting behavior is concentrated in the early rainy season, whereas the period of fruit scarcity follows the peak hunting season. Although the months with most frequent tool-assisted hunting were relatively low in fruit abundance (table 17.2), other months had equally scarce or lower availability of fruits for Fongoli chimpanzees. January's average fruit availability was similar to June and July, when 81% of all recorded tool-assisted hunting occurred, while August and September were the months characterized by greatest fruit scarcity at Fongoli. No records of tool-assisted hunting have been recorded in the month of January (n = 11 years, 2006–2017), while the months of August (n = 24 total tool-assisted hunts) and September (n = 10 total tool-assisted hunts) comprise 7% of all tool-assisted hunting recorded between January 2006 and September 2017.

Tool-assisted hunting by chimpanzees at other sites has been observed only rarely (three times at Mahale, Tanzania: Nakamura and Itoh 2008). However, hunting and meat-eating at other sites indicate patterns exhibited by forest-dwelling chimpanzees that could inform our understanding of the seasonal tool-assisted hunting at Fongoli. At Ngogo, Uganda, Mitani and Watts (2001) found that chimpanzees hunted primarily during times of food abundance and concluded that findings supported the male social bonding hypothesis, as they were more likely to hunt when accompanied by other males. These authors rejected energetic or reproductive explanations for chimpanzee hunting behavior (Mitani and Watts 2001). However, Tennie, O'Malley, and Gilby (2014) concluded that even meat scraps provided beneficial nutrients to chimpanzees, albeit with risks involved. The tool-assisted hunting of *Galago* prey by Fongoli chimpanzees is almost certainly less energetically expensive and physically less risky than chasing down most other types of vertebrate prey, and also allows low-ranking individuals access to meat, but this fails to explain the seasonal emphasis of this behavior at Fongoli. In their review, Tennie, O'Malley, and Gilby (2014) characterize vertebrate prey as a resource that is available year-round, but this may not be the case if *Galago* nesting behavior varies according to season (see below). Additionally, rather than examining the availability of all fruits for Fongoli chimpanzees, consideration of the top foods in the diet may better explain the pattern of tool-assisted hunting at Fongoli.

Neither the necessity hypothesis nor the relative profitability hypothesis are supported for Fongoli chimpanzee termite fishing. It is surprising that these chimpanzees spend so much time consuming a resource with apparently very little gain. However, *Macrotermes* nests are abundant at Fongoli, more so than any other site with comparable data, thus, supporting the opportunity hypothesis. Previous research indicated a positive correlation of termite fishing and mean temperatures, which suggested that during times of intense heat this stationary activity using little energy would be valuable in coping in this environment, in addition to nutrient and protein intake (Bogart and Pruetz 2011).

Chimpanzees at five long-term research sites use tools to access termites: Fongoli, Senegal; Bossou, Guinea (rare); Goualougo, Republic of Congo; Gombe, Tanzania, and Mahale, Tanzania (reviewed in Pruetz et al. in press). Sanz and Morgan (2013) provide a compilation of the data available at most of these sites to offer insight into foraging hypotheses. Results from research at Mahale, Gombe, and Goualougo support the opportunity hypothesis (Sanz and Morgan 2013), while Bossou termite fishing better fits the necessity hypothesis (Yamakoshi 1998). Notably, interpretations of the chimpanzees' patterns at Gombe and Mahale were inferred using rainfall and insect activity, not fruit abundance. Termite fishing by the Fongoli chimpanzees follows the pattern found at most other sites of supporting the opportunity hypothesis. However, indirect data gathered from two Central African sites indicate termite fishing there fits under the necessity hypothesis. Chimpanzee termite fishing increased during periods of low fruit abundance at La Belgique, Cameroon (Deblauwe 2009), and Ndoki Forest, Republic of Congo (combining data from Hashimoto et al. 2003 and Suzuki, Kuroda, and Nishihara 1995). Of further note, chimpanzees at Semliki, Uganda, do not termite fish despite the relative abundance of *Macrotermes* mounds (Webster et al. 2014), and nor do chimpanzees at Gashaka, Nigeria (Buba et al. 2016), suggesting more than just opportunity may influence this tool use behavior.

The relative profitability hypothesis has been less researched. However, termite fishing by chimpanzees at Goualougo (Sanz and Morgan 2013), Gombe (O'Malley and Power 2014), and La Belgique (Deblauwe and Janssens 2008) provide data to suggest this hypothesis. For example, O'Malley and Power (2014) discussed that *Macrotermes* termites at Gombe are nutritionally important for specific minerals, fat, and protein when accounting for total daily intake (see Rothman et al. 2014 for nutritional review). Here, we used metabolizable energy ingested per unit of time (kJ $g{\cdot}hr^{-1}$) for termite soldiers and workers, and did not consider essential micronutrients, which may explain the discrepancies between Fongoli and these other sites regarding termite eating. More research on daily and seasonal macro- and micro-nutrient intake would elucidate the relative profitability hypothesis regarding termite fishing in the future.

Baobab cracking occurred during most months of the year, presumably because this desiccated ripe fruit preserves for several months in the semiarid environment at Fongoli (Lindshield et al. 2017). Cracking was concentrated during the months of the baobab fresh fruiting season when other types of fruit were relatively scarce. Several lines of evidence support the necessity hypothesis for baobab cracking, although other hypotheses cannot be excluded. While Fongoli chimpanzees were more likely to encounter baobab fruit during the peak months of fruit production, thus increasing encounter opportunity, overall fruit availability was concomitantly low. Baobab unripe fruit was about average in profitability, which lends support for the opportunity hypothesis. However, the relatively profitability hypothesis was consistent with our findings on ripe baobab fruit. Scarcity is the strongest explanation for baobab cracking at Fongoli overall, the relative profitability hypothesis is supported for ripe baobab fruit, and the opportunity hypothesis was supported by the seasonal nature of baobab cracking and relative profitability of unripe fruit.

Baobab fruit cracking facilitates access to an energy- and nutrient-dense food (Lindshield et al. 2017), and our time-allocation findings show that cracking is an efficient feeding strategy. This behavior was first reported in unhabituated chimpanzees at Assirik in Niokolo Koba National Park (Baldwin 1979) and later at Fongoli (Gašperšič 2008). Fongoli and Assirik chimpanzees are separated by a distance of about 50 km, and the social transmission of this behavior is theoretically possible via dispersal corridors in the region. Although anvil use at Fongoli usually involved unripe and ripe baobab fruit, immature chimpanzees have cracked open *Strychnos spinosa* fruits, and an edentulous and elderly male opened *Saba senegalensis* fruits using anvils (Pruetz, pers. observation). While we showed that fruit cracking is not essential for accessing baobab pulp and seed, in many cases, anvil use could be more efficient than biting or prying open fruits with teeth and hands alone. Moreover, anvil use could minimize tooth wear and damage during the life course of an individual (Boesch and Boesch 1982).

Future Directions

Chimpanzees in Senegal live at the northern edge of the species' geographical range, and they cope annually with extreme pressures such as high heat, relatively scarce and scattered food resources, and water scarcity. Using savanna chimpanzees in a relational form of a referential model (sensu Moore 1996), where they are compared to forest-dwelling chimpanzees, allows us to identify potential selective pressures that could have shaped hominin adaptation in an open, hot, and dry environment.

Alternative explanations for the seasonal tool-assisted hunting at Fongoli include the social hypothesis, which would predict that increased feeding competition in the larger parties typical of the early rainy season at Fongoli drives the behavior (Pruetz and Bertolani 2009). This allows females and other subordinate individuals, such as adolescents, to increase their access to a highly prized food source (vertebrate meat), as we have proposed previously (Pruetz and Bertolani 2007). Predictions stemming from the opportunity hypothesis would entail opportunistic encounter rates by Fongoli chimpanzees with *Galago* prey. This research is currently ongoing at Fongoli and entails study of *Galago* availability and cavity use over the annual cycle. Similarly, data on the nutritional profitability of *Galago* prey compared to other foods at Fongoli would inform the relative profitability hypothesis.

Future research on the energetics of certain behaviors, such as termite fishing and baobab cracking, that would examine energy expenditure and gain would elucidate the relative profitability hypothesis regarding foraging behaviors. The data presented here on termite fishing indicates Fongoli chimpanzees gain very little profit in termite fishing; however, our data set is limited and we propose further research into this area in conjunction with more detailed nutritional analyses regarding micronutrients to inform the relative profitability hypothesis. Furthermore, possible studies on thermoregulation would be useful to examine questions on how different environments impact feeding choices.

Acknowledgments

We are thankful for the research contributions of Paco Bertolani, Jessica Rothman, Sylvia Ortmann, Anna Olson, Mody, Waly, and Mboule Camara, Dondo Kante, Jacques Keita, Ulises Villalobos Flores, Sally MacDonald, Caroline Melia, Andrea Quinn, James Quinn, Kaleigh Reyes, El Hadji Sakho, Michel Sadiakho, Kristina Walkup, and Erin Wessling. Financial support for this study was provided by the Leakey Foundation, National Geographic Society, National Science Foundation, Primate Conservation Inc., Rufford Small Grant Foundation, Iowa State University, and Wenner-Gren Foundation for Anthropological Research. We are grateful to our colleagues at the Direction des Eaux et Forêts, Chasses et de la Conservation des Sols, Republique du Sénégal for authorizing this research.

References

Altmann, J. 1974. "Observational study of behavior—sampling methods." *Behaviour* 49 (3–4): 227–67.

Baldwin, P. J. 1979. "The natural history of the chimpanzee (*Pan troglodytes verus*) at Mt. Assirik, Senegal." PhD diss., Stirling University.

Bertolani, P., and J. D. Pruetz. 2011. "Seed reingestion in savannah chimpanzees (*Pan troglodytes verus*) at Fongoli, Senegal." *International Journal of Primatology* 32: 1123–32.

Boesch, C., and H. Boesch. 1982. "Optimisation of nut-cracking with natural hammers by wild chimpanzees." *Behaviour* 83 (3–4): 265–86.

Boesch, C., and H. Boesch-Achermann. 2000. *The Chimpanzees of the Taï Forest: Behavioural Ecology and Evolution*. Oxford: Oxford University Press.

Bogart, S. L. 2009. "Behavioral ecology of savanna chimpanzees (*Pan troglodytes verus*) with respect to insectivory at Fongoli, Senegal." PhD diss., Iowa State University.

Bogart, S. L., and J. D. Pruetz. 2008. "Ecological context of chimpanzee (*Pan troglodytes verus*) termite fishing at Fongoli, Senegal." *American Journal of Primatology* 70: 605–12.

Bogart, S. L., and J. D. Pruetz. 2011. "Insectivory of savanna chimpanzees (*Pan troglodytes verus*) at Fongoli, Senegal." *American Journal of Physical Anthropology* 145: 11–20.

Buba, U., V. Sommer, G. Jesus, and A. Pascual-Garrido. 2016. "Constructing social identity? Potential non-nutritional functions of chimpanzee insectivory." *PeerJ Preprints* 4: e1845v1.

Conklin, N. L., and R. W. Wrangham. 1994. "The value of figs to a hind-gut fermenting frugivore—a nutritional analysis." *Biochemical Systematics and Ecology* 22 (2): 137–51.

Deblauwe, I. 2009. "Temporal variation in insect-eating by chimpanzees and gorillas in southeast Cameroon: Extension of niche differentiation." *International Journal of Primatology* 30: 229–52.

Deblauwe, I., and G. P. J. Janssens. 2008. "New insights in insect prey choice by chimpanzees and gorillas in Southeast Cameroon: The role of nutritional value." *American Journal of Physical Anthropology* 135: 42–55.

Fahy, G. E., M. Richards, J. Riedel, J.-J. Hublin, and C. Boesch. 2013. "Stable isotope evidence of meat eating and hunting specialization in adult male chimpanzees." *Proceedings of the National Academy of Sciences* 110 (15): 5829–33.

Fox, E. A., A. F. Sitompul, and C. P. van Schaik. 1999. "Intelligent tool use in wild Sumatran orangutans." In *The Mentality of Gorillas and Orangutans*, edited by S. T. Parker, R. W. Mitchell, and H. L. Miles, 99–116. Cambridge: Cambridge University Press.

Gašperšič, M. 2008. "Fongoli chimpanzees (*Pan troglodytes verus*) and baobab (*Adansonia digitata*) fruits: Modeling the evolution of hominin material culture on percussive technology." PhD diss., Ljubljana Graduate School of the Humanities.

Goodall, J. 1986. *The Chimpanzees of Gombe: Patterns of Behavior*. Cambridge, MA: Belknap Press of Harvard University Press.

Gruber, T. 2013. "Historical hypotheses of chimpanzee tool use behavior in relation to natural and human-induced changes in an East African rain forest." *Revue de Primatologie* 5: 1–16.

Hashimoto, C., S. Suzuki, Y. Takenoshita, J. Yamagiwa, A. K. Basabose, and T. Furuichi. 2003. "How fruit abundance affects the chimpanzee party size: A comparison between four study sites." *Primates* 44: 77–81.

Hicks, T. C., H. S. Kühl, C. Boesch, P. Dieguez, A. Emmanuel Ayimisin, R. Martin Fernandez, D. Barubiyo Zungawa, M. Kambere, J. Swinkels, S. B. J. Menken, J. Hart, R. Mundry, and P. Roessingh. 2019. "Specialized chimpanzee technology in Northern DR Congo: The Bili-Uéré behavioral realm." *Folia Primatologica* 90 (1): 3–64.

Humle, T., F. Maisels, J. F. Oates, A. Plumptre, and E. A. Williamson. 2016. "*Pan troglodytes* (errata version published in 2018)." *IUCN Red List of Threatened Species 2016* e.T15933A129038584.

Koops, K., T. Furuichi, and C. Hashimoto. 2015. "Chimpanzees and bonobos differ in intrinsic motivation for tool use." *Scientific Reports* 5: 11356.

Koops, K., W. C. McGrew, and T. Matsuzawa. 2013. "Ecology of culture: Do environmental factors influence foraging tool use in wild chimpanzees (*Pan troglodytes verus*)?" *Animal Behaviour* 85: 175–85.

Koops, K., C. Schöning, W. C. McGrew, and T. Matsuzawa. 2015. "Chimpanzees prey on army ants at Seringbara, Nimba Mountains, Guinea: Predation patterns and tool use characteristics." *American Journal of Primatology* 77: 319–29.

Koops, K., E. Visalberghi, and C. P. van Schaik. 2014. "The ecology of primate culture." *Biology Letters* 10 (11): 20140508.

Lambert, J. E. 2007. "Seasonality, fallback strategies, and natural selection: A chimpanzee and cercopithecoid model for interpreting the evolution of hominin diet." In *Evolution of the Human Diet: The Known, the Unknown, and the Unknowable*, edited by P. S. Unger, 324–43. New York: Oxford University Press.

Lindshield, S. M. 2014. "Multilevel analysis of the foraging decisions of western chimpanzees (*Pan troglodytes verus*) and resource scarcity in a savanna environment at Fongoli, Senegal." PhD diss., Iowa State University.

Lindshield, S. M., B. J. Danielson, J. M. Rothman, and J. D. Pruetz. 2017. "Feeding in fear? How adult male western chimpanzees (*Pan troglodytes verus*) adjust to predation and savanna habitat pressures." *American Journal of Physical Anthropology* 163: 480–96.

Luncz, L. V., and C. Boesch. 2015. "The extent of cultural variation between adjacent chimpanzee (*Pan troglodytes verus*) communities; A microecological approach." *American Journal of Physical Anthropology* 156: 67–75.

Marchant, L., and W. C. McGrew. 2005. "Percussive technology: Chimpanzee baobab smashing and the evolutionary modeling of hominin knapping." In *Stone Knapping: The Necessary Conditions for a Uniquely Hominin Behaviour*, edited by V. Roux and B. Bril, 341–50. Cambridge: Cambridge University Press.

McGrew, W. C. 1979. "Evolutionary implications of sex differences in chimpanzee predation and tool use." In *The Great Apes*, edited by D. A. Hamburg and E. R. McCown, 441–63. Menlo Park, CA: Benjamin/Cummings.

Mitani, J., and D. Watts. 2001. "Why do chimpanzees hunt and share meat?" *Animal Behaviour* 61: 915–24.

Moore, J. 1996. "Savanna chimpanzees, referential models and the last common ancestor." In *Great Ape Societies*, edited by W. C. McGrew, L. F. Marchant, T. Nishida, 275–92. Cambridge: Cambridge University Press.

Nakamura, M., and N. Itoh. 2008. "Hunting with tools by Mahale chimpanzees." *Pan Africa News* 15: 3–6.

Ogawa, H., G. Idani, J. Moore, L. Pintea, and A. Hernandez-Aguilar. 2007. "Sleeping parties and nest distribution of chimpanzees in savanna woodland, Ugalla, Tanzania." *International Journal of Primatology* 28: 1397–1412.

O'Malley, R. C. 2011. "Environmental, nutritional, and social aspects of insectivory by Gombe Chimpanzees." PhD diss., University of Southern California.

O'Malley, R. C, and M. L. Power. 2014. "The energetic and nutritional yields from insectivory for Kasekela chimpanzees." *Journal of Human Evolution* 71: 46–58.

Pickering, T. R. 2013. *Rough and Tumble: Aggression, Hunting and Human Evolution*. Berkeley: University of California Press.

Pickering, T. R., and M. Dominguez-Rodrigo. 2010. "Chimpanzee referents and the emergence of human hunting." *Open Anthropology Journal* 3: 107–13.

Pruetz, J. D. 2006. "Feeding ecology of savanna chimpanzees (*Pan troglodytes verus*) at Fongoli, Senegal." In *Feeding Ecology in Apes and Other Primates*, edited by G. Hohmann, M. M. Robbins, and C. Boesch, 161–82. Cambridge: Cambridge University Press.

Pruetz, J. D. 2007. "Evidence of cave use by savanna chimpanzees (*Pan troglodytes verus*) at Fongoli, Senegal: Implications for thermoregulatory behavior." *Primates* 48: 316–19.

Pruetz, J. D., and P. Bertolani. 2007. "Savanna chimpanzees (*Pan troglodytes verus*) hunt with tools." *Current Biology* 17: 1–6.

Pruetz, J. D., and P. Bertolani. 2009. "Chimpanzee (*Pan troglodytes verus*) behavioral responses to stresses associated with living in a savanna-mosaic environment: Implications for hominin adaptations to open habitats." *PaleoAnthropology*: 252–62.

Pruetz, J. D., P. Bertolani, K. Boyer Ontl, S. Lindshield, M. Shelley, and E. G. Wessling. 2015. "New evidence on the tool-assisted hunting exhibited by chimpanzees (*Pan troglodytes verus*) in a savanna habitat at Fongoli, Sénégal." *Royal Society Open Science* 2: 140507.

Pruetz, J. D., S. L. Bogart, S. M. Lindshield, and K. R. Walkup. In press. "Feeding related tool use in primates—A systematic overview." In *Primate Diet and Nutrition: Needing, Finding, and Using Food*, edited by J. E. Lambert and J. Rothman. Chicago: University of Chicago Press.

Pruetz, J. D., and D. Kante. 2010. "Successful return of a wild infant chimpanzee (*Pan troglodytes verus*) to its natal group after capture by poachers." *African Primates* 7: 35–41.

Pruetz, J. D., and T. C. LaDuke. 2010. "Reaction to fire by savanna chimpanzees (*Pan troglodytes verus*) at Fongoli, Senegal: Conceptualization of 'fire behavior' and the case for a chimpanzee model." *American Journal of Physical Anthropology* 141: 646–50.

Rothman, J. M., D. Raubenheimer, M. A. H. Bryer, M. Takahashi, and C. C. Gilbert. 2014. "Nutritional contributions of insect in primate diets: Implications for primate evolution." *Journal of Human Evolution* 71: 59–69.

Russak, S. M. and W. C. McGrew. 2008. "Chimpanzees as fauna: Comparisons of sympatric large mammals across long-term study sites." *American Journal of Primatology* 70: 402–9.

Rutz, C., and J. J. H. St. Clair. 2012. "The evolutionary origins and ecological context of tool use in New Caledonian crows." *Behavioural Processes* 89 (2): 153–65.

Sanz, C. M., and D. M. Morgan. 2013. "Ecological and social correlates of chimpanzee tool use." *Philosophical Transactions of the Royal Society of London B* 368: 20120416.

Shumaker, R. W., K. R. Walkup, and B. B. Beck. 2011. *Animal Tool Behavior: The Use and Manufacture of Tools by Animals*. Baltimore: Johns Hopkins University Press.

Skinner, M. F., and J. D. Pruetz. 2012. "Reconstruction of periodicity of repetitive linear enamel hypopolasia from perikymata counts on imbricational enamel among dry-adapted chimpanzees (*Pan troglodytes verus*) from Fongoli, Senegal." *American Journal of Physical Anthropology* 149: 468–82.

Spagnoletti, N., E. Visalberghi, M. P. Verderane, P. Iza, and D. Fragaszy. 2012. "Stone tool use in wild bearded capuchin monkeys, *Cebus libidinosus*. Is it a strategy to overcome food scarcity?" *Animal Behaviour* 83: 1285–94.

Sponheimer, M., J. E. Loudon, D. Codron, M. E. Howells, J. D. Pruetz, J. Codron, D. J. de Ruiter, and J. A. Lee-Thorp. 2006. "Do 'savanna' chimpanzees consume C4 resources?" *Journal of Human Evolution* 51: 128–33.

Stanford, C. B. 1998. *Chimpanzee and Red Colobus: The Ecology of Predator and Prey*. Cambridge, MA: Harvard University Press.

Stanford, C. B. 1999. *The Hunting Apes: Meat Eating and the Origins of Human Behavior*. Princeton, NJ: Princeton University Press.

Suzuki, S., S. Kuroda, and T. Nishihara. 1995. "Tool-set for termite-fishing by chimpanzees in the Ndoki Forest, Congo." *Behaviour* 132: 219–35.

Tennie, C., R. C. O'Malley, I. C. Gilby. 2014. "Why do chimpanzees hunt? Considering the benefits and costs of acquiring and consuming vertebrate versus invertebrate prey." *Journal of Human Evolution* 71: 38–45.

Watts, D. P., and J. C. Mitani. 2002. "Hunting behavior of chimpanzees at Ngogo, Kibale National Park, Uganda." *International Journal of Primatology* 23 (1): 1–12.

Webster, T. H., W. C. McGrew, L. F. Marchant, C. L. R. Payne, and K. D. Hunt. 2014. "Selective insectivory at Toro-Semliki, Uganda: Comparative analyses suggest no 'savanna' chimpanzee pattern." *Journal of Human Evolution* 71: 20–27.

Wessling, E. G., H. S. Kühl, R. Mundry, T. Deschner, and J. D. Pruetz. 2018. "The costs of living at the edge: Seasonal stress in wild savanna-dwelling chimpanzees." *Journal of Human Evolution* 121: 1–11.

White, T. D., S. H. Ambrose, G. Suwa, D. F. Su, D. DeGusta, R. L. Bernor, J.-R. Boisserie, M. Brunet, E. Delson, S. Frost, N. Garcia, I. X. Giaourtsakis, Y. Haile-Selassie, F. C. Howell, T. Lehmann, A. Likius, C. Pehlevan, H. Saegusa, G. Semprebon, M. Teaford, and E. Vrba. 2009. "Macrovertebrate paleontology and the Pliocene habitat of *Ardipithecus ramidus*." *Science* 326 (5949): 67–93.

Yamakoshi, G. 1998. "Dietary responses to fruit scarcity of wild chimpanzees at Bossou Guinea: Possible implications for ecological importance of tool use." *American Journal of Physical Anthropology* 106: 283–95.

18

Cultural Transmission in Dispersing Primates

LYDIA V. LUNCZ AND ERICA VAN DE WAAL

Introduction

Culture plays a big part in our human lives, often unconsciously influencing our daily behavior. Our ability to learn socially from each other leads to transmission of knowledge and the establishment of population-specific behavioral patterns (Boyd and Richerson 2005). What food we prefer, our body language, the concept of beauty or justice, and each interpersonal interaction is profoundly affected by our broader social surroundings and what we are exposed to during our upbringing. The physical aspect of our culture manifests itself in objects and artifacts that differ between social groups, countries, and continents. The earliest indications of our material culture are stone tools associated with early hominins that have been dated to 3.3 million years ago (Harmand et al. 2015). This timing has raised the possibility of cultural evolutionary roots shared with other primates and even more distantly related taxa. The evolution of our cultural abilities has long received extensive research interest with interdisciplinary efforts combining the fields of archaeology, anthropology, and psychology to understand the driving factors behind the origin of our cultural abilities and their development.

Culture was long considered to be a uniquely human trait. In recent years a focus on animals has increased and it now is apparent that other species exhibit culturally influenced behaviors (e.g., communication, foraging style, tool use, or social convention). The understanding of these cultural traits in wild animals has advanced significantly through contributions from research on primates (Boesch 2003; Dindo, Whiten, and de Waal 2009; Luncz, Mundry, and Boesch 2012; Panger et al. 2002; Santorelli et al. 2011; Schofield et al. 2017; van Schaik et al. 2003; Whiten et al. 1999), cetaceans (Whitehead and Rendell 2014), birds (Aplin et al. 2015), and fish (Laland, Atton, and Webster 2011). The cultural abilities of primates have received increased attention ever since the early 1960s when Japanese macaques (*Macaca fuscata*) were observed to invent a new behavior that spread throughout the group (Kawai 1963). Researchers observed how a young monkey on Koshima Island started to wash provisioned sweet potatoes in water before eating it. The other group members started to copy this invention until sometime later the entire troop on Koshima was known to wash their sweet potatoes before consumption.

Song sequences of whales and birds have also been linked to socially transmitted behavior (Jenkins 1978; Rendell and Whitehead 2001), creating local and population-specific song patterns that are distinguishable between social groups. In humpback whales (*Megaptera novaeangliae*), for example, groups that live in different areas have very distinct songs but, intriguingly, migrating individuals were observed to introduce new song patterns into their adopted groups (Noad et al. 2000). In contrast, dispersing primates were observed to adopt the traditions of their new groups (Luncz and Boesch 2014; Matsuzawa and Yamakoshi 1998) and therefore have not only been observed to invent new behaviors but also have social learning mechanisms that enable the spread of these behaviors.

Behaviors that differ between social groups have provided the principal initial means used to detect and study culture in wild animals. A landmark study in the late 1990s investigated the behavioral differences seen at six long-term field sites and provided the first systematic evidence for cultural diversity between chimpanzee (*Pan troglodytes*) groups across the African continent (Whiten et al. 1999). Since then, with the emergence of multiple habituated populations, research has advanced to report cultural differences even between directly neighboring groups. This finding was valuable as ecology and genetics could largely be excluded as explanations for observed diversity, enabling us to focus on underlying social effects that shape behavior (Tennie, Hopper, and van Schaik, chapter 19 this volume). Several different behaviors between social groups in several primate species were studied in more detail, including chimpanzees (*Pan troglodytes verus*) (Koops et al. 2015; Luncz and Boesch 2014; Luncz, Mundry, and Boesch 2012), orangutans (*Pongo* sp.) (Fox et al. 2004), and vervet monkeys (*Chlorocebus pygerythrus*) (van de Waal 2018).

However, despite long-term field research investigating cultural behavior in wild primates, empirical insight into how diversity amongst populations emerges and how it is maintained remains rare. Dispersing individuals between neighboring groups have recently added valuable information that helped researchers to understand cultural abilities in chimpanzees (Luncz and Boesch 2014; Luncz, Wittig, and Boesch 2015) and vervet monkeys (van de Waal, Borgeaud, and Whiten 2013), and comparable studies of between-group transfers of captive chimpanzees have also been undertaken (Vale et al. 2017). Nonetheless, relatively little is known about the social learning processes during the dispersal of wild primates, and each species shows different patterns depending on demography and life history. Depending on the primate species, males or females disperse (Strier, Lee, and Ives 2014). This dynamic has allowed us to compare the effect of an individual's sex on social behavior during migrating events. Only recently has habituation of multiple neighboring groups allowed research into behavioral changes after dispersal. Comparisons of behavior including immigrants and local group members are now possible before and after dispersal.

Female chimpanzees, for example, generally leave their native community at the onset of sexual maturity, at an age when their cultural repertoire is fully formed and they reliably perform the behavior of their native group. At this age, around 13 or 14 years old, they integrate into a new group. Detailed data on dispersal patterns are absent as the lack of multiple habituated neighboring groups in the past has hindered direct observations of transferring individuals and the effect they might have on migrants and other group members. The dispersal of monkey species is only slightly better understood. Research with multiple neighboring groups has been ongoing for decades as habituation is generally faster than in chimpanzees. In vervet monkeys, for example, the males disperse, and, unlike chimpanzees, they do so multiple times during their lives (Cheney and Seyfarth 1983).

Two different research approaches have been used to investigate cultural transmission in dispersing wild primates: natural observations and field experiments. These methods are complementary but each has unique advantages and limitations (for additional discussion of methods used to study primate social learning see Hopper and Carter, chapter 7 this volume). Both the observational and the experimental approaches have added valuable information to understanding the cultural abilities of primates in the wild. In this chapter, we compare dispersal patterns of wild primates using information from natural observations as well as field experiments.

Natural observation of undisturbed wild animal populations is invaluable as a foundational approach when studying cultural abilities and mechanisms of social learning that have evolved under natural conditions. Observing natural behavior in the wild is a necessary first step in understanding the full behavioral capacity of wild animals, which can then inform the design of controlled experiments targeting the underlying mechanisms of specific behavior. Performing experiments ignorant of a species' natural history makes little sense and certainly limits the interpretation of observed responses during testing.

In the Taï National Park, Côte d'Ivoire, three neighboring groups of chimpanzees (*Pan troglodytes verus*) have been habituated to the presence of humans. This offers the unique opportunity to investigate culturally influenced behavior by simultaneously minimizing the effect of ecological and genetic influence. Evidence for behavioral differences between three groups (North, Middle, and South groups) has been reported (Boesch 2003), and includes group-specific styles of foraging (tool use) and communication. Taï chimpanzees are well known for their nut-cracking behavior: they crack five different nut species, though one of them, the *Coula edulis* nut, stands out as the main energy source from November until March. *Coula* nuts are rather soft compared to the other four nut species in the Taï forest. This allows the chimpanzees a choice between stone or wooden tools with which to crack them open, and this, in turn, provides a possibility for differences in tool-type choice emerging between the groups. Additionally, tool selection is easily measurable: tools create long-lasting evidence in the chimpanzees' home range. This allows investigations into tool selection patterns and environmental assessment to answer questions about the social impact of tool choice.

Many captive experiments have revealed the scope of species' ability to learn socially (e.g., Bonnie et al. 2007; Laland, Atton, and Webster 2011; Whiten, Horner, and de Waal 2005, see Hopper and Carter, chapter 7 this volume for a review). However, these experiments are often performed with captive groups that may lack natural social systems and individuals that do not face the same challenges as their wild conspecific counterparts. Thus, field experiments are essential for revealing the frequency and type of social learning used by wild primates. Recently, experiments on social learning have been successfully conducted with chimpanzees in the wild (e.g., Gruber et al. 2011).

Vervet monkeys are an ideal species for field experiments as unlike many primates they are not neophobic, and the fact that they are not endangered and have a pest status in South Africa provides researchers with easier access and ethical approval for experimental work and captures. Additionally, their group size, matrilineal social structure, linear hierarchy, and the multiple dispersals of males, make vervets perfect models to investigate social learning experimentally, especially the between-group transfer of traits. Field experiments on up to six groups of habituated wild vervet monkeys at two different field sites, the Loskop Dam Nature Reserve and the Inkawu Vervet Project (IVP) in Mawana Game Reserve, have confirmed the importance of social learning for wild primates. Some of these experiments used two-action tasks (i.e., boxes that can be opened with two different actions to retrieve a small food reward) to test the fidelity of solution copying (van de Waal, Claidière,

TABLE 18.1. Life history comparisons between dispersal in two primate species: Chimpanzees and vervets.

Life History Facts	*Chimpanzee (Pan troglodytes verus)*	*Vervet (Chlorocebus pygerythrus)*
Dispersing sex	Females	Males
Age at first dispersal	Sexual maturity (~12–14)	Sexual maturity (~4–6)
Number of dispersals per lifetime	Once per lifetime?	~ Once per year
Number of observed dispersals	1 immigrant (Diva) from Middle group immigrating into South group; 7 immigrants from SoS group, immigrating into South group	15 immigrations observed in four groups during one mating season of the food coloring experiment
Time until observed behavioral changes	One nut-cracking season for Diva (~4 months); maximum 43 days for the SoS immigrants	Instantly, as soon as the males were not outranked and could choose the color of the food to eat

and Whiten 2015; van de Waal et al. 2010); the transmission of food handling techniques of sandy grapes (van de Waal, Bshary, and Whiten 2014; van de Waal et al. 2012); and the social learning of arbitrary food preference using differently colored edible corn (van de Waal, Borgeaud, and Whiten 2013).

Chimpanzees and vervet monkeys remain the only two primate species where the effects of dispersal on their cultural behavior have been studied in detail (see table 18.1 for main life history differences between chimpanzees and vervets). We therefore compare the dispersal behavior of these two primates, which have been studied by ourselves and our colleagues over long periods. We provide an overview of the main cultural differences known between neighboring primate groups within the same population. Furthermore, we summarize similarities and differences of cultural transmission pattern during immigration events in chimpanzees and vervet monkeys. Finally, we place the cultural transmission behavior of primates in a larger context through comparisons with birds and cetaceans, to evaluate general patterns in migrating animals.

Approach

The field sites at which we conduct our research support long-term projects with many years of observations. The detailed knowledge about the natural behavior of multiple neighboring groups makes them ideal candidates to compare the social parameters between species that might influence cultural transmission.

The chimpanzee study site is located in the Taï National Park in the far west of Côte d'Ivoire. Behavioral observations have been continually ongoing since the early 1980s with the North group (Boesch and Boesch-Achermann 2000). In 2000, the habituation of the South group was completed, followed by the successful habituation of the East group in 2007. All groups have been under observation and followed daily since their habituation, creating a long-term database of behavioral observations. Group sizes have been variable throughout the last decades but during the time of data collection for the observations reported here, North group had around 20 members, South group 45, and East group 50. The size of their home range was estimated according to group size, and varied between 25 and 35 km^2 (fig. 18.1).

Data on nut cracking and tool selection was collected between November 2007 and April 2014. We recorded nut-cracking behavior using focal and scan sampling methods (Altmann 1974) and focused on material and size selection, efficiency of opening nuts, and social behavior related to nut cracking. In order to investigate potential changes in tool selection over time within one group, we extracted the physical properties of selected tools over the last 30 years in the North group from the Taï Chimpanzee Project's long-term database. Using archaeological methods, we additionally investigated previous tool preferences in immigrating females (Luncz, Wittig, and Boesch 2015). By recovering tool remains at abandoned nut-cracking sites, we reconstructed tool selection patterns of the groups the females belonged to prior to their emigration. By comparing these patterns with direct observations after immigration, we were able to investigate cultural transmission patterns in tool selection of migrating chimpanzees (Luncz, Wittig, and Boesch 2015).

The Inkawu Vervet Project (IVP) is located in Mawana Game Reserve, in KwaZulu Natal, South Africa. IVP started in 2010 with the aim of building a long-term experimental field site. Experiments are currently conducted on six main groups of vervet monkeys. Three groups are fully habituated (Ankhase, Baie Dankie, and Noha) and researchers have worked with them since 2010. Another two groups (Lemon Tree and Kubu) have been habituated well enough since 2012 to allow for focal sampling on all the adults. The sixth group, Crossing, has been followed since 2014 but, to date, only scan data has been collected. All home ranges of the groups overlap with at least two other groups (fig. 18.1). Home range sizes are approximately 1.6 km^2 and the size of four groups that took part in the experiment varied between 23 and 45 individuals.

A large-scale field experiment was conducted on these four vervet groups (van de Waal, Borgeaud, and Whiten 2013). We exposed whole groups to two colors of corn (dyed pink and blue). The monkeys were trained to avoid one

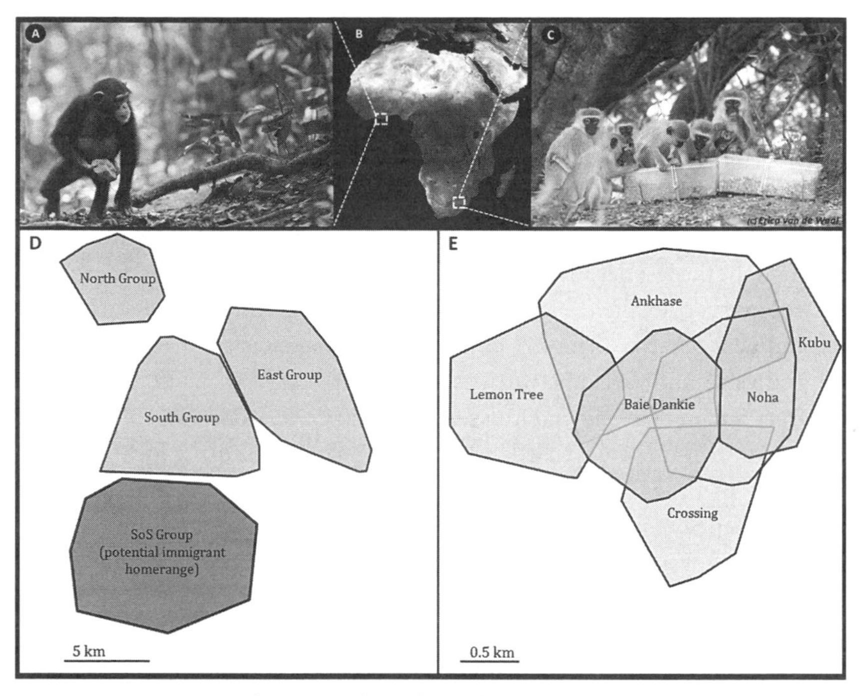

FIGURE 18.1. Home ranges of primates under study and the behavior that was investigated for the comparative approach between chimpanzees and vervet monkeys. A) *Coula edulis* nut cracking on three neighboring chimpanzee groups in the Taï forest. B) Locations of field sites studying cultural transmission in chimpanzees (Côte d'Ivoire) and vervet monkeys (South Africa). C) Colored corn experiments in six neighboring vervet groups in the Mawana Game Reserve. D) Home ranges of neighboring chimpanzee groups. E) Home ranges of neighboring vervet groups.

color that was made bitter using locally available aloe. Which color was the palatable one was counterbalanced across groups. We then offered both colors of corn without any bitter treatment to test whether a new generation (26 infants) would learn the arbitrarily assigned feeding preference of their mothers and groupmates. This experimental phase took place during the mating season when males disperse between the groups, so both dispersing males and the new crop of infants were naive to the earlier training of group members. In our study populations, adult males disperse on average once a year (mean duration of stay of adult males in 2012–2015 = 283 days, Sobrino, pers. comm.).

Results

Here, we review data on tool behavior in wild chimpanzees in Côte d'Ivoire and corn foraging data in wild vervet monkeys in South Africa. Both chimpanzees and vervet monkeys displayed behaviors that differed between neighboring groups. Each approach, field experiments with vervet monkeys and observational data with chimpanzees, independently found that migrating individuals conformed to the behavior of local group members after dispersal, despite the fact that migrants previously performed a different variant of the behavior.

Specifically, we found that chimpanzees belonging to different social groups showed very specific behavioral variants spanning all aspects of life, including tool use, communication, foraging, and hunting (Luncz and Boesch 2015). Diversity within a social group was found to be low; across several domains, an individual's behavior matched the behavior of its social group (Luncz and Boesch 2014). For details of behavioral differences between Taï chimpanzee groups, see table 18.2 (as in Luncz and Boesch 2015).

Strikingly, despite frequent female migrations between the Taï chimpanzees groups, the groups differ in their preferred hammer-type selection for cracking *Coula edulis* nuts. Differences are seen in tool material and size choices (Luncz, Mundry, and Boesch 2012). At the beginning of the season, when nuts are still hard, the majority of individuals prefer stone hammers, regardless of which group they belong to. However, when the dry season advances, and nuts become easier to crack, two of the groups, North and East, increasingly switch to wooden hammers. In contrast, South group members prefer stone tools throughout the entire nut-cracking season.

Selection patterns within a group have remained similar over decades, despite immigrating females and a complete turnover of member individuals (Luncz and Boesch 2014). Migration of individuals is rare and has only been observed a few times during the 34 years that the site has been studied. In 2009, a habituated female from the rapidly disappearing Middle group (not shown on the map due to their disappearance) integrated into South group. We compared the tool selection of her first nut-cracking season in her new group with her tool selection one year later, when she had become a full member of the group. In her first year, there was a significant difference in her tool-selection patterns: she used more wooden tools than her South group peers. In the second year after her immigration, this difference had disappeared. Our data suggest that she adopted the tool-selection pattern of her new group despite previous use of an alternative tool type. Archaeological and observational methods also show that several immigrants took up the tool selection pattern of their new group (Luncz, Wittig, and Boesch 2015). By recreating tool-selection patterns from abandoned nut-cracking sites, we constructed previous tool choices of seven immigrating females before they migrated. This pattern was then compared to observed tool selection after immigration. The results showed that immigrants took up the tool-selection pattern of their new group within a few weeks of being exposed to the new behaviors.

More recent studies have shown that the neighboring groups also differ in their nut-cracking efficiency (Luncz et al. 2018). In general, we found that stone tools are more efficient than wooden tools. Nut-cracking techniques differed as well. This leaves room to speculate that transferring females must encounter cases where their original tool selection has led to different foraging efficiencies than that of their adopted group. Despite the fact that potentially advantageous personal knowledge is present, they still adopted the tool pattern of their new groups. Similar results of potential conformist learning mechanisms have been found in an experimental setting in captive animals (Haun, Rekers, and Tomasello 2012; Kendal, Coolen, and Laland 2004; Whiten, Horner, and de Waal 2005).

Our data demonstrate that transferring female chimpanzees conform to the tool-selection pattern of their new community. Differences in the efficiency of tool properties suggest that these immigrants are exposed to costs and benefits in personal foraging success. Nevertheless, conformity is observed repeatedly during migrating events in chimpanzees as well as other animal species (for example, vervet monkeys and birds; Aplin et al. 2015; van de Waal, Borgeaud, and Whiten 2013), highlighting the importance of group belonging in wild animals.

From the food-coloring experiment run with vervet monkeys, we found that all but one of the 26 infants learned the arbitrary, experimentally seeded local norm (i.e., the option that was most common in their natal group: van de Waal, Borgeaud, and Whiten 2013). The one exception was the son of the lowest-ranking female in the largest group. He copied his mother who ate the previously distasteful color, since the other color was monopolized by dominants. Similarly, an experiment with captive chimpanzees revealed that subjects would conform to the group preference unless they were unable due to a lack of access to resources (Hopper et al. 2011). More excitingly, we were also able to test 15 dispersing males who entered our experimental groups, 14 of which adopted the local color preference. Ten of these males came from groups trained in the conflicting color preference (the other five came from groups outside of our study population) and nine of the ten abandoned their initial preference to conform to the local norm. The dispersing males conformed directly after observing their new group preference. These findings reveal the potency of conformity in wild primates, as these immigrant males, despite their conflicting knowledge about what food was good to eat, conformed to their new group preference. This reflects another example of

TABLE 18.2. Behavioral differences between three habituated groups of chimpanzees at Taï National Park in Côte d´Ivoire.

Community	*North*	*South*	*East*
A: Tool use			
- All tool use occurrence frequency	+++	+	++
- Rubbing one's back on vegetation after rain	+	+	-
- Leaf sponging to drink	Leaf	Leaf/ bark	Leaf
- Container use for drinking[1]	-	+	-
Nut cracking:			
- Hammer material (over nut season)	Stone→wood	Stone	Stone→wood
- Average wooden hammer size	Small	Large	Small→large
- Wood hammer transport in mouth	+	-	+
- Nut cracking in trees	Stone/wood (♀♂)	Stone (♀)	Stone/wood (♀)
B: Foraging			
- *Thoracotermes* mound pounding	+	-	-
- *Diospyros mannii* fruit consumption	Seed swallow	Seed spit	Seed swallow
- *Dorylus spp.* ant nest raid, hand depth	Arm	Wrist	?
- Frequency of insect consumption	+++	+	+
- *Treculia Africana* hard-shelled fruits:	+++	+	+
a- State of fruit when eaten	Decomposed/fresh	Fresh	Fresh
b- Location of fruit pounding	On ground/in trees	In trees	In trees
- Parts of *Haloplegia azurea* eaten	Leaves	Leaves/stem	Stem
- Part of *Strychnos aculeata* eaten	Decomposed seeds	Flesh in fresh fruit	?
C: Social interaction and communication			
- Day nest constructed for play start	-	+	+
- Leaf held in mouth for play start	+	-	?
- Day nest constructed for courtship	-	+	-
- Knuckle knock courtship	+	-	-
- Leaf clipping before drumming[2]	Rip leaf to pieces	Rip leaf to pieces	Rip leaf off stem
- Buttress drumming	♀♂	♀♂	♂
- Rain dance	Slow/calm	Aggressive	?
D: Hunting behavior			
- Vocalization during hunt	-	-	+
- Prey captured most often	B	B	B/P
- Monkey skull pound open	+	-	-
- Presence of ♀-hunters	+	+	-
- Meat-sharing	Hunters	Hierarchy	Hunters

+ = Present (frequency of observed behavior: += once a week, ++= at least twice a week, +++= daily)

- = Absent

? = Lack of observation time

B = *Piliocolobus badius*

P = *Colobus polycomos*

[1]Container use for drinking: In South-group females (N=4) have been observed to drink out of broken shells of the *Strychnos* fruits that filled up with rain water. Once the water was drunk they did not refill the container.

[2]Leaf clipping: Before buttress drumming and display, male chimpanzees break a leafy stem from surrounding vegetation and, without ingesting the leaves, rip bites off the leaves which makes loud ripping sounds. In South and North-group, chimps rip multiple pieces off one leaf, whereas in East-group the leaf is ripped off the stem once, then they go on to ripping off the next leaf.

strong conformity of foraging choice, reported in an experiment with palatable and unpalatable food on dyads of captive rats (*Rattus norvegicus*, Galef and Whiskin 2008).

A follow-up test from the colored corn experiment took place after two fissions occurred in one of our experimental groups of vervet monkeys (van de Waal, van Schaik, and Whiten 2017). We were able to observe twice how three of the lowest-ranked females left their original group. These females formed small splinter groups by themselves. Due to their low rank the corn originally marked as edible, which continued to be preferred by high-rankers, was often inaccessible to them so they often had to sample the alternative color of corn that now tasted the same. Despite their knowledge that both colors were now palatable, once they were in their small fission groups, these females expressed a 100% preference for the corn color that was tasty from the beginning in their original group and that had become the main food source for higher ranking individuals. This thus revealed an unexpected resilience and conformity to a learned and arbitrary food preference and echoed previous experiments with primates that showed the impact of rank on experimental performances (Cronin et al. 2014; Drea and Wallen 1999).

Through multiple field experiments with these vervet monkeys we have discovered several of the social learning strategies used by this species. Infants selectively copy their mothers (van de Waal, Borgeaud, and Whiten 2013; van de Waal, Bshary, and Whiten 2014), group members later copy the philopatric sex (females) and not the dispersing sex (males) (van de Waal et al. 2010) and do so without exhibiting a dominance-bias (i.e., they copy all females and not only the dominant one) (Botting et al. 2018). These strategies support the pattern of dispersing individuals conforming to the preferences exhibited by the majority of individuals in their new group. Analyses on intergroup behavioral variation in universal behaviors (like diet, gestural communication, social network, and conflict management) across our studied groups highlight the potential breadth of traditions within a single population (van de Waal 2018) and confirm that the social learning capacities of vervets in experimental tests can also be found under natural conditions.

Discussion

Despite substantial differences in their social life, immigration has been found to trigger conformity to new local behaviors in both chimpanzees and vervet monkeys. In both cases, immigrants were observed to adopt the behavior of their new group despite differing previous knowledge. Stable diversity between primate groups permits the study of cultural transmission during dispersal between social groups, whether the diversity is induced by researchers or naturally present in the study groups.

Immigration has proven to be a useful feature for the investigation of cultural transmission in nonhuman animals. Behavioral differences in their natural environment can tell us about the circumstances driving diversity, but face limitations in controlling for subtle environmental components. Field experiments can be set up in a more controlled natural group dynamic, ruling out potential ecological explanations for the conformity observed. This approach is however still limited by the fact that the trait transmitted socially is artificial compared to what may be the natural use and scope of conformity in a given species. However, together, the two methodological approaches build a complementary view to understanding the role of cultural transmission in wild primates (Hopper and Carter, chapter 7 this volume).

In addition to observations in the wild, research focusing on the cultural abilities of primates has been transferred into a captive setting. This approach has provided information on possible underlying social learning mechanisms ranging from stimulus enhancement to imitation, emulation and even teaching that can promote cultural diversity; a topic that observations in the wild have not yet been able to address. Several social learning strategies have also been proposed in order to understand when and from whom individuals learn socially (Laland 2004). For example, learning biases toward high-ranking individuals have been found in some studies on captive chimpanzees (Horner et al. 2010; Kendal et al. 2015; but not in all: Watson et al. 2015) and not in capuchin monkeys (*Cebus apella*, Dindo, Whiten, and de Waal 2009).

Another social learning bias common in human primates is copying the majority of group members, termed conformity. Conformity has been defined as abandoning personal preferences or behaviors to match alternatives exhibited by a majority of others (Whiten and van de Waal 2017). Our examples in immigrant chimpanzees (Luncz and Boesch 2014; Luncz, Wittig, and Boesch 2015) and vervets (van de Waal, Borgeaud, and Whiten 2013) demonstrate the existence of conformity in wild primates, and an elegant experiment on wild great tits revealed that even birds may conform to the local feeding habits when migrating to a new population (Aplin et al. 2015). Conformity has also been tested for in captive studies, sometimes reporting contrary results (Vale et al. 2017). Some conditions like insecurity, stress, and the lack of group belonging are not re-creatable in a captive setting. If conformity happens mainly in immigration events, in order to match the behavior of locals or to help integration in new groups, it will be very difficult to create this complex context in captivity.

Social psychologists distinguished two important different functions of conformity called informational and normative conformity (Claidière and Whiten 2012; Deutsch and Gerard 1955). Informational conformity is defined as serving the function of providing truthful information about the environment. In the context of the vervet monkey experiment this concerns which of two food options is the best to consume. In contrast, normative conformity was defined as serving purely social functions, such as simply being more like others and strengthening bonds, or adhering to societal norms (often implying punishment if one does not follow the norm). Observed uniformity of feeding choice within splinter groups of vervet monkeys has been described as "social" conformity; a different term was used because normative conformity is often taken to require a recognition that norms should be followed. By this term the authors hypothesized that individuals act like others not to gain information, but instead with a social aim of "being like others" (van de Waal, van Schaik, and Whiten 2017, see also Hopper et al. 2011). Such a disposition may need no underlying complex cognition, but merely the following of a motivational rule like to preferentially act as the majority do; such conformity could have been selected to assist intragroup integration, and favor behavior such as coordinated traveling and foraging.

Given the importance for social bonding in group-living animals (see Wittig et al., chapter 5 this volume), such social conformity may turn out to be a significant phenomenon in nature (de Waal and Luttrell 1986; Silk 2007; Silk, Alberts, and Altmann 2003). This might be especially true for immigrants that must integrate into new groups, such as the chimpanzee females and vervet males we describe in this chapter. Vocal convergence has been described among associating individuals: birds (Lachlan, Janik, and Slater 2004; Catchpole and Slater 2008), whales (Garland et al. 2011), and primates (Townsend, Watson, and Slocombe, chapter 11 this volume; Watson et al. 2015), and social conformity could be relevant in explaining these findings. Indeed, as hypothesized by the bonding and identification-based observational learning (BIOL) model (de Waal 2001), researchers found that primates behaved more affectionately to others if they matched their own behavior (Nielsen 2009; Paukner et al. 2005, 2009). Group cohesion might also benefit from social conformity by motivating group members to travel and forage together, consequently reducing predation risks and ranging conflicts. A potential cost of this strategy is that it increases within-group competition over food, but it might be outweighed in many species by the benefit of staying in the safety of the group.

Observed behavioral flexibility of immigrants has been proposed to be driven by informational and/or social conformity, leading to conformist behavior within one social group. There must be a very strong driving factor that makes immigrants adopt a new (in some cases less efficient) behavior. The unfamiliar, new surrounding might push them to seek information quickly ("copy if uncertain"). This may be crucial knowledge for the survival of the immigrant. Possibly the best method to fill in that information gap is to copy successful group members, who seem to do especially well in that specific home range. Conforming to the new group's behavior, in the sense of abandoning personal knowledge to match the behavior displayed by the majority of the new group, can therefore be a safe way for immigrants to gain missing information.

Conclusion

Our data suggest that cultural transmission in nonhuman primates is driven by conformist tendencies of dispersing individuals, highlighting the previously underestimated importance of group belonging triggering conformity in wild primates. Most interestingly, conformity seems to persistently reappear in studies of immigration in the wild, including birds, monkeys, and apes. As in the well-worn expression "When in Rome, do as the Romans do," the reoccurrence of similar mechanisms across animal species urges us to place chimpanzee behavior in a conceptual context with animals with similar social structures. Especially around research topics with such limited data available, comparative research, including captive work, will bring us closer to understanding the complexity of the emergence of cultural diversity in wild animals.

Our work additionally highlights the urgent necessity to protect wild primate populations and their habitat. The diversity of primate behavior is at risk with the ever-growing destruction of their ecosystems (Chapman et al., chapter 25 this volume; Hartel et al., chapter 26 this volume; Morgan et al., chapter 27 this volume). Only through the protection of multiple populations will it be possible to understand the multifaceted behavioral diversity between wild primate groups. Increased focus on directly neighboring groups will allow us to further our understanding of dispersal and cultural learning in wild primates. With diminishing primate populations, the opportunity to study the cultural abilities of our primate cousins might forever disappear.

Future Directions

Observing primate dispersal while it is happening has been rare in behavioral research. By putting GPS collars on the dispersing sex of a study population, we might finally understand more of the dispersal patterns of wild primates.

What do they do between the time leaving one group and joining another? How many dispersals are done jointly between individuals of the same origin group? This would open a new window to explore primate sociality.

Likewise, the current examples of conformity in immigrant primates are restricted to the two examples on chimpanzees and vervets that we detailed. We urge other primatologists to investigate whether similar patterns occur in their study species. With more studies on this topic, we hope to understand more of the evolution of conformity and culture, and the origins that enabled humans to become such a uniquely cultural species.

Acknowledgments

Field work in Côte d´Ivoire has been sponsored by the Max Planck Society. We thank the "Ministère de l'enseignement supérieure et de la recherche scientifique," and the OIPR (Office Ivorien des Parcs et Réserves) for granting permission to conduct research in Côte d'Ivoire and the Taï National Park. We further acknowledge the "Centre Suisse de la Recherche Scientifique" in Abidjan for their collaboration and support. For the field work in South Africa, we thank Kerneels van der Walt for permission to conduct the study on his land and the whole IVP team for the help with data collection. L. V. L. was supported by the Leverhulme Trust (Early Career Fellowship) and E. W. is grateful for the support of grants by the Swiss National Science Foundation (31003A_159587 and PP03P3_170624) and the Society in Science–Branco Weiss Fellowship.

References

Altmann, J. 1974. "Observational study of behavior: Sampling methods." *Behaviour* 49 (3/4):227–67.

Aplin, L. M., D. R. Farine, J. Morand-Ferron, A. Cockburn, A. Thornton, and B. C. Sheldon. 2015. "Experimentally induced innovations lead to persistent culture via conformity in wild birds." *Nature* 518 (7540): 538–41.

Boesch, C. 2003. "Is culture a golden barrier between human and chimpanzee?" *Evolutionary Anthropology* 12 (2): 82–91.

Boesch, C., and H. Boesch-Achermann. 2000. *The Chimpanzees of the Taï Forest: Behavioural Ecology and Evolution*. Oxford: Oxford University Press.

Bonnie, K. E, V. Horner, A. Whiten, and F. B. M de Waal. 2007. "Spread of arbitrary conventions among chimpanzees: A controlled experiment." *Proceedings of the Royal Society of London B* 274 (1608): 367–72.

Botting, J., A. Whiten, M. Grampp, and E. van de Waal. 2018. "Field experiments with wild primates reveal no consistent dominance-based bias in social learning." *Animal Behaviour* 136: 1–12.

Boyd, R., and P. J. Richerson. 2005. *The Origin and Evolution of Cultures*. Oxford: Oxford University Press.

Catchpole, C. K., and P. J. B. Slater. 2008. "Bird song: Biological themes and variation." Cambridge: Cambridge University Press.

Cheney, D. L., and R. M. Seyfarth. 1983. "Nonrandom dispersal in free-ranging vervet monkeys: Social and genetic consequences." *American Naturalist* 122 (3): 392–412.

Claidière, N., and A. Whiten. 2012. "Integrating the study of conformity and culture in humans and nonhuman animals." *Psychological Bulletin* 138 (1): 126–45.

Cronin, K. A., B. A. Pieper, E. J. C. van Leeuwen, R. Mundry, and D. B. M. Haun. 2014. "Problem solving in the presence of others: How rank and relationship quality impact resource acquisition in chimpanzees (Pan Troglodytes)." *PLoS One* 9 (4): e93204.

Deutsch, M., and H. B. Gerard. 1955. "A study of normative and informational social influences upon individual judgment." *Journal of Abnormal and Social Psychology* 51 (3): 629–36.

de Waal, F. B. M. 2001. *The Ape and the Sushi Master: Cultural Reflections of a Primatologist*. New York: Basic Books.

de Waal, F. B. M., and L. M. Luttrell. 1986. "The similarity principle underlying social bonding among female rhesus monkeys." *Folia Primatologica* 46 (4): 215–34.

Dindo, M., A. Whiten, and F. B. M. de Waal. 2009. "In-group conformity sustains different foraging traditions in capuchin monkeys (*Cebus apella*)." *PLoS One* 4 (11): e7858.

Drea, C., and K. Wallen. 1999. "Low-status monkeys 'play dumb' when learning in mixed social groups." *Proceedings of the National Academy of Sciences* 96 (22): 12965–69.

Fox, E. A., C. P. van Schaik, A. Sitompul, and D. N. Wright. 2004. "Intra- and interpopulational differences in orangutan (*Pongo pygmaeus*) activity and diet: Implications for the invention of tool use." *American Journal of Physical Anthropology* 125 (2): 162–74.

Galef, B. G., and E. E. Whiskin. 2008. "'Conformity' in Norway rats?" *Animal Behaviour* 75 (6): 2035–39.

Garland, E. C., A. W. Goldizen, M. L. Rekdahl, R. Constantine, C. Garrigue, N. Daeschler Hauser, M. M. Poole, J. Robbins, and M. J. Noad. 2011. "Dynamic horizontal cultural transmission of humpback whale song at the ocean basin scale." *Current Biology* 21 (8): 687–91.

Gruber, T., M. N. Muller, V. Reynolds, R. Wrangham, and K. Zuberbühler. 2011. "Community-specific evaluation of tool affordances in wild chimpanzees." *Scientific Reports* 1 (November): 128.

Harmand, S., J. E. Lewis, C. S. Feibel, C. J. Lepre, S. Prat, A. Lenoble, X. Boës, R. L. Quinn, M. Brenet, A. Arroyo, N. Taylor, S. Clément, G. Daver, J.-P. Brugal, L. Leakey, R. A. Mortlock, J. D. Wright, S. Lokorodi, C. Kirwa, D. V. Kent, and H. Roche. 2015. "3.3-million-year-old stone tools from Lomekwi 3, West Turkana, Kenya." *Nature* 521 (7552): 310–15.

Haun, D. B. M., Y. Rekers, and M. Tomasello. 2012. "Majority-biased transmission in chimpanzees and human children, but not orangutans." *Current Biology* 22 (8): 727–31.

Hopper, L. M., S. J. Schapiro, S. P. Lambeth, and S. F. Brosnan. 2011. "Chimpanzees' socially maintained food preferences indicate both conservatism and conformity." *Animal Behaviour* 81 (6): 1195–1202.

Horner, V., D. Proctor, K. E. Bonnie, A. Whiten, and F. B. M. de Waal. 2010. "Prestige affects cultural learning in chimpanzees." *PLoS One* 5 (5): e10625.

Jenkins, P. F. 1978. "Cultural transmission of song patterns and dialect development in a free-living bird population." *Animal Behaviour* 26, part 1 (February): 50–78.

Kawai, M. 1963. "On the newly-acquired behaviors of the natural troop of Japanese monkeys on Koshima Island." *Primates* 4 (1): 113–15.

Kendal, R. L., I. Coolen, and K. N. Laland. 2004. "The role of conformity in foraging when personal and social information conflict." *Behavioral Ecology* 15 (2): 269–77.

Kendal, R., L. M. Hopper, A. Whiten, S. F. Brosnan, S. P. Lambeth, S. J. Schapiro, and W. Hoppitt. 2015. "Chimpanzees copy dominant and knowledgeable individuals: Implications for cultural diversity." *Evolution and Human Behavior* 36 (1): 65–72.

Koops, K., C. Schöning, M. Isaji, and C. Hashimoto. 2015. "Cultural differences in ant-dipping tool length between neighbouring chimpanzee communities at Kalinzu, Uganda." *Scientific Reports* 5 (July): 12456.

Lachlan, R. F., V. M. Janik, and P. J. B. Slater. 2004. "The evolution of conformity-enforcing behaviour in cultural communication systems." *Animal Behaviour* 68 (3): 561–70.

Laland, K. N. 2004. "Social learning strategies." *Animal Learning & Behavior* 32 (1): 4–14.

Laland, K. N., N. Atton, and M. M. Webster. 2011. "From fish to fashion: Experimental and theoretical insights into the evolution of culture." *Philosophical Transactions of the Royal Society of London B* 366 (1567): 958–68.

Luncz, L. V., and C. Boesch. 2014. "Tradition over trend: Neighboring chimpanzee communities maintain differences in cultural behavior despite frequent immigration of adult females." *American Journal of Primatology* 76 (7): 649–57.

Luncz, L. V., and C. Boesch. 2015. "The extent of cultural variation between adjacent chimpanzee (*Pan troglodytes verus*) communities; A microecological approach." *American Journal of Physical Anthropology* 156 (1): 67–75.

Luncz, L. V., R. Mundry, and C. Boesch. 2012. "Evidence for cultural differences between neighboring chimpanzee communities." *Current Biology* 22 (10): 922–26.

Luncz, L. V., G. Sirianni, R. Mundry, and C. Boesch. 2018. "Costly culture: Differences in nut-cracking efficiency between wild chimpanzee groups." *Animal Behaviour* 137: 63–73.

Luncz, L. V., R. M. Wittig, and C. Boesch. 2015. "Primate archaeology reveals cultural transmission in wild chimpanzees (*Pan troglodytes verus*)." *Philosophical Transactions of the Royal Society of London B* 370 (1682): 20140348.

Matsuzawa, T., and G. Yamakoshi. 1998. "Comparison of chimpanzee material culture between Bossou and Nimba, West Africa." In *Reaching into Thought: The Minds of the Great Apes*, edited by A. E. Russon, K. A. Bard, and S. T. Parker, 211–32. Cambridge: Cambridge University Press.

Nielsen, M. 2009. "The imitative behaviour of children and chimpanzees: A window on the transmission of cultural traditions." *Revue de Primatologie* 1 (April).

Noad, M. J., D. H. Cato, M. M. Bryden, M.-N. Jenner, and K. C. S. Jenner. 2000. "Cultural revolution in whale songs." *Nature* 408 (6812): 537. https://doi.org/10.1038/35046199.

Panger, M. A., S. Perry, L. Rose, J. Gros-Louis, E. Vogel, K. C. Mackinnon, and M. Baker. 2002. "Cross-site differences in foraging behavior of white-faced capuchins (*Cebus capucinus*)." *American Journal of Physical Anthropology* 119 (1): 52–66.

Paukner, A., J. R. Anderson, E. Borelli, E. Visalberghi, and P. F. Ferrari. 2005. "Macaques (*Macaca nemestrina*) recognize when they are being imitated." *Biology Letters* 1 (2): 219–22.

Paukner, A., S. J. Suomi, E. Visalberghi, and P. F. Ferrari. 2009. "Capuchin monkeys display affiliation toward humans who imitate them." *Science* 325 (5942): 880–83.

Rendell, L., and H. Whitehead. 2001. "Culture in whales and dolphins." *Behavioral and Brain Sciences* 24 (2): 309–24.

Santorelli, C. J., C. M. Schaffner, C. J. Campbell, H. Notman, M. S. Pavelka, J. A. Weghorst, and F. Aureli. 2011. "Traditions in spider monkeys are biased towards the social domain." *PLoS One* 6 (2): e16863.

Schofield, D. P., W. C. McGrew, A. Takahashi, and S. Hirata. 2017. "Cumulative culture in nonhumans: Overlooked findings from Japanese monkeys?" *Primates* (December): 1–10.

Silk, J. B. 2007. "Social components of fitness in primate groups." *Science* 317 (5843): 1347–51.

Silk, J. B., S. C. Alberts, and J. Altmann. 2003. "Social bonds of female baboons enhance infant survival." *Science* 302 (5648): 1231–34.

Strier, K. B., P. C. Lee, and A. R. Ives. 2014. "Behavioral flexibility and the evolution of primate social states." *PLoS One* 9 (12): e114099.

Vale, G. L., S. J. Davis, E. van de Waal, S. J. Schapiro, S. P. Lambeth, and A. Whiten. 2017. "Lack of conformity to new local dietary preferences in migrating captive chimpanzees." *Animal Behaviour* 124 (Supplement C): 135–44.

van de Waal, E. 2018. "On the neglected behavioural variation among neighbouring primate groups." *Ethology* 124 (12): 845–54.

van de Waal, E., C. Borgeaud, and A. Whiten. 2013. "Potent social learning and conformity shape a wild primates' foraging decisions." *Science* 340 (6131): 483–85.

van de Waal, E., R. Bshary, and A. Whiten. 2014. "Wild vervet monkey infants acquire the food-processing variants of their mothers." *Animal Behaviour* 90 (Supplement C): 41–45.

van de Waal, E., N. Claidière, and A. Whiten. 2015. "Wild vervet monkeys copy alternative methods for opening an artificial fruit." *Animal Cognition* 18 (3): 617–27.

van de Waal, E., M. Krützen, J. Hula, J. Goudet, and R. Bshary. 2012. "Similarity in food cleaning techniques within matrilines in wild vervet monkeys." *PLoS One* 7 (4): e35694.

van de Waal, E., N. Renevey, C. M. Favre, and R. Bshary. 2010. "Selective attention to philopatric models causes directed social learning in wild vervet monkeys." *Proceedings of the Royal Society of London B* 277 (1691): 2105–11.

van de Waal, E., C. P. van Schaik, and A. Whiten. 2017. "Resilience of experimentally seeded dietary traditions in wild vervets: Evidence from group fissions." *American Journal of Primatology* 79 (10): e22687.

van Schaik, C. P., M. Ancrenaz, G. Borgen, B. Galdikas, C. D. Knott, I. Singleton, A. Suzuki, S. Suci Utami, and M. Merrill. 2003. "Orangutan cultures and the evolution of material culture." *Science* 299 (5603): 102–5.

Watson, S. K., S. W. Townsend, A. M. Schel, C. Wilke, E. K. Wallace, L. Cheng, V. West, and K. E. Slocombe. 2015. "Vocal learning in the functionally referential food grunts of chimpanzees." *Current Biology* 25 (4): 495–99.

Whitehead, H., and L. Rendell. 2014. *The Cultural Lives of Whales and Dolphins*. Chicago: University of Chicago Press.

Whiten, A., J. Goodall, W. C. McGrew, T. Nishida, V. Reynolds, Y. Sugiyama, C. E. G. Tutin, R. W. Wrangham, and C. Boesch. 1999. "Cultures in chimpanzees." *Nature* 399 (6737): 682–85.

Whiten, A., V. Horner, and F. B. M. de Waal. 2005. "Conformity to cultural norms of tool use in chimpanzees." *Nature* 437 (7059): 737–40.

Whiten, A., and E. van de Waal. 2017. "Social learning, culture and the 'socio-cultural brain' of human and non-human primates." *Neuroscience & Biobehavioral Reviews* 82: 58–75.

19

On the Origin of Cumulative Culture: Consideration of the Role of Copying in Culture-Dependent Traits and a Reappraisal of the Zone of Latent Solutions Hypothesis

CLAUDIO TENNIE, LYDIA M. HOPPER, AND CAREL P. VAN SCHAIK

Knowledge is the only good that increases when you share it.
MARIE VON EBNER-ESCHENBACH

Introduction

In 2009, Tennie and colleagues published their controversial Zone of Latent Solutions (ZLS) hypothesis, which argued that certain species' cultural behaviors represent traits that each member of the species could, in principle, independently produce (Tennie, Call, and Tomasello 2009). Thus, the observed variance across wild chimpanzee (*Pan troglodytes*) groups represents differential expression of a subset of all potential behaviors that lie within that species' ZLS. The theory did not discount the importance of observational learning in facilitating the emergence of group-level patterns of latent solutions. Rather, Tennie et al. (2009) argued that non-copying observational learning variants alone (see table 19.1) would suffice in encouraging widespread and stable individual expression of latent solutions within a population, resulting in a patchy geographic distribution of various behavioral traits that all fall within the species' ZLS.

This hypothesis clearly distinguished modern human culture, along with its necessary underlying observational learning mechanisms, from that of chimpanzees' and many other species' cultures. It argued that only humans could copy traits beyond their ZLS. This hypothesis, perhaps unsurprisingly, was strongly criticized. Indeed, one of the coauthors of this chapter, Hopper, was a critic of the original ZLS hypothesis, and by coauthoring this chapter, our aim is to find common ground, address earlier misconceptions, and redress the original framing of the hypothesis.

This chapter has two main aims. First, we provide an update for the ZLS

TABLE 19.1. Glossary of terms referenced

Copying observational learning	Culture-dependent traits (CDTs) require observational learning mechanisms that enable the transmission of the form of a behavior or artifact (or artifact production). Therefore, CDTs require form-copying learning mechanisms, and in this chapter we use the umbrella term "copying observational learning." Note that it is still unclear which observational learning mechanisms (Whiten et al. 2004) fall under this umbrella term. Also note that copying observational learning does not necessarily lead to CDTs (e.g., the copying may remain inside the Zone of Latent Solutions [ZLS]). We use the term observational, rather than social, learning to more intuitively include learning forms that do not require agents (e.g., various forms of emulation).
Non-copying observational learning	Non-copying observational learning types do not transmit a behavior's or artifact's form. Instead, they guide the learner's behavior and learning (e.g., to targets), and/or may increase the individual's general motivation to act. Sometimes such forms of observational learning are labeled as low-fidelity social learning (e.g., Whiten et al. 2009). We believe this term is potentially misleading—there is no fidelity in form transmission (i.e., no form copying) in such learning. To avoid the resulting confusion, in this chapter we make the distinction between copying and non-copying forms of observational learning, rather than between high- and low-fidelity forms.
Culture	The presence of an innovation (can be a latent solution) at time t must have a positive influence—via observational learning—on the likelihood of the presence of this innovation at time *t*+1. That is, any form of observational learning—even happening a single time—will count as culture (compare with Neadle, Allritz, and Tennie 2017).
Culture-dependent traits	The presence and observation of the form of at least one trait at time *t* is required for the presence of the same form of at least one trait at time *t*+1 (approximations of form also count: thus we would allow for the evolution of the trait by copying error alone).
Cumulative culture	The form of at least one trait at time *t* must have an influence—via observational learning—on the (somewhat different) form of at least one trait at time t+1. Can lead to culture-dependent traits or may stay within the ZLS ("gray" zone of cumulative culture). May or may not be open-ended (clear evidence for the latter currently only in humans). May or may not increase efficiency of the trait.
Latent solution	A trait that can be expressed by several culturally unconnected (i.e., naive) non-enculturated and non-deprived individuals within their lifetime, without the need to observe the trait in another individual and/or the resulting products, and also without the need to observe similar traits/products to the one in question (though there may be a need to individually innovate precursors to the trait before the trait is also shown). Non-copying observational learning and copying within the ZLS can lead to socially mediated serial reinnovation of latent solutions within populations.
Open-ended cumulative culture	Open-ended cumulative culture (Tennie, Caldwell, and Dean 2018) describes much of modern human technology (e.g., we might always envision a faster rocket, a higher tower) and where new combinations, sometimes via donated culture (Pradhan, Tennie, and van Schaik 2012), allow us to reach ever-higher outcomes (if there is a limit, then we have not reached it yet).
Asymptotic cumulative culture	Asymptotic cumulative culture is one in which social learning channels and leads to an optimum (total or local; compare to endnote in Tennie, Call, and Tomasello 2009); note that there is evidence for asymptotic cumulative culture in non-human animals (Claidière et al. 2014; Sasaki and Biro 2017).

hypothesis, in part by clarifying misconceptions and by refining the tenets of the ZLS approach (even further clarifications and responses can be found in Tennie et al. in revision). Second, we evaluate how well the ZLS account has stood the test of time over the ten years since Tennie et al. (2009) was published. In this way, our aim is to put the culture of chimpanzees in context, by comparing the observational learning mechanisms that underlie it, to that of humans. Our focus will therefore be on transmission, though this means that we will also talk about innovation (latent solutions are much closer to innovation than to transmission).

CUMULATIVE CULTURE AND CULTURE-DEPENDENT TRAITS

Whatever medium (paper, screen, hologram) you are reading this from, it owes its existence to a prolonged process of copying (and inescapable copying error), drift, and selection. It is a product of cumulative culture, for which earlier variants of a trait spread socially by *copying* observational learning mechanisms and then become the basis of further changes that spread again (compare with the "ratchet effect"; Tomasello, Kruger, and Ratner 1993). No "culturally naive" human raised alone (on a "cultural island") would have been able to produce anything like this device; it is therefore not only an example of cumulative culture, but also a "culture-dependent" trait (CDT; Reindl et al. 2017). CDTs can be objects, such as tools, including their production methods (our foci in this chapter), but may also include social behaviors based on actions, such as some dances, rituals, and gestures, or social behaviors based on copying of sounds, such as vocal language. In humans, many of these can only be understood as being products of a history of consecutive modification.

Given the enormous importance of cumulative culture for our species, many have asked whether nonhuman animals (henceforth animals) are also capable of showing cumulative culture. Currently, the most widely used definition of cumulative culture is concerned with traits (behaviors and/or their resulting products) that no individual can independently reinnovate (by "independent" we mean independent of cultural access to the trait or its cumulative cultural "neighbors"; see Boyd and Richerson 1996; Muthukrishna and Henrich 2016). This definition targets an *outcome* of the cumulative cultural process—the resulting traits—instead of the *process* or mechanisms leading to those traits. A lack of individual reinnovation may derive from the traits becoming too complex and/or too arbitrary over time (e.g., specific actions and sequences within rituals; Legare and Nielsen 2015). However, while we typically consider only these late outcomes of cumulative culture, earlier outcomes of the cumulative cultural process may remain within the reach of individuals (i.e., within their ZLS; see Pradhan, Tennie, and van Schaik 2012). Thus, a lack of evidence for a species having CDTs does not preclude evidence for (early) cumulative culture. How, then, can we identify this earlier, or "gray," zone of cumulative culture?

If each innovation within a species' ZLS emerges with a certain probability, we are asking about the likelihood of another *conditional* innovation (i.e., one *dependent* on the presence of an earlier innovation, emerging within the lifetime of a single individual). For instance, let us assume some animals spontaneously innovate the technique of cracking objects using unmodified stones (innovation 1), but, occasionally, one individual then innovates the use of a leaf sheath to more securely grip the stone (innovation 2). Intuitively, this would feel like cumulative culture. However, while this second innovation may arise rarely, because both innovation 1 *and* innovation 2 occurred within an individual's lifetime, *both* innovations, *even in combination*, are within the ZLS of these animals and, therefore, do not represent a CDT (as per the definition of Reindl et al. 2017; and see table 19.1 for definitions used in this chapter). This example represents the gray zone of cumulative culture. Clearly, when additional conditional innovations are made, the probability of them still being all sequentially innovated by a single individual during its own lifetime becomes vanishingly small. At what point, then, shall we speak of CDTs? In the following, we will describe how we can tackle this question. Also, we will consider what factors allows us, but not chimpanzees and many other species, to escape our own ZLS to generate CDTs.

Over the past two decades, and using indirect evidence for observational learning, "cultures" (sensu Neadle et al. 2017) have been inferred for various animals, including nonhuman great apes (henceforth great apes; Robbins et al. 2016; van Schaik et al. 2003; Whiten et al. 1999, see also Hopper and Carter, chapter 7 this volume; Luncz and van de Waal, chapter 18 this volume), monkeys (Leca, Gunst, and Huffman 2007; Perry et al. 2003; Santorelli et al. 2011), birds (Hunt and Gray 2003, see also Hopper and Carter, chapter 7 this volume; Luncz and van de Waal, chapter 18 this volume) and cetaceans (Whitehead and Rendell 2014). Specifically considering the case of great apes, inferred cultural behaviors appear in the form of simple technologies, comfort behaviors, social conventions, and signal variants (see Whiten 2011 for a review for chimpanzees; van Schaik et al. 2009, for orangutans, *Pongo* sp.).

When we focus on technology or material culture, with one possible exception (Tasmanians after they became isolated from the Australian mainland; McGrew 1987), even the least technologically sophisticated mobile human foragers have larger tool repertoires than any other primate, including chimpanzees (for additional discussions of chimpanzee tool use, see Luncz and van de Waal, chapter 18 this volume; Mann, Stanton, and Murray, chapter 3 this volume; Pruetz, Bogart, and Lindshield, chapter 17 this volume). Moreover, although humans underwent a long period of increasing cultural complexity, culminating in a cultural explosion within roughly the past 100,000 years (Hoffecker 2017), no cultural "big bang" has been documented for chimpanzees' and other great apes' behavior or material culture (or indeed any other nonhuman animal). Therefore, we cannot currently infer cumulative cultural processes for nonhuman primates, which makes it unlikely that great apes possess CDTs. For example, judging from excavated tools, the basic technique of nut cracking has seemingly remained static (here, and elsewhere, we mean *variable stasis*—around an unchanging mean) in chimpanzees for at least four thousand years (Mercader et al. 2007). Similarly, indirect evidence for a likely stasis in tool techniques also exists for monkey stone tool behavior (long-tailed macaques, *Macaca fascicularis*, hammering: from Carpenter's initial, admittedly not very detailed, report in 1887 to Gumert, Kluck, and Malaivijitnond 2009; capuchins, *Cebus* sp.: Haslam et al. 2016). Of course, even longer periods of stasis (hundreds of thousands of years in each case) characterizes the archaeological record of our early stone-tool making ancestors, suggesting that early hominin stone tool types, too, do not represent CDTs (Tennie et al. 2016, 2017; compare also to a narrower, and thus less-flexible, more genetically based adaptation argument by Corbey et al. 2016).

THE CASE FOR LATENT SOLUTIONS

A major challenge for evolutionary anthropology is to explain why modern human tool culture advanced so rapidly, becoming greatly elaborated and consisting of a blinding variety of CDTs, especially those that are "open-ended" (see table 19.1; e.g., Boyd and Richerson 1996; Tennie, Call, and Tomasello 2009; Tomasello 1999), while great ape tool cultures seemingly have not (e.g., Dean et al. 2014; Tennie, Call, and Tomasello 2009).

Based on theoretical considerations, and what was known at the time about modern great ape tool use, Tennie et al. (2009) argued that modern ape tool cultures lack CDTs and instead consist of traits from within their ZLS (see also Tennie and Hedwig 2009). To explain the difference between human and other great ape culture, Tennie et al. assumed that great apes do not have access to the rich group-level "collective brain/collective intelligence" as humans do (Henrich 2015; Henrich and Tennie 2017; Muthukrishna and

Henrich 2016; Ridley 2010; Tennie and Over 2012; Tomasello 1999). This is not to deny that chimpanzees can, and do, use some variant(s) of observational learning. However, observational learning per se does not automatically lead to cumulative culture or CDTs (at least not of an open-ended type): *copying* observational learning mechanisms are required. Note that this learning must also be discriminatory to some degree; "blind mimicry" (of either actions or results) does not lead to the original, *directed* ratchet effect (see Tomasello, Kruger, and Ratner 1993). In particular, observational learning powerful enough to allow individuals to copy traits outside their ZLS is required. As we argue below, such copying mechanisms might be absent or rare in great apes and other primates (see also Lewis and Laland 2012; Tennie, Call, and Tomasello 2009). Alternatively, great ape copying might exist—or even be common—but with only limited reach and bandwidth (and in this case these skills may fall short of escaping their ZLS).

The original ZLS hypothesis assumed that every naive member of a taxon is bound by its ZLS and able to reinnovate only "cultural variants" that already are or can be shown by other members of its taxon (Tennie, Call, and Tomasello 2009; cf. Bandini and Tennie 2017). Non-copying observational learning mechanisms alone can lead to an illusion of a spread/copy of these latent solutions within a population (as their frequencies may increase to population-wide levels), even though the actual form of the behavior does not spread due to a lack of copying (box 19.1). Non-copying variants of observational learning here merely result in an increase in frequency (and stabilization) of the respective form. The ZLS hypothesis proposes that the observed uniformity (variable sameness around a mean) derives from an ontogenetic interplay between environmental channeling (e.g., Koops, Visalberghi, and van Schaik 2014) and evolved anatomy, motivation, and cognitive skills.

Uniformity within ape culture may thus be best described as repetitions of form without replications of form, or as "socially-mediated serial reinnovations" (Bandini and Tennie 2017). This view is quite consistent with long-term studies of skill acquisition by immature chimpanzees and orangutans (Lonsdorf 2006; Schuppli et al. 2016), in which immatures peer at feeding and tool use events of others but then subsequently practice/develop the skill individually until they master it themselves. Moreover, the kinds of non-copying observational learning mechanisms that can fuel this system may be widespread in the animal kingdom (e.g., Hoppitt and Laland 2013). And so, what makes chimpanzees special is likely not their observational learning skills (as is currently often assumed), but, rather, the wide reach of their individual innovative skills (i.e., the relatively large size of their ZLS as compared

BOX 19.1. How Population Level Differences Can Arise via the Latent Solution Approach, without Copying of Traits

Let us assume that the use of a stone as a hammer is within the reach of individuals of a species: naive subjects, when tested, can and do spontaneously use stone hammers to crack nuts (i.e., nut cracking is a latent solution; see prediction 1). Consider two geographically (or socially) separate populations, both living in similar environments and being genetically indistinguishable: population 1 and population 2. Given the widespread use of diverse non-copying observational learning mechanisms in the animal kingdom (and the close link between individual ability and observational learning, Reader and Laland 2002), we can further assume that, if an individual in one population is able to start using a stone as a hammer, that her initial use of this stone hammer increases the *frequency* of stone hammer usage in her whole population. This can happen, for example, by others becoming more interested in the stones (i.e., stimulus enhancement). By having their attention drawn to the stone, and exploring the options afforded it, other individuals will be more likely to reinnovate stone hammer use themselves. Several non-copying observational learning mechanisms and factors can also work *together* here (plus even copying that remains inside the Zone of Latent Solutions). One outcome can be that the whole of population 1 shows the stone-hammer behavior, while population 2 does not. Why, though, would we not expect population 2 to show this behavior, at least eventually? There may be several reasons for this. In this particular example, we already have excluded subtle genetic and ecological differences (and their interplay), but even then other potential factors remain, of which we shall consider a few here. It might be that the species is generally neophobic, and so initial seeding of the behavior is unlikely in general; making it unlikely that population 2 repeats what population 1 did. Population 2 might also have solved the same problem in a *different* way, and where their solution became prevalent in an analogous way as stone hammer use did in population 1 (but where both are latent solutions). Tennie, Call, and Tomasello (2009) described this potential between-population variation as the outcome of cultural founder effects. If a population has an alternative solution, they might be "reluctant" to engage with the environment in ways not expressed by its members (e.g., because they might conform to majorities, or are conservative in their behavior, or show functional fixedness, Brosnan and Hopper 2014). While we do not know the mechanism, we do know that once the range of latent solutions seems more settled in wild populations, new innovations rarely stick (Nishida, Matsusaka, and McGrew 2009), though they *sometimes* still do occur and then, sometimes, increase in frequency (Hobaiter et al. 2014).

FIGURE 19.1. An illustration of the illusion of uniformity due to copying. In this illustration two different chimpanzees (one in the wild, one in captivity) independently arrived at a similar behavioral form individually (here, stick tool use to access food in small cavities). Many thanks to the artist William Daniel Snyder.

to most other species). Nonhuman great apes are great evolved *innovators*, but they are not accomplished *copiers* (see fig. 19.1).

AUTO-ENCULTURATION OF EARLY HOMININS

A recent, intriguing, and logically consistent hypothesis argues that increased rates of environmental change in the *relevant* environment (the *Environment of Evolutionary Adaptedness*) of both humans and great apes should have led *both* to be better at observational learning today than we can assume the last common ancestor was (Henrich 2015). Yet even (unenculturated) modern great apes, while adept at various non-copying observational learning, show very little spontaneous evidence of copying observational learning—at least not regarding copying outside their ZLS (see below). Given general phylogenetic arguments, it is therefore parsimonious to assume that our last common ancestor also lacked copying observational learning mechanisms that allowed them to copy traits beyond their ZLS (Tennie et al. 2016, 2017). Of course, at some point in time our ancestors must have become able to copy what they could not have (re)innovated on their own. But the evolution of these copying skills is unlikely to have happened in a vacuum; what we may envision is an automatic coevolution of observational learning skills, individual learning skills and of the culture(s) that resulted (compare Henrich 2015; Lewis and Laland 2012; Pradhan, Tennie, and van Schaik 2012; Sterelny 2012). Eventually (and likely with ups and downs) this process will have led into a gray zone of cumulative culture, and then to CDTs. What might have facilitated these shifts?

In addressing this, data on human-enculturated apes are highly relevant.

Human-enculturated apes show more varied behavior than their non-enculturated cousins and they even show very clear (yet crude) action *copying* skills—perhaps especially when specifically trained to do so by humans (e.g., Custance, Whiten, and Bard 1995; Miles, Mitchell, and Harper 1996; Russon and Galdikas 1993). Therefore, the genetic *potential* of apes seemingly includes the capacity to develop action copying skills, probably resulting from changes in the brain (Bard and Hopkins 2017; Pope et al. 2018), especially given that such structures seem to be absent in non-enculturated, untrained great apes (Hecht et al. 2012). Copying of actions has been proposed as a major driver of cultural evolution (Tomasello 1999). Since a similar *potential* to copy actions can be reasonably expected in the last common ancestor, it could therefore have allowed for an (initially slow) autocatalytic evolution of cultural processes and cultural change (Pradhan, Tennie, and van Schaik 2012), leading to the gradual, but clear, manifestation of (action) copying skills at some point in our lineage. In a similar vein, other copying skills (goal copying, environmental results copying) could also have become enhanced (with emergent properties of their own; compare Acerbi and Tennie 2016). This might mean that humans today are biologically highly prepared for copying (Tomasello 1999). But, in addition, or alternatively, these copying skills themselves could have become partly or wholly dependent on culture itself; they could have been (and could continue to be) socially learned during this process ("observational learning of observational learning"; see Mesoudi et al. 2016; cf. Heyes 2012, 2018 including comments and response).

Approach: Revising the ZLS

A criticism of the original ZLS hypothesis (Tennie, Call, and Tomasello 2009) is that it lacked precision or "operational criteria" (Schofield et al. 2017). In the following, we reformulate the ZLS hypothesis more precisely, in stochastic terms. For this, we take advantage of the fact that the ZLS hypothesis makes two testable claims. Importantly, to ensure ecologically valid data when testing these claims, we propose that only non-enculturated and/or untrained subjects of the same species (or even of the same subspecies, if possible) be regarded as valid, relevant test cases. We stipulate this because (i) the ZLS hypothesis is concerned primarily with explaining modern ape culture and (ii) human enculturation does not occur in wild ape settings, so human-enculturated subjects are ecologically invalid to test the ZLS hypothesis. We are also not interested in the results of deprived subjects, for the same reason (more detail on subject selection in Henrich and Tennie 2017).

ZLS PREDICTION 1: IF A SPECIES CANNOT BE SOCIALLY ENCOURAGED (MEDIATED) TO SHOW A BEHAVIOR, IT ALSO CANNOT INNOVATE (OR REINNOVATE) THE BEHAVIOR INDIVIDUALLY

For a species, such as (likely) chimpanzees, that is bound within its ZLS, if the probability, *p*, of acquiring a particular trait x for individual *i* through individual innovation (i.e., P_{innov} [x, *i*]), is effectively zero (≈ 0), then no other individual of the same species (*j*, *k*, etc.) should have a meaningfully higher chance to acquire trait x, even if aided by observational learning opportunities (i.e., P_{acq} [x, *j*] ≈ *p* ≈ 0). Thus, individuals innovate and reinnovate (including via socially mediated serial reinnovation) traits *within* their ZLS, but not traits *outside* their ZLS. In contrast, for a species able to produce and sustain CDTs, such as modern humans, once traits beyond their ZLS are innovated (again, with likelihood *p*), other naive members of that same species should be able to observationally learn these CDTs. For such species, when a naive individual's P_{innov} [x, *i*] = *p*, it is instead true that P_{acq} [x, *j*]>> *p* (i.e., individuals should have a realistic chance to acquire a trait through copying observational learning, basically bypassing the sequence of innovations that led to the current state).

The value of P_{acq} for the non-innovating acquirer will depend on the effectiveness of observational learning. In the absence of any observational learning, the acquisition probability should be the same as for P_{innov}. Non-copying variants of observational learning, such as stimulus or local enhancement, and copying observational learning that remain limited to copying within a species' ZLS (a possibility we return to below), can raise this probability (i.e., P_{acq} [x, *j*]> *p*) to some extent by providing various constellations of factors that favor individual reinnovation. However, only powerful variants of copying observational learning may increase this probability in a substantial way, maybe even to a point close to 1 (i.e., P_{acq} [x, *j*]>> *p*).

The argument presented here is that unenculturated great apes lack the required copying observational learning mechanisms (including copying variants of teaching) and this is why their observational learning appears insufficient to lead to CDTs on a population level, not because P_{innov} is close to zero. Thus, although both modern humans *and* apes have a ZLS (Reindl et al. 2016), only modern humans have the required copying mechanisms and can culturally generate and maintain traits *outside* their ZLS, in a Vygotskian fashion (Reindl, Bandini, and Tennie 2018, see also Tennie et al. in revision).

Evaluating ZLS Prediction 1

Testing prediction 1 requires that we operationalize *p*. To do so, we must make an inference based on the number of observations of the behavior in question. It follows that the more we see a particular trait—across culturally independent, naive individuals, within and between populations, in the field or in captivity—the higher the estimated likelihood of innovation. Counting innovations in this way also allows us to use field data, which confer three key advantages: (i) full ecological validity, (ii) extended time frames, and (iii) potentially large sample sizes. However, when we aim to experimentally examine the occurrence probability of target traits, more often than not this means testing in captive situations, which are usually much more restricted in time *and* in sample size. Balancing such limitations, however, is that the likelihood of innovation in captivity is increased relative to the wild because of the so-called "captivity" or "spare-time" effect (Benson-Amram, Weldele, and Holekamp 2013; van Schaik et al. 2016) and because we can provide all the required materials needed to encourage individual innovation (simulating the effect of various non-copying observational learning processes during socially mediated individual reinnovation events).

We believe data from both wild and captive individuals can be framed under a common operationalization: if any target trait was performed at least twice by an innovator (inferred or directly observed) *and* at least once in one or more individuals from one or more culturally unconnected populations (so as to preclude influences of copying) it is within the ZLS of the species/subspecies. The trait is then regarded as a latent solution where P_{innov} [x, *i*]> 0. In contrast, if the trait has never been observed, or (to be conservative) only ever has been observed (even repeatedly) in a *single* individual, it is to be regarded (for the time being, pending further testing) as being *beyond* the species' ZLS (P_{innov} [x, *i*] ≈ 0). The same is true for a behavior that was observed once each in two individuals. If the behavior was observed once each in three culturally unconnected individuals, it is to be regarded as within the ZLS of the species/subspecies.

One way to test the first prediction is to train "demonstrator" individuals to acquire a trait—or to show an advanced technological solution—that has never been innovated before by any known individual of that species (an effective $P_{innov} \approx 0$) and then to allow observers to watch the trained demonstrator. While children have been shown to copy what they cannot innovate (Reindl et al. 2017), current data for unenculturated, untrained great apes support our prediction that if a behavior is beyond the species' ZLS, other members of the population should not be able to acquire this trait through observational learning. For example, when a male chimpanzee was trained to perform an action with an estimated $P_{innov} \approx 0$ (a "praying arm" action that never appeared in baseline observations), a second chimpanzee failed to copy this novel food-rewarded action (Tennie, Call, and Tomasello 2012). Furthermore, an earlier study by Tomasello et al. (1997) produced similar findings using female chimpanzee demonstrators and 18 observers: the trained novel food-rewarded actions were never copied. Similar data exist for bonobos (*Pan paniscus*, Clay and Tennie 2017). In these studies, $P_{innov} \approx 0$ and, in accordance with prediction 1, there was zero success (i.e., close to *p*) in observational learning of a trait beyond the ZLS. Using human demonstrators (who need little training) produces similar results (Tennie, Call, and Tomasello 2009).

However, when traits *are* within a species' ZLS, the *appearance* of copying can be frequent. Indeed, copying may actually be involved sometimes, though only copying of traits within that species' ZLS. Key examples with chimpanzees come from studies using so-called two-target tasks, such as moving a door to the left (trait 1) or the right (trait 2) to gain access to a reward (more traits may be possible, but typically there are two). In *baseline* conditions of these tasks, apes generally produce both solutions' action patterns spontaneously (e.g., push door right *and* push door left; e.g., Hopper et al. 2008; Kendall et al. 2015; Tennie, Call, and Tomasello 2006). In line with our definitions, *both* action traits are thus within the species' ZLS and so, any production of these actions by naive individuals in experimental conditions (after observations of others) may be copying within the ZLS, or even simply be evidence of reinnovation of these traits—merely mediated/catalyzed by non-copying observational learning.

When we look for specific variants of copying observational learning (even copying is split into sub-mechanisms), one that is rare or absent is again action copying. Indeed, with certain two-target tasks chimpanzees also reproduced the trait even when no social (action) demonstration was shown, but simply the target object movement (via a so-called ghost display; see Hopper 2010). The fact that this happens more reliably when social demonstrations are provided (instead of "ghosts"; Hopper et al. 2007) does not mean that action copying is responsible. Rather, it could be that traits within a species' ZLS are more strongly elicited by a social than a non-social model (Hopper et al. 2015a), perhaps due to how the information is encoded by the observer (Howard et al. 2017). For species bound by the ZLS, non-copying observational learning encourages a homogenous response at the group level as to which of the two traits they produced in such two-target tasks (see also box 19.1).

One potential exception to prediction 1 is the "Pan-pipes" apparatus; a variant of two-target tasks for which a stick tool could be used for either a "lift" or "poke" trait to release a reward from the apparatus (Hopper et al. 2007, 2015a; Whiten, Horner, and de Waal 2005). In all the published baseline conditions in which the Pan-pipes was provided, only one trait, poke, was spontaneously innovated by chimpanzees without social information, and even then it was only discovered by one chimpanzee of the more than 50 tested in this situation—across two different research facilities (Hopper et al. 2007; Whiten, Horner, and de Waal 2005). As is fairly typical for these tasks, in the observational learning experimental conditions, trained demonstrators performed one of the two traits to their respective social groups, with one group being seeded with the poke and one with the lift trait (Hopper et al. 2007; Whiten, Horner, and de Waal 2005). The demonstrated trait dominated the respective group's approach to the box. Thus, the Pan-pipe studies were argued to have identified the successful *copying* of at least one (the lift) trait (which never spontaneously appeared in the baseline conditions), and which was therefore inferred to have been outside the ZLS of chimpanzees (see also Hopper et al. 2015a). However, this conclusion was premature. It ignored the fact that *both* traits (poke and lift) appeared in *trait*-naive subjects; namely in subjects in groups seeded with the *alternative* trait. In other words, the lift trait was innovated by subjects in the poke-seeded test groups, despite never having seen that trait demonstrated (Hopper et al. 2007; Whiten, Horner, and de Waal 2005). Importantly, in an additional (non-published) study with chimpanzees in another location that were given the Pan-pipe apparatus without social demonstrations (baseline), *both* the lift and poke traits appeared spontaneously (C. Hrubesch, pers. comm., November 16, 2017). Similar spontaneous discovery of both traits has also been shown by human children presented with the Pan-pipes (Hopper et al. 2010). Neither trait, therefore, should be considered as beyond the chimpanzee—or human—ZLS.

In sum, we know of no two-target task that, to date, has successfully tested apes' copying of traits beyond their ZLS (at least not when testing non-enculturated, untrained apes). Challenging this conclusion, a recent study with chimpanzees purportedly identified the cumulation of a technological trait beyond the level attainable by naive chimpanzees (i.e., a claim of CDT-copying in chimpanzees; Vale et al. 2017). However, while it is true that the chimpanzees' performance in the test conditions that allowed for the possibility of observational learning somewhat exceeded the apes' performance in conditions in which observational learning was not possible (i.e., baseline), their performance in both conditions was similar, and high. Furthermore, the baseline condition was both underpowered *and* different in important respects from the observational learning conditions. In order to evaluate claims for or against the existence and observational learning of CDTs, just as in evaluating whether a trait is a latent solution or not, the baseline condition must instead be the experimental focus. It must match the conditions of testing for observational learning as closely as possible in method (see e.g., Hopper et al. 2015a). And, as a rule of thumb, we propose that baseline conditions should contain at least as many subjects as all observational learning conditions combined, and that these control subjects be tested for at least twice the amount of time as test subjects. We add this requirement because any social learning may very likely mediate, and thus speed up, the individual expression of behaviors. Speeding up expression, however, would not mean that social learning is *necessary* for the trait to occur and/or that copying social learning mediates it. For example, Collias and Collias (1964) tested for the spontaneous manufacture of a weaverbird (*Textor cuculatus*) nest by a naive male (in essence, a baseline condition). The bird did make a wild-type ("uniform") nest in the end—but it simply took him *longer* to do so than non-naive birds.

Tennie, Call, and Tomasello (2009) proposed that great apes lack CDTs because they do not imitate, as indicated by their relative lack of action copying, while humans are adept at it (see also Haun and Rapold 2009; Tomasello 1999). The presence of CDTs was thus linked to the ability to imitate novel actions or novel action sequences (so-called production imitation: Byrne and Russon 1998). More recently, and challenging these earlier predictions, end-state emulation (where subjects individually recreate an environmental end result) has been shown to be sufficient to allow CDTs to be copied in human children (Reindl et al. 2017) and adults (Caldwell and Millen 2008, 2009; Caldwell et al. 2012; Zwirner and Thornton 2015). Therefore, in theory, individual reengineering of artifacts/results might also allow *other* species to pass on CDTs, including primates (Fragaszy et al. 2013; Meulman et al. 2012). It may, however, again be that the emulation prowess of apes may be insufficient to transmit CDTs (see experiment in Tennie, Call, and Tomasello 2009, which potentially allowed for such learning to no effect). Additionally, the apparent lack of action copying among wild apes may prevent the emergent boost of transmission fidelity that is enabled by combining the copying of several types of information at once (Acerbi and Tennie 2016). It may also be that several mechanisms might currently hide under the label of end-state emulation: one may be an insightful object-guided copying mechanism (close reengineering; seemingly used by humans). Another might involve much less copying and be more individually-based product copying, which requires the learner to be near ready to produce the artifact in question on his/her own (sensu Tennie, Call, and Tomasello 2009). Lastly, more fine-grained emulation forms also exist, namely so-called object movement reenactment

(Custance, Whiten, and Fredman 1999; Whiten et al. 2004), and *their* role in ape culture (and perhaps in passing on of ape CDTs) needs to be examined more in future. There is the possibility that object movement reenactment may underlie the target matching in two-target tasks that is so frequently observed in studies on great apes (described above).

CDTs may derive their cultural dependency from inherent unlikeliness and/or from overall hierarchical complexity (compare Byrne 2007), and/or from linear/sequential complexity (e.g., dance steps). Concerning the latter, we also need to create tasks that test for the transmission of culture-dependent *arbitrary sequences* in great apes (compare Ghirlanda, Lind, and Enquist 2017), where each step on its own may be a part of their ZLS, but in combination they exceed it. But to datc, studies have reported either limited or no copying by great apes of even short sequences (Call and Tomasello 1995; Dean et al. 2012; Whiten 1998—and note that enculturated apes were tested in this study). Concerning the former, we will next look at *individual* innovation of even seemingly complex behavior in chimpanzees.

ZLS PREDICTION 2: LATENT SOLUTIONS HAVE A SUBSTANTIAL LIKELIHOOD BOTH TO BE INNOVATED IN NAIVE INDIVIDUALS AND TO INCREASE IN FREQUENCY (AND STABILIZE) THROUGH SOCIALLY MEDIATED REINNOVATION

For a behavior *within* the ZLS, the probability of acquiring a trait through individual innovation is above 0; thus, when properly tested in baselines, all naturally observed (i.e., not human-trained) traits of a given species should reappear in naive individuals. In addition, the probability of showing the trait after the encouraging influence of some form of observational learning should be >0 (i.e., $0 \ll P_{innov}[x, i] \ll 1$, and $P_{innov}[x, i] < P_{acq}[x, j]$). We include the condition $P_{innov}[x, i] \ll 1$ in order to exclude hard-wired behaviors, which will be (nearly) invariably shown by all members of a population or species in a given environment, or even across environments. These behaviors would have a likelihood of 1 or close to 1, and thus this would leave no real frequency gap that could be filled by social learning (i.e., it would lead to a ceiling effect). These behaviors would *still* represent latent solutions (as they do not require social learning to appear)—but they simply do not lend themselves to testing in this way (for further discussion see Tennie et al. in revision).

For species that are restricted to their ZLS, the story should end here. For modern humans, in stark contrast, even if $P_{acq}[x, j]$ is $\gg 0$ (i.e., if observational learning allows for a *spread* of the trait, via powerful copying), the probability of independent innovation by naive individuals (P_{innov} [x, *i*]) can still be very close to zero.

Evaluating Prediction 2

Experimental data currently available largely support prediction 2. For example, Price et al. (2009) found there was a low, but consistent, level of success by naive chimpanzees in baseline conditions to combine or extend plastic tools designed especially to facilitate combinatory responses—but when observational learning was made possible, the behavior in question showed the predicted increase in frequency. In another study, subjects spontaneously used water to raise objects within a tube (i.e., it was a trait within their ZLS; Hanus et al. 2011) but more instances of this behavior were performed by other individuals after observational learning was made possible (noteworthy even in situations that disabled action copying; Tennie, Call, and Tomasello 2010). These studies may thus serve as *experimental models* of how the ZLS hypothesis envisions socially mediated serial reinnovation in the wild to explain the observed differences in frequencies between groups, i.e., their cultures (see also box 19.1).

Considering the data from wild animals, if a behavior appears in several culturally unconnected populations, this fulfills the criterion for a suspected latent solution. This follows from the powerful prediction given above: when properly tested in baselines, all cultural behaviors of a given species in the wild should reappear in naive individuals. With this we can also test for the reinnovation of traits in captivity that appear in only a single population in the wild. These tests have been labeled "latent solution experiments" (Tennie and Hedwig 2009), and have also been adapted to test culturally naive wild populations (Gruber et al. 2009).

Several latent solution experiments have provided evidence of wild cultural behaviors reappearing in naive apes who lacked observational learning opportunities for the target traits (Bandini and Tennie 2017; Huffman and Hirata 2004; Menzel et al. 2013; Tennie et al. 2008). For example, one of the few suspected cases of gorilla (*Gorilla gorilla*) culture (a particular form of nettle feeding) was observed to emerge spontaneously in naive captive gorillas (Tennie and Hedwig 2009; Tennie et al. 2008; see also Masi 2011), although Byrne, Hobaiter, and Klailova (2011) have questioned this conclusion. Similarly, leaf swallowing (a proposed self-medication behavior) has been observed in naive apes tested in more than one captive facility, revealing it to be within the ZLS of chimpanzees and bonobos (Huffman and Hirata 2004; Menzel et al. 2013). Crucially, tool-use behaviors also reappear in naive apes:

naive, captive chimpanzees were reported to spontaneously use the detailed action form underlying wild algae scooping behavior (Bandini and Tennie 2017). These studies support our prediction that the *frequency* of an ape latent solution may be influenced by *non-copying* observational learning, but the *form* derives most likely via individual reinnovation, and is not passed via *copying* variants of observational learning (box 19.1).

The systematic latent solution experiments project is still ongoing—in fact, it has just begun. However, of highest priority are those cases where the suspected cultural variants seem rare and/or locally restricted, such as producing a brush tool *prior to use* in chimpanzees—which, to date, has been described for only one single wild population (Sanz and Morgan 2007), but has now also been reported for a captive population of chimpanzees (Hopper et al. 2015b). Considering orangutans (van Schaik et al. 2009), only one behavior has been, so far, identified as being similarly locally restricted: tree-hole tool use. However, this behavioral form is commonly reinnovated by captive, naive orangutans (i.e., has been shown to be a latent solution, Lehner, Burkart, and van Schaik 2010).

Behaviors that are more widespread, i.e., that appear across culturally unconnected populations, also require testing (even though they can already be inferred to have had more than one single innovator and therefore to likely be latent solutions; a ZLS variant of the logic of the "method of exclusion"). For example, chimpanzee nut-cracking is present in different populations across many hundreds of kilometers from each other (Morgan and Abwe 2006), and also appears to be a latent solution for capuchin monkeys (naive subjects reinnovated it; Visalberghi 1987). Similarly, orangutans have been recorded performing tool-assisted seed-extraction from *Neesia* fruits (van Schaik et al. 2009) in multiple, geographically distant (i.e., only indirectly unconnected) populations. However, only by experimentally testing these behaviors can we fully exclude the remaining theoretical possibility that *all* these occurrences go back in time to a single innovation, carried over across all past population drifts (the "Diffusion from Unitary Origin" [null] model; Whiten et al. 2001).

The *form* of wild cultural traits themselves can also appear cumulative, although looks can be deceiving (Collias and Collias 1964; Tennie and Hedwig 2009). Examples of such behaviors proposed as candidates for cumulative culture include tool *sets*, in which two or more different tools are used in succession. For example, chimpanzees perforate an ant nest with one tool and then extract ants with a second tool type (Sanz, Schöning, and Morgan 2009; see also Sanz, Call, and Morgan 2009) and chimpanzees harvest honey from arboreal bees' nests by using one stick to provide access to the nest and then a second to extract the honey (Boesch, Head, and Robbins 2009).

Given the hierarchical dependency of one (or more) tools on others, these behaviors may represent the apes' gray zone of cumulative culture and they deserve testing via latent solution experiments to check whether they reach culture-dependent status or remain within the ZLS (Bernstein-Kurtycz et al. in press).

With this, we conclude the part of our chapter in which we more clearly define the tenets of the ZLS. We will now move to addressing two major misconceptions of the ZLS, regarding ontogenetic influences on the ZLS and also regarding the case of social culture (can it fall under the ZLS, too?). Additional discussion of misconceptions and further details on the ZLS account can be found elsewhere (Tennie et al. in revision).

Addressing Misconceptions

ONTOGENETIC INFLUENCES ON THE ZLS

Behavior may also depend on development, and so development plays a role in the latent solutions account. For example, there may be sensitive periods in which latent solutions are *more easily* developed (including by being socially mediated). Earlier we excluded enculturated and trained animals from our account, but what if, during development, it turns out to matter which kinds of latent solutions and resulting states of the environment individual great apes have previously had access to? Somewhat consistent with such a view, there are reports that are suggestive of *wild-born* captive-housed chimpanzees appearing to be more adept at using tools (Brent, Bloomsmith, and Fisher 1995) and making sleeping nests (Fultz et al. 2013) than their captive-born counterparts. Similarly, but possibly via different mechanisms, human-enculturated chimpanzees may also show a greater propensity for innovating behaviors within their ZLS (and perhaps even beyond) than non-enculturated chimpanzees. Conversely, but also supporting the importance of ontogeny, *deprived* apes show a limited skill set (Menzel, Davenport, and Rogers 1970).

An additional developmental factor that needs to be considered, when understanding what traits within an animal's ZLS might be expressed, is physical maturity. Nut cracking with hammers, for example, requires some minimal strength and manual coordination that younger chimpanzees lack (Inoue-Nakamura and Matsuzawa 1997; see also De Resende, Ottoni, and Fragaszy 2008 for comparable reports for capuchin monkeys' tool-use development). Therefore, not observing the complete trait in young subjects may merely indicate that before a certain age individuals cannot physically or mentally perform the trait, rather than it being outside that species' ZLS (compare

Schuppli et al. 2016). Likewise, testing subjects when it is *too late* to do so (because of a decrease in motivation over ontogenetic time, because of cognitive degradation, or because of earlier experience blocking expressions of target behavior) may also mean the trait is not expressed by that individual, despite it being within the species' ZLS (Lacreuse et al. 2014; box 19.1).

DOES THE ZLS ALSO APPLY TO TRAITS *OUTSIDE* THE TECHNOLOGICAL DOMAIN?

The ZLS account is compatible with the general account provided by Byrne and colleagues on the gestural communicative repertoire of great apes—the general conclusion (in our words) of their findings being that the *gestural* repertoire of great apes is also *not* culture dependent (overview in Byrne et al. 2017). In essence, this means that ape *gestures* can also be regarded as latent solutions (see Hobaiter, chapter 10 this volume for a more detailed discussion of ape gestural communication).

However, one animal culture domain in which the main tenet of the ZLS account is not (always) applicable is vocal culture (see Hobaiter, chapter 10 this volume; Townsend, Watson, and Slocombe, chapter 11 this volume for a more detailed discussion of ape vocal communication). Some animal vocalizations are most likely (at least partly) copied (e.g., in some cetaceans and birds), *and* have led to CDTs (e.g., there is little chance that a specific humpback whale song will ever be spontaneously reinnovated *in the same form* by another naive whale; Garland et al. 2011; see also Tennie, Caldwell, and Dean 2018).

Conclusion

We believe that the ZLS hypothesis (Tennie, Call, and Tomasello 2009) has not aged poorly, in particular for the cases of tool use in great apes. Here, we have largely focused on the case of chimpanzees (for whom we know the most): for chimpanzee tool use, many of the predictions of the ZLS hypothesis have not (yet?) been disproved. Given how easy it should be, in theory, to refute the main tenets of the ZLS hypothesis (for example by showing copying outside the ZLS), it seems noteworthy that this has not yet convincingly happened—and even the evidence for copying *within* the ZLS is not strong at the moment (at least not for action copying). Thus, it may be time to change the null hypothesis for great ape (and possibly other animal) tool use: their tool cultures may all be based on socially mediated individual reinnovations of latent solutions, and not on copying.

Acknowledgments

Many thanks for Elisa Bandini for providing helpful comments for several earlier versions of the manuscript. We also thank Damien Neadle, William Snyder, Steve Ross, Rachel Harrison, Alba Motes Rodrigo, and Eva Reindl for helpful comments on (or references for) earlier versions of the manuscript, and Matthias Allritz and Thibaud Gruber for requesting a few of the above clarifications of the original ZLS account. Many thanks to Christine Hrubesch for sharing data, and to William Daniel Snyder for creating and making accessible figure 19.1. At the time of writing, L. M. H. was supported by the Leo S. Guthman Fund. This project has received funding from the European Research Council (ERC) under the European Union's Horizon 2020 research and innovation program (grant agreement 714658; STONECULT project). This project was also financially supported by the Institutional Strategy of the University of Tübingen (Deutsche Forschungsgemeinschaft, ZUK 63).

References

Acerbi, A., and C. Tennie. 2016. "The role of redundant information in cultural transmission and cultural stabilization." *Journal of Comparative Psychology* 130 (2): 62–70.

Bandini, E., and C. Tennie. 2017. "Spontaneous reoccurrence of 'scooping,' a wild tool-use behavior, in naïve chimpanzees." *PeerJ* 5: e3814.

Bard, K., and W. D. Hopkins. 2017. "Early socio-emotional intervention mediates long-term effects of atypical rearing on structural co-variation in gray matter in adult chimpanzees." *Psychological Science* 29 (4): 594–603.

Benson-Amram, S., M. L. Weldele, and K. E. Holekamp. 2013. "A comparison of innovative problem-solving abilities between wild and captive spotted hyenas, *Crocuta crocuta*." *Animal Behaviour* 85 (2): 349–56.

Bernstein-Kurtycz, L. M., L. M. Hopper, S. R. Ross, and C. Tennie. In press. "Zoo-housed chimpanzees can spontaneously use tool sets but perseverate on previously successful tool-use methods." *Animal Behavior and Cognition*.

Boesch, C., J. Head, and M. M. Robbins. 2009. "Complex tool sets for honey extraction among chimpanzees in Loango National Park, Gabon." *Journal of Human Evolution* 56 (6): 560–69.

Boyd, R., and P. J. Richerson. 1996. "Why culture is common, but cultural evolution is rare." *Proceedings of the British Academy* 88: 77–93.

Brent, L., M. A. Bloomsmith, and S. D. Fisher. 1995. "Factors determining tool-using ability in two captive chimpanzee (*Pan troglodytes*) colonies." *Primates* 36 (2): 265–74.

Brosnan, S. F., and L. M. Hopper. 2014. "Psychological limits on animal innovation." *Animal Behaviour* 92: 325–32.

Byrne, R. W. 2007. "Culture in great apes: Using intricate complexity in feeding skills to trace the evolutionary origin of human technical prowess." *Philosophical Transactions of the Royal Society of London B* 362 (1480): 577–85.

Byrne, R. W., E. Cartmill, E. Genty, K. E. Graham, C. Hobaiter, and J. Tanner. 2017. "Great ape gestures: Intentional communication with a rich set of innate signals." *Animal Cognition* 20 (4): 1–15.

Byrne, R. W., C. Hobaiter, and M. Klailova. 2011. "Local traditions in gorilla manual skill: Evidence for observational learning of behavioral organization." *Animal Cognition* 14 (5): 683–93.

Byrne, R. W., and A. E. Russon. 1998. "Learning by imitation: A hierarchical approach." *Behavioral and Brain Sciences* 21 (5): 667–84.

Caldwell, C. A., and A. E. Millen. 2008. "Studying cumulative cultural evolution in the laboratory." *Philosophical Transactions of the Royal Society of London B* 363 (1509): 3529–39.

Caldwell, C. A., and A. E. Millen. 2009. "Social learning mechanisms and cumulative cultural evolution: Is imitation necessary?" *Psychological Science* 20 (12): 1478–83.

Caldwell, C. A., K. Schillinger, C. L. Evans, and L. M. Hopper. 2012. "End state copying by humans (*Homo sapiens*): Implications for a comparative perspective on cumulative culture." *Journal of Comparative Psychology* 126 (2): 161–69.

Call, J., and M. Tomasello. 1995. "Use of social information in the problem solving of orangutans (*Pongo pygmaeus*) and human children (*Homo sapiens*)." *Journal of Comparative Psychology* 109 (3): 308–20.

Claidière, N., K. Smith, S. Kirby, and J. Fagot. 2014. "Cultural evolution of systematically structured behavior in a non-human primate." *Proceedings of the Royal Society of London B* 281 (1797): 20141541.

Clay, Z., and C. Tennie. 2017. "Is overimitation a uniquely human phenomenon? Insights from human children as compared to bonobos." *Child Development* 89 (5): 1535–44.

Collias, E. C., and N. E. Collias. 1964. "The development of nest-building behavior in a weaverbird." *Auk* 81 (1): 42–52.

Corbey, R., A. Jagich, K. Vaesen, and M. Collard. 2016. "The Acheulean handaxe: More like a bird's song than a Beatles' tune?" *Evolutionary Anthropology* 25 (1): 6–19.

Custance, D. M., A. Whiten, and K. A. Bard. 1995. "Can young chimpanzees (*Pan troglodytes*) imitate arbitrary actions? Hayes & Hayes (1952) revisited." *Behaviour* 132 (11): 837–59.

Custance, D. M., A. Whiten, and T. Fredman. 1999. "Social learning of an artificial fruit task in capuchin monkeys (*Cebus apella*)." *Journal of Comparative Psychology* 113 (1): 13–23.

Dean, L. G., R. L. Kendal, S. J. Schapiro, B. Thierry, and K. N. Laland. 2012. "Identification of the social and cognitive processes underlying human cumulative culture." *Science* 335 (6072): 1114–18.

Dean, L. G., G. L. Vale, K. N. Laland, E. Flynn, and R. L. Kendal. 2014. "Human cumulative culture: A comparative perspective." *Biological Reviews* 89 (2): 284–301.

De Resende, B. D., E. B. Ottoni, and D. M. Fragaszy. 2008. "Ontogeny of manipulative behavior and nut-cracking in young tufted capuchin monkeys (*Cebus apella*): A perception–action perspective." *Developmental Science* 11 (6): 828–40.

Fragaszy, D. M., D. Biro, Y. Eshchar, T. Humle, P. Izar, B. Resende, and E. Visalberghi. 2013. "The fourth dimension of tool use: Temporally enduring artifacts aid primates learning to use tools." *Philosophical Transactions of the Royal Society of London B* 368 (1630): 20120410.

Fultz, A., L. Brent, S. D. Breaux, and A. P. Grand. 2013. "An evaluation of nest-building behavior by sanctuary chimpanzees with access to forested habitats." *Folia Primatologica* 84 (6): 405–20.

Garland, E. C., A. W. Goldizen, M. L. Rekdahl, R. Constantine, C. Garrigue, N. D. Hauser, M. M. Poole, J. Robbins, and M. J. Noad. 2011. "Dynamic horizontal cultural transmission of humpback whale song at the ocean basin scale." *Current Biology* 21 (8): 687–91.

Ghirlanda, S., J. Lind, and M. Enquist. 2017. "Memory for stimulus sequences: A divide between humans and other animals?" *Royal Society Open Science* 4 (6): 161011.

Gruber, T., M. N. Muller, P. Strimling, R. Wrangham, and K. Zuberbühler. 2009. "Wild chimpanzees rely on cultural knowledge to solve an experimental honey acquisition task." *Current Biology* 19 (21): 1806–10.

Gumert, M. D., M. Kluck, and S. Malaivijitnond. 2009. "The physical characteristics and usage patterns of stone axe and pounding hammers used by long-tailed macaques in the Andaman Sea Region of Thailand." *American Journal of Primatology* 71 (7): 594–608.

Hanus, D., N. Mendes, C. Tennie, and J. Call. 2011. "Comparing the performances of apes (*Gorilla gorilla, Pan troglodytes, Pongo pygmaeus*) and human children (*Homo sapiens*) in the floating peanut task." *PLoS One* 6 (6): e19555.

Haslam, M., L. V. Luncz, R. A. Staff, F. Bradshaw, E. Ottoni, and T. Falótico. 2016. "Pre-Columbian monkey tools." *Current Biology* 26 (13): 521–22.

Haun, D. B., and C. J. Rapold. 2009. "Variation in memory for body movements across cultures." *Current Biology* 19 (23): R1068–69.

Hecht, E. E., D. A. Gutman, T. M. Preuss, M. M. Sanchez, L. A. Parr, and J. K. Rilling. 2012. "Process versus product in social learning: Comparative diffusion tensor imaging of neural systems for action execution–observation matching in macaques, chimpanzees, and humans." *Cerebral Cortex* 23 (5): 1014–24.

Henrich, J. 2015. *The Secret of Our Success: How Culture Is Driving Human Evolution, Domesticating Our Species, and Making Us Smarter*. Princeton, NJ: Princeton University Press.

Henrich, J., and C. Tennie. 2017. "Cultural evolution in chimpanzees and humans." In *Chimpanzees and Human Evolution*, edited by M. N. Muller, R. W. Wrangham, and D. R. Pilbeam, 645–702. Cambridge, MA: Harvard University Press.

Heyes, C. 2012. "Grist and mills: On the cultural origins of cultural learning." *Philosophical Transactions of the Royal Society of London B* 367 (1599): 2181–91.

Heyes, C. 2018. "Précis of cognitive gadgets: The cultural evolution of thinking." *Behavioral and Brain Sciences*: 1–57.

Hobaiter, C., T. Poisot, K. Zuberbühler, W. Hoppitt, and T. Gruber. 2014. "Social network analysis shows direct evidence for social transmission of tool use in wild chimpanzees." *PLoS Biology* 12 (9): e1001960.

Hoffecker, J. F. 2017. *Modern Humans: Their African Origin and Global Dispersal*. New York: Columbia University Press.

Hopper, L. M. 2010. "'Ghost' experiments and the dissection of social learning in humans and animals." *Biological Reviews* 85: 685–701.

Hopper, L. M., E. G. Flynn, L. A. N. Wood, and A. Whiten. 2010. "Observational learning of tool use in children: Investigating cultural spread through diffusion chains and learning mechanisms through ghost displays." *Journal of Experimental Child Psychology* 106 (1): 82–97.

Hopper, L. M., S. P. Lambeth, S. J. Schapiro, and A. Whiten. 2008. "Observational learning in chimpanzees and children studied through 'ghost' conditions." *Proceedings of the Royal Society of London B* 275 (1636): 835–40.

Hopper, L. M., S. P. Lambeth, S. J. Schapiro, and A. Whiten. 2015a. "The importance of witnessed agency in chimpanzee social learning of tool use." *Behavioural Processes* 112: 120–29.

Hopper, L. M., A. Spiteri, S. P. Lambeth, S. J. Schapiro, V. Horner, and A. Whiten. 2007. "Experimental studies of traditions and underlying transmission processes in chimpanzees." *Animal Behaviour* 73 (6): 1021–32.

Hopper, L. M., C. Tennie, S. R. Ross, and E. V. Lonsdorf. 2015b. "Chimpanzees create and modify probe tools functionally: A study with zoo-housed chimpanzees." *American Journal of Primatology* 77 (2): 162–70.

Hoppitt, W., and K. N. Laland. 2013. *Social Learning: An Introduction to Mechanisms, Methods, and Models*. Princeton, NJ: Princeton University Press.

Howard, L. H., K. E. Wagner, A. L. Woodward, S. R. Ross, and L. M. Hopper. 2017. "Social models enhance apes' memory for novel events." *Scientific Reports* 7: 40926.

Huffman, M. A., and S. Hirata. 2004. "An experimental study of leaf swallowing in captive chimpanzees: Insights into the origin of a self-medicative behavior and the role of social learning." *Primates* 45 (2): 113–18.

Hunt, G. R., and R. D. Gray. 2003. "Diversification and cumulative evolution in New Caledonian crow tool manufacture." *Proceedings of the Royal Society of London B* 270 (1517): 867–74.

Inoue-Nakamura, N., and T. Matsuzawa. 1997. "Development of stone tool use by wild chimpanzees (*Pan troglodytes*)." *Journal of Comparative Psychology* 111 (2): 159–73.

Kendall, R., L. M. Hopper, A. Whiten, S. F. Brosnan, S. P. Lambeth, S. J. Schapiro, and W. Hoppitt. 2015. "Chimpanzees copy dominant and knowledgeable individuals: Implications for cultural diversity." *Evolution and Human Behavior* 36 (1): 65–72.

Koops, K., E. Visalberghi, and C. P. van Schaik. 2014. "The ecology of primate material culture." *Biology Letters* 10 (11): 20140508.

Lacreuse, A., J. L. Russel, W. D. Hopkins, and J. G. Herndon. 2014. "Cognitive and motor aging in female chimpanzees." *Neurobiology of Aging* 35 (3): 623–32.

Leca, J.-B., N. Gunst, and M. A. Huffman. 2007. "Japanese macaque cultures: Inter- and intra-troop behavioral variability of stone handling patterns across 10 troops." *Behaviour* 144 (3): 251–81.

Legare, C. H., and M. Nielsen. 2015. "Imitation and innovation: The dual engines of cultural learning." *Trends in Cognitive Sciences* 19 (11): 688–99.

Lehner, S., J. M. Burkart, and C. P. van Schaik. 2010. "An evaluation of the geographic method for recognizing innovations in nature, using zoo orangutans." *Primates* 51 (2): 101–18.

Lewis, H. M., and K. N. Laland. 2012. "Transmission fidelity is the key to the build-up of cumulative culture." *Philosophical Transactions of the Royal Society of London B* 367 (1599): 2171–80.

Lonsdorf, E. V. 2006. "What is the role of mothers in the acquisition of termite-fishing behaviors in wild chimpanzees (*Pan troglodytes schweinfurthii*)?" *Animal Cognition* 9 (1): 36–46.

Masi, S. 2011. "Differences in gorilla nettle-feeding between captivity and the wild: Local traditions, species typical behaviors or merely the result of nutritional deficiencies?" *Animal Cognition* 14 (6): 921.

McGrew, W. C. 1987. "Tools to get food: The subsistants of Tasmanian aborigines and Tanzanian chimpanzees compared." *Journal of Anthropological Research* 43 (3): 247–58.

Menzel, C., A. Fowler, C. Tennie, and J. Call. 2013. "Leaf surface roughness elicits leaf swallowing behavior in captive chimpanzees (*Pan troglodytes*) and bonobos (*P. paniscus*), but not in gorillas (*Gorilla gorilla*) or orangutans (*Pongo abelii*)." *International Journal of Primatology* 34 (3): 533–53.

Menzel, E. W., Jr., R. K. Davenport, and C. M. Rogers. 1970. "The development of tool using in wild-born and restriction-reared chimpanzees." *Folia Primatologica* 12 (4): 273–83.

Mercader, J., H. Barton, J. Gillespie, J. Harris, S. Kuhn, R. Tyler, and C. Boesch. 2007. "4,300-year-old chimpanzee sites and the origins of percussive stone technology." *Proceedings of the National Academy of Sciences* 104 (9): 3043–48.

Mesoudi, A., L. Chang, S. R. X. Dall, and A. Thornton. 2016. "The evolution of individual and cultural variation in social learning." *Trends in Ecology & Evolution* 31 (3): 215–25.

Meulman, E. J. M., C. M. Sanz, E. Visalberghi, and C. P. van Schaik. 2012. "The role of terrestriality in promoting primate technology." *Evolutionary Anthropology* 21: 58–68.

Miles, H. L., R. W. Mitchell, and S. E. Harper. 1996. "Simon says: The development of imitation in an enculturated orangutan." In *Reaching into Thought: The Minds of the Great Apes*, edited by A. E. Russon, K. A. Bard, and S. T. Parker, 278–99. Cambridge: Cambridge University Press.

Morgan, B. J., and E. E. Abwe. 2006. "Chimpanzees use stone hammers in Cameroon." *Current Biology* 16 (16): R632–33.

Muthukrishna, M., and J. Henrich. 2016. "Innovation in the collective brain." *Philosophical Transactions of the Royal Society of London B* 371 (1690): 20150192.

Neadle, D., M. Allritz, and C. Tennie. 2017. "Food cleaning in gorillas: Social learning is a possibility but not a necessity." *PLoS One* 12 (12): e0188866.

Nishida, T., T. Matsusaka, and W. C. McGrew. 2009. "Emergence, propagation or disappearance of novel behavioral patterns in the habituated chimpanzees of Mahale: A review." *Primates* 50 (1): 23–36.

Perry, S., M. Baker, L. Fedigan, J. Gros-Louis, K. Jack, K. C. MacKinnon, J. H. Manson, M. Panger, K. Pyle, and L. Rose. 2003. "Social conventions in wild white-faced capuchin monkeys: Evidence for traditions in a neotropical primate." *Current Anthropology* 44 (2): 241–68.

Pope, S. M., J. P. Taglialatela, S. A. Skiba, and W. D. Hopkins. 2018. "Changes in frontoparietotemporal connectivity following do-as-I-do imitation training in chimpanzees (*Pan troglodytes*)." *Journal of Cognitive Neuroscience* 30 (3): 421–31.

Pradhan, G. R., C. Tennie, and C. P. van Schaik. 2012. "Social organization and the evolution of cumulative technology in apes and hominins." *Journal of Human Evolution* 63 (1): 180–90.

Price, E. E., S. P. Lambeth, S. J. Schapiro, and A. Whiten. 2009. "A potent effect of observational learning on chimpanzee tool construction." *Proceedings of the Royal Society of London B* 276 (1671): 3377–83.

Reader, S. M., and K. N. Laland. 2002. "Social intelligence, innovation, and enhanced brain size in primates." *Proceedings of the National Academy of Sciences* 99 (7): 4436–41.

Reindl, E., I. A. Apperly, S. R. Beck, and C. Tennie. 2017. "Young children copy cumulative technological design in the absence of action information." *Scientific Reports* 7: 1788.

Reindl, E., E. Bandini, and C. Tennie. 2018. "The zone of latent solutions and its relation to the classics: Vygotsky and Köhler." In *Evolution of Primate Social Cognition*, edited by L. D. Di Paolo, F Di Vincenzo, and F. De Petrillo. Dordrecht: Springer.

Reindl, E., S. R. Beck, I. A. Apperly, and C. Tennie. 2016. "Young children spontaneously invent wild great apes' tool-use behaviors." *Proceedings of the Royal Society of London B* 283 (1825): 20152402.

Ridley, M. 2010. *The Rational Optimist: How Prosperity Evolves*. Harper/HarperCollins Publishers.

Robbins, M. M., C. Ando, K. A. Fawcett, C. C. Grueter, D. Hedwig, Y. Iwata, J. L. Lodwick, S. Masi, R. Salmi, T. S. Stoinski, A. Todd, V. Vercellio, and J. Yamagiwa. 2016. "Behavioral variation in gorillas: Evidence of potential cultural traits." *PLoS One* 11(9): e0160483.

Russon, A. E., and B. M. F. Galdikas. 1993. "Imitation in free-ranging rehabilitant orangutans (*Pongo pygmaeus*)." *Journal of Comparative Psychology* 107: 147.

Santorelli, C. J., C. M. Schaffner, C. J. Campbell, H. Notman, M. S. Pavelka, J. A. Weghorst, and F. Aureli. 2011. "Traditions in spider monkeys are biased towards the social domain." *PLoS One* 6: e16863.

Sanz, C. M., J. Call, and D. Morgan. 2009. "Design complexity in termite-fishing tools of chimpanzees (*Pan troglodytes*)." *Biology Letters* 5: 293–96.

Sanz, C. M., and D. B. Morgan. 2007. "Chimpanzee tool technology in the Goulougo Triangle, Republic of Congo." *Journal of Human Evolution* 52 (4): 420–33.

Sanz, C. M., C. Schöning, and D. B. Morgan. 2009. "Chimpanzees prey on army ants with specialized tool set." *American Journal of Primatology* 72 (1): 17–24.

Sasaki, T., and D. Biro. 2017. "Cumulative culture can emerge from collective intelligence in animal groups." *Nature Communications* 8: 15049.

Schofield, D. P., W. C. McGrew, A. Takahashi, and S. Hirata. 2017. "Cumulative culture in nonhumans: Overlooked findings from Japanese monkeys?" *Primates* 59 (2): 113–22.

Schuppli, C., S. I. F. Forss, E. J. M. Meulman, N. Zweifel, K. C. Lee, E. Rukmana, E. R. Vogel, M. A. van Noordwijk, and C. P. van Schaik. 2016. "Development of foraging skills in two orangutan populations: Needing to learn or needing to grow?" *Frontiers in Zoology* 13 (43): 1–17.

Sterelny, Kim. 2012. *The Evolved Apprentice*. Cambridge, MA: MIT Press.

Tennie, C., E. Bandini, C. P. van Schaik, and L. M. Hopper. In revision. "The Zone of Latent Solutions and its relevance to understanding ape cultures." *Biology and Philosophy*.

Tennie, C., D. R. Braun, L. S. Premo, and S. P. McPherron. 2016. "The island test for cumulative culture in the paleolithic." In *The Nature of Culture*, edited by M. N. Haidle, N. J. Conard, and M. Bolus, 121–33. Dordrecht: Springer.

Tennie, C., C. Caldwell, and L. G. Dean. 2018. "Cumulative culture." In *International Encyclopedia of Anthropology*, edited by H. Callan. Oxford: Wiley-Blackwell.

Tennie, C., J. Call, and M. Tomasello. 2006. "Push or pull: Imitation vs. emulation in great apes and human children." *Ethology* 112 (12): 1159–69.

Tennie, C., J. Call, and M. Tomasello. 2009. "Ratcheting up the ratchet: On the evolution of cumulative culture." *Philosophical Transactions of the Royal Society of London B* 364: 2405–15.

Tennie, C., J. Call, and M. Tomasello. 2010. "Evidence for emulation in chimpanzees in social settings using the floating peanut task." *PLoS One* 5: e10544.

Tennie, C., J. Call, and M. Tomasello. 2012. "Untrained chimpanzees (*Pan troglodytes schweinfurthii*) fail to imitate novel actions." *PLoS One* 7: e41548.

Tennie, C., and D. Hedwig. 2009. "How latent solution experiments can help to study differences between human culture and primate traditions." In *Primatology: Theories, Methods and Research*, edited by E. Potocki and J. Krasinski, 95–112. New York: Nova Publishers.

Tennie, C., D. Hedwig, J. Call, and M. Tomasello. 2008. "An experimental study of nettle feeding in captive gorillas." *American Journal of Primatology* 70: 584–93.

Tennie, C., and H. Over. 2012. "Cultural intelligence is key to explaining human tool use." *Behavioral and Brain Sciences* 35 (4): 242–43.

Tennie, C., L. S. Premo, D. R. Braun, and S. P. McPherron. 2017. "Resetting the null hypothesis: Early stone tools and cultural transmission." *Current Anthropology* 58 (5): 652–54.

Tomasello, M. 1999. *The Cultural Origins of Human Cognition*. Cambridge, MA: Harvard University Press.

Tomasello, M., J. Call, J. Warren, G. T. Frost, M. Carpenter, and K. Nagell. 1997. "The ontogeny of chimpanzee gestural signals: A comparison across groups and generations." *Evolution of Communication* 1 (2): 223–53.

Tomasello, M., A. C. Kruger, and H. H. Ratner. 1993. "Cultural learning." *Behavioral and Brain Sciences* 16 (3): 495–552.

Vale, G. L., S. J. Davis, S. P. Lambeth, S. J. Schapiro, and A. Whiten. 2017. "Acquisition of a socially learned tool use sequence in chimpanzees: Implications for cumulative culture." *Evolution and Human Behavior* 38 (5): 635–44.

van Schaik, C. P., M. Ancrenaz, G. Borgen, B. Galdikas, C. D. Knott, I. Singleton, A. Suzuki, S. S. Utami, and M. Merrill. 2003. "Orangutan cultures and the evolution of material culture." *Science* 299 (5603): 102–5.

van Schaik, C. P., M. Ancrenaz, R. Djojoasmoro, C. D. Knott, H. C. Morrogh-Bernard, N. O. Kisar, S. S. Utami Atmoko, and M. A. van Noordwijk. 2009. "Orangutan cultures revisited." In *Orangutans: Geographic Variation, Behavioral Ecology and Conservation*, edited by S. A. Wich, S. S. Utami Atmoko, T. Mitra Setia, and C. P. van Schaik, 299–309. New York: Oxford University Press.

van Schaik, C. P., J. M. Burkart, L. Damerius, S. I. F. Forss, K. Koops, M. A. van Noordwijk, and C. A. Schuppli. 2016. "The reluctant innovator: Orangutans and the phylogeny of creativity." *Philosophical Transactions of the Royal Society of London B* 371 (1690): 20150183.

Visalberghi, E. 1987. "Acquisition of nut-cracking behavior by two capuchin monkeys (*Cebus apella*)." *Folia Primatologica* 49 (3–4): 168–81.

Whitehead, H., and L. Rendell. 2014. *The Cultural Lives of Whales and Dolphins*. Chicago: University of Chicago Press.

Whiten, A. 1998. "Imitation of the sequential structure of actions by chimpanzees (*Pan troglodytes*)." *Journal of Comparative Psychology* 112 (3): 270–81.

Whiten, A. 2011. "The scope of culture in chimpanzees, humans and ancestral apes." *Philosophical Transactions of the Royal Society of London B* 366 (1567): 997–1007.

Whiten, A., J. Goodall, W. C. McGrew, T. Nishida, V. Reynolds, Y. Sugiyama, C. E. G. Tutin, R. W. Wrangham, and C. Boesch. 1999. "Cultures in chimpanzees." *Nature* 399 (6737): 682–85.

Whiten, A., J. Goodall, W. C. McGrew, T. Nishida, V. Reynolds, Y. Sugiyama, C. E. G. Tutin, R. W. Wrangham, and C. Boesch. 2001. "Charting cultural variation in chimpanzees." *Behaviour* 138, no. 11 (2001): 1481–1516.

Whiten, A., V. Horner, and F. B. M. de Waal. 2005. "Conformity to cultural norms of tool use in chimpanzees." *Nature* 437: 737–40.

Whiten, A., V. Horner, C. A. Litchfield, and S. Marshall-Pescini. 2004. "How do apes ape?" *Learning & Behavior* 32 (1): 36–52.

Whiten, A., N. McGuigan, S. Marshall-Pescini, and L. M. Hopper. 2009. "Emulation, imitation, over-imitation and the scope of culture for child and chimpanzee." *Philosophical Transactions of the Royal Society of London B* 364: 2417–28.

Zwirner, E., and A. Thornton. 2015. "Cognitive requirements of cumulative culture: Teaching is useful but not essential." *Scientific Reports* 5: 16781.

20

Cognitive Control and Metacognition in Chimpanzees

MICHAEL J. BERAN, BONNIE M. PERDUE,
AND AUDREY E. PARRISH

Introduction

The cognitive revolution in psychology emerged initially within human psychological studies, but very quickly came to be embraced by some who worked with nonhuman animals (Honig and Fetterman 1992; Hulse, Fowler, and Honig 1978; Roitblat, Bever, and Terrace 1984). Through innovative new techniques, researchers began to uncover a variety of capacities in a wide range of species that begged an explanation other than one grounded in stimulus-response relations. These included representational capacities such as categorization, numerical cognition, episodic memory, planning abilities, and abstract modes of learning about relations among stimuli. Many of these discoveries occurred with nonhuman primates, leading to new conceptualization of the mental lives of those in this taxa (e.g., Maestripieri 2003).

A new phase in understanding animal minds then emerged, with the idea that animals were not only cognitive, but perhaps also *metacognitive*. Metacognition involves the control processes and monitoring processes that reflect upon internal states of knowing and not knowing, rather than external cues that elicit responses based on the property of those cues. Human metacognition is held to be one of our most defining intellectual features (Benjamin, Bjork, and Schwartz 1998; Flavell 1979; Koriat and Goldsmith 1996; Metcalfe and Kober 2005; Nelson 1992), and so the search for such capacities in nonhuman animals was controversial, but also influential on new thoughts about the depth of the mental lives of animals (Carruthers 2008; Crystal 2014; Crystal and Foote 2009; Hampton 2009; Jozefowiez, Staddon, and Cerutti 2009; Kornell 2009, 2014; Le Pelley 2012; Smith 2009; Smith et al. 2008). This area of research has become an important component to better understanding the emergence of human metacognition through evolution and development (see Beran et al. 2012). Work with nonhuman primates has largely dominated this effort.

THE UNCERTAINTY RESPONSE PARADIGM

The earliest work, with macaque monkeys, made use of a so-called uncertainty response (UR) added as an extra response option to a variety of existing comparative cognition tasks. This allowed monkeys that were working on computer tasks to integrate a UR into their typical classification responses. For example, monkeys might learn to classify a group of pixelated boxes as being sparse or dense, or they could choose to escape making the classification by selecting the UR (Smith et al. 1997). Similarly, monkeys could be trained to judge the sameness or difference of two stimuli, or escape making that judgment via the UR (Shields, Smith, and Washburn 1997). These early studies showed that monkeys chose the UR on exactly those trial levels for which errors were most likely to occur, and they did this across a variety of discrimination and memory tasks. In other tasks, rhesus monkeys showed that they would choose to take tests or not take tests based on memory strength (e.g., Fujita 2009; Hampton 2001). Again, monkeys chose memory tests they were likely to pass, and avoided those they were less likely to pass, and, they performed better when they chose to take tests compared to when they were forced to take them (also see Basile et al. 2015; Templer and Hampton 2012).

INFORMATION-SEEKING PARADIGM

As these UR-paradigm studies began appearing in the literature, another approach was being developed with apes and children that emphasized organisms' natural proclivity to search out and obtain preferred foods. This approach has had a strong impact on thinking about animal metacognition. In the first experiment of their seminal paper, Call and Carpenter (2001) tested three chimpanzees (*Pan troglodytes*) and three orangutans (*Pongo pygmaeus*) at the Yerkes National Primate Research Center, USA. The apparatus consisted of PVC tubes in which food could be placed. These tubes either were open in such a way that subjects could easily see the contents, or involved an occluder that allowed the experimenter to hide a piece of food without the subject being able to see it. There were three combinations of tubes used in testing trials: both tubes were open, both were closed, or one was open and one was closed. The baiting of the foods was either visible to the subject or took place behind a cardboard occluder between the experimenter and subject, and various combinations of open and closed tubes were presented to subjects. Call and Carpenter also varied whether the subject could select a tube immediately after baiting, or whether a 5-second delay occurred before the selection. The main variable of interest was whether subjects chose a tube first or chose to look into the tubes to determine their contents before making a selection. The researchers also identified several potential strategies depending on the configuration of the tubes (open versus closed): efficient (searching until food was seen, or making a selection on a trial with an empty open tube and a closed tube, or searching two closed tubes), insufficient, or excessive.

Chimpanzees and orangutans performed similarly on the task, with the general pattern of looking more on trials in which the baiting had occurred behind the occluder. On these unseen trials, subjects were more likely to engage in an efficient search pattern than the excessive or insufficient search patterns. Even when the searches were excessive, they were usually still systematic in that they were not repetitive. The apes also looked more on trials with a delay before the selection could be made as compared to those in which the selection was immediately following the baiting. Overall this pattern of responding suggested that the subjects used the information about what they had or had not seen and responded in an efficient way to the task. There were some limitations to the design of the first experiment and a second experiment was included to explore these potential confounds. Specifically, there was no cost to looking and it required little effort, so this might have increased the apes' likelihood to perform excessive searches. The closed tubes may have also introduced confusion and accounted for part of the observed response patterns in the first experiment.

In their second experiment, Call and Carpenter (2001) introduced a modified version of the previous task and increased the sample size to include eleven chimpanzees. The number of tubes was increased to three to increase the cost of looking and all of the tubes were now open. Also, there was a five-second delay on all trials before a response could be made and they varied whether the subjects observed the baiting process or whether this took place behind an occluder (i.e., seen versus unseen). They found that chimpanzees selected the correct tube significantly more often than chance, and importantly, looked significantly more often in the hidden condition than in the visible one. These findings were critical in establishing that chimpanzees appear to respond adaptively to what they do or do not know.

Subsequent work with a variety of species using a creative mixture of such information-seeking tasks has consistently shown that apes tend to look for further information when they need it prior to making choices, and respond more directly when they do not, whereas evidence from monkeys (*Cebus apella*, *Macaca mulatta*), pigeons (*Columba livia*), and rats (*Rattus norvegicus*) is more mixed (e.g., Beran and Smith 2011; Beran, Smith, and Perdue 2013; Call 2010; Castro and Wasserman 2013; Hampton, Zivin, and Murray

2004; Iwasaki, Watanabe, and Fujita 2013; Kirk, McMillan, and Roberts 2014; Marsh and MacDonald 2012a, 2012b; Roberts et al. 2009; Suda-King 2008; Suda-King et al. 2013). The use of information-seeking tasks has therefore highlighted what could be seen as a strong argument that at least the great apes control and monitor their memory and knowledge states, and then integrate simple information-seeking behaviors into their responding when it is necessary.

EXPLORING THE MECHANISMS OF METACOGNITION

This information-seeking paradigm and other metacognition tasks have stimulated a great deal of debate on the nature of metacognition in nonhuman animals and the interpretation of the findings from such studies (e.g., Carruthers 2008; Hampton 2009; Kornell 2014; Smith et al. 2008), and some concerns have been raised about this approach. In particular, alternative explanations for the information-seeking task described above might involve first-order mental states regarding beliefs about food. If the food is not seen, there are no such beliefs about food and the animal will engage in searching behavior until food has been seen (Carruthers 2008). Another nonmetacognitive explanation is that animals' responses are shaped by a priority to look before choosing and an assumption that one should go where something appetitizing is located rather than randomly searching through space (Crystal and Foote 2011). Together, these principles could explain the pattern of responding observed in the information-seeking studies without necessarily evoking a need for a metacognitive account in which an organism reflects on its own knowledge states or memory. We adapted the information-seeking task to address these concerns from a methodological perspective.

Beran, Smith, and Perdue (2013) assessed metacognition in three language-trained chimpanzees, living at the Language Research Center, USA, who demonstrated proficiency in using symbols (lexigrams) as representations of food items. As such, we were able to require responses about the *identity* of food items, rather than the location of food items, as done in the previous studies. This change allowed for these alternative explanations to be controlled for because the subjects always had a belief about the food and knew its location, but in these studies they had to name it rather than locate it. A similar pattern of responding as found in the earlier studies would provide further support that the pattern of responses seems to be supported by metacognitive processes.

In the first study, a container was baited with a food item that the chimpanzee could name with its lexigram and the baiting took place out of view of the subject. Then an experimenter (experimenter 1) approached the chimpanzee's enclosure and either showed the contents of the container or tilted it away such that the contents remained unseen. Next experimenter 1 moved the container to the floor near an adjacent enclosure. A second experimenter (experimenter 2), who had no knowledge about whether the contents had been seen or not or what was in the container, then entered the area and stood near the entrance to the second enclosure. The chimpanzees had to move through space into a new enclosure. When they entered the new space, there was a lexigram board located near the enclosure entrance and experimenter 2 was standing near the board to read any responses made by the chimpanzee. The chimpanzees could immediately attempt to name what was hidden in the container or they could walk in a different direction toward where the container was sitting on the floor. If they got close to the container, they could peer inside to see the contents and then return and name the item on the lexigram board. If the item was correctly named, they would receive the food. The chimpanzees essentially had two choices when entering the second enclosure—go directly to the keyboard and attempt to name the item, or first go to look at the contents of the container and then name the item. If the chimpanzees were monitoring their knowledge state regarding the contents of the container, they should name the item immediately on the trials in which they were shown the contents (visible trials) and look into the container on trials before attempting to name the objects that were unseen (nonvisible). All chimpanzees were significantly more likely to look first on nonvisible trials and name first on visible trials, a pattern consistent with a metacognitive interpretation.

To complete this task, it is possible the chimpanzees may have used a rule such as "if see food, name item" rather than metacognitive monitoring of their knowledge. To address this, in a second experiment food was seen on every trial. In this version, experimenter 1 brought in two unique containers, each with a different food inside. The contents of one container were shown to the chimpanzees and the contents of the other were not, and then one of those containers was moved to the second enclosure. Therefore, the chimpanzees saw a food item on every trial, but it was not always the one that was moved and made available to them in the second enclosure. Now, they could not rely on a strategy of naming what they saw, or looking when they had no name, because every trial gave them a food they could name. But, that was not enough information—it also mattered which container was the one they could try to get the food from in the second location. Even with this change, the chimpanzees named the item in the container more often in the condition where the food was shown to them and then moved compared to when it was shown to them, but the other container was moved. Together these results provided evidence that chimpanzees named things when they knew where those things were, and what they were, but otherwise they sought more information to solve the task.

CONFIDENCE MOVEMENTS IN CHIMPANZEES

In more recent work, a new way to test metacognition in chimpanzees relied upon the idea that chimpanzees would naturally try to maximize food intake by moving toward potential food sources (Beran et al. 2015). Confidence in their own memory and knowledge states could be assessed in chimpanzees via a "confidence movement" paradigm. Here, chimpanzees completed a variety of computerized tasks designed to assess memory *and* the animals' confidence in those memory judgments by requiring subjects to move through space to retrieve a food reward from a separate location that would only be delivered if the subject completed the task correctly. Crucially, the computer tasks did not generate auditory feedback based on the accuracy of their responding until after a delay period had passed. Furthermore, the food rewards for correct responses were delivered at a secondary location some distance away from the computer apparatus. These two task features required the chimpanzees to move through space prior to external cues regarding the correctness of their responding in order to retrieve a food reward, which was otherwise lost and could not be recovered if the chimpanzees failed to move to the reward delivery-site. Thus, the chimpanzees had to evaluate the accuracy of their response and whether they should move away from the computer to retrieve a reward. The three chimpanzees that were tested demonstrated varying degrees of metacognitive monitoring in this task, moving to the reward location significantly more often following correct trials than incorrect trials. We interpreted these spontaneous confidence movements as evidence of metacognition in which chimpanzees monitored memory states and varying degrees of knowledge in these cognitive tasks, and appropriately responded on this information prior to external cuing of their performance. Here, we discuss this paradigm and the associated results in detail before describing new studies with one chimpanzee that highlight crucial aspects of the generalizability of this finding.

All chimpanzees in the Beran et al. (2015) study lived at the Language Research Center, USA, and had a long history of engaging with computerized cognition tasks including tests of memory (e.g., Beran and Rumbaugh 2001; Beran and Washburn 2002; Beran et al. 2004). Crucially, they always had received immediate auditory feedback on the accuracy of their responses (melodic tone for correct trials and buzz tone for incorrect trials) as well as immediate delivery of a food reward following correct responses at the computer apparatus. In the confidence movement study, however, food delivery and auditory feedback were delayed for a few seconds and food rewards for correct trials were delivered at a secondary location distant from the computer apparatus. Thus, if the chimpanzees correctly completed a computer trial, they could move to an adjacent enclosure with the food dispenser *prior to* receiving external feedback (i.e., the computer-generated auditory cues) about their responses. If the chimpanzees moved to the food-delivery site following a correct trial, an automated dispenser delivered a food reward into the adjacent enclosure that could be retrieved and immediately consumed by the subject. If a chimpanzee failed to move to the adjacent cage following a correct trial (or moved to this location too long after the auditory feedback), the food reward was dispensed but fell away from the enclosure and was lost. Alternatively, incorrect trials were not rewarded, thus moving to the food-dispenser was not beneficial on these trials and was ultimately costly in terms of the time expenditure of moving back and forth between sites. Another key aspect to this design was that the chimpanzees were not trained to move to this secondary location, so we captured spontaneous movements based on the animals' confidence in their own knowledge states. The idea was simple—to see whether the chimpanzees would go early when they knew answers, and wait to see how they performed first when they did not.

In the first experiment, two chimpanzees named Lana (female, 44 years old) and Sherman (male, 41 years old) were tested on a visual matching-to-sample (MTS) task. As noted earlier, these chimpanzees were language-trained using the Language Research Center's lexigram system, learning to associate geometric symbols with a variety of corresponding items in their environment (e.g., food, items, places, people; Rumbaugh 1977; Savage-Rumbaugh 1986). In the MTS task, the chimpanzees matched a photograph of a food object or item (e.g., banana) to one of four lexigrams representing the photographed item. In this test, we sometimes gave them photographs of items for which they knew the correct lexigram symbol, but other times we showed them photographs of things they did not have lexigrams for, so that we could create objectively easy and hard trials to determine how that affected movements to the dispenser.

Both chimpanzees were significantly more likely to move to the food-delivery site early (before the auditory cue) on correct trials than incorrect trials. Because Sherman and Lana had a long history with these lexigram symbols, and the chimpanzees used some symbols more often than others, there was a concern that they could be responding on the basis of previous associations and reinforcement history with the better-known symbols rather than on the basis of metacognitive monitoring. Thus, in experiment 2, arbitrary stimuli (clip-art images) were used in the MTS task and the task also incorporated a variable delay interval between the sample presentation and choice options. This delayed MTS (DMTS) task created possible instances of forgetting for the chimpanzees, with delay intervals lasting 1, 4, 7, 10, or 13 seconds. The prediction was that longer delays should lead to lower matching accuracy on the DMTS task as well as fewer instances of early movements to the food-delivery site. Lana, Sherman, and an additional chimpanzee, Mercury (a 27-year-old male), were tested. Mercury performed at chance level (25%) across all forgetting intervals in the DMTS task and he always moved to the dispenser *after* the auditory cue was signaled. Thus, his results indicated no evidence of metacognitive monitoring or adequate memory for the samples in the DMTS task. However, Sherman and Lana continued to perform well on the DMTS task across all forgetting intervals. Importantly, both chimpanzees were significantly more likely to move to the dispenser early (before auditory feedback) on correct trials than incorrect trials.

An important new kind of trial was introduced in experiment 3. "No Sample" trials occurred when the chimpanzees were not shown a sample image but still forced to select among four match images, one of which was randomly assigned as the "correct" response. These trials had the shortest delay—1 second. If chimpanzees were using the delay length as a cue to whether to move early or not, this would create a real problem, because the delay was short, but they also had not seen a sample image and so would have to guess. The prediction was that longer delay intervals should lead to lower task accuracy as well as fewer early movements. Mercury again performed at chance levels across the DMTS task and always moved to the food dispenser location after auditory feedback. Sherman and Lana, on the other hand, continued to perform well in the DMTS task although longer delays created more memory errors, as we had predicted. As before, both chimpanzees were significantly more likely to move to the food delivery location prior to auditory feedback on correct trials than incorrect trials. For No Sample trials, both chimpanzees performed at chance levels as expected and, importantly, did not move to the dispenser early on these trials as often as they did on correct trials where samples had been presented. These results indicated that, for Lana and Sherman, the signal to move early was not simply tied to delay length, as the shortest delay length (No Sample trials) did not produce the most movements. Instead these movements were contingent upon memory accuracy, which was contingent upon forgetting interval.

Because Mercury performed poorly across all previous experiments, some additional variations of the task were presented to him. What led to a fascinating immediate change in his behavior was eliminating the delay between the auditory feedback and reward delivery. With the delay between feedback and food delivery removed, Mercury was unable to wait until after the auditory feedback to be able to catch the food reward from the dispenser. Now, Mercury was significantly more likely to move to the dispenser on correct trials than incorrect trials and did not move to the dispenser on the random No Sample trials. Mercury's results in the final experiment therefore aligned with the findings from Sherman and Lana and highlighted the need to consider all aspects of an experimental design and how they might interact with motivation of individual subjects.

This confidence movement paradigm provides an additional tool for studying metacognition in nonhuman species as it requires little to no training in that animals are naturally motivated to seek out and maximize food intake. Thus, spontaneous instances of metacognitive monitoring can be viewed that reflect an individual's confidence in their memory, performance, or knowledge states. These confidence judgments are easily dissociated from feedback, if those movements occur prior to external cuing. Finally, this paradigm can be manipulated depending upon the needs of the individual (e.g., Mercury) or the species under investigation. Because verbal reports of confidence in nonhuman animals and young children are impossible to solicit, paradigms that eliminate this reporting and instead rely upon natural behavioral repertoires are key to best illustrating capacities for control and monitoring. And, once such movements are in place, it is possible to provide tests of generalization to see whether such confidence occurs in new task settings.

Approach: Assessing Generalization of Confidence Movements in a Chimpanzee

Chimpanzee Sherman was re-presented with the testing situation in which the dispenser was located some distance from his computer apparatus, and auditory feedback to indicate correctness and incorrectness of responding also was delayed. This occurred more than a year after Sherman completed the study described earlier, and so we were interested in the degree to which his confidence movements would reemerge in a number of transfer tests, unlike those he originally was given in this context. These transfer tests are considered a crucial aspect of any study of metacognition in nonhuman species, as they allow investigators to determine the degree to which the responses that are considered to reflect metacognitive control or monitoring will occur to new stimuli, new tests, and thus are not constrained to being used only in trained contexts. In addition, tests of confidence and information-seeking behavior in nonhuman species have been particularly impressive for their generalization by species such as rhesus monkeys to new tasks, and even qualitatively different kinds of task (e.g., Kornell, Son, and Terrace 2007; Morgan et al. 2014). For this reason, it is important to determine whether the confidence movements we outlined earlier would occur in tests other than memory tasks. If so, they could be considered to reflect more than metamemory assessments, but perhaps also metacognitive judgments of perceptual discrimination abilities, for example.

Thus, Sherman performed a quantity judgment task, in which he had to select the array containing the larger number of two digital dot sets onscreen. This task assessed relative quantity judgments, and the confidence that Sherman might show in those judgments. A second experiment presented Sherman with a task that induced cognitive bias—a decoy task. In this task, a third stimulus operates to selectively enhance a bias toward another stimulus, and in some cases this evokes objective choice errors. The question here is whether, in this state of cognitive bias, Sherman still exhibited confidence in his choices in the way that humans often show high confidence even in cases of poor performance when they operate under perceptual or cognitive biases.

GENERAL METHODS

The experimental setup and basic procedures were identical to the Beran et al. (2015) study described before. All computer programs were written in Visual Basic 6.0. Trials were presented on a personal computer with a 17-inch monitor. Sherman responded by manipulating a digital joystick with his hand, to control an onscreen cursor and bring it into contact with stimuli onscreen. Attached to the computer was a universal feeder that dispensed food items (sugar-free candies) down a tube into a port in an adjacent enclosure. This port allowed Sherman to reach in and catch a falling reward, but if he was not in place to do that, the reward fell into a funnel that routed it out of the enclosure so that he could not obtain it (fig. 20.1).

For both computer tasks that we will describe, during formal testing sessions, Sherman had to complete the trial in one location, and he could move at any time to the other location where the reward dispenser was located. For these tasks, feedback was delayed by 4 seconds after each completed trial. If he was incorrect, Sherman heard a buzz tone, and the next trial was immediately presented on the computer screen. Thus, there was no timeout period for incorrect trials, allowing him to start immediately on the next trial if he was in

FIGURE 20.1. Demonstration of the general testing situation. Sherman worked in one location on a computer task, with a dispenser located in another location (top panel). The white arrow indicates the location in space where Sherman had to be past when feedback was given in order to score that trial as an early move. The middle panel shows Sherman moving, and the bottom panel shows him at the dispenser.

front of the computer apparatus. If he was correct, a melodic tone was played, and then 3 seconds after that, a reward was dispensed in the adjacent enclosure. This meant that on correct trials, Sherman would have to move quickly from the computer apparatus to the dispenser area if he was to catch the reward in time. If, however, he moved over early, before any feedback from the computer, he could do so in a slower manner, with plenty of time to reach the dispenser location. During these tests, there was no experimenter near the computer apparatus or the dispenser. The experimenter remained away from these areas so as to have no influence on Sherman's possible movement patterns during trials. Instead, the experimenter sat in an area of the laboratory from which he could record Sherman's position at the time that the feedback (buzz tone or melodic tone) was presented. If Sherman was still in the enclosure with the computer apparatus, this was recorded as no early movement. If he had moved into the enclosure with the dispenser before feedback was given, this was recorded as an early movement. Trials were video-recorded also in a way that allowed for determining Sherman's position when feedback was given (fig. 20.1). Sherman determined the pace of testing on his own, as he made the choice of when to engage with the computer apparatus.

Experiment 1: Relative Quantity Judgment Task

In this task, Sherman was presented with two dot arrays, one at the left center of the screen, and one at the right center, and he had to choose the array with more dots to be rewarded (fig. 20.2). This was a new task for him, and so first he was trained on the task rules and was given immediate feedback after responses. He was given food rewards by the experimenter (and thus did not need to move through space to retrieve the reward), and he experienced an eight second timeout period for incorrect responses during which the screen remained blank. On these training trials, one array contained 4 to 20 dots onscreen (randomly selected across trials), and the other array contained three times that number. The side with the larger number of dots was randomly determined on each trial. These were easy discriminations to allow him to learn the rule for choosing. Sherman completed three sessions of 60, 80, and 80 trials in this training phase. He was correct on 44/60 trials in session 1 (73%), 62/80 trials in session 2 (78%), and 72/80 trials in session 3 (90%). He then moved to the test phase.

In the test phase, the delayed feedback contingencies were in place and rewards were delivered at the distant food delivery site as described above. The task itself was made more difficult. On 1/3 of the trials, one set contained 4 to 20 dots, and the other set contained double that number. On 1/3 of trials, one set contained 12 to 20 dots, and the other contained 21 to 29 dots, and on 1/3 of trials, one array contained 18 to 20 dots and the other contained 21 to 23 dots. Thus, a wide range of trial difficulties were presented in this phase. Sherman completed four sessions of 30 trials each in this experiment.

All analyses were conducted using VassarStats.net. Overall, Sherman was correct on 62.5% of the trials, indicating that this was not an easy task for him. However, this was a performance level that exceeded the 50% chance level, $p < 0.01$, binomial test. With regard to his early movements, Sherman moved early to the dispenser on 53.3% of his correctly completed trials, and on only 31.1% of his incorrectly completed trials. This was a significant difference, χ^2 ($N = 120$) = 4.12, $p = 0.042$. Thus, Sherman demonstrated overall another example of earlier movements on trials that he correctly completed compared to those that he did not complete correctly. This is a direct extension of the previous work but in a domain different than the original study assessing confidence judgments on a memory task.

Despite the overall effect, though, it was clear that on a session-by-session basis Sherman's performance was highly variable. Figure 20.3 shows that for the first session, Sherman did not differentially move early depending on outcome (16/30 early moves overall). In session 2, although Sherman moved early

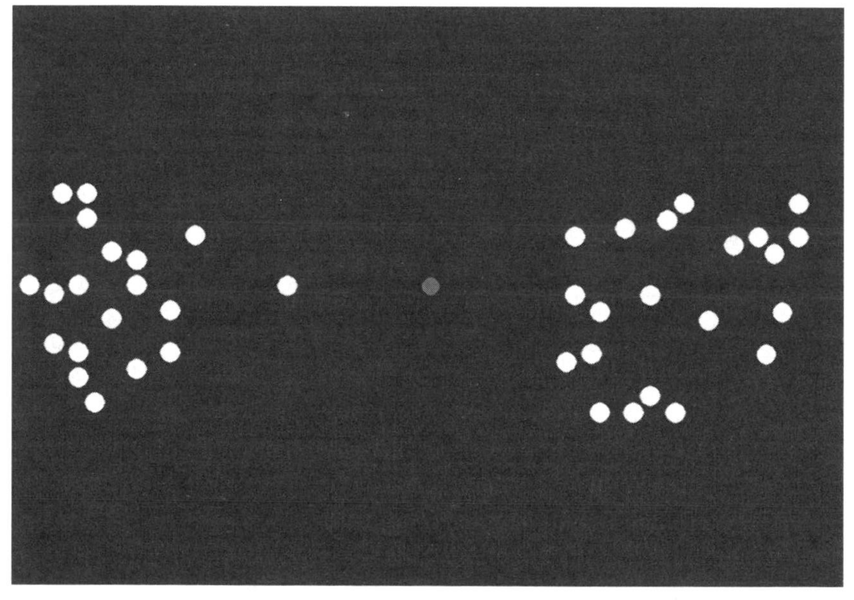

FIGURE 20.2. An example trial in the dot quantity discrimination task. The dot at center was the cursor, and was colored red. Sherman moved it left or right to indicate which of the sets contained more dots overall.

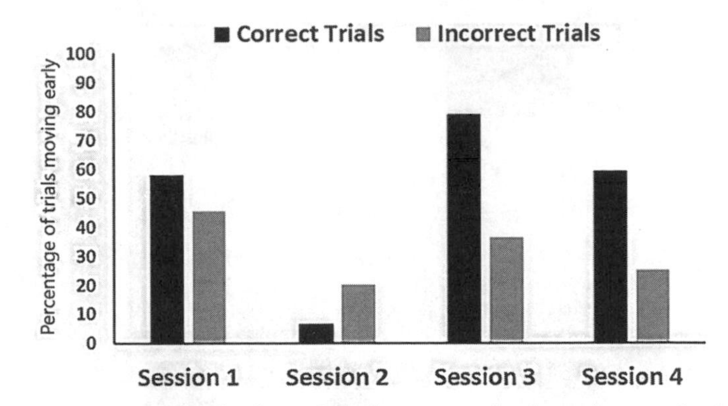

FIGURE 20.3. The percentage of trials on which Sherman moved early as a function of trial outcome (correct or incorrect) and session number.

on a greater percentage of *incorrect trials*, this is misleading in some sense because in that session he only went early on 4 of 30 trials (once when correct, three times when incorrect). In session 3 (19 of 30 early moves) and session 4 (15 of 30 early moves) is where the effect finally emerged more clearly, with a greater percentage of those moves occurring for correct trials compared to incorrect trials. Thus, it did take Sherman some time to establish the pattern of early moves occurring more often when he was correct. Of course, trials were all variable, so he was not learning when to move based on specific dot sets that were presented, but he likely did learn to better anticipate food reward as he gained more experience with this primary quantity discrimination task.

Experiment 2: Decoy Task

During training sessions, Sherman saw two rectangles on the screen during each trial (fig. 20.4). The orientation of each rectangle was manipulated so that one rectangle was presented with a longer horizontal plane than vertical plane (landscape). The other rectangle was presented in the opposite orientation (portrait). In half of the trials, the computer program created the first stimulus rectangle with a width that was randomly chosen from the range of 120 to 169 pixels and a height randomly chosen from a range of 60 to 109 pixels (landscape). On the other half of the trials, the program created a rectangle with a height that was randomly chosen from the range of 120 to 169 pixels and a width randomly chosen from a range of 60 to 109 pixels (portrait). In both cases, the second rectangle was created with a height that was determined by multiplying the width of the first rectangle by the value [1 + (Level * 0.05)] and a width that was determined by multiplying the height of

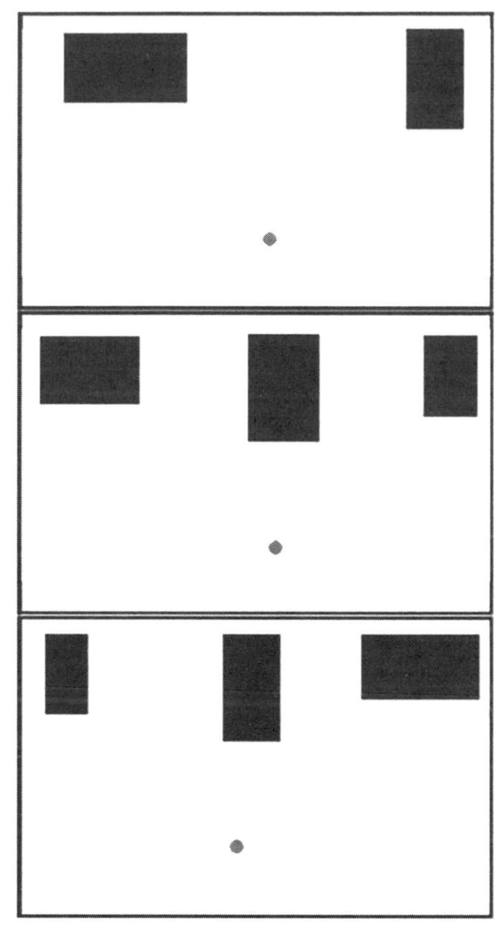

FIGURE 20.4. The decoy task. The top panel shows a baseline trial with no decoy stimulus. The correct choice is at left. The middle panel shows a congruent trial in which the decoy (smallest) rectangle is in the same orientation as the largest rectangle (in the center). The bottom panel shows an incongruent trial where the decoy rectangle is in the same orientation as the smaller of the other two rectangles (the correct choice is at right).

the first rectangle by the value [1 + (Level * 0.05)]. With this design, Level was an objective value for trial difficulty as it indicated the degree of difference in area between the two rectangles. This approach also allowed us to present Sherman with one rectangle that was wider than it was tall (landscape) and another that was taller than it was wide (portrait). Sherman had to select the larger (by area) rectangle to obtain a reward. On each trial, whether the larger rectangle was tall (portrait) or wide (landscape) was randomly determined.

Sherman first completed three training sessions of 95, 100, and 100 trials to teach him the rule to choose the larger rectangle. In these sessions, there was no delayed feedback, and he received food rewards immediately at the computer for correctly completed trials, and an 8 second timeout for incorrect trials during which time the computer screen was blank. In session 1, he completed 20 trials with Level set to 15, and he was correct on all 20 of those trials (100%). Note that these were trials in which the difference in rectangles was very large. He then completed 25 trials at Level 10, and was correct on 22 of those 25 trials (88%). He then completed 50 trials at Level 8, and he was correct on 36 of those 50 trials (72%). In session 2 and session 3, Level was randomly selected on each trial from the range 1 (most difficult trials) to 8 (least difficult trials). Thus, in these sessions, Sherman experienced the full range of trial difficulties in discriminating between the two rectangles. However, there was no delay in the presentation of auditory feedback, and Sherman did not have to move to a new location to retrieve food rewards.

In the test phase, two new aspects were added: a decoy stimulus and a delay before the presentation of auditory feedback. In terms of the task, the same trial types were present as in training, with just two rectangles onscreen. These were called baseline trials. In other trials, a third rectangle was introduced, and it was the decoy (fig. 20.4). The decoy was always the smallest rectangle on the screen in such trials; it always matched the orientation of one of the other two rectangles; and it was placed onscreen next to the other rectangle in that same orientation. Two conditions were presented with decoy stimuli. In the congruent condition, the decoy rectangle had the same orientation as the larger of the other two rectangles on the screen. In the incongruent condition, the decoy rectangle had the same orientation as the smaller of the other two rectangles on the screen. The decoy rectangle was drawn with a width and a height that were only 75% that of the other rectangle with the same orientation. If the decoy impacted Sherman's performance, then it should improve in the congruent condition, presumably because the decoy attracts Sherman to the truly larger of the other two rectangles because the decoy shares an orientation (and physical appearance) with it. At the same time, performance should suffer in the incongruent condition because the decoy attracts attention to the smaller (incorrect) of the other two rectangles because the decoy shares orientation features with it. This is the result we reported previously with monkeys (Parrish et al. 2015). The key question here is whether there is any evidence that Sherman not only falls prey to the decoy effect, but that he is seemingly not aware of that susceptibility. Rather, he should experience high confidence (and thus move to the reward dispenser early) on trials for which his performance is particularly poor in making the primary discrimination of largest rectangle because of the decoy effect (as is the case in incongruent decoy trials).

Thus, during test sessions, the two largest rectangles onscreen were drawn using the equation above, with Level randomly determined on each trial. Easier discriminations were those in which Level was set at 6, 7, or 8, and harder discriminations were those where Level was set at 1, 2, or 3. Baseline trials were presented with probability 0.50, and congruent and incongruent trials were presented with probability .25 each, randomly determined on each trial. Sherman completed 180 trials, with 5 sessions of 30 trials, one session of 25 trials, and a final session in which he finished the 5 trials that were not completed in his sixth session. From this design, Sherman was presented with objectively more difficult or less difficult primary discriminations between the two largest rectangles, and the program also manipulated whether the decoy stimulus (when present) would operate to aid discrimination or impede discrimination, if Sherman fell prey to a decoy effect.

We separated performance into those trials that were objectively easier because the rectangles varied more in size and those that were harder due to closer size of those rectangles. We also separated performance for baseline trials and for the congruent and incongruent test conditions. Figure 20.5 shows the results. For easier discriminations, there was no effect of the presence of a decoy stimulus, $p > 0.10$, Fisher's Exact test, although numerically this did decrease Sherman's percentage of correct choices in the incongruent condition compared to the congruent condition. For these trials, Sherman moved to the dispenser area significantly more often on correctly completed trials than incorrectly completed trials, $p < 0.001$, Fisher's Exact test. He did not, however, show any difference in his early movements for the different trial types that were correctly completed, all $p > 0.10$, Fisher's Exact test. Data again were analyzed using the VassarStats.net website.

The story was very different, however, for the more difficult discriminations. First, Sherman showed a clear decoy effect. He performed significantly better on congruent decoy trials than on baseline trials, $p = 0.03$, Fisher's Exact test. He did not perform significantly poorer on incongruent trials compared to baseline trials, $p = 0.10$, Fisher's Exact test, but he was numerically poorer. And, despite only selecting the decoy stimulus itself in 5 of 91 trials, he was significantly better on congruent trials than incongruent trials, $p < 0.001$, Fisher's exact test.

Perhaps the most interesting behavioral pattern comes from comparing early moves for correct and incorrect trials. Overall, there was no difference in early moves for correct and incorrect trials, $p = 0.12$, Fisher's Exact test. However, it seemed that this was because of different response patterns depending on the type of decoy stimulus presented. To examine this, we conducted a loglinear analysis to see whether there was an effect of trial type and

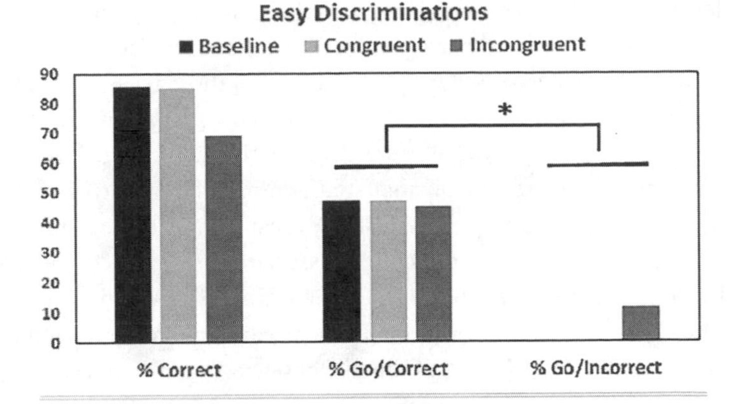

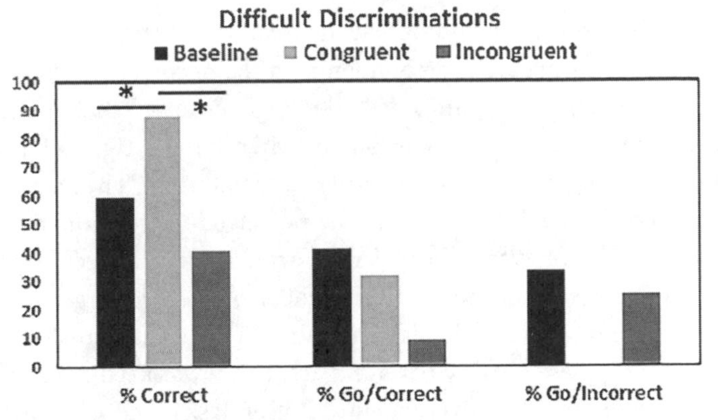

FIGURE 20.5. Chimpanzee Sherman's performance and early movements in the decoy task. The data are separated into easier and harder discrimination trials, as determined by the Level of trials. Asterisks indicate significant differences between selected trial types or response outcomes.

outcome (correctness or incorrectness) on early movements. This was confirmed, χ^2 (N = 90) = 20.02, p < 0.001. These findings indicate that although Sherman was more likely to move early on correctly completed congruent trials, he moved early more often on *incorrectly* completed incongruent trials than on correctly completed trials in this condition. This would seem to indicate that Sherman felt more confident when he was wrong than when he was correct when a decoy was present to potentially lead him to choose the wrong matching-orientation rectangle.

Discussion

Chimpanzee Sherman's performance in these generalization tasks provided two important results. First, for perceptual discrimination tasks, in which psychophysical relations were assessed in the judgments Sherman made, his confidence movements were similar to what he showed in metamemory tests (Beran et al. 2015). Sherman performed a difficult quantity comparison task, and he moved to the dispenser early more often on those trials he completed correctly than on trials he did not complete correctly. This confirms that his movements to where he expects reward to be given are based on internal signals of confidence, presumably that are the result of monitoring trial difficulty when making a primary choice. Unlike escape or uncertainty responses, which allow animals to avoid difficult trials, Sherman had to complete each trial. He could then wait for feedback, which functionally still allowed him to obtain all earned rewards, or he could move early to ease some of the motor demands on quickly getting to the dispenser in time when feedback was given that a trial was correct.

Perhaps the more novel finding was the clear demonstration of a dissociation between performance and confidence in the decoy task. Although this is a single chimpanzee in a single test, the results showed that Sherman felt more confident when he was wrong than when he was right, and this occurred when he was under a cognitive/perceptual illusion. The decoy stimuli, though rarely selected, influenced how he perceived the relation of the other rectangles in the task. When those decoys acted against correct responding, by steering Sherman more toward the smaller of the two target stimuli, he seemingly was not aware of this fallacy in his choice behavior. He moved to the dispenser on these illusory trials, expecting a reward. This is compelling because many of the criticisms of metacognition tasks used with animals allow those animals to readily track reinforcement history to responses in the presence of certain stimuli. This concern emerged early in this area of study, and has been addressed in a number of ways. For example, monkeys can be given runs of trials to complete, and then only summary feedback on overall performance, so that they cannot map specific responses to specific stimuli and the resulting reward or punishment that comes from those responses. In these cases, monkeys can learn to perform tasks, and will still opt out of trials that present exactly those stimuli that they are at greatest risk of making errors with (Couchman et al. 2010; Smith et al. 2006). Other approaches have forced monkeys to make responses first, and then wager on those responses with high or low bets (Kornell et al. 2007), thereby forcing animals to complete trials rather than escape from potentially aversive stimuli that become so through negative feedback.

In the case of Sherman with the decoy task, it appears that he was not aware that the decoy was leading him astray in his choice behavior, because he moved to collect an expected reward in incorrect probe trials. This is another instance, therefore, of dissociation of outcome and metacognitive judgment, and it is one that matches many of the reports of metacognition in humans (e.g., Levin et al. 2000; Rhodes and Castel 2008; Roediger and DeSoto 2014). This result, and other recent new approaches to studying things such as fluency and its effects of metacognitive responding, are promising. For example, Ferrigno, Kornell, and Cantlon (2017) reported that monkeys showed no difference in trials where they had to remember stimuli with high fluency (dark contrasting colors that were easy to see) compared to those with low fluency (lighter colors that were more difficult to see). However, although performance was the same, wagering responses to those stimuli varied in some cases, with the monkeys wagering less confidently after experiences low fluency trials. As with chimpanzee Sherman, the monkeys' confidence did not match their performance. Whether such effects can be seen in other metacognitive monitoring tasks or tasks of information-seeking behavior remain to be seen. But, the work with Sherman makes clear that confidence movements are generalizable, and dissociable from performance outcomes in ways that are interesting psychologically.

Future Directions

Researchers have been studying metacognition in animals for just over 20 years, and much progress has been made. This is certainly true in terms of the diversity of paradigms that have been developed to assess metacognition. And, from these studies we have progressed the degree to which we can argue that other primates (and some non-primate species) monitor their cognitive processes such as perception and memory, and they also monitor how they control guided information seeking to aid in correct responding. We also have learned some of the limitations experienced by other species when it comes to metacognitive control.

Less progress has been made in terms of broad species assessments, given that even within the primates, such research has been conducted using multiple tasks with only rhesus macaques, capuchin monkeys, and the great apes. This is a shortcoming in terms of building a phylogenetic map of different capacities and limitations in different primate species and other non-primate species. For example, the consistent failures or half-successes of capuchin monkeys relative to rhesus monkeys (e.g., Beran, Perdue, and Smith 2014; Perdue et al. 2015) begs for additional species comparisons. In addition, some of the tests used with apes, such as the metacognitive-movement task we described here, have yet to be given to other species. And, many tasks have been developed and used in laboratory settings, some of which could be adapted for use in zoos or other settings, potentially facilitating a richer comparative approach. Knowing whether zoo apes or monkeys might also show confidence movements, or make other responses to uncertainty or to gather information, would help researchers understand the generalized capacities for metacognition in these species. For example, a recent study with semi-wild rhesus monkeys replicated reports from laboratory tests of guided information-seeking behavior in monkeys. In that study (Rosati and Santos 2016), monkeys observed a human appearing to hide a food reward in an apparatus that contained one or two tubes. As in the laboratory studies, the monkeys tended to search the correct location when they observed this baiting event. However, if their view was occluded, they would instead look more often first to find the food before making a choice. They would peer into a center location that allowed them to check both potential hiding spots when their view had been occluded but they did not try to look from this location if information seeking was impossible. As with laboratory studies (e.g., Call 2010; Call and Carpenter 2001; Marsh and MacDonald 2012a), this result emerged without need of training animals what to do, and so this is a nice example of how those working with animals in more naturalistic settings can adapt and use laboratory tests to provide an account of how well those tests generalize to other groups of animals with alternative histories and living conditions. Alternatively, more laboratory-based tests are needed that better mirror the natural challenges faced by these animals in terms of confidence assessments and metacognitive monitoring.

Finally, it remains to be seen whether paradigms can be developed that truly require self-reflective processes to accompany the control and monitoring processes we have described. For adult humans, metacognition often is defined as having not only the quality of control and monitoring, but also the awareness of oneself as being in those reflective states as they are occurring (e.g., Carruthers 2008). This is not a definition of metacognition promoted by everyone (e.g., Proust 2013), but it remains the case that the study of metacognition is important, in part, because it offers the chance to better understand the conscious experiences of other animals, and perhaps most likely in the case of chimpanzees and other apes. Therefore, it is important to continue to consider experimental designs that might push farther the extent to which we can confidently assert that animals are aware of themselves as being in states of evaluating their own cognitive faculties. This will not be easy, and perhaps may not be possible, but the effort is required because the results may be crucial for approaching how best to care for other species and how to understand more fully their psychology.

Acknowledgments

Correspondence concerning this chapter can be addressed to Michael Beran, Language Research Center, Georgia State University, University Plaza, Atlanta, GA 30302. Email to mberan1@gsu. edu. This research was supported by NIH grants HD061455 and HD060563 and by NSF grants BCS1552405 and BCS0956993. The authors dedicate this chapter to Duane M. Rumbaugh (1929–2017), who worked with chimpanzee Sherman and many others, teaching them to use symbols and studying how they learned and communicated. Each of us had the honor to serve as the Duane M. Rumbaugh fellow at Georgia State University, and Duane taught us so much about the importance of valuing and learning from chimpanzees. We also dedicate this to the chimpanzees Panzee (1985–2014), Mercury (1986–2016), Lana (1970–2017), and Sherman (1973–2018). We worked with them (and played with them) for so many years, and without their willingness to engage in our scientific questions, we would know so much less about chimpanzee cognition and metacognition.

References

Basile, B. M., G. R. Schroeder, E. K. Brown, V. L. Templer, and R. R. Hampton. 2015. "Evaluation of seven hypotheses for metamemory performance in rhesus monkeys." *Journal of Experimental Psychology: General* 144: 85–102.

Benjamin, A. S., R. A. Bjork, and B. L. Schwartz. 1998. "The mismeasure of memory: When retrieval fluency is misleading as a metacognitive index." *Journal of Experimental Psychology: General* 127: 55–68.

Beran, M. J., J. Brandl, J. Perner, and J. Proust. 2012. *Foundations of Metacognition*. Oxford: Oxford University Press.

Beran, M. J., J. L. Pate, D. A. Washburn, and D. M. Rumbaugh. 2004. "Sequential responding and planning in chimpanzees (*Pan troglodytes*) and rhesus macaques (*Macaca mulatta*)." *Journal of Experimental Psychology: Animal Behavior Processes* 30: 203–12.

Beran, M. J., B. M. Perdue, S. E. Futch, J. D. Smith, T. A. Evans, and A. E. Parrish. 2015. "Go when you know: Chimpanzees' confidence movements reflect their responses in a computerized memory task." *Cognition* 142: 236–46.

Beran, M. J., B. M. Perdue, and J. D. Smith. 2014. "What are my chances? Closing the gap in uncertainty monitoring between rhesus monkeys (*Macaca mulatta*) and capuchin monkeys (*Cebus apella*)." *Journal of Experimental Psychology: Animal Learning and Cognition* 40: 303–16.

Beran, M. J., and D. M. Rumbaugh. 2001. "'Constructive' enumeration by chimpanzees (*Pan troglodytes*) on a computerized task." *Animal Cognition* 4: 81–89.

Beran, M. J., and J. D. Smith. 2011. "Information seeking by rhesus monkeys (*Macaca mulatta*) and capuchin monkeys (*Cebus apella*)." *Cognition* 120: 90–105.

Beran, M. J., J. D. Smith, and B. M. Perdue. 2013. "Language-trained chimpanzees (*Pan troglodytes*) name what they have seen but look first at what they have not seen." *Psychological Science* 24: 660–66.

Beran, M. J., and D. A. Washburn. 2002. "Chimpanzee responding during matching-to-sample: Control by exclusion." *Journal of the Experimental Analysis of Behavior* 78: 497–508.

Call, J. 2010. "Do apes know that they could be wrong?" *Animal Cognition* 13: 689–700.

Call, J., and M. Carpenter. 2001. "Do apes and children know what they have seen?" *Animal Cognition* 4: 207–20.

Carruthers, P. 2008. "Meta-cognition in animals: A skeptical look." *Mind and Language* 23: 58–89.

Castro, L., and E. A. Wasserman. 2013. "Information-seeking behavior: Exploring metacognitive control in pigeons." *Animal Cognition* 16: 241–54.

Couchman, J. J., M. V. C. Coutinho, M. J. Beran, and J. D. Smith. 2010. "Beyond stimulus cues and reinforcement signals: A new approach to animal metacognition." *Journal of Comparative Psychology* 124: 256–68.

Crystal, J. D. 2014. "Where is the skepticism in animal metacognition?" *Journal of Comparative Psychology* 128: 152–54.

Crystal, J. D., and A. L. Foote. 2009. "Metacognition in animals: Trends and challenges." *Comparative Cognition and Behavior Reviews* 4: 54–55.

Crystal, J. D., and A. L. Foote. 2011. "Evaluating information-seeking approaches to metacognition." *Current Zoology* 57: 531–42.

Ferrigno, S., N. Kornell, and J. F. Cantlon. 2017. "A metacognitive illusion in monkeys." *Proceedings of the Royal Society of London B* 284: 20171541.

Flavell, J. H. 1979. "Metacognition and cognitive monitoring: A new area of cognitive-developmental inquiry." *American Psychologist* 34: 906–11.

Fujita, K. 2009. "Metamemory in tufted capuchin monkeys (*Cebus apella*)." *Animal Cognition* 12: 575–85.

Hampton, R. R. 2001. "Rhesus monkeys know when they remember." *Proceedings of the National Academy of Sciences* 98: 5359–62.

Hampton, R. R. 2009. "Multiple demonstrations of metacognition in nonhumans: Converging evidence or multiple mechanisms?" *Comparative Cognition and Behavior Reviews* 4: 17–28.

Hampton, R. R., A. Zivin, and E. A. Murray. 2004. "Rhesus monkeys (*Macaca mulatta*) discriminate between knowing and not knowing and collect information as needed before acting." *Animal Cognition* 7: 239–46.

Honig, W. K., and J. G. Fetterman. 1992. *Cognitive Aspects of Stimulus Control*. Hillsdale, NJ: Lawrence Erlbaum Associates.

Hulse, S. H., H. Fowler, and W. K. Honig. 1978. *Cognitive Processes in Animal Behavior*. Hillsdale, NJ: Lawrence Erlbaum Associates.

Iwasaki, S., S. Watanabe, and K. Fujita. 2013. "Do pigeons (*Columba livia*) seek information when they have insufficient knowledge?" *Animal Cognition* 13: 211–21.

Jozefowiez, J., J. E. R. Staddon, and D. T. Cerutti. 2009. "Metacognition in animals: How do we know that they know?" *Comparative Cognition and Behavior Reviews* 4: 29–39.

Kirk, C. R., N. McMillan, and W. A. Roberts. 2014. "Rats respond for information: Metacognition in a rodent?" *Journal of Experimental Psychology: Animal Learning and Cognition* 40: 249–59.

Koriat, A., and M. Goldsmith. 1996. "Monitoring and control processes in the strategic regulation of memory accuracy." *Psychological Review* 103: 490–517.

Kornell, N. 2009. "Metacognition in humans and animals." *Current Directions in Psychological Science* 18: 11–15.

Kornell, N. 2014. "Where is the 'meta' in animal metacognition?" *Journal of Comparative Psychology* 128: 143–49.

Kornell, N., L. K. Son, and H. S. Terrace. 2007. "Transfer of metacognitive skills and hint seeking in monkeys." *Psychological Science* 18: 64–71.

Le Pelley, M. E. 2012. "Metacognitive monkeys or associative animals? Simple reinforcement learning explains uncertainty in nonhuman animals." *Journal of Experimental Psychology: Learning, Memory, and Cognition* 38: 686–708.

Levin, D. T., N. Momen, S. B. Drivdahl, and D. J. Simons. 2000. "Change blindness blindness: The metacognitive error of overestimating change-detection ability." *Visual Cognition* 7: 397–412.

Maestripieri, D. 2003. *Primate Psychology*. Cambridge, MA: Harvard University Press.

Marsh, H. L., and S. E. MacDonald. 2012a. "Information seeking by orangutans: A generalized search strategy?" *Animal Cognition* 15: 293–304.

Marsh, H. L., and S. E. MacDonald. 2012b. "Orangutans (*Pongo abelii*) 'play the odds': Information-seeking strategies in relation to cost, risk, and benefit." *Journal of Comparative Psychology* 126: 263–78.

Metcalfe, J., and H. Kober. 2005. "Self-reflective consciousness and the projectable self." In *The Missing Link in Cognition: Origins of Self-Reflective Consciousness*, edited by H. S. Terrace and J. Metcalfe. New York: Oxford University Press.

Morgan, G., N. Kornell, T. Kornblum, and H. S. Terrace. 2014. "Retrospective and prospective metacognitive judgments in rhesus macaques (*Macaca mulatta*)." *Animal Cognition* 17: 249–57.

Nelson, T. O. 1992. *Metacognition: Core Readings*. Toronto: Allyn and Bacon.

Parrish, A. E., T. A. Evans, and M. J. Beran. 2015. "Rhesus macaques (*Macaca mulatta*) exhibit the decoy effect in a perceptual discrimination task." *Attention, Perception, & Psychophysics* 77: 1715–25.

Perdue, B. M., B. A. Church, J. D. Smith, and M. J. Beran. 2015. "Exploring potential mechanisms underlying the lack of uncertainty monitoring in capuchin monkeys." *International Journal of Comparative Psychology* 28.

Proust, J. 2013. *The Philosophy of Metacognition: Mental Agency and Self-Awareness*. Oxford: Oxford University Press.

Rhodes, M. G., and A. D. Castel. 2008. "Memory predictions are influenced by perceptual information: Evidence for metacognitive illusions. *Journal of Experimental Psychology: General* 137: 615–25.

Roberts, W. A., M. C. Feeney, N. McMillan, K. MacPherson, E. Musolino, and M. Petter. 2009. "Do pigeons (*Columba livia*) study for a test?" *Journal of Experimental Psychology: Animal Behavior Processes* 35: 129–42.

Roediger, H. L., and K. A. DeSoto. 2014. "Confidence and memory: Assessing positive and negative correlations." *Memory* 22: 76–91.

Roitblat, H. L., T. G. Bever, and H. S. Terrace. 1984. *Animal Cognition*. Hillsdale, NJ: Lawrence Erlbaum Associates.

Rosati, A. G., and L. A. Santos. 2016. "Spontaneous metacognition in rhesus monkeys." *Psychological Science* 27: 1181–91.

Rumbaugh, D. M. 1977. *Language Learning by a Chimpanzee: The Lana Project*. New York: Academic Press.

Savage-Rumbaugh, E. S. 1986. *Ape Language: From Conditioned Response to Symbol.* New York: Columbia University Press.

Shields, W. E., J. D. Smith, and D. A. Washburn. 1997. "Uncertain response by humans and rhesus monkeys (*Macaca mulatta*) in a psychophysical same-different task." *Journal of Experimental Psychology: General* 126: 147–64.

Smith, J. D. 2009. "The study of animal metacognition." *Trends in Cognitive Sciences* 13: 389–96.

Smith, J. D., M. J. Beran, M. V. C. Coutinho, and J. J. Couchman. 2008. "The comparative study of metacognition: Sharper paradigms, safer inferences." *Psychonomic Bulletin and Review* 15: 679–91.

Smith, J. D., M. J. Beran, J. S. Redford, and D. A. Washburn. 2006. "Dissociating uncertainty responses and reinforcement signals in the comparative study of uncertainty monitoring." *Journal of Experimental Psychology: General* 135: 282–97.

Smith, J. D., W. E. Shields, J. Schull, and D. A. Washburn. 1997. "The uncertain response in humans and animals." *Cognition* 62: 75–97.

Suda-King, C. 2008. "Do orangutans (*Pongo pygmaeus*) know when they do not remember?" *Animal Cognition* 11: 21–42.

Suda-King, C., A. E. Bania, E. E. Stromberg, and F. Subiaul. 2013. "Gorillas' use of the escape response in object choice memory tests." *Animal Cognition* 16: 65–84.

Templer, V. L., and R. R. Hampton. 2012. "Rhesus monkeys (*Macaca mulatta*) show robust evidence for memory awareness across multiple generalization tests." *Animal Cognition* 15: 409–19.

PART SEVEN

Caring for Chimpanzees

21

Chimpanzees in US Zoos, Sanctuaries, and Research Facilities: A Survey-Based Comparison of Atypical Behaviors

MOLLIE A. BOOMSMITH, ANDREA W. CLAY, STEPHEN R. ROSS, SUSAN P. LAMBETH, CORRINE K. LUTZ, SARAH D. BREAUX, RHONDA PIETSCH, AMY FULTZ, MICHAEL L. LAMMEY, SARAH L. JACOBSON, AND JAINE E. PERLMAN

Introduction

ANIMAL WELFARE AND THE ASSESSMENT OF ATYPICAL BEHAVIOR

Behavioral aspects of chimpanzee welfare have been actively studied for more than three decades, with empirical research focusing on social housing, environmental enrichment, facility design, and animal training methods (Bloomsmith and Else 2005; Ross, chapter 24 this volume). This work has been conducted in research facilities (Baker and Easley 1996; Bloomsmith, Alford, and Maple 1988; Brent 1992), zoos (Herrelko, Buchannan-Smith, and Vick 2015; Hopper, Freeman, and Ross 2016; Ross and Lukas 2006; Ross and Shender 2016), and sanctuaries (Fultz et al. 2013, 2016; Kalcher et al. 2008; Kalcher-Sommersguter et al. 2013). The scientific study of chimpanzee (*Pan troglodytes*) welfare has long included the evaluation of atypical or abnormal behavior as one category of well-being measures (Walsh, Bramblett, and Alford 1982). Abnormal behaviors are defined as those that are not typical of the species in the wild, or that occur at very different frequencies in captivity than in the wild (Erwin and Deni 1979). Abnormal behaviors are generally regarded as indicators of poor welfare because of their association with animals living in restrictive environments, deficiencies in the animals' social environment, and insufficient management practices (Baker 1997; Brent 1992; Davenport and Rogers 1970; Davenport, Menzel, and Rogers 1966; Maki, Fritz, and England 1993; Turner, Davenport, and Rogers 1969), but they may not always be directly linked to welfare (Hopper, Freeman, and Ross 2016; Mason and Latham 2004). It is critical that we understand factors underlying atypical behaviors so we can properly interpret which may be indicators of poor welfare, and which may not.

The aim of this project was to contribute to a better understanding of abnormal behavior by describing the characteristics of the chimpanzees performing them using a very large and diverse sample of individuals living in three different captive settings (in research facilities, zoos, and sanctuaries in the United States). Using a written survey method, we were able to characterize the prevalence of seven types of abnormal behavior in relation to the chimpanzees' early rearing, group size, sex, and age and how this related to the setting in which the chimpanzees lived.

THE BEHAVIORS EXAMINED

A wide diversity of abnormal behaviors has been documented among captive chimpanzees (Jacobson, Ross, and Bloomsmith 2016; Nash et al. 1999; Walsh, Bramblett, and Alford 1982). The survey used in the current study included questions about the prevalence of seven different types of abnormal behaviors in individual chimpanzees: coprophagy (consuming feces), hair-plucking (the removal of hair from one's own body or from that of a companion), repeated regurgitation and reingestion (RandR) of food, stereotyped rocking (repeated back-and-forth movements, generally while seated), self-directed abnormal behaviors (e.g., bizarre posturing), self-injurious behavior (SIB), and a category of other stereotyped behaviors (e.g., pacing). This list includes abnormal behaviors with varying degrees of validated association with welfare. For instance, coprophagy, or seed reingestion from feces, has been reported in multiple wild communities of chimpanzees and may be an adaptive strategy to maximize nutritional intake in certain conditions (Bertolani and Pruetz 2011; Payne, Webster, and Hunt 2008), though it appears to be much more pervasive in captive conditions (Hopper, Freeman, and Ross 2016). In contrast, self-injurious behavior can cause serious physical harm and is likely the most concerning behavioral problem observed in nonhuman primates (Novak et al. 2012).

COMPARISONS OF ABNORMAL BEHAVIOR ACROSS FACILITY TYPE

Chimpanzees living under human care in the United States live in four major settings: research institutions, sanctuaries, zoos, and privately-owned facilities (e.g., pets, performers), and within these categories there is a broad diversity of housing and care. For many years the largest proportion of this chimpanzee population has lived in research settings, but with the ongoing retirement of chimpanzees to sanctuaries, these two populations (sanctuary and research facility) are now fairly equivalent in size (Hirata et al., chapter 9 this volume). Zoos hold smaller numbers of chimpanzees than sanctuaries or research facilities, and private individuals own the smallest portion of the population (ChimpCARE 2018). Our study compared the profile of the abnormal behaviors described above across research facilities, zoos, and sanctuaries. Other recent work on abnormal behavior has focused on chimpanzees living in just one setting: Birkett and Newton-Fisher (2011) and Jacobson, Ross, and Bloomsmith (2016) on zoo chimpanzees; Bloomsmith et al. (2019) on chimpanzees in the research environment; and Kalcher-Sommersguter et al. (2013) on chimpanzees in a sanctuary. Our survey approach facilitated the involvement of many facilities and allowed the application of consistent methods to data gathering, thus permitting more direct comparisons within a very large sample. In addition, this study collated data from multiple facilities within each housing category, which provides a broader scope than some previous studies carried out at a single facility.

Our intention with this study was to document the occurrence of abnormal behaviors across a diverse population of chimpanzees. Since our sample included a very large number of subjects that live in a wide variety of environmental conditions this allowed a more robust assessment of the influential factors than can be completed in a single type of setting. Although any behavioral differences that might be associated with chimpanzees living in varying facility types would be noteworthy, we also must keep in mind that such differences could be due to intrinsic characteristics of the subpopulation (e.g., age, sex) and/or past experiences (e.g., early rearing conditions, research history) of the animals, rather than being directly associated with the setting in which the chimpanzees currently lived when we conducted our survey. Our analysis allowed for inspection of some of these factors, age and sex for example, but others, such as research history, exposure to humans, personality, past housing conditions, and past social experiences (e.g., size and stability of social groups over the course of the individual's lifetime), were not assessed here although they may influence some of the effects we found for facility, age, sex, early rearing history, and current group size. For example, most chimpanzees that live in sanctuaries originated elsewhere (some are wild-born, and most came from research facilities or were personal pets or entertainers), and their behavior at the sanctuary is influenced both by their past and by their present environments.

Our current understanding of many abnormal behaviors in primates is that once they are established, they tend to be maintained over time, and it is often difficult to eliminate them despite active attempts at treatment by changing the physical, social, or internal physiological environment (Novak et al. 2012). Some stereotypies, for example, are understood to change over time, coming under "central control" following a high number of repetitions. Central control is a form of automatic processing that requires minimal cognitive processing or need for sensory feedback, and accounts for some stereotypies becoming more difficult to interrupt or change with environmental alteration (Mason and Latham 2004). This process may also disassociate the stereotyped behavior from current welfare status such that an animal observed pacing cannot be assumed to do so because of a deficit in his or her current environment or because he or she is currently experiencing distress. Therefore, not all stereotypies are considered good indicators of current stress or frustration (Mason and Latham 2004). Because of these characteristics of some abnormal behaviors, it may not be reasonable to expect that changing an animal's environment will eliminate their abnormal behavior expression. Seen another way, the current living situation of a chimpanzee may not be the best (and is certainly not the only) predictor of that individual's abnormal behavior profile. Similarly, an individual's current abnormal behavior profile may not be the best, and certainly is not the only, indicator of the individual's welfare in his or her current living situation.

IMPACT OF FACTORS TO BE EVALUATED

Our study focused on four major factors to determine their influence on abnormal behavior.

1) Rearing history: The single, best-established factor that influences abnormal behavior in chimpanzees is the social context in which young chimpanzees develop. Male chimpanzees from research facilities who were not mother-reared are more likely to show abnormal behavior (of any type) than mother-reared males (Bloomsmith et al. 2019) and similar results have been found for chimpanzees in zoos (Jacobson, Ross, and Bloomsmith 2016; Martin 2002) and sanctuaries (Freeman and Ross 2014). In addition to the differential impact of being mother- or human-reared, being wild-born (WB) and brought into captivity as a young animal may also be influential.

2) Sex: Although there have been reports of male biases for abnormal behavior in certain primates (Capitanio 1986; Gottlieb, Capitanio, and McCowen 2013), this may not be the case for chimpanzees. However, female chimpanzees are more likely to show coprophagy than males (Bloomsmith et al. 2019; Jacobson, Ross, and Bloomsmith 2016; Kalcher-Sommersguter et al.

2013; Nash et al. 1999) and this may be related to a female bias for social learning, especially during early development (Lonsdorf 2005).

3) Age: We examined age to identify possible impacts on abnormal behavior, as there has been limited analysis in the past exploring the influence of chimpanzees' age and the expression of abnormal behaviors.

4) Social group size: Scientists who study chimpanzee behavior in the field, zoo, and research environments agree that social housing is a key contributor to chimpanzee welfare (AZA Ape TAG 2010; Bloomsmith and Else 2005; Bloomsmith and Baker 2001; Pruetz and McGrew 2001; Ross, chapter 24 this volume), but little has been done to document its effect on the expression of abnormal behavior. In a recent conference presentation, abnormal behavior was reported higher in chimpanzees housed in larger groups (>7 group members), although most other behaviors were unaffected by group size (Reamer et al. 2016). Certainly more data are needed to explore how group size (and the additional space larger groups generally live in) may impact abnormal behavior.

Approach

We sent a written questionnaire to the following facilities housing chimpanzees in the United States: 1) All (six) research facilities that were accredited by the Association for Assessment and Accreditation of Laboratory Animal Care International (AAALAC) (Yerkes National Primate Research Center; Southwest National Primate Research Center; Keeling Center, MD Anderson Cancer Center; New Iberia Research Center; Alamogordo Primate Foundation; Language Research Center, Georgia State University); 2) all (33) zoos accredited by the Association of Zoos and Aquariums (AZA) and also holding chimpanzees; and 3) two chimpanzee sanctuaries (Chimp Haven and The Center for Great Apes) that were accredited by the Global Federation of Animal Sanctuaries (Chimp Haven is also AAALAC accredited).

We sent out the questionnaires between 2015 and 2017 in a staggered manner and each facility completed the information for each chimpanzee in their care at that time and who had been housed at that facility for at least two years. In the questionnaire, we asked each facility to score each chimpanzee as either having been observed, at least once in the past two years, engaging in an abnormal behavior, or not having been observed engaging in that behavior (see Bloomsmith et al. 2019 for the entire questionnaire). The prevalence of these behaviors was by our definition inherently conservative: exhibiting a behavior at least once in a two-year period.

There was no overlap in the population and so no individual chimpanzee was assessed twice. It should be noted that subsets of the data have been used in other analyses: Bloomsmith et al. (2019) focused only on the chimpanzees living in research facilities and Jacobson, Ross, and Bloomsmith (2016) focused on zoo chimpanzees.

DEFINITION OF VARIABLES

We defined "early rearing" as the predominant type of social setting during the subject's first 12 months of life. For example, mother-reared (MR) subjects were those that spent more than six of their first 12 months with their mother. Similarly, those identified as not-mother-reared (NOTMR) spent more than six of their first 12 months in a nursery being raised by humans, in a human home, or other human setting, and may have lived with other chimpanzee infants. Wild-born (WB) chimpanzees were those who were trapped and brought into captivity as youngsters. We know little about their duration of maternal rearing, although they certainly had some before being forcibly separated. As such, we categorized WB chimpanzees separately here. Those with unknown (UNK) rearing histories in captive settings were eliminated from statistical comparisons that included early rearing as an independent variable.

SUBJECT DEMOGRAPHICS

Our survey captured 1,122 subjects, 62.5% from primate research facilities, 18.2% from zoos, and 19.3% from sanctuaries. Female subjects made up 55.5% of the sample, males 44.5%. It is notable that this study includes the largest number of chimpanzees living under human care ever assessed from a welfare perspective. We categorized the subjects as "adult" if they were 12 to 39 years of age at the time of the survey (76.2% of the sample), "immature" if they were 11 or younger (10.0%), and "elderly" if they were 40 or older (13.8%). The mean age of our sample was 26.9 years (SD = 11.4), with a range of two to 77 years. We categorized group size as pairs (2 animals), small groups (3–7 animals), and large groups (8 or more animals). There were 86 subjects (7.7%) in pairs (we included the one singly-housed subject here, so he could be included in all of the other analyses; he showed no abnormal behaviors), 602 in small groups (53.7%), and 434 (38.7%) in large groups. MR subjects made up 35.3% of the sample; 8.9% were WB; NOTMR made up 41.6% of our sample; and unknown rearing (UNK) was reported for 14.2% of the sample. Table 21.1 provides age, rearing, and sex demographics for the sample.

TABLE 21.1. Age, sex, and rearing demographics for the chimpanzees reported in our survey.

	Sex		*Rearing*				
Age Category	*Female*	*Male*	*MR*	*NOTMR*	*UNK*	*WB*	*SUBTOTAL*
Immature (≤11 years)	66	46	61	51	0	0	112
Adult (12–39 years)	465	390	324	394	131	6	855
Elderly (≥40 years)	92	63	11	22	28	94	155
SUBTOTAL	623	499	396	467	159	100	TOTAL: 1122

Note: MR = mother reared, NOTMR = not mother reared but born in captivity, UNK = unknown origin, and WB = wild born.

DATA ANALYSIS

Based on the information collected for this data set, we compiled a list of five categorical predictors (age, sex, rearing history, current type of facility, and group size). We also included the interaction of sex × rearing due to previous results indicating this interaction could be significant (Bloomsmith et al. 2019). Predictors were entered into Forward:Wald logistic regression analyses (conducted with IBM SPSS Statistical Software version 24) for each abnormal behavior surveyed: coprophagy, RandR, stereotypic rocking, stereotypy other than rocking, hair-plucking, self-directed abnormal behavior, and SIB with wounding. Reference categories for categorical variables were as follows: 1) sex = female; 2) rearing = MR; 3) facility type = research facility; 4) age = adult; and 5) group size = small groups. Tests of significance compare test groups (non-reference groups) to the reference group, but do not compare test groups to each other (if there is more than one), so, for example, we assessed significant differences between WB and MR or between NOTMR and MR, but not between WB and NOTMR. Reference groups were chosen either because they represented the "standard" against which we wanted to compare the other groups (e.g., MR as standard versus NOTMR or WB), or because they represented the majority of cases (e.g., there were more animals living in small groups than in pairs or large groups). We excluded animals with UNK rearing history (n = 159) from this regression analysis, so our sample size for these analyses was n = 963.

For each dependent measure, we tested the five predictor variables for multicollinearity by calculating tolerance and variance inflation factors (VIF) in a linear regression. Tolerance below 0.4 and VIF above 2.5 are indicators of high collinearity (Allison 2012). We found that none of the predictors scored below 0.4 tolerance or above 2.5 VIF for any of the dependent measures we assessed. We entered the full set of possible predictors, including the sex by rearing interaction, into a Forward:Wald logistic regression for each dependent variable, set to add and retain only predictors with significant β coefficients ($p < 0.05$) and to report each step in which significant predictors were added (and resulted in a significant model). The significant model with the most predictors that also passed the Hosmer-Lemeshow (HL) goodness-of-fit test ($p > 0.05$) is reported here. We do not report predictors that we removed due to insignificant β coefficients.

Results and Discussion

PREVALENCE OF ABNORMAL BEHAVIORS IN COMPLETE SAMPLE

Prior to running logistic regression analyses, we calculated the prevalence of each abnormal behavior in each type of setting, research facility, zoo, and sanctuary within the full sample (N = 1,122) (table 21.1). The most prevalent abnormal behavior in research facilities was rocking, in zoos the most prevalent was coprophagy, and in sanctuaries it was hair-plucking (fig. 21.1). Over the entire data set, we found that 45.9% of the chimpanzees displayed some type of abnormal behavior in the two-year period prior to survey completion, and that 8.9% of them showed three or more abnormal behaviors. When we excluded coprophagy, 38.1% showed some kind of abnormal behavior and 6.2% showed three or more.

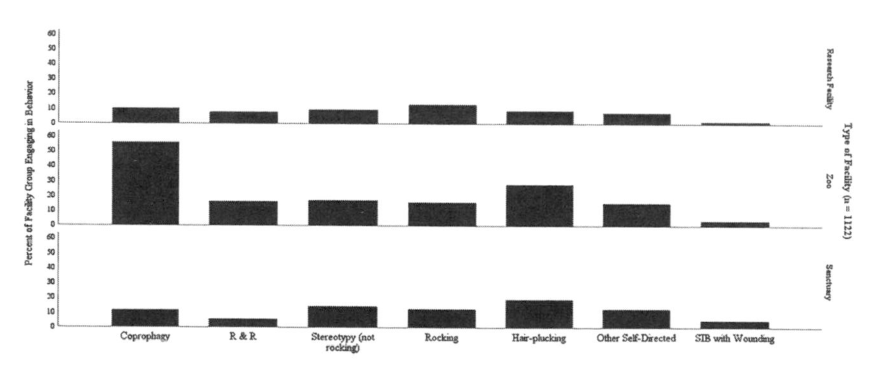

FIGURE 21.1. Prevalence of abnormal behaviors in each type of facility (percent of sample exhibiting the behavior). R & R, repeated regurgitation and reingestion; SIB, self-injurious behavior.

Coprophagy

The regression analysis for coprophagy resulted in an equation that retained rearing, sex, and facility type as significant predictors. Age was initially retained, but the model including age failed the goodness-of-fit test; the model with rearing, sex and facility type passed (H.L. $\chi^2[6] = 12.55$, $p = 0.051$). The model was significant ($\chi^2[5] = 229.63$, $p < 0.001$), had a −2 log likelihood of 735.26, and pseudo-R^2 range of 21.2% (CoxandSnell) to 33.5% (Nagelkerke). Based on coefficient values and significance, MR individuals were significantly more likely to engage in coprophagy as compared to NOTMR, females were significantly more likely than males to engage in coprophagy, and chimpanzees in zoos were significantly more likely to engage in coprophagy than those in research facilities (fig. 21.2a).

Coprophagy was associated with early rearing, sex, and facility type, but not with group size or age. In line with previous reports (Hopper, Freeman, and Ross 2016; Jacobson, Ross, and Bloomsmith 2016; Nash et al. 1999), MR subjects had higher probabilities than NOTMR individuals to engage in coprophagy. Some have argued that coprophagy in chimpanzees should not be classified as indicating poor welfare because it is associated with the desirable social condition of mother rearing and with desirable social interactions (Hopper, Freeman, and Ross 2016; Jacobson, Ross, and Bloomsmith 2016). Nonetheless, coprophagy is not a *desirable* behavior. It is much less frequent in the wild than in captive conditions and so should not be interpreted as desirable because it is "natural." Coprophagy may have negative health consequences by contributing to parasitic and bacterial disease transmission through a fecal/oral route of exposure as occurs between companion animals and owners (Esch and Petersen 2013) or due to the reuse of sewage water for crop irrigation (Beuchat and Ryu 1997). Increased focus on dietary improvements and feeding enrichment strategies is needed to attempt to minimize coprophagy. Additionally, other factors related to the type of facility in which chimpanzees live (e.g., being restricted to outside portions of enclosures, see below) should be more fully investigated for their impact on this behavior. If this behavior is chiefly socially learned, we might be able to "interrupt" this transmission through some creative management techniques.

The 100 WB chimpanzees were almost exclusively elderly, and so have spent more years in captivity than most NOTMR or MR chimpanzees. The WB chimpanzees had at least some early mother rearing (perhaps more than the 6-month minimum we applied for defining MR) and this social experience may facilitate coprophagy. Some may have had other experience with coprophagy prior to capture that could have been accentuated once in captivity and confronted with a very different diet. Females were more likely than males to show coprophagy,

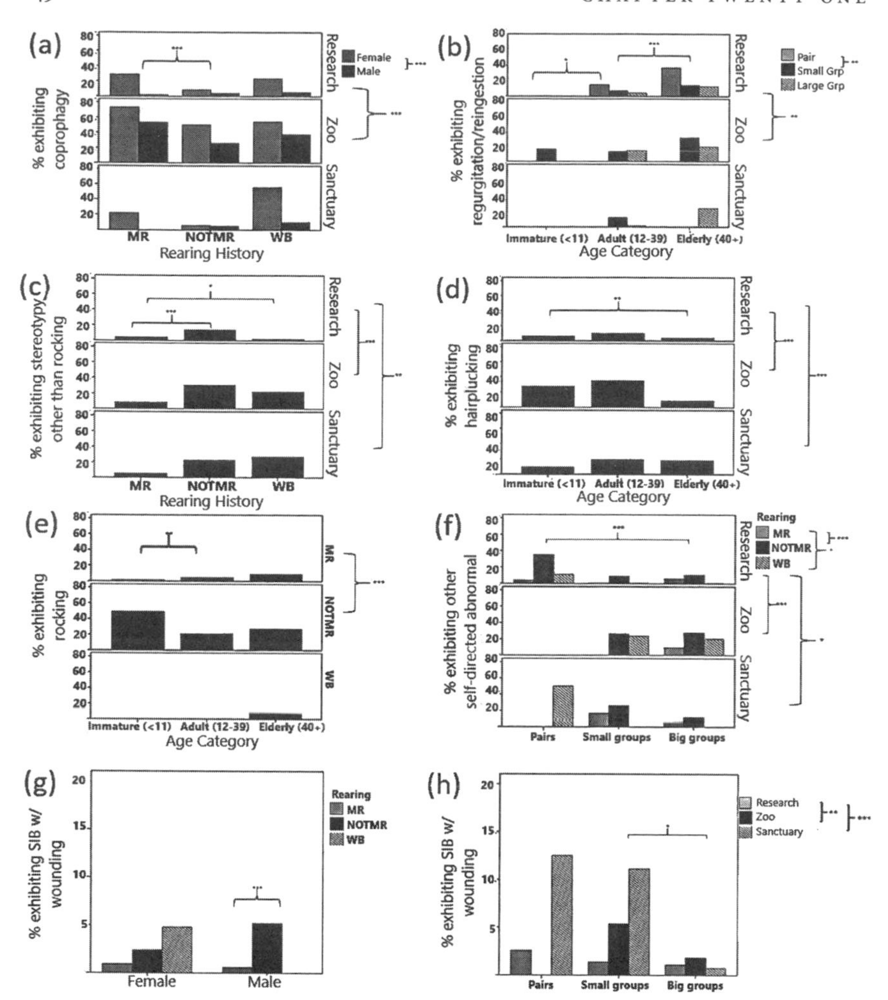

FIGURE 21.2. Prevalence of abnormal behaviors based on significant, retained predictors (for all figures, * indicates a predictor with a p < .05, ** indicates a p < .01, and *** indicates a p < .001). BIG GRP, big group; MR, mother reared; NOTMR, not mother reared but born in captivity; SIB, self-injurious behavior; SM GRP, small group; and WB, wild born.

as has been shown previously (Bloomsmith et al. 2019; Jacobson, Ross, and Bloomsmith 2016; Kalcher-Sommersguter et al. 2013; Nash et al. 1999).

Chimpanzees in zoos had a higher probability of engaging in coprophagy than animals in research facilities although we cannot elucidate the cause for this. While one might hypothesize that the potentially negative effects of the presence of zoo visitors (particularly large crowds of noisy zoo visitors) could contribute to the expression of this behavior, several published studies did not find such an effect (Chamove, Hosey, and Schaetzel 1988; Fernandez et al. 2009; Hosey 2005). In another study, chimpanzees showed behavioral changes when larger crowds of zoo visitors were present, but abnormal behaviors (fecal-related behaviors, self-plucking of hair, stereotypic behavior) did not vary with crowd sizes (Wood 1998). Bonnie, Ang, and Ross (2016) demonstrated that even as crowd size increased, chimpanzees did not avoid use of the areas of their exhibit closest to zoo visitors and crowd size had no effect on the frequency of most behaviors analyzed, including abnormal behavior. These studies indicate that abnormal behaviors, such as coprophagy, are unlikely to be accentuated due to the effects of zoo visitors.

Another general difference between zoos and the other facility types is that some zoos restrict chimpanzees to some portions of their enclosures (often the outdoor portions) for most of the day where zoo visitors can best see them. Research facilities and sanctuaries typically restrict chimpanzees during much shorter periods of time such as during cleaning routines. The effects of this management strategy have not been adequately addressed in the literature, but chimpanzees with outdoor access showed less abnormal behavior (e.g., coprophagy, RandR) than those living only indoors (Baker and Ross 1998) and restriction to only indoor spaces increased abnormal behavior in chimpanzees in a research facility (Bloomsmith, Lambeth, and Haberstroh 1999). Perhaps zoo chimpanzees have a similar response when restricted to portions of their enclosures over longer periods of time. Another possibility for differences in coprophagy across facilities is variation in feeding routines (Baker and Easley 1996; Bloomsmith, Alford, and Maple 1988; Bloomstrand et al. 1986; Fritz et al. 1992; Martin 2002; Struck et al. 2007).

Lastly, if WB individuals are potentially a source for increased coprophagy levels, the higher number of individuals remaining in their matrilineal groups and/or being raised by their mothers in zoos may "preserve" this behavior better than for those in other settings where a smaller proportion of the population is MR. Indeed, the behavior may then spread further, from zoo to zoo, due to recommendations made by the Species Survival Plan for the transfer of animals within the accredited zoo population. These ideas about differences across facilities are speculative, but they underscore the importance of comparing welfare across different types of facilities with varying management methods to better understand and care for chimpanzees.

Regurgitation/Reingestion

The regression analyses for regurgitation/reingestion (RandR) resulted in an equation that retained age, facility type, and group size as significant predictors. This was the model with the most significant predictors and also passed the HL test for goodness-of-fit (H.L. $\chi^2[6] = 5.75$, $p = 0.452$). The model was significant ($\chi^2[6] = 45.05$, $p < 0.001$), had a −2 log likelihood of 543.77, and pseudo-R^2 range of 4.6% (CoxandSnell) to 10.0% (Nagelkerke). Based on coefficient values and significance, adult chimpanzees were significantly more likely to engage in RandR than immature individuals, elderly chimpanzees were more likely than adults, chimpanzees in zoos were more likely than those in research facilities, and chimpanzees in pairs were more likely than those in small groups (fig. 21.2b).

Age, facility type, and group size were associated with RandR. Advancing age increased the probability that chimpanzee showed this behavior. The lack of influence of early rearing on RandR is consistent with other findings (Jacobson, Ross, and Bloomsmith 2016; Nash et al. 1999) though it differs from what has been found in gorillas (Gould and Bres 1986). Management procedures, such as feeding enrichment can modify RandR behaviors. Enhancements for chimpanzees including supplying substrates such as hay for foraging opportunities, feeding a larger number of meals, providing a greater diversity of fresh produce, using feeding puzzles that require more work to obtain food, and even the use of positive reinforcement training techniques to reward other behaviors, can all suppress RandR (Baker and Easley 1996; Bloomsmith, Alford, and Maple 1988; Bloomstrand et al. 1986; Morgan, Howell, and Fritz 1993; Struck et al. 2007). For example, a biscuit-free gorilla diet with low caloric density has greatly reduced RandR in gorillas (Less, Kuhar, and Lukas 2014). It is feasible that widespread improvements in modern management and care programs have had positive influences on younger chimpanzees, therefore reducing their incidence of RandR, but that among older chimpanzees who established this behavior some years ago, it is not eliminated by the improvements that began later in their lives. Latham and Mason (2008) describe that in some cases, restrictive environments can lead to lifelong changes in central nervous system functioning and behavioral control processes, which may cause individuals who have developed abnormal behaviors to be unable to cease those behaviors even when the environment is improved. Additionally, enjoyment of RandR has been reported by humans who engage in this activity (Mayes et al. 1988), and chimpanzees could have a similar, self-reinforcing experience.

Chimpanzees in zoos had a higher probability of engaging in RandR than those in research facilities, and the factors discussed above related to coprophagy may also be relevant here. There was a higher prevalence of RandR among those subjects living in pairs than among those living in small groups. The vast majority of subjects that lived in pairs were housed in research facilities (90.1%), and figure 21.2 illustrates that the pair-housed, elderly subjects in research facilities accounted for a large portion of the RandR. These data may indicate that pair housing is not stimulating enough to avoid the expression of RandR and that complex social groupings are more desirable for supporting species-typical behavior. It is interesting that two seemingly similar behaviors, RandR and coprophagy, seem to be differentially socially influenced. There is more RandR in small social groups but there is no impact of early rearing, while coprophagy is not influenced by group size but is affected by early rearing.

Stereotypy Other than Rocking

The regression analysis for stereotypic behaviors other than rocking resulted in an equation that retained rearing and facility as significant predictors. This was the model with the most significant predictors that also passed the HL test for goodness-of-fit (H.L. $\chi^2[5] = 4.09$, $p = 0.537$). Coefficients for group size and age were also significant when added to the model but resulted in a poor goodness-of-fit (H.L. $p < 0.05$). The model including rearing and facility was significant ($\chi^2[4] = 51.06$, $p < 0.001$), had a −2 log likelihood of 657.39, and pseudo-R2 range of 5.2% (CoxandSnell) to 9.9% (Nagelkerke). Based on coefficient values and significance, NOTMR and WB animals were significantly more likely to engage in stereotypies other than rocking than MR, and chimpanzees living in zoos or sanctuaries were more likely than those in research facilities (fig. 21.2c).

These types of stereotypies (e.g., pacing, repeated hand movements) were associated with early rearing and facility type. Our findings confirmed that NOTMR chimpanzees and WB chimpanzees were more likely than those raised by their mothers to show these stereotypies. Since stereotypies are associated with socially restrictive early rearing across nonhuman primate species (Berkson 1968; Erwin and Deni 1979; Gottlieb, Capitanio, and McCown 2013; Martin 2002; Mason 1968), it is not surprising that NOTMR chimpanzees have a higher prevalence of this behavior. But similar prevalence among WB animals is a surprising finding and may be due to historical animal management practices and/or the experiences young chimpanzees had when they were captured in the wild. Stereotyped behaviors may help animals cope with suboptimal environmental conditions (Mason and Latham 2004), such as nursery-rearing and social restriction (Martin 2002). Once stereotypies have developed, they may persist as a response to encountered stressors, even if the environment is generally appropriate (e.g., anticipation of desired event such as food delivery in Waitt and Buchanan-Smith 2001, or brief separation from a groupmate in Suomi 1991).

Chimpanzees in research facilities showed reduced prevalence of these behaviors when compared to those in zoos and sanctuaries. Since so many of the animals in sanctuaries originated from research facilities, why would they be more likely to show stereotypy when in a sanctuary? Perhaps the selection criteria applied to choose the first chimpanzees that were sent to sanctuary could have resulted in a higher number of chimpanzees with a tendency to express stereotyped behavior. However, this would not explain the boost in likelihood associated with living in a zoo. Likely, there is some other factor at work here that we have not yet evaluated, such as the varied origins of zoo chimpanzees (Jacobson, Ross, and Bloomsmith 2016). Behaviors included in this category may have different etiologies, possibly clouding the interpretation of these data. For example, pacing may be associated with one set of factors, while repetitive hand movements may be associated with others.

Hair-Plucking

The regression analyses for hair-plucking resulted in an equation that retained age and facility as significant predictors. This was the model with the most significant predictors and also passed the HL test for goodness-of-fit (H.L. $\chi^2[4] = 4.30$, $p = 0.367$). The model was significant ($\chi^2[4] = 56.50$, $p < 0.001$), had a −2 log likelihood of 752.59, and pseudo-R^2 range of 5.7% (Coxand-Snell) to 10.0% (Nagelkerke). Based on coefficient values and significance, adult chimpanzees were significantly more likely to hair pluck than elderly individuals, and chimpanzees living in zoos or sanctuaries were more likely than those in research facilities to perform hair-plucking (fig. 21.2d).

The probability of hair-plucking was associated with age and facility type. Elderly subjects had a lower prevalence than adults. Additionally, hair-plucking was more common in zoo and sanctuary-living chimpanzees than in research chimpanzees. It is interesting that hair-plucking was not related to the social variables we tested—early rearing or social group size. There have been earlier indications that this behavior may be socially learned (Nash et al. 1999), but it did not show a positive relationship to mother rearing (although it was most prevalent in MR) or to living in large groups, both of which would be expected to accompany a socially facilitated behavior. Hair-plucking could still be influenced by social learning even if it is not accentuated by mother rearing, and this has been identified in gorillas (Less, Kuhar, and Lukas 2013). Vazquez and Lutz (2017) reported alopecia in chimpanzees was not related to group size, corroborating this finding. Recent studies of macaque alopecia report some relationships with stress (Novak et al. 2014) and also individual temperament (Coleman et al. 2017).

However, in our survey, hair-plucking included both hair-plucking oneself and hair-plucking of others, and these behaviors could have different etiologies and factors for expression. For example, hair-plucking of others may be more related to social issues, while self-directed hair-plucking may be more related to stress, as has been found in bonobos (Brand and Marchant 2015, 2018). In humans, trichotillomania, or hair-plucking of self, is theorized to be a condition related to poor impulse control, anxiety disorder (Grant et al. 2017), and/or stress; different "subtypes" of hair-pulling have been assessed (Lochner, Seedat, and Stein 2010) and many etiological theories have been put forward (for a review, see Duke et al. 2010). Chimpanzees who hair-pluck others during social grooming may have a different motivation from those who hair-pluck themselves, and thus the influence of some of the predictors may have been obscured.

Rocking

The regression analyses for rocking resulted in an equation that retained age and rearing as significant predictors. This was the model with the most significant predictors and also passed the HL test for goodness-of-fit (H.L. $\chi^2[3] = 5.60$, $p = 0.133$). The model was significant ($\chi^2[4] = 92.38$, $p < 0.001$), had a −2 log likelihood of 699.04, and pseudo-R^2 range of 9.1% (CoxandSnell) to 16.3% (Nagelkerke). Based on coefficient values and significance, immature chimpanzees were significantly more likely to rock than adults, and NOTMR chimpanzees were significantly more likely than MR individuals (fig. 21.2e).

Rocking was influenced by early rearing and age, but unlike all other behaviors surveyed, it was not influenced by facility type. As was the case for stereotypies other than rocking, NOTMR chimpanzees were more likely to rock than MR. Stereotypies are associated with socially restrictive early rearing (Berkson 1968; Erwin and Deni 1979; Gottlieb, Capitanio, and McCowen 2013; Martin 2002; Mason 1968) and with current social environment (Pazol and Bloomsmith 1993) across nonhuman primate species. Rocking was more prevalent in the immature chimpanzees in this sample than in the adults. Stereotyped rocking typically begins at a young age, and may persist into adolescence and adulthood, being displayed when the chimpanzees are frightened, distressed, or apprehensive (Davenport and Menzel 1963; Davenport and Rogers 1970). In these cases, rocking seems to continue helping the chimpanzees to cope with suboptimal environmental conditions (Mason and Latham 2004). This change with age seems to be consistent with our data as older NOTMR chimpanzees were less likely than younger NOTMR to display rocking. Our findings are similar to other studies (Pazol and Bloomsmith 1993; Spijkerman et al. 1994) with no sex difference in stereotyped rocking reported, while contradicting a male bias for rocking found by Nash et al. (1999).

Self-Directed Abnormal Behavior

The regression analyses for self-directed abnormal behavior resulted in a model that retained rearing, facility type and group size as significant predictors. This was the model with the most significant predictors and also passed the HL test for goodness-of-fit (H.L. $\chi^2[7] = 10.85$, $p = 0.145$). The model was significant ($\chi 2[6] = 66.55$, $p < 0.001$), had a −2 log likelihood of 558.24, and pseudo-R^2 range of 6.7% (CoxandSnell) to 14.0% (Nagelkerke). Based on coefficient values and significance, NOTMR and WB chimpanzees were more likely to engage in self-directed abnormal behavior than MR individuals, chimpanzees in zoos and sanctuaries were more likely than those in research facilities, and chimpanzees in pairs were more likely than those in small groups (fig. 21.2f).

Early rearing, facility type, and group size were associated with the expression of self-directed abnormal behaviors such as self-clasping (in which the chimpanzees embraces their own body with their arms) and bizarre posturing (such as sitting stationary with one arm raised in the air). NOTMR and WB animals had a higher prevalence of these behaviors than MR subjects, similar to the pattern found for stereotyped behaviors other than rocking. Again, this effect of a restrictive early social life is consistent with other studies of chimpanzees and of other nonhuman primates (Kalcher-Sommersguter et al. 2013; Lutz, Well, and Novak 2003; Novak et al. 2012), although no earlier studies were able to test for the effects of being WB. We also found that zoo and sanctuary chimpanzees had a higher likelihood of showing self-directed abnormal behaviors than those currently living in research facilities. Chimpanzees living in pairs were more likely to display self-directed abnormal behavior than those in small groups. The highest prevalence (50%) was among sanctuary-living pairs, but there were only eight subjects in this category. Similar to our finding regarding RandR, it may be that pair housing is not stimulating or challenging enough to avoid the expression of these behaviors, so boredom may be one cause of these behaviors, in contrast to some other abnormal behaviors. While Reamer et al. (2016) found higher levels of abnormal behavior in chimpanzees living in larger groups, our finding contradicts that result for this particular abnormal behavior.

Self-Injurious Behavior with Wounding

Twenty-three chimpanzees were reported to engage in self-injurious behavior (SIB). The regression analyses for self-injurious behavior with wounding resulted in an equation that retained the sex × rearing interaction, facility type, and group size as significant predictors. This was the model with the most significant predictors and also passed the HL test for goodness-of-fit (H.L. $\chi^2[7] = 11.18$, $p = 0.131$). The model was significant ($\chi^2[6] = 27.95$, $p < 0.001$), had a −2 log likelihood of 189.29, and pseudo-R^2 range of 2.9% (CoxandSnell) to 14.2% (Nagelkerke). Based on coefficient values and significance, NOTMR males were more likely to engage in self-injurious behavior with wounding as compared to MR males, chimpanzees in zoos and sanctuaries were more likely than those in research facilities, and chimpanzees in small groups were more likely than those in big groups (figures 21.2g and 21.2h).

We found that early rearing interacted with sex, and that facility type and group size were associated with SIB. NOTMR male chimpanzees were more likely than MR males to engage in SIB (see also Fritz 1986). This confirms findings in other nonhuman primates of a strong relationship between SIB and early life stressors (e.g., early removal from mother) (Novak et al. 2012). Subjects living in larger groups had a reduced prevalence of SIB when compared to smaller groups, supporting the idea that the social environment is important in both preventing and minimizing SIB (Novak 2003). However, we have not established a causal relationship between group size and SIB. In this case it may be that chimpanzees with better social skills resulting from enhanced social backgrounds are also those who can successfully live in larger groups, thus indirectly resulting in reduced likelihoods of showing SIB. Subjects in zoos and sanctuaries had a higher probability of displaying SIB than those in research facilities, and, as for other behaviors we have assessed, there are a number of possible explanations for facility-based differences including the possibility that such behaviors are more so a product of past environments than current settings.

SIB is a very concerning problem among primates other than chimpanzees (Novak et al. 2012). Throughout our entire sample, 2.4% of the animals had exhibited a self-inflicted wound in the prior two years. These could have included both significant wounding that required veterinary intervention as well as more minor wounds that did not require clinical treatment. In all facility types, self-injurious behavior was the rarest abnormal behavior reported. Consistent with published literature, among male chimpanzees, NOTMR have a higher probability of engaging in SIB compared to MR animals. SIB in other primates has been shown to have a strong relationship to early life stressors such as early removal from mother and individual housing (Gottlieb, Capitanio, and McCowen 2013; Novak 2003; Novak et al. 2012). SIB appears to function to reduce arousal, and unfortunately this may contribute to it often being resistant to treatment (Novak 2003). However, it is important to note that a comprehensive and intensive treatment that used a combination of social, training, enrichment, and drug interventions was successful in treating one extreme case of chimpanzee SIB (Bourgeois, Vazquez, and Brasky 2007).

For the analysis of SIB, we had relatively few cases to analyze, which impacted the ability to measure the effects of multiple influencing factors. We did not distinguish between different intensities of SIB with wounding, so a bite-inflicted wound and picking at the skin were scored similarly. This is perhaps another category of behavior that could be better understood as multiple categories, with differing etiologies. Like trichotillomania, dermatillomania (skin picking) in humans has been theorized to occur due to obsessive-compulsive disorder, impulse control issues, and/or anxiety, while self-harming such as cutting or otherwise injuring oneself is, for humans, considered to be a method for coping with intense anger, pain, or frustration (Neziroglu, Rabinowitz, and Jacofsky 2008; McKay and Andover 2012). More in-depth research is called for to better understand causes of SIB in chimpanzees.

CONCLUSIONS ABOUT EARLY REARING AND WILD-BORN INFLUENCE

It is notable that chimpanzees' early rearing had such substantial effects on the later expression of abnormal behavior since we operationalized it as the first year of life. That first year had profound and lasting influences on multiple measures of welfare. This emphasizes the importance of early mother rearing. Truly understanding the welfare of an individual chimpanzee must include assessing the animal's history *and* the current circumstances, both of which impact the animal's present behavior. While we are aware of the negative consequences of a lack of mother rearing, characterization of the early experiences of WB chimpanzees is more ambiguous and much less studied. WB chimpanzees likely had some period of appropriate mother rearing but also likely experienced a possibly sudden and traumatic separation from their mother early in life. Our study is one of the first of captive chimpanzee welfare that has an adequate sample to test for impacts of being wild born. WB chimpanzees engaged in all of the abnormal behaviors surveyed, so they are not buffered from developing abnormal behavior once moved into captive

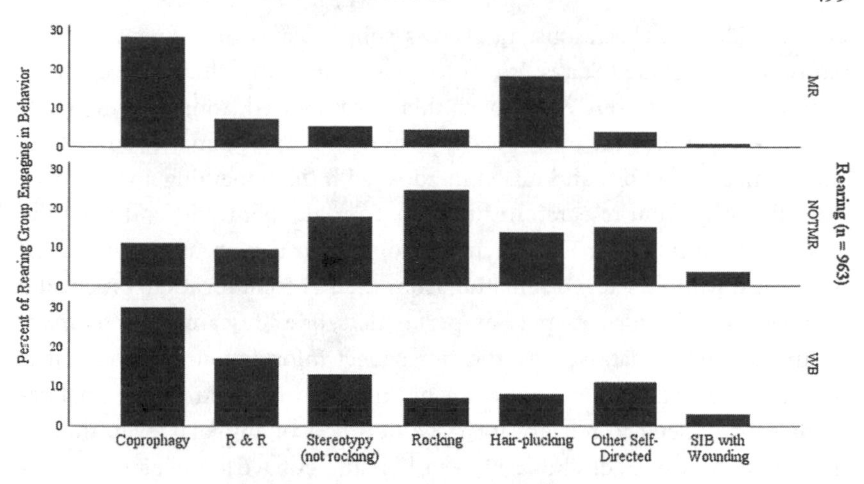

FIGURE 21.3. Prevalence of abnormal behaviors per early rearing history (percent of sample exhibiting the behavior). MR, mother reared; NOTMR, not mother reared but born in captivity; R & R, repeated regurgitation and reingestion; SIB, self-injurious behavior; and WB, wild born.

environments as some other species seem to be (Latham and Mason 2008). Previous research has revealed that WB chimpanzees living in a research environment showed less stereotypic behavior than their counterparts born in the research facility (Davenport and Menzel 1963) and that zoo-living chimpanzees who originated from the wild were much less likely to show coprophagy than zoo-living chimpanzees who originated from research facilities (Jacobson, Ross, and Bloomsmith 2016). In the current study, however, WB chimpanzees were similar to NOTMR in the prevalence of stereotypies other than rocking and self-directed abnormal behaviors, and a substantial proportion of WB displayed coprophagy (fig. 21.3). It is clear from these results that WB individuals may have propensities toward developing certain abnormal behaviors, so it is sound practice to consider them separately from NOTMR or MR animals whenever possible.

POPULATION OVERLAP AND INFLUENCE OF ORIGIN AND ENCLOSURE TYPE

What we provide here is a broad snapshot of what and where abnormal behaviors are expressed in a number of captive chimpanzee settings. Although we categorize the subpopulations based on their current location, these subpopulations do not operate in isolation from each other, and formative experiences that may influence the expression of abnormal behavior in one setting may well have taken place when that chimpanzee lived in a different location. There has been substantial cross-subpopulation movement of chimpanzees in the United States. While the vast majority of chimpanzees living in research settings were either born there or imported from the wild, there are some who were previously pets or performers. Approximately 60% of the zoo animals were born and raised in zoos, with the remaining 40% coming from the wild, from research institutions, from the pet trade, and from the entertainment industry. The sanctuary population is even more extreme in this regard with very few originating from their current location; practically all are either wild-born, ex-pets, ex-performers, or ex-research chimpanzees.

In the current data set, we did not collect information on origin (it is sometimes unavailable and may not be indicated in electronic animal records so is difficult to retrieve for large numbers of animals), even though that would have been desirable. Two earlier studies have found effects of origin on abnormal behaviors. Jacobson, Ross, and Bloomsmith (2016) found that 15 zoo-living chimpanzees who originated from research facilities had a greater likelihood to display coprophagy, but there were no origin impacts on other abnormal behaviors. Nash et al. (1999) reported that of the 11 abnormal behaviors with sufficient data for analysis of origin, only coprophagy was significantly affected such that those who came from zoos and later moved to research facilities had increased likelihoods of showing it. It is interesting that only coprophagy was affected by origin in both studies, and that the two results are contradictory (zoo origin had higher levels in one, while research origin had higher levels in the other). Nonetheless, we suggest further study of this potentially influential factor.

There is inherent variability among the physical caging-structures at the institutions. While all chimpanzees in this study live in settings that are accredited by their respective industry agencies, there are considerable differences within each setting in terms of the environmental conditions (size and complexity) in which they live. Enclosure designs also vary *within* a single facility. In addition, there are a number of types of chimpanzee enclosures that are used by all three categories of facilities (e.g., dome-type chimpanzee enclosures exist in all). Because of this variability and overlap, it is not appropriate to associate one type of environment with one facility type and future research should assess the impact of housing type as well as facility type.

CONCLUSIONS ABOUT ABNORMAL BEHAVIOR AND THE VALENCE OF CHIMPANZEE WELFARE

Abnormal behaviors are often interpreted as evidence of overall poor welfare since they are associated with conditions such as early removal from mother, individual housing, and repeated stressful events. Our findings confirmed previous work indicating that MR chimpanzees are more prone to engage in coprophagy than those not raised by their mothers. None of the other abnormal behaviors we studied were significantly and positively associated with mother rearing. RandR was not associated with any desirable condition that we can identify at this time, so it should be interpreted within the framework of poor welfare. This corroborates conclusions by Hill (2009) that RandR in gorillas is indicative of suboptimal welfare and that there are potentially negative health consequences to this behavior. The relationships between hair-plucking, alopecia, and welfare in other nonhuman primates are somewhat complex and inconsistent, and our study does not clearly reveal the welfare valence of hair-plucking for chimpanzees. We suggest that delineating the potential differences between hair-plucking in the context of social grooming and hair-plucking from oneself may allow us to better discern the welfare implications of these behaviors.

The other abnormal behaviors we studied (stereotyped rocking, stereotypies other than rocking, self-directed abnormal behavior, SIB) should be interpreted as evidence of negative welfare either historically or currently because they are all related to conditions that are not species-typical (e.g., NOTMR, smaller group size). Stereotypies in other species, including humans, have an arousal-reducing function and help animals cope with suboptimal conditions (Mason and Latham 2004), so this is likely in chimpanzees as well. SIB with the potential for creating wounds has an additional, obvious negative impact on welfare. Ideally, all of these behaviors should be discouraged by replacing them with species-normative behaviors.

Future Directions

Here, we have reported on the prevalence of abnormal behavior in chimpanzees living in US zoos, research facilities, and sanctuaries via analyses of the largest and most diverse sample assessed in a study of chimpanzee welfare. The profile of abnormal behavior prevalence varied across the different settings in which chimpanzees currently live, and we found that age, sex, early rearing history, and current social group size all were associated with the prevalence of some abnormal behaviors. We found that each type of abnormal behavior was correlated with multiple factors, which corroborates earlier work showing the complex etiology of chimpanzee abnormal behavior (Nash et al. 1999). Our comparative approach allowed us to broadly characterize the circumstances in which these behaviors are being expressed among and across the diverse set of subjects that were evaluated. Three of the factors analyzed were essentially immutable characteristics of the individual chimpanzees: their sex, age, and early rearing history. Two of the factors relate to the current environment in which the chimpanzees live—the type of facility and the size of their social groups. Clearly, we found that both types of variables are associated with the prevalence of abnormal behavior in chimpanzees.

Despite the strength of our approach, it has disadvantages as well. It is important to keep in mind that the measures of prevalence of abnormal behavior we used do not reflect the frequency, duration, or intensity of these behaviors, which are also important to consider when evaluating welfare (Novak et al. 2012). We also remain mindful of the caveats we described earlier about behavioral differences across facility types that could be due to characteristics of the subpopulations (e.g., age distribution) and/or past experiences (e.g., complexity of past social groups, research history), rather than reflecting current care or housing type. We also note that most chimpanzees living in sanctuaries have spent the majority of their lives in other facility types, and that some living in zoos and research facilities were also from other sources, and this may impact the current expression of behavior. Because contemporary abnormal behavior of chimpanzees is related to both their current and past experiences (as well as other, non-experiential factors), and because some abnormal behaviors come under central control such that they are difficult to modify (Mason and Latham 2004), it is not appropriate to attribute abnormal behavior solely to the current environment in which a chimpanzee lives. Furthermore, these data are not causal; as with all correlation-based analyses, we cannot infer cause from our results. Finally, we wish to acknowledge that it would be valuable to have more precise information regarding observational methods, the amount of time observing each individual, and staff training in behavior at each facility and whether and how these differences impacted the reporting of behaviors, especially infrequently occurring abnormal behaviors (see Jacobson, Ross, and Bloomsmith 2016 for a consideration of these limitations).

Our survey-based study is a productive step in understanding abnormal behaviors in chimpanzees, and it provides general results that encourage future research to investigate topics in more detail than was possible here. For example, studies of the impact of genetic relatedness, features of animal facilities, animal management systems, and more detailed social histories might all be informative and will require more granular and experimental approaches. It would be ideal to follow chimpanzees as they move from one type of facility to another and to measure within-subject behavioral changes associated with those moves (although even this approach would still potentially have an age confound); that approach would be a useful complement to the cross-sectional comparisons of the current study. Additionally, we suggest that further exploration of the different types of SIB and hair-plucking and their potentially different etiologies would allow a better understanding of chimpanzee welfare and lead to more appropriate strategies for reducing these behaviors.

The measurable impacts of early rearing, age, and sex all indicate that there are some chimpanzees within the current US population who have propensities to show abnormal behaviors, possibly no matter where they live or in what size social group. Knowing the complexity of past and present factors that influence abnormal behavior should moderate expectations about eliminating abnormal behavior by placing chimpanzees in different environments. Indeed, it is not the case that moving chimpanzees to sanctuary eliminates abnormal behavior, and that should not be the single metric applied to judge the welfare value of a chimpanzee environment. Comprehensive welfare assessment should include a wide variety of behavioral measures (e.g., species-typical behaviors, social behaviors, abnormal behaviors), health measures, measures of affective state (e.g., cognitive bias assessment), and physiological measures (e.g., cortisol, heart rate). In addition, as we have demonstrated, the presence of abnormal behavior in a sanctuary, a research facility, or a zoo does not mean that the animals in that facility are currently experiencing poor welfare; any number of factors, past and present, may influence the expression of abnormal behaviors. The focus of this study was on abnormal behavior, but we are currently evaluating other information collected on this same sample of chimpanzees, including assessing species-typical behaviors and information on housing conditions and animal training.

It is our hope that the findings from our project will inform future improvements in behavioral management needed to address existing abnormal behaviors of chimpanzees living under human care, and will also identify aspects of chimpanzee behavior and welfare that most need future research. This survey has identified common areas of concern and has created opportunities for people to collaborate across different types of chimpanzee facilities to better promote chimpanzee welfare.

Acknowledgments

This project was supported by the NIH/OD Cooperative Agreement U42-OD 011197 to the Michale E. Keeling Center for Comparative Medicine and Research. The Southwest National Primate Research Center resources are supported by NIH grant P51-OD011133 from the Office of Research Infrastructure

Programs/Office of the Director. The views and opinions expressed in this publication represent the authors' views alone, and do not express or imply the views, endorsement, or financial support of the Federal government or any of its agencies, including the National Institutes of Health, unless otherwise stated by an authorized representative thereof. The authors thank the husbandry, veterinary, and behavioral management staff for their care of the chimpanzees at each facility.

References

Allison, P. 2012. "When can you safely ignore multicollinearity?" *Statistical Horizons* 5 (1): 1–2.

AZA Ape TAG. 2010. "Chimpanzee (*Pan troglodytes*) care manual." Silver Spring, MD: Association of Zoos and Aquariums.

Baker, K. C. 1997. "Straw and forage material ameliorate abnormal behaviors in adult chimpanzees." *Zoo Biology* 16 (3): 225–36.

Baker, K. C., and S. P. Easley. 1996. "An analysis of regurgitation and reingestion in captive chimpanzees." *Applied Animal Behaviour Science* 49 (4): 403–15.

Baker, K. C., and S. K. Ross. 1998. "Outdoor access: The behavioral benefits to chimpanzees." *American Journal of Primatology* 45 (2): 166.

Berkson, G. 1968. "Development of abnormal stereotyped behaviors." *Developmental Psychobiology* 1 (2): 118–32.

Bertolani, P., and J. D. Pruetz. 2011. "Seed reingestion in savannah chimpanzees (*Pan troglodytes verus*) at Fongoli, Senegal." *International Journal of Primatology* 32: 1123.

Beuchat, L. R., and J. H. Ryu. 1997. "Produce handling and processing practices." *Emerging Infectious Diseases* 3 (4): 459.

Birkett, L. P., and N. E. Newton-Fisher. 2011. "How abnormal is the behaviour of captive, zoo-living chimpanzees?" *PLoS One* 6 (6): e20101.

Bloomsmith, M. A., P. L. Alford, and T. L. Maple. 1988. "Successful feeding enrichment for captive chimpanzees." *American Journal of Primatology* 16 (1): 155–64.

Bloomsmith, M. A, and K. Baker. 2001. "Social management of captive chimpanzees." In *Special Topics in Primatology: The Care and Management of Captive Chimpanzees*, vol. 2, edited by L. Brent, 205–41. San Antonio: American Society of Primatologists.

Bloomsmith, M. A., A. Clay, S. P. Lambeth, C. Lutz, S. Breaux, M. Lammey, A. Franklin, K. Neu, J. E. Perlman, L. Reamer, M. C. Mareno, S. J. Schapiro, M. Vazquez, and S. Bourgeois. 2019. "A survey of behavioral indices of welfare in research chimpanzees (*Pan troglodytes*) in the United States." *Journal of the American Association for Laboratory Animal Science* 58 (2):160–77.

Bloomsmith, M. A., and J. G. Else. 2005. "The behavioral management of chimpanzees in biomedical research facilities: The state of the science." *ILAR Journal* 46 (2): 192–201.

Bloomsmith, M. A., S. P. Lambeth, and M. D. Haberstroh. 1999. "Chimpanzee use of enclosures." *American Journal of Primatology* 49 (1): 36.

Bloomstrand, M., K. Riddle, P. Alford, and T. L. Maple. 1986. "Objective evaluation of a behavioral enrichment device for captive chimpanzees (*Pan troglodytes*)." *Zoo Biology* 5 (3): 293–300.

Bonnie, K. E., M. Y. L. Ang, and S. R. Ross. 2016. "Effects of crowd size on exhibit use by and behavior of chimpanzee (*Pan troglodytes*) and Western lowland gorillas (*Gorilla gorilla*) at a zoo." *Applied Animal Behaviour Science* 178: 102–10.

Bourgeois, S. R., M. Vazquez, and K. Brasky. 2007. "Combination therapy reduces self-injurious behavior in a chimpanzee (*Pan troglodytes*): A case report." *Journal of Applied Animal Welfare Science* 10 (2): 123–40.

Brand, C. M., and L. F. Marchant. 2015. "Hair plucking in captive bonobos (*Pan paniscus*)." *Applied Animal Behaviour Science* 171: 192–96.

Brand, C. M., and L. F. Marchant. 2018. "Prevalence and characteristics of hair plucking in captive bonobos (*Pan paniscus*) in North America." *American Journal of Primatology* 80 (4): e22751.

Brent, L. 1992. "Woodchip bedding as enrichment for captive chimpanzees in an outdoor enclosure." *Animal Welfare* 1 (3): 161–70.

Capitanio, J. P. 1986. "Behavioural pathology." In *Comparative Primate Biology: Behaviour, Conservation, and Ecology*, edited by G. Mitchell and J. Erwin, 2A:411–54. New York: Liss.

Chamove, A. S., G. R. Hosey, and P. Schaetzel. 1998. "Visitors excite primates in zoos." *Zoo Biology* 7 (4): 359–69.

ChimpCARE. 2018. "Chimpanzees in the U.S." http://www.chimpcare.org/map.

Coleman, K., C. K. Lutz, J. M. Worlein, D. H. Gottlieb, E. Peterson, G. H. Lee, N. D. Robertson, K. Rosenberg, M. T. Menard, and M. A. Novak. 2017. "The correlation between alopecia and temperament in rhesus macaques (*Macaca mulatta*) at four primate centers." *American Journal of Primatology* 79 (1): 1–10.

Davenport, R. K., and E. W. Menzel. 1963. "Stereotyped behaviour of the infant chimpanzee." *Archives of General Psychiatry* 8: 99–104.

Davenport, R. K., E. W. Menzel, and C. M. Rogers. 1966. "Effects of severe isolation on 'normal' juvenile chimpanzees." *Archives of General Psychiatry* 14: 134–38.

Davenport, R. K., and C. M. Rogers. 1970. "Differential rearing of the chimpanzee: A project survey." In *The Chimpanzee*, vol. 3, *Immunology, Infections, Hormones, Anatomy, and Behavior of Chimpanzees*, edited by G. H. Bourne, 337–60. Baltimore: University Park Press.

Duke, D. C., M. L. Keeley, G. R. Geffken, and E. A. Storch. 2010. "Trichotillomania: A current review." *Clinical Psychology Review* 30: 181–93.

Erwin, J., and R. Deni. 1979. "Strangers in a strange land: Abnormal behaviors or abnormal environments? In *Captivity and Behavior: Primates in Breeding Colonies, Laboratories, and Zoos*, edited by J. Erwin, T. L. Maple, and G. Mitchell, 1–28. New York: Van Nostrand Reinhold.

Esch, K. J., and C. A. Petersen. 2013. "Transmission and epidemiology of zoonotic protozoal diseases of companion animals." *Clinical Microbiology Review* 26 (1): 58–85.

Fernandez, E. J., M. A. Tamborski, S. R. Pickens, and W. Timberlake. 2009. "Animal-visitor interactions in the modern zoo: Conflicts and interventions." *Applied Animal Behaviour Science* 120 (1–2): 1–8.

Freeman, H. D., and S. R. Ross. 2014. "The impact of atypical early histories on pet or performer chimpanzees." *PeerJ* 2: e579.

Fritz, J. 1986. "Resocialization of asocial chimpanzees." In *Primates: The Road to Self-Sustaining Populations*, edited by K. Benirschke, 351–60. New York: Springer.

Fritz, J., S. Maki, L. T. Nash, Y. Martin, and M. Matevi. 1992. "The relationship between forage material and levels of coprophagy in captive chimpanzees (*Pan troglodytes*)." *Zoo Biology* 11 (5): 313–18.

Fultz, A., L. Brent, S. D. Breaux, and A. P. Grand. 2013. "An evaluation of nest-building behavior by sanctuary chimpanzees with access to forested habitats." *Folia Primatologica* 84 (6): 405–20.

Fultz, A., R. Jackson-Jewett, K. Taylor, C. Philipp, A. Yanagi, E. Loeser, and L. Case. 2016. "The effects of dietary changes on chimpanzee behavior." *PeerJ Preprints* 4: e1793v1.

Gottlieb, D. H., J. P. Capitanio, and B. McCowan. 2013. "Risk factors for stereotypic behavior and self-biting in rhesus macaques (*Macaca mulatta*): Animal's history, current environment, and personality." *American Journal of Primatology* 75 (10): 995–1008.

Gould, E., and M. Bres. 1986. "Regurgitation and reingestion in captive gorillas: Description and intervention." *Zoo Biology* 5: 241–50.

Grant, J. E., S. A. Redden, E. W. Leppink, and S. R. Chamberlain. 2017. "Trichotillomania and co-occuring anxiety." *Comprehensive Psychiatry* 72: 1–5.

Herrelko, E. S., H. M. Buchanan-Smith, and S. J. Vick. 2015. "Perception of available space during chimpanzee introductions: Number of accessible areas is more important than enclosure size." *Zoo Biology* 34 (5): 397–405.

Hill, S. P. 2009. "Do gorillas regurgitate potentially-injurious stomach acid during 'regurgitation and reingestion'?" *Animal Welfare* 18: 123–27.

Hopper, L. M., H. D. Freeman, and S. R. Ross. 2016. "Reconsidering coprophagy as an indicator of negative welfare for captive chimpanzees." *Applied Animal Behaviour Science* 176: 112–19.

Hosey, G. R. 2005. "How does the zoo environment affect the behaviour of captive primates?" *Applied Animal Behaviour Science* 90 (2): 107–29.

Jacobson, S. L., S. R. Ross, and M. A. Bloomsmith. 2016. "Characterizing abnormal behavior in a large population of zoo-housed chimpanzees: Prevalence and potential influencing factors." *PeerJ* 4: e2225.

Kalcher, E., C. Franzi, K. Crailsheim, and S. Preuschoft. 2008. "Differential onset of infantile deprivation produces distinctive long-term effects in adult ex-laboratory chimpanzees (Pan troglodytes)." *Developmental Psychobiology* 50 (8): 777–88.

Kalcher-Sommersguter, E., C. Franz-Schaider, S. Preuschoft, and K. Crailsheim. 2013. "Long-term evaluation of abnormal behavior in adult ex-laboratory chimpanzees (*Pan troglodytes*) following re-socialization." *Behavioral Science* 3: 99–119.

Latham, N. R., and G. J. Mason. 2008. "Material deprivation and the development of stereotypic behavior." *Applied Animal Behaviour Science* 110 (1–2): 84–108.

Less, E. H., R. Bergl, R. Ball, P. M. Dennis, C. W. Kuhar, S. R. Lavin, M. A. Raghanti, J. Wensvoort, M. A. Willis, and K. E. Lukas. 2014. "Implementing a low-starch biscuit-free diet in zoo gorillas: The impact on behavior." *Zoo Biology* 33: 63–73.

Less, E. H., C. W. Kuhar, and K. E. Lukas. 2013. "Assessing the prevalence and characteristics of hair-plucking behaviour in captive Western lowland gorillas (*Gorilla gorilla gorilla*)." *Animal Welfare* 22 (2): 175–83.

Lochner, C., S. Seedat, and D. J. Stein. 2010. "Chronic hair-pulling: Phenomenology-based subtypes." *Journal of Anxiety Disorders* 24 (2): 196–202.

Lonsdorf, E. V. 2005. "Sex differences in the development of termite-fishing skills in the wild chimpanzees, *Pan troglodytes schweinfurthii*, of Gombe National Park, Tanzania." *Animal Behaviour* 70: 673–83.

Lutz, C., A. Well, and M. Novak. 2003. "Stereotypic and self-injurious behavior in rhesus macaques: A survey and retrospective analysis of environment and early experience." *American Journal of Primatology* 60 (1): 1–15.

Maki, S., J. Fritz, and N. England. 1993. "An assessment of early differential rearing conditions on later behavioral-development in captive chimpanzees." *Infant Behavior & Development* 16 (3): 373–81.

Martin, J. E. 2002. "Early life experiences: Activity levels and abnormal behaviours in resocialised chimpanzees." *Animal Welfare* 11 (4): 419–36.

Mason, G. J., and N. R. Latham. 2004. "Can't stop, won't stop: Is stereotypy a reliable animal welfare indicator?" *Animal Welfare* 13: S57–69.

Mason, W. A. 1968. "Early social deprivation in the nonhuman primates: Implications for human behavior." *Environmental Influences*: 70–101.

Mayes, S. D., F. J. Humphrey, H. A. Handford, and J. F. Mitchell. 1988. "Rumination disorder: Differential diagnosis." *Journal of the American Academy of Child & Adolescent Psychiatry* 27 (3): 300–302.

McKay, D., and M. Andover. 2012. "Should non-suicidal self-injury be a putative obsessive-compulsive related condition? A critical appraisal." *Behavioral Modifications* 36 (1): 3–17.

Morgan, L., S. M. Howell, and J. Fritz. 1993. "Regurgitation and reingestion in a captive chimpanzee (*Pan troglodytes*)." *Lab Animal* 22 (8): 42–45.

Nash, L. T., J. Fritz, P. A. Alford, and L. Brent. 1999. "Variables influencing the origins of diverse abnormal behaviors in a large sample of captive chimpanzees (*Pan troglodytes*)." *American Journal of Primatology* 48: 15–29.

Neziroglu, F., D. Rabinowitz, and M. Jacofsky. 2008. "Skin picking phenomenology and severity comparison." *Primary Care Companion Journal of Clinical Psychiatry* 10 (4): 306–12.

Novak, M. A. 2003. "Self-injurious behavior in rhesus monkeys: New insights into its etiology, physiology, and treatment." *American Journal of Primatology* 59 (1): 3–19.

Novak, M. A., A. F. Hamel, K. Coleman, C. K. Lutz, J. Worlein, M. Menard, A. Ryan, K. Rosenberg, and J. S. Meyer. 2014. "Hair loss and hypothalamic–pituitary–adrenocortical axis activity in captive rhesus macaques (*Macaca mulatta*)." *Journal of the American Association for Laboratory Animal Science* 53 (3): 261–66.

Novak, M. A., B. J. Kelly, K. Bayne, and J. S. Meyer. 2012. "Behavioral disorders of nonhuman primates." In *Nonhuman Primates in Biomedical Research: Biology and Management*, edited by C. R. Abee, K. Mansfield, S. Tardif, and T. Morris, 177–96. Oxford: Academic Press, Elsevier.

Payne, C. L. R., T. H. Webster, and K. D. Hunt. 2008. "Coprophagy by the semi-habituated chimpanzees of Semliki, Uganda." *Pan Africa News* 15 (2): 29–32.

Pazol, K. A., and M. A. Bloomsmith. 1993. "The development of stereotyped body rocking in chimpanzees (*Pan Troglodytes*) reared in a variety of nursery settings." *Animal Welfare* 2 (2): 113–29.

Pruetz, J. D. E., and W. C. McGrew. 2001. "What does a chimpanzee need? Using behavior to guide the care and management of captive populations." In *Special Topics in Primatology*, vol. 2, *The Care and Management of Captive Chimpanzees*, edited by L. Brent, 17–37. San Antonio: American Society of Primatologists.

Reamer, L. A., C. F. Talbot, L. M. Hopper, M. C. Mareno, K. Hall, S. F. Brosnan, S. P. Lambeth, and S. J. Schapiro. 2016. "The effects of group size on the behavior of captive chimpanzee (*Pan troglodytes*)." Abstract presented at the Joint Meeting of the International Primate Society and American Society of Primatologists, Chicago, IL, August, 21–27, 2016.

Ross, S. R., and K. E. Lukas. 2006. "Use of space in a non-naturalistic environment by chimpanzees (*Pan troglodytes*) and lowland gorillas (*Gorilla gorilla gorilla*)." *Applied Animal Behaviour Science* 96 (1–2): 143–52.

Ross, S. R., and M. A. Shender. 2016. "Daily travel distances of zoo-housed chimpanzees and gorillas: Implications for welfare assessments and space requirements." *Primates* 57 (3): 395–401.

Spijkerman, R. P., J. Dienske, J. A. R. A. M. Vanhooff, and W. Jens. 1994. "Causes of body rocking in chimpanzees (*Pan troglodytes*)." *Animal Welfare* 3 (3): 193–211.

Struck, K., E. N. Videan, J. Fritz, and J. Murphy. 2007. "Attempting to reduce regurgitation and reingestion in a captive chimpanzee through increased feeding opportunities: A case study." *Lab Animal* 36 (1): 35–38.

Suomi, S. J. 1991. "Early stress and adult emotional reactivity in rhesus monkeys." In *The Childhood Environment and Adult Disease*, edited by R. B. Gregory and J. Whelan, 171–83, discussion 183–88. Chichester: John Wiley & Sons.

Turner, C. H., R. K. Davenport, and C. M. Rogers. 1969. "The effect of early deprivation on the social behavior of adolescent chimpanzees." *American Journal of Psychiatry* 125 (11): 1531–36.

Vazquez, M., and C. K. Lutz. 2017. "Activity budget and alopecia in captive chimpanzees (*Pan troglodytes*): Impact of sex, age, and group size." Abstract presented at the Meeting of the American Society of Primatologists, Washington, DC, August, 25–28, 2017.

Waitt, C., and H. M. Buchanan-Smith. 2001. "What time is feeding? How delays and anticipation of feeding schedules affect stump-tailed macaque behavior." *Applied Animal Behaviour Science* 75: 75–85.

Walsh, S., C. A. Bramblett, and P. L. Alford. 1982. "A vocabulary of abnormal behaviors in restrictively reared chimpanzees." *American Journal of Primatology* 3 (1–4): 315–19.

Wood, W. 1998. "Interactions among environmental enrichment, viewing crowds and zoo chimpanzees (*Pan troglodytes*)." *Zoo Biology* 17 (3): 211–30.

22

When Is "Natural" Better? The Welfare Implications of Limiting Reproduction in Captive Chimpanzees

KATHERINE A. CRONIN AND STEPHEN R. ROSS

Introduction

Modern zoos often prioritize recreating the natural environment for an animal as much as possible to promote positive experiences for guests (Davey, Henzi, and Higgins 2005; Finlay, James, and Maple 1988; Ross et al. 2012), to maintain viable zoo populations long-term (Gillman et al. 2017; Rabin 2003; Seidensticker and Forthman 1998), and to facilitate good individual animal welfare (Brando and Buchanan-Smith 2018; Fàbregas, Guillén-Salazar, and Garcés-Narro 2012; Hancocks 1980). The presumed positive relationship between environments that socially and physically approximate the wild and animal welfare largely stems from the perspective that an animal is likely to experience poor welfare if they are unable to express motivations that have been favored by natural selection (Duncan and Petherick 1991; Fraser and Duncan 1998). By recreating a natural environment, people are more likely to provide outlets for motivations that have been favored by natural selection, minimizing frustration and, by extension, increasing the probability that animals in their care experience good welfare.

In most cases, opting for a natural environment increases the probability that managers are providing animals with social and physical characteristics necessary to express motivations and avoid frustrations. This is especially true in cases where the links between specific behavioral expressions and welfare are not yet well understood. However, in cases where managers have identified behaviors that encourage positive welfare outcomes and are able to create environments that promote such behaviors, even in "unnatural" ways, the link between naturalism and welfare may be rendered moot (Markowitz 1975, 1982). For instance, a species that is highly motivated to express extractive foraging behavior may benefit from the opportunity to engage with a plastic food puzzle very unlike what would be present in nature but which

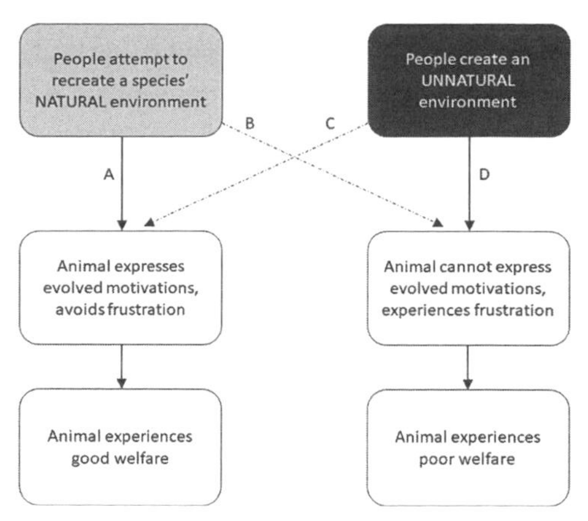

FIGURE 22.1. Schematic representation of when and why recreating a species-typical environment often leads to better animal welfare. Here we consider two starting points, one in which managers are able to create a captive environment that closely replicates the environment in which much of the species evolutionary history has occurred ("NATURAL" starting point, light gray box), and one in which managers create an environment different from the one in which much of the species evolutionary history has occurred ("UNNATURAL" starting point, dark gray box). The desired endpoint is to achieve good animal welfare. For simplification we assume welfare is solely predicted by presence or absence of frustration. In the case of the NATURAL starting point, because the animal is living in an environment that replicates the conditions under which its motivations evolved, the animal is able to express motivations, avoid frustration, and experience positive welfare (path A). However, it is possible that people are unable to successfully recreate essential components of the environment related to evolved motivations, leading to poor welfare (path B). In the case of the UNNATURAL starting point, it is possible that people will have the knowledge and skills to create environments that do not match the conditions under which motivations evolved but still elicit important, highly motivated behaviors (path C). For most species, however, this would be difficult to accomplish with certainty simply due to the lack of scientific knowledge of the full set of species-specific motivations. It is more likely that an animal living in an environment that does not replicate the conditions under which its motivations evolved will be unable to express important motivations, and will experience frustration and poor welfare (path D).

elicits the same behavior. Figure 22.1 illustrates the potential links between naturalistic environments and the promotion of animal welfare.

A "natural environment" for chimpanzees (*Pan troglodytes*) would largely recreate both the physical and social features of the environment that chimpanzees would experience in their natural habitat. Often the discussion of naturalism in captive environments focuses on whether aspects of the physical environment offer the animals the necessary resources to meet their biological needs (e.g., Hancocks 1980; Ross et al. 2009b, 2011). The social environment has also been considered, with the bulk of the discussion in the zoo and laboratory community focusing on the importance of providing social housing for species, such as chimpanzees, that typically live in social groups (e.g., Novak and Suomi 1988; Olsson and Westlund 2007; Ross et al. 2009a, see also Ross, chapter 24 this volume). However, the *composition* of those social groups has not been the focus of much research. In the context of their wild counterparts, chimpanzees housed in captivity live in social groups that often differ in composition from the wild in several ways. For instance, relatedness within the group may be quite different from what is typically seen in the wild, for example with matrilines remaining together (although there is variability in female dispersal; see review in Knott and Harwell, chapter 1 this volume; Mann, Stanton, and Murray, chapter 3 this volume), and sex ratios may also differ from patterns observed in the wild. However, one pervasive difference between the composition of wild and captive social groups is the number of young individuals present. In zoos, laboratories, and sanctuaries, breeding is restricted by management practices, resulting in social groups that have few (or no) young individuals present.

In the wild, adult female chimpanzees are almost always somewhere in the cycle of breeding, pregnancy, and lactation, and a large proportion of their activity budget is devoted to reproduction and rearing (Goodall 1986). Consequently, the presence of infants in wild chimpanzee social groups is ubiquitous for virtually all group members, though the frequency of interactions with infants clearly varies based on factors such as sex and rank (Markham et al. 2014; Nishida 1970; Pusey 1990). Within zoos accredited by the Association of Zoos and Aquariums (AZA), however, chimpanzee reproduction is controlled through hormonal birth control for females and/or vasectomies for males to maintain a population size determined by zoo capacity while minimizing the loss of genetic diversity (McAuliffe 2018). Only a small number of chimpanzees are permitted to reproduce and opportunities to rear offspring are rare compared to the relatively unrestricted opportunities in the wild. For instance, over a 20-year period (1995–2014), only 4% (range 1.7% to 8.1%, SEM 0.4%) of potentially breeding females produced offspring each year in AZA-accredited zoos (Ross 2015). Opportunities to rear young in the sanctuary community are even rarer where reproduction is virtually absent. Nearly 900 chimpanzees live in sanctuaries accredited by the Pan African Sanctuary Alliance (Cronin, West, and Ross 2016), and in these environments, the minimum standards stipulate that animals are not permitted to reproduce except when an active release program is in place (Pan African Sanctuary Alliance 2016).

The controlled reproduction of chimpanzees in zoos and sanctuaries generates very different social environments for chimpanzees in captivity compared to the wild; environments that one may easily argue to be "non-natural." Specifically, as a result of the differing reproductive rates, social groups in captivity contain fewer young individuals, a mismatch that could indicate a welfare concern (Brando and Buchanan-Smith 2018; Hancocks 1980; Lund 2006; Seidensticker and Forthman 1998). Furthermore, while there is no empirical work we are aware of demonstrating a positive effect of the presence of young individuals on the welfare of groupmates, it could be argued that the experience of rearing and interacting with young provides welfare benefits that have been well-established through other lines of work. For example, close social bonds have a positive effect on health and welfare (e.g., Hennessy, Kaiser, and Sachser 2009; Seyfarth and Cheney 2012; Silk et al. 2010), as do variable and dynamic environments (e.g., Bracke and Hopster 2006; Lukas, Hoff, and Maple 2003). Thus, evaluation of the relationship between the presence (or absence) of young in a group and the welfare of other group members is a relevant, as-of-yet unaddressed question.

Chimpanzees have been the focus of a great deal of welfare-related research, and several factors that affect their welfare in captivity are well established (see Bloomsmith et al., chapter 21 this volume; Ross, chapter 24 this volume). The approach of using natural behavior patterns seen in wild chimpanzees to guide the care and management of captive chimpanzees has been widely espoused (Pruetz and McGrew 2001). For instance, three-dimensional physical environments are constructed to encourage naturalistic locomotor patterns and enrichment tasks are designed to mimic naturally occurring foraging challenges. Similarly, a common tactic in improving the welfare of captive chimpanzees is to set behavior "targets," where managers aspire to create conditions in which the proportion of time captive chimpanzees spend in various behaviors matches that recorded in wild chimpanzees. While there is relatively broad consensus to, for instance, increase captive chimpanzee activity and foraging rates that differ quite broadly between wild and captive chimpanzees, there has been far less attention to whether and how managers should aspire to match other behaviors. In that light, virtually no scientific work has been done on the question of how controlling reproduction, and therefore substantively reducing opportunities for infant rearing and interaction with young, impacts the welfare of chimpanzees in human care.

In a recent study we began to empirically address the question of whether the opportunity to rear offspring was associated with differences in welfare in adult female chimpanzees living in captivity (Cronin, West, and Ross 2016). We made use of the existing variation in the reproductive experiences of chimpanzees at Chimfunshi Wildlife Orphanage Trust in Zambia and categorized adult females by whether they currently did or did not have dependent young (n = 15 and 28, respectively). We considered several behavioral measures of compromised welfare such as stereotypies, abnormal posture, and self-directed behaviors (e.g., Llorente et al. 2015; Pomerantz and Terkel 2009), as well as behavioral measures of positive welfare such as play (e.g., Held and Špinka 2011). We also considered how females responded to a potentially stressful social challenge, gathering in a small space for daily feeding, under the assumption that individuals experiencing compromised welfare are poorly equipped to handle additional stressors (e.g., Boccia, Laudenslager, and Reite 1995; Moberg 2000). In sum, we found no differences in any welfare indicators, positive or negative, between females who currently did and females who did not have dependent young, providing no support for the idea that the opportunity to rear young is an important influence on captive chimpanzees' welfare.

Our previous study (Cronin, West, and Ross 2016) was limited by specifically examining adult females with or without dependent, biological offspring, but not considering whether the presence of *any* young in the group, regardless of kinship, impacted the welfare of group members. In our previous study of the Chimfunshi chimpanzees, the adult females lived across four different social groups, each of which had some dependent young present (range in dependent young was 3–8 per group). In accredited zoos, where breeding is purposefully limited, there are several social groups without any infants or juveniles present. A chimpanzee social group lacking infants and juveniles creates a mismatch from the natural chimpanzee environment in which reproduction is more frequent and the demographic composition of groups includes younger individuals who may provide stimulation for other group members. Here, we empirically address whether the presence of young (related or otherwise) in a social group is associated with measurable indicators of welfare, focusing our analyses on the population of chimpanzees living in zoos in the United States accredited by the AZA where breeding is restricted.

Approach

We analyzed whether the presence of young in a social group of captive chimpanzees was associated with differences in welfare. We did so by extracting the prevalence of abnormal behavior across chimpanzees (age range 2–78 years, N = 207) living in 26 AZA-accredited zoos in the United States. These data were gathered as part of a previously conducted survey of curators, managers, and zookeepers who regularly worked with the chimpanzees and were responsible for their care (see Jacobson, Ross, and Bloomsmith 2016, for details). Respondents were provided with a list of abnormal behaviors (table 22.1) and asked to document which of the behaviors each chimpanzee in their care displayed at least once in the previous two-year period. We focused on

TABLE 22.1. The definition of abnormal behaviors provided in the survey.

Behavior	*Definition*
Hair pluck	Pulling out hair on self or another
Rock	Repetitive and sustained swaying movement without piloerection
Regurgitation and reingestion	The deliberate regurgitation of food and subsequent consumption of the food
Self-injurious behavior	Biting, picking, or scratching at own body to cause injury
Pacing	Locomoting repetitively along the same path with no clear objective
Other	Any other behavior deemed abnormal, space to describe

abnormal behavior as our primary indicator of welfare because it is a widely accepted and easily measurable indictor. Coprophagy, which historically has been included in lists of abnormal behavior, was not included in our analysis due to the growing consensus that its prevalence is not reliably related to welfare in chimpanzees (Cronin, West, and Ross 2016; Hopper, Freeman, and Ross 2016; Jacobson, Ross, and Bloomsmith 2016; Nash et al. 1999, although see Bloomsmith et al., chapter 21 this volume).

Jacobson and colleagues (2016) compared several logistic regression models considering combinations of the factors *rearing type* (mother- or non-mother-reared), *origin* (zoo, laboratory, or wild caught), and *sex* to determine which combination of historical factors best predicted whether or not a chimpanzee performed any abnormal behaviors. Using an information-theoretic approach and, in particular, model comparisons based on Akaike information criterion (AIC) values, the authors determined that a model including all three variables best predicted the occurrence of abnormal behavior and that not being mother-reared was the best predictor of abnormal behavior. Here, in order to test the hypothesis that the presence of young in the group is associated with differences in welfare, we consider whether the addition of a categorical predictor variable that indicates whether or not dependent young were present in the social group better explains the probability of abnormal behavioral expression.

Dependent young were defined as any individual younger than seven years of age (Cronin, West, and Ross 2016; Nishida 2011). We queried the North American Regional Studbook for Chimpanzees (Ross 2015) to determine whether the social group contained individuals younger than seven years at the time the survey was administered to staff. Demographic data regarding the population was extracted from the studbook using the software Poplink version 2.4 (Faust et al. 2012).

We conducted logistic regressions in R version 3.4.1 (R Core Team 2014) using the glm function in base R. In addition, we used the packages "Amelia" (Honaker, King, and Blackwell 2011) and "AICcmodavg" (Mazerolle and Mazerolle 2017) to check for missing variables and to compare AIC values, respectively. Specifically, we compared corrected AIC values (AICc) given that the value of n/number of fitted parameters was less than 4 (Symonds and Moussalli 2011). Results were plotted using the packages "ggplot2" and "RColorBrewer" (Neuwirth 2011; Wickham, Chang, and Wickham 2013). The reference variable for sex was female, for rearing was mother-reared, for the presence of young was young absent, and for origin was wild. Individuals with an unknown rearing history (n = 42) were excluded, resulting in a sample of 165 chimpanzees living in 26 institutions, of which 38% were male and 62% were female, a sex ratio representative of the complete AZA population (Ross 2015).

Results

In order to test whether the presence of dependent young was associated with improved welfare, as indicated through a decrease in the prevalence of abnormal behaviors, we included whether or not each individual was in a group that contained dependent young in the logistic regression model that had proven most explanatory in a recent analysis of predictors of abnormal behavior (Jacobson, Ross, and Bloomsmith 2016). Full model results are shown in table 22.2 and reconfirm that rearing condition is a significant influence on the prevalence of abnormal behaviors. The AICc of the model including the presence of dependent young was greater than the AICc obtained from the original model, indicating that the inclusion of the predictor indicating whether young were present did not improve the model fit (table 22.3).

TABLE 22.2. Logistic regression model results for chimpanzee abnormal behavior with predictor variables and constant.

			95% CI for odds ratio		
	β (SE)	*P*	*Lower*	*Odds ratio*	*Upper*
Constant	−0.51 (0.38)	0.18			
Young present (yes)	0.21 (0.34)	0.53	0.63	1.24	2.42
Rearing (non-mother)	1.13 (0.44)	< 0.01	1.33	3.10	7.50
Sex (male)	−0.08 (0.81)	0.81	0.47	0.92	1.82
Origin (lab)	1.46 (0.90)	0.10	0.85	4.32	33.27
Origin (private)	−0.51 (0.78)	0.51	0.13	0.60	2.80
Origin (zoo)	−0.09 (0.44)	0.84	0.39	0.91	2.19

TABLE 22.3. Logistic regression models compared to predict the occurrence of abnormal behavior in chimpanzees.

Model	*k*	*AICc*	*Delta i*	*ER*
Origin + Rearing + Sex	7	274.17	0	
Origin + Rearing + Sex + Young Present	8	276.16	1.99	2.7

Note: k = number of fitted parameters including the intercept; AICc = second-order Akaike's information criterion; Delta i = change in AIC, ER = evidence ratio, which provides a measure of how much more likely the best model (the model with lowest AICc) is than the comparison model.

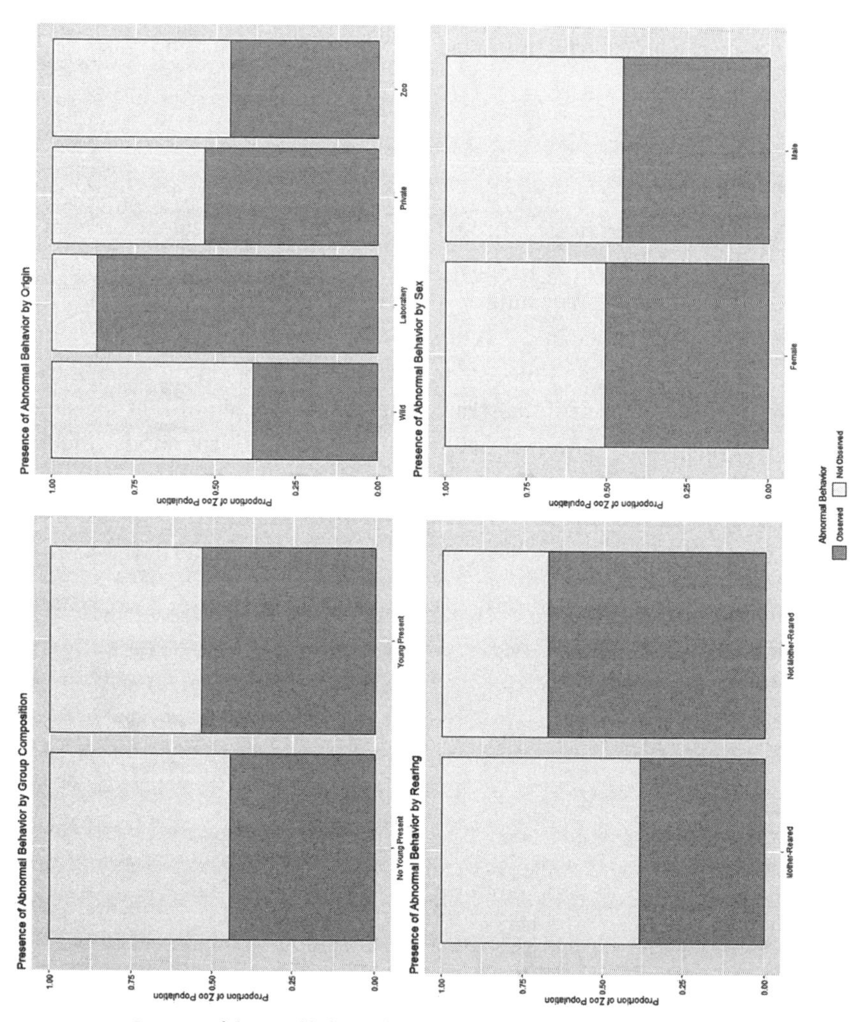

FIGURE 22.2. Presence of abnormal behavior by each predictor variable.

Furthermore, likelihood ratio tests using the ANOVA function and chi-square distribution to compare the full model including the presence of dependent young with a reduced model excluding that predictor indicated no significant difference ($\chi^2 1 = 0.391$, $p = 0.532$). The proportion of the population that did and did not show abnormal behavior, grouped by each predictor variable, is shown in figure 22.2.

Discussion

Minimizing the mismatch between natural, wild conditions and captive conditions is widely thought to be a powerful strategy to promote positive animal welfare. This is likely because recreating conditions that match the environment in which much of the species' evolutionary history occurred is one way to safeguard against frustrations that stem from an inability to perform behaviors that have been under the pressure of natural selection. In a previous study, we began to investigate the relationship between breeding and welfare by showing that female chimpanzees in a sanctuary setting did not show differences in behavioral indicators of welfare dependent upon whether they currently were rearing young (Cronin, West, and Ross 2016). In sum, we found no differences in any welfare indicators, positive or negative, between females who currently did and did not have dependent young, providing no support for the idea that the opportunity to rear young was an important influence on welfare. Here we take a second step in that investigation by analyzing whether the presence of young in a social group is associated with behavioral indicators of welfare (abnormal behaviors) for others in the group. Through our present analysis, we find no evidence that the presence of young in the group is related to this measure of welfare, suggesting that this specific mismatch between wild (natural) and captive conditions (presence of youngsters) does not compromise captive chimpanzees' welfare.

The primary limitation of our current analysis is the coarseness of the survey data. Here, we analyzed whether an individual was reported to have been observed performing one of the abnormal behaviors listed in table 22.1. The data quality is dependent on the degree to which the respondents were able to observe the chimpanzees and the degree of familiarity they have with the behaviors. Broadly speaking, these should be conservative estimates of the behavioral frequencies but there is no reason to think the reports would be differentially reported between our experimental groups. Furthermore, we acknowledge that observation of abnormal behavior is just one way to evaluate welfare. For a more complete understanding one would consider additional behaviors that may be indicative of welfare, such as play (e.g., Held and

Špinka 2011; Oliveira et al. 2010), alongside hormonal data and indications of mental state or emotional well-being (Botreau et al. 2007; Hill and Broom 2009; Mason and Mendl 1993).

While our data suggest there is little evidence for negative welfare consequences to limited breeding in captive chimpanzees, we acknowledge that alternative explanations for these findings are possible. For instance, it is possible that behavioral benefits of rearing and exposure to infants are buffered to some degree by the inherent stresses of motherhood. Studies of wild chimpanzees demonstrate that mothers with dependent offspring are typically less gregarious (Goodall 1986; Murray et al. 2014) in part perhaps to guard against infant vulnerability (Otali and Gilchrist 2006) or the costs of traveling with an infant (Wrangham 2000). Indeed, mothers with dependent offspring seem to face a unique set of social and ecological challenges compared to other females, and that may be exacerbated by factors such as rank (Markham et al. 2014). An understanding of the degree to which these maternal-related stressors are relevant in captive settings may inform how best to characterize both positive and negative welfare-related aspects of reproduction for this species.

Considering the reproductive lives of captive chimpanzees in the context of their wild counterparts highlights many differences between the two groups. Wild chimpanzee females reproduce earlier and more often than captive chimpanzee females (Goodall 1986; Nishida 2011; Ross 2015). Female chimpanzees rearing young often separate from their social groups for weeks at a time (Goodall 1986; Nishida 2011) whereas captive female chimpanzees remain in relatively close proximity to their group throughout parturition and rearing. Wild chimpanzee males and females live in groups with more young chimpanzees present, and have more opportunities for interacting with young, than captive chimpanzees (Markham et al. 2014; Pusey 1990; Ross 2015). How this suite of differences positively or negatively affects the overall welfare of chimpanzees living in captivity is probably an intractable question, given the number of dimensions in which their lives differ and the difficulty of quantifying overall welfare. However, for at least two of these differences—opportunities to actively rear young (Cronin, West, and Ross 2016) and the presence of youngsters in the group (this analysis)—there seems no measurable differences in behavioral indicators of welfare between what might be considered the "natural" condition and that which exists in captive settings.

The question of whether limitation on breeding negatively impacts welfare is a contemporary question with far-reaching implications (Cronin, West, and Ross 2016; Penfold et al. 2014; Powell and Ardaiolo 2016). For instance, gauging whether there is a welfare benefit to providing the opportunity to breed and rear offspring helps inform decisions associated with the use of euthanasia of healthy individuals as a means of population management, a current topic of discussion (Cohen and Fennell 2016; Lacy 1991; Powell and Ardaiolo 2016). Determining the welfare consequences of reproductive-related behaviors is an urgent need in order to scientifically support or oppose such controversial practices. This study tackled how reproductive management, and the option to breed and rear offspring, potentially impacts the welfare of chimpanzees. It is notable (but not conclusive) that consistent results have now been reported for investigations focusing on sanctuary and zoo chimpanzees, for males and females, and with different methodologies. However, we agree that mismatches between wild and captive conditions call for increased attention and acknowledge that one must consider null effects with caution. We invite additional empirical investigation into this question so that a larger body of evidence can be referenced to determine how managed breeding may or may not impact the quality of lives of chimpanzees in human care.

Future Directions

The relationship between the "naturalness" of an animal's social and physical environment and the animal's welfare is a complicated one. We generally advocate that the most straightforward path to good welfare, especially in the absence of clear predictors, is to at least functionally replicate aspects of wild environments. However, empirical studies, such as those conducted here, can help us identify cases where unnatural conditions seem not to directly and negatively affect welfare. In the case of chimpanzees, we are gaining confidence in the finding that one aspect of a natural life, the opportunity to breed and rear young, seems not to be a necessary element to be able to achieve good welfare. However, we are only scraping the surface of this complex topic and invite additional empirical investigation into the effects of managed breeding on individual welfare. Here we considered whether there was a measurable, population-level difference in welfare that corresponded to the presence of young in the social groups of chimpanzees housed in accredited North American zoos. While there was no effect on welfare at the population level, there may be individual differences in whether chimpanzees benefit from young in the group that correspond to individual temperaments, past experience, or life history stages that were not captured in these results. Considering whether and how the presence of young differentially impacts individuals would be important in future work, ideally by pursuing a longitudinal approach that allows one to track how welfare may change in response to different social environments, and by integrating multiple welfare indicators.

Acknowledgments

We thank the organizers and contributors to the Chimpanzees in Context symposium for thought-provoking discussions, Sarah L. Jacobson for advice on data analyses, and the Leo S. Guthman Fund for supporting the research.

References

Boccia, M. L., M. L. Laudenslager, and M. L. Reite. 1995. "Individual differences in macaques' responses to stressors based on social and physiological factors: Implications for primate welfare and research outcomes." *Laboratory Animals* 29 (3): 250–57.

Botreau, R., M. Bonde, A. Butterworth, P. Perny, M. B. M. Bracke, J. Capdeville, and I. Veissier. 2007. "Aggregation of measures to produce an overall assessment of animal welfare. Part 1: A review of existing methods." *Animal* 1 (8): 1179–87.

Bracke, M. B., and H. Hopster. 2006. "Assessing the importance of natural behavior for animal welfare." *Journal of Agricultural and Environmental Ethics* 19 (1): 77–89.

Brando, S., and H. M. Buchanan-Smith. 2018. "The 24/7 approach to promoting optimal welfare for captive wild animals." *Behavioural Processes* 156: 83–95.

Cohen, E., and D. Fennell. 2016. "The elimination of Marius, the giraffe: Humanitarian act or callous management decision?" *Tourism Recreation Research* 41 (2): 168–76.

Cronin, K. A., V. West, and S. R. Ross. 2016. "Investigating the relationship between welfare and rearing young in captive chimpanzees (*Pan troglodytes*)." *Applied Animal Behaviour Science* 181: 166–72.

Davey, G., P. Henzi, and L. Higgins. 2005. "The influence of environmental enrichment on Chinese visitor behavior." *Journal of Applied Animal Welfare Science* 8 (2): 131–40.

Duncan, I. J., and J. C. Petherick. 1991. "The implications of cognitive processes for animal welfare." *Journal of Animal Science* 69 (12): 5017–22.

Fàbregas, M. C., F. Guillén-Salazar, and C. Garcés-Narro. 2012. "Do naturalistic enclosures provide suitable environments for zoo animals?" *Zoo Biology* 31 (3): 362–73.

Faust, L. J., Y. M. Bergstrom, S. D. Thompson, and L. Bier. 2012. PopLink (Version 2.4). Chicago: Lincoln Park Zoo.

Finlay, T., L. R. James, and T. L. Maple. 1988. "People's perceptions of animals: The influence of zoo environment." *Environment and Behavior* 20 (4): 508–28.

Fraser, D., and I. J. Duncan. 1998. " 'Pleasures,' 'pains' and animal welfare: Toward a natural history of affect." *Animal Welfare* 7 (4): 383–96.

Gillman, S. J., K. Ziegler-Meeks, C. Eager, T. A. Tenhundfeld, W. Shaffstall, M. J. Stearns, and A. E. Crosier. 2017. "Impact of mimicking natural dispersion on breeding success of captive North American Cheetahs (*Acinonyx jubatus*)." *Zoo Biology* 36 (5): 332–40.

Goodall, J. 1986. *The Chimpanzees of Gombe: Patterns of Behavior*. Cambridge, MA: Belknap Press of Harvard University Press.

Hancocks, D. 1980. "Bringing nature into the zoo: Inexpensive solutions for zoo environments." *International Journal for the Study of Animal Problems* 1 (3): 170–77.

Held, S. D., and M. Špinka. 2011. "Animal play and animal welfare." *Animal Behaviour* 81 (5): 891–99.

Hennessy, M. B., S. Kaiser, and N. Sachser. 2009. "Social buffering of the stress response: Diversity, mechanisms, and functions." *Frontiers in Neuroendocrinology* 30 (4): 470–82.

Hill, S. P., and D. M. Broom. 2009. "Measuring zoo animal welfare: Theory and practice." *Zoo Biology* 28 (6): 531–44.

Honaker, J., G. King, and M. Blackwell. 2011. "Amelia II: A program for missing data." *Journal of Statistical Software* 45 (7): 1–47.

Hopper, L. M., H. D. Freeman, and S. R. Ross. 2016. "Reconsidering coprophagy as an indicator of negative welfare for captive chimpanzees." *Applied Animal Behaviour Science* 176: 112–19.

Jacobson, S. L., S. R. Ross, and M. A. Bloomsmith. 2016. "Characterizing abnormal behavior in a large population of zoo-housed chimpanzees: Prevalence and potential influencing factors." *PeerJ* 4: e2225.

Lacy, R. C. 1991. "Zoos and the surplus problem: An alternative solution." *Zoo Biology* 10 (4): 293–97.

Llorente, M., D. Riba, S. Ballesta, O. Feliu, and C. Rostán. 2015. "Rehabilitation and socialization of chimpanzees (*Pan troglodytes*) used for entertainment and as pets: An 8-year study at Fundació Mona." *International Journal of Primatology* 36 (3): 605–24.

Lukas, K. E., M. P. Hoff, and T. L. Maple. 2003. "Gorilla behavior in response to systematic alternation between zoo enclosures." *Applied Animal Behaviour Science* 81 (4): 367–86.

Lund, V. 2006. "Natural living—A precondition for animal welfare in organic farming." *Livestock Science* 100 (2): 71–83.

Markham, A. C., R. M. Santymire, E. V. Lonsdorf, M. R. Heintz, I. Lipende, and C. M. Murray. 2014. "Rank effects on social stress in lactating chimpanzees." *Animal Behaviour* 87: 195–202.

Markowitz, H. 1975. "In defense of unnatural acts between consenting animals." In *Proceedings of the 51st Annual American Association of Zoological Parks and Aquariums (AAZPA) Conference, Calgary, Canada*, 103–6. Topeka: Hills-Riviana.

Markowitz, H. 1982. "Behavioral enrichment in the zoo." New York: Van Nostrand Reinhold.

Mason, G., and M. Mendl. 1993. "Why is there no simple way of measuring animal welfare?" *Animal Welfare* 2 (4): 301–19.

Mazerolle, M. J., and M. M. J. Mazerolle. 2017. Package 'AICcmodavg.'

McAuliffe, J. 2018. *Population Analysis and Breeding and Transfer Plan for the Chimpanzee (Pan troglodytes) Green Species Survival Program (trademark)*. Chicago: Lincoln Park Zoo.

Moberg, G. P. 2000. "Biological response to stress: Implications for animal welfare." In *The Biology of Animal Stress: Basic Principles and Implications for Animal Welfare*, edited by G. P. Moberg and J. A. Mench, 1–21. Wallingford: CABI Publishing.

Murray, C. M., E. V. Lonsdorf, M. A. Stanton, K. R. Wellens, J. A. Miller, J. Goodall, and A. E. Pusey. 2014. "Early social exposure in wild chimpanzees: Mothers with sons are more gregarious than mothers with daughters." *Proceedings of the National Academy of Sciences* 111 (51): 18189–94.

Nash, L. T., J. Fritz, P. A. Alford, and L. Brent. 1999. "Variables influencing the origins of diverse abnormal behaviors in a large sample of captive chimpanzees (*Pan troglodytes*)." *American Journal of Primatology* 48 (1): 15–29.

Neuwirth, E. 2011. RColorBrewer: Colorbrewer palettes. R Package Version 1.2.

Nishida, T. 1970. "Social behavior and relationship among wild chimpanzees of the Mahali Mountains." *Primates* 11 (1): 47–87.

Nishida, T. 2011. *Chimpanzees of the Lakeshore: Natural History and Culture at Mahale*. Cambridge: Cambridge University Press.

Novak, M. A., and S. J. Suomi. 1988. "Psychological well-being of primates in captivity." *American Psychologist* 43 (10): 765–73.

Oliveira, A. F. S., A. O. Rossi, L. F. R. Silva, M. C. Lau, and R. E. Barreto. 2010. "Play behaviour in nonhuman animals and the animal welfare issue." *Journal of Ethology* 28 (1): 1–5.

Olsson, I. A. S., and K. Westlund. 2007. "More than numbers matter: The effect of social factors on behaviour and welfare of laboratory rodents and non-human primates." *Applied Animal Behaviour Science* 103 (3): 229–54.

Otali, E., and J. S. Gilchrist. 2006. "Why chimpanzee (*Pan troglodytes schweinfurthii*) mothers are less gregarious than nonmothers and males: The infant safety hypothesis." *Behavioral Ecology and Sociobiology* 59 (4): 561–70.

Pan African Sanctuary Alliance. 2016. *Pan African Sanctuary Alliance Operations Manual.*

Penfold, L. M., D. Powell, K. Traylor-Holzer, and C. S. Asa. 2014. "'Use it or lose it': Characterization, implications and mitigation of female infertility in captive wildlife." *Zoo Biology* 33: 20–28.

Pomerantz, O., and J. Terkel. 2009. "Effects of positive reinforcement training techniques on the psychological welfare of zoo-housed chimpanzees (*Pan troglodytes*)." *American Journal of Primatology* 71 (8): 687–95.

Powell, D. M., and M. Ardaiolo. 2016. "Survey of US zoo and aquarium animal care staff attitudes regarding humane euthanasia for population management." *Zoo Biology* 35 (3): 187–200.

Pruetz, J. D., and W. C. McGrew. 2001. "What does a chimpanzee need? Using natural behavior to guide the care and management of captive populations." *Care and Management of Captive Chimpanzees*: 17–37.

Pusey, A. E. 1990. "Behavioural changes at adolescence in chimpanzees." *Behaviour* 115 (3): 203–46.

Rabin, L. A. 2003. "Maintaining behavioural diversity in captivity for conservation: Natural behaviour management." *Animal Welfare* 12 (1): 85–94.

R Core Team. 2014. *R: A Language and Environment for Statistical Computing.* Vienna: The R Foundation. http://www.R-project.org.

Ross, S. R. 2015. "Chimpanzee (*Pan troglodytes*) North American regional studbook." Chicago: Lincoln Park Zoo.

Ross, S. R., M. A. Bloomsmith, T. L. Bettinger, and K. E. Wagner. 2009a. "The influence of captive adolescent male chimpanzees on wounding: Management and welfare implications." *Zoo Biology* 28 (6): 623–34.

Ross, S. R., L. M. Melber, K. L. Gillespie, and K. E. Lukas. 2012. "The impact of a modern, naturalistic exhibit design on visitor behavior: A cross-facility comparison." *Visitor Studies* 15 (1): 3–15.

Ross, S. R., S. J. Schapiro, J. Hau, and K. E. Lukas. 2009b. "Space use as an indicator of enclosure appropriateness: A novel measure of captive animal welfare." *Applied Animal Behaviour Science* 121 (1): 42–50.

Ross, S. R., K. E. Wagner, S. J. Schapiro, J. Hau, and K. E. Lukas. 2011. "Transfer and acclimatization effects on the behavior of two species of African great ape (*Pan troglodytes* and *Gorilla gorilla gorilla*) moved to a novel and naturalistic zoo environment." *International Journal of Primatology* 32: 99–117.

Seidensticker, J., and D. Forthman. 1998. "Evolution, ecology, and enrichment: Basic considerations for wild animals in zoos." In *Second Nature: Environmental Enrichment for Captive Animals*, edited by D. J. Shepherdson, J. D. Mellen, and M. Hutchins, 15–29. Washington, DC: Smithsonian Institution Press.

Seyfarth, R. M., and D. L. Cheney. 2012. "The evolutionary origins of friendship." *Annual Review of Psychology* 63: 153–77.

Silk, J. B., J. C. Beehner, T. J. Bergman, C. Crockford, A. L. Engh, L. R. Moscovice, and D. L. Cheney. 2010. "Strong and consistent social bonds enhance the longevity of female baboons." *Current Biology* 20 (15): 1359–61.

Symonds, M. R., and A. Moussalli. 2011. "A brief guide to model selection, multimodel inference and model averaging in behavioural ecology using Akaike's information criterion." *Behavioral Ecology and Sociobiology* 65 (1): 13–21.

Wickham, H., W. Chang, and M. H. Wickham. 2013. Package 'ggplot2.' *Computer Software Manual*. R Package Version 0.9.3.1.

Wrangham, R. W. 2000. "Why are male chimpanzees more gregarious than mothers? A scramble competition hypothesis." In *Primate Males: Causes and Consequences of Variation in Group Composition*, edited by P. M. Kappeler, 248–58. Cambridge: Cambridge University Press.

23

How Chimpanzee Personality and Video Studies Can Inform Management and Care of the Species: A Case Study

ELIZABETH S. HERRELKO, SARAH-JANE VICK, AND HANNAH M. BUCHANAN-SMITH

Introduction

Sociality is a common feature across primate species. As humans, we spend the majority of our lives maneuvering within or between groups of individuals (Carstensen 1991) and our success within personal and professional endeavors is often reliant upon how well we interact with others (Ferris et al. 2000; Offer and Fischer 2018; Wolff and Moser 2009). This concept is not unique to humans or even nonhuman primate species (referred to as primates hereafter); it is an important factor for the care and welfare of all group-living animals. Appropriately, social interaction is a focal point of federal regulation in the United States (USDA 1991) as well as a core component of animal welfare theory: Five Domains model (which updated the Five Freedoms, FAWC 1979, to incorporate positive welfare states; Mellor 2016; Mellor and Beausoleil 2015). Zoos, sanctuaries, and laboratories work to meet the social needs of primates in many ways, including collaborations within and across institutions to encourage appropriate species-typical behaviors (Ross, chapter 24 this volume). The gold standard for captive animal behavior has long been to compare activities to wild populations (Cronin and Ross, chapter 22 this volume; Fraser and Broom 1990; Hediger 1969), while recognizing that there are both limitations and advantages of captive care (e.g., Veasey 2017). Direct comparisons with the wild are not always an appropriate measuring stick, especially for activity budgets (Howell and Cheyne 2019; Veasey, Waran, and Young 1996), but each species' natural history, particularly in terms of sociality, must be the benchmark given evolved capacities.

Chimpanzees (*Pan troglodytes*) in the wild live in social communities of over one hundred individuals and regularly practice fission-fusion (Nishida 1979) by breaking apart into smaller social groups or parties within their fluid society (Aureli et al. 2008). They dynamically interact with individuals within their own communities (Aureli et al. 2008; Kummer 1971), but when faced with outside individuals or groups, interactions can be aggressive in nature (Wilson and Wrangham 2003). Permanently migrating from one group to another, generally carried out by females when they become sexually mature (Nishida et al. 2003), can be a challenging process. To become part of another group's social group is not easy, as coalitions between females tend to be stable over time and males frequently change coalitions in order to increase their rank (de Waal 1984). For chimpanzees in captivity, the success of merging groups is dependent not only on the behavior of the animals themselves, but also on the facilitation process by animal care staff.

The psychological well-being of animals relies heavily on the ability to meet each species' social needs—including introducing or removing individuals within a group (Visalberghi and Anderson 1993)—as well as the individual needs of animals (Coleman 2017). To increase success, we look to previous studies and behavior models to guide our management decisions and care of the species. Pervasive individual differences, potentially stemming from rearing and atypical life histories, determine how individual animals respond to social and environmental challenges (Carere and Maestripieri 2013; Freeman and Ross 2014; Réale et al. 2007; Suomi 1997). It is through these differences that we have been able to explain behavior in nonhuman primates for many decades (Alford et al. 1995; Freeman and Gosling 2010; Gold and Maple 1994; Goodall 1971; Schel et al. 2013; Yerkes 1939).

Personality serves as a heuristic for understanding the actions of others and can be applied to understand individual variation in a range of species (Gosling 2001). Personality profiles of primates can be measured through human-provided ratings of traits (e.g., Capitanio 1999; Freeman et al. 2013; King and Figueredo 1997; Stevenson-Hinde and Zunz 1978), behavioral expression of cortisol profiles (e.g., Capitanio et al. 2005), or from behavioral test battery data (e.g., Massen et al. 2013; Uher, Asendorpf, and Call 2008). Using instruments such as the Hominoid Personality Questionnaire (Weiss 2017), in which human caretakers score animals' personality using a series of adjective descriptors ("traits"), produces ratings that are consistent across raters (King and Figueredo 1997; Weiss et al. 2009), over time (King, Weiss, and Sisco 2008) and, in the case of at least four of the same six constructs, in chimpanzees across different habitats and rater nationalities (King, Weiss, and Farmer 2005; King, Weiss, and Sisco 2008; Weiss, King, and Hopkins 2007; Weiss et al. 2009). Such data have revealed that, similar to humans, chimpanzees share the "big five" personality factors of Extraversion, Neuroticism, Openness, Agreeableness, and Conscientiousness, but also have a distinct Dominance trait (King and Figueredo 1997).

The success of personality research with nonhuman animals stems from human-based studies. Researchers have found success in using the Five Factor Model (Goldberg 1990) to predict the outcomes of group performance in the human business world. For example, ideal groups included a combination of individuals with a mixture of high and low scores on Extraversion and Neuroticism and at least a few individuals with high scores on Conscientiousness, Agreeableness, and Openness (Neuman, Wagner, and Christiansen 1999). Based on the Five Factor Model, human personality studies served as the starting point to assess personality for many nonhuman primate studies (Gold and Maple 1994; King and Figueredo 1997). While studies indicate that the measures of personality factors in primates are reliable, repeatable, and generalizable, the validity of these measures can be demonstrated by an association with outcomes, such as behavioral differences across a variety of contexts or situations (Freeman et al. 2013; King and Weiss 2011). A relationship between traits and observable behavior exists for chimpanzees and other primates. For example, chimpanzees who were rated high in the factor Neuroticism were associated with increased rates of self-directed behaviors (SDBs; e.g., Baker and Aureli 1997) and those rated higher in Openness showed higher voluntary participation in cognitive training activities (Herrelko, Vick, and Buchanan-Smith 2012, see also Hopper et al. 2014; Morton, Lee, and Buchanan-Smith 2013; Pederson, King, and Landau 2005). Personality ratings are likely to be similar among chimpanzees who choose to spend time together (Massen and Koski 2014; Morton et al. 2015) and they correlate with longer-term outcomes such as happiness and longevity, although the precise mechanisms are not fully understood. For example, higher Extraversion ratings predicted keeper-evaluated happiness in orangutans (*Pongo* sp., Weiss, Adams, and King 2011) and longevity in captive gorillas (*Gorilla gorilla gorilla*), findings that might be due to higher rates of social affiliation providing a social buffer against stressors (Weiss et al. 2013). It has also been shown that chimpanzees who experience atypical rearing histories are rated lower on Extraversion than chimpanzees raised with their mothers in a social setting (Freeman, Weiss, and Ross 2016). In this way, primates' personality ratings can be useful metrics for aiding primate care and assessing welfare (Coleman 2017).

While primate personality ratings have proven valuable for a number of welfare and husbandry metrics, such as training outcomes (Reamer et al. 2014), abnormal behavior prevalence (Gottleib, Capitanio, and McCowan 2013), and longevity (Altschul et al. 2018; Weiss et al. 2011), their role in predicting the success of captive group introductions has yet to be investigated. This is surprising as individual variation among chimpanzees is also evident in their responses to social contexts; these responses often have immediate social outcomes that are likely to influence fitness. Individual differences have been shown to impact hunting initiation in the wild (Gilby, Eberly, and Wrangham 2008); influence the likelihood of collaboration under experimental conditions that require collective action for success (Schneider, Melis, and Tomasello 2012); correlate with differing responses to experimentally induced inequity (Brosnan et al. 2015); and underpin differential responses to outgroup encounters in captivity (Brent, Kessel, and Barrera 1997). Therefore, here, we examine chimpanzee personality in relation to a novel and high-stakes social challenge: the integration of two established social groups. Specifically, we use a case study of the introduction of two chimpanzee groups at a zoo and evaluate the relative importance of chimpanzee personality and introduction methodology (e.g., video introductions and visual access periods) on introduction outcomes.

Introducing unfamiliar individuals can be stressful and challenging for chimpanzees; inter-group interactions are complex and multifaceted (Brent, Kessel, and Barrera 1997; Clark 2010; Schel et al. 2013; Seres, Aureli, and de Waal 2001). The introduction process generally starts by providing visual access before proceeding to physical introductions (often in dyads) where successful pairings are separated from their original groups to gradually create a new social group (Brent, Kessel, and Barrera 1997; Fritz and Howell 2001; Schel et al. 2013; Seres, Aureli, and de Waal 2001). Although familiarity is an important factor in reducing aggressive interactions between chimpanzees (Fritz and Fritz 1979), prior exposure is not always required for an introduction to be successful (Brent, Kessel, and Barrera 1997).

Captive chimpanzee introductions have the potential, much like inter-group interactions in the wild, to be volatile; they can be difficult and complex events to manage (Brent, Kessel, and Barrera 1997) and further research is required to refine introduction methods and improve welfare. While moving primates between institutions to form new groups is a widespread practice in zoos, sanctuaries, and laboratories, introducing unfamiliar conspecifics can still have psychological and physical risks (Joint Working Group on Refinement 2009; Schapiro et al. 2012; Yamanashi et al. 2016). The topic of chimpanzee introductions is prevalent within conference presentations and industry publications discussing a range of topics including the influence of rearing conditions on introduction success (Brent 2001); the importance of certain demographics of successful outcomes, specifically valuing a combination of previous familiarity between individuals, females with high reproductive value (no offspring and cycle regularly), and established males being younger than the male being introduced (Stevens and van Elsacker 2005);

how husbandry training can benefit introductions (Whittaker 2006); the introduction process itself (AZA Ape TAG 2010); behavior during the introduction process (Bloomsmith 2015; Bloomsmith et al. 1998, 1999; Clay et al. 2015); and considerations for individuals with atypical behavior profiles (Seres 2008; Tresz 2011). However, there is a paucity of peer-reviewed literature on this topic (i.e., Bashaw, Gullott, and Gill, 2009; Brent, Kessel, and Barrera 1997; Fritz and Howell 2001; Herrelko, Buchanan-Smith, and Vick 2015; Mcdonald 1994; Noon 1991; Seres, Aureli, and de Waal 2001; van Hooff 1973).

To help fill the gap of systematic, species-specific knowledge, we are reliant upon what we know about other primate species. First-hand accounts of introductions and recommended techniques are described for several species: macaques (*Macaca* sp., McGrew 2017; Truelove et al. 2017), marmosets (*Callithrix jacchus*, Majolo, Buchanan-Smith, and Morris 2003), woolly monkeys (*Lagothrix lagotricha*, Barnes and Cronin 2012), orangutans (Hamburger 1988), and mixed-species exhibits (Gentry and Margulis 2008). Systematic research is primarily available from data on rhesus macaques (*M. mulatta*) housed in laboratory environments. For example, opportunities for social interactions elicit more species-typical behavior than inanimate environmental enrichment (Schapiro et al. 1996). Incremental introductions combined with an enclosure designed to allow visual and physical separation of individuals helps to reduce wounding in rhesus macaques (Westergaard et al. 1999). Agonistic interactions initially occur as rhesus macaques establish rank but dissipate as new groups settle and become organized (Bernstein, Gordon, and Rose 1974). The physiological effects of group formation (a stressor) can be modulated by the addition of a companion in rhesus macaques (Gust et al. 1996). For an extensive list of mammal introduction publications, including a framework for the introduction and socialization process, see Powell (2010).

There is limited empirical evidence exploring the behavioral predictors of chimpanzee introduction outcomes (Brent, Kessel, and Barrera 1997), particularly for high-risk mergers of larger groups. The introductions of larger groups mainly composed of adults could potentially result in substantial injuries or in worst-case scenarios, the isolation of individuals from social groups or even fatalities (van Hooff 1973). When considering predictors of success for chimpanzee introductions, in addition to personality ratings, the chimpanzees' sex and rank differences are also important factors. Males are more aggressive than females (Alford et al. 1995; Brent, Kessel, and Barrera 1997) and high-ranking individuals exhibit more aggression than low-ranking individuals (Muller and Wrangham 2004).

A visual access, non-contact period between unfamiliar chimpanzees could potentially predict compatibility and be a useful assessment tool (Bloomsmith et al. 1998; Brent, Kessel, and Barrera 1997), allowing for both physical assessments (e.g., age and sex) and the opportunity for limited social interactions during an introduction. Therefore, we examined whether personality traits or behavioral responses to both visual only (video) and non-contact encounters predicted the social outcomes of introductions. Video technology has been successfully used in research, as enrichment, and to aid training with chimpanzees in a number of facilities (Bloomsmith, Keeling, and Lambeth 1990; Bloomsmith and Lambeth 2000; Bloomsmith et al. 2000; Hirata 2007; Menzel, Savage-Rumbaugh, and Lawson 1985; Perlman et al. 2010; Price et al. 2009) and has informally been used as "video dating" for gorillas (P. Patterson, pers. comm. 2006) and chimpanzees (S. Ross, pers. comm. 2018). Despite interest in the technology from researchers, keepers, and chimpanzees, video has not yet been systematically tested as a potential tool to enhance animal management in the creation of new groups, although evidence for its efficacy as providing social information is mixed (Hopper, Lambeth, and Schapiro 2012).

In this study, we evaluated chimpanzees' personality profiles, and their behavior during visual access periods and video introductions, as potential predictors of introduction outcomes during the merger of two large social groups in a zoo setting. A primary yet challenging goal in introductions is to be able to gauge cues (vocal and non-vocal) from the chimpanzees and react appropriately to provide a suitable environment in which they can develop dynamic social groups (Seres 2008). With potential variability in process and outcome for each chimpanzee introduction, developing models to gauge the likelihood of initial success for specific combinations may help reduce risk and lead to improvements in welfare by guiding future introductions (e.g., highlighting likely aggressive, affiliative, or neutral combinations based on statistical models). Finally, transfers and introductions happen within a variety of species across a range of settings; although chimpanzees are the focus of this chapter, this study also has the potential to inform our understanding of individual and social factors in introductions more broadly.

Approach

STUDY ANIMALS

We studied 22 chimpanzees, ranging in age from 11 to 49 years old, that included a group of resident chimpanzees living at the Royal Zoological Society of Scotland's Edinburgh Zoo in Scotland (n = 5 males, n = 6 females) and a group of former laboratory-housed chimpanzees being transferred from Beekse Bergen Safaripark (BBS) in the Netherlands to Edinburgh Zoo (n = 6 males, n = 5 females). The groups were housed in the Edinburgh Zoo, within Budongo Trail, an exhibit dedicated to chimpanzee husbandry and research. For further details on study animals and housing (see Herrelko, Vick, and Buchanan-Smith 2012).

APPARATUS

We used two cameras (Sanyo Xacti HD700 and Panasonic SDR-8W21) to record footage and an Apple MacBook MB062LL/B to edit (QuickTime Pro) and play the videos on a 19″ open-frame flat-screen monitor (Elo 1939L) mounted on a steel L-shaped cart with a protective Perspex barrier (12 mm). We recorded behavior on check sheets with the use of an audible beeper to indicate intervals for instantaneous scan sampling.

PERSONALITY AND RANK ASSESSMENT

Prior to the introduction process, we calculated personality profiles for each chimpanzee using the Hominoid Personality Questionnaire, an assessment of personality factors (table 23.1). Six staff members (Edinburgh Zoo, n = 4; BBS, n = 2), who each worked with the chimpanzees for at least two years, provided the chimpanzee personality ratings. We calculated rank through staff assessments of dominance behaviors (table 23.2; Herrelko 2011).

VIDEO INTRODUCTIONS

Prior to and during the physical introductions, we presented all chimpanzees with video clips of the unfamiliar individuals they would meet during the introduction process. A whistle indicated the start of each session when chimpanzees could choose to watch the monitor in an off-exhibit area of their enclosure. Access to the viewing area was not restricted, chimpanzees could choose to come and go throughout the sessions. Seven "video introduction" sessions, spaced throughout the physical introduction process, were run, each of which included one five-minute video clip for each unfamiliar chimpanzee and one for the overall group. The viewing opportunity for each group totaled 9 hours and 20 minutes. As introductions progressed and a new, larger group of chimpanzees formed (ultimately to include all 22 individuals), we updated video clips to show footage of the unfamiliar chimpanzees in their new group. When the newly formed group included 12 or more individuals, we reduced the clip length to 2.5 minutes per chimpanzee to ensure session length did not exceed one hour. We recorded each chimpanzee's responses to

TABLE 23.1. Definitions of personality traits, behavioral responses, and introduction outcomes. Asterisks (*) indicate the measures included in regression models. Carets (^) indicate variables omitted from analysis due to a low Intraclass Correlation score.

Measures	*Definition*
Personality Factors	
Dominance*	*Positive Loading:* Bullying, decisive, dominant, independent, intelligent, manipulative, persistent, stingy, and persistent
	Negative Loading: Anxious, cautious, dependent, fearful, submissive, timid, vulnerable
Extraversion	*Positive Loading:* Active, affectionate, friendly^, imitative, playful, sociable
	Negative Loading: Depressed, individualistic^, lazy^, solitary
Conscientiousness*	*Positive Loading:* Predictable^
	Negative Loading: Aggressive, clumsy^, defiant, disorganized, distractible, erratic, impulsive^, irritable^, jealous, quitting, thoughtless^
Agreeableness	*Positive Loading:* Conventional, gentle, helpful^, protective, sensitive^, sympathetic
Neuroticism	*Positive Loading:* Autistic, excitable
	Negative Loading: Cool, stable, unemotional
Openness	*Positive Loading:* Curious, innovative, inquisitive, inventive
Responses to Videos	
Watching	Level of interest (estimated %) determined by head orientation towards monitor
Self-directed behaviors	Rates per minute for: rub, scratch, self-groom, and yawn
Affiliation*	Rates per minute for: Approach, follow and smell with touching, embrace, hug, groom, mouth, pat, play, play invite, copulate, display, inspect genitals, mount, and kiss
Aggression*	Rates per minute for: Display behaviors including sway and pant hoot, chase, hit, bite, charge (chimps or objects), stomp, foot, fight, kick, or jump over
Not present	Not physically in the introduction area (estimated %)
Introduction Outcomes and Response to Visual Contact	
Affiliative behavior*	See affiliative behavior above (estimated %)
Aggressive behavior*	See aggressive behavior above (estimated %)
Neutral	No overt social behaviors exhibited (estimated %)

Note: Personality definitions include example traits used in this study, following Weiss et al. (2009); affiliative and aggressive behavior definitions follow those from Brent, Kessel, and Barrera (1997).

the videos in real time using 30-second point sampling of behavioral states and all occurrences of behavioral events (table 23.1). Throughout the video introduction sessions, the keepers maintained their usual routine, which included passing by the off-exhibit area. We made no extra efforts to interact with the chimpanzees when the videos were playing and there were no additional food provisions, other than any items leftover from a previous feed.

INTRODUCTION PROCESS

The introduction process occurred in three stages: auditory/olfactory access, visual access, and physical introductions. Following the arrival of the chimpanzees

TABLE 23.2. Dominance ranks for the Edinburgh and Beekse Bergen chimpanzee groups prior to the introduction process based on keeper and researcher assessment, in order of rank (parentheses: sex and age, in years, at start of introductions).

Rank	*Edinburgh Group*	*Beekse Bergen Group*
High	1—Qafzeh (male, 17)	1—Claus (male, 16)
	2—David (male, 35)	2—Paul (male, 16)
	3—Louis (male, 34)	3—Eva (female, 29)
	4—Emma (female, 28)	4—Pearl (female, 41)
Medium	5—Kindia (male, 13)	5—Rene (male, 17)
	6—Lucy (female, 33)	6—Bram (male, 23)
	7—Liberius (male, 11)	7—Frek (male, 16)
Low	8—Kilimi (female, 17)	8—Edith (female, 13)
	9—Lyndsey (female, 25)	9—Heleen (female, 18)
	10—Ricky (male, 49)	10—Sophie (female, 28)
	11—Cindy (female, 46)	11—Lianne (female, 21)

Note: For more information on ranking methodology, see Herrelko (2011).

from BBS to Edinburgh Zoo, animal care staff housed the groups in separate areas of Budongo Trail; they could hear and smell, but not see, each other. After six days, the chimpanzees had visual access to each other with a 2 m wide enclosure serving as a barrier between the groups. One week later, the physical introductions began: animal care staff briefly separated one or more chimpanzees from each group and allowed them to have limited physical contact with each other through mesh fencing. If aggression was low, staff opened the door separating the chimpanzees so they could share the same space and interact without any barriers. The duration of each stage varied based on chimpanzee behavior and progress from one stage to another was dependent upon evaluation by animal staff. Following each successful introduction, newly introduced individuals joined the new, larger group of chimpanzees. The introduction process took just over 3.5 months (Schel et al. 2013). Due to the dynamic and complex nature of each introduction (e.g., multiple individuals quickly moving and interacting in a smaller, controlled area) and difficulty filming in this location, it was not possible to record detailed observations. Animal staff, following each encounter, estimated the percentage of time spent exhibiting social behaviors in each possible dyad and evaluated the introduction outcome. Social behavior categories that identify the physical introduction outcomes are aggressive, affiliative, and neutral (table 23.1).

DATA ANALYSES

We analyzed all data using SPSS v16. When evaluating data from the video introduction sessions, we estimated percentages of time for behavioral states and calculated rates per minute for behavior events. The personality analyses included 44 traits after the exclusion of 10 traits with low (below zero; Weiss et al. 2011) intraclass correlation coefficients (ICC; Shrout and Fleiss 1979) between raters: clumsy, friendly, helpful, impulsive, individualistic, irritable, lazy, predictable, sensitive, and thoughtless. We also used ICC to assess reliability of the behavioral estimates between four raters (animal and research staff involved in the introductions) during the physical introductions and excluded one category (submissive behavior) due to low reliability.

Following Weiss et al. (2009), to generate personality profiles, we unit-weighted variable scores from each rater and listed them as positive or negative in accordance with the defined loadings for each personality factor (table 23.1). After omitting the variables that did not reach suitable inter-observer reliability, we averaged the remaining scores to create six factor scores for each chimpanzee/rater combination and calculated the average personality factor scores for each chimpanzee. To simplify interpretation, we converted unit-weighted factor scores into T scores (mean = 50, standard deviation = 10, following King, Weiss, and Farmer 2005; Weiss et al. 2009). The resulting profiles are shown in figure 23.1.

We assessed behavior during all phases of introductions with descriptive statistics and correlated personality ratings with the chimpanzees' sex,

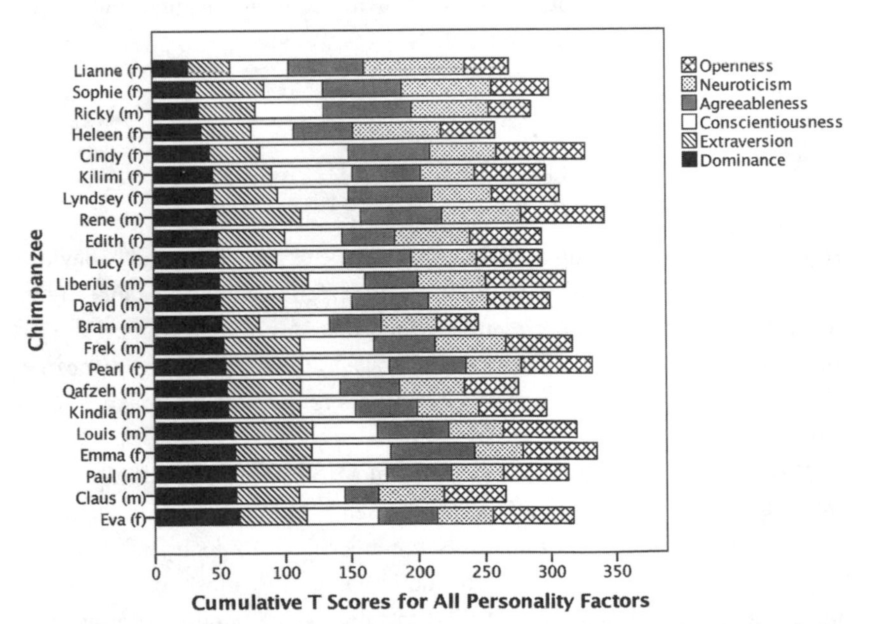

FIGURE 23.1. Chimpanzee personality profiles in order of the Dominance factor, from lowest to highest (top to bottom) with sex listed in parentheses. Different colors and patterns represent the *T*-score for all six factors (as defined in the legend).

rank, and time spent watching the videos. Regression models (Method: Enter, Field 2009) enabled us to examine the relationship between personality or behavioral measures and social outcomes during the physical introductions, categorized as aggressive, affiliative, and neutral. We performed analyses for all individuals in relation to sex (male and female), rank (high and low), and familiarity (in- and out-groups). The variance inflation factor and tolerance scores indicated multicollinearity between variables. We omitted any highly correlated ($r > 0.90$) variables and non-normal data, as well as any models with heteroscedasticity and non-linearity. Each condition included two predictors due to sample size restrictions and correlations between variables. Dominance significantly correlated with all personality traits except for Conscientiousness, therefore we only included Dominance and Conscientiousness in our models. Similarly, rates of aggressive and affiliative behavior correlated with estimated percentage of time spent watching videos and rates of SDBs; aggression and affiliation served as predictors because these measures were also available in the visual contact phase. When examining relative personality, following Morton et al. (2015), we calculated the absolute difference between personality factor scores for each dyad occurrence during the introduction process and conducted regression analyses with bootstrapping to compare each factor to the social outcomes of the introductions. The results report only significant models ($p < 0.05$) with non-significant variables removed from the final model reported.

Results

THE CHIMPANZEES' BEHAVIOR

Across all phases of the introduction process, we observed neutral behaviors most frequently (mean ± SE = 47.90% ± 2.45%), followed by avoidance of the introduction area (i.e., "not present"), which was only possible during the visual access period (mean ± SE = 22.61% ± 2.54%), affiliative behaviors (mean ± SE = 11.12% ± 1.38%), and aggressive behaviors toward others (mean ± SE = 9.78% ± 1.45%).

Visual Access Period: During the visual access period, the chimpanzees avoided the introduction area (i.e., not present) and remained in adjoining enclosure areas about half of the time (mean ± SE = 42.77% ± 5.48%). When they chose to be present in the introduction area, they most commonly exhibited neutral behaviors (mean ± SE = 27.74% ± 4.79%), followed by aggressive (mean ± SE = 24.62% ± 5.93%) and affiliative (mean ± SE = 3.87% ± 1.47%) behaviors toward others. When examining these behaviors according to sex, rank, and personality, only "absence from the introduction area" yielded significant results: females, low-ranked individuals, and those rated low in Dominance and high in Neuroticism were more likely to avoid the introduction area (sex: $r_{pb} = -0.45$, p = 0.037; rank: $r = 0.58$, p = 0.004; Dominance: $r = -0.68$, p = 0.001; Neuroticism: $r = 0.58$, p = 0.005).

Video Introductions: The chimpanzees watched an average of 12% of the video introduction clips (mean ± SE = 12.04% ± 1.45%). When videos were playing, mean rates per minute were low for SDBs and social behaviors: scratching (mean ± SE = 0.06 ± 0.01), rubbing (mean ± SE = 0.01 ± 0.002), self-grooming (mean ± SE = 0.02 ± 0.004), yawning (mean ± SE = 0.01 ± 0.002), and exhibiting aggressive (mean ± SE = 0.01 ± 0.002) and affiliative (mean ± SE = 0.01 ± 0.002) behaviors toward others in proximity or toward the monitor. When examining these behaviors in relation to sex, rank, and personality, only aggressive behavior yielded significant results: individuals low in Agreeableness were more likely to exhibit aggressive behavior during the video introductions ($r = -0.048$, p = 0.025).

DID THE CHIMPANZEES' PRE-INTRODUCTION BEHAVIOR OR PERSONALITIES PREDICT THEIR RESPONSE TO INTRODUCTIONS?

We used multiple regressions to investigate whether (1) behaviors exhibited during the visual access period, (2) behaviors exhibited during video introductions, or (3) personality profiles could significantly predict the way the chimpanzees would behave during the physical introductions.

Visual Access Period: Analyses of the chimpanzees' behavior during the visual access period did not yield significant models for predicting chimpanzee behavior during the physical introductions.

Video Introductions: Higher rates of affiliation performed when the chimpanzees were observing the video footage were associated with higher rates of aggression toward all individuals during the subsequent physical introductions, but there were no significant behavioral models when analyzed by sex, rank, or familiarity (table 23.3). In contrast, lower rates of affiliative behavior performed while the chimpanzees were observing the videos were associated with higher rates of neutral behavior toward all individuals later during the physical introductions. When separated by sex, rank, and familiarity, lower rates of affiliative behavior during the video introductions were associated with higher rates of neutral behavior toward high-ranking individuals during the physical introductions; higher rates of aggressive behavior during the video introductions were associated with higher rates of neutral

TABLE 23.3. Significant models of video introductions predicting behaviors during the physical introductions. Models, in bold, indicate to whom the behavior is directed. The model applies to behavior exhibited by all individuals, unless noted.

Behavior Predicted	*Models*	*B*	*SE B*	*β*	*T*	*Sig*
Aggressive	**All Individuals**	**$\Delta R^2 = 0.20$, $F_{1,20} = 6.39$, $p = 0.020$**				
	Constant	7.21	1.65		4.38	<0.0001
	Affiliative	74.28	29.39	0.49	2.53	0.02
Neutral	**All Individuals**	**$\Delta R^2 = 0.25$, $F_{1,20} = 8.03$, $p = 0.010$**				
	Constant	52.63	2.98		19.52	<0.0001
	Affiliative	−136.46	48.15	−0.54	−2.83	0.01
	Males	**$\Delta R^2 = 0.21$, $F_{1,20} = 6.54$, $p = 0.019$**				
	Constant	56.54	2.96		19.12	<0.0001
	Aggressive	47.52	18.58	0.50	2.56	0.019
	High Ranks	**$\Delta R^2 = 0.24$, $F_{1,20} = 7.71$, $p = 0.012$**				
	Constant	64.60	4.14		15.58	<0.0001
	Affiliative	−205.55	74.01	−0.53	−2.78	0.012
	Low Ranks	**$\Delta R^2 = 0.34$, $F_{1,20} = 5.83$, $p = 0.012$**				
	Constant	67.02	2.82		23.81	<0.0001
	Affiliative	358.37	136.48	0.49	2.63	0.018
	Aggressive	88.08	36.57	0.45	2.41	0.028

behavior toward males during introductions; and both higher rates of affiliative and aggressive behavior during video introductions were associated with higher rates of neutral behavior toward low-ranking individuals. No significant models were found when data were split according to familiarity.

Personality Profiles: Higher ratings of the trait Dominance were associated with higher aggression toward all individuals during the physical introductions (table 23.4). When separated by sex, rank, and familiarity, higher Dominance ratings were associated with higher aggression toward males during the physical introductions but rank and familiarity analyses yielded no significant effects. We did not find significant personality predictors of neutral behavior toward all individuals, but when separated by sex, rank, and familiarity, higher Conscientiousness was associated with higher rates of neutral behavior toward males, high-ranked chimpanzees, as well as both in- and out-group individuals during the physical interactions. No significant personality-based models predicted affiliative behavior during physical introductions. When chimpanzees were introduced that had similar Dominance or Conscientiousness we saw higher rates of aggression toward each other during introductions. Dyads of chimpanzees with different Conscientiousness scores introduced to one another were associated with a higher occurrence of neutral behaviors.

Discussion

Personality profiles and behaviors exhibited during video introductions predicted the chimpanzees' aggressive and neutral, but not affiliative, behavior during physical introductions. Neither sex, rank, nor familiarity predicted behavioral outcomes during the introduction. This suggests that the compatibility of personalities associated with affiliative outcomes might reside at the individual level (e.g., Massen and Koski 2014) rather than by sex, rank, or familiarity.

TABLE 23.4. Significant models predicting behaviors during the physical introductions based on personality factors. Models are listed by the category of individuals they met during the introductions.

Behavior Predicted	*Models*	*B*	*SE B*	*β*	*T*	*Sig*
Aggressive	**All Individuals**	**$\Delta R^2 = 0.29, F_{1,20} = 9.63, p = 0.006$**				
	Constant	−9.59	6.36		−1.51	0.147
	Dominance	0.39	0.12	0.57	3.10	0.006
	Males	**$\Delta R^2 = 0.22, F_{1,20} = 6.90, p = 0.016$**				
	Constant	−6.47	5.41		−1.19	0.246
	Dominance	0.28	0.11	0.51	2.63	0.016
	All Dyads	**$\Delta R^2 = 0.01, F_{1,417} = 4.66, p = 0.031$**				
	Constant	9.06	1.07		8.48	0.000
	Dominance	−0.16	0.08	−0.11	−2.16	0.031
	Homophily					
		$\Delta R^2 = 0.01, F_{1,417} = 6.13, p = 0.014$				
	Constant	9.57	1.15		8.34	0.000
	Conscientiousness	−0.21	0.08	−0.12	−2.48	0.014
	Homophily					
Neutral	**Males**	**$\Delta R^2 = 0.47, F_{1,20} = 19.70, p = 0.001$**				
	Constant	6.87	11.88		0.58	0.570
	Conscientiousness	1.05	0.24	0.70	4.44	<0.0001
	High Ranks	**$\Delta R^2 = 0.31, F_{1,20} = 10.22, p = 0.005$**				
	Constant	7.53	15.93		0.47	0.641
	Conscientiousness	1.01	0.32	0.58	3.20	0.005
	In-Group	**$\Delta R^2 = 0.34, F_{1,20} = 11.70, p = 0.003$**				
	Constant	18.32	12.99		1.41	0.170
	Conscientiousness	0.88	0.26	0.61	3.42	0.003
	Out-Group	**$\Delta R^2 = 0.35, F_{1,20} = 12.35, p = 0.002$**				
	Constant	11.73	14.87		0.79	0.440
	Conscientiousness	1.04	0.29	0.61	3.52	0.002
	All Dyads	**$\Delta R^2 = 0.01, F_{1,417} = 5.34, p = 0.021$**				
	Constant	57.76	2.22		26.043	0.000
	Conscientiousness	0.377	0.163	0.112	2.311	0.210
	Homophily					

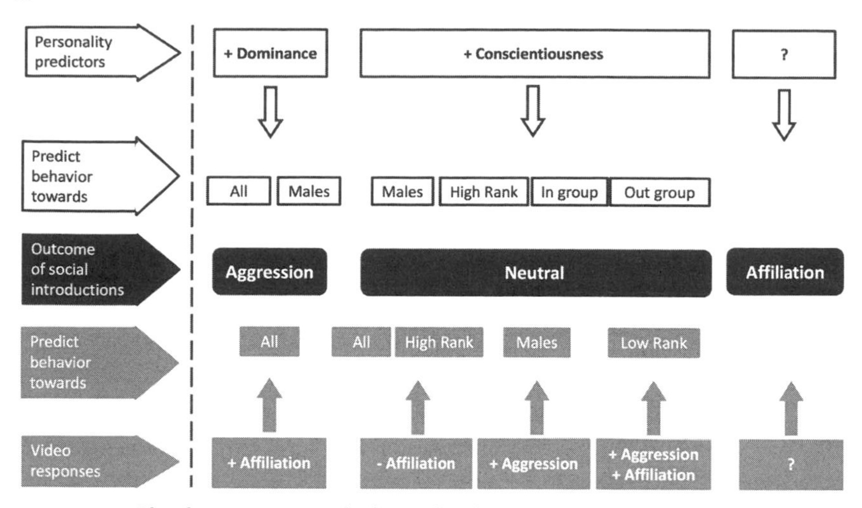

FIGURE 23.2. Flowchart summarizing the factors that play a role in predicting the outcomes of social introductions.

Although the chimpanzees watched the videos for only 12% of the time they were provided on average (i.e., just over an hour), their interest in watching videos was positively associated with both affiliative and aggressive, but not neutral introduction outcomes. Overall, the chimpanzees' personality traits better predicted introduction outcomes (see fig. 23.2 for a flowchart of overarching patterns in predicting social behaviors) compared to the other factors (i.e., behavior, sex, and rank). Specifically, chimpanzees rated highly on the trait Conscientiousness were more likely to show neutral behavior during introductions. This is not surprising given the role that personality plays in self-regulation and predicting anger in humans. Conscientiousness in humans negatively correlates with anger and mediates the relationship or progression between anger and aggression (Jensen-Campbell et al. 2007). When considering relative personality, those with similar Conscientiousness scores were more likely to be aggressive to each other whereas those differing in Conscientiousness were more likely to be neutral, suggesting that there is less uncertainty when interacting with similar personalities (Massen and Koski 2014) and more unknown factors when dealing with varying personalities.

Chimpanzees rated highly on the factor Dominance were more aggressive toward all individuals, particularly males and out-group individuals, supporting previous research indicating that Dominance and Excitability as traits were positively related to agonistic or aggressive behavior (Murray 2011; Pederson, King, and Landau 2005). Sex, rank, and group status were also important within these personality models, reflecting the complexity of chimpanzee social dynamics (e.g., de Waal 1982; Goodall 1986). However, the concept of homophily, the tendancy for individuals who interact to be similar, simplifies a complex dynamic showing that like attracts like and aggression is more likely to occur between those with similar Dominance scores.

The generally neutral outcomes we recorded during the introduction process, however, are not congruent with chimpanzee social dynamics, in terms of typical responses to contact with unfamiliar individuals (e.g., Baker and Aureli 1997; Watts et al. 2006). This suggests either considerable behavioral flexibility in coping with challenges, or that the introductions were not overly arousing for these particular chimpanzees. Given that primates employ strategies to mediate the stressors of increases in social density, including a decrease in social interactions when managing short-term conflict (Judge and de Waal 1993), the former is more likely. The importance of Conscientiousness indicates that self-regulation is central to managing social challenge (e.g., Calkins and Fox 2002). Although these data cannot identify whether the neutral outcomes were due to emotion regulation, avoidance, or a lack of arousal (i.e., we did not collect physiological markers), describing chimpanzee introduction events as calm or not arousing would be unusual (e.g., de Waal 1982; Fritz and Howell 2001; Seres, Aureli, and de Waal 2001). However, the predominance of neutral outcomes overall concurs with longer-term evidence that the integration of these groups was successful, with low levels of aggression. Relationship formation was gradual and out-group interactions remained low and stable even after a year (Schel et al. 2013), likely facilitated by the design of the enclosure, which allowed subgroups to choose to occupy different areas available (e.g., Herrelko, Buchanan-Smith, and Vick 2015).

Evidence on the importance of a gradual introduction process, which provides the opportunity for chimpanzees to gain experience with each other in a protected and controlled environment, is inconsistent. Familiarity is an important factor in reducing aggressive interactions between chimpanzees (Fritz and Fritz 1979), but research also shows that prior exposure is not always required for an introduction to be successful (Brent, Kessel, and Barrera 1997). While there is no evidence to support gradual, over direct, introductions in monkey species, Bernstein (1991, as cited in Brent, Kessel, and Barrera 1997) noted that a visual access, non-contact period could potentially predict compatibility. When gradually building up to physical contact in chimpanzee introductions, researchers have seen social behaviors change over time; compared to baseline levels, agonism increased during the visual access period and did not further increase during the physical introductions, whereas passive behaviors did the opposite and decreased during visual access and increased during physical introductions (Bloomsmith et al. 1998). Similarly, in the current study, the chimpanzees' behavior during the visual access period did not predict introduction outcomes. The physical constraints may provide an explanation: during the visual access period, individuals could avoid any encounters, as indicated by the relatively high percentage of time absent from the area.

With gradual introductions that occur in steps (Alford et al. 1995; Brent, Kessel, and Barrera 1997; Fritz and Fritz 1979; Fritz and Howell 2001; McDonald 1994; Noon 1991; Seres, Aureli, and de Waal 2001), the way in which each chimpanzee behaves during the first stages of the introduction process will likely play a role in the keepers' overall perceptions of each individual. Nevertheless, without any significant models predicting future introduction outcomes, should the behaviors exhibited during the visual access period factor into keeper decisions?

Despite evidence that gradual introductions that facilitate familiarity are not necessarily needed for integration success (Brent, Kessel, and Barrera 1997), the decreased aggression reported over time (Bloomsmith et al. 1998) suggests that the early visual access periods (i.e., protected contact) might aid as an outlet from some initial aggression; chimpanzees can challenge each other without risk of physical repercussions. Compared to the physical introductions, the visual access phase had 152% more aggressive behaviors, 65% less affiliative behaviors, and 42% less neutral behaviors. The higher percentage of aggression is likely related to the nature of the visual access period where the closest access the groups had to each other was through two layers of steel mesh that were two meters apart. Within the protected environment, the chimpanzees had room to show their strength without having to prove it in a physical fight. This concept of false bravado was also evident during some of the dyad and small group introductions when individuals transitioned from protected contact to full contact. For example, Liberius, an 11-year-old male from the Edinburgh group, was difficult to integrate into the new, larger group of chimpanzees (ultimately to include all 22 individuals), likely due to his inexperience as the only juvenile. During the two aborted attempts to introduce him to unfamiliar females, he was initially paired with his mother and then with another familiar female. Both times, he exhibited aggression toward the unfamiliar individuals while in protected contact; however, once they had full access to each other, his aggression toward the new chimpanzees was non-existent. Interestingly, he redirected his aggression to his in-group partners, with few repercussions.

Although the visual access period could not predict behaviors during the physical introductions, behaviors seen during this time should nonetheless be used to inform overall decisions because how each chimpanzee behaves contributes to their behavioral repertoires in terms of this unique situation. Behaviors exhibited during a visual access period might inform concepts not examined in this study (e.g., showing weakness or stability of coalitions within the original groups); however, without predictive power, we suggest behaviors exhibited during isolated visual access periods do not serve as the primary evaluation tool during introductions or play a dominance role in management decisions. Further study is required to examine the potentially informative nature of the specific behaviors and interactions exhibited during visual access periods.

Overall, the prevalence of neutral outcomes indicates that individuals generally coped with this social challenge, and that they regulated or avoided increased arousal during the process. Personality predicted short-term outcomes during social introductions that are likely to impact longer-term relationship formation, a topic that warrants systematic data collection. Individual variations in social approach and avoidance tendencies likely lead to significant differences in social buffering from stressors and long-term fitness outcomes (Capitanio 1999; Réale et al. 2007; Weiss et al. 2013). For managed populations, personality assessments could be applied more broadly to inform decisions and potentially reduce risks for animals and staff.

Applications and Future Directions

When talking to animal care staff about their animals, the topic of personality or individual differences inevitably comes up. Although animal management is primarily evidence based, it also values the subjective opinions of those who work closely with the animals. By asking staff to complete personality assessments for individual animals, researchers can quantify the subjective opinions of those who know the animals best and test the results against patterns of behavior to predict future behavior (King and Weiss 2011). Whether this applies to the prediction of introduction behaviors or for other zoo management scenarios, we are carrying out the process of taking what we think and turning it into what we know (C. Saffoe, pers. comm. 2015).

Zoo-based research is on the rise (Loh et al. 2018). While a great benefit for our field, this sometimes involves an increased workload for animal care teams; as active participants in the zoo community, staff receive several survey-based requests each year (B. Malinsky, pers. comm. 2015). For this study, we required at least two people to complete the questionnaire for each chimpanzee: in total, 44 completed surveys. With positive interest from the staff, and a few reminders, we received 66 completed surveys (four sets for the Edinburgh group and two sets for the BBS group). Knowing this was an extra activity on top of everyday responsibilities and sometimes challenging to complete, given the number of chimpanzees involved, we reinforced survey participation with known motivators. Developing personality questionnaires that are shorter and less taxing for staff to complete (especially when completing questionnaires for multiple animals) may help such efforts (e.g., Hopper, Cronin, and Ross 2018).

When using an individualized management approach (Seres, Aureli, and de Waal 2001), animal staff rely on their knowledge of each chimpanzee, including details about their personalities and social tendencies when interacting with others. Although animal staff are experts in chimpanzee behavior, their expertise is typically limited to their resident group. In the majority of our experiences, animal staff eagerly share information about each transferred animal and in many circumstances, spend time with the animals and their new care staff once they reach the new destination. To refine the process and guarantee the transfer to specific knowledge, we recommend incorporating low-cost and relatively brief, quantifiable measures of behavior patterns, specifically personality profiles, into cross-institutional transfer documentation.

Even without calculating the specific degree to which different individuals will likely behave based on video introductions or personality profiles, we anticipate that these models could be used informally to understand general patterns. The specific questionnaire used in this case study is freely available online (http://extras.springer.com/2011/978-1-4614-0175-9), but if formal personality profiles cannot be completed (due to time restrictions or limited access to a specialist to run the ICC and factor analysis), we recommend an alternative process to still benefit from these data. We encourage animal care staff to (1) familiarize themselves with the personality factors and traits defining each factor, and mentally categorize each individual as high, medium, or low in each category; (2) discuss how they categorized each individual with their team to create a general agreement of informal ratings; (3) use fig. 23.2 to consider how these factors are likely to shape the outcome of each specific introduction planned and use this evaluation to inform the process and associated risk assessments.

With the understanding that staff time is a valuable resource in every zoo, sanctuary, and laboratory, we consider assessments of individual differences to be an important part of the management process. Given that personality is a relatively stable construct (Caspi, Roberts, and Shiner 2005; Hampson and Goldberg 2006), the time taken to complete personality profiles has the potential for a long-lasting impact on animal management strategies that aim to promote welfare across an individual's lifespan (Brando and Buchanan-Smith 2018). When paired with data from other approaches, such as social network analyses and demographic changes over time, techniques for managing so-

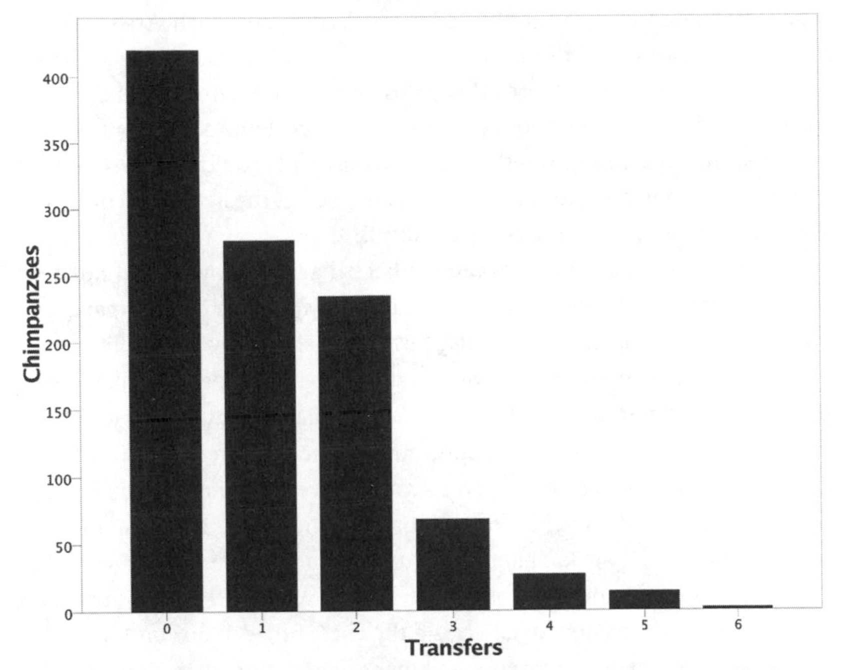

FIGURE 23.3. The distribution of chimpanzees from AZA and EAZA institutions organized by the number of institutional transfers experienced (compiled from Ross 2014; Carlsen 2015).

cial groups with varying, atypical life histories, coping strategies, exhibit usage (e.g., what makes a good exhibit animal and how that can impact collection plans), long-term introduction successes, social partner pairing prior to transport, and so on, the potential for personality profiles to positively impact animal welfare and management practices is considerable.

Personality measures have great potential as a tool in captive management and conservation when applications are evidence based (Coleman 2017; Watters and Powell 2012). The North American and European studbook populations represent 1,041 chimpanzees from AZA and EAZA institutions, the majority of which have experienced one or more institutional transfers (fig. 23.3, Carlsen 2015; Ross 2015). Animal shipments are costly in terms of staff time and resources and are potentially stressful for the animals in transit (Schapiro et al. 2012; Yamanashi et al. 2016). The planning and implementation of a move includes many individuals from population management experts, to registrars, animal keepers, managers, veterinarians, and live-animal transport specialists. Once on site, introductions are not always successful and there are risks of injury and incompatibility across individuals. In addition to the scenarios highlighting potential areas where personality profiles could help animal welfare and management, zoological organizations (e.g., Association of Zoos and Aquariums, British and Irish Association of Zoos and Aquariums, and European Association of Zoos and Aquaria) and sanctuaries (where animals are often transferred when retired from the entertainment industry or laboratory environments, e.g., Fultz 2017) would benefit from research examining whether these assessments could positively inform sociality decisions prior to transport (e.g., to predict the likelihood of success in introductions or breeding compatibility).

Given that personality ratings can predict behavior in novel social situations for captive primates (rhesus macaques: Capitanio 1999; captive chimpanzee introductions: this chapter), it could also be beneficial to apply personality framework to in situ research and/or conservation. For many decades, personality terminology has been used to describe wild chimpanzees and understand their behavior (e.g., Goodall 1986). When groups in the wild are competing for resources, cost/benefit analysis plays a role in the decision to physically compete or not (numerical assessments: Wilson, Britton, and Franks 2002; asymmetries in ownership: Davies 1978; "value of the future": Enquist and Leimar 1990). Personalities of the individuals involved may play a role in these decisions; quantifiable personality research in situ could thus help broaden our understanding of the relationship between proximate functions and ultimate outcomes.

Acknowledgments

This study would not have been possible without the collaborative team effort from the University of Stirling, the Royal Zoological Society of Scotland, Burning Gold Productions, and Alex Weiss from the University of Edinburgh. We are very grateful to all the staff who facilitated our research in Budongo Trail (Edinburgh Zoo, Royal Zoological Society of Scotland) and Beekse Bergen Safaripark (the Netherlands), the Behaviour and Evolution Research Group at the University of Stirling who provided helpful expertise and advice, Blake Morton for statistical advice, and Lydia Hopper and Steve Ross for organizing the Chimpanzees in Context conference and accompanying book, as well as providing valuable edits for this chapter. E. S. H. was supported by the David Bohnett Foundation and the Smithsonian's National Zoological Park during the final stages of manuscript preparation.

References

Alford, P. L., M. A. Bloomsmith, M. E. Keeling, and T. F. Beck. 1995. "Wounding aggression during the formation and maintenance of captive, multimale chimpanzee groups." *Zoo Biology* 14 (4): 347–59.

Altschul, D. M., W. D. Hopkins, E. S. Herrelko, M. Inoue-Murayama, T. Matsuzawa, J. E. King, S. R. Ross, and A. Weiss. 2018. "Personality links with lifespan in chimpanzees." *eLife* 7: e33781.

Aureli, F., C. M. Schaffner, C. Boesch, S. K. Bearder, J. Call, C. A. Chapman, R. Connor, A. Di Fiore, R. I. M. Dunbar, S. P. Henzi, K. Holekamp, A. H. Korstjens, R. Layton, P. Lee, J. Lehmann, J. H. Manson, G. Ramos-Fernandez, K. B. Strier, and C. P. van Schaik. 2008. "Fission-fusion dynamics: New research frameworks." *Current Anthropology* 49 (4): 627–54.

AZA Ape TAG. 2010. *Chimpanzee (Pan troglodytes) Care Manual.* Silver Spring, MD: Association of Zoos and Aquariums.

Baker, K. C., and F. Aureli. 1997. "Behavioural indicators of anxiety: An empirical test in chimpanzees." *Behaviour* 134 (13): 1031–50.

Barnes, H. A., and A. Cronin. 2012. "Hand-rearing and reintroduction of woolly monkey *Lagothrix lagotricha* at Monkey World Ape Rescue Centre, UK." *International Zoo Yearbook* 46: 164–74.

Bashaw, M. J., R. L. Gullott, and E. C. Gill. 2009. "What defines successful integration into a social group for hand-reared chimpanzee infants?" *Primates* 51 (2): 139–47.

Bernstein, I. S., T. P. Gordon, and R. M. Rose. 1974. "Aggression and social controls in rhesus monkey (*Macaca mulatta*) groups revealed in group formation studies." *Folia Primatologica* 21: 81–107.

Bloomsmith, M. A. 2015. "Recent findings in chimpanzee welfare." *American Journal of Primatology* 77 (S1): 82.

Bloomsmith, M. A., K. C. Baker, S. K. Ross, and S. P. Lambeth. 1998. "Enlarging chimpanzee social groups: The behavioral course of introductions." *American Journal of Primatology* 45: 171.

Bloomsmith, M. A., K. C. Baker, S. K. Ross, and S. P. Lambeth. 1999. "Chimpanzee behavior during the process of social introductions." In *Annual Conference Proceedings*, 270–73. Silver Spring, MD: Association of Zoos and Aquariums.

Bloomsmith, M. A., M. E. Keeling, and S. P. Lambeth. 1990. "Videotapes: Environmental enrichment for singly housed chimpanzees." *Laboratory Animals* 19: 42–46.

Bloomsmith, M. A., and S. P. Lambeth. 2000. "Videotapes as enrichment for captive chimpanzees (*Pan troglodytes*)." *Zoo Biology* 19 (6): 541–51.

Bloomsmith, M. A., S. P. Lambeth, J. E. Perlman, M. A. Hook, and S. J. Schapiro. 2000. "Control over videotape enrichment for socially housed chimpanzees." *American Journal of Primatology* 51: 41–45.

Brando, S., and H. M. Buchanan-Smith. 2018. "The 24/7 approach to promoting optimal welfare for captive wild animals." *Behavioural Processes* 156: 83–95.

Brent, L. 2001. "The influence of rearing condition on chimpanzee introductions." In *The Apes: Challenges for the 21st Century, Conference Proceedings*, 103–4. Brookfield, IL: Chicago Zoological Society.

Brent, L., A. L. Kessel, and H. Barrera. 1997. "Evaluation of introduction procedures in captive chimpanzees." *Zoo Biology* 16 (4): 335–42.

Brosnan, S. F., L. M. Hopper, S. Richey, H. D. Freeman, C. F. Talbot, S. D. Gosling, S. P. Lambeth, and S. J. Schapiro. 2015. "Personality influences responses to inequity and contrast in chimpanzees." *Animal Behaviour* 101: 75–87.

Calkins, S. D., and N. A. Fox. 2002. "Self-regulatory processes in early personality development: A multilevel approach to the study of childhood social withdrawal and aggression." *Development and Psychopathology* 14 (3): 477–98.

Capitanio, J. P. 1999. "Personality dimensions in adult male rhesus macaques: Prediction of behaviors across time and situation." *American Journal of Primatology* 47 (4): 299–320.

Capitanio, J. P., S. P. Mendoza, W. A. Mason, and N. Maninger. 2005. "Rearing environment and hypothalamic-pituitary-adrenal regulation in young rhesus monkeys (*Macaca mulatta*)." *Developmental Psychobiology* 46 (4): 318–30.

Carere, C., and D. Maestripieri. 2013. *Animal Personalities: Behavior, Physiology, and Evolution*. Chicago: University of Chicago Press.

Carlsen, F. 2015. *European Studbook for the Chimpanzee Pan troglodytes: 1st Edition of Joint EEP Studbook 2014*. Frederiksberg, Denmark: Copenhagen Zoo. https://www.zoo.dk/files/stambog_chimpanser_zoo_2014.pdf.

Carstensen, L. L. 1991. "Selectivity theory: Social activity in life-span context." *Annual Review of Gerontology and Geriatrics* 11: 195–217.

Caspi, A., B. W. Roberts, and R. L. Shiner. 2005. "Personality development: Stability and change." *Annual Review of Psychology* 56: 453–84.

Clark, F. E. 2010. "Space to choose: Network analysis of social preferences in a captive chimpanzee community, and implications for management." *American Journal of Primatology* 73 (8): 748–57.

Clay, A. W., M. A. Bloomsmith, A. Franklin, K. Neu, and J. E. Perlman. 2015. "Captive chimpanzee (*Pan troglodytes*) behavior during socialization procedures." *American Journal of Primatology* 77: 88.

Coleman, K. 2017. "Individual differences in temperament and behavioral management." In *Handbook of Primate Behavioral Management*, edited by S. J. Schapiro, 95–113. Boca Raton: CRC Press.

Davies, N. B. 1978. "Territorial defense in the speckled wood butterfly (*Pararge Aegeria*): The resident always wins." *Animal Behaviour* 26: 138–47.

de Waal, F. B. M. (1982) 2007. *Chimpanzee Politics: Power and Sex among Apes*. 25th anniversary edition. Baltimore: Johns Hopkins University Press.

de Waal, F. B. M. 1984. "Sex differences in the formation of coalitions among chimpanzees." *Ethology and Sociobiology* 5: 239–55.

Enquist, M., and O. Leimar. 1990. "The evolution of fatal fighting." *Animal Behaviour* 39 (1): 1–9.

FAWC. 1979. Farm Animal Welfare Council press statement.

Ferris, G. R., P. L. Perrewé, W. P. Anthony, and D. C. Gilmore. 2000. "Political skill at work." *Organizational Dynamics* 28 (4): 52–37.

Field, A. 2009. *Discovering Statistics Using SPSS*. London: SAGE Publications.

Fraser, A. F., and D. M. Broom. 1990. *Farm Animal Behaviour and Animal Welfare*. London: Baillere Tindall.

Freeman, H. D., S. F. Brosnan, L. M. Hopper, S. P. Lambeth, S. J. Schapiro, and S. D. Gosling. 2013. "Developing a comprehensive and comparative questionnaire for measuring personality in chimpanzees using a simultaneous top-down/bottom-up design." *American Journal of Primatology* 75 (10): 1042–53.

Freeman, H. D., and S. D. Gosling. 2010. "Personality in nonhuman primates: A review and evaluation of past research." *American Journal of Primatology* 72 (8): 653–71.

Freeman, H. D., and S. R. Ross. 2014. "The impact of atypical early histories on pet or performer chimpanzees." *PeerJ* 2: e579.

Freeman, H. D., A. Weiss, and S. R. Ross. 2016. "Atypical early histories predict lower extraversion in captive chimpanzees." *Developmental Psychobiology* 58 (4): 519–27.

Fritz, J., and S. Howell. 2001. "Captive chimpanzee social group formation." In *Special Topics in Primatology*, vol. 2, *The Care and Management of Captive Chimpanzees*, 173–204. San Antonio: American Society of Primatologists.

Fritz, P., and J. Fritz. 1979. "Resocialization of chimpanzees." *Journal of Medical Primatology* 8: 202–21.

Fultz, A. 2017. "A guide for modern sanctuaries with examples from a captive chimpanzee sanctuary." *Animals Studies Journal* 6 (2): 9–29.

Gentry, L., and S. W. Margulis. 2008. "Behavioral effects of introducing pied tamarin (*Saguinus bicolor*) to black howler monkey (*Alouatta caraya*) and white-faced saki (*Pithecia pithecia*) in a zoological park." *American Journal of Primatology* 70: 1–5.

Gilby, I. C., L. E. Eberly, and R. W. Wrangham. 2008. "Economic profitability of social predation among wild chimpanzees: Individual variation promotes cooperation." *Animal Behaviour* 75 (2): 351–60.

Gold, K. C., and T. L. Maple. 1994. "Personality assessment in the gorilla and its utility as a management tool." *Zoo Biology* 13 (5): 509–22.

Goldberg, L. R. 1990. "An alternative 'description of personality': The big-five structure." *Journal of Personality and Social Psychology* 59: 1216–29.

Goodall, J. (1971) 1999. *In the Shadow of Man*. Rev. ed. London: Phoenix.

Goodall, J. 1986. *The Chimpanzees of Gombe: Patterns of Behavior*. Cambridge, MA: Belknap Press of Harvard University Press.

Gosling, S. D. 2001. "From mice to men: What can we learn about personality from animal research?" *Psychological Bulletin* 127: 45–86.

Gottlieb, D. H., J. P. Capitanio, and B. McCowan. 2013. "Risk factors for stereotypic behavior and self-biting in rhesus macaques (*Macaca mulatta*): Animal's history, current environment, and personality." *American Journal of Primatology* 75 (10): 995–1008.

Gust, D. A., T. P. Gordon, A. R. Brodie, and H. M. McClure. 1996. "Effects of companions in modulating stress associated with new group formation in juvenile rhesus macaques." *Physiology & Behavior* 59 (4–5): 941–45.

Hamburger, L. 1988. "Introduction of two young orang-utans *Pongo pygmaeus* into an established family group." *International Zoo Yearbook* 27: 273–78.

Hampson, S. E., and L. R. Goldberg. 2006. "A first large-cohort study of personality-trait stability over the 40 years between elementary school and midlife." *Journal of Personality and Social Psychology* 91 (4): 763–79.

Heidiger, H. 1969. *Man and Animal in the Zoo*. London: Routledge and Kegan Paul.

Herrelko, E. S. 2011. "An assessment of the development of a cognitive research programme and introductions in zoo-housed chimpanzees." PhD diss., University of Stirling.

Herrelko, E. S., H. M. Buchanan-Smith, and S. J. Vick. 2015. "Perception of available space during chimpanzee introductions: Number of accessible areas is more important than enclosure size." *Zoo Biology* 34 (5): 397–405.

Herrelko, E. S., S.-J. Vick, and H. M. Buchanan-Smith. 2012. "Cognitive research in zoo-housed chimpanzees: Influence of personality and impact on welfare." *American Journal of Primatology* 74 (9): 828–40.

Hirata, S. 2007. "A note on the responses of chimpanzees (*Pan troglodytes*) to live self-images on television monitors." *Behavioural Processes* 75: 85–90.

Hopper, L. M., K. A. Cronin, and S. R. Ross, 2018. "A multi-institutional assessment of a short-form personality questionnaire for use with macaques." *Zoo Biology* 37 (5): 281–89.

Hopper, L. M., S. P. Lambeth, and S. J. Schapiro. 2012. "An evaluation of the efficacy of video displays for use with chimpanzees (*Pan troglodytes*)." *American Journal of Primatology* 74 (5): 442–49.

Hopper, L. M., S. A. Price, H. D. Freeman, S. P. Lambeth, S. J. Schapiro, and R. L. Kendal. 2014. "Influence of personality, age, sex, and estrous state on chimpanzee problem-solving success." *Animal Cognition* 17 (4): 835–47.

Howell, C. P., and S. M. Cheyne. 2019. "Complexities of using wild versus captive activity budget comparisons for assessing captive primate welfare." *Journal of Applied Animal Welfare Science* 22 (1): 78–96.

Jensen-Campbell, L. A., J. M. Knack, A. M. Waldrip, and S. D. Campbell. 2007. "Do Big Five personality traits associated with self-control influence the regulation of anger and aggression?" *Journal of Research in Personality* 41 (2): 403–24.

Joint Working Group on Refinement. 2009. "Refinements in husbandry, care and common procedures for non-human primates: Ninth report of the BVAAWF/FRAME/RSPCA/UFAW joint working group on refinement." *Laboratory Animals* 43: S1:1–47.

Judge, P. G., and F. B. M. de Waal. 1993. "Conflict avoidance among rhesus monkeys: Coping with short-term crowding." *Animal Behaviour* 46: 221–32.

King, J. E., and A. J. Figueredo. 1997. "The five-factor model plus dominance in chimpanzee personality." *Journal of Research in Personality* 31 (2): 257–71.

King, J. E., and A. Weiss. 2011. "Personality from the perspective of a primatologist." In *Personality and Temperament in Nonhuman Primates*, edited by A. Weiss, J. King, and L. Murray, 77–99. New York: Springer.

King, J. E., A. Weiss, and K. H. Farmer. 2005. "A chimpanzee (*Pan troglodytes*) analogue of cross-national generalization of personality structure: Zoological parks and an African sanctuary." *Journal of Personality* 73 (2): 389–410.

King, J. E., A. Weiss, and M. M. Sisco. 2008. "Aping humans: Age and sex effects in chimpanzee (*Pan troglodytes*) and human (*Homo sapiens*) personality." *Journal of Comparative Psychology* 122 (4): 418–27.

Kummer, H. 1971. *Primate Societies: Group Techniques of Ecological Adaptation*. Chicago: Aldine.

Loh, T. L., E. R. Larson, S. R. David, L. S. de Souza, R. Gericke, M. Gryzbek, A. S. Kough, P. W. Willink, and C. R. Knapp. 2018. "Quantifying the contribution of zoos and aquariums to peer-reviewed scientific research." *FACETS* 3: 287–99.

Majolo, B., H. M. Buchanan-Smith, and K. Morris. 2003. "Factors affecting the successful pairing of unfamiliar common marmoset (*Callithrix jacchus*) females: Preliminary results." *Animal Welfare* 12 (3): 327–38.

Massen, J. J. M., A. Antonides, A. M. K. Arnold, T. Bionda, and S. E. Koski. 2013. "A behavioral view on chimpanzee personality: Exploration tendency, persistence, boldness, and tool-orientation measured with group experiments." *American Journal of Primatology* 75: 947–58.

Massen, J. J. M., and S. E. Koski. 2014. "Chimps of a feather sit together: Chimpanzee friendships are based on homophily in personality." *Evolution and Human Behavior* 35 (1): 1–8.

Mcdonald, S. 1994. "The Detroit Zoo chimpanzees *Pan troglodytes*: Exhibit design, group composition and the process of group formation." *International Zoo Yearbook* 33 (1): 235–47.

McGrew, K. 2017. "Pairing strategies for cynomologus macaques." In *Handbook of Primate Behavioral Management*, edited by S. J. Schapiro, 255–64. Boca Raton: CRC Press.

Mellor, D. J. 2016. "Updating animal welfare thinking: Moving beyond the 'five freedoms' towards 'a life worth living.'" *Animals* 6 (3): 21–41.

Mellor, D. J., and N. J. Beausoleil. 2015. "Extending the 'Five Domains' model for animal welfare assessment to incorporate positive welfare states." *Animal Welfare* 24 (3): 241–53.

Menzel, E. W., E. S. Savage-Rumbaugh, and J. Lawson. 1985. "Chimpanzee (*Pan troglodytes*) spatial problem solving with the use of mirrors and televised equivalents of mirrors." *Journal of Comparative Psychology* 99 (2): 211–17.

Morton, F. B., P. C. Lee, and H. M. Buchanan-Smith. 2013. "Taking personality selection bias seriously in animal cognition research: A case study in capuchin monkeys (*Sapajus apella*)." *Animal Cognition* 16 (4): 677–84.

Morton, F. B., A. Weiss, H. M. Buchanan-Smith, and P. C. Lee. 2015. "Capuchin monkeys with similar personalities have higher-quality relationships independent of age, sex, kinship and rank." *Animal Behaviour* 105: 163–71.

Muller, M. N., and R. W. Wrangham. 2004. "Dominance, aggression and testosterone in wild chimpanzees: A test of the 'challenge hypothesis.'" *Animal Behaviour* 67 (1): 113–23.

Murray, L. 2011. "Predicting primate behavior from personality ratings." In *Personality and Temperament in Nonhuman Primates*, edited by A. Weiss, J. King, and L. Murray, 129–67. New York: Springer.

Neuman, G. A., S. H. Wagner, and N. D. Christiansen. 1999. "The relationship between work-team personality composition and the job performance of teams." *Group & Organization Management* 24: 28–45.

Nishida, T. 1979. "The social structure of chimpanzees of the Mahale Mountains." In *The Great Apes*, edited by D. A. Hamburg and E. R. McCown, 72–121. Menlo Park, CA: Benjamin/Cummings.

Nishida, T., N. Corp, M. Hamai, T. Hasegawa, M. Hiraiwa-Hasegawa, K. Hosaka, K. D. Hunt, N. Itoh, K. Kawanaka, A. Matsumoto-Oda, J. C. Mitani, M. Nakamura, K. Norikoshi, T. Sakamaki, L. Turner, S. Uehara, and K. Zamma. 2003. "Demography, female life history, and reproductive profiles among the chimpanzees of Mahale." *American Journal of Primatology* 59 (3): 99–121.

Noon, C. 1991. "Resocialization of a group of ex-laboratory chimpanzees, *Pan troglodytes*." *Journal of Medical Primatology* 20: 375–81.

Offer, S., and C. S. Fischer. 2018. "Difficult people: Who is perceived to be demanding in personal networks and why are they there?" *American Sociological Review* 83 (1): 111–42.

Pederson, A. K., J. E. King, and V. I. Landau. 2005. "Chimpanzee (*Pan troglodytes*) personality predicts behavior." *Journal of Research in Personality* 39 (5): 534–49.

Perlman, J. E., V. Horner, M. A. Bloomsmith, S. P. Lambeth, and S. J. Schapiro. 2010. "Positive reinforcement training, social learning, and chimpanzee welfare." In *The Mind of the Chimpanzee: Ecological and Experimental Perspectives*, edited by E. V. Lonsdorf, S. R. Ross, and T. Matsuzawa, 320–31. Chicago: University of Chicago Press.

Powell, D. M. 2010. "A framework for introduction and socialization processes for mammals." In *Wild Mammals in Captivity: Principles and Techniques*, 2nd ed., edited by D. G. Kleiman, K. V. Thompson, and C. K. Baer, 49–61. Chicago: University of Chicago Press.

Price, E. E., S. P. Lambeth, S. J. Schapiro, and A. Whiten. 2009. "A potent effect of observational learning on chimpanzee tool construction." *Proceedings of the Royal Society of London B* 276 (1671): 3377–83.

Réale, D., S. M. Reader, D. Sol, P. T. McDougall, and N. J. Dingemanse. 2007. "Integrating animal temperament within ecology and evolution." *Biological Reviews* 82 (2): 291–318.

Reamer, L. A., R. L. Haller, E. J. Thiele, H. D. Freeman, S. P. Lambeth, and S. J. Schapiro. 2014. "Factors affecting initial training success of blood glucose testing in captive chimpanzees (*Pan troglodytes*)." *Zoo Biology* 33 (3): 212–20.

Ross, S. R. 2015. *North American Regional Chimpanzee Studbook (Pan troglodytes)*. Silver Spring, MD: Association of Zoos and Aquariums.

Schapiro, S. J., M. A. Bloomsmith, L. M. Porter, and S. A. Suarez. 1996. "Enrichment effects on rhesus monkeys successively housed singly, in pairs, and in groups." *Applied Animal Behaviour Science* 48 (3–4): 159–71.

Schapiro, S. J., S. P. Lambeth, K. Rosenmaj Jacobsen, L. E. Williams, B. N. Nehete, and P. N. Nehete. 2012. "Physiological and welfare consequences of transport, relocation, and acclimatization of chimpanzees (*Pan troglodytes*)." *Applied Animal Behaviour Science* 137 (3–4): 183–93.

Schel, A. M., B. Rawlings, N. Claidière, C. Wilke, J. Wathan, J. Richardson, S. Pearson, E. S. Herrelko, A. Whiten, and K. Slocombe. 2013. "Network analysis of social changes in a captive chimpanzee community following the successful integration of two adult groups." *American Journal of Primatology* 75 (3): 254–66.

Schneider, A.-C., A. P. Melis, and M. Tomasello. 2012. "How chimpanzees solve collective action problems." *Proceedings of the Royal Society of London B* 279 (1749): 4946–54.

Seres, M. 2008. "Chimpanzee (*Pan troglodytes*) introductions and group formations in captivity: Unique individuals, damaged minds—All learning to get along." In *The 3rd HOPE International Symposium*. Tama Zoo and University of Tokyo.

Seres, M., F. Aureli, and F. B. M. de Waal. 2001. "Successful formation of a large chimpanzee group out of two preexisting subgroups." *Zoo Biology* 20 (6): 501–15.

Shrout, P. E., and J. L. Fleiss. 1979. "Intraclass correlations: Uses in assessing rater reliability." *Psychological Bulletin* 86 (2): 420–28. https://doi.org/10.1037/0033-2909.86.2.420.

Stevens, J., and L. van Elsacker. 2005. "The successful integration of an adult male chimpanzee in a complex group." In *Proceedings of the 7th Annual Symposium on Zoo Research*, edited by A. Nicklin, 165–74. London: British & Irish Association of Zoos & Aquariums.

Stevenson-Hinde, J., and M. Zunz. 1978. "Subjective assessment of individual rhesus monkeys." *Primates* 19: 473–82.

Suomi, S. J. 1997. "Early determinants of behaviour: Evidence from primate studies." *British Medical Bulletin* 53 (1): 170–84.

Tresz, H. 2011. "Successful introductions of individually housed chimpanzees into group settings." *International Zoo News* 58 (5): 300–314.

Truelove, M. A., A. L. Martin, J. E. Perlman, J. S. Wood, and M. A. Bloomsmith. 2017. "Pair housing of macaques: A review of partner selection, introduction techniques, monitoring for compatibility, and methods for long-term maintenance of pairs." *American Journal of Primatology* 79 (1): 1–15.

Uher, J., J. B. Asendorpf, and J. Call. 2008. "Personality in the behaviour of great apes: Temporal stability, cross-situational consistency and coherence in response." *Animal Behaviour* 75: 99–112.

USDA. 1991. Title 9, Animal and Animal Products. Part 3, Standards. Subpart D, Specifications for the Humane Handling, Care, Treatment, and Transportation of Nonhuman Primates, Section 3.81 (Environment enhancement to promote psychological well-being), 9 CFR 3.81.

Van Hooff, J. A. R. A. M. 1973. "A structural analysis of the social behaviour of a semi-captive group of chimpanzees." In *Social Communication and Movement, Studies of Interaction and Expression in Man and Chimpanzee*, edited by M. von Cranach and I. Vine, 75–162. London: Academic Press.

Veasey, J. S. 2017. "In pursuit of peak animal welfare; The need to prioritize the meaningful over the measurable." *Zoo Biology* 36 (6): 413–25.

Veasey, J. S., N. K. Waran, and R. J. Young. 1996. "On comparing the behaviour of zoo housed animals with wild conspecifics as a welfare indicator." *Animal Welfare* 5: 13–24.

Visalberghi, E., and J. R. Anderson. 1993. "Reasons and risks associated with manipulating captive primates' social environments." *Animal Welfare* 2: 3–15.

Watters, J. V., and D. M. Powell. 2012. "Measuring animal personality for use in population management in zoos: Suggested methods and rationale." *Zoo Biology* 31 (1): 1–12.

Watts, D. P., M. Muller, S. J. Amsler, G. Mbabazi, and J. C. Mitani. 2006. "Lethal intergroup aggression by chimpanzees in Kibale National Park, Uganda." *American Journal of Primatology* 68 (2): 161–80.

Weiss, A. 2017. "Exploring factor space (and other adventures) with the hominoid personality questionnaire." In *Personality in Nonhuman Animals*, edited by J. Vonk, A. Weiss, and S. Kuczaj. Cham: Springer.

Weiss, A., M. J. Adams, and J. E. King. 2011. "Happy orang-utans live longer lives." *Biology Letters* 7 (6): 872–74.

Weiss, A., M. J. Adams, A. Widdig, and M. S. Gerald. 2011. "Rhesus macaques (*Macaca mulatta*) as living fossils of hominoid personality and subjective well-being." *Journal of Comparative Psychology* 125: 72–83.

Weiss, A., M. C. Gartner, K. C. Gold, and T. S. Stoinski. 2013. "Extraversion predicts longer survival in gorillas: An 18-year longitudinal study." *Proceedings of the Royal Society of London B* 280 (1752): 20122231.

Weiss, A., M. Inoue-Murayama, K-W. Hong, E. Inoue, T. Udono, T. Ochiai, T. Matsuzawa, S. Hirata, and J. E. King. 2009. "Assessing chimpanzee personality and subjective well-being in Japan." *American Journal of Primatology* 71 (4): 283–92.

Weiss, A., J. E. King, and W. D. Hopkins. 2007. "A cross-setting study of chimpanzee (*Pan troglodytes*) personality structure and development: Zoological parks and Yerkes National Primate Research Center." *American Journal of Primatology* 69 (11): 1264–77.

Westergaard, G. C., M. K. Izard, J. H. Drake, S. J. Suomi, and J. D. Higley. 1999. "Rhesus macaque (*Macaca mulatta*) group formation and housing: Wounding and reproduction in a specific pathogen free (SPF) colony." *American Journal of Primatology* 49 (4): 339–47.

Whittaker, M. 2006. "Managing monkey behavior: Advancing the social management of old world monkeys." In *AZA Regional Conference Proceedings*. Chicago: AZA Annual Conference.

Wilson, M. L., N. F. Britton, and N. R. Franks. 2002. "Chimpanzees and the mathematics of battle." *Proceedings of the Royal Society of London B* 269 (1496): 1107–12.

Wilson, M. L., and R. W. Wrangham. 2003. "Intergroup relations in chimpanzees." *Annual Review of Anthropology* 32 (1): 363–92.

Wolff, H.-G., and K. Moser. 2009. "Effects of networking on career success: A longitudinal study." *Journal of Applied Psychology* 94 (1): 196–206.

Yamanashi, Y., M. Teramoto, N. Morimura, S. Hirata, M. Inoue-Murayama, and G. Idani. 2016. "Effects of relocation and individual and environmental factors on the long-term stress levels in captive chimpanzees (*Pan troglodytes*): Monitoring hair cortisol and behaviours." *PLoS One* 11 (7): e0160029.

Yerkes, R. M. 1939. "The life history and personality of the chimpanzee." *American Naturalist* 73 (745): 97–112.

24

Chimpanzee Welfare in the Context of Science, Policy, and Practice

STEPHEN R. ROSS

Introduction: The Context of Chimpanzee Welfare

Assessing and addressing the welfare considerations of chimpanzees (*Pan troglodytes*) is a daunting task. This is, in part, because virtually everything about chimpanzees is complicated. They live in complicated social groups, splitting off and coming back together in various patterns. In the wild, their surroundings are complex and multi-dimensional, and their ability to adapt from chilly mountainsides to arid desert caves only further complicates the goal of characterizing (and replicating) their environmental needs in a captive setting. And of course, chimpanzees are a species renowned for their intellect, problem-solving abilities, and emotional complexity. Indeed, *complicated* may be an understatement.

The closest model to such a comprehensive review of chimpanzee welfare issues might be Pruetz and McGrew's (2001) chapter entitled "What a Chimpanzee Needs," and more broadly, the entire volume in which that chapter appeared: Linda Brent's *Care and Management of Captive Chimpanzees* (2001), which remains essential reading for captive chimpanzee managers almost 20 years after its publication. Pruetz and McGrew (2001) espouse what's been known as a "natural-living" approach, in which the gold standard for the living conditions of captive animals is to match, as closely as possible, that which they experience as wild animals. As such, their recommendations for chimpanzee captive care are based on how wild chimpanzees socialize, travel, and eat. While critics of a natural-living approach question whether the full range of wild chimpanzee behavior is inherently necessary for their well-being (see Cronin and Ross, chapter 22 this volume; Else 2013), it's undoubtedly an important starting point for welfare considerations, especially in cases where very little is known about a given species. But after decades of studying chimpanzees in captive situations, there is now a strong foundational set of literature but one that has not been recently reviewed (Bloomsmith and Else 2005). As such, this chapter diverges from a focus on what is known about wild chimpanzees to make recommendations for captive chimpanzee care, to focus instead on what has already been revealed through studies of captive chimpanzees in order to further progress their care and management (see also Bloomsmith et al., chapter 21 this volume).

In completing this review, several other bits of information have been incorporated that will hopefully provide relevant context with which to interpret the literature. When applicable, the relevant regulatory standards or guidelines that relate to specific areas of interest will be referenced (though admittedly, this will primarily be focused on the United States), and there will be an emphasis on the degree to which such scientific findings are being implemented at this time. The intention here is that both of these will assist those seeking to implement these scientific findings that would be more difficult to interpret in isolation. Finally, there will be an overt promotion for a "bias-for-action" approach (Keeling, Alford, and Bloomsmith 1991): considering all the available scientific evidence at hand, but advocating for policy and management advancement even in the absence of such quantitative evidence.

EXPERT ADVICE

To help guide the prioritization of welfare-related issues for chimpanzees, input from a group of 20 experienced professionals in the field of captive chimpanzee care, working in zoos, sanctuaries, and research centers around the world, was gathered. These experts were requested to rank a list of welfare considerations from those they consider most important for chimpanzee welfare to that of least importance. While this is hardly a rigorously derived finding, the experts' responses provided assurance that this chapter's focus was maintained on the most relevant and impactful areas of research. Figure 24.1 displays the results of the expert poll (note that some topics are subsumed into broader categories for the purposes of this chapter's organization).

The Social Life

REGULATORY CONTEXT AND CURRENT PRACTICES

Overwhelmingly, the experts I surveyed agreed that there is nothing as important to chimpanzees as the companionship of other chimpanzees (fig. 24.1). Given this, it is unsurprising that there exists a range of guidelines that have made recommendations for minimum and target group sizes for captive

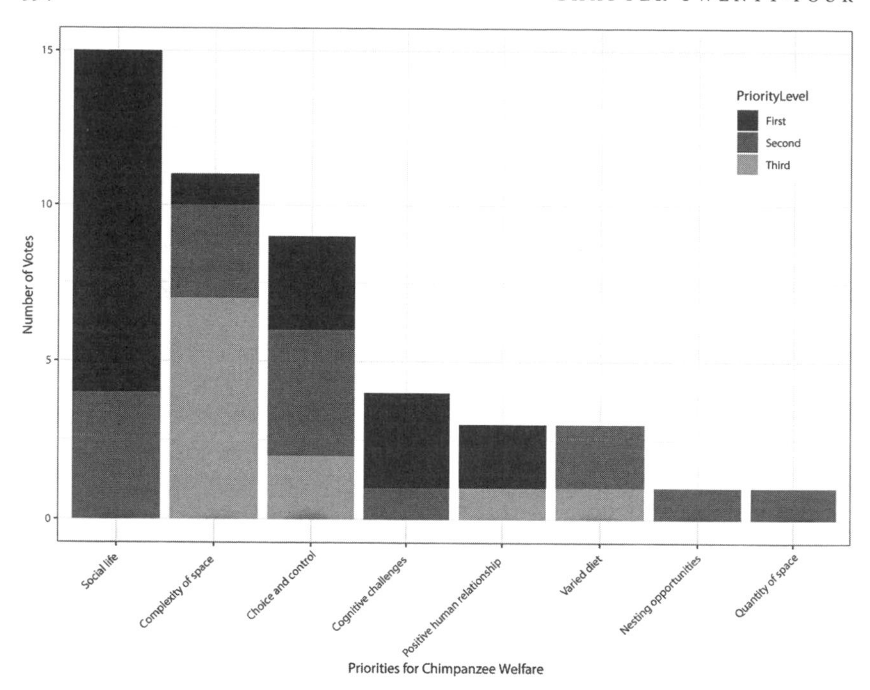

FIGURE 24.1. The result of the poll of experts (n = 20) who rated and characterized priorities for chimpanzee welfare. The stacked bars represent the number of first, second, and third place votes for each category.

chimpanzees. Many are roughly similar to those made by the National Institutes of Health (NIH) Working Group that developed the definitions of ethologically appropriate environments and which stated, "Chimpanzees must have the opportunity to live in sufficiently large, complex, multi-male, multi-female social groupings, ideally consisting of at least 7 individuals. Pairs, trios, and even small groups of 4 to 6 individuals do not provide the social complexity required to meet the social needs of this cognitively advanced species" (Dpcpsi.nih.gov 2018). There remains a gap, however, between those guidelines and the status of chimpanzee social groups in the various facilities in which they are housed globally, although there are signs that this is improving. Over a decade ago, Bloomsmith and Else (2005) reported that of 1,226 laboratory-housed chimpanzees living in six US facilities, 7.6% were individually housed. There has been significant improvement in this area more recently: Bloomsmith et al. (2019) reported only one individually housed chimpanzee living in laboratories (n = 701 chimpanzees), and although 51.9% still live in small groups, the mean group size is 4.6 individuals. In accredited US zoos, the mean group size has increased from 4.5 in 2001 to over 7.1 in 2013

(Ross unpublished data) and, at the time of writing, group sizes range from three to 17 chimpanzees (n = 251). The largest group sizes are likely found in sanctuaries, where several groups are composed of over 20 chimpanzees within US sanctuaries (Fultz, pers. comm.) and beyond the US, the size of chimpanzee groups in range-country sanctuaries can exceed 40 individuals (e.g., Chimfunshi Wildlife Orphanage Trust, Zambia: Cronin et al. 2014).

A LACK OF SOCIAL OPPORTUNITIES HAS NEGATIVE EFFECTS IN THE DEVELOPMENTAL PERIOD

The growth in group sizes for captive chimpanzees likely incurs a range of behavioral benefits though frankly there is relatively little empirical evidence that can definitively support that claim. More robust is the knowledge base behind the negative effects of *reduced* social opportunities, perhaps most so in a chimpanzee's developmental period. More than fifty years ago, scientists first investigated the effects of severe sensory and social isolation on nursery-reared chimpanzees. These experiences had profound and lasting negative effects, including the development of high levels of abnormal behavior indicative of distress and dramatic deficits in both social and cognitive abilities ability (e.g., Berkson and Mason 1964; Davenport, Rogers, and Rumbaugh 1973; Menzel, Davenport, and Rogers 1963). The consequences of these practices diminished chimpanzees' ability to appropriately navigate social groups later in life, including negative effects on their successful copulation and infant-rearing. Later studies evaluated less-severe rearing conditions, but still found substantive impacts of even moderate maternal deprivation.

Broadly speaking, chimpanzees raised in nursery settings without their mother, tend to display more behavioral deficits than those reared by their mother (e.g., Bloomsmith and Haberstroh 1995; Brent 1995; Maki, Fritz, and England 1993; Spijkerman et al. 1997) and these effects can be expressed even many decades later (Clay, Bard, and Bloomsmith 2017). While Bloomsmith et al. (2006) reported that more modern and progressive nursery-rearing practices may have even fewer negative effects than one would assume, there is broad consensus that mother rearing is unequivocally the best outcome whenever possible for young chimpanzees and their welfare (Ape TAG 2010).

The evaluation of how an appropriate social grouping (or lack thereof) potentially impacts the welfare of captive chimpanzees was extended in a recent set of studies that took a holistic approach to evaluating the importance of mother rearing. Dissatisfied with the typical approach of classifying chimpanzees' rearing on a binary scale (mother-reared versus nursery-reared) and seeking to extend the discussion of social impacts beyond rearing years, scientists with Lincoln Park Zoo's Project ChimpCARE conducted a multidisciplinary assessment of chimpanzees' early rearing experiences. They developed a continuous metric—the CHI (Chimpanzee-Human Interaction) Index—that captured the degree to which chimpanzees were exposed to both species-typical (chimpanzee) social interactions and atypical interactions with humans (often from being owned as pets, or in the entertainment trade). This comprehensive project found that chimpanzees without sufficient exposure to other chimpanzees when young tended to show lower rates of grooming and sexual behaviors later in life, even when housed in appropriate physical and social conditions (Freeman and Ross 2014). The chimpanzees' behavioral responses were mirrored by personality ratings that revealed individuals who had had more exposure to humans early in their lives scored lower on a measure of Extraversion than chimpanzees who had experienced a more naturalistic upbringing (Freeman, Weiss, and Ross 2016). An atypical rearing impacted not only the chimpanzees' behavioral well-being, but also their physiological well-being: chimpanzees that spent most of their time around humans rather than other chimpanzees when younger, expressed chronically higher stress hormone levels (cortisol), even decades after their lives as pets and performers (Jacobson et al. 2017).

SOCIAL HOUSING EFFECTS OUTSIDE THE DEVELOPMENTAL PERIOD

Given the highly social nature of chimpanzees, it is difficult to imagine that housing them individually would be without negative consequences. Brent (2001) provides a thorough review of the effects of individual housing on behavior, reproduction, maternal behavior, and cognition, as well as detailing the degree to which management and enrichment strategies might help buffer potential negative consequences. Brent also authored a study in which the behavior of adult chimpanzees was monitored following their move to individual housing for a research protocol lasting five weeks. In spite of the short-term nature of this housing protocol, these chimpanzees exhibited increased rates of stereotyped behaviors as well as decreases in environmental manipulations and vocalizations (Brent, Lee, and Eichberg 1989). Baker and Easley (1996) found no differences in rates of abnormal behavior when comparing chimpanzees living alone versus those in pairs or trios, but the severity of the abnormal behavior was greater in those that were solitary. More recent research showed that larger groups (over 7) may confer some behavioral benefits, such as increased activity and affiliation (Neal Webb, Hau, and Schapiro 2019).

WELFARE CONSEQUENCES AND CONSIDERATIONS OF SOCIAL HOUSING

While housing chimpanzees with other chimpanzees can result in dynamic and enriching environments, group-housing will be beneficial for the animals only if compatible groups of appropriate composition can be formed, and maintained over time (Herrelko, Vick, and Buchanan-Smith, chapter 23 this volume). Of course, with large, dynamic social groupings come challenges, and many of these may result in negative influences on welfare (e.g., aggression) that may influence short-term welfare as well as more long-term effects should these challenges be chronically present (e.g., an unstable group with constant tension and fighting). Importantly, Baker et al. (2000) demonstrated that chimpanzees living in large groups, while incurring the highest level of *minor* wounding, were no more likely to incur serious wounding levels than those individuals living in smaller social groups in smaller spaces.

In addition to the number of individuals, managers need to consider the composition of the group in terms of ages, sex, and other individual factors. For instance, while the presence of infants, adolescents, and other immature individuals is ubiquitous in wild chimpanzee groups, limited breeding in managed populations makes this much rarer in captivity. One study that sought evidence for welfare benefits of the presence of youngsters found no such support for that assertion (Cronin, West, and Ross 2016, see also Cronin and Ross, chapter 22 this volume). The maintenance of a successful and strategic breeding program, and the addition of young chimpanzees to social groups, has a range of demographic benefits to the population management, but there is no evidence to suggest that breeding should be done simply to infer a welfare benefit to individuals. In terms of sex ratio, many of the contemporary guidelines advocate for maintaining chimpanzees in a multi-male/multi-female social group to facilitate appropriate social behaviors, despite the historical apprehension about how males (and young males in particular) incite unrest, aggression, and wounding in the group. A multi-institutional study of chimpanzee social groups in zoos tackled that question and examined wounding rates across different types of groups (Ross et al. 2009). They found no effect on wounding related to the number of males in the group and in fact found that uni-male groups were the ones with the most wounding, supporting an earlier study with the same findings (Williams et al. 2010). To allow young males in the group to experience adolescence in a multi-male group and become part of a strong core of males, now seems the best path forward for group stability.

FUTURE DIRECTIONS

It is a bit surprising, given the clarity with which chimpanzee experts identified the social dimension as the most important to welfare, that there is not more direct evidence for the benefits of social housing on chimpanzee welfare outside the developmental period. This is likely due in part to the improvements in both zoos and research centers in augmenting group sizes beyond solitary housing and pairs and trios, meaning that the behavior of solitarily-housed chimpanzees cannot be studied. Nonetheless, there is substantial opportunity to delineate the potential benefits of varying group sizes and to provide empirical backing to recommendations and guidelines that propose appropriate group sizes. Of course, such evaluations are rife with potential difficulties as investigators must at least attempt to isolate the effects of group size with factors such as individual personalities, hierarchies, enclosure size, housing elements, and group cohesion. Clearly there is a lack of evidence that points specifically to optimum group sizes, and therefore little direction by which to guide how many chimpanzees should live together. A bias-for-action approach would suggest that managers should aim for maximizing group size within the spatial constraints of the enclosure; that is, to make the group as large and dynamic as possible without tipping the balance of social density to a situation of crowding that may lead to negative welfare effects. As such, the factors of group size and physical space are inextricably intertwined (Christman and Leone 2007) and, accordingly, I turn focus to what we know about the importance of available space for captive chimpanzees.

Space

REGULATORY CONTEXT AND CURRENT PRACTICES

From a regulatory standpoint, one of the most contentious issues is the quantification of appropriate space to provide captive animals. Despite all we know about wild chimpanzee ranging patterns, the minimum allowable floor space for an individually housed chimpanzee in the United States is just 15 square feet (4.6 m^2) (NRC 1996). As such, space requirements were at the center of deliberations about "ethologically appropriate environments" (EAE) as defined by the NIH's Council of Councils Working Group on the Use of Chimpanzees in NIH-supported Research (2013). From reviewing the literature on space use in captive chimpanzees, and after extensive discussion with chimpanzee experts, the group recommended a density of 1,000 ft^2 (304.8 m^2) per individual, but the NIH chose not to approve this recommendation

TABLE 24.1. Comparison of select standards and guidelines for necessary space for chimpanzees

	Standards and Recommendations for Chimpanzees (square feet of floor space)					
Number of Chimpanzees	*USDA Min. Standard*[1]	*Switzerland Min. Requirement*[2]	*NIH EAE Rec.*[3]	*AZA Care Manual Rec.*[4]	*GFAS Rec.*[5]	*PASA Operations Manual*[6]
5	60	1,098	1,250	2,000	5,000	13,450
10	120	1,959	2,500	7,000	6,250	26,910
15	180	2,820	3,750	12,000	11,250	40,350

[1]The United States Department of Agriculture (USDA) is the governmental agency charged with setting minimum standards for animal housing in the United States. [2]The Switzerland (2001) Minimum Requirements for the Keeping of Wild Animals is an example of European standards. [3]The National Institutes of Health (NIH)'s Council of Councils convened a working group to make recommendations to guide the development of ethologically appropriate environments (EAE) for federally owned and supported laboratory-housed chimpanzees. [4]The Association of Zoos and Aquariums (AZA) is the primary accrediting body for zoos in the United States and has published a set of recommendations for care of chimpanzees. [5]The Global Federation of Animal Sanctuaries (GFAS) has published recommendations to guide care of chimpanzees in sanctuary settings. [6]The Pan African Sanctuary Alliance (PASA) is the accrediting body for African sanctuaries housing chimpanzees and other primates and has published their care guidelines as an Operations Manual.

based on "concerns about the scientific basis for this recommendation and the expected costs of implementing it" (Else 2013). Ultimately, the NIH recommended that the minimum density of the primary living space of captive chimpanzees should be 250 ft^2 (76.2 m^2) per chimpanzee, a quarter of the original recommendation.

Table 24.1 shows the recommendations from various guidelines proposing appropriate enclosure size for socially housed chimpanzees. The disparity across these guidelines demonstrates the lack of uniformity around this issue. In practice, a relatively recent survey of chimpanzee enclosures in American zoos accredited by the Association of Zoos and Aquariums revealed that, on average, chimpanzees were housed in over 35,000 square feet of space (10,668.0 m^2), including both indoor and outdoor areas (Ross, unpublished data). Kansas City Zoo, USA, had the largest outdoor yard of the zoos surveyed, measuring approximately 3 acres (40,538.4m^2), for a single group of chimpanzees that is comprised of 12 individuals at the time of writing. Sanctuary spaces can be even larger: the largest forested habitat at Chimp Haven, USA, is approximately 5 acres (20,234.3 m^2) in size, while the four habitats at the Chimfunshi Wildlife Orphanage Trust in Zambia range from 50 to 200 acres (200,000–800,000 m^2) (van Leeuwen, Cronin, and Haun 2014). Spaces in laboratory environments are quite variable in scope. Traditional indoor-outdoor runs, which are typically used to house individuals or pairs of chimpanzees, often measure approximately 200 square feet (18.6 m^2) but there are also larger, open-topped corrals in laboratory settings that measure 4,300 square feet (400 m^2) and house larger groups of over 10 individuals (e.g., Kendal et al. 2015).

HOW MUCH SPACE IS ENOUGH?

To inform a discussion of space requirements for captive chimpanzees' well-being, one can turn to what is known about how chimpanzee behave in different spaces and how they utilize their environments. Early investigations regarding how space affects the behavior of captive chimpanzees focused primarily on animal density and crowding. Chimpanzees living in crowded conditions have been reported to adopt tension-reduction tactics in which affiliation rates are elevated in response to the potential for increased aggressive encounters (e.g., Nieuwenhuijsen and de Waal 1982), conflict avoidance strategies in which both aggressive and affiliative behaviors are reduced (e.g., Videan and Fritz 2007), or decreases in overall activity and affiliation (e.g., Aureli and de Waal 1997). The range of behavioral adaptations to crowding emphasizes the adaptable nature of this species but there are also studies in which the chimpanzees increase abnormal and anxiety-related behaviors in response to increased animal density (Duncan et al. 2013; Ross et al. 2010). As such it's difficult to conclude that smaller spaces (and therefore higher animal densities) do not have a negative effect on chimpanzee behavior and welfare.

QUALITY/COMPLEXITY OF SPACE

Studies of crowding, such as those described above, typically examine chimpanzees living in short-term housing not intended for long-term use. To extend the line of study surrounding the effects of space, several studies have compared chimpanzees living in larger, but often qualitatively different spaces (e.g., Clarke, Juno, and Maple 1982; Goff et al. 1994; Jensvold et al. 2001). In these cases, the behavioral changes following transfers were broadly characterized as positive and beneficial for the chimpanzees, even despite what short-term stress might come as the result of the transport process. Yamanashi et al. (2016) reported that while relocation of chimpanzees can result in some stress, other factors, such as those related to aggression (rearing, group type, sex), may be a more important consideration to welfare. Another study looked at the impacts of chimpanzee moves between laboratory settings and found only relatively short-lasting effects (up to 12 weeks) even after very long moves (Schapiro et al. 2012). Ross et al. (2011b) observed chimpanzees moved from a smaller, hardscape zoo exhibit to a larger, more naturalistic exhibit and found the chimpanzees showed decreases in the frequency of abnormal behaviors and visual monitoring of humans as well as evidence of a transition period in which anxiety-related behaviors decreased over the course of a year after transfer. Similarly, Neal Webb, Hau, and Schapiro (2018) found that laboratory-housed chimpanzees that were moved from a smaller area (below EAE guidelines, i.e., <76.2 m^2/chimpanzee) to a larger enclosure (above EAE guidelines) exhibited more locomotion and higher behavioral diversity scores. Importantly, this transfer was between environments that could reasonably be considered of equal environmental complexity. When chimpanzees were moved from a smaller (an arguably more complex) environment to a larger (and arguably less complex) environment, positive behavioral changes were evident as well—increases were observed in the chimpanzees' foraging and behavioral diversity, as well as decrease in their self-scratching behavior. The latter result was perhaps the more surprising of the findings; *quantity* of space seemed to have a greater influence on the chimpanzees than *quality* of space, counter to most assertions in this area.

In combination, there does seem to be at least some modest evidence of behavioral benefits to increasing space for captive chimpanzees, though Reamer et al. (2015) failed to detect behavioral differences between chimpanzees housed in differently sized spaces. A growing, though conservative, consensus is that the relationship between available space and behavioral benefits may be asymptotic (Reamer et al. 2015; Ross et al. 2011a). That is, as space increases grow larger and larger, the potential for behavioral benefit grows increasing less impactful. However, it's important to note that there is no evidence for negative effects of "too much space" nor any indication of the point at which increases in space do not result in any worthwhile benefits.

Another form of assessing the impact of improved physical space is by measuring space use and preferences for particular environmental elements. Ross et al. (2009) reported that chimpanzees expressed significant preferences to be around the meshed walls and especially corners of their enclosure, and that they generally avoided open spaces. Unsurprisingly, several studies have identified chimpanzee preferences for elevated areas (e.g., Ross and Lukas 2006; Ross et al. 2009; Traylor-Holzer and Fritz 1985). When transferred to a new, larger, and more naturalistic indoor-outdoor space, chimpanzees maintained or strengthened many of these environmental preferences and importantly, distributed their space use proportionately to what was available (Ross et al. 2011b). This suggests a "build it and they will come" paradigm; though chimpanzees' space use is relatively selective, they may expand their use of space to be proportionate to that which is available.

Outdoor areas in particular confer complexity in ways that are difficult to replicate in controlled, indoor spaces, and, as such, the provision of outdoor areas seems particularly important for chimpanzees. The evidence for the importance of outdoor spaces comes from much of the literature described previously, in which chimpanzees are moved from an indoor-only area to one in which outdoor space is also available (e.g., Jensvold et al. 2001). While the behavioral changes from these housing shifts are evident, it is difficult to distinguish the degree to which such changes were specifically the result of access to outdoors as compared to other changes in the quality and quantity of space. Ross et al. (2011a) did not find clear environmental preferences for an outdoor yard but conferred three potential benefits of such provision: the increase in available space, the creation of distinct areas, and inherently higher environmental complexity.

FUTURE DIRECTIONS

When faced with choosing between an expansive but barren environment and a dynamic but more intimate space, we might expect a chimpanzee to select the latter. However, these assertions are, in part, based on the false dichotomy that pits quality versus quantity—that managers must select one over the other—when in fact many of the largest enclosures are among the most complex as well. With large enclosures, come the inherent benefits of providing spatial choice and control as chimpanzees are able to select from a variety of physical and social "niches." Providing large spaces need not come at the expense of providing complexity and in most cases, the very largest spaces are inherently the most dynamic. More space gives the opportunity for more choice and in the next section, I transition to the importance of choice and control and how they are linked to captive chimpanzee welfare.

Opportunities for Choice and Control

One of the most prominent concepts in captive animal management and welfare research today is the importance of providing opportunities for choice and control. There is abundant research on humans that demonstrates the attraction to having agency over one's own environment and so it is not surprising that such effects are evident with nonhuman animals as well. Examples of passive choices that can be provided to chimpanzees are social choices (who to spend time with and when—the larger the social group the more choices that are afforded); spatial choices (where to spend time—large, dynamic enclosures offer more choices); and dietary choices (what to eat and when to eat).

These examples demonstrate that while social, spatial, and dietary considerations are important in and unto themselves, much of their inherent value comes from the degree of choice they imbue. More difficult at times is encouraging chimpanzees to fully exploit the choices provided to them. For example, and as discussed earlier, chimpanzees can be highly selective in their spatial use of enclosures. One study of zoo-housed chimpanzees living in a dynamic and naturalistic, indoor-outdoor facility found that they spent half of their time in just 3.2% of the space available (Ross et al. 2011b). Should these results lead us to believe that chimpanzees need only 3.2% of available space to be happy, or might they benefit simply from having the choice or opportunity to explore areas they only rarely visit? One of the few studies to empirically evaluate the value of choice was conducted at Lincoln Park Zoo, USA, and focused on chimpanzees' choice to access outdoor areas (Kurtycz, Wagner, and Ross 2014). Investigators collected data on the apes in two conditions: when they were restricted to an indoor enclosure (no choice) and when they were permitted free access to an adjacent outdoor yard (choice). Importantly, observations were recorded only when the apes remained indoors (irrespective of whether they had the option to be outside), thereby isolating the potential impacts of choice from the potential impacts of using the outdoor space. When afforded the choice to go outside, the chimpanzees demonstrated more frequent social and self-directed behaviors and lower levels of inactivity, which was interpreted as a positive response, including increased arousal, to the availability of environmental choice.

CONTROL

Like the effects of choice, there are not many experimental evaluations to directly demonstrate how the provision of control may have welfare benefits for chimpanzees. In a study focusing on the effects of providing control via a joystick-controlled computer task, Baker et al. (2001) found that chimpanzees with control of the game did not differ in their behavior from those who had access to the same display but without any ability to control the task. However, a later phase to the study showed that subjects with prior control of the tasks displayed lower frequencies of anxiety-related behavior experiencing a mild stressor. In a parallel study Bloomsmith et al. (2000) showed that chimpanzees who had control over video programming using a manual switchbox were more social and playful than chimpanzees without such control. Likewise, chimpanzees that were relegated to other's choices in video content tended to demonstrate higher rates of anxiety-related scratching behavior.

FUTURE DIRECTIONS

Captive chimpanzees, like any species living under human care, have relatively few choices to make in their lives. Social groups are relatively stagnant, physical environments often lack substantial dynamicity, and decisions about which activities are available and what foods they can eat, are made by human caregivers. This intrinsic state of reliance emphasizes the effort that managers must make to provide some of that control and those choices back to the chimpanzees whenever possible (Morimura 2006; Schapiro and Lambeth 2007). While this is straightforward conceptually, there are a myriad of practical challenges to studying choice and control. These include the management outcomes from providing choices that animals do not exercise and the scientific challenge of isolating the effects of choice from the potential benefits of providing them. Very few intervention strategies are flexible and responsive enough to facilitate adaptive choice opportunities but one such stimulus is human interaction. In the next section I explore the importance of those relationships and the opportunities and risks associated with them.

Relationships with Humans

Even in the least intensive management scenarios, where chimpanzees are given numerous choices and expansive environments, the lives of these individuals are inextricably tied to, and reliant upon, the humans that care for them. As such, the importance of a positive caretaker-chimpanzee bond is likely very influential, as has been demonstrated in many other managed species (Hemsworth, Barnett, and Coleman 1993; Hosey and Melfi 2014). Earlier, we described the often very negative impact that humans can have on chimpanzees when they are reared without appropriate exposure to conspecifics, but here we examine the influence of humans later in life and over three levels of "intensity" characterizing these interactions as (i) direct formalized interactions, such as part of positive-reinforcement training, (ii) direct informal interactions, such as spontaneous and unstructured play, and (iii) passive presence, such as the effects of unfamiliar humans (e.g., zoo visitors).

DIRECT AND TARGETED INTERACTIONS: THE IMPACT OF HUSBANDRY TRAINING

While the layperson may assume that zookeepers and caregivers routinely interact with chimpanzees through bouts of wrestling and direct contact, the fact is that for virtually all professional organizations, the most direct interaction between humans and chimpanzees comes as part of either research tasks or positive-reinforcement training (PRT)—both of which involve having the two species separated by a safe barrier, typically cage mesh. Despite the fact that physical contact is often an unnecessary aspect of these interactions, these represent the most intensive interactions with chimpanzees and the most overt potential influence of humans on chimpanzee behavior and welfare. In terms of the scope of these types of interactions, it is likely quite broad. Virtually all accredited zoos and research settings employ a regular PRT program as part of standard husbandry practices, and the prevalence of such programs is growing in sanctuaries.

Given its prevalence in various captive settings, there is substantial interest in determining whether and how training techniques affect the welfare of chimpanzees (see Perlman et al. 2010 for review). Pomerantz and Terkel (2009) reported a significant reduction in abnormal and stress-related behaviors (as well as increases in affiliation) following implementation of the training program in a zoo setting, which seemed to be especially impactful for low-ranking members of the group. The results of training programs may be best demonstrated in evaluating the potential ameliorating effects on the stress of husbandry and veterinary activities. Chimpanzees trained for voluntary cooperation for intramuscular injections of anesthesia (Lambeth et al. 2006) or blood draws (Perlman et al. 2010) showed reduced physiological measures of stress compared to those who did not participate in such training. Chimpanzees who participated in PRT were found to spend less time inactive or in solitary activities during training sessions. Likewise, these individuals showed improved social interactions outside of training sessions, such as increases in play behavior (Bloomsmith et al. 1993, 1999; Laule and Whittaker 2007).

INFORMAL INTERACTIONS

When housed in captive settings, chimpanzees may be influenced by humans in ways that are less direct or intentional than PRT sessions. Human interaction as a form of enrichment for chimpanzees is purported to have a positive impact but there are mixed results from the few studies investigating such interactions. Baker (2004) assessed the effects of positive human interaction (an additional 10 minutes per day of relaxed treat feeding and playing) on the behavior of adult lab-housed chimpanzees. When observations were conducted outside of these interaction times, they found that chimpanzees groomed each other more, showed lower levels of oral abnormal behaviors, and were less reactive to the displays of neighboring chimpanzees. The results suggest relatively small increases in simple, unstructured affiliations between humans and chimpanzees could confer benefits for these chimpanzees and should be implemented as part of regular behavioral management. The degree to which these informal interactions provides benefits that are measurably different than formal PRT sessions remains murky. In one attempt to compare these types of interactions, Bloomsmith et al. (1999) found that chimpanzees that participated in PRT training expressed decreased abnormal behaviors during the intervention compared to no such decrease for chimpanzees engaged in less formal interactions. However, in times outside the interventions, the results were the opposite, with only the non-training group showing lower rates of abnormal behavior. But one cannot assume that all positively-intentioned interactions result in positive outcomes for chimpanzees. Chelluri, Ross, and Wagner (2013) examined the potential impact of informal and unscheduled interactions between caregivers and zoo-housed chimpanzees, including play, spontaneous feeding, and other positive vocal and visual interactions performed through a mesh barrier. They found that in times that involved such interactions chimpanzees demonstrated ten times higher agonism and 50% lower prosocial interactions with other chimpanzees than in times in which no such interactions with caregivers took place. Though there was also an 18% decrease in self-directed behaviors, these well-intentioned interactions seem to produce effects that are at the least arousing and possibly disruptive.

PASSIVE INTERACTIONS

Chimpanzees may also be affected simply by the presence of humans, familiar or unfamiliar. Lambeth, Bloomsmith, and Alford (1997) observed significantly more wounding in laboratory-housed chimpanzees on weekdays than on weekends, a trend they attributed to greater staff presence and activity during the work week (i.e., familiar humans, see also Williams et al. 2010 for similar results). Considering unfamiliar humans, Maki, Alford, and Bramblett (1987) showed that the presence of unfamiliar humans was correlated with agonism in laboratory-housed chimpanzees. Zoos offer an obvious venue to study the effect of unfamiliar humans given the regular presence of visitors; however, the data, to date, are mixed. While Wood (1998) reported that zoo-housed chimpanzees show lower frequencies of species-typical feeding, playing, and grooming during busy weekends than weekdays, more recently, Bonnie, Ang, and Ross (2016) used a long-term data set to assess visitor effects on behavior and space use of zoo-housed chimpanzees and found that the chimpanzees' space use seemed completely unaffected by crowd size (i.e., they did not avoid crowds, despite visitor density). Additionally, Bonnie, Ang, and Ross reported that crowd size had no effect on the chimpanzees' behavior, demonstrating that, at least in this modern facility, the presence of unfamiliar humans had little effect on chimpanzee welfare. Exploring these questions in a sanctuary setting, where one might assume that chimpanzees are less acclimated to the presence of unfamiliar humans compared to zoos, Hansen et al. (2020) found that while the chimpanzees exhibited some small behavioral changes when unfamiliar humans were present, these were not likely associated with compromised welfare but more likely associated with the fact that the groups received food during these times.

FUTURE DIRECTIONS

There continues to be interest and study of human-animal relationships and their importance to the welfare of captive animals. While much of this research has not been conducted with chimpanzees, the outcomes of these studies should be considered. If one takes an absolute naturalistic approach to the utility of human interactions as influences on chimpanzee behavior, it may be to aspire to reduce or even eliminate them altogether. However, this approach likely underestimates the degree to which human must be a part of the lives of captive chimpanzees. While these interactions should, as much as possible, be the result of free-choice on the part of the chimpanzees, we also cannot ignore that some interactions cannot be eliminated in most settings. As such, continued study of which interactions have effects, and how best to optimize those interactions when they happen, is a worthy course of study. In addition, we need to consider the intrinsic factors that may influence human interactions such as rearing history, age, sex, and past interactions with humans. We should also explore our potential to act as a form of enrichment—we have the potential to have the desired characteristics of the very best enrichment: flexible, responsive, and engaging. In the next section I will focus on (nonhuman) enrichment and its effects on chimpanzee welfare.

Environmental Enrichment

REGULATORY CONTEXT AND CURRENT PRACTICES

In the United States (and other regions worldwide), facilities housing chimpanzees (and other primates) are required to develop, document, and adhere to a plan for environmental enhancement (USDA). While obviously well-intentioned, directives have left much open to interpretation in terms of what could or should be provided in order to best benefit captive species. Baker (2007) surveyed 22 research facilities housing primates (four with chimpanzees) and determined a huge growth in formalized enrichment programs over the past several decades including an almost ubiquitous presence of a full-time position dedicated to enrichment initiatives. There has been similar growth in the zoo and sanctuary communities, though in many cases such programs are less-formalized.

SCIENTIFIC EVALUATION OF ENRICHMENT PRACTICES

The well-established cognitive abilities of chimpanzees have motivated managers to devise enrichment devices that challenge their problem-solving abilities (i.e., "goal oriented" enrichment; reviewed in table 24.2). While such tasks are rarely replicates of actions required in natural, wild conditions, one may presume that the aims of the tasks may functionally replicate complex choices that wild chimpanzees face regularly (Clark 2017). Some tasks more directly relate to natural chimpanzee behaviors, such as extractive tool-use, and artificial representations of termite mounds are almost ubiquitous in chimpanzee enclosures these days. Such enrichment devices have proven relatively successful in reducing inactivity and increasing time foraging, though one must also consider the potential for negative consequences of cognitive tasks that are "too challenging" (Leavens et al. 2001; Wagner, Hopper, and Ross 2016; Yamanashi and Matsuzawa 2010).

In addition to goal-oriented enrichment, aimed to provide cognitive challenges for chimpanzees, a range of "manipulatable" and/or "destructible" enrichment items have been introduced and evaluated with chimpanzees (as reviewed in table 24.2). For example, Videan et al. (2005) demonstrated that destructible materials were used more by chimpanzees than non-destructible items. This is supported by other studies that have indicated that items like large sheaths of wrapping paper were more attractive than indestructible rubber kong toys (Pruetz and Bloomsmith 1992) or rigid plastic balls, which may not maintain the long-term attention of chimpanzees (Bloomsmith et al. 1990).

Several studies have also investigated the impact of adding sensory elements to the environment, such as lights and smells (reviewed across many taxa by Wells 2009). While these are relatively passive interventions—chimpanzees need not actively engage with them—there have been indications that the novelty of different sensory inputs may positively impact their behavior (as reviewed in table 24.2).

FUTURE DIRECTIONS

As evidenced by table 24.2, the impacts of environmental enrichment on chimpanzee behavior and welfare are well studied. Nonetheless, there remain questions deserving of empirical attention in order to progress this field. While we are relatively well informed about the short-term value of a variety of interventions, evidence suggests these effects are short-lived in many cases

TABLE 24.2. Published studies in which particular enrichment interventions have been studied with chimpanzees

Enrichment intervention	*Effect of enrichment*	*Number of subjects*	*Citation*
Goal-Oriented Enrichment			
Puzzleboard	Decreased aggressive, affiliative, inactive, and self-directed behavior while the device was in use. Females used more than males.	29	Brent and Eichberg (1991)
Fishing/tool-use device	Reduced inactivity by 52% and increased foraging up to 31%. Dominant individuals showed more change than subordinates.	6	Celli et al. (2003)
Tube-feeder apparatus	Chimpanzees chose to use a variety of methods to access the reward (juice) in a tube feeder.	4	Morimura (2004b)
Opaque tube maze	Overall low use of device. Rough scratching increased when device was present but decreased when it was used.	6	Clark and Smith (2013)
Artificial termite mound	Chimpanzees used a variety of tools with the mound. Younger individuals spent more time at the mound and used previously used tools.	10	Nash (1982)
Grass foraging device	Subjects used the container for 4% of time and only increased foraging when it contained additional food items. There was no evidence of habituation.	n/a	Lambeth and Bloomsmith (1994)
Beverage-serving device	Daily feeding duration increased.	5	Morimura (2004a)
Tokens	Increased rate of travel compared to baseline periods without the token exchange. Increase in locomotion was not dependent on participation in token exchange study. Activity returned to baseline within 2 hours of testing.	6	Hopper et al. (2016)
Food puzzle	Reduction in abnormal behaviors and inactivity. Aggression was induced by presence of puzzle.	18	Maki et al. (1989)
Food puzzle	Food puzzle lowered levels of agonism and abnormal behavior, and increased grooming behaviors. Individual variation was high.	20	Bloomstrand et al. (1986)
Multiple	This paper reports on a number of studies including exposure to videos and tool tasks.	n/a	Morimura (2006)
Cognitive testing via touchscreens	Repeated interest and willingness to participate in touchscreen tasks suggested that the research was enriching.	11	Herrelko, Vick, and Buchanan-Smith (2012)

(*continues*)

TABLE 24.2. (continued)

Enrichment intervention	Effect of enrichment	Number of subjects	Citation
Manipulatable and Destructible Enrichment			
Multiple manipulatable items	Destructible items used more than other types of item and were used mostly by subadults. Mother-reared, and those with indoor/outdoor access, used enrichment more than human-reared chimpanzees or those confined to indoors.	75	Videan et al. (2005)
Novel plastic toys	Novel objects decreased inactivity and self-grooming and almost eliminated abnormal behaviors. Manipulation of toys decreased over time.	4	Paquette and Prescott (1988)
Paper and rubber Kong® toys	Paper was used 27% of the time and Kong® toy was used 10% of the time but use declined over first hour of exposure.	22	Pruetz and Bloomsmith (1992)
Rigid plastic balls	Balls were used 2.5–7.1% of the time.	16	Bloomsmith et al. (1990)
Televisions, plastic balls, mirrors	No age, sex, or social group differences in use, but the chimpanzees used the television the most, then plastic balls, and mirrors were used least.	20	Brent and Stone (1996)
Various manipulatable toys	Use of toys was higher when provided one at a time. Toy use was not related to age or level of abnormal behavior.	9	Brent and Stone (1998)
Shredded paper	Decreased affiliative behavior, including clinging and tandem-walking, with the provision of shredded paper.	5	Kessel, Brent, and Walljasper (1995)
Straw and forage	Subjects spent 8.7% of observed time manipulating the straw and abnormal behaviors were reduced. No evidence of habituation was found over 9 weeks. Males foraged more than females.	13	Baker (1997)
Woodchip bedding	Subjected engaged with woodchips for 20.5% of time, spending more time in the morning than afternoon. Abnormal behavior was reduced.	16	Brent (1992)
Uprooted trees	Use of uprooted trees up to 41.9% during first days and later dropped to 3.5%. Juveniles used them more than adults.	28	Maki and Bloomsmith (1989)
Passive and Sensory Enrichment			
Essential oils	Peppermint oils increased the chimpanzees' activity.	18	Struthers and Campbell (1996)

TABLE 24.2. *(continued)*

Enrichment intervention	*Effect of enrichment*	*Number of subjects*	*Citation*
Scented cloths	No difference in the chimpanzees' use of scented and unscented cloths.	n/a	Ostrower and Brent (1997)
Mirror	When mirror gave visual access to neighboring animals, subjects used it 30% of time, resulting in increased sexual and agonistic behaviors, and decreased affiliation. When the mirror showed an empty room, agonism decreased compared to times when subjects could see neighbors with the mirror.	28	Lambeth and Bloomsmith (1992)
Videotapes	No change in behavior but viewing videotapes occupied a significant portion of activity budget.	10	Bloomsmith and Lambeth (2000)
Videotapes	Chimpanzees watched videos of species-relevant behavior as much as human-centric programing. Singly housed chimpanzees watched videotapes longer than socially housed animals.	10	Bloomsmith, Keeling, and Lambeth (1990)
Various	Novel enrichment (combination of browse, video, ice blocks, tool device, burlap sacks, mirror) was used more earlier in the day. Inactivity and abnormal behaviors increased the longer enrichment was present.	11	Wood (1998)
Goldfish in a plastic aquarium	Chimpanzees spent 3.5% of their time directing attention to the fish but novelty wore off quickly.	7	Kessel and Brent (1996)
Music	Chimpanzees were more interested in video than in audio stimuli, but audio effects persisted longer.	3	Sauquet and Llorente (2014)
Music	Instrumental music was more effective at increasing affiliation. Vocal music and slower tempos decreased agonistic behavior, especially males.	57	Videan et al. (2007)
Music	Decreased aggressive (especially for all-male groups) and active (especially for mixed-sex groups) behaviors and some changes were evident after music had ceased.	57	Fritz, Roeder, and Nelson (2003)
Music	The genre of music did not matter, but chimpanzees showed fewer active social behaviors when music was playing.	18	Wallace et al. (2017)
Colored light	Green light may have a calming effect on the behavior of captive chimpanzees.	6	Fritz, Howell, and Schwandt (1997)

and, as such, programs that are designed to foster long-term influence on behavior and welfare must be designed and importantly, evaluated.

Technology-based enrichment, not only the use of touchscreens and computers, but also adaptive and interactive physical devices, such as automated feeders and animal-controlled devices, are increasingly considered for use with captive animals (Clay et al. 2011; Egelkamp and Ross 2019; see also Martin and Adachi, chapter 8 this volume). The majority of enrichment endeavors often utilize some form of food-based incentive. Given that obesity, diabetes, and heart disease are all priority considerations in captive management of chimpanzees, further work should be done to design, implement, and evaluate enrichment for which chimpanzees are motivated without the caloric downsides and/or that increases their activity (e.g., Hopper, Shender, and Ross 2016). This is especially true for female chimpanzees, who seem disproportionally affected, in terms of body weight, by food-based enrichment programs (Brent, Bloomsmith, and Fisher 1995). Beyond the context of food-based enrichment, a healthy diet is key to chimpanzees' well-being and the next section focuses on the research related to captive chimpanzee diet.

Diet

REGULATORY CONTEXT

The importance of a varied and appropriate diet was emphasized in my survey of chimpanzee welfare experts (fig. 24.1). Collectively, they ranked it as the fourth most important consideration on the list ahead of topics that have received more research attention, such as caretaker relationships, cognitive challenge, and quantity of space. Unfortunately, there is very little literature that clearly demonstrates the welfare benefits of different diets. The AZA Chimpanzee Care Manual (Ape TAG 2010) reflects that gap in the literature, with a relatively sparse set of recommendations, including those that point to both wild chimpanzee diets and human nutritional requirements as starting points to consider. In terms of complete, published captive chimpanzee diets, there appears to be just one (Howell and Fritz 1999).

CONSIDERATION RELATING TO DIETS

Careful consideration of the nutritional composition of diets and food enrichment is necessary to counter problems with obesity. For example, diets containing a high proportion of fat and sugar, such as the "peanut butter bombs" described by Borman, Gratton-Fabbri, and Fritz (2000), may prove to be very popular with chimpanzees, but are not likely to be considered part of a healthy diet. The AZA Chimpanzee Care Manual recommends dietary fiber as very important for captive chimpanzees (Ape TAG 2010) but unfortunately the chimpanzees themselves may choose differently when given the choice. Remis (2002) tested zoo-housed chimpanzees to determine that when provided the choice, they preferred foods high in non-starch sugars and high sugar-to-fiber ratios, and low in dietary fiber (see also Huskisson et al. 2020). Nonetheless, while high-fiber foods may not be what chimpanzees want, they can still be related to improved welfare. One study has shown that providing fibrous browse was successful in reducing regurgitation and reingestion behavior in chimpanzees that were otherwise unresponsive to previous alterations to their diet (Struck et al. 2007).

Beyond *what* is fed to chimpanzees, it is also important to consider *when* and *how* they are fed. For example, while there is some evidence of pre-feeding agonism in captive chimpanzees (Howell et al. 1993), feeding chimpanzees on a more unpredictable schedule can lead to increases in species-typical behavior patterns and may confer welfare benefits (Bloomsmith and Lambeth 1995). This follows with the extant literature on how predictability is related to captive animal welfare (reviewed in Bassett and Buchanan-Smith 2007). Furthermore, cooperative feeding techniques, such as those described in Bloomsmith et al. (1994) in which a dominant individual is rewarded for not stealing subordinate's food, may be particularly useful in reducing aggression and ensuring all group members get their required diet. Considering *how* they are fed, it is important to maximize the proportion of time chimpanzees dedicate to foraging and food manipulation. To that end Bloomsmith Alford, and Maple (1988) demonstrated the effectiveness of a non-competitive feeding enrichment strategy that both reduced agonism and lowered rates of abnormal behaviors. Similarly, Yamanashi and Hayashi (2011) used a series of cognitive experiments as a form of food-based enrichment for chimpanzees and demonstrated that such interventions could create activity budgets (feeding and resting times) that are equivalent to those demonstrated by wild chimpanzees. Another relevant factor associated with the presentation of diets is the degree of novelty in food items. Visalberghi et al. (2002) examined responses to novel food items and found a high degree of inter-individual variation in terms of the willingness to consume novel foods. However, the authors concluded that the presentation of such new items was a simple means to provide enrichment and improve well-being for captive chimpanzees.

FUTURE DIRECTIONS

There is little question that there is a critical connection between the diets we provide to captive chimpanzees and their physical health. Being able to replicate the complexity and richness of diets that are consumed by great apes is a daunting task and there is increasing reason to think that inadequacies in captive diets may be related to everything from cardiac disease (McManamon and Lowenstine 2011) to the expression of oral stereotypies (Lukas 1999). Nonetheless, there are virtually no substantive empirical assessments of the effects of diet changes on captive chimpanzees, despite some promising findings in closely related species (e.g., *Gorilla gorilla*, Less et al. 2014).

Final Conclusions

The measurement, assessment, and improvement of welfare for captive chimpanzees has been the focus of dozens of studies and yet it seems we have barely scratched the surface. Chimpanzees are incredibly complicated, but it may be their inherent adaptability that has clouded our progress in determining how best to advance their care. Faced even with difficult and inappropriate social and physical conditions, chimpanzees often demonstrate an amazing resiliency, showing very few obvious signs of distress. But care should be taken to not mistake this lack of overt signals as an indication that conditions have been optimized for chimpanzees. The bias-for-action approach espoused throughout this chapter was raised in the context of improving chimpanzee care by Dr. Michale Keeling in 1991, where he described the traditional rate of change in care practices as "glacial." With this review of the current state of science related to chimpanzee, it may be time again to reinvigorate those efforts and continue to build upon the foundation of research described here. The bias-for-action approach compels a continued effort to raise the bar, even without clear, demonstrated evidence that such change will result in any observable effects. The interpretation of the collective evidence here is unequivocally to push toward more space, bigger groups, and more choices. There seems little if any indication that advancement in these areas will result in negative effects for captive chimpanzees, yet the potential benefits may be great for them. We should rely heavily on science to guide the changes we make, and to assess the effects of such changes. But equally so, we should not buffer our urgency in the face of missing data and continue to innovate and progress care practices on behalf of our closest living relatives.

Acknowledgments

Special thanks to Anne Kwiatt for constructing the extensive table on chimpanzee enrichment and to the Leo S. Guthman Fund for their support of the Lester Fisher Center for the Study and Conservation of Apes.

References

Ape TAG. 2010. *Chimpanzee (Pan troglodytes) Care Manual*. Silver Spring, MD: Association of Zoos and Aquariums.

Aureli, F., and F. B. M. de Waal. 1997. "Inhibition of social behavior in chimpanzees under high-density conditions." *American Journal of Primatology* 41 (3): 213–28.

Baker, K. C. 1997. "Straw and forage material ameliorate abnormal behaviors in adult chimpanzees." *Zoo Biology* 16 (3): 225–36.

Baker, K. C. 2004. "Benefits of positive human interaction for socially-housed chimpanzees." *Animal Welfare* 13 (2): 239–45.

Baker, K. C. 2007. "Enrichment and primate centers: Closing the gap between research and practice." *Journal of Applied Animal Welfare Science* 10 (1): 49–54.

Baker, K. C., M. L. Bloomsmith, S. Ross, S. Lambeth, and P. Noble. 2001. "Control vs. passive exposure to joystick-controlled computer tasks intended as enrichment for chimpanzees (*Pan troglodytes*)." *American Journal of Primatology* 54 (1): 64.

Baker, K. C., and S. P. Easley. 1996. "An analysis of regurgitation and reingestion in captive chimpanzees." *Applied Animal Behaviour Science* 49 (4): 403–15.

Baker, K. C., M. Seres, F. Aureli, and F. B. M. de Waal. 2000. "Injury risks among chimpanzees in three housing conditions." *American Journal of Primatology* 51 (3): 161–75.

Bassett, L., and H. M. Buchanan-Smith. 2007. "Effects of predictability on the welfare of captive animals." *Applied Animal Behaviour Science* 102 (3): 223–45

Berkson, G., and W. A. Mason. 1964. "Stereotyped behaviors of chimpanzees: Relation to general arousal and alternative activities." *Perceptual and Motor Skills* 19 (2): 635–52.

Bloomsmith, M. A., P. L. Alford, and T. L. Maple. 1988. "Successful feeding enrichment for captive chimpanzees." *American Journal of Primatology* 16 (2): 155–64.

Bloomsmith, M. A., K. C. Baker, S. P. Lambeth, S. K. Ross, and S. J. Schapiro. 2000. "Is giving chimpanzees control over environmental enrichment a good idea." In *The Apes: Challenges for the 21st Century, Conference Proceedings*, 88–89. Brookfield, IL: Chicago Zoological Society.

Bloomsmith, M. A., K. C. Baker, S. K. Ross, and S. P. Lambeth. 1999. "Comparing animal training to non-training human interaction as environmental enrichment for chimpanzees." *American Journal of Primatology* 49 (1): 35–36.

Bloomsmith, M. A., K. C. Baker, S. R. Ross, and S. P. Lambeth. 2006. "Early rearing conditions and captive chimpanzee behavior: Some surprising findings." In *Nursery Rearing of Nonhuman Primates in the 21st Century*, edited by G. P. Sackett, G. Ruppenthal, and K. Elias, 289–312. Boston: Springer.

Bloomsmith, M. A., A. Clay, S. Lambeth, C. Lutz, S. Breaux, M. Lammey, A. Franklin, K. Neu, J. Perlman, L. Reamer, M. C. Mareno, S. J. Schapiro, M. Vazquez, and S. Bourgeois. 2019. "A survey of behavioral indices of welfare in research chimpanzees (*Pan troglodytes*) in the United States." *Journal of the American Association for Laboratory Animal Science* 58 (2): 160–77.

Bloomsmith, M. A., and J. G. Else. 2005. "Behavioral management of chimpanzees in biomedical research facilities: The state of the science." *ILAR Journal* 46 (2): 192–201.

Bloomsmith, M. A., T. W. Finlay, J. J. Merhalski, and T. L. Maple. 1990. "Rigid plastic balls as enrichment devices for captive chimpanzees." *Laboratory Animal Science* 40 (3): 319–22.

Bloomsmith, M. A., and M. D. Haberstroh. 1995. "Effect of early social experience on the expression of abnormal behavior among juvenile chimpanzees." *American Journal of Primatology* 36 (11).

Bloomsmith, M. A., M. E. Keeling, and S. P. Lambeth. 1990. "Videotapes: Environmental enrichment for singly housed chimpanzees." *Lab Animal* 19 (1): 42–46.

Bloomsmith, M. A., and S. P. Lambeth. 1995. "Effects of predictable versus unpredictable feeding schedules on chimpanzee behavior." *Applied Animal Behaviour Science* 44 (1): 65–74.

Bloomsmith, M. A., and S. P. Lambeth. 2000. "Videotapes as enrichment for captive chimpanzees." *Zoo Biology* 19 (6): 541–51.

Bloomsmith, M. A., S. P. Lambeth, G. E. Laule, and R. H. Thurston. 1993. "Training as environmental enrichment for chimpanzees." *American Journal of Primatology* 30: 299.

Bloomsmith, M. A., G. E. Laule, P. L. Alford, and R. H. Thurston. 1994. "Using training to moderate chimpanzee aggression during feeding." *Zoo Biology* 13 (6): 557–66.

Bloomstrand, M., K. Riddle, P. Alford, and T. L. Maple. 1986. "Objective evaluation of a behavioral enrichment device for captive chimpanzees (*Pan troglodytes*)." *Zoo Biology* 5 (3): 293–300.

Bonnie, K. E., M. Y. Ang, and S. R. Ross. 2016. "Effects of crowd size on exhibit use by and behavior of chimpanzees (*Pan troglodytes*) and Western lowland gorillas (*Gorilla gorilla*) at a zoo." *Applied Animal Behaviour Science* 178: 102–10.

Borman, R., L. Gratton-Fabbri, and J. Fritz. 2000. "Peanut butter bombs: An enrichment device for captive chimpanzees." *Newsletter of the Primate Foundation of Arizona* 8 (4): 4–6.

Brent, L. 1992. "Woodchip bedding as enrichment for captive chimpanzees in an outdoor enclosure." *Animal Welfare* 1 (3): 161–70.

Brent, L. 1995. "Feeding enrichment and body weight in captive chimpanzees." *Journal of Medical Primatology* 24 (1): 12–16.

Brent, L. 2001. "Care and management of captive chimpanzees." Paper presented at the American Society of Primatologist Conference, San Antonio, TX.

Brent, L., M. A. Bloomsmith, and S. D. Fisher. 1995. "Factors determining tool-using ability in two captive chimpanzee (*Pan troglodytes*) colonies." *Primates* 36 (2): 265–74.

Brent, L., and J. W. Eichberg. 1991. "Primate puzzleboard: A simple environmental enrichment device for captive chimpanzees." *Zoo Biology* 10: 353–60.

Brent, L., D. R. Lee, and J. W. Eichberg. 1989. "The effects of single caging on chimpanzee behavior." *Laboratory Animal Science* 31: 81–85.

Brent, L., and A. M. Stone. 1996. "Long term use of televisions, balls, and mirrors as enrichment for paired and singly housed chimpanzees." *American Journal of Primatology* 39 (2): 139–45.

Brent, L., and A. Stone. 1998. "Destructible toys as enrichment for captive chimpanzees." *Journal of Applied Animal Welfare Science* 1 (1): 5–14.

Celli, M. L., M. Tomonaga, T. Udono, M. Teramoto, and N. Kunimaru. 2003. "Tool use task as environmental enrichment for captive chimpanzees." *Applied Animal Behaviour Science* 81 (2): 171–82.

Chelluri, G. I., S. R. Ross, and K. E. Wagner. 2013. "Behavioral correlates and welfare implications of informal interactions between caretakers and zoo-housed chimpanzees and gorillas." *Applied Animal Behaviour Science* 147 (3): 306–15.

Christman, M. C., and E. H. Leone. 2007. "Statistical aspects of the analysis of group size effects in confined animals." *Applied Animal Behaviour Science* 103 (3): 265–83.

Clark, F. E. 2017. "Cognitive enrichment and welfare: Current approaches and future directions." *Animal Behavior and Cognition* 4 (1): 52–71.

Clark, F. E., and L. J. Smith. 2013. "Effect of a cognitive challenge device containing food and non-food rewards on chimpanzee well-being." *American Journal of Primatology* 75: 807–16.

Clarke, A. S., C. J. Juno, and T. L. Maple. 1982. "Behavioral effects of a change in the physical environment: A pilot study of captive chimpanzees." *Zoo Biology* 1 (4): 371–80.

Clay, A. W., K. A. Bard, and M. A. Bloomsmith. 2017. "Effects of sex and early rearing condition on adult behavior, health, and well-being in captive chimpanzees (*Pan troglodytes*)." *Behavioural Processes* 156: 58–76.

Clay, A. W., B. M. Perdue, D. E. Gaalema, F. L. Dolins, and M. A. Bloomsmith. 2011. "The use of technology to enhance zoological parks." *Zoo Biology* 30 (5): 487–97.

Council of Councils. 2013. *Council of Councils Working Group on the Use of Chimpanzees in NIH-Supported Research Report*. Bethesda, MD: National Institutes of Health. https://dpcpsi.nih.gov/sites/default/files/FNL_Report_WG_Chimpanzees_0.pdf.

Cronin, K. A., E. J. van Leeuwen, V. Vreeman, and D. B. Haun. 2014. "Population-level variability in the social climates of four chimpanzee societies." *Evolution and Human Behavior* 35 (5): 389–96.

Cronin, K. A., V. West, and S. R. Ross. 2016. "Investigating the relationship between welfare and rearing young in captive chimpanzees (*Pan troglodytes*)." *Applied Animal Behaviour Science* 181: 166–72.

Davenport, R. K., C. M. Rogers, and D. M. Rumbaugh. 1973. "Long-term cognitive deficits in chimpanzees associated with early impoverished rearing." *Developmental Psychology* 9 (3): 343.

Duncan, L. M., M. A. Jones, M. van Lierop, and N. Pillay. 2013. "Chimpanzees use multiple strategies to limit aggression and stress during spatial density changes." *Applied Animal Behaviour Science* 147 (1): 159–71.

Egelkamp, C. L., and S. R. Ross. 2019. "A review of zoo-based cognitive research using touch-screen interfaces." *Zoo Biology* 38 (2): 220–35.

Else, J. G. 2013. "A review of literature and animal welfare/ regulatory requirements and guidance pertaining to the space density needs of captive research chimpanzees." Unpublished manuscript.

Freeman, H. D., and S. R. Ross. 2014. "The impact of atypical early histories on pet or performer chimpanzees." *PeerJ* 2: e579.

Freeman, H. D., A. Weiss, and S. R. Ross. 2016. "Atypical early histories predict lower extraversion in captive chimpanzees." *Developmental Psychobiology* 58 (4): 519–27.

Fritz, J., S. M. Howell, and M. L. Schwandt. 1997. "Colored light as environmental enrichment for captive chimpanzees (*Pan troglodytes*)." *Laboratory Primate Newsletter* 36: 1–4.

Fritz, J., E. Roeder, and C. Nelson. 2003. "A stereo music system as environmental enrichment for captive chimpanzees." *Lab Animal* 32 (10): 31.

Goff, C., S. M. Howell, J. Fritz, and B. Nankivell. 1994. "Space use and proximity of captive chimpanzee (*Pan troglodytes*) mother/offspring pairs." *Zoo Biology* 13 (1): 61–68.

Hansen, B. K., A. L. Fultz, L. M. Hopper, and S. R. Ross. 2020. "Understanding the effects of public programs on sanctuary-housed chimpanzees." *Anthrozoös* 33 (4): 481–95.

Hemsworth, P. H., J. L. Barnett, and G. J. Coleman. 1993. "The human-animal relationship in agriculture and its consequences for the animal." *Animal Welfare* 2 (1): 33–51.

Herrelko, E. S., S. J. Vick, and H. M. Buchanan-Smith. 2012. "Cognitive research in zoo-housed chimpanzees: Influence of personality and impact on welfare." *American Journal of Primatology* 74 (9): 828–40.

Hopper, L. M., M. A. Shender, and S. R. Ross. 2016. "Behavioral research as physical enrichment for captive chimpanzees." *Zoo Biology* 35 (4): 293–97.

Hosey, G., and V. Melfi. 2014. "Human-animal interactions, relationships and bonds: A review and analysis of the literature." *International Journal of Comparative Psychology* 27 (1).

Howell, S., and J. Fritz. 1999. "The nuts and bolts of captive chimpanzee diets and food as enrichment: A survey." *Journal of Applied Animal Welfare Science* 2 (3): 205–15.

Howell, S. M., M. Matevia, J. Fritz, L. Nash, and S. Maki. 1993. "Pre-feeding agonism and seasonality in captive groups of chimpanzees (*Pan troglodytes*)." *Animal Welfare* 2 (2): 153–63.

Huskisson, S. M., S. L. Jacobson, C. L. Egelkamp, S. R. Ross, and L. M. Hopper. 2020. "Using a touchscreen paradigm to evaluate food preferences and response to novel photographic stimuli of food in three primate species (*Gorilla gorilla gorilla*, *Pan troglodytes*, and *Macaca fuscata*)." *International Journal of Primatology*, online first.

Jacobson, S. L., H. D. Freeman, R. M. Santymire, and S. R. Ross. 2017. "Atypical experiences of captive chimpanzees (*Pan troglodytes*) are associated with higher hair cortisol concentrations as adults." *Royal Society Open Science* 4 (12): 170932.

Jensvold, M. L. A., C. M. Sanz, R. S. Fouts, and D. H. Fouts. 2001. "Effect of enclosure size and complexity on the behaviors of captive chimpanzees (*Pan troglodytes*)." *Journal of Applied Animal Welfare Science* 4 (1): 53–69.

Keeling, M. E., P. L. Alford, and M. A. Bloomsmith. 1991. "Decision analysis for developing programs of psychological well-being: A bias-for-action approach." In *Through the Looking Glass: Issues of Psychological Well-Being in Captive Nonhuman Primates*, edited by M. A. Novak and A. J. Peto, 57–68. Washington, DC: American Psychological Association.

Kendal, R., L. M. Hopper, A. Whiten, S. F. Brosnan, S. P. Lambeth, S. J. Schapiro, and W. Hoppitt. 2015. "Chimpanzees copy dominant and knowledgeable individuals: Implications for cultural diversity." *Evolution and Human Behavior* 36 (1): 65–72.

Kessel, A., and L. Brent. 1996. "Goldfish as enrichment for singly housed chimpanzees." *Animal Technology* 47: 1–8.

Kessel, A. L., L. Brent, and T. Walljasper. 1995. "Shredded paper as enrichment for infant chimpanzees." *Laboratory Primate Newsletter* 34: 4–4.

Kurtycz, L. M., K. E. Wagner, and S. R. Ross. 2014. "The choice to access outdoor areas affects the behavior of great apes." *Journal of Applied Animal Welfare Science* 17 (3): 185–97.

Lambeth, S. P., and M. A. Bloomsmith. 1992. "Mirrors as enrichment for captive chimpanzees (*Pan troglodytes*)." *Laboratory Animal Science* 42 (3): 261–66.

Lambeth, S. P., and M. A. Bloomsmith. 1994. "A grass foraging device for captive chimpanzees (*Pan troglodytes*)." *Animal Welfare* 3 (1): 13–24.

Lambeth, S. P., M. A. Bloomsmith, and P. L. Alford. 1997. "Effects of human activity on chimpanzee wounding." *Zoo Biology* 16 (4): 327–33.

Lambeth, S. P., J. Hau, J. E. Perlman, M. Martino, and S. J. Schapiro. 2006. "Positive reinforcement training affects hematologic and serum chemistry values in captive chimpanzees (*Pan troglodytes*)." *American Journal of Primatology* 68 (3): 245–56.

Laule, G., and M. Whittaker. 2007. "Enhancing nonhuman primate care and welfare through the use of positive reinforcement training." *Journal of Applied Animal Welfare Science* 10 (1): 31–38.

Leavens, D. A., F. Aureli, W. D. Hopkins, and C. W. Hyatt. 2001. "Effects of cognitive challenge on self-directed behaviors by chimpanzees (*Pan troglodytes*)." *American Journal of Primatology* 55 (1): 1–14.

Less, E. H., K. E. Lukas, R. Bergl, R. Ball, C. W. Kuhar, S. R. Lavin, M. A. Raghanti, J. Wensvoort, M. A. Willis, and P. M. Dennis. 2014. "Implementing a low-starch biscuit-free diet in zoo gorillas: The impact on health." *Zoo Biology* 33 (1): 74–80.

Lukas, K. E. 1999. "A review of nutritional and motivational factors contributing to the performance of regurgitation and reingestion in captive lowland gorillas (*Gorilla gorilla gorilla*)." *Applied Animal Behaviour Science* 63 (3): 237–49.

Maki, S., P. L. Alford, M. A. Bloomsmith, and J. Franklin. 1989. "Food puzzle device simulating termite fishing for captive chimpanzees (*Pan troglodytes*)." *American Journal of Primatology* Supplement 1: 71–78.

Maki, S., P. L. Alford, and C. Bramblett. 1987. "The effects of unfamiliar humans on aggression in captive chimpanzee groups." *American Journal of Primatology* 12 (3): 358.

Maki, S., and M. A. Bloomsmith. 1989. "Uprooted trees facilitate the psychological well-being of captive chimpanzees." *Zoo Biology* 8: 79–87.

Maki, S., J. Fritz, and N. England. 1993. "An assessment of early differential rearing conditions on later behavioral development in captive chimpanzees." *Infant Behavior & Development* 16 (3): 373–81.

McManamon, R., and L. J. Lowenstine. 2011. "Cardiovascular disease in great apes." *Zoo and Wild Animal Medicine* 7: 408–15.

Menzel, E. W., R. K. Davenport, and C. M. Rogers. 1963. "Effects of environmental restriction upon the chimpanzee's responsiveness in novel situations." *Journal of Comparative and Physiological Psychology* 56 (2): 329.

Morimura, N. 2004a. "A note on effects of a daylong feeding enrichment program for chimpanzees (*Pan troglodytes*)." *Applied Animal Behaviour Science* 106 (1): 178–83.

Morimura, N. 2004b. "A note on enrichment for spontaneous tool use by chimpanzees (*Pan troglodytes*)." *Applied Animal Behaviour Science* 82: 241–47.

Morimura, N. 2006. "Cognitive enrichment in chimpanzees: An approach of welfare entailing an animal's entire resources." In *Cognitive Development in Chimpanzees*, edited by T. Matsuzawa, M. Tomonaga, and M. Tanaka, 368–91. Tokyo: Springer.

Nash, V. J. 1982. "Tool use by captive chimpanzees at an artificial termite mound." *Zoo Biology* 1 (3): 211–221.

National Research Council. 1996. *Guide for the Care and Use of Laboratory Animals*. 7th ed. Washington, DC: National Academies Press.

Neal Webb, S. J., J. Hau, and S. J. Schapiro. 2018. "Captive chimpanzee (*Pan troglodytes*) behavior as a function of space per animal and enclosure type." *American Journal of Primatology* 80 (3): e22749.

Neal Webb, S. J., J. Hau, and S. J. Schapiro. 2019. "Does group size matter? Captive chimpanzee (*Pan troglodytes*) behavior as a function of group size and composition." *American Journal of Primatology* 81 (1): e22947.

Nieuwenhuijsen, K., and F. B. M. de Waal. 1982. "Effects of spatial crowding on social behavior in a chimpanzee colony." *Zoo Biology* 1 (1): 5–28.

Ostrower, S., and L. Brent. 1997. "Olfactory enrichment for captive chimpanzees: Response to different odors." *Laboratory Primate Newsletter* 36 (1): 8–10.

Paquette, D., and J. Prescott. 1988. "Use of novel objects to enhance environments of captive chimpanzees." *Zoo Biology* 7 (1): 15–23.

Perlman, J. E., V. Horner, M. A. Bloomsmith, S. P. Lambeth, and S. J. Schapiro. 2010. "Positive reinforcement training, social learning, and chimpanzee welfare." In *The Mind of the Chimpanzee: Ecological and Experimental Perspectives*, edited by E. V. Lonsdorf, S. R. Ross, and T. Matsuzawa, 320–31. Chicago: University of Chicago Press.

Pomerantz, O., and J. Terkel. 2009. "Effects of positive reinforcement training techniques on the psychological welfare of zoo-housed chimpanzees (*Pan troglodytes*)." *American Journal of Primatology* 71 (8): 687–95.

Pruetz, J. D., and M. A. Bloomsmith. 1992. "Comparing two manipulable objects as enrichment for captive chimpanzees." *Animal Welfare* 1 (2): 127–37.

Pruetz, J. D., and W. C. McGrew. 2001. "What does a chimpanzee need? Using natural behavior to guide the care and management of captive populations." In *Care and Management of Captive Chimpanzees*, edited by L. Brent, 17–37. San Antonio: CRC Press, American Society of Primatologists.

Reamer, L., C. F. Talbot, L. M. Hopper, M. C. Mareno, K. Hall, S. F. Brosnan, S. P. Lambeth, and S. J. Schapiro. 2015. "Assessing quantity of space for captive chimpanzee welfare." *American Journal of Primatology* 77: 84–85.

Remis, M. J. 2002. "Food preferences among captive western gorillas (*Gorilla gorilla gorilla*) and chimpanzees (*Pan troglodytes*)." *International Journal of Primatology* 23 (2): 231–49.

Ross, S. R., M. A. Bloomsmith, T. L. Bettinger, and K. E. Wagner. 2009. "The influence of captive adolescent male chimpanzees on wounding: Management and welfare implications." *Zoo Biology* 28 (6): 623–34.

Ross, S. R., S. Calcutt, S. J. Schapiro, and J. Hau. 2011a. "Space use selectivity by chimpanzees and gorillas in an indoor–outdoor enclosure." *American Journal of Primatology* 73 (2): 197–208.

Ross, S. R., and K. E. Lukas. 2006. "Use of space in a non-naturalistic environment by chimpanzees (*Pan troglodytes*) and lowland gorillas (*Gorilla gorilla gorilla*)." *Applied Animal Behaviour Science* 96 (1): 143–52.

Ross, S. R., S. J. Schapiro, J. Hau, and K. E. Lukas. 2009. "Space use as an indicator of enclosure appropriateness: A novel measure of captive animal welfare." *Applied Animal Behaviour Science* 121 (1): 42–50.

Ross, S. R., K. E. Wagner, S. J. Schapiro, and J. Hau. 2010. "Ape behavior in two alternating environments: Comparing exhibit and short-term holding areas." *American Journal of Primatology* 72 (11): 951–59.

Ross, S. R., K. E. Wagner, S. J. Schapiro, J. Hau, and K. E. Lukas. 2011b. "Transfer and acclimatization effects on the behavior of two species of African great ape (*Pan troglodytes* and *Gorilla gorilla gorilla*) moved to a novel and naturalistic zoo environment." *International Journal of Primatology* 32 (1): 99–117.

Sauquet, T., and M. Llorente. 2014. "Do chimpanzees like videos and music? Sensorial stimulation and its impact on the welfare of chimpanzees at the Fundació Mona (Girona, Spain)." *Folia Primatologica* 85 (1): 55.

Schapiro, S. J., and S. P. Lambeth. 2007. "Control, choice, and assessments of the value of behavioral management to nonhuman primates in captivity." *Journal of Applied Animal Welfare Science* 10 (1): 39–47.

Schapiro, S. J., S. P. Lambeth, K. R. Jacobsen, L. E. Williams, B. N. Nehete, and P. N. Nehete. 2012. "Physiological and welfare consequences of transport, relocation, and acclimatization of chimpanzees (*Pan troglodytes*)." *Applied Animal Behaviour Science* 137 (3): 183–93.

Spijkerman, R. P., J. A. Van Hooff, H. Dienske, and W. Jens. 1997. "Differences in subadult behaviors of chimpanzees living in peer groups and in a family group." *International Journal of Primatology* 18 (3): 439–54.

Struck, K., E. N. Videan, J. Fritz, and J. Murphy. 2007. "Attempting to reduce regurgitation and reingestion in a captive chimpanzee through increased feeding opportunities: A case study." *Lab Animal* 36 (1): 35.

Struthers, E., and J. Campbell. 1996. "Scent-specific behavioral response to olfactory enrichment in captive chimpanzees (*Pan troglodytes*)." Paper presented at the XVth Congress of the International Primatological Society Proceedings, Madison, WI.

Traylor-Holzer, K., and P. Fritz. 1985. "Utilization of space by adult and juvenile groups of captive chimpanzees (*Pan troglodytes*)." *Zoo Biology* 4 (2): 115–27.

van Leeuwen, E. J., K. A. Cronin, and D. B. Haun. 2014. "A group-specific arbitrary tradition in chimpanzees (*Pan troglodytes*)." *Animal Cognition* 17 (6): 1421–25.

Videan, E. N., and J. Fritz. 2007. "Effects of short and long-term changes in spatial density on the social behavior of captive chimpanzees (*Pan troglodytes*)." *Applied Animal Behaviour Science* 102 (1): 95–105.

Videan, E. N., J. Fritz, S. Howell, and J. Murphy. 2007. "Effects of two types and two genre of music on behavior in captive chimpanzees (*Pan troglodytes*)." *Journal of the American Association for Laboratory Animal Science* 46 (1): 66–70.

Videan, E. N., J. Fritz, M. L. Schwandt, H. F. Smith, and S. Howell. 2005. "Controllability in environmental enrichment for captive chimpanzees (*Pan troglodytes*)." *Journal of Applied Animal Welfare Science* 8 (2): 117–30.

Visalberghi, E., M. Myowa Yamakoshi, S. Hirata, and T. Matsuzawa. 2002. "Responses to novel foods in captive chimpanzees." *Zoo Biology* 21 (6): 539–48.

Wagner, K. E., L. M. Hopper, and S. R. Ross. 2016. "Asymmetries in the production of self-directed behavior by chimpanzees and gorillas during a computerized cognitive test." *Animal Cognition* 19 (2): 343–50.

Wallace, E. K., D. Altschul, K. Koèrfer, B. Benti, A. Kaeser, S. P. Lambeth, B. M. Waller, and K. Slocombe. 2017. "Is music enriching for group-housed captive chimpanzees (*Pan troglodytes*)?" *PLoS One* 12 (3): e0172672.

Wells, D. L. 2009. "Sensory stimulation as environmental enrichment for captive animals: A review." *Applied Animal Behaviour Science* 118 (1): 1–11.

Williams, R. C., L. T. Nash, C. J. Scarry, E. N. Videan, and J. Fritz. 2010. "Factors affecting wounding aggression in a colony of captive chimpanzees (*Pan troglodytes*)." *Zoo Biology* 29 (3): 351–64.

Wood, W. 1998. "Interactions among environmental enrichment, viewing crowds, and zoo chimpanzees (*Pan troglodytes*)." *Zoo Biology* 17: 211–30.

Yamanashi, Y., and M. Hayashi. 2011. "Assessing the effects of cognitive experiments on the welfare of captive chimpanzees (*Pan troglodytes*) by direct comparison of activity budget between wild and captive chimpanzees." *American Journal of Primatology* 73 (12): 1231–38.

Yamanashi, Y., and T. Matsuzawa. 2010. "Emotional consequences when chimpanzees (*Pan troglodytes*) face challenges: Individual differences in self-directed behaviours during cognitive tasks." *Animal Welfare* 19 (1): 25–30.

Yamanashi, Y., M. Teramoto, N. Morimura, S. Hirata, M. Inoue-Murayama, and G. I. Idani. 2016. "Effects of relocation and individual and environmental factors on the long-term stress levels in captive chimpanzees (*Pan troglodytes*): Monitoring hair cortisol and behaviors." *PLoS One* 11 (7): e0160029.

PART EIGHT

Conserving Chimpanzees

25

Chimpanzee Conservation: What We Know, What We Do Not Know, and Ways Forward

COLIN A. CHAPMAN, KIM VALENTA, SARAH BORTOLAMIOL, SAM K. MUGUME, AND MENG YAO

Introduction

Biodiversity is being lost at an accelerating rate, with current extinction rates approximately 1,000 times higher than background rates observed in the fossil record (Pimm et al. 2014). Recent estimates suggest that 11,000–58,000 species are lost annually and that extant vertebrate species have declined in abundance by approximately 25% since 1970 (Dirzo et al. 2014). Humans are clearly responsible for this accelerating loss of biodiversity by causing habitat conversion, climate change, the spread of exotic species, and wildlife overexploitation (Dirzo et al. 2014; Laurance et al. 2012; Ripple et al. 2015). This has led to almost 50% of the world's primate species being at risk of extinction (Estrada 2013; Estrada et al. 2017; Mittermeier et al. 2009), and 14.5% being critically endangered (IUCN Redlist database 2020).

The endangerment of chimpanzees (*Pan troglodytes*) is particularly important because of their iconic status and phylogenetic closeness to humans. It is estimated that chimpanzee populations have experienced a significant reduction in the past 20 to 30 years and the overall population reduction over three generations is estimated to exceed 50% (Walsh et al. 2003), hence qualifying this taxon for endangered status. It is officially estimated that only half a million chimpanzees exist in the wild, with 65,000 in West Africa, 9,000 in Nigeria and Cameroon, 140,000 in Central Africa, and 256,000 in the Democratic Republic of Congo above the Congo River and in East Africa (Humle et al. 2016). However, other estimates suggest much lower numbers (Oates 1996; Sop et al. 2015) and the level of endangerment and the nature of the threats differ among regions and across subspecies (Kühl et al. 2017).

Chimpanzees occur at low densities and they are very difficult to habituate and follow for scientific observations due to their fission-fusion social system and large home range (Bertolani and Boesch 2008; Boesch and

Boesch-Achermann 2000). As a result, to evaluate conservation threats to chimpanzees it is useful to take a comparative approach and contrast them to other diurnal primates that inhabit the same forest. These other primates are easier to study and census, thus there is much more data on their response to adverse conditions people create. Therefore, in our evaluation of chimpanzee conservation, we will draw on a number of examples of how monkeys respond to threats that similarly affect chimpanzees. However, it is important to acknowledge that chimpanzees represent something special. They can be considered to fill important roles as umbrella, flagship, or phylogenetically important species (Hartel et al., chapter 26 this volume; Wrangham et al. 2008). As such, many would argue that they deserve special attention. Directing conservation efforts to chimpanzees can make great advances at fostering public awareness and raising funding for conservation. Chimpanzee conservation efforts must include a myriad of activities, including protecting their habitat, decreasing bushmeat hunting, and improving park-people interactions, thus conserving chimpanzees conserves their habitats and the plants and animals therein.

The objective of our review is to document current threats to chimpanzee populations in such a way as to illustrate what the scientific and conservation communities know and what they do not know. In doing so, we hope to illustrate the way forward for both communities. There are likely thousands of publications, government documents, reports from NGOs, and theses published every few years on chimpanzee abundance and distribution. Therefore, this is not meant to be an exhaustive review, but rather we focus on major issues. We concentrate our evaluation on the populations in East Africa, particularly Uganda, as this is an area we work in and know well; however, we make comparisons with threats facing other chimpanzee populations, as well as other African primate species, to put chimpanzees' conservation concerns in context.

Deforestation

Threats to chimpanzees come in many forms, but likely one of the most significant is habitat loss (Morgan et al., chapter 27 this volume). Between 2000 and 2012, it is estimated that 2.3 million km2 of forest was lost globally, and in the tropics rates of loss increased by 2,101 km^2 per year (Hansen et al. 2013). To put this in perspective, the global annual loss is approximately the size of Mexico (1.96 million km^2). The loss is greatest in South America and Africa. However, estimates of deforestation vary greatly. For example, a recent study estimates a 62% acceleration in deforestation in the humid tropics between

TABLE 25.1. Landsat estimates of forest area (106 ha) in 1990, 2000, and 2010 for different countries in Tropical Africa, with data from Kim, Sexton, and Townshend (2015)

Country	*1990*	*2000*	*2010*
Cameroon	20.32	20.21	19.88
Congo	23.88	23.66	23.43
Democratic Republic of Congo	153.23	152.2	147.93
Equatorial Guinea	2.59	2.56	2.54
Gabon	23.38	22.92	22.99
Liberia	7.46	7.27	7.23
Madagascar	8.93	8.55	7.58
Sierra Leone	3.76	3.70	3.53
Total	**243.55**	**241.07**	**235.11**

1990 and 2000 (Kim, Sexton, and Townshend 2015) that dramatically contradicts the 25% reduction reported by the Food and Agriculture Organization of the United Nations (FAO 2010). This is regrettable as conservation biologists do not really know the magnitude of the problem they are dealing with and thus have a difficult time estimating the impact on chimpanzee populations. Furthermore, in such circumstances policy makers have the option of selecting the lower deforestation estimates when they establish policy if that suits their needs. By one estimate Africa shows the largest accelerating rate of loss (table 25.1) (Kim, Sexton, and Townshend 2015), which is alarming when one is considering chimpanzee conservation. The largest loss, and the largest increasing rate of forest loss, is in the Democratic Republic of Congo. This loss will be accentuated by the discovery of oil reserves in the Virunga National Park and the desire to exploit this resource (Gouby 2015). The oil industry opens up roads that are used to extract bushmeat and the oil workers hunt to feed themselves (Wilkie 2000). Per-country deforestation rates are a function both of the pace of logging and conversion to agriculture and of how much forest remains. For example, in Uganda, closed-canopy tropical forest once covered 20% of the country's land area, but deforestation reduced this to just 3% by 1990 (Howard et al. 2000), thus in terms of km2 there is not much left to lose. However, Uganda still lost 18% of its remaining forest between 1990 and 2000 (Howard et al. 2000) and the most recent estimate suggests that the annual rate of loss of tropical high forest is 7% (Pomeroy and Tushabe 2004). Very soon, Uganda will have little or no forests left that could support chimpanzee outside of the three forested National Parks (Chapman, Lawes, and Eeley 2006; Chapman et al. 2013). Deforestation rates in Uganda, however, pale in comparison to other African countries that harbor chimpanzees. The worlds' highest deforestation range, according to the FAO (2010), is Nigeria and it is estimated that between 2000 and 2005, Nigeria lost 55.7% of its primary forest, while the rural poor saw few advances in the quality of life. One should not simply think of these values as a strict loss in forest; rather, they also represent a fragmentation and the genetic isolation of populations. From that perspective, deforestation has resulted in the fragmentation of 58% of the subtropical forests and 46% of tropical forests (Chapman et al. 2007; Estrada et al. 2017; Haddad et al. 2015).

These statistics represent the general loss of forest; however, chimpanzees are a very flexible genus, being found in woodland and riverine forests to dense closed high-canopy forests (Hockings et al. 2015; see also Pruetz, Bogart, and Lindshield, chapter 17 this volume). Thus, the overall decline in suitable habitat, not just forest, would be a more accurate metric. A recent study predicted the distribution of suitable environmental conditions for chimpanzees in the 1990s and 2000s and demonstrated that the area of suitable habitat declined by 207,827 km^2, from 2,015,480 to 1,807,653 km^2 (i.e., an area approximately the size of the US state of Kansas or just larger than the country of Senegal) (Junker et al. 2012). This represents a 10.3% decline in a decade, and the authors of the study conclude that this represents a dramatic decline in suitable environmental conditions and call for an immediate increase in conservation efforts. In addition to this, the number of suitable areas did not decline significantly, suggesting that the size of suitable areas has shrunk. This raises the serious question of whether existing populations are large enough to maintain viable populations in the long term. A second study predicted that by 2030 only 10% of the current African great ape habitat will remain (Nelleman and Newton 2002).

In general, chimpanzee habitat loss corresponds with an increase in agriculture in tropical countries, which globally expanded by 48,000 km^2 per year between 1999 and 2008 (Phalan et al. 2013). One estimate suggests that approximately 1 billion ha of additional agricultural land, primarily in developing countries, will need to be converted to agriculture by 2050 to meet the demands of the growing human population—an area larger than Canada (Laurance, Sayer, and Cassman 2014). Ultimately, these changes are driven by increased human population size and consumption rates (Crist, Mora, and Engelman 2017). The UN Population Division estimates that the world's population is expected to rise from 7 billion in 2011 to 9 billion in 2050. Making the situation more dire for chimpanzees is the fact that, in primate range countries in Africa, human population density in 1950 was 8 people/km^2, while in 2010 it had increased to 35 people/km^2 (Estrada 2013) and in some protected parks harboring chimpanzees, human population density neighboring the park exceeds 400 people/km^2 (Hartter et al. 2015; MacKenzie and Hartter 2013). Also, as prosperity in tropical countries increases, there is a tendency for people to want to eat higher on the food chain (e.g., cattle), which demands greater land conversion.

Chimpanzees are a charismatic species that can be a rallying point for conservation effects and conserving their populations is the morally correct thing to do. As a result, it is an urgent imperative that we attempt to reverse a number of these trends and given that many populations may be approaching the point where they are no longer viable, the time for action is now.

Bushmeat

The bushmeat trade is a large commercial and local industry that is decimating many animal populations, particularly in West and Central Africa (Fa, Peres, and Meeuwig 2002; Walsh et al. 2003). The need to understand the bushmeat trade is partially created by the fact that a large proportion of the remaining chimpanzees are not in protected areas—approximately 45–81% of West African chimpanzees are not in parks or reserves (Kormos et al. 2003). However, hunting is frequently a serious problem in protected areas as well (Refisch and Koné 2005). For example, in Budongo Forest Reserve, Uganda, 21% of chimpanzees suffer from limb injuries caused by hunting snares (Byrne and Stokes 2002) and in Kibale National Park, Uganda, 31% of identified chimpanzees over the age of two exhibited limb disability (Cibot et al. 2016; Hartel et al., chapter 26 this volume). This is not an issue localized to Uganda: Quiatt, Reynolds, and Stokes (2002) documented that 32 of 422 chimpanzees (7.6%) in 10 different communities across Africa had limb disabilities likely resulting from snares.

There are a number of single-case market studies (Covey and McGraw 2014; Martin 1983), and hunting and the sale of ape meat have been reported from Nigeria (McFarland 1994), Central African Republic (Goldsmith 1995), Democratic Republic of Congo (Basabose, Mbake, and Yamagiva 1995), Gabon (Harcourt 1980), and Equatorial Guinea (Fa et al. 1995). This is not just a local trade; it is an international trade. Chaber et al. (2010) report that 273 tonnes of bushmeat are confiscated annually at the Charles de Gaulle airport in Paris, France. Assuming that the weight of the average cow is 625 kg (weight of an average Canadian cow, but an average Ankole cow in Uganda weighs 485 kg), this would mean that the equivalent of 440 cows were confiscated. This does not include the bushmeat that was not detected, the importation into other airports in France or worldwide, or the bushmeat that may come in through other routes, such as via shipping or over land. However, despite this sort of information on the extent of general bushmeat hunting, there is little large-scale evidence or quantification of bushmeat hunting impacts on populations of chimpanzees, and most studies simply examine the quantity of all bushmeat and do not distinguish between animal species (reviewed by Taylor et al. 2015).

While not informing chimpanzee endangerment, some of these bushmeat studies reveal some alarming statistics. For example, it has been estimated that four million metric tons of bushmeat were extracted each year from the Congo basin alone (equivalent to approximately 4,500,000 cows, Fa and Brown 2009). The rate of extraction was estimated to be increasing by 90,000 tonnes each year in 2002 and the rate of increase may have gone up (Fa, Peres, and Meeuwig 2002). In a study considering bushmeat on a species-by-species basics, Kano and Asato (1994) estimated that in the Mataba River region of northeastern Republic of Congo, 0.02 chimpanzees were killed annually per km^2. This represents an annual offtake of 5–7%, which, given chimpanzee's slow life history strategy, is unsustainable. Two country-wide surveys have been done in Côte d'Ivoire over a decade apart (1989–1990 and 2007) in national parks and classified forests, and on Mount Kope. These surveys documented a 90% decline in chimpanzee nest encounter rates over 17 years and attributed the decline to the 50% increase in the human population (Campbell et al. 2008). Distressingly, this study illustrates that even the national parks that are intended to be refuges for chimpanzees and other animals are not functioning as safe havens, and species in parks can still be driven to extinction through hunting (see also Laurance et al. 2012; McGraw 2005; Oates et al. 2000). Such trade from parks is very common in many regions. In a global analysis of 60 parks, Laurance et al. (2012) documented that researchers consider only approximately half of all reserves to have been effective over the last 20–30 years, while the remainder of the reserves are experiencing an alarming erosion of biodiversity, which includes the loss of primate species. This phenomenon is poignantly illustrated by a park-wide survey in Taï National Park, Côte d'Ivoire: regardless of primate species, density was 100 times higher near the protected research station and tourism site than in the remainder of the park (N'Goran et al. 2012). Similarly, in Moukalaba Doudou National Park, Gabon, surveys demonstrated that ape nest density (distinguishing chimpanzee and gorilla nests is not possible) was threes time lower at the park borders near human population centers, as compared to the park interior (Kuehl et al. 2009).

Associated with the bushmeat trade is the lucrative illegal trade of chimpanzees to discreditable zoos (often privately owned or in developing countries) and private owners. A baby chimpanzee can fetch $12,500 US, and often more (Shukman and Piranty 2017), and China has been singled out as the main destination for many of these illegally trafficked apes (Stiles et al. 2013). But the cost to the population is much larger than the one animal that makes it to these markets—hunters will typically shoot as many adults as possible to prevent adults from interfering with the capture of the baby and to get animals for bushmeat. An inquiry by the British Broadcasting Corporation estimated that for every infant captured, 10 adults would be killed (Shukman and Piranty 2017).

Disease

A further threat to chimpanzees is from disease, particularly Ebola (Walsh et al. 2007). However, surprisingly little is known about Ebola's impact on chimpanzees at the population level. This is partially due to the fact that researchers working on behavior and ecology have not been closely connected with researchers in the medical and veterinary fields, though this is changing rapidly (Goldberg, Paige, and Chapman 2012; Goldberg et al. 2008; Leendertz et al. 2006b; Rouquet et al. 2005). It is also due to the fact that chimpanzee populations typically inhabit remote forest regions, where road access is limited at best and researchers often lack access to those regions and are thereby unable to document disease outbreaks. Thus, Ebola outbreaks are often inferred from either low chimpanzee densities or declines in densities (Huijbregts et al. 2003; Walsh et al. 2003). The most dramatic case where the decline was estimated occurred in the Lossi Sanctuary in northwest Republic of Congo, where the chimpanzee population declined by more than 80% (Bermejo et al. 2006; Leroy et al. 2004). Another well-documented case involved a habituated chimpanzee community in Taï National Park, Côte d'Ivoire, where 11 of 43 (26%) members of one group disappeared in 1994, and where Ebola was confirmed as the cause (Boesch 2008; Formenty et al. 1999).

Diseases other than Ebola may also threaten chimpanzee populations (Knott and Harwell, chapter 1 this volume). Anthrax has been documented to have killed at least six individuals in Taï National Park (Leendertz et al. 2004) and at least three individuals in Cameroon (Leendertz et al. 2006a). Respiratory diseases have also caused deaths of chimpanzees at Kibale National Park, Uganda (Scully et al. 2018), Taï National Park (Kondgen et al. 2008), and Gombe National Park, Tanzania (Pusey, Wilson, and Collins 2008) and some of these infections are believed to have been transmitted to chimpanzees from people and vise versa. For example, molecular and epidemiologic analyses demonstrated that the outbreak of a respiratory disease in the chimpanzees of

Kibale National Park in 2013, which killed an infant chimpanzee, was consistent with the common cold in humans (Scully et al. 2018). However, to date, peaks in respiratory cases in local clinics have not been shown to correspond to peaks in similar symptoms in chimpanzees (Chapman and Melissa Emery Thompson, unpublished data).

There is a detailed record of the causes of deaths of chimpanzees at Gombe over 40 years and analyses indicate that disease accounted for 58% of the 86 deaths where the cause of death was known (Lonsdorf et al. 2018; Pusey, Wilson, and Collins 2008; Williams et al. 2008). Furthermore, major epidemics accounted for 50% of these disease-related deaths, attributed to a polio-like disease, mange, and respiratory diseases (Pusey, Wilson, and Collins 2008).

What seems clear to us is that disease, particularly Ebola and Anthrax, can play a major role influencing the size of chimpanzee populations. However, we know relatively little about its overall impact at the population level. This calls for closer collaboration between ecologists and veterinary scientists (Leendertz et al. 2006b) and large-scale monitoring schemes (Leendertz et al. 2006b; Leroy et al. 2004)—unfortunately there does not seem to be the political will to pay for such monitoring. These actions are definitely needed if we are to be able to make informed conservation plans for chimpanzees and decrease the risk of transmission of Ebola to the human population. The link between human health and conservation warrants further investigation as this may prove to be a win-win situation (Kirumira et al. 2019).

Climate Change

Thus far, deforestation and habitat loss seem to be the major threats to chimpanzees; however, another potential threat to chimpanzee populations comes from global climate change. Admittedly this risk is one that is very difficult to evaluate, but here we present data with which we can speculate on the potential outcomes of this global phenomenon. The Intergovernmental Panel on Climate Change (IPCC) estimates that the earth warmed by 0.85°C (0.65 to 1.06°C) between 1880 to 2012 (IPCC 2014) and temperature is projected to increase by 1.5°C by 2100 (IPCC 2014). Given where primates occur, estimates suggest that they will experience 10% greater warming than this global average, and some primate species will experience a 50% greater temperature increase for every 1°C of global warming (Graham, Matthews, and Turner 2016). Primates will also face changes in rainfall. This is because rising temperature alters global patterns of circulation, which affects rainfall patterns; however, changes will not occur uniformly around the globe (Graham, Matthews, and Turner 2016). Precipitation changes will likely be quite varied across the areas occupied by primates (i.e., from >7.5% increases per °C of global warming to >7.5% decreases) (Graham, Matthews, and Turner 2016). Furthermore, there will be "climate change hotspots" and if these areas contain endangered species, the consequences could be very severe and even result in extinctions. Projections vary considerably; however, considering moderate greenhouse gas emission estimates, it is estimated that 75% of all tropical forests present in 2000 will experience temperatures that are higher than the temperatures that presently support closed canopy forest by 2100 (Wright, Muller-Landau, and Schipper 2009).

For Africa, climate change projections are that the rainforest regions will become 3 to 4°C hotter over the next century under the most likely emission scenarios (Malhi et al. 2013; Zelazowski et al. 2011). This will lead to the retreat of forest in some areas, to be replaced by woodland or savannah. With regard to how rainfall patterns will be altered in Africa with climate change, the picture is less clear. It seems likely that East African forests will become wetter. Climate models for West Africa and the Congo Basin produce conflicting results; some suggest more rain, while others suggest less (Zelazowski et al. 2011). However, as recently as 3,000 years ago, there was a substantial retreat of both of these forest types (Oslisly et al. 2013), thus a climate-change induced forest retreat is certainly a possibility (Malhi et al. 2013).

The distribution of suitable chimpanzee habitat has been modeled for Cameroon and Nigeria under three different climate change scenarios for the years 2020, 2050, and 2080 (Clee et al. 2015). The availability of suitable habitat in northwest Cameroon and Eastern Nigeria is predicted to remain largely unchanged through 2080; however, in central Cameroon, habitat is predicted to decline dramatically over the coming century. This must be taken seriously in conservation planning, because the population in Central Cameroon represents half of the population of the chimpanzee subspecies *Pan troglodytes ellioti*, and this region also experiences high levels of hunting.

Lehmann, Korstjens, and Dunbar (2010) constructed a simulation model based on chimpanzee time budgets and how they would be altered by rising temperatures and changing rainfall patterns. The authors noted that climate variability would play a particularly important role in the degree of change in ape population size and distribution. Unfortunately, few climate change models consider such variability, yet intra-annual weather variability can also strongly impact species behavior and survival. The effect of such changes on chimpanzee populations will be driven partly by such direct effects of climate change (e.g., responses to temperature, disappearing habitat), but it has become clear in recent years that the indirect effects, mediated via species interactions, could be pronounced and have very significant impacts

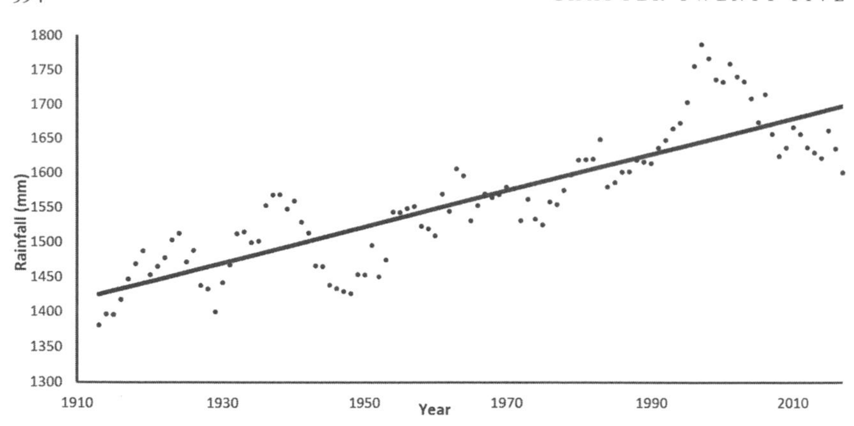

FIGURE 25.1. The 10 year running average of annual rainfall (mm) at Makerere University Biological Field Station, Kibale National Park, Uganda from 1903 to 2014.

on populations as well (Angert, LaDeau, and Ostfeld 2013). This is one area where research on other primates can inform us of what to expect chimpanzees will experience, such as our work on the red colobus (*Procolobus rufomitratus*) of Kibale National Park, Uganda.

It is relatively easy to imagine that in areas becoming hotter and drier, food trees will die and chimpanzees will die along with them or need to move. This is supported by data from Amboseli National Park, Kenya, where the average daily maximum temperature increased by 0.275°C per year between 1976 and 2000, which is an order of magnitude greater than that predicted by climate change models (Altmann, Alberts, and Roy 2002). This change corresponded to a dramatic loss of tree and shrub cover (Altmann, Alberts, and Roy 2002) and may have driven the concomitant decline in local vervet monkey populations (Struhsaker 1973, 1976). What happens in situations where the climate gets wetter is much less clear, but potentially equally negative for chimpanzees and resultant population stability. For example, Kibale National Park, Uganda, has experienced climate change well above the global average. The area receives 300 mm more rainfall/year than in 1900 and the average maximum monthly temperature has increased by 4.4 C° in the last 40 years (fig. 25.1). Corresponding with this change in climate, we have documented the cessation of fruiting of a number of plant species (Chapman et al. 2005), meaning that there is less fruit available for the frugivores, such as chimpanzees. One example of the localized effects of climate change on fruiting trees that primates rely on for food is that of *Trilepisium madagascariense* (formerly *Bosqueia phoberos*). This species has stopped fruiting at a site to the north of the park (Kanyawara) but continues to fruit at a site to the south

(Dura River), which is drier because of a natural north-south decline in rainfall associated with a decline in elevation.

In Kibale National Park there have also been changes in the quality of the leaf resources that correspond with changing temperature and precipitation. Greenhouse experiments indicate that elevated temperature, rainfall, and CO^2 to levels predicted by climate change models will impact the nutritional composition of leaves (Robinson, Ryan, and Newman 2012; Stiling and Cornelissen 2007; Zvereva and Kozlov 2006). Rothman et al. (2015) show a general increase in fiber and tannins and a decline in protein compared to data collected 15 and 30 years previously. This study examined leaves that were important in the diet of red colobus; however, this may also apply to the terrestrial herbaceous vegetation that is often considered an important fallback food for chimpanzees (Lambert 2007; Malenky and Wrangham 1994; Marshall et al. 2009; Wrangham et al. 1991). A decline in the quality of fallback foods that are eaten when more-preferred foods are not available could have serious impacts on a chimpanzee population.

The wetter conditions that East African forests will experience, and that West and Central African forests may experience (model predictions vary), are likely to create conditions where diseases become more prevalent. Connections between climate and disease are well established in the human medical literature, with specific diseases occurring during certain seasons, or erupting in association with specific unseasonable weather conditions. For example, in sub-Saharan Africa, meningococcal meningitis epidemics erupt during the hot dry season and subside soon after the onset of the rains (Patz et al. 1996). Climate change can affect disease transmission by influencing the ecology of hosts and vectors, or by causing resource shifts that stress the animal's physiology, making them more susceptible to infection (Haines and Patz 2004). For example, heavy rains are associated with outbreaks of waterborne diseases in humans. In the United States, 68% of waterborne disease outbreaks, such as *Giardia* and *Cryptosporidium* (both of which infect chimpanzees) were preceded by precipitation events above the 80th percentile (Hunter 2003). One reason that wetter conditions promote disease is that they facilitate the survival of infective stage parasitic larvae and eggs. For example, in an experimental study Larsen and Roepstorff (1999) demonstrated a reduction in the number of pig parasite eggs recovered in hot, dry months compared to wetter months. A study in Kibale National Park on black-and-white colobus (*Colobus guereza*) supports the idea that wetter conditions promote parasitism (Chapman et al. 2010). The study demonstrated that groups in wetter habitats (e.g., wet valley bottoms) had elevated gastrointestinal infections, as compared to groups in the same region that lived in drier areas (e.g., hilltops). Also, in Kibale National Park there is a north-south decline in rainfall associated with decreasing elevation and as predicted groups living in the north had elevated gastrointestinal infections compared to groups in the drier south (Chapman et al. 2010). The population effects of such changes in parasite infections remain to be evaluated.

Genetic Viability

Knowledge of patterns of chimpanzee population genetic diversity can provide pivotal insights into the evolutionary history, population structure, mating system, demographic dynamics, and population viability, which should be used to inform conservation design and recovery efforts. High-coverage genome sequencing of chimpanzees supports two distinct lineages, each comprising two genetic groups: Central/Eastern and Nigeria-Cameroon/Western chimpanzees (Prado-Martinez et al. 2013). The Central chimpanzee shows the highest genetic diversity and largest effective population size (N_e), whereas the Western chimpanzee shows the lowest level of genetic diversity and smallest N_e among all populations (Prado-Martinez et al. 2013).

Despite the wide distribution and relatively large total population size, many of the remaining chimpanzee populations are small and dispersed. This situation is particularly severe in Senegal, Mali, the Cabinda enclave of Angola, Equatorial Guinea, and Sudan. Traditionally, the 50/500 rule has been applied to gauge the minimal viable population size of endangered species, i.e., $N_e = 50$ as the minimum population size for avoiding inbreeding depression, and $N_e = 500$ for retaining evolutionary potential. Accumulating evidence suggests that these thresholds are too low and a 100/1,000 rule might be more appropriate for the purpose of maintaining population viability in the short term, as well as the long term (Frankham, Bradshaw, and Brook 2014). It is worth noting that N_e of a wild population is often only a small fraction of the census population size, as censuses include immatures and non-breeding individuals (Frankham 1995; Palstra and Ruzzante 2008). Additionally, high variance in sex ratio and reproductive success can lead to a large further reduction of the ratio of N_e to census population size (Frankham 1995; Luikart et al. 2010). Taken together, very few extant wild chimpanzee populations fulfill the minimum population rules for conservation.

Human activity and habitat modification often increase genetic structure and limit gene flow within and among habitat areas, leading to genetic isolation and inbreeding within small populations (Knight, Chapman, and Hale 2016). High degrees of inbreeding are frequently correlated with low genetic diversity, impaired resistance to disease and environmental stress, and reduced growth rate and reproductive success, all of which hamper individual fitness and population viability (Keller and Waller 2002; Knight, Chapman, and Hale 2016). Therefore, maintaining ecological corridors is critical to ensure continued chimpanzee dispersal between habitat pockets and combat genetic erosion caused by population isolation and inbreeding (Basabose, Mbake, and Yamagiva 2015). However, genetic information of many local chimpanzee populations remains uninvestigated or sporadic at best, posing significant challenges for science-guided management. Future conservation efforts should stress genetic monitoring of chimpanzee populations to understand population dynamics and the impacts of anthropogenic disturbance and environmental variables on population demographic and genetic patterns (Schwartz, Luikart, and Waples 2007). Populations with high genetic diversity (i.e., genetic reservoirs) can potentially be of importance for genetic rescue of small, inbred populations threatened with extinction (Whiteley et al. 2015).

The genetic viability of populations should also be used by conservation managers to identify conservation priorities in situations where difficult choices have to be made, which, sadly, is often the case. If conservation managers were to follow these rules of thumb for the minimum viable population size for avoiding inbreeding depression and retaining evolutionary potential as around 1,000 breeding individuals and it was assumed that only one-third of the animals recorded in a typical census were breeding (due to the proportion of immature individuals and the sometimes small proportion of males that sire offspring), this would mean that populations of greater than 3,000 individuals should be conservation priorities. For East African chimpanzees, this would mean emphasizing conservation efforts in the Democratic Republic of Congo and potentially Uganda (table 25.2; the Central African Republic is data deficient). Let us consider Uganda in more detail (table 25.3). Of the 26 forests for which we have census data, only three (Budongo, Bugoma, and Kibale) are estimated to have 1,000 individuals, and no area has the 3,000 individuals needed to maintain the minimum population size for avoiding inbreeding depression and retaining evolutionary potential (the Bugoma Forest Reserve has experienced intense pressure recently and the previous estimates now likely overestimate what remains, C. Chapman and P. Omeja, pers. observations, May 2017). How this information is used should be debated; however, it does mean that the protection of Budongo and Kibale should be a Ugandan priority. The situation elsewhere in Africa is not unlike that in Uganda. For example, Campbell et al. (2008) report a 90% decline in the chimpanzee populations of Côte d'Ivoire between 1990 and 2007 (see also N'Goran et al. 2013).

TABLE 25.2. Estimates of eastern chimpanzee populations in each country in which they occur (adapted from Plumptre et al. 2010).

Country	Country Size (km2)	Chimpanzee Population
Burundi	27,834	450
Central African Republic	622,984	?
Democratic Republic of Congo	2,345,409	42,798
Rwanda	26,338	275
Sudan	1,886,068	?
Tanzania	947,303	2,750
Uganda	241,038	5,000
Total	**6,096,974**	**51,273**

TABLE 25.3. Estimates of the population size of chimpanzees in Uganda (NP = National Park, FR = Forest Reserve)

Area	Population Estimate	Survey Year	Source
Budongo NP	500–1,000	1999–2002	Plumptre, Cox, and Mugume (2003)
Bugoma FR	500–1,000	1999–2002	Plumptre , Cox, and Mugume (2003)
Bugoma–Budongo Corridor	50–100	1999–2002	Plumptre, Cox, and Mugume (2003)
Buhungiro	Extirpated	2008–2010	Koojo (2016)
Bulindi	260	2006	McLennan (2008)
Bwindi Impenetrable NP	100–300	1999–2002	Plumptre, Cox, and Mugume (2003)
Echuya FR	Extirpated	1999–2002	Plumptre, Cox, and Mugume (2003)
Ibambaro FR	Extirpated	1999–2002	Plumptre, Cox, and Mugume (2003)
Itwara FR	100–300	1999–2002	Plumptre, Cox, and Mugume (2003)
	34	2008–2010	Koojo (2016)
Kagombe FR	100–300	1999–2002	Plumptre, Cox, and Mugume (2003)
Kagorra region	<50	1999–2002	Plumptre, Cox, and Mugume (2003)
Kalinzu FR	100–300	2006	Plumptre et al. (2008)
	220	1999–2002	Plumptre, Cox, and Mugume (2003)
Kasato FR	<50	1999–2002	Plumptre, Cox, and Mugume (2003)
Kasyoha-Kitomi FR	300–500	2006	Plumptre et al. (2008c)
	370	1999–2002	Plumptre, Cox, and Mugume (2003)
Kibale NP	500–1,000	2015	Sop et al. (2015)
	1,298	1999–2002	Plumptre, Cox, and Mugume (2003)
	921	2005	Wanyama et al. (2009), Wanyama (2005)
	1,931*	No date	Plumptre and Cox (2006)
Kibego FR	<50	1999–2002	Plumptre, Cox, and Mugume (2003)
	Present	2008–2010	Koojo (2016)
Kitechura FR	Extirpated	1999–2002	Plumptre, Cox, and Mugume (2003)
Kyambura Wildlife Reserve	50–100	1999–2002	Plumptre, Cox, and Mugume (2003)
Maramagambo	100–300	1999–2002	Plumptre, Cox, and Mugume (2003)
Matiri FR	Extirpated	1999–2002	Plumptre, Cox, and Mugume (2003)
	Present	2008–2010	Koojo (2016)
Muhangi FR	<50	1999–2002	Plumptre, Cox, and Mugume (2003)

TABLE 25.3. *(continued)*

Area	*Population Estimate*	*Survey Year*	*Source*
Otzi FR	<50	1999–2002	Plumptre, Cox, and Mugume (2003)
Rwenzori Mountains	300–500	1999–2002	Plumptre, Cox, and Mugume (2003)
NP	466	2012	S. M. Koojo, unpublished data
South Bugoma	<50	1999–2002	Plumptre, Cox, and Mugume (2003)
Toro-Semliki Wildlife Reserve	50–100	2010–2011	Samson and Hunt (2012)
Wambabya FR	100–300	1999–2002	Plumptre, Cox, and Mugume (2003)

Note: Adapted from Sop et al. (2015) with additions from a variety of sources.

*Calculated.

Future Directions

There are a number of excellent studies that discuss means of reducing logging, or lessening its impact (Bicknell and Peres 2010; Morgan 2007; Morgan et al. 2017; Putz, Dykstra, and Heinrich 2000), and there are many excellent texts that discuss means to decrease the risk of disease transmission from humans to chimpanzees (Boesch 2008; Cranfield 2008; Leendertz et al. 2016; Pusey, Wilson, and Collins 2008; Rwego et al. 2008; Wallis and Lee 1999; Woodford, Butynski, and Karesh 2002). Thus, we are not going to review what has already been so well reviewed. Rather, we would like to present novel ways forward.

It should be emphasized that a large number of review studies and the recommendations from many long-term researchers point to the importance of improved and more extensive law enforcement (Struhsaker, Struhsaker, and Siex 2005; Tranquilli et al. 2012, 2014). For chimpanzees, particular attention should be paid to snare removal, which takes different search strategies than regular patrols that are searching for poachers (Muller and Wrangham 2000; Quiatt, Reynolds, and Stokes 2002). In addition, long-term research sites and tourist establishments have proven effective at reducing poaching and should be encouraged (Sandbrook and Semple 2006; Sarkar et al. accepted with revision); however, this development must be done in a fashion that minimizes the risk of disease transmission.

Education and public outreach constitute a conservation strategy that has been employed for decades. The assumption behind this is that if conservation biologists can illustrate to the community the value of a protected area, they will not exploit and harm its resources. Unfortunately, contrary to expectations, studies in Africa have demonstrated that community outreach programs designed to promote positive community attitudes through education are seldom associated with successful conservation outcomes (Struhsaker, Struhsaker, and Siex

2005). In fact, a detailed study of protected areas in Uganda found no evidence that such programs promoted positive community attitudes toward parks (Mugisha and Jacobson 2004). It is our opinion that these negative results do not mean that this approach should be abandoned, but rather we should learn from past experiences and make the approach more effective. In fact, there is a resurgence of the application of this approach (Padua 2010; Savage et al. 2010) and its careful long-term evaluation (Jacobson 2010; Kuhar et al. 2010), some of the original problems of such programs are being addressed (Kasenene and Ross 2008; Struhsaker, Struhsaker, and Siex 2005), and new refined approaches that deal with chimpanzees appear promising (Leeds et al. 2017). Since education often targets the young, the impact of such programs will be seen only after many years (Chapman, Struhsaker, and Lambert 2005; Struhsaker, Struhsaker, and Siex 2005). Also new outreach approaches should be investigated (Leeds et al. 2017).

We have initiated one such new local outreach approach involving the union of the provision of health care and conservation; namely the delivery of health care to local communities bordering Ugandan national parks through a mobile health clinic system and the establishment of a large permanent clinic (Chapman et al. 2015). The mobile health clinic is a means to reach many people; in fact, it is estimated that in its first year of operation it helped 1,000 patients a month and delivered conservation outreach information to 10,000 people a month (Kirumira et al. 2019). These examples suggest that if the goal is to conserve chimpanzees, the local communities' livelihoods must be considered so that they perceive receiving benefits from the protected areas and thus are encouraged to conserve the system and its chimpanzees. This also calls for strategic intervention in other livelihood activities for local communities so that pressure on chimpanzee habitat can be reduced.

Lastly, a number of researchers have questioned the model typically used by large governmental or non-governmental organizations to fund conservation projects (Oates 1999; Struhsaker, Struhsaker, and Siex 2005; Terborgh 1999). Typically, groups like the World Bank give large sums of money to the central government over a relatively short period of time (e.g., 5 years) and after this short period funding stops altogether. We suggest that changes should be made in funding strategies. A recent study used an evaluation of 90 "success stories" provided by conservation scientists and practitioners and explored characteristics of the projects that were "perceived" successful (Chapman et al. 2016). The conservation community viewed successful projects to most often be long-term, small spatial scale, and relatively low budget, and involving a protectionist approach alone or in combination with another approach. This suggests that extending funding over longer periods of time and investing in long-term projects would help make conservation gains.

To reduce or remove current threats across the whole range of chimpanzees will require huge efforts on a very large scale and very significant funding. It will require that international and national agencies gain the cooperation of local people, alternative sources of income and protein be found, and a great deal of effort be placed on education and outreach, with novel approaches being attempted and evaluated, increased efforts toward law enforcement, genetic monitoring of existing populations with corridors being established when possible, and protection of large areas where large populations are still found (i.e., Democratic Republic of Congo). In reality, it is unlikely that a project of such a magnitude will be initiated, but attempts must be made on whatever scale is possible.

Acknowledgments

Funding was provided by Canada Research Chairs Program, Wildlife Conservation Society, Natural Science and Engineering Research Council of Canada, National Geographic, and Fonds Québécois de la Recherche sur la Nature et les Technologies. We thank Aaron Sandel and Sophie Muset for helpful comments on this manuscript.

References

Altmann, J., S. C. Alberts, and S. B. Roy. 2002. "Dramatic change in local climate patterns in Amboseli Basin, Kenya." *African Journal of Ecology* 40: 248–51.

Angert, A. L., S. L. LaDeau, and R. S. Ostfeld. 2013. "Climate change and species interactions: Ways forward." *Annals of the New York Academy of Sciences* 1297: 1–7.

Basabose, A. K., E. Inoue, S. Kamungu, B. Murhabale, E.-F. Akomo-Okoue, and J. Yamagiwa. 2015. "Estimation of chimpanzee community size and genetic diversity in Kahuzi-Biega National Park, Democratic Republic of Congo." *American Journal of Primatology* 77: 1015–25.

Basabose, A. K., S. Mbake, and J. Yamagiva. 1995. "Research and conservation of eastern lowland gorillas in the Kahuzi-Biega National Park, Zaire." *Gorilla Conservation News* 9: 11–12.

Bermejo, M., J. D. Rodriguez-Teijeiro, G. Illera, A. Barroso, C. Vila, and P. D. Walsh. 2006. "Ebola outbreak killed 5000 gorillas." *Science* 314: 1564–64.

Bertolani, P., and C. Boesch. 2008. "Habituation of wild chimpanzees (*Pan troglodytes*) of the south group at Taï Forest, Côte d'Ivoire: Empirical measure of progress." *Folia Primatologica* 79: 162–71.

Bicknell, J., and C. A. Peres. 2010. "Vertebrate population responses to reduced-impact logging in a neotropical forest." *Forest Ecology and Management* 259: 2267–75.

Boesch, C. 2008. "Why do chimpanzees die in the forest? The challenges of understanding and controlling for wild ape health." *American Journal of Primatology* 70: 722–26.

Boesch, C., and H. Boesch-Achermann. 2000. *The Chimpanzees of the Taï Forest: Behavioural Ecology and Evolution*. New York: Oxford University Press.

Byrne, R. W., and E. J. Stokes. 2002. "Effects of manual disability on feeding skills in gorillas and chimpanzees." *International Journal of Primatology* 23: 539–54.

Campbell, G., H. Kuehl, N. G. P. Kouame, and C. Boesch. 2008. "Alarming decline in West African chimpanzees in Côte d'Ivoire." *Current Biology* 18: R903–4.

Chaber, A.-L., S. Sallebone-Webb, Y. Lignereux, A. A. Cunningham, and J. M. Rowcliffe. 2010. "The scale of illegal meat importation from Africa to Europe via Paris." *Conservation Letters* 3: 317–23.

Chapman, C. A., L. J. Chapman, T. T. Struhsaker, A. E. Zanne, C. J. Clark, and J. R. Poulsen. 2005. "A long-term evaluation of fruiting phenology: Importance of climate change." *Journal of Tropical Ecology* 21: 31–45.

Chapman, C. A., A. DeLuycker, R. A. Reyna-Hurtado, J. C. Serio-Silva, T. B. Smith, K. B. Strier, and T. L. Goldberg. 2016. "Safeguarding biodiversity: What works according to the conservation community." *Oryx* 50: 302–7.

Chapman, C. A., R. R. Ghai, A. L. Jacob, S. M. Koojo, R. Reyna-Hurtado, J. M. Rothman, D. Twinomugisha, M. D. Wasserman, and T. L. Goldberg. 2013. "Going, going, gone: A 15-year history of the decline of primates in forest fragments near Kibale National Park, Uganda." In *Primates in Fragments: Complexity and Resilience*, edited by L. K. Marsh and C. A. Chapman. New York: Springer.

Chapman, C. A., M. J. Lawes, and H. A. C. Eeley. 2006. "What hope for African primate diversity?" *African Journal of Ecology* 44: 1–18.

Chapman, C. A., L. Naughton-Treves, M. J. Lawes, M. D. Wasserman, and T. R. Gillespie. 2007. "The conservation value of forest fragments: Explanations for population declines of the colobus of Western Uganda." *International Journal of Primatology* 23: 513–78.

Chapman, C. A., M. L. Speirs, S. A. M. Hodder, and J. M. Rothman. 2010. "Colobus monkey parasite infections in wet and dry habitats: Implications for climate change." *African Journal of Ecology* 48: 555–58.

Chapman, C. A., T. T. Struhsaker, and J. E. Lambert. 2005. "Thirty years of research in Kibale National Park, Uganda, reveals a complex picture for conservation." *International Journal of Primatology* 26: 539–55.

Chapman, C. A., B. van Bavel, C. Boodman, R. R. Ghai, J. F. Gogarten, J. Hartter, L. E. Mechak, P. A. Omeja, S. Poonawala, D. Tuli, and T. L. Goldberg. 2015. "Providing health care to promote people-park relations." *Oryx* 49 (4): 636–42.

Cibot, M., S. Krief, J. Philippon, P. Couchoud, A. Seguya, and E. Pouydebat. 2016. "Feeding consequences of hand and foot disability in wild adult chimpanzees (*Pan troglodytes schweinfurthii*)." *International Journal of Primatology* 37: 479–94.

Clee, P. R. S., E. E. Abwe, R. D. Ambahe, N. M. Anthony, R. Fotso, S. Locatelli, F. Maisels, M. W. Mitchell, B. J. Morgan, A. A. Pokempner, and M. K. Gonder. 2015. "Chimpanzee population structure in Cameroon and Nigeria is associated with habitat variation that may be lost under climate change." *BMC Evolutionary Biology* 15: 1–13.

Covey, R., and W. S. McGraw. 2014. "Monkeys in a West African bushmeat market: Implications for cercopithecid conservation in eastern Liberia." *Tropical Conservation Science* 7: 115–25.

Cranfield, M. R. 2008. "Mountain gorilla research: The risk of disease transmission relative to the benefit from the perspective of ecosystem health." *American Journal of Primatology* 70: 751–54.

Crist, E., C. Mora, and R. Engelman. 2017. "The interaction of human population, food production, and biodiversity protection." *Science* 356: 260–64.

Dirzo, R., H. S. Young, M. Galetti, G. Ceballos, N. J. B. Isaac, and B. Collen. 2014. "Defaunation in the anthropocene." *Science* 345: 401–6.

Estrada, A. 2013. "Socioeconomic context of primate conservation: Population, poverty, global economic demands, and sustainable land use." *American Journal of Primatology* 75: 30–45.

Estrada, A., P. A. Garber, A. B. Rylands, C. Roos, E. Fernandez-Duque, A. Di Fiore, K. A.-I. Nekaris, V. Nijman, E. W. Heymann, and J. E. Lambert. 2017. "Impending extinction crisis of the world's primates: Why primates matter." *Science Advances* 3: e1600946.

Fa, J. E., and D. Brown. 2009. "Impacts of hunting on mammals in African tropical moist forests: A review and synthesis." *Mammal Review* 39: 231–64.

Fa, J. E., J. Juste, J. del Val, and J. Castroviejo. 1995. "Impact of market hunting on mammal species in Equatorial Guinea." *Conservation Biology* 9: 1107–15.

Fa, J. E., C. A. Peres, and J. Meeuwig. 2002. "Bushmeat exploitation in tropical forests: An intercontinental comparison." *Conservation Biology* 16: 232–37.

FAO. 2010. *Global Forest Resource Assessment 2010*. Rome: Food and Agriculture Organization of the United Nations.

Formenty, P., C. Boesch, M. Wyers, C. Steiner, F. Donati, F. Dind, F. Walker, and B. L. Geunno. 1999. "Ebola virus outbreaks among wild chimpanzees living in a rain forest of Côte d'Ivoire." *Journal of Infectious Diseases* 179: S120–26.

Frankham, R. 1995. "Effective population size/adult population size ratios in wildlife: A review." *Genetical Research* 66: 95–107.

Frankham, R., C. J. A. Bradshaw, and B. W. Brook. 2014. "Genetics in conservation management: Revised recommendations for the 50/500 rules, red list criteria and population viability analyses." *Biological Conservation* 170: 56–63.

Goldberg, T. L., T. R. Gillespie, I. B. Rwego, E. E. Estoff, and C. A. Chapman. 2008. "Forest fragmentation as cause of bacterial transmission among primates, humans, and livestock, Uganda." *Emerging Infectious Diseases* 14: 1375–82.

Goldberg, T. L., S. Paige, and C. A. Chapman. 2012. "The Kibale EcoHealth Project: Exploring connections among human health, animal health, and landscape dynamics in western Uganda." In *Conservation Medicine: Applied Cases of Ecological Health*, edited by A. A. Aguirre and P. Daszak. Oxford: Oxford University Press.

Goldsmith, M. L. 1995. "Ranging and grouping patterns of western lowland gorillas (*Gorilla g. gorilla*) in the Central African Republic." *Gorilla Conservation News* 9: 5–6.

Gouby, M. 2015. "Democratic Republic of Congo wants to open up Virunga National Park to oil exploration." *Guardian*, May 6.

Graham, T. L., H. D. Matthews, and S. E. Turner. 2016. "A global-scale evaluation of primate exposure and vulnerability to climate change." *International Journal of Primatology* 37: 158–74.

Haddad, N. M., L. A. Brudvig, J. Clobert, K. F. Davies, A. Gonzalez, R. D. Holt, T. E. Lovejoy, J. O. Sexton, M. P. Austin, and C. D. Collins. 2015. "Habitat fragmentation and its lasting impact on Earth's ecosystems." *Science Advances* 1: e1500052.

Haines, A., and J. A. Patz. 2004. "Health effects of climate change." *Journal of the American Medical Association* 291: 99–103.

Hansen, M. C., P. V. Potapov, R. Moore, M. Hancher, S. A. Turubanova, A. Tyukavina, D. Thau, S. V. Stehman, S. J. Goetz, T. R. Loveland, A. Kommareddy, A. Egorov, L. Chini, C. O. Justice, and J. R. G. Townshend. 2013. "High-resolution global maps of 21st-century forest cover change." *Science* 342: 850–53.

Harcourt, A. H. 1980. "Gorilla-eaters of Gabon." *Oryx* 15: 248–51.

Hartter, J., S. J. Ryan, C. A. MacKenzie, A. Goldman, N. Dowhaniuk, M. Palace, J. E. Diem, and C. A. Chapman. 2015. "Now there is no land: A story of ethnic migration in a protected area landscape in western Uganda." *Population and Environment* 36: 452–79.

Hockings, K. J., M. R. McLennan, S. Carvalho, M. Ancrenaz, R. Bobe, R. W. Byrne, R. I. Dunbar, T. Matsuzawa, W. C. McGrew, and E. A. Williamson. 2015. "Apes in the anthropocene: Flexibility and survival." *Trends in Ecology & Evolution* 30: 215–22.

Howard, P. C., T. R. B. Davenport, F. W. Kigenyi, P. Viskanic, M. C. Balzer, C. J. Dickinson, J. S. Lwanga, R. A. Matthews, and E. Mupada. 2000. "Protected area planning in the tropics: Uganda's national system of forest nature reserves." *Conservation Biology* 14: 858–75.

Huijbregts, B., P. De Wachter, L. S. N. Obiang, and M. E. Akou. 2003. "Ebola and the decline of gorilla *Gorilla gorilla* and chimpanzee *Pan troglodytes* populations in Minkebe Forest, North-Eastern Gabon." *Oryx* 37: 437–43.

Humle, T., F. Maisels, J. F. Oates, A. Plumptre, and E. A. Williamson. 2016. "*Pan troglodytes* (errata version published in 2018)." *IUCN Red List of Threatened Species 2016*: e.T15933A129038584.

Hunter, P. R. 2003. "Climate change and waterborne and vector-borne disease." *Journal of Applied Microbiology* 94: 37S–46S.

IPCC. 2014. *Climate Change 2014: Synthesis Report. Contribution of Working Groups I, II and III to the Fifth Assessment Report of the Intergovernmental Panel on Climate Change*. Edited by Core Writing Team, R. K. Pachauri, and L. A. Meyer. Geneva: IPPC.

IUCN Redlist database. 2020. https://www.iucnredlist.org/search/stats?taxonomies=100091&searchType=species.

Jacobson, S. K. 2010. "Effective primate conservation education: Gaps and opportunities." *American Journal of Primatology* 72: 414–19.

Junker, J., S. Blake, C. Boesch, G. Campbell, L. du Toit, C. Duvall, A. Ekobo, G. Etoga, A. Galat-Luong, and J. Gamys. 2012. "Recent decline in suitable environmental conditions for African great apes." *Diversity and Distributions* 18: 1077–91.

Kano, T., and R. Asato. 1994. "Hunting pressure on chimpanzees and gorillas in the Mataba River area of northeastern Congo." *African Study Monographs* 15 (3): 143–62.

Kasenene, J. M., and E. A. Ross. 2008. "Community benefits from long-term research programs: A case study from Kibale National Park, Uganda." In *Science and Conservation in African Forests: The Benefits of Long-Term Research*, edited by R. W. Wrangham and E. A. Ross, 99–114. Cambridge: Cambridge University Press.

Keller, L. F., and D. M. Waller. 2002. "Inbreeding effects in wild populations." *Trends in Ecology & Evolution* 17: 230–41.

Kim, D. H., J. O. Sexton, and J. R. Townshend. 2015. "Accelerated deforestation in the humid tropics from the 1990s to the 2000s." *Geophysical Research Letters* 42: 3495–3501.

Kirumira, D., D. Baranga, J. Hartter, K. Valenta, C. Tumwesigye, W. Kagoro, and C. A. Chapman. 2019. "Evaluating a union between health care and conservation: A mobile clinic improves park-people relations, yet poaching increases." *Conservation and Society* 17 (1): 51–62.

Knight, A., H. M. Chapman, and M. Hale. 2016. "Habitat fragmentation and its implications for endangered chimpanzee *Pan troglodytes* conservation." *Oryx* 50: 533–36.

Kondgen, S., H. Kuhl, P. K. N'Goran, P. D. Walsh, S. Schenk, N. Ernst, R. Biek, P. Formenty, K. Maetz-Rensing, B. Schweiger, S. Junglen, H. Ellerbrok, A. Nitsche, T. Briese, W. I. Lipkin, G. Pauli, C. Boesch, and F. H. Leendertz. 2008. "Pandemic human viruses cause decline of endangered great apes." *Current Biology* 18: 260–64.

Koojo, S. M. 2016. "Influence of human activities on the status of wildlife in four central forest reserves in Uganda." PhD diss., Makerere University.

Kormos, R., C. Boesch, M. I. Bakarr, and T. M. Butynski. 2003. *West African Chimpanzees: Status Survey and Conservation Action Plan*. Gland, Switzerland: IUCN/SSC Primate Specialist Group.

Kuehl, H. S., C. Nzeingui, S. L. D. Yeno, B. Huijbregts, C. Boesch, and P. D. Walsh. 2009. "Discriminating between village and commercial hunting of apes." *Biological Conservation* 142: 1500–1506.

Kuhar, C. W., T. L. Bettinger, K. Lehnhardt, O. Tracy, and D. Cox. 2010. "Evaluating for long-term impact of an environmental education program at the Kalinzu Forest Reserve, Uganda." *American Journal of Primatology* 72: 407–13.

Kühl, H. S., T. Sop, E. A. Williamson, R. Mundry, D. Brugière, G. Campbell, H. Cohen, E. Danquah, L. Ginn, and I. Herbinger. 2017. "The critically endangered western chimpanzee declines by 80%." *American Journal of Primatology* 79: e22681.

Lambert, J. E. 2007. "Seasonality, fallback strategies, and natural selection: A chimpanzee and cercopithecoid model for interpreting the evolution of hominin diet." In *Evolution of Human Diet: The Known, the Unknown, and the Unknowable*, edited by P. S. Ungar, 324–43. Oxford: Oxford University Press.

Larsen, M. N., and A. Roepstorff. 1999. "Seasonal variation in development and survival of *Ascaris suum* and *Trichuris suis* eggs on pastures." *Parasitology* 119 (2): 209–20.

Laurance, W. F., D. Carolina Useche, J. Rendeiro, M. Kalka, C. J. A. Bradshaw, S. P. Sloan, S. G. Laurance, M. Campbell, K. Abernethy, A. Alvarez, V. Arroyo-Rodriguez, P. Ashton, J. Benítez-Malvido, A. Blom, K. S. Bobo, C. H. Cannon, M. Cao, R. Carroll, C. Chapman, R. Coates, C. Cords, D. Danielsen, B. D. Dijn, E. Dinerstein, M. A. Donnelly, D. Edwards, F. Edwards, N. Farwig, P. Fashing, P.-M. Forget, M. Foster, G. Gale, D. Harris, R. Harrison, H. Hart, S. Karpanty, W. J. Kress, J. Krishnaswamy, W. Logsdon, J. Lovett, W. Magnusson, F. Maisels, A. R. Marshall, D. McClearn, D. Mudappa, M. R. Nielsen, R. Pearson, N. Pitman, J. V. D. Ploeg, A. Plumptre, J. Poulsen, M. Quesada, H. Rainey, D. Robinson, C. Roetgers, F. Rovero, F. Scatena, C. Schulze, D. Sheil, T. Struhsaker, J. Terborgh, D. Thomas, R. Timm, J. N. Urbina-Cardona, K. Vasudevan, S. J. Wright, J. C. Arias-G., L. Arroyo, M. Ashton, P. Auzel, D. Babaasa, F. Babweteera, P. Baker, O. Banki, M. Bass, I. Bila-Isia, S. Blake, W. Brockelman, N. Brokaw, C. A. Brühl, S. Bunyavejchewin, J.-T. Chao, J. Chave, R. Chellam, C. J. Clark, J. Clavijo, R. Congdon, R. Corlett, H. S. Dattaraja, C. Dave, G. Davies, B. de Mello Beisiegel, R. N. P. Silva, A. Di Fiore, A. Diesmos, R. Dirzo, D. Doran-Sheehy, M. Eaton, L. Emmons, A. Estrada, C. Ewango, L. Fedigan, F. Feer, B. Fruth, J. G. Willis, U. Goodale, S. Goodman, J. C. Guix, P. Guthiga, W. Haber, K. Hamer, I. Herbinger, J. Jane Hill, Z. Huang, I. F. Sun, K. Ickes, A. Itoh, N. Ivanauskas, B. Jackes, J. Janovec, D. H. Janzen, M. Jiangming, C. Jin, T. Jones, H. Justiniano, E. Kalko, A. Kasangaki, T. Killeen, H.-b. King, E. Klop, C. Knott, I. Koné, E. Kudavidanage, J. L. S. Ribeiro, J. Lattke, R. Laval, R. Lawton, M. Leal, M. Leighton, M. Lentino, C. Leonel, J. Lindsell, L. L. Ling-Ling, K. E. Linsenmair, E. Losos, A. Lugo, J. Lwanga, A. L. Mack, M. Martins, W. S. McGraw, R. McNab, L. Montag, J. M. Thompson, J. Nabe-Nielsen, M. Nakagawa, S. Nepal, M. Norconk, V. Novotny, S. O'Donnell, M. Opiang, P. Ouboter, K. Parker, N. Parthasarathy, K. Pisciotta, D. Prawiradilaga, C. Pringle, S. Rajathurai, U. Reichard, G. Reinartz, K. Renton, G. Reynolds, V. Reynolds, E. Riley, M.-O. Rödel, J. Rothman, P. Round, S. Sakai, T. Sanaiotti, T. Savini, G. Schaab, J. Seidensticker, A. Siaka, M. R. Silman, T. B. Smith, S. S. de Almeida, N. Sodhi, C. Stanford, K. Stewart, E. Stokes, K. E. Stoner, R. Sukumar, M. Surbeck, M. Tobler, T. Tscharntke, A. Turkalo, G. Umapathy, M. Weerd, J. V. Vega Rivera, M. Venkataraman, L. Venn, C. Verea, C. V. de Castilho, M. Waltert, B. Wang, S. Watts, W. Weber, P. West, D. Whitacre, K. Whitney, D. Wilkie, S. Williams, D. D. Wright, P. Wright, L. Xiankai, P. Yonzon, and R. Zamzani. 2012. "Averting biodiversity collapse in tropical forest protected areas." *Nature* 489: 290–94.

Laurance, W. F., J. Sayer, and K. G. Cassman. 2014. "Agriculture expansion and its impacts on tropical nature." *Trends in Ecology & Evolution* 29: 107–16.

Leeds, A., K. E. Lukas, C. J. Kendall, M. A. Salvin, E. A. Ross, M. M. Robbins, C. van Weeghel, and R. A. Bergl. 2017. "Evaluating the effect of a year-long film focused environmental education program on Ugandan student knowledge of and attitudes toward great apes." *American Journal of Primatology* 79 (8): e22673.

Leendertz, F. H., H. Ellerbrok, C. Boesch, E. Couacy-Hymann, K. Matz-Rensing, R. Hakenbeck, C. Bergmann, P. Abaza, S. Junglen, Y. Moeblus, L. Vigilant, P. Formenty, and G. Pauli. 2004. "Anthrax kills wild chimpanzees in a tropical rainforest." *Nature* 430: 451–52.

Leendertz, F. H., F. Lankester, P. Guislain, C. Neel, O. Drori, J. Dupain, S. Speede, P. Reed, N. Wolfe, S. Loul, E. Mpoudi-Ngole, M. Peeters, C. Boesch, G. Pauli, H. Ellerbrok, and E. M. Leroy. 2006a. "Anthrax in Western and Central African great apes." *American Journal of Primatology* 68: 928–33.

Leendertz, F. H., G. Pauli, K. Maetz-Rensing, W. Boardman, C. Nunn, H. Ellerbrok, S. A. Jensen, S. Junglen, and C. Boesch. 2006b. "Pathogens as drivers of population declines: The importance of systematic monitoring in great apes and other threatened mammals." *Biological Conservation* 131: 325–37.

Leendertz, S. A. J., S. A. Wich, M. Ancrenaz, R. A. Bergl, M. K. Gonder, T. Humle, and F. H. Leendertz. 2016. "Ebola in great apes–Current knowledge, possibilities for vaccination, and implications for conservation and human health." *Mammal Review* 47 (2): 98–111.

Lehmann, J., A. H. Korstjens, and R. I. M. Dunbar. 2010. "Apes in a changing world—The effects of global warming on the behaviour and distribution of African apes." *Journal of Biogeography* 37: 2217–31.

Leroy, E. M., P. Rouguet, P. Formenty, S. Souquiere, A. Kilbourne, J.-M. Forment, M. Bermejo, S. Smit, W. Karesh, R. Swanepoel, S. R. Zaki, and P. E. Rollin. 2004. "Multiple Ebola virus transmission events and rapid decline of central African wildlife." *Science* 303: 387–90.

Lonsdorf, E. V., T. R. Gillespie, T. M. Wolf, I. Lipende, J. Raphael, J. Bakuza, C. M. Murray, M. L. Wilson, S. Kamenya, and D. Mjungu. 2018. "Socioecological correlates of clinical signs in two communities of wild chimpanzees (*Pan troglodytes*) at Gombe National Park, Tanzania." *American Journal of Primatology* 80 (1): e22562.

Luikart, G., N. Ryman, D. A. Tallmon, M. K. Schwartz, and F. W. Allendorf. 2010. "Estimation of census and effective population sizes: The increasing usefulness of DNA-based approaches." *Conservation Genetics* 11: 355–73.

MacKenzie, C., and J. Hartter. 2013. "Demand and proximity: Drivers of illegal forest resource extraction." *Oryx* 47: 288–97.

Malenky, R. K., and R. W. Wrangham. 1994. "A quantitative comparison of terrestrial herbaceous food consumption by *Pan paniscus* in the Lomako Forest Zaire, and *Pan troglodytes* in the Kibale Forest, Uganda." *American Journal of Primatology* 32: 1–12.

Malhi, Y., S. Adu-Bredu, R. A. Asare, S. L. Lewis, and P. Mayaux. 2013. "The past, present and future of Africa's rainforests." *Philosophical Transactions of the Royal Society of London B* 368: 20120312.

Marshall, A. J., C. M. Boyko, K. L. Feilen, R. H. Boyko, and M. Leighton. 2009. "Defining fallback foods and assessing their importance in primate ecology and evolution." *American Journal of Physical Anthropology* 140 (4): 603–14.

Martin, G. 1983. "Bushmeat in Nigeria as a natural resource with environmental implications." *Environmental Conservation* 10: 125–32.

McFarland, K. 1994. "Update on gorillas in Cross River State, Nigeria." *Gorilla Conservation News* 8: 13–14.

McGraw, W. S. 2005. "Update on the search for Miss Waldron's red colobus monkey." *International Journal of Primatology* 26: 605–19.

McLennan, M. R. 2008. "Beleaguered chimpanzees in the agricultural district of Hoima, Western Uganda." *Primate Conservation* 23: 45–54.

Mittermeier, R. A., J. Wallis, A. B. Rylands, J. Ganzhorn, J. F. Oates, E. A. Williamson, E. Palacious, E. Heymann, M. C. M. Jierulff, Y. Long, J. Saupriatna, C. Roos, S. Walker, L. Cortes-Ortiz, and C. Schwitzer. 2009. "Primates in peril: The world's 25 most endangered primates 2008–2010." *Primate Conservation* 24: 1–57.

Morgan, D. 2007. *Best Practice Guidelines for Reducing the Impact of Commercial Logging on Great Apes in Western Equatorial Africa*. Gland, Switzerland: IUCN.

Morgan, D., R. Mundry, C. Sanz, C. E. Ayina, S. Strindberg, E. Lonsdorf, and H. S. Kühl. 2017. "African apes coexisting with logging: Comparing chimpanzee (*Pan troglodytes troglodytes*) and gorilla (*Gorilla gorilla gorilla*) resource needs and responses to forestry activities." *Biological Conservation* 218: 277–86.

Mugisha, A. R., and S. K. Jacobson. 2004. "Threat reduction assessment of conventional and community-based conservation approaches to managing protected areas in Uganda." *Environmental Conservation* 31: 233–41.

Muller, M. N., and R. W. Wrangham. 2000. "The knuckle-walking wounded." *Natural History* 109: 44.

Nelleman, C., and A. Newton. 2002. *The Great Apes: The Road Ahead*. Arendal: UNEP.

N'Goran, P. K., C. Boesch, R. Mundry, E. K. N'Goran, I. Herbinger, F. E. Yapi, and H. S. Kuhl. 2012. "Hunting, law enforcement, and African primate conservation." *Conservation Biology* 26: 565–71.

N'Goran, P. K., C. Y. Kouakou, E. K. N'Goran, S. Konaté, I. Herbinger, F. A. Yapi, H. S. Kühl, and C. Boesch. 2013. "Chimpanzee conservation status in the world heritage site Taï National Park, Côte d'Ivoire." *International Journal of Innovation and Applied Studies* 3: 326–36.

Oates, J. F. 1996. *African Primates: Status Survey and Conservation Action Plan*. Rev. ed. Gland, Switzerland: IUCN.

Oates, J. F. 1999. *Myth and Reality in the Rain Forest*. Berkeley: University of California Press.

Oates, J. F., M. Abedi-Lartey, W. S. McGraw, and T. T. Struhsaker. 2000. "Extinction of a West African red colobus monkey." *Conservation Biology* 14: 1526–32.

Oslisly, R., L. White, I. Bentaleb, C. Favier, M. Fortungne, J.-F. Gillet, and D. Sebag. 2013. "Climatic and cultural changes in west Congo Basin forests over the past 5000 years." *Philosophical Transactions of the Royal Society of London B* 368: 20120304.

Padua, S. M. 2010. "Primate conservation: Integrating communities through environmental education programs." *American Journal of Primatology* 72: 450–53.

Palstra, F. P., and D. E. Ruzzante. 2008. "Genetic estimates of contemporary effective population size: What can they tell us about the importance of genetic stochasticity for wild population persistence?" *Molecular Ecology* 17: 3428–47.

Patz, J. A., P. R. Epetein, T. A. Burke, and J. M. Balbus. 1996. "Global climate change and emerging infectious diseases." *Journal of the American Medical Association* 275: 217–23.

Phalan, B., M. Bertzky, S. H. M. Butchart, P. F. Donald, J. P. W. Scharlemann, A. Stattersfield, and A. Balmford. 2013. "Crop expansion and conservation priorities in tropical countries." *PLoS One* 8: e51759.

Pimm, S. L., C. N. Jenkins, R. Abell, T. M. Brooks, J. L. Gittleman, L. N. Joppa, P. H. Raven, C. M. Roberts, and J. O. Sexton. 2014. "The biodiversity of species and their rates of extinction, distribution, and protection." *Science* 344: 1246752.

Plumptre, A. J., and D. Cox. 2006. "Counting primates for conservation: Primate survey in Uganda." *Primates* 47: 65–73.

Plumptre, A. J., D. Cox, and S. Mugume. 2003. "The status of chimpanzees in Uganda." *Albertine Rift Technical Report Series* 2.

Plumptre, A. J., S. Nampindo, N. Mutungire, M. Gonya, and T. Akuguzibwe. 2008. *Surveys of Chimpanzees and Other Large Mammals in Uganda's Forest Reserves in the Greater Virunga Landscape.* Kampala, Uganda: Wildlife Conservation Society (WCS).

Plumptre, A. J., R. Rose, G. Nangendo, E. A. Williamson, K. Didier, J. Hart, F. Mulindahabi, C. Hicks, B. Griffin, H. Ogawa, S. Nixon, L. Pintea, A. Vosper, M. McClennan, F. Amsini, A. McNeilage, J. R. Makana, M. Kanamori, A. Hernandez, A. Piel, F. Stewart, J. Moore, K. Zamma, M. Nakamura, S. Kamenya, G. Idani, T. Sakamaki, M. Yoshikawa, D. Greer, S. Tranquilli, R. Beyers, T. Furuichi, C. Hashimoto, and E. Bennet. 2010. *Eastern Chimpanzee (*Pan troglodytes schweinfurthii*): Status Survey and Conservation Action Plan 2010-2020.* Gland, Switzerland: IUCN.

Pomeroy, D., and H. Tushabe. 2004. "The state of Uganda's biodiversity." National Biodiversity Data Bank. Makerere University, Kampala.

Prado-Martinez, J., P. H. Sudmant, J. M. Kidd, H. Li, J. L. Kelley, B. Lorente-Galdos, K. R. Veeramah, A. E. Woerner, T. D. O'Connor, G. Santpere, A. Cagan, C. Theunert, F. Casals, H. Laayouni, K. Munch, A. Hobolth, A. E. Halager, M. Malig, J. Hernandez-Rodriguez, I. Hernando-Herraez, K. Prufer, M. Pybus, L. Johnstone, M. Lachmann, C. Alkan, D. Twigg, N. Petit, C. Baker, F. Hormozdiari, M. Fernandez-Callejo, M. Dabad, M. L. Wilson, L. Stevison, C. Camprubi, T. Carvalho, A. Ruiz-Herrera, L. Vives, M. Mele, T. Abello, I. Kondova, R. E. Bontrop, A. Pusey, F. Lankester, J. A. Kiyang, R. A. Bergl, E. Lonsdorf, S. Myers, M. Ventura, P. Gagneux, D. Comas, H. Siegismund, J. Blanc, L. Agueda-Calpena, M. Gut, L. Fulton, S. A. Tishkoff, J. C. Mullikin, R. K. Wilson, I. G. Gut, M. K. Gonder, O. A. Ryder, B. H. Hahn, A. Navarro, J. M. Akey, J. Bertranpetit, D. Reich, T. Mailund, M. H. Schierup, C. Hvilsom, A. M. Andres, J. D. Wall, C. D. Bustamante, M. F. Hammer, E. E. Eichler, and T. Marques-Bonet. 2013. "Great ape genetic diversity and population history." *Nature* 499: 471–75.

Pusey, A. E., M. L. Wilson, and D. A. Collins. 2008. "Human impacts, disease risk, and population dynamics in the chimpanzees of Gombe National Park, Tanzania." *American Journal of Primatology* 70: 738–44.

Putz, F. E., D. P. Dykstra, and R. Heinrich. 2000. "Why poor logging practices persist in the tropics." *Conservation Biology* 14: 505–8.

Quiatt, D., V. Reynolds, and E. J. Stokes. 2002. "Snare injuries to chimpanzees (*Pan troglodytes*) at 10 study sites in East and West Africa." *African Journal of Ecology* 40: 303–5.

Refisch, J., and I. Koné. 2005. "Impact of commercial hunting on monkey populations in the Taï Region, Côte d'Ivoire." *Biotropica* 37: 136–44.

Ripple, W. J., T. M. Newsome, C. Wolf, R. Dirzo, K. T. Everatt, M. Galetti, M. W. Hayward, G. I. Kerley, T. Levi, and P. A. Lindsey. 2015. "Collapse of the world's largest herbivores." *Science Advances* 1: e1400103.

Robinson, E. A., G. D. Ryan, and J. A. Newman. 2012. "A meta-analytical review of the effects of elevated CO2 on plant-arthropod interactions highlights the importance of interacting environmental and biological variables." *New Phytologist* 194: 321–36.

Rothman, J. M., C. A. Chapman, T. T. Struhsaker, D. Raubenheimer, D. Twinomugisha, and P. G. Waterman. 2015. "Cascading effects of global change: Decline in nutritional quality of tropical leaves." *Ecology* 96: 873–78.

Rouquet, P., J. M. Froment, M. Bermejo, A. Kilbourn, W. Karesh, P. Reed, B. Kumulungui, P. Yaba, A. Delicat, P. E. Rollin, and E. M. Leroy. 2005. "Wild animal mortality monitoring and human Ebola outbreaks, Gabon and Republic of Congo, 2001–2003." *Emerging Infectious Diseases* 11: 283–90.

Rwego, I. B., G. Isabirye-Basuta, T. R. Gillespie, and T. L. Goldberg. 2008. "Gastrointestinal bacterial transmission among humans, mountain gorillas, and livestock in Bwindi Impenetrable National Park, Uganda." *Conservation Biology* 22: 1600–1607.

Samson, D. R., and K. D. Hunt. 2012. "A thermodynamic comparison of arboreal and terrestrial sleeping sites for dry-habitat chimpanzees (*Pan troglodytes schweinfurthii*) at the Toro-Semliki Wildlife Reserve, Uganda." *American Journal of Primatology* 74: 811–18.

Sandbrook, C., and S. Semple. 2006. "The rules and the reality of mountain gorilla *Gorilla beringei beringei* tracking: How close do tourists get?" *Oryx* 40: 428–S33.

Sarkar, D., C. A. Chapman, K. Valenta, S. Angom, W. Kagoro, and R. Sengupta. Accepted with revision. "Research stations as instrument of conservation: Value expressed by community perceptions." *Conservation Letters.*

Savage, A., R. Guillen, I. Lamilla, and L. Soto. 2010. "Developing an effective community conservation program for cotton-top tamarins (*Saguinus oedipus*) in Colombia." *American Journal of Primatology* 72: 379–90.

Schwartz, M. K., G. Luikart, and R. S. Waples. 2007. "Genetic monitoring as a promising tool for conservation and management." *Trends in Ecology & Evolution* 22: 25–33.

Scully, E. J., S. Basnet, R. W. Wrangham, M. N. Muller, E. Otali, D. Hyeroba, K. A. Grindle, T. E. Pappas, M. E. Thompson, and Z. Machanda. 2018. "Lethal respiratory disease associated with human rhinovirus C in wild chimpanzees, Uganda, 2013." *Emerging Infectious Diseases* 24: 267.

Shukman, D., and S. Piranty. 2017. "The secret trade in baby chimps." *BBC News: Science and Environment* 30 January 2017 https://www.bbc.co.uk/news/resources/idt-5e8c4bac-c236-4cd9-bacc-db96d733f6cf.

Sop, T., S. M. Cheyne, F. Maisels, S. A. Wich, and E. A. Williamson. 2015. "Abundance Annex: Ape population abundance estimates." In *State of the Apes*, edited by A. Lanjouw, H. Rainer, and A. White. Arcus Foundation. Cambridge: Cambridge University Press.

Stiles, D., I. Redmond, D. Cress, C. Nellemann, and R. K. Formo, eds. 2013. *Stolen Apes—The Illicit Trade in Chimpanzees, Gorillas, Bonobos and Orangutans. A Rapid Response Assessment.* Arendal: UNEP, GRID-Arendal.

Stiling, P., and T. Cornelissen. 2007. "How does elevated carbon dioxide (CO_2) affect plant-animal interactions? A field experiment and meta-analysis of CO_2-mediated changes on plant chemistry and herbivore performance." *Global Change Biology* 13: 1823–42.

Struhsaker, T. T. 1973. "A recensus of vervet monkeys in the Masai-Amboseli Game Reserve, Kenya." *Ecology* 54: 930–32.

Struhsaker, T. T. 1976. "A further decline in numbers of Amboseli vervet monkeys." *Biotropica* 8: 211–14.

Struhsaker, T. T., P. J. Struhsaker, and K. S. Siex. 2005. "Conserving Africa's rain forests: Problems in protected areas and possible solutions." *Biological Conservation* 123: 45–54.

Taylor, G., J. P. W. Scharlemann, M. Rowcliffe, N. Kümpel, M. B. J. Harfoot, J. E. Fa, R. Melisch, E. J. Milner-Gulland, S. Bhagwat, K. A. Abernethy, A. S. Ajonina, L. Albrechtsen,

S. Allebone-Webb, E. Brown, D. Brugiere, C. Clark, M. Colello, G. Cowlishaw, D. Crookes, E. De Merode, J. Dupain, T. East, D. Edderai, P. Elkan, D. Gill, E. Greengrass, C. Hodgkinson, O. Ilambu, P. Jeanmart, J. Juste, J. M. Linder, D. W. Macdonald, A. J. Noss, P. U. Okorie, V. J. J. Okouyi, S. Pailer, P. R. Poulsen, M. Riddell, J. Schleicher, B. Schulte-Herbruggen, M. Starkey, N. van Vliet, C. Whitham, A. S. Willcox, D. S. Wilkie, J. H. Wright, and L. M. Coad. 2015. "Synthesising bushmeat research effort in West and Central Africa: A new regional database." *Biological Conservation* 181: 199–205.

Terborgh, J. 1999. *Requiem for Nature*. Washington, DC: Island Press.

Tranquilli, S., M. Abedi-Lartey, K. Abernethy, F. Amsini, A. Asamoah, C. Balangtaa, S. Blake, E. Bouanga, T. Breuer, T. M. Brncic, G. Campbell, R. Chancellor, C. A. Chapman, T. R. Davenport, A. Dunn, J. Dupain, A. Ekobo, M. Eno-Nku, G. Etoga, T. Furuichi, S. Gatti, A. Ghiurghi, C. Hashimoto, J. Hart, J. Head, M. Hega, I. Herbinger, T. C. Hicks, L. H. Holbech, B. Huijbregts, H. S. Kühl, I. Imong, S. L. Yeno, J. Linder, P. Marshall, P. M. Lero, D. Morgan, L. Mubalama, P. K. N'Goran, A. Nicholas, S. Nixon, E. Normand, L. Nziguyimpa, Z. Nzooh-Dongmo, R. Ofori-Amanfo, B. G. Ogunjemite, C. A. Petre, H. J. Rainey, S. Regnaut, O. Robinson, A. Rundus, C. M. Sanz, D. T. Okon, A. Todd, Y. Warren, and V. Sommer. 2014. "Protected areas in tropical Africa: Assessing threats and conservation activities." *PLoS One* 9: e114154.

Tranquilli, S., M. Abedi-Lartey, F. Amsini, L. Arranz, A. Asamoah, O. Babafemi, N. Barakabuye, G. Campbell, R. Chancellor, T. R. B. Davenport, A. Dunn, J. Dupain, C. Ellis, G. Etoga, T. Furuichi, S. Gatti, A. Ghiurghi, E. Greengrass, C. Hashimoto, J. Hart, I. Herbinger, T. C. Hicks, L. H. Holbech, B. Huijbregts, I. Imong, N. Kumpel, F. Maisels, P. Marshall, S. Nixon, E. Normand, L. Nziguyimpa, Z. Nzooh-Dogmo, D. T. Okon, A. Plumptre, A. Rundus, J. Sunderland-Groves, A. Todd, Y. Warren, R. Mundry, C. Boesch, and H. Kuehl. 2012. "Lack of conservation effort rapidly increases African great ape extinction risk." *Conservation Letters* 5: 48–55.

Wallis, J., and D. R. Lee. 1999. "Primate conservation: The prevention of disease transmission." *International Journal of Primatology* 20: 803–26.

Walsh, P. D., K. A. Abernethy, M. Bermejo, R. Beyersk, P. De Wachter, M. E. Akou, B. Huljbregis, D. I. Mambounga, A. K. Toham, A. M. Kilbourn, S. A. Lahm, S. Latour, F. Maisels, C. Mbina, Y. Mihindou, S. N. Obiang, E. N. Effa, M. P. Starkey, P. Telfer, M. Thibault, C. E. G. Tutin, L. J. T. White, and D. S. Wilkie. 2003. "Catastrophic ape decline in Western Equatorial Africa." *Nature* 422: 611–14.

Walsh, P. D., T. Breuer, C. Sanz, D. Morgan, and D. Doran-Sheehy. 2007. "Natural history miscellany—Potential for ebola transmission between gorilla and chimpanzee social groups." *American Naturalist* 169: 684–89.

Wanyama, F. 2005. "Ground census of mammals in Kibale National Park Uganda: Unpublished report." Ugandan Wildlife Authority. Kampala, Uganda.

Wanyama, F., R. Muhabwe, A. J. Plumptre, C. A. Chapman, and J. M. Rothman. 2009. "Censusing large mammals in Kibale National Park: Evaluation of the intensity of sampling required to determine change." *African Journal of Ecology* 48: 953–61.

Whiteley, A. R., S. W. Fitzpatrick, W. C. Funk, and D. A. Tallmon. 2015. "Genetic rescue to the rescue." *Trends in Ecology & Evolution* 30: 42–49.

Wilkie, D. S. 2000. "Roads, development, and conservation in the Congo Basin." *Conservation Biology* 14: 1614–22.

Williams, J. M., E. V. Lonsdorf, M. L. Wilson, J. Schumacher-Stankey, J. Goodall, and A. E. Pusey. 2008. "Causes of death in the Kasekela chimpanzees of Gombe National Park, Tanzania." *American Journal of Primatology* 70: 766–77.

Woodford, M. H., T. M. Butynski, and W. B. Karesh. 2002. "Habituating the great apes: The disease risks." *Oryx* 36: 153–60.

Wrangham, R. W., N. L. Conklin, C. A. Chapman, and K. Hunt. 1991. "The significance of fibrous foods for Kibale Forest Chimpanzees." *Philosophical Transactions of the Royal Society of London B* 334: 171–78.

Wrangham, R. W., G. Hagel, M. Leighton, A. J. Marshall, P. Waldau, and T. Nishida. 2008. "The great ape world heritage species project." In *Conservation in the 21st Century: Gorillas as a Case Study*, edited by T. S. Stoinski, H. D. Steklis, and P. T. Mehlman, 282–95. Boston: Springer.

Wright, S. J., H. C. Muller-Landau, and J. Schipper. 2009. "The future of tropical species on a warmer planet." *Conservation Biology* 23: 1418–26.

Zelazowski, P., Y. Malhi, C. Huntingord, S. Sitch, and J. B. Fisher. 2011. "Changes in the potential distribution of tropical forest on a warmer planet." *Philosophical Transactions of the Royal Society of London B* 369: 137–60.

Zvereva, E. L., and M. V. Kozlov. 2006. "Consequences of simultaneous elevation of carbon dioxide and temperature for plant-herbivore interactions: A meta-analysis." *Global Change Biology* 12: 27–41.

26

Holistic Approach for Conservation of Chimpanzees in Kibale National Park, Uganda

JESSICA A. HARTEL, EMILY OTALI, ZARIN MACHANDA, RICHARD W. WRANGHAM, AND ELIZABETH ROSS

Introduction

Human population growth coupled with crippling poverty in many rural areas across Africa has threatened the extinction and endangerment of many wildlife species (Estrada et al. 2017; Lamarque et al. 2009), especially those in small reserves (Brashares, Arcese, and Sam 2001). In particular, all four subspecies of chimpanzees (*Pan troglodytes* spp.), our closest living relatives, are classified as endangered (Humle et al. 2016) and many populations living in human-dominated landscapes are facing local extinction (Guinea: Hockings 2009; Nigeria: Greengrass 2009; Uganda: McLennan 2010). In the last century anthropogenic pressures—deforestation, bushmeat hunting, and disease transmission—have caused the confirmed or likely extinction of wild chimpanzees in four countries (Benin, Burkina Faso, Gambia, and Togo; Campbell and Houngbedji 2015; Ginn et al. 2013), and dramatic population decline in their remaining 21 range countries (IUCN: Humle et al. 2016; Eastern chimpanzees: Plumptre et al. 2010; Nigeria-Cameroon chimpanzees: Morgan et al. 2011; Western chimpanzees: Kühl et al. 2017; Kormos et al. 2003; central chimpanzees: Morgan et al., chapter 27 this volume). These pressures also threaten the health, longevity, and viability of many small unprotected local populations (Cowlishaw and Dunbar 2000; McCarthy et al. 2015; McLennan 2010). By 2030, predictions estimate that of the suitable African great ape habitat that currently remains, less than 10% will be relatively undisturbed by human infrastructural development (Nelleman and Newton 2002), thereby isolating most populations (Fitzgerald et al. 2018). While many species of animals and plants are endangered globally, the African great apes are particularly vulnerable because they have large home ranges and low population densities. The loss of apes is troubling because they are effectively "umbrella species" (Lambert 2011), meaning that ape conservation efforts also help to conserve many other sympatric species (Chapman et al., chapter 25 this volume). Such conservation activities are badly needed because, based on current and accelerating rates of bushmeat hunting and forest loss, it has been suggested that wild chimpanzees, gorillas, and bonobos all risk extinction by 2100 (Nishida et al. 2000).

While this biodiversity loss of chimpanzees and other great apes is of serious concern (Butchart et al. 2010), so too is the livelihood of local people. Unfortunately, the clash between the needs of rural populations and habitat protection has often been unsatisfactorily mitigated. During the last century over 14 million people have reportedly been displaced from their homes across Africa in the name of conservation (Dowie 2011). Tactics designed to protect wildlife areas included "fortress conservation," which can vilify traditional practices of land and animal use (Brockington 2004). Superficially, results can look good for conservation—over the past two decades the proportion of terrestrial land in Africa given to protected areas has grown from 9% to 15.5% (UNEP-WCMC 2019). But while legal protection to wildlife has increased (Bruner et al. 2001; Struhsaker, Struhsaker, and Siex 2005), the apparent success can be misleading as laws and policies may not actually be enforced. Many protected areas are merely "paper parks" (Blom, Yamindou, and Prins 2004) that exist on maps and in legislation, but in reality offer little to no protection. While these areas are officially designated as protected, conservation activities are often insufficient to halt degradation. How to implement successful conservation policies that are truly effective and sustainable in the long term is thus a critical question. During periods of active investment by relevant institutions, conservation initiatives, ecotourism, and educational outreach have all shown positive impacts on great ape biodiversity and species preservation (Tranquilli et al. 2011). However, such activities are vulnerable to failure when the initiative stops. For example, a conservation organization's presence in Marahoué National Park, Côte d'Ivoire, had elevated protection levels, but within six years of the conservation efforts ceasing, there was a 93% decrease in forest cover and an 82% reduction in the chimpanzee population (Campbell et al. 2008).

One solution may be the establishment of long-term (multi-decade) field sites, which have repeatedly been found to generate both direct and indirect conservation benefits in chimpanzees (Budongo, Uganda: Reynolds 2005; Gombe, Tanzania: Pusey et al. 2007; Kibale, Uganda: Wrangham and Ross 2008; Taï, Côte d'Ivoire: Campbell et al. 2011), other great ape species (gorillas: Williamson and Fawcett 2018), protected areas in general (Tranquilli et al. 2014), and even unprotected areas (Piel et al. 2015) such as logging concessions (Morgan et al. 2013, chapter 27 this volume). Long-term research sites create unique opportunities for on-the-ground conservation promotion, management, and sustainability because researchers stay long enough to build trusted relationships with key stake-holders, partly by providing practical help (Kibale, Uganda: Wrangham and Ross 2008). Even when researchers begin by focusing on purely scientific studies, they tend to incorporate conservation initiatives and/or educational outreach. Their local knowledge makes them well positioned to address issues of sustainability using appropriate approaches, while also being sensitive to the needs and struggles of peoples living sympatrically with wildlife. Results seem to be beneficial, such as in Taï National Park, Côte d'Ivoire, where a detailed study found that researcher presence positively affected biodiversity abundance relative to adjacent areas (Campbell et al. 2011), and when anti-poaching patrols were coordinated with local law enforcement, primate densities in particular were up to 100 times larger near research stations (N'Goran et al. 2012). While many studies have been published on the anthropogenic threats to chimpanzee conservation (see Chapman et al., chapter 25 this volume, for a review of these threats) and the actions of law enforcement to mitigate these threats in protected areas (Critchlow et al. 2015; Kablan et al. 2017), only a handful of research sites have published on their conservation and/or educational outreach initiatives (see Wrangham and Ross 2008 for a review of long-term chimpanzee research sites at Bossou, Budongo, Gombe, Kibale, Mahale, and Taï, and for the Virunga mountain gorillas). It is not clear whether this is due to a lack of establishment, data, priority, or something else.

The best-known and most long-term chimpanzee conservation and education initiatives come from Gombe Stream Research Center in Gombe National Park, Tanzania. Their community-centered conservation approach, Lake Tanganyika Catchment Reforestation and Education Project (TACARE), was established in 1994 to initially address deforestation outside the park. TACARE has contributed to positive attitudinal and behavioral changes (Anderson et al. 2004) as well as a reduction in the rate of forest degradation in and around Gombe (Pusey et al. 2007). However, to keep pace with the growing human population around the park (Anderson et al. 2004), more focused conservation strategies were identified, in collaboration with the Nature Conservancy, to incorporate satellite imagery in analyses and work more closely with local communities (Pusey et al. 2007). Over the years TACARE has grown their community-based conservation programs and is now the Jane Goodall Institute's flagship population, health, and environment project with six major initiatives that focus on community development, forestry, agriculture, health, land use planning with GIS, and the globally recognized environmental education for youth program, Roots and Shoots. However, we recognize that scaling community-based conservation projects to the level of the Jane Goodall Institute is challenging and requires more funding and management than most long-term research projects can afford.

Several other long-term chimpanzee research sites have reported a range of successful conservation education programs in/around their parks (see Wrangham and Ross 2008 for review). In West Africa, the Taï Chimpanzee Project in Taï National Park, Côte d'Ivoire, established the Wild Chimpanzee Foundation, which focuses on environmental education in local villages, wildlife and human encroachment field surveys, and supporting local development actions (Boesch et al. 2008b). Repeated exposure of their multimedia bushmeat awareness campaigns resulted in a significant (62%) decrease in bushmeat consumption in poverty-stricken rural households (Kouassi et al. 2017). In partnership with local teachers, their environmental education club PAN (peoples, animals, and nature) positively influenced children's attitudes toward nature and significantly increased their environmental knowledge (Borchers et al. 2014). The Bossou Chimpanzee Project in Guinea established environmental education and community development campaigns (Matsuzawa, Humle, and Sugiyama 2011), which included the Green Corridor Project to reconnect Bossou to the Nimba Mountains by replanting trees along a 4 km corridor (Matsuzawa and Kourouma 2008). While the corridor is nearly complete, the study chimpanzee population has sadly decreased to only seven as of March 2017 (Fitzgerald et al. 2018). In East Africa, the Mahale Mountains Chimpanzee Research Project in Mahale National Park, Tanzania, established the Mahale Wildlife Conservation Society, which in turn founded the *Pan African News* for researchers, built a nearby primary school and park visitor center, and guided student field trips in the park (Nishida and Nakamura 2008). The Budongo Conservation Field Station in the Budongo Forest Reserve, Uganda, has an active snare removal team, sustainable vermin control, selective forest harvesting programs, and local veterinary and education services (Asiimwe et al. 2016; Babweteera, Reynolds, and Züberbuhler 2008). The Ngogo Chimpanzee Project in Kibale National Park, Uganda, has an active snare removal team and works in collaboration with the North Carolina Zoo's UNITE for the Environment on conservation education initiatives in nine local schools.

From what has been learned at Gombe and other sites, a successful long-term future for conservation surely depends on capacity building and buy-in from people surrounding the protected areas (Shaffer et al. 2017). For this reason it has been suggested that future research and conservation initiatives need to integrate conservation biology with social action (Adams and Mulligan 2003; Lwanga and Isabirye-Basuta 2008) so as to build long-term relationships of trust with local communities (Breuer and Mavinga 2010; Collins and Goodall 2008; Kasenene and Ross 2008). Using a holistic approach to conservation, the Kibale Chimpanzee Project in Kibale National Park, Uganda, employs a conservation team who regularly patrol the park while working on conservation education initiatives in partnership with the Kasiisi Project in 16 local schools within 5 km of the national park. The success of this partnership is partly the result of the project founders, Drs. Richard Wrangham and Elizabeth Ross, whose marriage has fostered strong communication and collaboration between the two projects for the past 20 years. This chapter presents a case study of these two collaborative projects working in and around Kibale National Park to conserve chimpanzees.

THREATS TO CHIMPANZEES LIVING IN KIBALE NATIONAL PARK

Kibale National Park (hereafter refered to as Kibale) lies in the Eastern Afromontane region, one of the world's thirty-five biodiversity hotspots (Mittermeier et al. 2005; Wright 2005), and is a mid-altitude rainforest of 795 km^2 with the highest primate density and diversity in East Africa (Fashing and Cords 2000; Oates et al. 1990). Kibale is home to the largest known population of the eastern chimpanzee subspecies (*P. t. schweinfurthii*, Plumptre et al. 2010), and serves as a stronghold for many other threatened and endangered species, including 12 other primate species, elephants, golden cats, pangolins, and hundreds of other rare mammal, bird, reptile, amphibian, insect, and plant species (Plumptre et al. 2007). Chimpanzees living in Kibale have been designated by the IUCN as a high priority for conservation (Plumptre et al. 2010).

Though announced as a national park in 1993, Kibale faces numerous long-term threats. In 2015, the human population density around Kibale was estimated at 308 people/km^2 (MacKenzie et al. 2017a). Neighboring human populations are growing by 3.5% annually (Goldman et al. 2008; WPR 2017) and with 50% under the age of 15 (Bwambale 2012), the surrounding population is predicted to increase almost five-fold by 2050 (PRB 2017). Due to the low annual income in the region (Tumusiime and Vedeld 2015), wildlife is often viewed as a "free" resource tempting people to enter the park illegally to poach for bushmeat. Given the permeable nature of the park boundary, chimpanzees and other wildlife (typically elephants and baboons) sometimes range outside of the park when habitat and food sources are lost and/or seasonally variable. Resource-dense areas, such as a crop fields (mainly maize, fruits, sugar cane, etc.), located near the forest edge can become favorable foraging targets for chimpanzees. While chimpanzees are less destructive than many other wildlife (e.g., elephants, baboons, bush pigs, other monkeys; Naughton-Treves 1997; Tweheyo, Hill, and Obua 2005), at times, they do destroy villagers' crops (MacKenzie and Ahabyona 2012). Additionally, chimpanzees have injured and even killed children (Wrangham et al. 2000). Thus, the people living around Kibale have good reason to dislike the chimpanzees and farmers who are victims of these chimpanzee crop-foraging attacks may (over time) develop negative attitudes toward chimpanzees and their conservation (Garriga et al. 2017; McLennan and Hill 2012). People sometimes respond by setting snares in their gardens and/or attacking chimpanzees with dogs and spears when they encounter them.

Ultimately, it is the movement of both people and chimpanzees in and out of the park that fuels the human-chimpanzee conflict and directly threatens the survival of Kibale's chimpanzees. This manifests in three main ways, as chimpanzees (1) are caught in snares set for other bushmeat species in the park, (2) are injured or killed during crop foraging, and (3) catch human and domestic animal diseases (Parsons et al. 2014), including things like the common cold, which can lead to mortality (Scully et al. 2018). This chapter focuses predominantly on snaring and human-chimpanzee conflicts, rather than disease.

Though prohibited by law, snare traps made of wire or nylon are commonly set by hunters within the park's boundaries to catch small game (i.e., antelope, bushpigs). Snares act like landmines in the forest—they are cryptic, indiscriminate, and deadly. While not the intended target, chimpanzees are often accidental victims: their appendages become entangled and over time the snares cut deep into their flesh causing pain, infection, and permanent damage—often they lose hands and/or feet (fig. 26.1, Cohen 2010). Wrangham and Mugume (2000) estimated there were 15,000 snares set in Kibale at any given time, resulting in a 3.7% risk of a chimpanzee being snared each year. This anthropogenic pressure is too intense for the national wildlife authority to mitigate on their own, and as a result, approximately one-third of chimpanzees in Uganda have permanent snare injuries, ranging from missing/paralyzed digits to limb amputations (Plumptre et al. 2010).

Given that the threats to chimpanzees are so intertwined with human activity, for conservation efforts to succeed, local people must see value in chimpanzees and other wildlife despite the challenges they bring. Villagers' needs must therefore be incorporated into conservation agendas (Adams and Hutton 2007). This can be difficult because a fear of chimpanzees is instilled even in young children. In 2014, 62% of children's negative comments about chimpanzees cited their aggressive behavior (Elizabeth Ross, unpublished

FIGURE 26.1. Max (pictured left) lost both of his feet in two separate snare injuries, one at the age of five and the other at the age of eight, with the likely culprits being wire foot snares (pictured right).

data). Therefore, one of the keys to long-term conservation is to focus efforts toward young children. This is best achieved by improving academic and environmental standards (Fiallo and Jacobsen 1995; Kideghesho, Røskaft, and Kaltenborn 2007) since individuals who do not complete school tend to be poor and/or unemployed, and therefore more likely to poach wildlife (Knapp, Peace, and Bechtel 2017). Less than 60% of Ugandan children complete primary school (Mwesigwa 2015; UNPC 2017) and academic standards in rural schools, such as those surrounding Kibale, have tended to be low. Lacking the education and skills needed to enter the job market, local children turn to subsistence farming on progressively smaller patches of land as they transition to adulthood, and in tough times often depend on the forest and its resources to make ends meet. These ongoing community challenges both inside and out of the park must be continually assessed and addressed for local conservation strategies to work.

Our Holistic Approach: A Community Call to Action

Our NGOs, the Kibale Chimpanzee Project (KCP) and the Kasiisi Project, have developed collaborative initiatives that focus on the conservation of chimpanzees living in Kibale National Park. Established in 1987 by Richard Wrangham, KCP studies the Kanyawara chimpanzee community located on the northwest side of the park (Chapman and Wrangham 1993). After 10 years of long-term research, KCP, in partnership with the Uganda Wildlife Authority (UWA), established two key conservation-based projects in the area: the Kibale Snare Removal Program (KSRP) and the Kasiisi Project. Core aims of the KSRP and the Kasiisi Project partnership are to address the persistent threats to chimpanzee conservation in and around the park. The broader mission of both projects is conservation of Kibale and its wildlife residents. Together we use a multifaceted approach that includes on-the-ground conservation efforts within the park (KSRP) paired with community outreach and development outside the park (Kasiisi Project) to address persistent and evolving human-chimpanzee conflict issues. Both approaches work in concert with one another and are crucial to the conservation of chimpanzees and other wildlife at a local level. However, we do not stand alone. Our effectiveness depends on long-term partnerships with other conservation- and education-focused govermental organizations, NGOs, educators, and scientists at local and international levels (see acknowledgments).

In addition to our conservation efforts, our projects offer Ugandans highly desired, competitive salary positions with further education, health, and retirement benefits (KCP/KSRP: 21 personnel, Kasiisi Project: 18 personnel). Most employees live at the field station or in nearby communities along the park boundary. Because they share their experiences with family and friends, employees sensitize local communities to conservation issues and appear to help deter illegal activities in the park.

KIBALE SNARE REMOVAL PROGRAM: CONSERVATION VIA PARK MONITORING

As the conservation arm of KCP, the Kibale Snare Removal Program was founded in collaboration with the UWA in response to the high proportion of debilitating chimpanzee snare injuries observed in the park. KSRP has five main goals within and around Kibale: 1) conduct regular patrols to remove snares and apprehend poachers (when accompanied by UWA), 2) collect data on the occurrence and location of snares and other illegal activities, 3) collaborate with UWA to strengthen law enforcement, 4) provide data on snares and other illegal activities to park management, and 5) educate local communities to help curtail poaching.

Over the past 20 years, KSRP has grown from two employees to six, who deploy as two independent patrol teams that are often accompanied by UWA rangers. While the KSRP patrols cover roughly two-thirds of the park, UWA rangers conduct patrols throughout the entire park, and other snare removal teams (Ngogo and Sebitoli) cover their chimpanzee home ranges and beyond. These complementary and collaborative patrol efforts by all parties result in patrol coverage redundancy within the park, and therefore help strengthen our overall effectiveness as conservationist and wildlife protectors.

Like UWA rangers, KSRP immediately confiscates and destroys all encountered snares and other hunting evidence, such as nets and spears. When KSRP patrols are accompanied by UWA personnel, poachers are arrested. Rigorous data collection protocols and GPS technologies are used to document the location and relevant details of snares, other hunting/poaching events, additional illegal activities, and wildlife abundance measures. These data are used to evaluate the program's general effectiveness, improve its efficiency, and provide intelligence to the patrol teams regarding hotspots of snares and other poaching/illegal activities.

In past years, and as needed, KSRP has employed local village residents to serve as Community Liaisons to document and mitigate human-chimpanzee conflicts in communities where crop foraging and aggressive encounters were high. Liaisons engaged in open discussions with villagers about their feelings toward chimpanzees and the actions they took when faced with encounters. They advised about chimpanzee behavior and how villagers could avoid or reduce aggressive interactions. To further mitigate conflict, research project field assistants notify local farmers when chimpanzees are actively crop foraging, giving farmers the opportunity to encourage the chimpanzees to leave in non-aggressive ways.

KASIISI PROJECT: CONSERVATION VIA COMMUNITY DEVELOPMENT

As the community arm of KCP, the Kasiisi Project partners with with the Kibale Forest Schools Program (KFSP, Ugandan-registered NGO) to address the environmental impact of a rapidly increasing human population. The premise of the educational program is to enhance tolerance by highlighting the unique situation of living alongside chimpanzees and other endemic wildlife (Lee and Priston 2005). The project aims to foster this attitude in communities around Kibale. By linking our programs to research, school curricula, and community needs, we can change attitudes and behavior in ways that have powerful and practical conservation benefits. Our vision is to guide and nurture a generation of committed rural conservationists who are passionate about protecting the forest now and in the future by implementing sound environmental practices in their homes. This is especially important in the park-boundary communities where human-wildlife conflicts are often daily occurences. We address four main objectives known to significantly improve conservation outcomes: age, academic standards, environmental knowledge, and community trust and support.

Age: We target children in pre-adolescence, a time when they are most receptive to conservation messages (Chawla 2007), by working in 16 government primary schools located within 5 km of the northern and western boundary of Kibale, reaching over 11,000 children aged between six and 16.

Academic Standards: Our higher academic standards are achieved by utilizing a range of support strategies including improving infrastructure, teacher training, libraries and literacy, post-primary scholarships, clean water and sanitation, health education, school lunches, pre-schools, boarding facilities, and addressing the special needs of girls (e.g., sanitary pads, sex education). Prioritizing teachers is key to successful education outcomes. To incentivize teachers and combat absenteeism, we provide clean and safe classrooms, adequate latrines, and opportunities for advancing qualifications.

Environmental Knowledge: Increasing environmental knowledge is achieved by empowering children with accurate, exciting, and engaging information about their environment (specifically, Kibale and its chimpanzees). Conservation education programs for children are run primarily after school in Wildlife Clubs (WLCs) headed by teacher patrons. WLCs are familiar organizations that promote environmental knowledge and active conservation, and support the academic curriculum. We build passion for the environment through interesting, interactive, and comprehensive conservation education programs that prioritize hands-on engagement. Using brief instructional talks coupled with films, art/drama, debate, field trips, practical conservation activities, and interactive games, our programs build knowledge, stimulate empathy toward chimpanzees and other wildlife, and encourage changes in behavior. Programs are evaluated by measuring student attitudinal and behavioral changes using pre- and post-surveys, program participation and retention, and performance on standarized exams.

Community Trust and Support: As important centers of the community, schools offer regular access to parents, local government, and political leaders. All programs are designed and implemented in collaboration with parents, teachers, school management committees, district education and health departments, local clinics, village health teams, churches, and local political leaders. Regular community questionnaires keep the project updated on parents' opionions and requests. The project has a strong reputation for nonpartisanship, built on long-term reliable investment in areas people highly

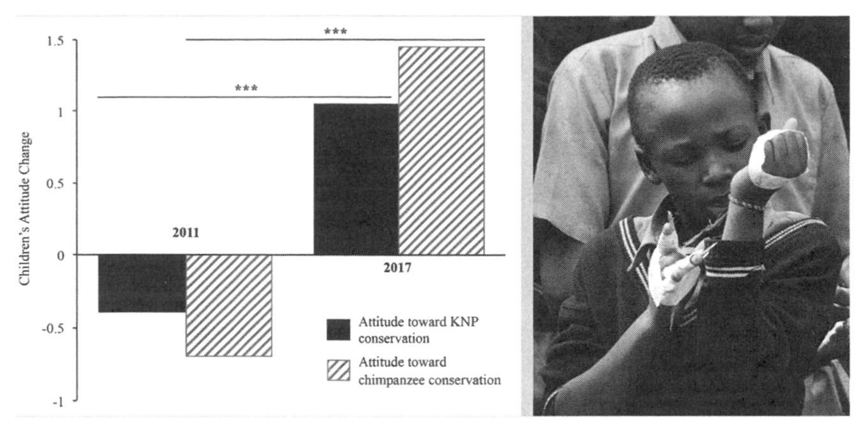

FIGURE 26.2. As the Kasiisi Project's conservation education programs have grown and improved, students' attitudes towards Kibale National Park and chimpanzees have gotten more positive (bar graph). In the Snare Care activity (pictured right) using tape and sticks to immobilize their fingers, wildlife club students get to experience some of the challenges chimpanzees face when they have paralyzed or missing digits/limbs from snares. *** $p < 0.05$.

value, such as their children's health and education. The schools where we work are key to building effective conservation networks between community, research, and local government, and our programs are crucial to molding young environmental stewards.

THE KSRP-KASIISI PROJECT PARTNERSHIP

KSRP and Kasiisi Project work collaboratively to integrate on-the-ground conservation efforts with educational outreach. To aid in this goal, the projects jointly employ a Conservation Education Liaison who works with both organizations to coordinate KSRP-led conservation education activities in Kasiisi Project schools and maintain an interactive connection between our programs. KSRP has a mini-documentary with local language subtitles (produced by Jane Goodall Institute) that is shown during conservation education presentations. Conservation films that specifically address regional threats have been found to positively influence student knowledge of and attitudes toward great apes (Leeds et al. 2017). Beyond presentations, KSRP employees engage WLC students and community members in chimpanzee-focused conservation activities and games. In 2014, KSRP implemented two interactive games, Snare Care and Ape Survivor, with WLC members. The goal was two-fold: to sensitize students to the anthropogenic dangers chimpanzees face in the forest and to help students empathize with the chimpanzees who have been permanently handicapped by snares. Snare Care (fig. 26.2) required students to identify a photographed snare hiddened in the forest. Failure to do so resulted in the student being "snared." Their new injury, which was constructed using tape and sticks, mimicked a Kanyawara chimpanzee's snare injury. Snared students then had to compete in a foraging task against non-snared students. Afterwards, KSRP team members and students discussed the snared individuals' challenges in the context of the chimpanzees living in Kibale and what the students could do to help. Both projects also regularly seek partnership opportunities with UWA and other local conservation-based organizations to implement new conservation education initiatives whenever possible.

Results

Both KSRP and Kasiisi Project initiatives have been ongoing since 1997, but data collection and reliability were not always consistent. The conservation data presented in this chapter were predominantly extracted from 2006 to 2017. Below we describe our findings regarding the demographics of chimpanzees affected by the snares, the efficacy of the Kibale Snare Removal Program, and the results of the Kasiisi Project, our conservation education and community outreach program.

CHIMPANZEE SNARE DEMOGRAPHICS

Community size (excluding infants) as of December 2017 was 42 chimpanzees with a cumulative total of 453 possible years exposed to snare risks. Approximately 28% of the currently-living Kanyawara chimpanzees (past the age of infancy) have been snared at least once (four individuals were snared twice), resulting in a 4.0% risk of being snared per year. Among those, 57% have suffered a permanent snare injury, including the young male Max who lost both of his feet in two separate snaring events (fig. 26.1). However, when accounting for the entire known history of the Kanyawara chimpanzee community (from 1987 to present, N = 107 chimpanzees with 1,225 possible years of snare risk), the percentage is much higher. Approximately 45% of the past and present community members (past the age of infancy) have been snared at least once (seven individuals snared twice or thrice), resulting in a 4.7% cumulative risk of being snared per year. Of those, 88% suffered a permanent snare injury. Two individuals in the study population are known to have died directly from snare wounds.

The data show no significant difference between the proportion of male (N = 22) and female (N = 27) snare victims ($x^2[1] = 0.5$, ns, data analyzed in

SPSS v.23). However, there was a significant interaction between snarings and chimpanzee age ($x^2[2] = 10.7$, $p < 0.01$). Juveniles (4.0 to 9.9 years) were the most at-risk age category for being snared (59%, N = 20) followed by adults (15.0 years and older, 26%, N = 9) and then subadults (10.0 to 14.9 years, 15%, N = 5). There have been no known infant snaring events at Kanyawara. The mean snare age was 12.4 years (±6.8 SD, N = 32). There was no significant interaction between sex and age ($x^2[2] = 1.1$, ns).

KIBALE SNARE REMOVAL PROGRAM

The number of permanent snare injuries has declined in recent years due to the implementation of veterinarian-supervised snare interventions starting in 2006. Emergency protocols were developed in partnership with local veterinarians, the Jane Goodall Institute in Uganda, and UWA to improve response time and reduce snare severity, infection, and permanent damage. Sine 2006, 64% of attempted interventions (N = 11 attempts out of 15 snare events) have resulted in successful snare removal. Additionally, in all of these success cases, the digit/limb was saved from amputation. However, social, behavioral, and environmental conditions sometimes make intervention attempts impossible (27% of cases).

Chimpanzee Snaring Interval

The inter-snare interval (ISI) refers to the length of time elapsed between observed chimpanzee snare injuries at Kanyawara. Prior to KSRP (from 1990 to 1997), the ISI averaged 8.9 months (±2.2 SE, N = 7) compared to 13.0 months (±3.0 SE, N = 19) after KSRP began to be present.

Poacher Activity

Since 1997, KSRP has worked closely with UWA to deploy more than 5,000 KSRP patrols removing over 8,500 snares from the park and assisted UWA in the arrest of countless poachers. On these patrols, significantly more neck snares (68%) were removed compared to foot snares (32%, $x^2[1] = 326.8$, $p < 0.0001$) and significantly more wire snares (89%) were removed compared to nylon snares (11%, $x^2[1] = 2052.1$, $p < 0.0001$) from 2014 to 2017. For chimpanzees, foot snares are notoriously more dangerous because of their trigger/spring mechanism that causes the wire/nylon to instantly tighten, embed, lock, and restrict blood supply to extremites. The savagery of the snare is typically exacerbated by the newly entangled chimpanzee's startled reaction as

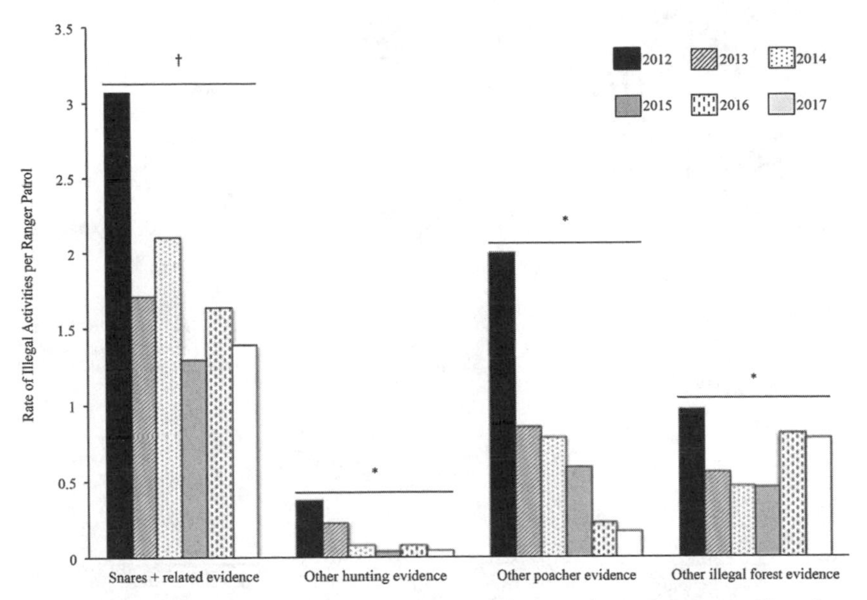

FIGURE 26.3. Over a five-year period (2012–2017), KSRP has observed a steady decline in all poacher-related activities per patrol (*$p < 0.05$; †approaching significance).

she or he struggles to break free, which only causes the snare to tighten even more and the likelihood of permanent injury to increase (fig. 26.1). Over the life of the program, KSRP patrols have resulted in the removal of 44.4 snares/month (on average) from Kibale. In 2014, when the program expanded to include two independent snare removal teams, the monthly snare removal average significantly increased from 39.2 (±3.1 SE) snares per month to 60.2 (±4.8 SE) snares per month ($t[192] = 3.5$, $p < 0.001$). While increased patrol effort and manpower over the life of the program may predict an increase in snare detection and removal per patrol, instead a significantly steady decline has been observed in all poacher-related activities from 2012 to 2017 ($F[1,4] = 12.4$, $p < 0.05$). Specifically, the number of snare ($F[1,4] = 5.9$, $p = 0.07$), non-snare hunting ($F[1,4] = 11.0$, $p < 0.05$), and non-hunting poacher ($F[1,4] = 17.7$, $p < 0.05$) activities per patrol have declined over recent years (fig. 26.3).

Data from 2009 to 2017 show that the type of illegal activity significantly influenced its distance from the park boundary ($F[3] = 566.8$, $p < 0.0001$). Snares tended to be more densely concentrated near the park boundary and outside of the Kanyawara chimpanzee home range (fig. 26.4). The average distance of snares from the park boundary was 0.9 km (±15 SE). Snares were significantly closer to the park boundary compared to other non-snare hunting (avg. 2.6 km ±104.2 SE; $t[5417] = -16.9$, $p < 0.0001$) and non-hunting

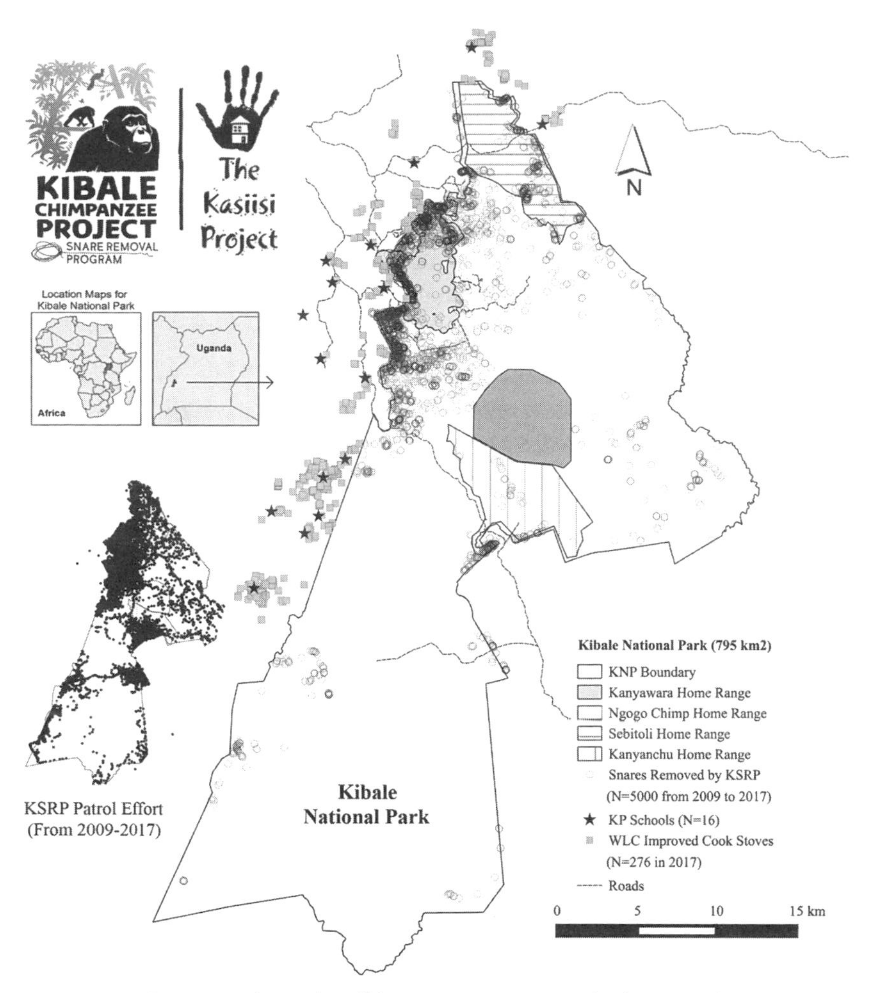

FIGURE 26.4. Larger map depicts the collaborative conservation work of KSRP (with UWA support) and Kasiisi Project both inside and outside of KNP. Snares occur at higher densities along the boundary of KNP and outside of the Kanaywara chimpanzee home range. As a community development initiative, Kasiisi Project students built improved cook stoves for local people living near KNP. KNP boundary shapefile downloaded from Protected Planet (2019) with reported area of 795 km^2 provided by UWA. Chimpanzee homerange shapefiles provide by Paco Bertolani (pers. comm.) for Kanyawara or created using maps published in Amsler (2009) for Ngogo and CCB (2015) for Kanyanchu and Sebitoli.

poacher (avg. 11.5 km ±36.4 SE; $t[6645] = -7.5$, $p < 0.0001$) activities. However, snares were significantly further from the park boundary compared to other non-poaching illegal activities (avg. 0.3 km ±10.0 SE; $t[7957] = 29.9$, $p < 0.0001$). The majority (65%) of KSRP patrol effort was outside of the Kanyawara chimpanzee home range. However, effort alone did not account for the high proportion of illegal activity (82%) observed outside of the well-protected home range. This majority was significantly more than expected based on the percentage of patrol effort outside of the Kanyawara home range ($x^2[1] = 1278.0$, $p < 0.0001$). Only 15% of snare hunting, 7% of non-snare hunting, 18% of non-hunting poaching, and 22% of non-poaching illegal activities were observed within the home range.

Deforestation

In addition to snare removal, KSRP patrols help mitigate deforestation and resouce extraction in the park. Using MODIS data (NASA's Moderate Resolution Imaging Spectroradiometer data) to investigate changes in forest cover from 2000 to 2010, forest cover within the Kanyawara home range increased by +4.1% and by +0.7% for park areas within 5 km of Kanyawara. Conversely, forest cover in the remainder of (−2.7%) and outside (−3.1%) the national park declined during this same time period. Kibale has two other long-term chimpanzee field sites (Ngogo and Sebitoli) and one chimpanzee tourism site (Kanyanchu). Unlike Kanyawara, the other three long-term sites all showed forest cover decline: Ngogo (−3.8%), Sebitoli (−0.3%), and Kanyanchu (−6.2%, fig. 26.5). Simiar to the snare location pattern, and in support of the MODIS data, on-the-ground deforestation events documented by KSRP were concentrated more densely along the park boundary (avg. 0.3 km ±10 SE) and outside of the Kanyawara home range (77%).

KASIISI PROJECT

Academic Improvement

Academic performance is measured by government administered Primary Leaving Exams (PLE). Schools are divided into three types: 1) Kasiisi Project Core Schools—15+ year program members with full range of our academic programs, 2) Kasiisi Project Satellite Schools—fewer than eight-year program member with partial academic programs, and 3) non-Kasiisi Project Peer Schools. Data show improvement in PLE scores in all schools between 2004 and 2017 (lower scores are better, fig. 26.6), but Kasiisi Project schools significantly outperformed Peer Schools. In 2013 mean score for Kasiisi Project schools (16.02 ± 2.05) was significantly better than for peer schools (27.91 ± 1.55) (unpaired t-test: $t = 2.67$, df 26, $p < 0.01$). In Kabarole District, the most significant predictor of academic success in forest-edge schools was attending a Kasiisi Project school (MacKenzie et al. 2017b). Results from PLE scores in 2017 ranked five Kasiisi Project schools in the top 12 for the district despite

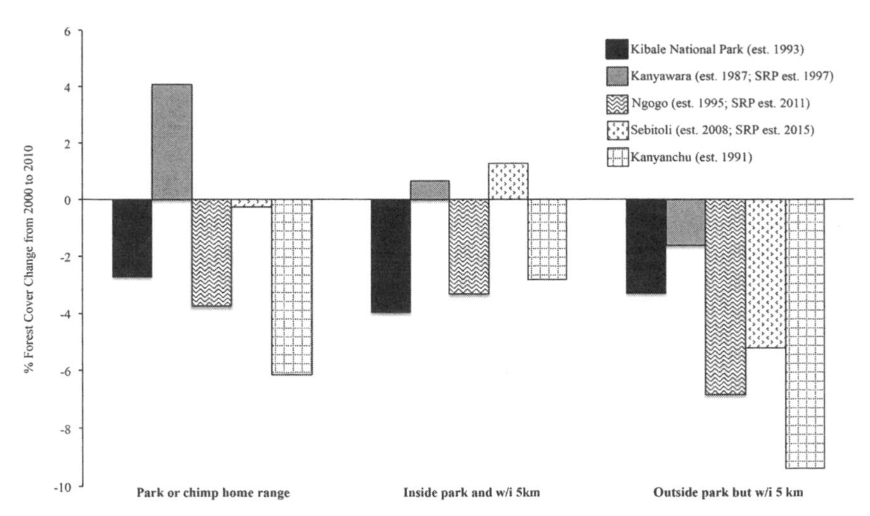

FIGURE 26.5. Kanyawara was the only area that showed an increase in tree cover over the 10-year period. It should be noted that while Kanyawara does have the longest running snare removal program of all the habituated field sites in Kibale, this area was selectively logged in the 1990s and forest regrowth may be contributing to an inflated value. Park or chimp home range represents the percent of forest cover change from 2000 to 2010 throughout KNP and also within the habituated chimpanzee community home ranges. Inside park and within 5 km represent boundary edges that are within 5 km of the park boundary or home range boundary and still fall within the park. Outside park but within 5 km represent boundary edges that are within 5 km of the park or home range boundary but are not inside the park boundary. Data from 2000 and 2010 MODIS satellite imagery.

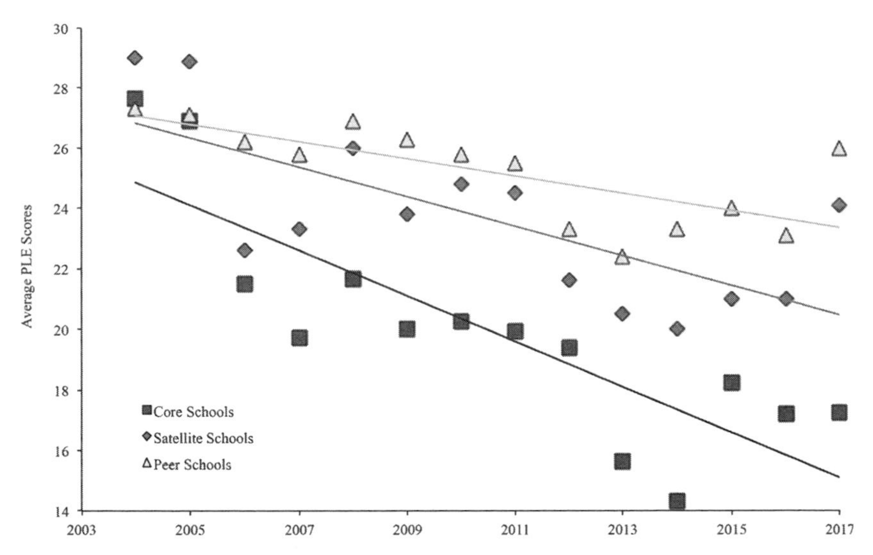

FIGURE 26.6. Kasiisi Project core schools (15+ year project members) outperformed both Kasiisi Project satellite schools (less than 8 year project members) and non-Kasiisi Project peer schools (no project membership). Lower Primary Leaving Exam (PLE) scores are better.

competition with private schools, which a) restrict enrollment and b) send poorly performing students to other PLE centers.

Conservation Education

Since 2007 we have collected data on schools' engagement in conservation programs, environmental knowledge, attitudes toward conservation, and rates of practical conservation activities in students, teachers, and parents.

When teachers were asked in 2008 what factors made teaching about the environment difficult, 60% cited lack of skills and knowledge, but by 2016 this dropped to only 10% of teachers. In 2016, teachers ranked Kasiisi Project conservation education programs highest in value to their schools and students (health ranked second).

In 2017, 14 Kasiisi Project/KFSP WLCs enrolled 680 children and conducted 135 conservation activities reaching 2,500 children. Enrollment and conservation activities in 2017 were up 40% from 2016 and 90% from 2015, indicating the schools' growing interest in good conservation education programs. In 2017, WLC members from Kasiisi Project schools scored significantly higher than non-members on environmental-based questions. Mean score for WLC members (5.44 ± 0.40) was higher than for non-WLC members (2.76 ± 0.68, unpaired t-test: $t = 3.3$, $df = 169$, $p < 0.001$). There was no difference in knowledge scores between members (2.65 ± 0.86) and non-members in control schools (3.05 ± 0.86, $t = 0.27$, $df = 34$, ns).

Mean children's attitude scores toward Kibale in Kasiisi Project schools were significantly higher in WLC members (1.27 ± 0.07) than non-members (0.89 ± 0.07, unpaired t-test: $t = 2.9$. $df = 279$, $p < 0.01$). Children's attitudes toward both Kibale and chimpanzees have significantly improved from 2011 to 2017 (fig. 26.2). Mean attitude scores toward Kibale were significantly higher in 2017 (1.41 ± 0.14) than 2011 (-0.13 ± 0.05, unpaired t-test: $t = 11$, $df = 791$, $p < 0.001$). Mean attitude scores toward chimpanzees were significantly higher in 2017 (1.14 ± 0.12) than 2011 (-0.13 ± 0.05, unpaired t-test: $t = 9.64$, $df = 755$, $p < 0.001$), a big change from 2011 when 66% responded with statements like "chimpanzees are dangerous" and "chimpanzees ruin our crops" (Glennon 2011).

Community Engagement and Trust

Our data show a growing engagement of parents in their children's conservation activities and increasing demands for accurate and practical environmental education.

In 2016, 65% knew the name of their child's WLC patron, 84% wanted more conservation education for their children, 87% wanted more engagement in their children's education, and 93% wanted their children to learn to conserve elephants.

In 2017, 276 families requested WLCs build fuel-efficient cookstoves in their homes (fig. 26.4), a 168% increase from 2016 (N = 103). In response to such community engagement requests, students from 14 Kasiisi Project schools completed over 390 practical conservation activities, including tree planting, building beehives and fuel-efficient stoves, planting sustainable gardens, and digging rubbish pits.

In 2017, 20% of parents ranked conservation as the most important program, even above those with more obvious practical impacts (i.e., 18% health, 15% sanitary pads, 14% latrines, 10% libraries, and 5% scholarships). Separate interviews with parents showed that 92% understood what the WLCs do, 89% had positive comments about it, and 75% agreed with conservation activities conducted at home as reported by their children.

Critical to gaining community support is trust—our most valuable asset and greatest achievement. We maintain a strong reputation for nonpartisanship, which shepherds our cooperation with parents, schools, and local government. Of 3,000 parents surveyed in 2016, 74% said that the Kasiisi Project could be trusted to be fair, a 164% increase since 2007. When asked for the best outcome from Kibale becoming a national park, 33% of parents said that it brought our programs to their communities, whereas 34% said increased income and employment.

Discussion

Due to the support and long-term presence of KCP, for over 20 years KSRP and Kasiisi Project have been active and collaborative players aiming to help conserve Kibale National Park. While both projects address chimpanzee conservation, each takes its own approach. KSRP uses on-the-ground tactics to promote chimpanzee conservation and preserve the integrity of Kibale in real-time, whereas Kasiisi Project uses educational tactics to sensitize young people and increase the likelihood they will adopt conservation practices as adults in the future. Simply put, KSRP functions as the immediate bandaid to poaching, while Kasiisi Project aims to be the future cure.

KSRP has been successful over the years in helping UWA to reduce the overall snare density in the park, particularly at Kanyawara, while concurrently sensitizing local communities and schools to snare and other poaching-related wildlife injuries and causalities. As a result of this collaborative effort, the snare probability for chimpanzees and other wildlife in Kibale has declined along with other poacher threats. While cumulative snaring rates at Kanyawara exceed the estimated rate of chimpanzee snare injuries for Uganda (33%; Plumptre et al. 2010), rates among presently-living chimpanzees are lower (28%). Declining snare rates at Kanyawara suggest that the consistent presence of KSRP and UWA patrols may be acting as a deterent to local poachers within the Kanyawara home range. Contrary to the canopy cover declines observed in other areas of Kibale, the forest cover at Kanyawara has shown signs of regrowth following its logging in 1995. On all accounts, Kanyawara appears to be better protected than other areas within the park—an effect that is likely the combination of UWA-supported KSRP patrols plus many field assistants, students, long-term researchers, and trail cutters who also frequent the area. As a further deterrent, Kanyawara is home to the Makerere University Biological Field Station and an UWA outpost. For poachers, Kanyawara is more risky than other areas in the park. However, the snare and other illegal activity (i.e., firewood and forest product collection) pressure along the park boundary and outside of the Kanyawara home range remains intense and future efforts should concentrate more heavily in these areas.

The Kasiisi Project has worked diligently to make their rural forest schools academically, professionally, and practically competitive with private schools. Kasiisi Project schools have consistently outranked many private schools in the region, and have had many pupils receive academic scholarships and go on to graduate from local and international universities, including Harvard University. Some scholars have even returned to Kasiisi Project as teachers or administrators. Through their popular WLCs and conservation education programs, children's attitudes toward chimpanzees and other wildlife have improved, inspiring some to join UWA and work in conservation. Taï's educational outreach program reported similar attitudinal and behavioral shifts in 16 nearby villages due to the implementation of an interactive play about a chimpanzee's family whose mother is killed by a poacher. The chimpanzee family mourned the loss of their mother, exhibiting many human-like behaviors. The most common message retained by villagers four months after the play was "The chimpanzee is like a human," which reportedly led to reduced bushmeat consumption and a more positive perception of chimpanzees (Boesch et al. 2008b). The Kasiisi Project also regularly uses interactive student-based performance dramas with empathetic undertones to effectively communicate complex and culturally sensitive conservation messages to both children and adults.

Kasiisi Project has also given students the tools to implement practical conservation strategies at home through their fuel-efficient cookstoves and other sustainability programs. However, it is difficult to measure the true effectiveness of these conservation education programs in real time. It may be a decade before Kasiisi Project students are in decision-making positions. Gradually, adults have recognized the value that Kasiisi Project brings to the community and are slowly becoming more open-minded and willing to change. In the meantime, Kasiisi Project is striving to empower students and give them the tools they need to succeed both personally and professionally, so unlike their parents, they will not have to depend on the forest for resources.

Long-term conservation can be enhanced by active conservation measures paired with a well-educated population with good environmental knowledge (Fiallo and Jacobson 1995; Kideghesho, Røskaft, and Kaltenborn 2007). By linking our programs to long-term research, school curricula, and community needs, we can galvanize today's youth and change attitudes and behavior in ways that have powerful and practical conservation benefits. In the past 20 years, in an area of forest with KSRP's active conservation program that is bordered by six Kasiisi Project school communities, the park boundary has not been eroded. In that time there have been many signs of success. There has been a 43% increase in the local chimpanzee community (from 42 individuals to 60; Muller and Wrangham 2014); populations of wild pigs, duiker, and monkeys have grown; and buffalo have returned to the forest swamps. Furthermore, the average time interval between snaring incidents in chimpanzees has risen from 8.9 to 13.0 months, and despite a 3-fold rise in elephant traps, relative abundance of elephants (measured by tracks crossing a set of trails walked twice a month for a year) have increased tenfold (Chapman, pers. comm.; Omeja et al. 2016; Wrangham, pers. comm.). On a small scale, what we are doing appears to be working for now.

Future Directions

While these programs have made measurable progress in and around Kibale, chimpanzees and the other wildlife living under their umbrella remain under siege from the expanding human population. Manpower has been key to increasing program success. In recent years, Ngogo and Sebitoli also established snare removal programs in and around their long-term field sites in Kibale, which has resulted in patrol coverage for approximately two-thirds of the park in addition to regular UWA patrols that strive to cover the entire park. Given that snare density is highest within 1 km of the park boundary, establishing additional patrol teams who could exclusively cover the park boundary zones in their entirety could further assist UWA's existing efforts to discourage poachers from entering the park while increasing snare extraction rates. However, funding and program management can be a limiting factor. Other long-term field sites should work in collaboration with their national wildlife authority to establish their own conservation patrol teams who are focused on the anthropogenic threats most common in their region. Field sites in close proximity to one another should collaborate to maximize the coverage and effectiveness of their patrols.

In addition to scheduling regular patrols, we must be innovative in our approach to keep up with the compounding poacher presence. Alternative methods of poacher detection using drones (Bondi et al. 2018; Wich 2015) and camera traps (Hossain et al. 2016; Widness and Aronsen 2018) have been used in the past, but a growing area of interest is the use of conservation dogs (Hurt and Smith 2009). For some time now, detection dogs have been used to reliably locate the feces of many elusive species (i.e., black bear: Wasser et al. 2004; whales: Rolland et al. 2006), including primates (i.e., cross-river gorilla: Arandjelovic et al. 2014), with most studies reporting higher success rates in dog-directed surveys than human-directed efforts. Conservation dogs have been used to detect both biological scents (i.e., animal and human scents, including poachers) and non-biological scents (i.e., landmines, flammable products, hazardous chemicals) with a high degree of confidence (reviewed in Browne, Stafford, and Fordham 2006). Given their consistent success across diverse habitats (Leigh and Dominick 2015), this has led to the integration of dogs in conservation practices that extend beyond species and fecal detection (Hurt and Smith 2009). However, the data regarding snare detection is still spotty due to limited scent markers in the materials used. Working Dogs for Conservation is one organization that is currently marketing and testing the validity of using dogs to detect snares in savanna environments (Parker 2015). To date, however, there is no published information available on conservation dogs detecting snares in forested habitats, which is where most chimpanzee populations live. As more data become available and training techniques are perfected, this and other innovative approaches to conservation may revolutionize methods that wildlife authorities and conservation programs use to patrol protected areas while concurrently improving their effectiveness. Conservation dogs could also be an effective education tool in schools used to sensitize students to animals in general. However, this unorthodox method is not without challenges, including (but not limited to) extensive and continuous professional training of dogs and local handlers, mitigating zoonotic disease transmission threats, acquiring governmental clearance for use of dogs in a protected area, and providing preventative veterinary care, a balanced diet, sanitary and social housing with handlers, and armed protection for the dogs. Moving forward, these innovation approaches should be further explored and evaluated in the context of chimpanzee conservation to revise best practices when and where applicable.

At times, we have found our chimpanzee conservation efforts to be limited by veterinary availability and response time. Until recently, our project did not employ a full-time veterinarian, and at times there was no resident wildlife veterinarian in the region. Whenever a chimpanzee was snared, we would immediately send an emergency request to the Jane Goodall Institute in Uganda to deploy a veterinarian with a UWA mandate. Over the years, very few veterinarians in Uganda have been qualified and authorized to intervene in chimpanzee snare injuries, which can be problematic for these time-sensistive injuries if authorized veterinarians are not immediately available to respond. Employing a permanent veterinarian who specializes in chimpanzees and other wildlife would not only benefit long-term projects, but also benefit the park in general. A resident veterinarian could also respond to anthropogenic-induced emergencies in real time throughout the park, continually monitor the health of local chimpanzee populations, and educate the local communities about anthrozoonotic disease transmission (Lukasik 2002). A resident veterinarian would likely maximize the probability of snare removal intervention success and reduce the likelihood of permanent injury. The Budongo Conservation Field Station employs a resident chimpanzee veterinarian for this very purpose, which has reduced respiratory and gastrointestinal infections (Asiimwe et al. 2016), improved emergency response time in the field, and benefited other neighboring chimpanzee communities living in forest fragments and human-dominated landscapes (McLennan and Asiimwe 2016; McLennan et al. 2012). Veterinarian retention and long-term funding is undoubtedly challenging, but field sites with larger service areas or limited access to local wildlife veterinarians should consider whether the conservation benefits could outweigh the financial and management costs. It should be noted that deciding whether or not and how to intervene is fraught with philosophical, ethical, and logistical challenges. Prior to interventions, projects should discuss and develop a "standard protocol" for intervention, contingency plans, logistical preparation, and training opportunities with their veterinarian (see Lonsdorf et al. 2014 for intervention lessons learned and recommendations).

While Kasiisi Project programs primarily target children as the decision makers of tomorrow, today some of their parents are poaching duiker and clearing forests—hence the immediate need for KSRP. Unfortunately, there remains a prevailing perspective from the local people that the drawbacks of living close to the forest outweigh the benefits of such proximity (MacKenzie et al. 2017a). Time is of the essence and reaching beyond children to adults is urgent. Damerell, Howe, and Milner-Gulland (2013) found that educating children led to better conservation practices in their parents, suggesting that school-based programs can have an active reach beyond the children they target. In recent years, the Kasiisi Project has incorporated more forest schools along the northwestern boundary of the park into its program. The North Carolina Zoo's UNITE for the Environment program works in nine forest schools along the eastern boundary. Together they are implementing conservation education initiatives in schools along nearly half of the park's boundary. Long-term research programs should look for school and/or community-based NGOs they can partner with to develop and implement chimpanzee conservation initiatives near their study site. Researchers must invest in this effort and extend their involvement beyond the occasional talk or demonstration. Using a bottom-up approach, Kasiisi Project schools will continue to operate as the center of Kibale's rural communities to promote changing attitudes and practical conservation messages, starting with their students who then spread these values to their families and neighbors.

The KCP/KSRP-Kasiisi Project collaboration described in this chapter has been a voice of change in the region by providing desirable employment opportunities to local people while concurrently promoting chimpanzee research, conservation, and education initiatives. While there is still have a long way to go, improvements have been observed over the years in local people's actions and attitudes toward chimpanzees and the national park in general. We therefore recommend that other long-term field sites couple active, on-the-ground conservation strategies with consistent educational outreach in schools and local communities. A staying presence with regular programs is key to the success and local acceptance of these initiatives.

We extend a call to action to all chimpanzee researchers and long-term field sites to work together to develop and/or improve local conservation and education initiatives in their areas. We invite discussions and are happy to provide any resources for those interested in establishing programs and/or collaborating. After 20 years, we still remain committed to our mission and will continue to work with our local and international partners to conserve Kibale's chimpanzees and national park. Despite chimpanzees extreme behavioral flexibility in increasingly challenging anthropogenic environments (Hockings et al. 2009), the current trajectory is not sustainable. For the future of the species, it is imperative that we fight to keep chimpanzees in their own context—in viable habitats free of deadly anthropogenic threats.

Acknowledgments

We thank our field team, community educators, funders, and collaborators for their hard work, dedication, and generous support over the years. Without them, our work would not be possible.

KCP thanks the National Institutes of Health, National Science Foundation, Wenner-Gren Foundation, and Leakey Foundation for funding. KSRP thanks the Jane Goodall Institutes (JGI) in Austria, the Netherlands, and Switzerland for funding, and the North Carolina Zoo/UNITE for data collection devices and technical support. KCP and KSRP depend on long-term collaboration with the UWA and Makerere University Biological Field Station. Other important collaborators include the Uganda and USA JGI branches and the Ngogo and Sebitoli Chimpanzee Projects.

Kasiisi Project thanks the Disney Conservation Fund, Columbus Zoo, Cleveland Metroparks, National Geographic Soceity, Zoos and Aquariums Committed to Conservation, British and Foreign Schools Society, African Bird Club, and International Elephant Foundation for funding, and Francis Rwabuhinga, Gorret Nyabutono, Corey Dickinson, Zach Manta, Caroline Warne, and Sonya Kahlenberg for their hard work and contributions.

References

Adams, W. M., and J. Hutton. 2007. "People, parks and poverty: Political ecology and biodiversity conservation." *Conservation and Society* 5 (2): 147–83.

Adams, W. M., and M. Mulligan. 2003. *Decolonizing Nature: Strategies for Conservation in a Post-Colonial Era*. London: Earthscan Publications.

Amsler, S. J. 2009. "Ranging behavior and territoriality in chimpanzees at Ngogo, Kibale National Park, Uganda." PhD diss., University of Michigan.

Anderson, G. W., L. Gaffikin, L. Pintea, G. C. Kajembe, K. Yeboah, and B. J. Humplick. 2004. "Assessment of the Lake Tanganyika Catchment, Reforestation and Education (TACARE) project." Report to the US Agency for International Development and the Jane Goodall Institute. Washington, DC.

Arandjelovic, M., J. S. Head, C. Boesch, M. M. Robbins, and L. Vigilant. 2014. "Genetic inference of group dynamics and female kin structure in a western lowland gorilla population (*Gorilla gorilla gorilla*)." *Primate Biology* 1 (1): 29–38.

Asiimwe, C., G. Muhanguzi, E. Okwir, P. Okimat, A. W. Bugenyi, T. Mugabe, and F. Babweteera. 2016. "Transdisciplinary approach to solving conservation challenges: A case of Budongo Conservation Field Station, Uganda." *PeerJ Preprints* 4: e1848v1.

Babweteera, F., V. Reynolds, and K. Züberbuhler. 2008. "Conservation and research in the Budongo Forest Reserve, Masindi District, Western Uganda." In *Science and Conservation in African Forests: Benefits from Long-Term Research*, edited by R. W. Wrangham and E. A. Ross, 145–57. Cambridge: Cambridge University Press.

Blom, A., J. Yamindou, and H. H. T. Prins. 2004. "Status of the protected areas of the Central African Republic." *Biological Conservation* 118 (4): 479–87.

Boesch, C., H. Boesch, Z. Bertin, G. Bi, E. Normand, and I. Herbingee. 2008a. "The contribution of long-term research by the Taï Chimpanzee Project to conservation." In *Science and Conservation in African Forests: Benefits from Long-Term Research*, edited by R. W. Wrangham and E. A. Ross, 184–200. Cambridge: Cambridge University Press.

Boesch, C., C. Gnakouri, L. Marques, G. Nohon, I. Herbinger, F. Lauginie, H. Boesch, S. Kouamé, M. Traoré, and F. Akindes. 2008b. "Chimpanzee conservation and theatre: A case study of an awareness project around the Taï National Park, Côte d'Ivoire." In *Conservation in the 21st Century: Gorillas as a Case Study*, edited by T. S. Stoinski, H. D. Steklis, and P. T. Mehlman, 128–35. Boston: Springer.

Bondi, E., F. Fang, M. Hamilton, D. Kar, D. Dmello, J. Choi, R. Hannaford, A. Iyer, L. Joppa, M. Tambe, and R. Nevatia. 2018. "Spot poachers in action: Augmenting conservation drones with automatic detection in near real time." *Association for the Advancement of Artificial Intelligence*. https://www.cais.usc.edu/wp-content/uploads/2017/11/spot-camera-ready.pdf.

Borchers, C., C. Boesch, J. Riedel, H. Guilahoux, D. Ouattara, and C. Randler. 2014. "Environmental education in Côte d'Ivoire/West Africa: Extra-curricular primary school teaching shows positive impact on environmental knowledge and attitudes." *International Journal of Science Education, Part B* 4 (3): 240–59.

Brashares, J. S., P. Arcese, and M. K. Sam. 2001. "Human demography and reserve size predict wildlife extinction in West Africa." *Proceedings of the Royal Society of London B* 268 (1484): 2473–78.

Breuer, T., and F. B. Mavinga. 2010. "Education for the conservation of great apes and other wildlife in Northern Congo—The importance of nature clubs." *American Journal of Primatology* 72 (5): 454–61.

Brockington, D. 2004. "Community conservation, inequality and injustice: Myths of power in protected area management." *Conservation and Society* 2 (2): 411–32.

Browne, C., K. Stafford, and R. Fordham. 2006. "The use of scent-detection dogs." *Irish Veterinary Journal* 59 (2): 97–104.

Bruner, A. G., R. E. Gullison, R. E. Rice, and G. A. B. Da Fonseca. 2001. "Effectiveness of parks in protecting tropical biodiversity." *Science* 291 (5501): 125–28.

Butchart, S. H. M., M. Walpole, B. Collen, A. Van Strien, J. P. W. Scharlemann, R. E. A. Almond, J. E. M. Baillie, B. Bomhard, C. Brown, J. Bruno, K. E. Carpenter, G. M. Carr, J. Chanson, A. M. Chenery, J. Csirke, N. C. Davidson, F. Dentener, M. Foster, A. Galli, J. N. Galloway, P. Genovesi, R. D. Gregory, M. Hockings, V. Kapos, J. F. Lamarque, F. Leverington, J. Loh, M. A. McGeoch, L. McRae, A. Minasyan, M. Hernández Morcillo, T. E. E. Oldfield, D. Pauly, S. Qader, C. Revenga, J. R. Sauer, B. Skolnik, D. Spear, D. Stanwell-Smith, S. N. Stuart, A. Symes, M. Tierney, T. D. Tyrrell, J. C. Vie, and R. Watson. 2010. "Global biodiversity: Indicators of recent declines." *Science*: 1187512.

Bwambale, T. 2012. "Uganda has the youngest population in the world." *New Vision*, December 14. https://www.newvision.co.ug/new_vision/news/1311368/uganda-population-world.

Campbell, G., and M. Houngbedji. 2015. "Conservation status of the West African chimpanzee (*Pan troglodytes verus*) in Togo and Benin." Unpublished report to Primate Action Fund.

Campbell, G., H. Kuehl, A. Diarrassouba, P. K. N'Goran, and C. Boesch. 2011. "Long-term research sites as refugia for threatened and over-harvested species." *Biology Letters* 7 (5): 723–26.

Campbell, G., H. Kuehl, P. N'Goran Kouamé, and C. Boesch. 2008. "Alarming decline of West African chimpanzees in Côte d'Ivoire." *Current Biology* 18 (19): R903–4.

Chapman, C. A., and R. W. Wrangham. 1993. "Range use of the forest chimpanzees of Kibale: Implications for the understanding of chimpanzee social organization." *American Journal of Primatology* 3 (4): 263–73.

Chawla, L. 2007. "Childhood experiences associated with care for the natural world: A theoretical framework for empirical results." *Children Youth and Environments* 17 (4): 144–70.

Climate, Community, and Biodiversity (CCB). 2015. "Natural high forest rehabilitation project on degraded land of Kibale National Park." Project Design Document, V2.1.

Cohen, J. 2010. "A matter of life and limb." *Science* 328 (5974): 33.

Collins, A., and J. Goodall. 2008. "Long-term research and conservation in Gombe National Park, Tanzania." In *Science and Conservation in African Forests: Benefits from Long-Term Research*, edited by R. W. Wrangham and E. A. Ross, 158–72. Cambridge: Cambridge University Press.

Cowlishaw, G., and R. I. M. Dunbar. 2000. *Primate Conservation Biology*. Chicago: University of Chicago Press.

Critchlow, R., A. J. Plumptre, M. Driciru, A. Rwetsiba, E. J. Stokes, C. Tumwesigye, F. Wanyama, and C. M. Beale. 2015. "Spatiotemporal trends of illegal activities from ranger-collected data in a Ugandan national park." *Conservation Biology* 29 (5): 1458–70.

Damerell, P., C. Howe, and E. J. Milner-Gulland. 2013. "Child-orientated environmental education influences adult knowledge and household behaviour." *Environmental Research Letters* 8 (1): 015016.

Dowie, M. 2011. *Conservation Refugees: The Hundred-Year Conflict between Global Conservation and Native Peoples*. Cambridge, MA: MIT Press.

Estrada, A., P. A. Garber, A. B. Rylands, C. Roos, E. Fernandez-Duque, A. Di Fiore, K. A. I. Nekaris, V. Niijman, E. W. Heymann, J. E. Lambert, F. Rovero, C. Barelli, J. M. Setchell, T. R. Gillespie, R. A. Mittermeier, V. Verde Arregoitia, M. de Guinea, S. Gouveia, R. Dobrovolski, S. Shanee, N. Shanee, S. A. Boyle, A. Fuentes, K. C. MacKinnon, K. R. Amato, A. L. S. Meyer, S. Wich, R. W. Sussman, R. Pan, I. Kone, and B. Li. 2017. "Impending extinction crisis of the world's primates: Why primates matter." *Science Advances* 3: e1600946.

Fashing, P. J., and M. Cords. 2000. "Diurnal primate densities and biomass in the Kakamega Forest: An evaluation of census methods and a comparison with other forests." *American Journal of Primatology* 50 (2): 139–52.

Fiallo, E. A., and S. K. Jacobson. 1995. "Local communities and protected areas: Attitudes of rural residents towards conservation and Machalilla National Park, Ecuador." *Environmental Conservation* 22 (3): 241–49.

Fitzgerald, M., R. Coulson, A. M. Lawing, T. Matsuzawa, and K. Koops. 2018. "Modeling habitat suitability for chimpanzees (*Pan troglodytes verus*) in the Greater Nimba Landscape, Guinea, West Africa." *Primates*: 1–15.

Garriga, R. M., I. Marco, E. Casas-Díaz, B. Amarasekaran, and T. Humle. 2017. "Perceptions of challenges to subsistence agriculture, and crop foraging by wildlife and chimpanzees *Pan troglodytes verus* in unprotected areas in Sierra Leone." *Oryx* 52 (4): 761–74.

Ginn, L. P., J. Robison, I. Redmond, and K. A. I. Nekaris. 2013. "Strong evidence that the West African chimpanzee is extirpated from Burkna Faso." *Oryx* 47: 325–26.

Glennon, E. 2011. "The effect of conservation education programs on students' wildlife knowledge and attitudes toward conservation in primary schools Surrounding Kibale National Park, Uganda." Senior thesis, Bates College.

Goldman, A., J. N. Hartter, J. Southworth, and M. W. Binford. 2008. "The human landscape around the island park: Impacts and responses to Kibale National Park." In *Science and Conservation in African Forests: Benefits from Long-Term Research*, edited by R. W. Wrangham and E. A. Ross, 158–72. Cambridge: Cambridge University Press.

Greengrass, E. J. 2009. "Chimpanzees are close to extinction in southwest Nigeria." *Primate Conservation* 24: 77–83.

Hockings, K. J. 2009. "Living at the interface: Human–chimpanzee competition, coexistence and conflict in Africa." *Interaction Studies* 10 (2): 183–205.

Hossain, A. N. M., A. Barlow, C. Greenwood Barlow, A. J. Lynam, S. Chakma, and T. Savini. 2016. "Assessing the efficacy of camera trapping as a tool for increasing detection rates of wildlife crime in tropical protected areas." *Biological Conservation* 201: 314–19.

Humle, T., F. Maisels, J. F. Oates, A. Plumptre, and E. A. Williamson. 2016. "*Pan troglodytes* (errata version published in 2018)." *The IUCN Red List of Threatened Species 2016*: e.T15933A129038584.

Hurt, A., and D. A. Smith. 2009. "Conservation dogs." In *Canine Ergonomics: The Science of Working Dogs*, edited by W. S. Helton, 175–94. Boca Raton: CRC Press.

Kablan, Y. A., A. Diarrassouba, R. Mundry, G. Campbell, E. Normand, H. S. Kühl, I. Koné, and C. Boesch. 2017. "Effects of anti-poaching patrols on the distribution of large mammals in Taï National Park, Côte d'Ivoire." *Oryx* 53 (3): 469–78.

Kasenene, J. M., and E. A. Ross. 2008. "Community benefits from long-term research programs: A case study from Kibale National Park, Uganda." In *Science and Conservation in African Forests: Benefits from Long-Term Research*, edited by R. W. Wrangham and E. A. Ross, 158–72. Cambridge: Cambridge University Press.

Kideghesho, J. R., E. Røskaft, and B. P. Kaltenborn. 2007. "Factors influencing conservation attitudes of local people in Western Serengeti, Tanzania." *Biodiversity and Conservation* 16 (7): 2213–30.

Knapp, E. J., N. Peace, and L. Bechtel. 2017. "Poachers and poverty: Assessing objective and subjective measures of poverty among illegal hunters outside Ruaha National Park, Tanzania." *Conservation and Society* 15 (1): 24.

Kormos, R., C. Boesch, M. I. Bakarr, and T. M. Butynski. 2003. *West African Chimpanzees: Status Survey and Conservation Action Plan*. Gland, Switzerland: IUCN.

Kouassi, J. A. K., E. Normand, I. Koné, and C. Boesch. 2017. "Bushmeat consumption and environmental awareness in rural households: A case study around Taï National Park, Côte d'Ivoire." *Oryx* 53 (2): 293–99.

Kühl, H. S., T. Sop, E. A. Williamson, R. Mundry, D. Brugière, G. Campbell, H. Cohen, E. Danquah, L. Ginn, I. Herbinger, S. Jones, J. Junker, R. Kormos, C. Y. Kouakou, P. N'Goran, E. Normand, K. Shutt-Phillips, A. Tickle, E. Vendras, A. Welsh, E. G. Wessling, and C. Boesch. 2017. "The critically endangered western chimpanzee declines by 80%." *American Journal of Primatology* 79 (9): e22681.

Lamarque, F., J. Anderson, R. Fergusson, M. Lagrange, Y. Osei-Owusu, and L. Bakker. 2009. *Human-Wildlife Conflict in Africa: Causes, Consequences and Management Strategies*. FAO Forestry Paper no. 157. Rome: Food and Agriculture Organization of the United Nations.

Lambert, J. E. 2011. "Primate seed dispersers as umbrella species: A case study from Kibale National Park, Uganda, with implications for Afrotropical forest conservation." *American Journal of Primatology* 73 (1): 9–24.

Lee, P. C., and N. E. C. Priston. 2005. "Human attitudes to primates: Perceptions of pests, conflict and consequences for primate conservation." In *Commensalism and Conflict: The*

Human-Primate Interface, edited by J. D. Paterson and J. Wallis, 1–23. Norman, OK: American Society of Primatologists.

Leeds, A., K. E. Lukas, C. J. Kendall, M. A. Slavin, E. A. Ross, M. M. Robbins, D. van Weeghel, and R. A. Bergl. 2017. "Evaluating the effect of a year-long film focused environmental education program on Ugandan student knowledge of and attitudes toward great apes." *American Journal of Primatology* 79 (8): e22673.

Leigh, K. A., and M. Dominick. 2015. "An assessment of the effects of habitat structure on the scat finding performance of a wildlife detection dog." *Methods in Ecology and Evolution* 6 (7): 745–52.

Lonsdorf, E., D. Travis, R. Ssuna, E. Lantz, M. Wilson, K. Gamble, K. Terio, F. Leendertz, B. Ehlers, B. Keele, B. Hahn, T. Gillespie, J. Pond, J. Raphael, and A. Collins. 2014. "Field immobilization for treatment of an unknown illness in a wild chimpanzee (*Pan troglodytes schweinfurthii*) at Gombe National Park, Tanzania: Findings, challenges, and lessons learned." *Primates* 55 (1): 89–99.

Lukasik, M. 2002. "Establishing a long-term veterinary project for free-ranging chimpanzees in Tanzania." *Pan Africa News* 9 (2): 13–17.

Lwanga, J. S., and G. Isabirye-Basuta. 2008. "Long-term perspectives on forest conservation: Lessons from research in Kibale National Park." In *Science and Conservation in African Forests: Benefits from Long-Term Research*, edited by R. W. Wrangham and E. A. Ross, 158–72. Cambridge: Cambridge University Press.

MacKenzie, C. A., and P. Ahabyona. 2012. "Elephants in the garden: Financial and social costs of crop raiding." *Ecological Economics* 75: 72–82.

MacKenzie, C. A., S. P. Moffatt, J. Ogwang, P. Ahabyona, and R. R. Sengupta. 2017a. "Spatial and temporal patterns in primary school enrollment and exam achievement in rural Uganda." *Children's Geographies* 15 (3): 334–48.

MacKenzie, C. A., J. Salerno, J. Hartter, C. A. Chapman, R. R. Reyna, D. Mwesigye Tumusiime, and M. Drake. 2017b. "Changing perceptions of protected area benefits and problems around Kibale National Park, Uganda." *Journal of Environmental Management* 200: 217–28.

Matsuzawa, T., T. Humle, and Y. Sugiyama. 2011. *The Chimpanzees of Bossou and Nimba*. Tokyo: Springer.

Matsuzawa, T., and M. Kourouma. 2008. "The Green Corridor Project: Long-term research and conservation in Bossou, Guinea." In *Science and Conservation in African Forests: Benefits from Long-Term Research*, edited by R. W. Wrangham and E. A. Ross, 201–12. Cambridge: Cambridge University Press.

McCarthy, M. S., J. D. Lester, E. J. Howe, M. Arandjelovic, C. B. Stanford, and L. Vigilant. 2015. "Genetic censusing identifies an unexpectedly sizeable population of an endangered large mammal in a fragmented forest landscape." *BMC Ecology* 15 (1): 21.

McLennan, M. R. 2010. "Chimpanzee ecology and interactions with people in an unprotected human-dominated landscape at Bulindi, Western Uganda." PhD diss., Oxford Brookes University.

McLennan, M. R., and C. Asiimwe. 2016. "Cars kill chimpanzees: Case report of a wild chimpanzee killed on a road at Bulindi, Uganda." *Primates* 57 (3): 377–88.

McLennan, M. R., and C. M. Hill. 2012. "Troublesome neighbours: Changing attitudes towards chimpanzees (*Pan troglodytes*) in a human-dominated landscape in Uganda." *Journal for Nature Conservation* 20 (4): 219–27.

McLennan, M. R., D. Hyeroba, C. Asiimwe, V. Reynolds, and J. Wallis. 2012. "Chimpanzees in mantraps: Lethal crop protection and conservation in Uganda." *Oryx* 46 (4): 598–603.

Mittermeier, R. A., P. R. Gil, M. Hoffman, J. Pilgrim, T. Brooks, C. G. Mittermeier, J. Lamoreux, and G. A. B. da Fonseca. 2005. *Hotspots Revisited: Earth's Biologically Richest and Most Endangered Terrestrial Ecoregions*. Arlington, VA: Conservation International.

Morgan, B. J., A. Adeleke, T. Bassey, R. Bergl, A. Dunn, R. Fotso, E. Gadsby, K. Gonder, E. Greengrass, D. Koutou Koulagna, G. Mbah, A. Nicholas, J. Oates, F. Omeni, Y. Saidu, V. Sommer, J. Sunderland-Groves, J. Tiebou, and E. Williamson. 2011. "Regional action plan for the conservation of the Nigeria–Cameroon chimpanzee (*Pan troglodytes ellioti*)." IUCN/SSC Primate Specialist Group and Zoological Society of San Diego.

Morgan, D., C. Sanz, D. Greer, T. Rayden, F. Maisels, and E. Williamson. 2013. "Great apes and FSC: Implementing 'ape friendly' practices in Central Africa's logging concessions." Gland, Switzerland: IUCN/SSC Primate Specialist Group.

Muller, M. N., and R. W. Wrangham. 2014. "Mortality rates among Kanyawara chimpanzees." *Journal of Human Evolution* 66: 107–14.

Mwesigwa, A. 2015. "Uganda's success in universal primary education falling apart." *Guardian*, April 23.

Naughton-Treves, L. I. S. A. 1997. "Farming the forest edge: Vulnerable places and people around Kibale National Park, Uganda." *Geographical Review* 87 (1): 27–46.

Nelleman, C., and A. Newton. 2002. "Great apes—The road ahead. An analysis of great ape habitat, using GLOBIO methodology." Nairobi: UNEP.

N'Goran, P. K., C. Boesch, R. Mundry, E. K. N'Goran, I. Herbinger, F. A. Yapi, and H. S. Kuehl. 2012. "Hunting, law enforcement, and African primate conservation." *Conservation Biology* 26 (3): 565–71.

Nishida, T., and M. Nakamura. 2008. "Long-term research and conservation in the Mahale Mountains, Tanzania." In *Science and Conservation in African Forests: Benefits from Long-Term Research*, edited by R. W. Wrangham and E. A. Ross, 173–83. Cambridge: Cambridge University Press.

Nishida, T., R. W. Wrangham, J. H. Jones, A. Marshall, and J. Wakibara. 2000. "Do chimpanzees survive the 21st century." In *The Apes: Challenges for the 21st Century, Conference Proceedings*, edited by Brookfield Zoo, 43–51. Brookfield, IL: Chicago Zoological Society.

Oates, J. F., G. H. Whitesides, A. G. Davies, P. G. Waterman, S. M. Green, G. L. Dasilva, and S. Mole. 1990. "Determinants of variation in tropical forest primate biomass: New evidence from West Africa." *Ecology* 71 (1): 328–43.

Omeja, P. A., M. J. Lawes, A. Corriveau, K. Valenta, D. Sarkar, F. Pozzan Paim, and C. A. Chapman. 2016. "Recovery of tree and mammal communities during large-scale forest regeneration in Kibale National Park, Uganda." *Biotropica* 48 (6): 770–79.

Parker, M. 2015. "Assessment of detection and tracking dog programs in Africa. Working Dogs for Conservation." https://wd4c.org/

Parsons, M. B., T. R. Gillespie, E. V. Lonsdorf, D. Travis, I. Lipende, B. Gilagiza, S. Kamenya, L. Pintea, and G. M. Vazquez-Prokopec. 2014. "Global positioning system data-loggers: A tool to quantify fine-scale movement of domestic animals to evaluate potential for zoonotic transmission to an endangered wildlife population." *PLoS One* 9 (11): e110984.

Piel, A. K., A. Lenoel, C. Johnson, and F. A. Stewart. 2015. "Deterring poaching in western Tanzania: The presence of wildlife researchers." *Global Ecology and Conservation* 3: 188–99.

Plumptre, A. J., T. R. B. Davenport, M. Behangana, R. Kityo, G. Eilu, P. Ssegawa, C. Ewango, D. Meirte, C. Kahindo, M. Herremans, J. K. Peterhans, J. D. Pilgrim, M. Wilson, M. Languy, and D. Moyer. 2007. "The biodiversity of the Albertine rift." *Biological Conservation* 134 (2): 178–94.

Plumptre, A. J., R. Rose, G. Nangendo, E. A. Williamson, K. Didier, J. Hart, F. Mulindahabi, C. Hicks, B. Griffin, H. Ogawa, S. Nixon, L. Pintea, A. Vosper, M. McLennan, F. Amsini, A. McNeilage, J. R. Makana, M. Kanamori, A. Hernanex, A. Piel, F. Stewart, J. Moore, K. Zamma, M. Nakamura, S. Kamenya, G. Idani, T. Sakamaki, M. Yoshikawa, D. Greer, S. Tranquilli, R. Beyers, C. Hashimoto, T. Furuichi, and E. Bennett. 2010. "Eastern chimpanzee (*Pan troglodytes schweinfurthii*): Status survey and conservation action plan 2010–2020." Gland, Switzerland: IUCN.

Population Reference Bureau (PRB). 2017. "2017 World Population Data Sheet." https://www.prb.org/2017-world-population-data-sheet/.

Pusey, A. E., L. Pintea, M. L. Wilson, S. Kamenya, and J. Goodall. 2007. "The contribution of long-term research at Gombe National Park to chimpanzee conservation." *Conservation Biology* 21 (3): 623–34.

Reynolds, V. 2005. *The Chimpanzees of the Budongo Forest: Ecology, Behaviour and Conservation*. New York: Oxford University Press.

Rolland, R. M., P. K. Hamilton, S. D. Kraus, B. Davenport, R. M. Gillett, and S. K. Wasser. 2006. "Faecal sampling using detection dogs to study reproduction and health in north Atlanta right whales (*Eubalaena glacialis*)." *Journal of Cetacean Research and Management* 8 (2): 121–25.

Scully, E. J., S. Basnet, R. W. Wrangham, M. N. Muller, E. Otali, D. Hyeroba, K. A. Grindle, T. E. Pappas, M. Emery Thompson, Z. Machanda, K. E. Watters, A. C. Palmenberg, J. E. Gern, and T. L. Goldberg. 2018. "Lethal respiratory disease associated with human rhinovirus C in wild chimpanzees, Uganda, 2013." *Emerging Infectious Diseases* 24 (2): 267–74.

Shaffer, C. A., M. S. Milstein, C. Yukuma, E. Marawanaru, and P. Suse. 2017. "Sustainability and comanagement of subsistence hunting in an indigenous reserve in Guyana." *Conservation Biology* 31 (5): 1119–31.

Struhsaker, T. T., P. J. Struhsaker, and K. S. Siex. 2005. "Conserving Africa's rain forests: Problems in protected areas and possible solutions." *Biological Conservation* 123 (1): 45–54.

Tranquilli, S., M. Abedi-Lartey, K. Abernethy, F. Amsini, A. Asamoah, C. Balangtaa, S. Blake, E. Bouanga, T. Breuer, T. M. Brncic, G. Campbell, R. Chancellor, C. A. Chapman, T. R. B. Davenport, A. Dunn, J. Dupain, A. Ekobo, M. Eno-Nku, G. Etoga, T. Furuichi, S. Gatti, A. Ghiurghi, C. Hashimoto, J. A. Hart, J. Head, M. Hega, I. Herbinger, T. C. Hicks, L. H. Holbech, B. Huijbregts, H. S. Kuhl, I. Imong, S. Le-Duc Yeno, J. Linder, P. Marshall, P. Minasoma Lero, D. Morgan, L. Mubalama, P. K. N'Goran, A. Nicholas, S. Nixon, E. Normand, L. Nziguyimpa, Z. Nzooh-Dongmo, R. Ofori-Amanfo, B. G. Ogunjemite, C. A. Petre, H. J. Rainey, S. Regnaut, O. Robinso, A. Rundus, C. M. Sanz, D. Tiku Okon, A. Todd, Y. Warren, and V. Sommer. 2014. "Protected areas in tropical Africa: Assessing threats and conservation activities." *PLoS One* 9 (12): e114154.

Tranquilli, S., M. Abedi-Lartey, F. Amsini, L. Arranz, A. Asamoah, O. Babafemi, N. Barakabuye, G. Campbell, R. Chancellor, T. R. B. Davenport, A. Dunn, J. Dupain, C. Ellis, G. Etoga, T. Furuichi, S. Gatti, A. Ghiurghi, E. Greengrass, C. Hashimoto, J. Hart, I. Herbinger, T. C. Hicks, L. H. Holbech, B. Huijbregts, I. Imong, N. Kumpel, F. Maisels, P. Marshall, S. Nixon, E. Normand, L. Nziguyimpa, Z. Nzooh-Dogmo, D. Tiku Okon, A. Plumptre, A. Rundus, J. Sunderland-Groves, A. Todd, Y. Warren, R. Mundry, C. Boesch, and H. Kuehl. 2011. "Lack of conservation effort rapidly increases African great ape extinction risk." *Conservation Letters* 5 (1): 48–55.

Tumusiime, D. M., and P. Vedeld. 2015. "Can biodiversity conservation benefit local people? Costs and benefits at a strict protected area in Uganda." *Journal of Sustainable Forestry* 34 (8): 761–86.

Tweheyo, M., C. M. Hill, and J. Obua. 2005. "Patterns of crop raiding by primates around the Budongo Forest Reserve, Uganda." *Wildlife Biology* 11 (3): 237–47.

Uganda National Population Council (UNPC). 2017. "Transforming Uganda's economy: Opportunities to harness the demographic dividend for sustainable development." State of Uganda Population Report (SUPRE).

UNEP-WCMC and IUCN. 2019. "Protected planet: The World Database on Protected Areas (WDPA)." Cambridge: UNEP-WCMC and IUCN. www.protectedplanet.net.

Wasser, S. K., B. Davenport, E. R. Ramage, K. E. Hunt, M. Parker, C. Clark, and G. Stenhouse. 2004. "Scat detection dogs in wildlife research and management: Application to grizzly and black bears in the Yellowhead Ecosystem, Alberta, Canada." *Canadian Journal of Zoology* 82: 475–92.

Wich, S. A. 2015. "Drones and conservation." In *Drones and Aerial Observation: New Technologies for Property Rights, Human Rights, and Global Development*, 63–70. Washington, DC: New America.

Widness, J., and G. P. Aronsen. 2018. "Camera trap data on mammal presence, behaviour and poaching: A case study from Mainaro, Kibale National Park, Uganda." *African Journal of Ecology* 56: 383–89.

Williamson, E. A., and K. Fawcett. 2018. "Long-term research and conservation of the Virunga mountain gorillas." In *Science and Conservation in African Forests: The Benefits of Long-Term Research*, edited by R. W. Wrangham and E. A. Ross, 213–29. Cambridge: Cambridge University Press.

World Population Review (WPR). 2017. "Uganda population." Accessed December 20, 2017. http://worldpopulationreview.com/countries/uganda-population/.

Wrangham, R. W., and S. Mugume. 2000. "Snare removal program in Kibale National Park: A preliminary report." *Pan Africa News* 7 (2).

Wrangham, R. W., and E. A. Ross. 2008. *Science and Conservation in African Forests: The Benefits of Long-Term Research*. Cambridge: Cambridge University Press.

Wrangham, R. W., M. Wilson, B. Hare, and N. D. Wolfe. 2000. "Chimpanzee predation and the ecology of microbial exchange." *Microbial Ecology in Health and Disease* 12 (3): 186–88.

Wright, S. J. 2005. "Tropical forests in a changing environment." *Trends in Ecology & Evolution* 20 (10): 553–60.

27

Forest Certification and the High Conservation Value Concept: Protecting Great Apes in the Sangha Trinational Landscape in an Era of Industrial Logging

DAVID B. MORGAN, WILLIAM WINSTON, CREPIN EYANA AYINA, WEN MAYOUKOU, ERIC V. LONSDORF, AND CRICKETTE M. SANZ

Introduction

FORESTRY IN TROPICAL AFRICA: IMPACTS OF LOGGING ON GREAT APES

Ape responses to logging and environmental change are complex. Precisely defining the impact thresholds of forestry activities on chimpanzees (*Pan troglodytes*) and gorillas (*Gorilla gorilla*) is challenging due to the long life spans of these apes, their complex social systems, and relatively low densities. Reactions of apes to environmental disturbance have generally been studied in terms of before and after numeric assessments of abundances at particular sites. Conclusions suggest both *Pan* and *Gorilla* are able to persist in timber production forest with varying degrees of success in short time frames, but the long-term prospects of survival have not yet been assessed. Survival prospects do not reflect solely the impact of forestry practices, but also a myriad of indirect or collateral impacts, such as bushmeat hunting, further habitat degradation and land conversion, and introduced pathogens (Campbell et al. 2008; Wilkie et al. 2001, see also Chapman et al., chapter 25 this volume; Hartel et al., chapter 26 this volume). This makes it difficult to isolate specific responses to particular types of disturbance associated with logging practices.

Species responses to environmental disturbances vary depending on their resource use preferences. Chimpanzees and bonobos are considered ripe fruit specialists (Morgan and Sanz 2006; Newton-Fisher 1999), with a preference for primary forest (Furuichi, Inagaki, and Angoue-Ovono 1997; Tutin and Fernandez 1984). While some surveys of chimpanzees in logged and unlogged habitats have indicated a preference for less disturbed forests (Clark et al. 2009; Matthews and Matthews 2004; Poulsen, Clark and Bolker 2011; Stokes et al. 2010), other investigations provide no conclusive evidence of such preferences (Arnhem et al. 2008; Dupain et al. 2004; Plumptre and Reynolds 1994).

Beyond diet and habitat preferences, it is important to take into consideration the social disposition and ranging behavior of the African apes. Chimpanzees and bonobos have structured territorial ranges that effectively limit community members from spending considerable time outside of their home range; they rely heavily on the use of core areas. If a larger portion of the community range was disturbed, this could physically displace a group into the range of a neighboring community, which would result in social upheaval within the group and possibly lethal conflict between groups. Such a scenario has been suggested to have reduced chimpanzee densities subsequent to selective logging at Lopé, Gabon (White and Tutin 2001), and may have transpired elsewhere in the region.

In contrast to chimpanzees and bonobos, gorillas are not as territorial and do not have the same social limitations to their ranging. Gorillas are committed folivores who rely on a diet primarily consisting of herbaceous vegetation (Kuroda et al. 1996; Morgan and Sanz 2006; Tutin and Fernandez 1985, 1993), though some populations show considerably more frugivorous tendencies (Masi et al. 2009). Similar to chimpanzees, gorillas utilize a wide variety of habitat types such as primary forest, heavily inundated swamps, and secondary forests (Carroll 1988; Furuichi et al. 1997; Tutin and Fernandez 1984; Usongo 1998). This high degree of ecological flexibility has led some researchers to hypothesize that gorillas would find adequate resources in logged habitats, a view which has been supported by field surveys in some regions (Clark et al. 2009; Matthews and Matthews 2004; Stokes et al. 2010). Chimpanzee densities, however, were lower in forests logged 15 years prior to the survey in comparison to unlogged habitat (Stokes et al. 2010). If an initial reduction in chimpanzee densities occurs after logging, a time-lag response in species recovery may nonetheless follow.

Ecological impacts of timber exploitation and forest recovery potentially have long-term effects on foraging and nesting resources used by local ape populations. Since the 1960s, scientists have studied logging at Kibale National Park in Uganda, where the variation in disturbance between logging units has served as a natural experiment for examining the impact of timber removal on chimpanzee ecology and reproductive fitness. In one of the initial investigations of the impact of logging on apes, Skorupa (1988) found an inverse relationship between logging intensity and chimpanzee densities. Degree of habitat disturbance was indicated to be a potentially key factor driving the abundance of chimpanzees inhabiting post-logged forests. Subsequent research elucidated more specific influences, showing that female chimpanzees had lower reproductive success, with longer interbirth intervals and higher infant mortality, in areas with outtake rates of 17.0 m^3/ha (50.3% of basal area reduction) and 20.9 m^3/ha (46.6% basal area reduction) compared to females residing in less disturbed forests (Emery Thompson et al. 2007). The more intensive logging regimes may have reduced the food resource base for chimpanzees, with adjacent "refuge" areas potentially buffering from such impacts. Potts' (2011) research in Kibale indicates that the explanation may be more complex, as the actual impacts of logging on chimpanzee diets were low, even in cases where preferred food items were previously exploited. Overall chimpanzee abundance did not appear related to logging history, highlighting the fact that previously-logged forests, particularly those exploited only once or twice, may still retain important resource attributes for ape survival. However, it is important to reflect on the differences in spatial scales considered in these investigations as well as any possible indirect or secondary impacts that could be influencing ape densities. While timber exploitation has the potential to have both positive and negative effects on apes, the removal of timber in west and central Africa generally coincides with other potential impacts such as the growth of human population centers (Poulsen et al. 2009), and the rise of informal marketplaces related to non-timber forest products, such as the trade in bushmeat.

In regions such as northern Republic of Congo, the potential for such direct and indirect impacts is considerable considering the amount of continuous forest and wildlife inhabiting such regions (Blake et al. 2008; Strindberg et al. 2018). The forests surrounding the Trinational de la Sangha landscape were recently designated a United Nations Education, Science, and Cultural Organization World Heritage site (fig. 27.1). This landscape consists of four officially recognized national parks and one reserve spanning Cameroon, Central African Republic, and Republic of Congo. While the setting aside of these protected areas is important to conservation, they are surrounded by large logging concessions. The first small-scale inroads into this remote landscape were established in the early 1970s but it was not until the early 2000s (Laporte et al. 2007) that a surge in road construction occurred, forever changing the accessibility of these once "frontier" and "last of the wild" expanses (Bryant, Nielsen, and Tangley 1997; Sanderson et al. 2002). The Intact Forest Landscape (IFL) technique uses medium spatial resolution satellite information to map boundaries of contiguous forest mosaics and associated habitats with an area at least 500 km^2 (50,000 ha). These continuous areas are reasoned to signify the likely existence of stable flora and fauna due to the absence of overt anthropogenic disturbance (Potapov et al. 2008). From 2000 to 2013, a 77% reduction in IFLs occurred in this region, which is the most rapid

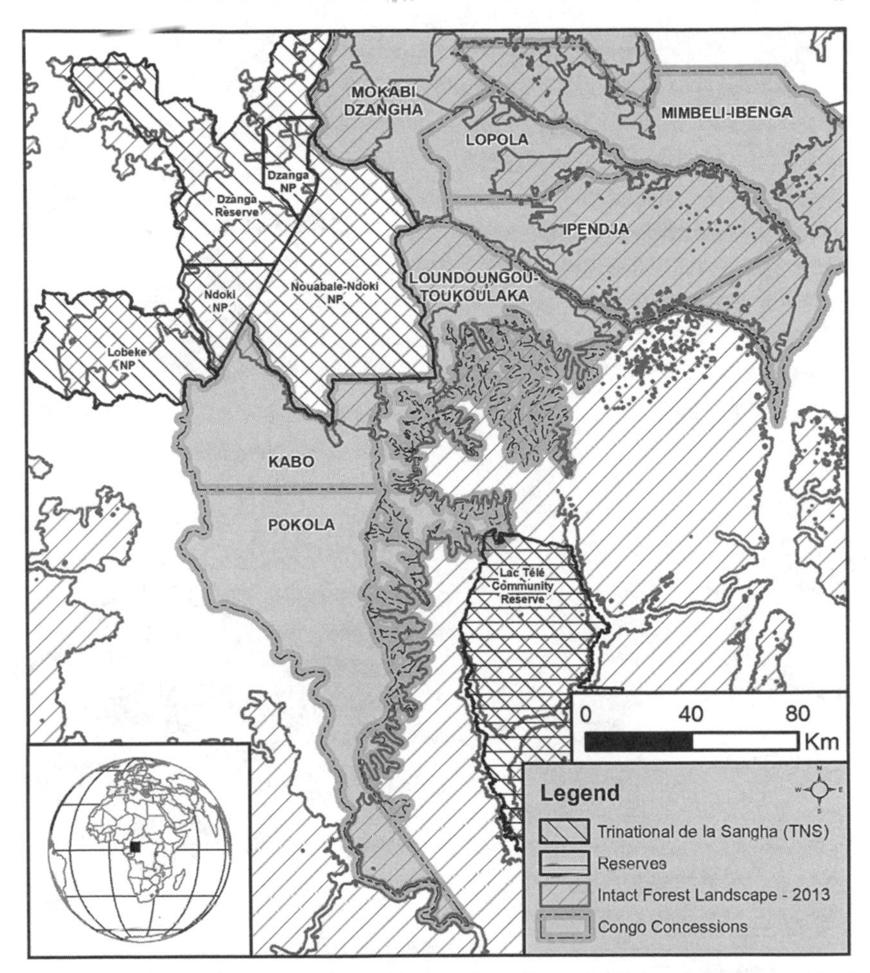

FIGURE 27.1. Protected areas of the Sangha Trinational Landscape and timber concessions contained important Intact Forest Landscape (IFL) as of 2013. Decline of Intact Forest Landscape (IFL) in the Kabo, Pokola, and Mimbeli-Ibenga concessions has been extensive.

change in all the tropics (Potapov et al. 2017). Most of this resulting change stemmed from the industrial timber industry.

Initial surveys in the logging concessions across the Sangha Trinational Landscape indicated that the Forest Stewardship Council (FSC) certification processes had produced positive results, which occurred over a five-year time frame, benefiting conservation of wildlife in the context of timber exploitation (Stokes et al. 2010). However, assessment of the full influence and any benefits accrued from certification necessitates a long-term approach. It also requires the establishment of an appropriate and timely monitoring regime to ensure that management practices are effective in their aims. There are valid

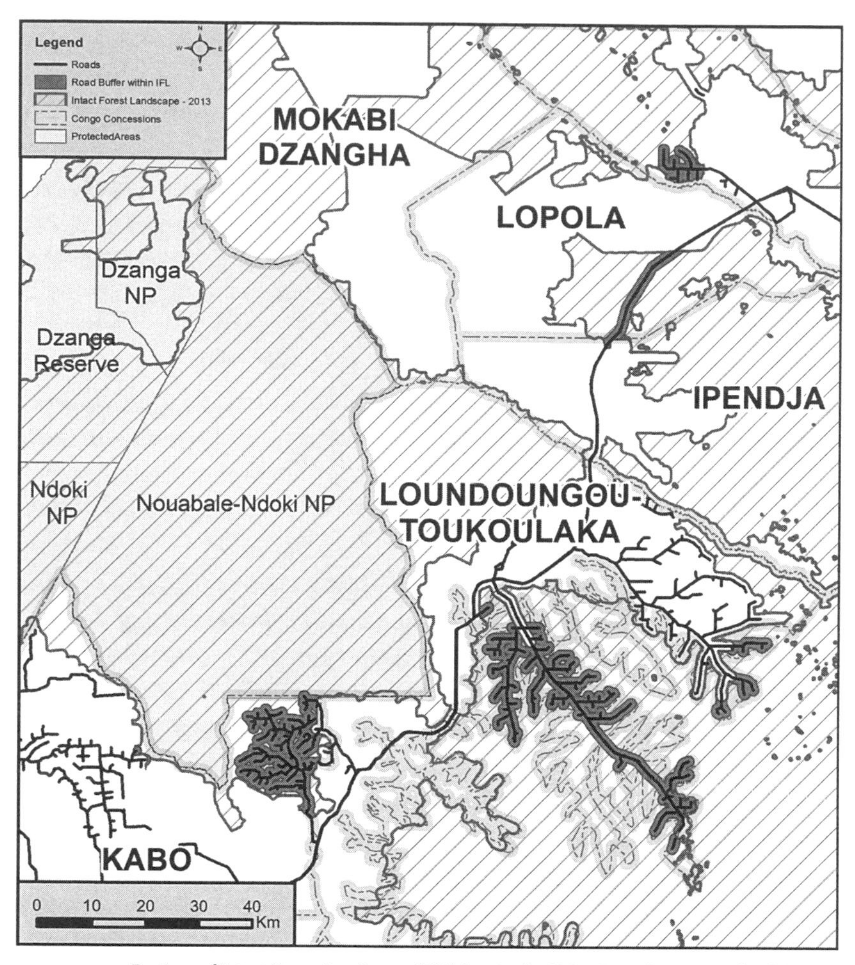

FIGURE 27.2. Regions of Intact Forest Landscape (IFL) loss in the Kabo, Loundougou-Toukoulaka, Lopola, and Mimbeli-Ibenga concessions. Different development strategies can be depicted based on the road placement and extent of area covered. Loss of forest canopy mostly coincided with proximity to roads and within mixed species forest.

concerns leveled at certified forestry operators and associated detrimental impacts within production zones, such as elevated levels of primary forest loss (Potapov et al. 2017) and deforestation (Brandt et al. 2016). In this chapter, we take a stepwise approach to identifying and assessing High Conservation Value Forests (HCVF) surrounding the Sangha Trinational Landscape. In the process, we update the IFL 2000–2013 inventory (Potapov et al. 2017) with a new estimate up to 2017 for the seven logging concessions neighboring the Sangha Trinational Landscape in Republic of Congo (fig. 27.2). At the landscape scale, we also provide an estimate of habitat suitability for chimpanzees and gorillas in the region based on satellite imagery analysis and species habitat use.

Advocates have argued for a more time sensitive and spatially explicit analytical research approach toward addressing anthropogenic disturbances that includes stakeholders and scientists working across disciplines (Bruenig 1996; Lindenmayer 2010; Nasi, Billand, and van Vliet 2012). Here, we provide a case study detailing such involvement including foresters, NGOs, and independent scientists. The HCVF approach is followed, placing emphasis on environmental values, which is at the core of principles of the forest certification standard (FSC 2009). Scaling down to the Forest Management Unit level, we estimate past and future impacts of primary roads in an active logging concession and relate development to deforestation. We use these results as a means to address the current debate surrounding the actual impact of logging IFL and influence on associated HCFVs within and outside these identified landscapes. Our overall aim is to improve conservation management policies to ensure the long-term survival of great apes and other wildlife in certified and non-certified logging concessions across the Congo Basin.

Approach

STUDY SITE

We analyzed seven logging concessions (Mokabi-Dzangha, leased by Rougier; Lopola, leased by BPL; Ipendja, leased by Thanry; Mimbeli-Ibenga, leased by SCTB; and Loundoungou-Toukoulaka, Kabo, Pokola, all leased by OLAM) along a vast stretch of lowland Guineo-Congolian forest that were allocated for timber harvesting. All but Mimbeli-Ibenga concessions included within this study were operated according to Management Plans that set regulations on timber regimes (Lescuyer et al. 2015). Various summary statistics were estimated at the concession level. Within the Kabo concession we focused on a Unite Forest Production (UFP) area (228.3 km^2) where active timber removal was taking place. The UFP3 consisted of three areas marked for exploitation from 2015 to 2019. The size and shape of each area was determined by timber inventories and the abundance of marketable timber species present (FAO 2016; Karsenty 2016).

Direct observational studies of gorilla and chimpanzee were carried out by the Goualougo Triangle Ape Project. The main study site and habituated ape groups are located in the unlogged IFL located between the Ndoki and

Goualougo Rivers. The Goualougo Triangle Ape Project was initiated in 1999 and is an ongoing project with studies located inside and outside of the Nouabalé-Ndoki National Park.

INTACT FOREST LANDSCAPE ACROSS THE TRINATIONAL DE LA SANGHA LANDSCAPE

We used the IFL inventory map (Potapov et al. 2008) to define "intactness" of the forest landscapes. This map is based on the extent of roads and settlements documented from Landsat images (of 30-m resolution) up to 2013 (Potapov et al. 2017). The regions highlighted are areas >500 km and >10 km wide that fall outside a 1-km buffer around such infrastructure (Potapov et al. 2008). However, new road networks have been established within the concessions surrounding the Sangha Trinational Landscape and within the IFL since 2013. Information on these new roads was provided by the timber operating company and projected in ArcGIS. We used satellite imagery to georeference roads and verify placement and subsequently updated the IFL inventories generated for areas of interest following Potapov et al. (2008).

PREDICTIVE CHIMPANZEE AND GORILLA HABITAT MODELING

The basis for our habitat suitability modeling was a Landsat ETM1 satellite image (collected on February 9, 2001) displayed in bands 4,5,3 (RGB) that overlaid the study area. Seventeen habitats were classified based on groundtruthed data within the image. Apes require two basic types of resources to persist on a landscape: nesting substrates and foraging resources. The model, therefore, requires estimates of availability of both of these resource types for each land cover type in the map. We ranked habitat types with a score of increasing preference from 0 to 10. Habitats were assessed separately in terms of quality for feeding and nesting, and later combined to produce a habitat landscape for each species.

Foraging preferences were based on our direct observations of ape foraging behavior in the Goualougo Triangle (unpublished data; Morgan et al. 2006), with consideration of reports of ape feeding ecology published from other sites (Doran et al. 2002; Morgan and Sanz 2006; Moutsambote et al. 1994; Remis et al. 2001; Rogers et al. 2004; Sabater-Pi 1979; Tutin and Fernandez 1985, 1993; Williamson et al. 1990). Suitability of habitat for nesting was based on studies of habitat use by gorillas and chimpanzees from line-transect

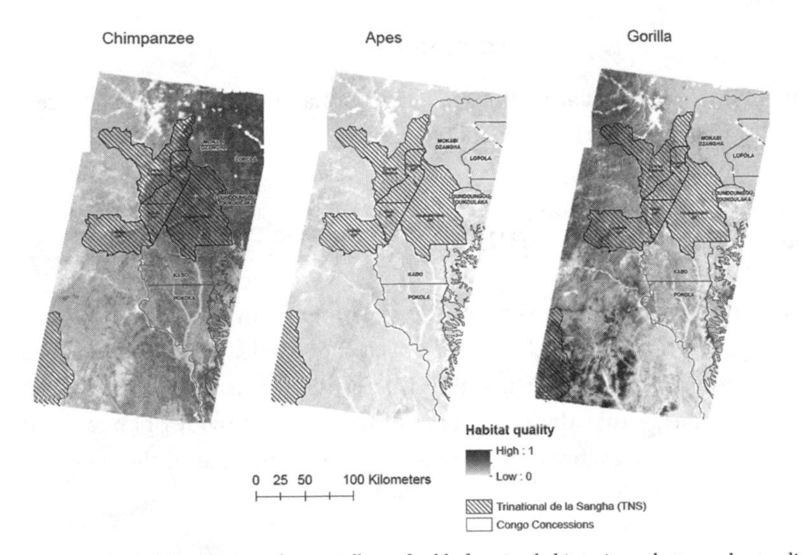

FIGURE 27.3. Spatial distribution of potentially-preferable foraging habitat gives a better understanding of distribution of potential High Conservation Value Forest (HCVF) for chimpanzees (left), great apes (center), and gorillas (right) across the Sangha Trinational area and adjacent forestry concessions. High quality habitat for gorillas and chimpanzees is indicated as increasing from gray to black. Less preferred habitat by species is indicated by lighter shading. Although habitat preferences overlap, there are distinct patterns in the spatial distribution of high quality habitat for gorillas and chimpanzees.

nest surveys and direct observations (Morgan et al. 2006). When scoring habitat preferences for each species, we also took into consideration the reports of ape nesting and habitat use from this region (Blake et al. 1995; Dupain et al. 2004; Fay et al. 1989; Furuichi et al. 1997; Matthews and Matthews 2004; Morgan 2001; Poulsen and Clark 2004; Tutin and Fernandez 1984; Tutin et al. 1995). Rather than base our ratings of preferred habitats on the Goualougo site, we attempted to estimate the likelihood that either species would select that habitat if its availability was equal to all other resources in an area. Apes use nesting habitat each evening and move into foraging habitats during the day, and their foraging distances, in combination with arrangement of different habitats, affect their likely persistence and abundance in a given landscape. Our model therefore also requires a typical foraging distance for chimpanzee and gorilla based on direct observations at Goualougo.

The first step of our model was to translate a land cover map (fig. 27.3) into a nesting suitability map and a forage resource availability map. Then, based on the amount and location of nesting and foraging resources, we calculated an overall habitat suitability score for chimpanzees and gorillas.

Nesting Suitability Map

The first step in calculating the ape habitat suitability score at each parcel was to identify the proportion of suitable ape nesting habitat (*HN*) in a parcel *x* as a function of land cover *j*:

(equation 1) $$\mathbf{HN_x} = \sum_{j=1}^{J} \mathbf{N_j P_{jx}}$$

where $N_j \in [0,1]$ represents compatibility of land cover *j* for nesting and p_{jx} is the proportion of parcel *x* that is covered by land cover *j*. This provides a landscape map of nesting suitability where $HN_x \in [0,1]$ (fig. 27.3). A score of 1 would indicate that the entire area of the parcel provides habitat suitable for nesting while a score of 0.2 would indicate 20 percent of the parcel's area provides suitable nesting habitat.

Forage Resource Map

We also calculated the proportion of suitable foraging habitat surrounding a parcel *x*, given by $HF_x \in [0,1]$. We assume that foraging frequency in parcel *m* declines exponentially with distance to a suitable nest site, and that apes forage in all directions with equal probability. Therefore, parcels farther away from nest parcel *x* contributed less to total resource availability than parcels nearby, and leads to the following prediction for the potential foraging resources available to ape nesting in parcel *x*, HF_x:

(equation 2) $$\mathbf{HF_x} = \frac{\sum_{m=1}^{M} \sum_{j=1}^{J} \mathbf{F_j\, p_{jm}\, e^{\frac{-D_{mx}}{\alpha}}}}{\sum_{m=1}^{M} \mathbf{e^{\frac{-D_{mx}}{\alpha}}}}$$

where p_{jm} is the proportion of parcels *m* in land cover *j*, D_{mx} is the Euclidean distance between parcels *m* and *x*, α is the expected foraging distance for the ape, and $F_j \in [0,1]$ represents relative amount of foraging resource in land cover *j*. The numerator is the distance-weighted resource summed across all *M* parcels. The denominator represents the maximum possible amount of forage in the landscape. This equation generates a distance-weighted proportion of habitat providing foraging resources within foraging range, normalized by the total forage available within that range (Winfree et al. 2005). Since ape abundance is potentially limited by both nesting and foraging resources, the overall suitability score on parcel *x* is simply the product of foraging and nesting such that $HS_x = HF_x HN_x \in [0,1]$.

DEFORESTATION

We used two measures of forest-cover loss. First, we estimated annual forest loss rates based on the examination of Landsat 8 data at a 30m spatial resolution from 2000 to 2016 (image GFG_10N_010) (Hansen et al. 2013). Cell counts from the global data set (cells = 30 m × 30 m) constituted the output.

Our second forest-cover loss analysis focused on the direct localized impact of road development and removal of trees. We estimated total area of forest lost due to land clearance for roads by placing a 20 m buffer around any new road. Estimated road width was 20 m based on Kleinschroth (2016). We modeled the removal of preferred ape foods, which were ranked according to frequency of use based on direct observations of feeding events by chimpanzees and gorillas, feeding traces encountered while following the apes, and fecal analysis in the Goualougo Triangle from 1999 to 2014 (Morgan and Sanz 2006). These included tree species within the fruit-bearing genus such as *Dialium*, *Chrysophyllum Greenwayodendron*, *Irvingia*, and *Pterocarpus*.

The influence of road development on ape resources was estimated by linking species-specific preferences to georeferenced commercial timber inventory data for this study area. For 2014 to 2017, the commercial timber inventory data corresponded with locations of new road construction within the UFP3 area. The inventory included tree stems with minimum diameter of exploitation ranging between 60 to 100 cm at breast height depending upon species (Congolaise Industrielle de Bois 2006). The spatially explicit inventory of individual trees surveyed in the UFP3 included stems from 40 different species of marketable and non-marketable trees. For this study, any tree stem falling within the pre-defined road buffer was considered removed during the development of the road. To assess potential carbon loss associated with forest disappearance, we used 395.7 Mg ha^{-1} as a reference, based on Lewis et al.'s (2013) calculation for mature forest in the Congo Basin.

Results

LANDSCAPE LEVEL ASSESSMENT OF INTACT FOREST LANDSCAPE

From 2013 to 2017, we estimate that 609 km^2 of intact forest was removed across five of the seven concessions inventoried in this study (fig. 27.2). There were differences in total amount of road-less area (km^2) being impacted between concessions during this time period (table 27.1). In Loundoungou-Toukoulaka, the reduction was equivalent to 358 km^2 over the last four years,

TABLE 27.1. Extent of road lengths and area reduction of Intact Forest Landscape (IFL) within timber concessions bordering the Sangha Trinational area.

Concession	*Area (km²)*	*Total Length of Roads (km)*	*Total IFL (km²)*	*Reduction in IFL 2013–2017 (km²)*
Pokola	5,277	784	656.48	0.07
Kabo	2,960	526	525.97	206.95
Ipendja	4,585	24	3,588.08	2.34
Lopola	1,970	36	746.32	40.72
Loundoungou-Toukoulaka	5,608	671	2,268.87	358.46
Mimbeli-Ibenga	6,716	160	2,811.48	42.00

Note: Concession database was obtained from the World Resource Institute (http://www.wri.org/resources/maps/republic-congo-logging-concessions-and-protected-areas). Logging road spatial database was provided by OLAM Ltd.

whereas the IFL loss in Kabo was 207 km². We found very little IFL remaining in the Pokola concession, which has experienced one or two cycles of timber exploitation throughout most of the area. IFL loss in Pokola was concentrated in inundated areas to the south of the concession where road placement was in close proximity to flooded forests. IFL loss in Lopola concession resulted from a long linear road that bisected the largest remaining extent of IFL in the concession area. While thc expansion into this road-less area was comparatively low, it fragmented the remaining IFL in the region. To the east, and in neighboring Mimbeli-Ibenga concession, 42 km² of IFL was lost as a result of logging activity.

PREDICTIVE CHIMPANZEE AND GORILLA HABITAT MODELING

The spatial distribution of preferred-foraging habitat gives a better understanding of the distribution of potential areas of High Conservation Value (HCV) for chimpanzees and gorillas across the Sangha Trinational Landscape and adjacent forestry concessions. Although the habitat preferences of these apes overlap, there are distinct patterns in the spatial distribution of high-quality habitat for chimpanzees and gorillas. Areas estimated to be high in habitat suitability for chimpanzees corresponded with IFL within the Nouabalé-Ndoki National Park in Republic of Congo and Ndoki National Park in Central African Republic (fig. 27.4). The highest average habitat suitability for gorillas was found in the Lobeke National Park (Cameroon) and the Dzanga Reserve (Central African Republic). Both of these areas had regions not included in the IFL inventory, as harvesting of timber occurred in these forests in the early 1970s. Logging has only recently been re-initiated in Dzanga Reserve.

DEFORESTATION IN THE KABO CONCESSION

Changing our focus from concession-scale analysis to the UFP3 area of interest, we found that IFL remained largely free of roads within the Kabo concession until 2014. Annual deforestation data from Hansen and colleagues (2013) indicated the forest of UFP3 was largely stable for the period spanning from 2001 to 2014, at which time forestry activities began to expand into this previously road-less area. From 2014 to 2016, forestry activities resulted in tree cover area loss ranging from 0.77 km²/year to 2.35 km²/year. The resulting forest-cover loss during this time period in the UFP3 totaled 4.92 km² of forest loss, most of which was spatially associated with logging roads (fig. 27.2). The forest removal was estimated to amount to 92,989.5 Mg AGB lost in the UFP3 from forestry activities during this period.

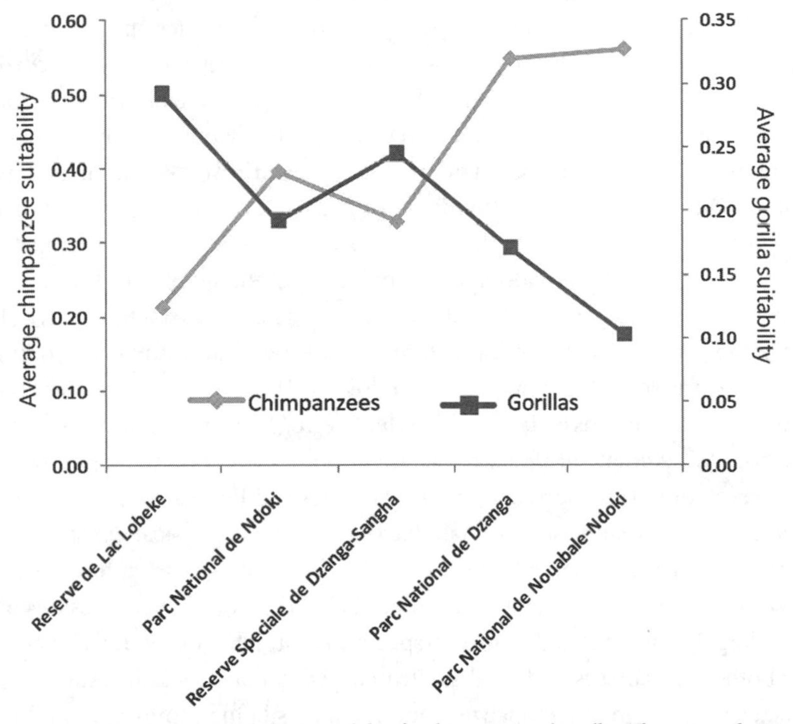

FIGURE 27.4. Protected average habitat suitability for chimpanzee and gorillas. The protected areas on the x-axis are arranged from west to east and show a chimpanzee-to-gorilla habitat gradient. The western parks protect relatively more gorillas while the eastern parks protect relatively more chimpanzees.

ROAD IMPACT ON GREAT APE RESOURCE BASE

The spatial distribution of deforestation obtained from global Landsat data largely aligned with the location of physical land clearing related to the establishment of primary and secondary roads. From 2014 to 2017, approximately 134.4 km of new roads were opened in UFP3. Forest removal resulting from direct impacts of road clearance was estimated at 2.69 km^2 of forest lost within this 227 km^2 area of the UFP3. Within 2017 alone, this likely resulted in the clearance of 199 individual tree stems representing 30 species of woody tree. The most common species removed was *Entandrophragma cylindricum*, followed by *Diospyros crassiflora* and *E. angolense*. Of the total estimated tree species removed during route placement, 11 species are included in the diet of chimpanzees and seven species are fed upon by gorillas.

Discussion

The forests of the Sangha Trinational Landscape are renowned for their high abundances of gorillas and chimpanzees. Through species-specific habitat modeling, we found that IFL in the Sangha Trinational Landscape corresponded with higher levels of suitable habitat for chimpanzees whereas non-IFL had elevated levels of predicted preferred gorilla habitat. Abundance estimates of both ape species based on nest surveys confirm these findings. They also provide further evidence that single or twice selectively logged forests may be altered at least initially to the benefit to gorillas and less so for sympatric chimpanzees. We also revealed that a significant percentage of road-less Sangha Trinational Landscape forest had been penetrated by extraction networks since the last assessment of intact forest, resulting in an estimated 650 km^2, or 6%, decrease of IFL since 2013. The loss of IFL was largely restricted to route construction associated with selective logging in dry mixed-species forests. Our estimate of IFL also included flooded and swamp forests, which are prevalent in the Loundoungou-Toukoulaka and Pokola concessions. Following the HCV approach, we tabulated the current physical extent of IFL in the region and also identified site-level HCVs and associated forestry impacts. Route construction patterns differed between adjacent concessions but were largely concentrated in mixed-species forest, which is preferred habitat for both chimpanzees and gorillas. Primary road construction resulted in a greater percentage of chimpanzee food resources being removed than those of gorillas. Flora regeneration likely depends on route type and associated damage, which could differentially impact chimpanzees and gorillas. Here we provide guidance on interpreting the HCV concept for other stakeholders interested in the use of this tool for conserving chimpanzees, gorillas, and other wildlife in active logging concessions.

As of 2017, road development in the Kabo and Loundoungou-Toukoulaka concessions was shown to be extensive. This is partially due to the large sizes of these concessions, which are 4,514 km^2, on average. However, the extent of forest impacted also varied with the different strategies of the operators. In all but one concession (Mokabi), the operators focused on road placement in pristine forests rather than returning to previously logged habitat. The greatest decline in IFL occurred in the Loundoungou-Toukoulaka concession, with an estimated 358 km^2 impacted since 2013. The road network in this concession traversed a previously road-less area to reach southern Annual Allowable Cut areas. The linearity of the main access road in this region was a reflection of the topography. The exploitable dry mature forest in this region was long and narrow extending into extensive flooded forests that buffered it on both sides. We documented a much denser road network for timber removal in the Kabo concession. The denser road network resulted in less impact on IFL with a total reduction of 206 km^2 over a three-year period. However, it did mark the arrival and expansion of exploitation into one of the last remaining expansive IFL within this concession. It also provided evidence of the operator's focus on timber removal, rather than construction of new extraction roads in this remote area.

INTACT FOREST LANDSCAPE AND HIGH CONSERVATION VALUE FORESTS

The boundaries of the Sangha Trinational protected area network were established in the late 1980s and early 1990s using threat indices that were developed as *ad hoc* functions of remoteness of the areas. The Parc National de Ndoki in Central African Republic and Nouabalé-Ndoki National Park in Republic of Congo were created largely encompassing IFL, which remains intact as of 2017. Our analysis verified there has been no illegal logging within the Nouabalé-Ndoki National Park, which is a concern for some other regions in the Republic of Congo, as estimates suggest that 70% of timber produced from Republic of Congo has been illegally sourced (Lawson 2014). Other protected areas within the Sangha Trinational Landscape (such as Lobeke National Park in Cameroon, Special Reserve de Dzanga Sangha and Parc National de Dzanga in Central African Republic) included once selectively logged forest within their boundaries. Our predictive modeling of suitable ape habitats indicated a high abundance of gorilla habitat within certain areas of Lobeke National Park and Special Reserve de Dzanga Sangha. The habitat disturbance associated with previous logging may have resulted in more open canopy forest, which is the preferred habitat of gorillas. While further monitoring is required, these results support a potential log-and-protect strategy that Rice, Gullison, and Reid (2001) proposed if stakeholders are interested in focusing conservation efforts on gorillas.

FSC Principle 9 requires that timber operators evaluate concessions to determine whether areas within the concession contain attributes that should be considered HCV. Concessions containing significant concentrations of great apes are considered to be HCV areas because all great ape species are listed as Endangered on the IUCN Red List of Threatened Species (IUCN 2016). Our coarse-grained regional scale analysis was a first step in the HCV assessment (fig. 27.5) and verified that the Sangha Trinational protected areas and bordering concessions qualify as an HCV 1. The suitability of the Sangha Trinational region to support high chimpanzee and gorilla populations is further validated by independent abundance estimates (Brncic, Maisels, and Strindberg 2018; Morgan et al. 2006; Stokes et al. 2010). Following FSC Principle 9, it is incumbent upon operators to consider the findings and guidance of independent scientists and NGOs when such information is available (FSC 2017). Large areas of IFL remain within northern Republic of Congo, but based on projections from road and settlement data, Potapov and colleagues (2017) estimated that the remaining IFL in the country will be lost within the next 60 years. Our updated estimates on the disappearance of IFL in this region support such claims. Further, we predict that all remaining IFL in the Kabo concession will be exploited in 2018 and 2019.

Most decisions about land allocations in the region have been made without detailed empirical data on species of concern. This information is now available from ongoing studies and can be used to better define "core areas" for conservation at large landscape scales (HCV 2), or for conserving forest attributes such as particular resources important to species-specific behaviors. Recent results indicate the number of behavioral repertoires documented across chimpanzee communities declines with increasing human disturbances, suggesting unique cultures of social groups are disappearing (Kuehl et al. 2019). Chimpanzees in this region are renowned for their tool-using propensities such as gathering honey from particular tree species or termites from earthen mounds (Sanz and Morgan 2009; Sanz, Morgan, and Gulick 2004). Protection measures for these particular resources should be appraised when planning exploitation of timber. Such protection measures have already been developed for identifying valued or "sacred" trees to local indigenous populations in some certified concessions (Hopkin 2007). Another potential avenue of

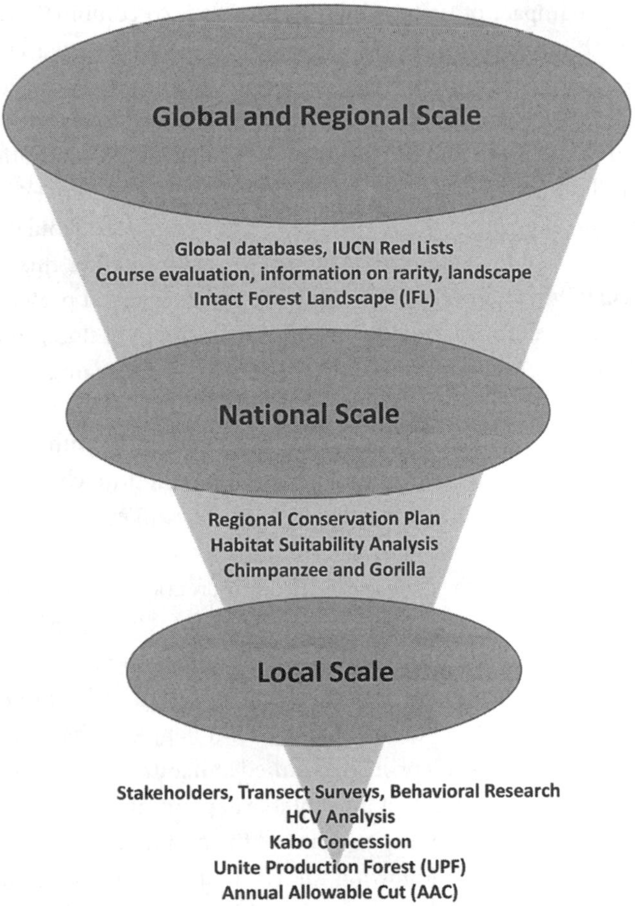

FIGURE 27.5. Following the High Conservation Value (HCV) approach, we show how different levels of information and data sources were used at various scales. At the regional level, coarse information, such as the Intact Forest Landscape (IFL) inventory, was used in identifying priority landscapes. Moving down to the national level and the Sangha Trinational protected area network, we found potential differences between apes in these varying conservation contexts using transect survey data and satellite imagery. Future potential sources of ape data could include regional distribution maps (see Junker et al. 2012; Maisels et al. submitted). At the local level, our HCV assessment utilized industry data (timber inventory and georeferenced spatial data on land-use and infrastructure such as roads, and annual allowable cut [AAC] areas) that was combined with information on chimpanzee and gorilla feeding ecology from direct observations of wild apes in this region. Critically, results from HCV value evaluation should be provided to local stakeholders (protected area managers, industry, and Forest Stewardship Council Regional Standard Committee).

expanding the impact of sustainability measures is to combine forest inventory data with information on the ecological needs of particular endangered species to generate quantitative parameters to inform selection of core areas or set-asides. For example, floristically interesting forests within remaining IFL located to the north and northeast of the Nouabalé-Ndoki National Park in the Loundoungou-Toukoulaka region deserve consideration (WWF 2017). We suggest an area within the IFL of the Loundoungou-Toukoulaka concession be inventoried for botanical attributes to verify whether this ecosystem qualifies as HCV 3 and potentially requiring environmental protection measures. Timber inventories provide an additional source of floral information typically lacking for the most at-risk regions, such as those outside protected areas where nearly 80% of chimpanzees in the region reside (Strindberg et al. 2018). Logging operators have already begun engaging with the academic community to enhance efforts to understand the ecological characteristics of production forests at a variety of spatial scales (Baccini et al. 2008; ter Steege 1998; Réjou-Méchain et al. 2008).

In most cases, outright protection through creation of key national protected areas is not possible. Approximately 14.9% of Republic of Congo's forests are IUCN Category 1 protected areas (Strindberg et al. 2018), which exceeds the Congo Basin average of 12.2% of forests designated as protected area across Central Africa (Mackinnon et al. 2016). It is urgent that a monitoring scheme following the HCV's principles aimed at habitat management in production forests be developed that is adaptable depending on local contexts. In fact, this is a unique opportunity to evaluate the cost and benefits of certified versus uncertified forestry following the HCV approach. Stakeholders concerned with impacts of exploitation on particular species or habitats should be able to predict potential changes associated with alteration in current forestry exploitation regimes. For example, the depletion of current marketable timber stocks and known natural tree growth trajectories in this region have led ecologists to recommend operators expand the diversity of timber species exploited and decrease the amount of canopy closure (Hall et al. 2003; Karsenty and Gourlet-Fleury 2006; Kleinschroth 2016). These prescriptions have been recommended as necessary to maintain timber production yields, but they also run counter to FSC timber certification standards. Greater emphasis on identifying and mitigating cumulative changes in quality and quantity of foraging and nesting resources over time and space is needed to ensure the long-term viability of chimpanzee and gorilla populations in these timber concessions. Given the accelerated loss of these forests, there is limited time to actualize such opportunities.

UNDERSTANDING THE IMPACT OF ROADS IN REMOTE FORESTS

The construction of roads and associated extraction routes have different short- and long-term consequences for continued exploitation of natural resources (Fuller et al. 2018) and resident wildlife populations (Laurance et al. 2006). As of 2017, an estimated 134 km of new road was opened and led to the immediate clearance of 2.69 km^2 of forest in the UFP3 area of the Kabo concession. By calculating total forest removed from an area of ape habitat, we provide a conservative estimate of potential direct impact on ape resources. In terms of ranging behavior, this represents an area equivalent to 15% of a chimpanzee "core area" of use based on estimates of community ranging made in the Goualougo Triangle (Morgan 2007). Not all route construction has the same spatial or floral impact. This study focused on primary roads, which result in the removal of more trees and the movement of greater amounts of soil than other routes such as secondary roads and skidder trails. Future studies on the immediate disturbance associated with primary, secondary, and skidder road development, followed by continuous monitoring of natural forest regeneration dynamics will provide valuable information on their implications for local apes.

Changes associated with road clearance had an immediate impact on apes: 35% of the removed trees were identified as food items of chimpanzees and 23% were included in gorilla diets. Research across the Sangha Trinational Landscape has shown that road construction associated with selectively logged forests has led to notable changes in tree species richness (Malcom and Ray 2000). The disturbance can lead to uniformity in the local floral assemblage and structurally altered habitats (Guariguata and Dupuy 1997). In the near term (<8 years), fruit-bearing tree resources removed will largely be replaced by non-woody vegetation at the ground level (Kleinschroth 2016; Malcom and Ray 2000). Such species may be favored by gorillas, but are typically less utilized by chimpanzees. This expansion of the foraging and nesting niches of gorillas likely has implications on local ranging patterns as well as population numbers. Forests logged at a greater intensity, with associated elevated levels of disturbance, could result in higher numbers of gorillas frequenting these areas after logging. Along primary logging roads, evidence from other sites suggests that the foraging prospects for chimpanzees may improve with time. Based on chronological surveys along primary logging roads in the region, the upper story of the forest will initially be typified by reduced canopy closure (Kleinschroth 2016) until fifteen years after exploitation, when pioneer tree species such as *Musanga cercopioides* grow and dominate the floral biomass and canopy. Chimpanzees in exploited forests of east Africa were found to incorporate the flowers and fruits of this tree species into their diets (Hashimoto 1995). Anecdotal observations in the Kabo logging concession also suggest chimpanzees make use of this new food and nesting resource. Stokes and colleagues (2010) found chimpanzee abundances in forests logged 15 years earlier were elevated and close to densities found in unlogged areas, indicating potential benefits of changing resources. This resource is temporary, however, and will last only three decades before a natural decline in abundance occurs as the forest transitions to include other tree species. An additional factor influencing the transition and ultimate legacy of roads is whether a road remains open after logging has ceased, either as a transport route for wood and commerce or for domestic travel. New routes provide easy and near complete access to once remote areas in concessions (Wilkie et al. 2001), which were previously buffered from threats such as hunting and pathogens of human origin known to be of threat to apes. Importantly, the vast majority (roughly 80%) of logging roads in northern Republic of Congo are temporary transport routes that are abandoned once logging has ceased, with inevitable natural forest regeneration to follow (Kleinschroth et al. 2015, 2016). Such road closures are mandatory with FSC certification as a measure to hinder illegal incursions or potential zoonotic disease transmission events. Most forests in this region have been exploited only once, and it remains unknown how these changes could influence great ape survival and fitness over longer time scales. An important avenue of future research will be to assess the dynamics of logging extraction routes and the transition of floristic communities influencing great ape populations.

Conservation Outlook for Great Apes

There are several reasons for cautious optimism when considering the conservation outlook of *Pan* and *Gorilla* in the context of timber exploitation in Africa. Until recently, logging disturbance across most of the region has been typified by relatively low rate and highly selective removal of a few timber species (Pérez et al. 2005). Two species (*Aucoumea klaineana* and *Entandrophragma cylindricum*) comprise 56% of the total timber production in the Congo Basin sub-region (Pérez et al. 2005). Neither of these species are of importance to great apes from a foraging or nesting perspective. Many of the concessions within the Republic of Congo and Gabon were located within IFL with low cumulative human influence. Therefore, remaining exploited forests in many of these concessions present an opportunity to assess and maintain identified natural attributes important to the survival of great apes. The relatively high abundance of great apes in this region is influenced by the distribution of closed or open canopy forest, which varies considerably over small spatial scales across this region. Understanding of species-specific preferences for particular habitat types and forest structures will continue to evolve with more complete information on ape ecology and be further enhanced by technological advances in remote sensing (Devos et al. 2008).

We previously identified the highest priority indicators for incorporation into Congo Basin FSC standards for promoting great ape conservation (Morgan and Sanz 2007; Morgan et al. 2013). In this study, we demonstrated how the HCV approach can be applied to conserving chimpanzee and gorillas in a timber production forest. From the landscape scale to the local area of interest, we identified habitat and resources impacted during exploitation that may pose a threat to great apes. This information is targeted for stakeholders and developers of High Conservation Value National Interpretations, which have already been drafted or finalized in five African countries (Republic of Congo, Democratic Republic of Congo, Cameroon, Gabon, and Liberia). We hope that the results presented will support discussions by a Standard Development Group on exploitation and preservation in IFL as well as inform reduced impact logging requirement of the Republic of Congo FSC indicators. The implications could be far reaching for remaining great ape populations.

Despite increasing human pressure and severe fragmentation of habitat, chimpanzees and gorillas defiantly persist (Bergl et al. 2012; Pusey et al. 2007; Robbins et al. 2011; Strindberg et al. 2018). Shifts in resource use and behavior observed across a continuum of human influence highlight the flexibility of apes in adapting to environmental changes and opportunities (Hockings, Anderson, and Matsuzawa 2009; Hockings et al. 2012; Meijaard et al. 2010; Morgan et al. 2017). Although these observations are encouraging, they also reflect the fact that apes are in crisis. The demise of our closest living relatives will be averted only by integrating ape resource needs into adaptive forestry management plans, which must become mandatory across the range states of African apes.

Acknowledgments

We are deeply appreciative of the opportunity to work in the Kabo Forestry Management Unit which is adjacent to the Nouabalé-Ndoki National Park. This work would not be possible without the continued support of the Ministère de l'Economie Forestière of the Government of the Republic of Congo and Wildlife Conservation Society's Congo Program. We thank OLAM Ltd.

and D. Paget, H. Ekani, and V. Istace for authorizing access to the inventory data. Special thanks are due to J. M. Fay, P. Telfer, P. Elkan, S. Elkan, B. Curran, M. Gately, E. Stokes, P. Ngouembe, D. Dos Santos, M. Ngangoue, Forrest Hogg, and Eric Arnhem. We also recognize the tireless dedication of C. Eyana-Ayina, S. Ndolo, A. Nzeheke, W. Mayoukou, M. Mguessa, I. Singono, and the Goualougo tracking team. Grateful acknowledgment of funding is due to the US Fish and Wildlife Service, Fondation pour le Tri-national de la Sangha, the Arcus Foundation, Cincinnati Zoo and Botanical Garden, Houston Zoo, Columbus Zoological Park, Indianapolis Zoo, the Associations of Zoos and Aquariuims, and Margot Marsh Biodiversity Fund.

References

Arnhem, E., J. Dupain, R. V. Drubbel, C. Devos, and M. Vercauteren. 2008. "Selective logging, habitat quality and home range use by sympatric gorillas and chimpanzees: A case study from an active logging concession in southeast Cameroon." *Folia Primatologica* 79 (1): 1–14.

Baccini, A., N. Laporte, S. J. Goetz, M. Sun, and H. Dong. 2008. "A first map of tropical Africa's above-ground biomass derived from satellite imagery." *Environmental Research Letters* 3 (4): 045011.

Bergl, R. A., Y. Warren, A. Nicholas, A. Dunn, I. Imong, J. L. Sunderland-Groves, and J. F. Oates. 2012. "Remote sensing analysis reveals habitat, dispersal corridors and expanded distribution for the critically endangered cross river gorilla *Gorilla gorilla diehli*." *Oryx* 46 (2): 278–89.

Blake, S., S. L. Deem, S. Stindberg, F. Maisels, L. Momont, I.-B. Isia, I. Douglas-Hamilton, W. B. Karesh, and M. D. Kock. 2008. "Roadless wilderness area determines forest elephant movements in the Congo Basin." *PLoS One* 3 (10): e3546.

Blake, S., E. Rogers, J. M. Fay, M. Ngangoue, and G. Ebeke. 1995. "Swamp gorillas in Northern Congo." *African Journal of Ecology* 33: 285–90.

Brandt, J. S., C. Nolte, and A. Agrawal. 2016. "Deforestation and timber production in Congo after implementation of sustainable forest management policy." *Land Use Policy* 52: 15–22.

Brncic, T., F. Maisels, and S. Strindberg. 2018. "Results of the 2016–2017 large mammal survey of the Ndoki-Likouala landscape." New York Wildlife Conservation Society.

Bruenig, E. F. 1996. *Conservation and Management of Tropical Rainforests: An Integrated Approach to Sustainability*. Wallingford: CABI Publishing.

Bryant, D., D. Nielsen, and L. Tangley. 1997. *The Last Frontier Forests. Ecosystems and Economies on the Edge*. Washington, DC: World Resources Institute.

Campbell, G., H. Kuehl, P. N. G. Kouame, and C. Boesch. 2008. "Alarming decline of West African chimpanzees in Côte d'Ivoire." *Current Biology* 18 (19): R903–4.

Carroll, R. W. 1988. "Relative density, range extension, and conservation potential of the lowland gorilla (*Gorilla gorilla gorilla*) in the Dzanga-Sangha region of Southwestern Central African Republic." *Mammalia* 52 (3): 309–23.

Clark, C. J., J. R. Poulsen, R. Malonga, and P. W. Elkan. 2009. "Logging concessions can extend the conservation estate for Central African tropical forests." *Conservation Biology* 23 (5): 1281–93.

Congolaise Industrielle des Bois (CIB) 2006. *Plan d'Amenagement de l'Unité Forestière d'Aménagement de Kabo (2005–2034)*. Brazzaville, Republic of the Congo: Ministry of Forest Economy.

Devos, C., C. Sanz, D. Morgan, J. R. Ononoga, N. Laporte, and M. C. Huynen. 2008. "Comparing ape densities and habitats in northern Congo: Surveys of sympatric gorillas and chimpanzees in the Odzala and Ndoki regions." *American Journal of Primatology* 70 (5): 439–51.

Doran, D. M., A. McNeilage, D. Greer, C. Bocian, P. Mehlman, and N. Shah. 2002. "Western lowland gorilla diet and resource availability: New evidence, cross-site comparisons, and reflections on indirect sampling." *American Journal of Primatology* 58: 91–116.

Dupain, J., P. Guislain, G. M. Nguenang, K. De Vleeschouwer, and L. Van Elsacker. 2004. "High chimpanzee and gorilla densities in a non-protected area on the northern periphery of the Dja Faunal Reserve, Cameroon." *Oryx* 38: 209–16.

Emery-Thompson, M., S. M. Kahlenberg, I. C. Gilby, and R. W. Wrangham. 2007. "Core area quality is associated with variance in reproductive success among female chimpanzees at Kibale National Park." *Animal Behaviour* 73: 501–12.

FAO. 2011. *State of the World's Forests*. Rome: Food and Agriculture Organization of the United Nations.

FAO. 2016. *Traceability: A Management Tool for Business and Governments*. Rome: Food and Agriculture Organization of the United Nations.

Fay, J. M., M. Agnagna, J. Moore, and R. Oko. 1989. "Gorillas (*Gorilla gorilla gorilla*) in the Likouala swamp forests of north central Congo: Preliminary data on populations and ecology." *International Journal of Primatology* 10: 477–86.

FSC. 2009. *Forest Management Evaluations Addendum–Forest Certification Reports*. FSC-STD-20-007a (V1-0) EN. Forestry Stewardship Council, A.C.

FSC. 2017. *FSC Facts and Figures*. Bonn, Germany: FSC International.

Fuller, T. L., T. P. Narins, J. Nackoney, T. C. Bonebrake, P. S. Clee, K. Morgan, A. Trochez, D. B. Mene, E. Boneweke, K. Y. Njabo, M. K. Gonder, M. Kahn, W. R. Allen, and T. B. Smith. 2018. "Assessing the impact of China's timber industry on Congo basin land use change." *Area* 51 (2): 340–49.

Furuichi, T., H. Inagaki, and S. Angoue-Ovono. 1997. "Population density of chimpanzees and gorillas in the Petit Loango Reserve, Gabon: Employing a new method to distinguish between nests of the two species." *International Journal of Primatology* 18 (6): 1029–46.

Guariguata, M. R., and J. M. Dupuy. 1997. "Forest regeneration in abandoned logging roads in Low Costa Rica." *Biotropica* 29: 15–28.

Hall, J. S., D. J. Harris, V. Medjibe, and P. M. S. Ashton. 2003. "The effects of selective logging on forest structure and tree species composition in a Central African forest: Implications for management of conservation areas." *Forest Ecology and Management* 183 (1–3): 249–64.

Hansen, M. C., P. V. Potapov, R. Moore, M. Hancher, S. A. Turubanova, A. Tyukavina, D. Thau, S. V. Stehman, S. J. Goetz, T. R. Loveland, A. Kommareddy, A. Egorov, L. Chini, C. O. Justice, and J. R. G. Townshend. 2013. "High-resolution global maps of 21st-century forest cover change." *Science* 342 (6160): 850–53.

Hashimoto, C. 1995. "Population census of the chimpanzees in the Kalinzu Forest, Uganda: Comparison between methods with nest counts." *Primates* 36 (4): 477–88.

Hockings, K. J., J. R. Anderson, and T. Matsuzawa. 2009. "Use of wild and cultivated foods by chimpanzees at Bossou, Republic of Guinea: Feeding dynamics in a human-Influenced environment." *American Journal of Primatology* 71 (8): 636–46.

Hockings, K. J., J. R. Anderson, and T. Matsuzawa. 2012. "Socioecological adaptations by chimpanzees, *Pan troglodytes verus*, inhabiting an anthropogenically impacted habitat." *Animal Behaviour* 83 (3): 801–10.

Hopkin, M. 2007. "Mark of respect." *Nature* 448: 402.

IUCN. 2016. Red List of Threatened Species. Gland, Switzerland: IUCN.

Junker, J., S. Blake, C. Boesch, G. Campbell, A. Dunn, L. du Toit, C. Duvall, A. Ekobo, G. Etoga, A. Galat-Luong, J. Gamys, J. Ganas-Swaray, S. Gatti, A. Ghiurghi, N. Granier, E. Greengrass, J. Hart, J. Head, I. Herbinger, T. C. Hicks, B. Huijbregts, I. S. Imong, N. Kuempel, S. Lahm, J. Lindsell, F. Maisels, M. McLennan, L. Martinez, B. Morgan, D. Morgan, F. Mulindahabi, R. Mundry, K. P. N'Goran, E. Normand, A. Ntongho, D. T. Okon, C. Petre, A. Plumptre, H. Rainey, S. Regnaut, C. Sanz, E. Stokes, A. Tondossama, S. Tranquilli, J. Sunderland-Groves, P. Walsh, Y. Warren, E. A. Williamson, and H. S. Kuehl. 2012. "Recent decline in suitable environmental conditions for African great apes." *Diversity and Distributions* 18: 1077–91.

Karsenty, A. 2016. "The contemporary forest concessions in West and Central Africa: Chronicle of a foretold decline?" *Forestry Policy and Institutions Working Paper No. 34*. Rome: Food and Agriculture Organization of the United Nations.

Karsenty, A., and S. Gourlet-Fleury. 2006. "Assessing sustainability of logging practices in the Congo Basin's managed forests: The issue of commercial species recovery." *Ecology and Society* 11 (1).

Kleinschroth, F. 2016. *Roads in the Rainforests: Legacy of Selective Logging in Central Africa*. PhD diss., Bangor University.

Kleinschroth, F., S. Gourlet-Fleury, P. Sist, F. Mortier, and J. R. Healey. 2015. "Legacy of logging roads in the Congo Basin: How persistent are the scars in forest cover?" *Ecosphere* 6 (4): 1–17.

Kleinschroth, F., J. R. Healey, P. Sist, F. Mortier, and S. Gourlet-Fleury. 2016. "How persistent are the impacts of logging roads on Central African forest vegetation?" *Journal of Applied Ecology* 53 (4): 1127–37.

Kuehl, H. S., C. Boesch, L. Kulik, F. Haas, M. Arandjelovic, P. Dieguez, G. Bocksberger, M. B. McElreath, A. Agbor, S. Angedakin, E. A. Ayimisin, E. Bailey, D. Barubiyo, M. Bessone, G. Brazzola, R. Chancellor, H. Cohen, C. Coupland, E. Danquah, T. Deschner, D. Dowd, A. Dunn, V. E. Egbe, H. Eshuis, A. Goedmakers, A. C. Granjon, J. Head, D. Hedwig, V. Hermans, I. Imong, K. J. Jeffery, S. Jones, J. Junker, P. Kadam, M. Kambere, M. Kambi, I. Kienast, D. Kujirakwinja, K. E. Langergraber, J. Lapuente, B. Larson, K. Lee, V. Leinert, M. Llana, G. Maretti, S. Marrocoli, R. Martin, T. J. Mbi, A. C. Meier, B. Morgan, D. Morgan, F. Mulindahabi, M. Murai, E. Neil, P. Niyigaba, L. J. Ormsby, R. Orume, L. Pacheco, A. Piel, J. Preece, S. Regnaut, A. Rundus, C. Sanz, J. van Schijndel, V. Sommer, F. Stewart, N. Tagg, E. Vendras, V. Vergnes, A. Welsh, E. G. Wessling, J. Willie, R. M. Wittig, Y. G. Yuh, K. Yurkiw, K. Zuberbuhler, and A. K. Kalan. 2019. "Human impact erodes chimpanzee behavioral diversity." *Science* 363 (6434): 1453–55.

Kuroda, S., T. Nishihara, S. Suzuki, and R. A. Oko. 1996. "Sympatric chimpanzees and gorillas in the Ndoki Forest, Congo." In *Great Ape Societies*, edited by W. C. McGrew, L. F. Marchant, and T. Nishida, 71–81. Cambridge: Cambridge University Press.

Laporte, N., J. A. Stabach, R. Grosch, T. S. Lin, and S. J. Goetz. 2007. "Expansion of industrial logging in Central Africa." *Science* 316: 1451.

Laurance, W. F., B. M. Croes, L. Tchignoumba, S. A. Lahm, A. Alonso, M. E. Lee, P. Campbell, and C. Ondzeano. 2006. "Impacts of roads and hunting on Central African rainforest mammals." *Conservation Biology* 20 (4): 1251–61.

Lawson, S. 2014. *Illegal Logging in the Democratic Republic of the Congo*. Energy, Environment and Resources EER PP 2014/03. London: Chatham House.

Lescuyer, G., M. N. Mvongo-Nkene, G. Monville, M. B. Elanga-Voundi, and T. Kakundika. 2015. "Promoting multiple-use forest management: Which trade-offs in the timber concessions of Central Africa?" *Forest Ecology and Management* 349: 20–28.

Lewis, S. L., B. Sonke, T. Sunderland, S. K. Begne, G. Lopez-Gonzalez, G. M. F. van der Heijden, O. L. Phillips, K. Affum-Baffoe, T. R. Baker, L. Banin, J. F. Bastin, H. Beeckman, P. Boeckx, J. Bogaert, C. De Canniere, E. Chezeaux, C. J. Clark, M. Collins, G. Djagbletey, M. N. K. Djuikouo, V. Droissart, J. L. Doucet, C. E. N. Ewango, S. Fauset, T. R. Feldpausch, E. G. Foli, J. F. Gillet, A. C. Hamilton, D. J. Harris, T. B. Hart, T. de Haulleville, A. Hladik, K. Hufkens, D. Huygens, P. Jeanmart, K. J. Jeffery, E. Kearsley, M. E. Leal, J. Lloyd, J. C. Lovett, J. R. Makana, Y. Malhi, A. R. Marshall, L. Ojo, K. S. H. Peh, G. Pickavance, J. R. Poulsen, J. M. Reitsma, D. Sheil, M. Simo, K. Steppe, H. E. Taedoumg, J. Talbot, J. R. D. Taplin, D. Taylor, S. C. Thomas, B. Toirambe, H. Verbeeck, J. Vleminckx, L. J. T. White, S. Willcock, H. Woell, and L. Zemagho. 2013. "Above-ground biomass and structure of 260 African tropical forests." *Philosophical Transactions of the Royal Society of London B* 368 (1625): 20120295.

Lindenmayer, D. B. 2010. "Landscape change and the science of biodiversity conservation in tropical forests: A view from the temperate world." *Biological Conservation* 143 (10): 2405–11.

Mackinnon, J., A. Aveling, R. C. D. Olivier, M. Murray, and C. Paolini. 2016. *Larger Than Elephants: Inputs for the Design of a Wildlife Conservation Strategy for Africa. A Regional Analysis*. Luxembourg: Publications Office of the European Union.

Malcolm, J. R., and J. C. Ray. 2000. "Influence of timber extraction routes on Central African small-mammal communities, forest structure, and tree diversity." *Conservation Biology* 14 (6): 1623–38.

Masi, S., C. Cipolletta, S. Ortmann, R. Mundry, and M. M. Robbins. 2009. "Does a more frugivorous diet lead to an increase in energy intake and energy expenditure? The case of the western lowland gorillas at Bai-Hokou, Central African Republic." *Folia Primatologica* 80 (6): 372–73.

Matthews, A., and A. Matthews. 2004. "Survey of gorillas (*Gorilla gorilla gorilla*) and chimpanzees (*Pan troglodytes troglodytes*) in southwestern Cameroon." *Primates* 45: 15–24.

Meijaard, E., G. Albar, Nardiyono, Y. Rayadin, M. Ancrenaz, and S. Spehar. 2010. "Unexpected ecological resilience in Bornean orangutans and implications for pulp and paper plantation management." *PLoS One* 50 (9): e12813.

Morgan, B. J. 2001. "Ecology of mammalian frugivores in the Reserve de Faune du Petit Loango, Gabon." PhD diss., University of Cambridge.

Morgan, D. B. 2007. "Socio-ecology of chimpanzees (*Pan troglodytes troglodytes*) in the Goualougo Triangle, Republic of Congo." PhD diss., University of Cambridge.

Morgan, D., R. Mundry, C. Sanz, C. Eyana Ayina, S. Strindberg, E. Lonsdorf, and H. S. Kühl. 2017. "African apes coexisting with logging: Comparing chimpanzee (*Pan troglodytes troglodytes*) and gorilla (*Gorilla gorilla gorilla*) resource needs and responses to forestry activities." *Biological Conservation* 218: 277–86.

Morgan, D., and C. Sanz. 2006. "Chimpanzee feeding ecology and comparisons with sympatric gorillas in the Goualougo Triangle, Republic of Congo." In *Primate Feeding Ecology in Apes and Other Primates: Ecological, Physiological, and Behavioural Aspects*, edited by G. Hohmann, M. Robbins, and C. Boesch, 97–122. Cambridge: Cambridge University Press.

Morgan, D., and C. Sanz. 2007. "Best practice guidelines for reducing the impact of commercial logging on wild apes in West Equatorial Africa." Gland, Switzerland: IUCN/SSC Primate Specialist Group.

Morgan, D., C. Sanz, D. Greer, T. Rayden, F. Maisels, and E. Williamson. 2013. "Great apes and FSC: Implementing 'ape friendly' practices in Central Africa's logging concessions." Gland, Switzerland: IUCN/SSC Primate Specialist Group.

Morgan, D., C. Sanz, J. R. Onononga, and S. Strindberg. 2006. "Ape abundance and habitat use in the Goualougo Triangle, Republic of Congo." *International Journal of Primatology* 27 (1): 147–79.

Moutsambote, J.-M., T. Yumoto, M. Mitani, T. Nishihara, S. Suzuki, and S. Kuroda. 1994. "Vegetation and plant list of species identified in the Nouabale-Ndoki Forest, Congo." *Tropics* 3 (3/4): 277–94.

Nasi, R., A. Billand, and N. van Vliet. 2012. "Managing for timber and biodiversity in the Congo Basin." *Forest Ecology and Management* 268: 103–11.

Newton-Fisher, N. E. 1999. "The diet of chimpanzees in the Budongo Forest Reserve, Uganda." *African Journal of Ecology* 37: 344–54.

Pérez, M. R., D. E. de Blas, R. Nasi, J. A. Sayer, M. Sassen, C. Angoue, N. Gami, O. Ndoye, G. Ngono, J. C. Nguinguiri, D. Nzala, B. Toirambe, and Y. Yalibanda. 2005. "Logging in the Congo Basin: A multi-country characterization of timber companies." *Forest Ecology and Management* 214 (1–3): 221–36.

Plumptre, A. J., and V. Reynolds. 1994. "The effect of selective logging on the primate populations in the Budongo Forest Reserve, Uganda." *Journal of Applied Ecology* 31: 631–41.

Potapov, P., M. C. Hansen, L. Laestadius, S. Turubanova, A. Yaroshenko, C. Thies, W. Smith, I. Zhuravleva, A. Komarova, S. Minnemeyer, and E. Esipova. 2017. "The last frontiers of wilderness: Tracking loss of intact forest landscapes from 2000 to 2013." *Science Advances* 3 (1): e1600821.

Potapov, P., A. Yaroshenko, S. Turubanova, M. Dubinin, L. Laestadius, C. Thies, D. Aksenov, A. Egorov, Y. Yesipova, I. Glushkov, M. Karpachevskiy, A. Kostikova, A. Manisha, E. Tsybikova, and I. Zhuravleva. 2008. "Mapping the world's intact forest landscapes by remote sensing." *Ecology and Society* 13 (2).

Potts, K. B. 2011. "The long-term impact of timber harvesting on the resource base of chimpanzees in Kibale National Park, Uganda." *Biotropica* 43 (2): 256–64.

Poulsen, J. R., and C. J. Clark. 2004. "Densities, distributions, and seasonal movements of gorillas and chimpanzees in swamp forest in Northern Congo." *International Journal of Primatology* 25: 285–306.

Poulsen, J. R., C. J. Clark, and B. M. Bolker. 2011. "Decoupling the effects of logging and hunting on an Afrotropical animal community." *Ecological Applications* 21 (5): 1819–36.

Poulsen, J. R., C. J. Clark, G. Mavah, and P. W. Elkan. 2009. "Bushmeat supply and consumption in a tropical logging concession in Northern Congo." *Conservation Biology* 23 (6): 1597–1608.

Pusey, A. E., L. Pintea, M. L. Wilson, S. Kamenya, and J. Goodall. 2007. "The contribution of long-term research at Gombe National Park to chimpanzee conservation." *Conservation Biology* 21 (3): 623–34.

Réjou-Méchain, M., R. Pelissier, S. Gourlet-Fleury, P. Couteron, R. Nasi, and J. D. Thompson. 2008. "Regional variation in tropical forest tree species composition in the Central African Republic: An assessment based on inventories by forest companies." *Journal of Tropical Ecology* 24: 663–74.

Remis, M. J., E. S. Dierenfeld, C. B. Mowry, and R. W. Carroll. 2001. "Nutritional aspects of western lowland gorilla (*Gorilla gorilla gorilla*) diet during seasons of fruit scarcity at Bai Houkou, Central African Republic." *International Journal of Primatology* 22 (5): 807–36.

Rice, R. E., R. E. Gullison, and J. W. Reid. 2001. "Sustainable forest management: A review of conventional wisdom." *Advances in Applied Biodiversity Science*. Washington, DC: Conservation International.

Robbins, M. M., M. Gray, K. A. Fawcett, F. B. Nutter, P. Uwingeli, I. Mburanumwe, E. Kagoda, A. Basabose, T. S. Stoinski, M. R. Cranfield, J. Byamukama, L. H. Spelman, and A. M. Robbins. 2011. "Extreme conservation leads to recovery of the Virunga mountain gorillas." *PLoS One* 6 (6): e19788.

Rogers, M. E., K. Abernethy, M. Bermejo, C. Cipolletta, D. Doran, K. McFarland, T. Nishihara, M. Remis, and C. E. G. Tutin. 2004. "Western gorilla diet: A synthesis from six sites." *American Journal of Primatology* 64 (2): 173–92.

Sabater-Pi, J. 1979. "Feeding behaviour and diet of chimpanzees (*Pan troglodytes troglodytes*) in the Okorobiko Mountains of Rio Muni (West Africa)." *Zeitschrift fuer Tierpsychologie* 50 (3): 265–81.

Sanderson, E. W., M. Jaiteh, M. A. Levy, K. H. Redford, A. V. Wannebo, and G. Woolmer. 2002. "The human footprint and the last of the wild." *Bioscience* 52 (10): 891–904.

Sanz, C. M., and D. B. Morgan. 2009. "Flexible and persistent tool-using strategies in honey-gathering by wild chimpanzees." *International Journal of Primatology* 30 (3): 411–27.

Sanz, C., D. Morgan, and S. Gulick. 2004. "New insights into chimpanzees, tools, and termites from the Congo basin." *American Naturalist* 164 (5): 567–81.

Skorupa, J. P. 1988. *The Effect of Selective Timber Harvesting on Rain-Forest Primates in Kibale Forest, Uganda*. PhD diss., University of California, Davis.

Stokes, E. J., S. Strindberg, P. C. Bakabana, P. W. Elkan, F. C. Iyenguet, B. Madzoke, G. A. F. Malanda, B. S. Mowawa, C. Moukoumbou, F. K. Ouakabadio, and H. J. Rainey. 2010. "Monitoring great ape and elephant abundance at large spatial scales: Measuring effectiveness of a conservation landscape." *PLoS One* 5 (4): e10294.

Strindberg, S., M. Maisels, L. Williamson, S. Blake, E. Stokes, R. Aba'a, G. Abitsi, A. Agbor, R. Ambahe, P. C. Bakabana, M. Bechem, A. Berlemont, P. Boundja, N. Bout, T. Breuer, G. Campbell, C. Inkamba, B. Bokoto de Semboli, P. de Wachter, A. Feistner, B. Fosso, R. Fotso, D. Greer, C. Iyenguet, K. Jeffery, F. Kiminou, H. Kühl, S. Latour, K. Pambou, C. Makoumbou, G. Malanda, R. Malonga, D. Morgan, G. Moukala, B. Mowawa, M. Murai, T. Nishihara, Z. Nzooh, L. Pintea, A. Amy Pokempner, H. Rainey, T. Rayden, H. Ruffler, C. Sanz, A. Todd, H. VanLeeuwe, A. Vosper, and D. Wilkie. 2018. "Guns, germs and trees: Determining density and distribution of gorillas and chimpanzees in Western Equatorial Africa." *Science Advances* 4 (4): eaar2964.

ter Steege, H. 1998. "The use of forest inventory data for a national protected area strategy in Guyana." *Biodiversity and Conservation* 7 (11): 1457–83.

Tutin, C. E. G., and M. Fernandez. 1984. "Nationwide census of gorilla and chimpanzee populations in Gabon." *American Journal of Primatology* 6: 313–36.

Tutin, C. E. G., and M. Fernandez. 1985. "Foods consumed by sympatric populations of *Gorilla g. gorilla* and *Pan t. troglodytes* in Gabon: Some preliminary data." *International Journal of Primatology* 6 (1): 27–42.

Tutin, C. E. G., and M. Fernandez. 1993. "Composition of the diet of chimpanzees and comparisons with that of sympatric lowland gorillas in the Lope Reserve, Gabon." *American Journal of Primatology* 30: 195–211.

Tutin, C. E. G., R. J. Parnell, L. J. T. White, and M. Fernandez. 1995. "Nest building by lowland gorillas in the Lopé Reserve, Gabon: Environmental influences and implications for censusing." *International Journal of Primatology* 16: 53.

Usongo, L. 1998. “Conservation status of primates in the proposed Lobeke Forest Reserve, South-East Cameroon.” *Primate Conservation* 18: 66–68.

White, L., and C. E. G. Tutin. 2001. “Why chimpanzees and gorillas respond differently to logging: A cautionary tale from Gabon.” In *African Rain Forest Ecology and Conservation: An Interdisciplinary Perspective*, edited by W. Weber, L. J. T. White, A. Vedder, and L. Naughton-Treves, 449–62. New Haven, CT: Yale University Press.

Wilkie, D. S., J. G. Sidle, G. C. Boundzanga, P. Auzel, and S. Blake. 2001. “Defaunation, not deforestation: Commercial logging and market hunting in Northern Congo.” In *The Cutting Edge: Conserving Wildlife in Logged Tropical Forest*, edited by R. A. Fimbel, A. Grajal, and J. G. Robinson, 375–99. New York: Columbia University Press.

Williamson, E. A., C. E. G. Tutin, M. E. Rogers, and M. Fernandez. 1990. “Composition of the diet of lowland gorillas at Lope in Gabon.” *American Journal of Primatology* 21: 265–77.

Winfree, R., J. Dushoff, E. E. Crone, C. B. Schultz, R. V. Budny, N. M. Williams, and C. Kremen. 2005. “Testing simple indices of habitat proximity.” *American Naturalist* 165 (6): 707–17.

World Wildlife Fund. 2017. “Position on IFL indicators for the Congo Basin.” World Wildlife Fund.

Contributors

Ikuma Adachi
Center for International Collaboration and Advanced Studies in Primatology
Primate Research Institute
Kyoto University
Japan
Matthias Allritz
School of Psychology and Neuroscience
University of St. Andrews
UK
Crepin Eyana Ayina
Goualougo Triangle Ape Project
Republic of Congo
Verena Behringer
Department of Primatology
Max Planck Institute for Evolutionary Anthropology
Germany
Michael J. Beran
Department of Psychology and Language Research Center
Georgia State University
USA
Mollie A. Bloomsmith
Yerkes National Primate Research Center
Emory University
USA
Stephanie L. Bogart
Department of Anthropology
University of Florida
USA

Sarah Bortolamiol
Department of Anthropology
McGill University
Canada
Sarah D. Breaux
New Iberia Research Center
USA
Sarah F. Brosnan
Departments of Psychology and Philosophy
Georgia State University
USA
Hannah M. Buchanan-Smith
Psychology Division, Faculty of Natural Sciences
University of Stirling
UK
Thomas Bugnyar
Department of Cognitive Biology
University of Vienna
Austria
Josep Call
School of Psychology and Neuroscience
University of St. Andrews
UK
Alecia J. Carter
Department of Anthropology
University College London
UK
Colin A. Chapman
Department of Anthropology
George Washington University
USA
Andrea W. Clay
Yerkes National Primate Research Center
Emory University
USA
Zanna Clay
Department of Psychology
Durham University
UK
Catherine Crockford
Department of Human Behavior, Ecology and Culture
Max Planck Institute for Evolutionary Anthropology
Germany

Katherine A. Cronin
Animal Welfare Science Program
Lincoln Park Zoo
USA
Tobias Deschner
Department of Primatology
Max Planck Institute for Evolutionary Anthropology
Germany
Shona Duguid
Warwick Business School
University of Warwick
UK
Robert E. Evans
Department of Ecology, Evolution, and Organismal Biology
Kennesaw State University
USA
Amy Fultz
Chimp Haven
USA
Jane Goodall
The Jane Goodall Institute
USA
Jessica A. Hartel
Department of Biology
University of North Georgia
USA
Faye S. Harwell
Department of Anthropology
Boston University
USA
Africa de las Heras
School of Psychology and Neuroscience
University of St. Andrews
UK
Elizabeth S. Herrelko
National Zoological Park
Smithsonian Institution
USA
Satoshi Hirata
Kumamoto Sanctuary
Kyoto University
Japan

Catherine Hobaiter
School of Psychology and Neuroscience
University of St. Andrews
UK
Gottfried Hohmann
Department of Primatology
Max Planck Institute for Evolutionary Anthropology
Germany
Lydia M. Hopper
Lester E. Fisher Center for the Study and Conservation of Apes
Lincoln Park Zoo
USA
Sarah L. Jacobson
CUNY Graduate Center
USA
Cheryl D. Knott
Department of Anthropology
Boston University
USA
Michael Lammey
Alamogordo Primate Facility
USA
Jack Lester
Department of Primatology
Max Planck Institute for Evolutionary Anthropology
Germany
Stacy Lindshield
Department of Anthropology
Purdue University
USA
Eric V. Lonsdorf
Institute on the Environment
University of Minnesota
USA
Lydia V. Luncz
Primate Models for Behavioral Evolution Lab, Institute of Cognitive and Evolutionary Anthropology
University of Oxford
UK
Corrine Lutz
Southwest National Primate Research Center
USA

Zarin Machanda
Department of Anthropology
Tufts University
USA
Janet Mann
Department of Biology and Department of Psychology
Georgetown University
USA
Christopher F. Martin
Indianapolis Zoo
USA
Jorg J. M. Massen
Animal Ecology Group
Utrecht University
The Netherlands
Wen Mayoukou
Goualougo Triangle Ape Project
Republic of Congo
Alexander Mielke
Department of Primatology
Max Planck Institute for Evolutionary Anthropology
Germany
David B. Morgan
Lester E. Fisher Center for the Study and Conservation of Apes
Lincoln Park Zoo
USA
Naruki Morimura
Kumamoto Sanctuary
Kyoto University
Japan
Sam Mugume
Makerere University Biological Field Station
Uganda
Carson M. Murray
Department of Anthropology, Center for the Advanced Study of Human Paleobiology
George Washington University
USA
Suska Nolte
School of Psychology and Neuroscience
University of St. Andrews
UK

Emily Otali
Kibale Chimpanzee Project
Makerere University Biology Field Station
Uganda
Audrey E. Parrish
Department of Psychology
The Citadel
USA
Susan (Lambeth) Pavonetti
Michale E. Keeling Center for Comparative Medicine and Research
The University of Texas MD Anderson Cancer Center
USA
Bonnie M. Perdue
Department of Psychology and Department of Neuroscience
Agnes Scott College
USA
Jaine Perlman
Yerkes National Primate Research Center
Emory University
USA
Rhonda Pietsch
Center for Great Apes
USA
Jill D. Pruetz
Department of Anthropology
Texas State University
USA
Stacy Rosenbaum
Department of Anthropology
University of Michigan
USA
Elizabeth Ross
Kasiisi Project
USA
Stephen R. Ross
Lester E. Fisher Center for the Study and Conservation of Apes
Lincoln Park Zoo
USA
Rachel Santymire
Davee Center for Epidemiology and Endocrinology
Lincoln Park Zoo
USA

Crickette M. Sanz
Department of Anthropology
Washington University in St. Louis
USA

Wouter A. A. Schaake
Department of Animal Ecology
Utrecht University
The Netherlands

Carel P. van Schaik
Department of Anthropology
University of Zurich
Switzerland

Natalie G. Schwob
Department of Psychology
Pennsylvania State University
USA

Sara A. Skiba
Department of Psychology
Georgia State University
USA

Katie E. Slocombe
Department of Psychology
University of York
UK

Margaret A. Stanton
Department of Psychology
Franklin and Marshall College
USA

Jeroen M. G. Stevens
Centre for Research and Conservation
Royal Zoological Society of Antwerp
Belgium

Tara Stoinski
Dian Fossey Gorilla Fund
USA

Jared P. Taglialatela
Department of Ecology, Evolution, and Organismal Biology
Kennesaw State University
USA

Claudio Tennie
Department for Early Prehistory and Quaternary Ecology
University of Tübingen
Germany

Simon W. Townsend
Department of Psychology
University of Warwick
UK
Gillian L. Vale
Michale E. Keeling Center for Comparative Medicine and Research
The University of Texas MD Anderson Cancer Center
USA
Kim Valenta
Department of Anthropology
McGill University
Canada
Sarah-Jane Vick
Psychology Division, Faculty of Natural Sciences
University of Stirling
UK
Erica van de Waal
Department of Ecology and Evolution
Université de Lausanne
Switzerland
Koshiro Watanuki
Primate Research Institute
Kyoto University
Japan
Stuart K. Watson
Department of Comparative Linguistics
University of Zurich
Switzerland
William Winston
Library Sciences
Washington University in St. Louis
USA
Roman M. Wittig
Department of Primatology
Max Planck Institute for Evolutionary Anthropology
Germany
Richard W. Wrangham
Department of Human Evolutionary Biology
Harvard University
USA

Shinya Yamamoto
Institute for Advanced Study
Kyoto University
Japan
Meng Yao
Institute of Ecology
Peking University
China

Index